Ecophysiology and Responses of Plants under Salt Stress

Parvaiz Ahmad · M.M. Azooz · M.N.V. Prasad
Editors

Ecophysiology and Responses of Plants under Salt Stress

Springer

Editors
Parvaiz Ahmad
Department of Botany, A.S. College
University of Kashmir
Srinagar, India

M.M. Azooz
Department of Botany
South Valley University
Qena, Egypt

M.N.V. Prasad
Department of Plant Sciences
University of Hyderabad
Hyderabad, India

ISBN 978-1-4614-4746-7 ISBN 978-1-4614-4747-4 (eBook)
DOI 10.1007/978-1-4614-4747-4
Springer New York Heidelberg Dordrecht London

Library of Congress Control Number: 2012947992

Printed on acid-free paper

Springer is part of Springer Science+Business Media (www.springer.com)

Foreword

Salinity imposes a serious problem to agriculture and biomass production. Several important crop species are sensitive to increased salinity which results in decreased photosynthetic performance and lower yield. Increased salinity of arable lands frequently is accompanying drought stress in arid and semiarid areas. On the other hand there are numerous species which not only tolerate increased salinity but they need salty soils and waters as their natural environment. Therefore, the understanding molecular basis of salt stress and the mechanisms of its tolerance is of utmost importance. The volume entitled "Ecophysiology and Responses of Plants under Salt Stress" gives up-dated information about recent achievements and future trends in research on salt stress in plants. International team of authors contributed eighteen chapters on various aspects of salt stress in plants, dealing both with the ecological and physiological responses, biochemical mechanisms, processes of salt stress damage and factors contributing to its defense and tolerance. The volume edited by the group of competent Editors brings together recent results in the broad area of salt stress in plants and gives good overview of the recent literature. In my opinion the volume is well prepared and necessary on the market and it may be recommended to researchers and students working in the field of salt stress in plants.

Krakow, Poland
Kazimierz Strzałka

Preface

Soil is the soul of life and is an invaluable natural resource. Healthy soil is essential for high-quality food production. Therefore, it is necessary to maintain soil fertility for sustainable crop production. Since the cultivable land is declining day by day because of urbanization, there is need to use the uncultivable land for food production to feed teeming population. Changing environment (biotic and abiotic stress) has a negative impact on the growth and development of the plants. Salinity stress is one of the major abiotic stresses that limit agricultural yield. Salinity is responsible for the induction of primary effects like ionic and osmotic stress, which in turn induce oxidative stress in plants.

In plants exposed to salt stress, all the major processes such as photosynthesis, protein synthesis, and energy and lipid metabolisms are affected. Reactive oxygen species (ROS) generated as a result of salt-imposed oxidative stress is highly deleterious for plants. ROS destroys the structure and functions of biomolecules such as membrane lipids, proteins and nucleic acids, and in higher concentrations causes death of the plant cells. Nonetheless, plants have mechanisms to counteract the deleterious effects of primary and secondary stresses through the generation of osmolytes or antioxidants.

This volume consists of 18 chapters which deal with the effect of salt stress on plants. Chapter 1 deals with the causes and types of salt stress and responses of plants. Chapter 2 describes how exogenous protectants help plants to withstand the negative effect of salt stress. Chapter 3 highlights the effect of salt stress on ion transport, water relations and oxidative damage in plants. Chapter 4 is about symbiotic coalition of lichens against salt stress. Chapter 5 deals with the changes in photosystem II under salt stress. Chapter 6 describes the effect of salt stress on the root system and its tolerance. Chapter 7 highlights the effect of salt stress on rice yield and the mechanism of salt tolerance in rice plants. Chapter 8 deals with aquaporins' activity, functions and their role in plant growth and development. Chapter 9 describes production of oil seed crops under salt stress and improving salt tolerance by different methods. Chapter 10 narrates the response of tomato plants to salinity and the role of ABA in plants under salt stress. Chapter 11 deals with changes of phenolic compound content in various plants under salt stress. Chapters 12 and 13 highlight the role of polyamines in plants under salt stress. Chapter 14 describes the physiological functions of jasmonates in relation to environmental stress in plants. Chapter 15 deals with the role of nitric oxide in osmoregulation, ion homeostasis and signalling in plants under salt stress. Chapters 16 and 17 highlight the role and metabolism of nutrients under salt stress. Chapter 18 describes a case study of Neretva river valley about soil and water management for sustained agriculture in alluvial plains and flood plains exposed to salinity.

In this volume, we have tried to provide the readers a background for understanding salt stress and tolerance mechanisms in plants. We are thankful to all the authors for their valuable contributions. We are also thankful to several colleagues who helped us directly or indirectly in completing this volume. We appreciate Hanna Smith (Associate Editor, Springer) and Margaret Burns (Developmental Editor, Springer) for their prompt help, suggestions and punctuality in publication of this collective volume.

Srinagar, India — Parvaiz Ahmad
Qena, Egypt — M.M. Azooz
Hyderabad, India — M.N.V. Prasad

Contents

Contributors

Parvaiz Ahmed Department of Botany, A. S. College, University of Kashmir, Srinagar, JK, India

Vera Alexieva Institute of Plant Physiology and Genetics, Bulgarian Academy of Sciences, Sofia, Bulgaria

M. Asgher Department of Botany, Aligarh Muslim University, Aligarh, UP, India

M.M. Azooz Department of Botany, Faculty of science, South Valley University, Qena, Egypt

Department of Biological Sciences, Faculty of Science, King Faisal University, Al-Hassa, Saudi Arabia

Renu Bhardwaj Department of Botanical and Environmental Sciences, Guru Nanak Dev University, Amritsar, PB, India

Venkatesh Bollina Saskatoon Research Centre, Agriculture and Agri-Food Canada, Saskatoon, SK, Canada

Joël Boustie Laboratory of Pharmacognosy and Mycology – UMR CNRS 6226 ISCR PNSCM, University of Rennes 1 (European University of Brittany), Rennes, France

Jeanne Cook-Moreau Department of Biochemistry and Molecular Genetics, University of Limoges (School of Medicine), Limoges, France

David Delmail Laboratory of Pharmacognosy and Mycology – UMR CNRS 6226 ISCR PNSCM, University of Rennes 1 (European University of Brittany), Rennes, France

Maduraimuthu Djanaguiraman 2004 Throckmorton Plant Sciences Center, Department of Agronomy, Kansas State University, Manhattan, KS, USA

Jhonathan E. Ephrath French Associates Institute for Agriculture and Biotechnology of Drylands, Ben-Gurion University of the Negev, Midreshet Ben-Gurion, Israel

Muhammad Farooq Institute of Plant Nutrition, Justus-Liebig-University, Giessen, Germany

Department of Agronomy, University of Agriculture, Faisalabad, Pakistan

The UWA Institute of Agriculture, The University of Western Australia, Crawley, WA, Australia

Masayuki Fujita Laboratory of Plant Stress Responses, Department of Applied Biological Science, Faculty of Agriculture, Kagawa University, Kagawa, Japan

Piotr Goliński Department of Chemistry, Poznań University of Life Sciences, Poznań, Poland

Martin Grube Institute of Plant Sciences, Karl-Franzens-Universität Graz, Graz, Austria

Asiya Hameed Department of Botany, Faculty of Science, Jamia Hamdard, New Delhi, India

Neha Handa Department of Botanical and Environmental Sciences, Guru Nanak Dev University, Amritsar, PB, India

Mirza Hasanuzzaman Laboratory of Plant Stress Responses, Department of Applied Biological Science, Faculty of Agriculture, Kagawa University, Kagawa, Japan

Department of Agronomy, Faculty of Agriculture, Sher-e-Bangla Agricultural University, Dhaka, Bangladesh

Jean-François Hausman Department EVA, Centre de Recherche Public – Gabriel Lippmann, Belvaux, Luxembourg

Noushina Iqbal Department of Botany, Aligarh Muslim University, Aligarh, UP, India

M. Iqbal R. Khan Department of Botany, Aligarh Muslim University, Aligarh, UP, India

Anjana Jajoo School of Life Science, Devi Ahilya University, Indore, MP, India

Sateesh Kagale Saskatoon Research Centre, Agriculture and Agri-Food Canada, Saskatoon, SK, Canada

Plant Biotechnology Institute, National Research Council, Saskatoon, SK, Canada

Mukesh Kanwar Department of Botanical and Environmental Sciences, Guru Nanak Dev University, Amritsar, PB, India

Dhriti Kapoor Department of Botanical and Environmental Sciences, Guru Nanak Dev University, Amritsar, PB, India

Zornitsa Katerova Institute of Plant Physiology and Genetics, Bulgarian Academy of Sciences, Sofia, Bulgaria

Harpreet Kaur Department of Botanical and Environmental Sciences, Guru Nanak Dev University, Amritsar, PB, India

Nafees A. Khan Department of Botany, Aligarh Muslim University, Aligarh, UP, India

Pascal Labrousse Laboratory of Botany and Cryptogamy – GRESE EA 4330, University of Limoges (Faculty of Pharmacy), Limoges, France

Isabelle Lefèvre Groupe de Recherche en Physiologie végétale (GRPV), Earth and Life Institute – Agronomy, Université catholique de Louvain, Louvain-la-Neuve, Belgium

Department EVA, Centre de Recherche Public – Gabriel Lippmann, Belvaux, Luxembourg

Stella Lovelli Department of Crop System, Forestry and Environmental Sciences, University of Basilicata, Potenza, Italy

Stanley Lutts Groupe de Recherche en Physiologie végétale (GRPV), Earth and Life Institute – Agronomy, Université catholique de Louvain, Louvain-la-Neuve, Belgium

Kenji Miura Faculty of Life and Environmental Sciences, University of Tsukuba, Tsukuba, Japan

Muneeb-u-Rehman Department of Medical Elementology and Toxicology, Faculty of Science, Jamia Hamdard, New Delhi, India

Małgorzata Muzolf-Panek Department of Food Quality Management, Poznań University of Life Sciences, Poznań, Poland

Kamrun Nahar Laboratory of Plant Stress Responses, Department of Applied Biological Science, Faculty of Agriculture, Kagawa University, Kagawa, Japan

Department of Agricultural Botany, Faculty of Agriculture, Sher-e-Bangla Agricultural University, Dhaka, Bangladesh

Naghabushana Nayidu Saskatoon Research Centre, Agriculture and Agri-Food Canada, Saskatoon, SK, Canada

Gabrijel Ondrašek Faculty of Agriculture, University of Zagreb, Zagreb, Croatia

Delphine Parrot Laboratory of Pharmacognosy and Mycology – UMR CNRS 6226 ISCR PNSCM, University of Rennes 1 (European University of Brittany), Rennes, France

Michele Perniola Department of Crop System, Forestry and Environmental Sciences, University of Basilicata, Potenza, Italy

Poonam Department of Botanical and Environmental Sciences, Guru Nanak Dev University, Amritsar, PB, India

Losanka P. Popova Bulgarian Academy of Sciences, Institute of Plant Physiology and Genetics, Sofia, Bulgaria

P.V. Vara Prasad 2004 Throckmorton Plant Sciences Center, Department of Agronomy, Kansas State University, Manhattan, KS, USA

Tabasum N. Qadri Department of Botany, Hamdard University, New-Delhi, India

Muriel Quinet Groupe de Recherche en Physiologie végétale (GRPV), Earth and Life Institute – Agronomy, Université catholique de Louvain, Louvain-la-Neuve, Belgium

Shimon Rachmilevitch French Associates Institute for Agriculture and Biotechnology of Drylands, Ben-Gurion University of the Negev, Midreshet Ben-Gurion, Israel

Saiema Rasool Department of Botany, Faculty of Science, Jamia Hamdard, New Delhi, India

Boris Rewald Institute of Forest Ecology, University of Natural Resources and Life Sciences (BOKU), Vienna, Austria

Davor Romić Faculty of Agriculture, University of Zagreb, Zagreb, Croatia

Marija Romić Faculty of Agriculture, University of Zagreb, Zagreb, Croatia

Sven Schubert Institute of Plant Nutrition, Justus-Liebig-University, Giessen, Germany

Antonio Scopa Department of Crop System, Forestry and Environmental Sciences, University of Basilicata, Potenza, Italy

Iskren Sergiev Institute of Plant Physiology and Genetics, Bulgarian Academy of Sciences, Sofia, Bulgaria

Indu Sharma Department of Botanical and Environmental Sciences, Guru Nanak Dev University, Amritsar, PB, India

Resham Sharma Department of Botanical and Environmental Sciences, Guru Nanak Dev University, Amritsar, PB, India

Oren Shelef French Associates Institute for Agriculture and Biotechnology of Drylands, Ben-Gurion University of the Negev, Midreshet Ben-Gurion, Israel

T.O. Siddiqi Department of Botany, Faculty of Science, Jamia Hamdard, New Delhi, India

Kadambot H.M. Siddique The UWA Institute of Agriculture, The University of Western Australia, Crawley, WA, Australia

Adriano Sofo Department of Crop System, Forestry and Environmental Sciences, University of Basilicata, Potenza, Italy

Dessislava Todorova Institute of Plant Physiology and Genetics, Bulgarian Academy of Sciences, Sofia, Bulgaria

Sophie Tomasi Laboratory of Pharmacognosy and Mycology – UMR CNRS 6226 ISCR PNSCM, University of Rennes 1 (European University of Brittany), Rennes, France

Agnieszka Waśkiewicz Department of Chemistry, Poznań University of Life Sciences, Poznań, Poland

Monika Zovko Faculty of Agriculture, University of Zagreb, Zagreb, Croatia

Chapter 1
Salt Stress: Causes, Types and Responses of Plants

Saiema Rasool, Asiya Hameed, M.M. Azooz, Muneeb-u-Rehman, T.O. Siddiqi, and Parvaiz Ahmad

1.1 Introduction

Abiotic factors like temperature, drought, salt stress, etc. results in the depletion of large number of food production in today's world and as a result these global changes have led to alarmist projections that seem to argue for additional strategies by which food supply can be guaranteed (Miflin 2000). Moreover, the increased productivity achieved in irrigated areas still do not benefit people because salinization following prolonged irrigation is unavoidable (Flowers and Yeo 1995; Postel 1999). These considerations have aroused strong interest in studying plant abiotic stress responses. Salinity has severely affected the agricultural productivity and the damaging effects of salt accumulation have influenced both ancient and modern civilizations. It is estimated that about 20% of the irrigated land in the present world is affected by salinity that is exclusively classified as arid and desert lands comprising

S. Rasool • A. Hameed • T.O. Siddiqi
Department of Botany, Faculty of Science, Jamia Hamdard, New Delhi 110062, India

M.M. Azooz
Department of Botany, Faculty of Science, South Valley University, 83523 Qena, Egypt

Department of Biological Sciences, Faculty of Science,
King Faisal University, Saudi Arabia

Muneeb-u-Rehman
Department of Medical Elementology and Toxicology, Faculty of Science,
Jamia Hamdard, Hamdard Nagar, New Delhi 110062, India

P. Ahmad (✉)
Department of Botany, A. S. College, University of Kashmir,
Srinagar 190008, India
e-mail: parvaizbot@yahoo.com; saimu083@yahoo.co.in

P. Ahmad et al. (eds.), *Ecophysiology and Responses of Plants under Salt Stress*,
DOI 10.1007/978-1-4614-4747-4_1, © Springer Science+Business Media, LLC 2013

25% of the total land of our planet (Yeo 1999). Saline soils with soluble salts affect plant growth at various stages leading to yield differences between crops and also the differences in their ion compositions at maturity. The loss of farmable land due to salinization directly affects the food requirement of the world population, which is projected to increase by 8.5 billion over the next 25 years. Since the cultivable land is declining day by day because of urbanization, there is need to use the uncultivable land for food production to feed teeming population. The maximum area of uncultivable land belongs to salt affected area. In developed and developing countries, the intensive use of agricultural practices has led to the degradations of farming land and water supplies. Preferably, use of crops that tolerate the high levels of salinity in the soils would be a practical contribution towards addressing the problem. Most crops tolerate salinity to a threshold level (Khan et al. 2006), a 'threshold' for a crop is defined as the value below which crop growth is generally not affected due to salinity.

According to Flowers et al. (1977) plants can be divided in to glycophytes and halophytes on the basis of their abilities to grow on different salt concentrations. Halophytes are the plants that grow and complete their life cycle on high concentration of salt, e.g. *Atriplex*, *Vesicaria*. Majority of the terrestrial plants including agricultural crops are glycophytic and cannot tolerate high concentration of salt. Plant growth and development is hampered due to salinity stress through: (1) low osmotic potential of soil solution (water stress), (2) nutritional imbalance, (3) specific ion effect (salt stress) or (4) a combination of these factors (Ashraf 1994). During the onset and development of salt stress within a plant, all the major processes such as photosynthesis, protein synthesis and energy and lipid metabolisms are affected.

Osmotic stress is caused due to the excess of Na^+ and Cl^- in the environment that decrease the osmotic potential of the soil solution and hence water uptake by the plant root. During osmotic stress plant also accumulate low molecular mass compounds known as compatible solutes or osmolytes like, proline, protein, mannitol, sorbitol, glycine betaine, etc. Salt induced osmotic stress is responsible for the oxidative stress caused by reactive oxygen species (ROS).

ROS, such as singlet oxygen ($_1O^2$), superoxide ions (O_2^-) and peroxides, the most widely distributed being hydrogen peroxide (H_2O_2) are toxic molecules (Apel and Hirt 2004; Triantaphylidès et al. 2008). ROS is capable of inducing damage to almost all cellular macromolecules including DNA (Jaleel et al. 2007a, b, c, 2008; Tuteja et al. 2009). ROS targets high-molecular mass molecules, such as membrane lipids or mitochondrial DNA, with the formation of lipid or nucleotide peroxides, especially at the level of thymine.

The harmful effect of ROS is primarily due to their ability to initiate a variety of autoxidative chain reactions on unsaturated fatty acids (Smirnoff 2000). Oxidative attack on proteins results in site specific amino acid modifications, fragmentation of the peptide chain, aggregation of cross linked reaction products and increased susceptibility to proteolysis (Ahmad et al. 2010b, c, 2011). ROS can also induce numerous lesions in DNA that cause deletions, mutations and other lethal genetic effects (Srivalli et al. 2003; Tuteja et al. 2009).

Table 1.1 Different classes of salinity

Salinity class	ECe range (dS/m)
Non-saline	0–2
Low salinity	2–4
Moderate salinity	4–8
High salinity	8–16
Severe salinity	16–32
Extreme salinity	>32

The toxic effects of ROS are counteracted by enzymatic as well as nonenzymatic antioxidative system such as: superoxide dismutase (SOD), catalase (CAT), ascorbate peroxidase (APX), glutathione reductase (GR), ascorbic acid (AsA), tocopherol, glutathione and phenolic compounds, etc. (Shi and Zhu 2008; Sharma and Dietz 2009; Ashraf 2009; Ahmad et al. 2008a, b, 2010a, b, c, 2011, 2012b). Normally, each cellular compartment contains more than one enzymatic activity that detoxifies a particular ROS.

1.2 Salinity in India and World

In the arid and semi-arid regions, low rainfall coupled with ambiguity of its occurrence has been the major limiting factors in crop production. This is mainly true for India because most of the agricultural productive regions lie in hyper-arid to sub-humid regions where evaporation far exceeds the rainfall. The data collected for salt affected areas in different states by Soil Resource Maps has been published by the NBBS and LUP, Nagpur (Bhargava 2005). The NBSS and LUP (National Bureau of Soil Survey and Land Use Planning) had mapped saline soils into six classes on the basis of ECe (electrical conductivity of the saturation extract) viz., slight, moderate, moderately strong, strong, severe and very severe (Table 1.1). While compiling the figures for salt affected soils the values for ECe ranging between 2–4 dS/m were discarded because they constitute non-saline class as per norms evolved by (Richards 1954). Similarly, Bureau categorized the soil sodicity classes in three classes on the basis of ESP (exchangeable sodium percentage) namely high (< 5), moderate (5–15) and strong (>15). All the soils with less than 5 ESP have been considered non-sodic or non-alkali. But all black soils or vertisols with >5 ESP have been considered sodic or alkali. The Rann of Kutchchh (Bangalore), which has an area of 2.1507 million ha and comprises a saline marsh, has been separated in the Maps published by the Bureau. The figures so extracted have been presented in Table 1.2.

At global scale more than 77 mha of land are salt affected and about 43 mha are attributed to secondary salinization as reported by FAO (2007). It is estimated that about one-third of the irrigated land in the major countries with irrigated agriculture is badly affected by salinity or is likely to be salinized in the near future.

Table 1.2 Distribution of saline/sodic soils in India (,000 ha)

State	Saline	Sodic	Total
Punjab	10.2	190.9	201.1
Haryana	175.2	255.7	430.9
Rajasthan	490.4	9.5	499.9
Dehli	21.2	Nil	21.2
Uttar Pradesh	128.9	3,394.1	3,523.0
Bihar	–	229.0	229.0
West Bengal	377.7	–	377.7
Maharashtra	317.3	114.6	431.9
Andhra Pradesh	312.7	217.2	529.9
Madhya Pradesh	1,743.4	26.6	1,770.0
Karnataka	100.0	10.0	110.0
Orissa	74.5	–	74.5
Tamil Nadu	50.0	46.2	96.2
Goa	1.0	–	1.0
A & N Islands	91.9	–	91.9
Karaikal	1.7	1.8	3.5
Pondicherry	1.5	1.9	3.4
Kerala	203.3	–	203.0
Jammu & Kashmir	40.0	20.2	60.0
Gujarat	3,598.4	846.2	4,444.6
Rann of Kutchchh	2,150.7	–	2,150.7
Total	9,889.7	5,363.7	15,253.4

Source: (Bhargava 2005)

Current estimates of the salt-affected soils as percent of the irrigated lands for different countries are: 27% for India, 28% for Pakistan, 13% for Israel, 20% for Australia, 15% for China, 50% for Iraq, and 30% for Egypt (Stockle 2001). In only New South Wales (NSW) state of Australia, salinity is estimated to affect 15% of the irrigated land.

There has been a sequential increase in the level and strength of salinity that affects soil. A number of major irrigation schemes throughout the world have suffered to some extent from the effect of salinity and/or sodocity. Many once-productive areas have become salt affected wastelands. Salt related problems occur within the boundaries of at least 75 countries (Szabolcs 1994). An alphabetical list of the countries with serious salinity problems include Australia, China, Egypt, India, Iraq, Mexico, Pakistan, the Soviet Union, Syria, Turkey and the United States (Rhoades 1998). It has been reported by different international organizations like ICID (International Commission on Irrigation and Drainage), UNEP (United Nations Environment Programme) and FAO (Food and Agriculture Organization) that salinization and water logging of the soil in arid and semi-arid regions are highly responsible for the loss of agricultural productivity on irrigated land. Therefore, the focus should be on urgent measures to combat desertification via modification of farming techniques and amelioration of salt affected and water logged soils with a resultant improvement in social and economic conditions of people dependent on agriculture (Rhoades 1998).

1.3 Causes and Types of Soil Salinity

Salinity has also been categorized as primary and secondary on the basis of their source of cause. Former occurs as natural salt in the landscape like salt marshes, salt lakes, tidal swamps or natural salts clads. While as latter results due to human activity such as urbanization and agriculture (irrigated and dry land). Following are the factors responsible for soil salinity:

For Primary salinity:

I. Weathering of rocks
II. Capillary rise from shallow brackish groundwater
III. Intrusion of sea water along the coast
IV. Salt laden sand blown by sea winds
V. Impeded drainage

Secondary salinization is due to human activities like:

I. Introduction of irrigation without proper drainage system
II. Industrial effluents
III. Overuse of fertilizers
IV. Removal of natural plant cover
V. Flooding with salt rich waters
VI. High water table and the use of poor quality ground water for irrigation

1.3.1 Types of Salinity

There are two types of salt affected soils:

(i) Sodic soils
(ii) Saline soils

The main differences between these two lies in the nature of anions and the pH of the soil. Studies demonstrate that carbonate or bicarbonate ions constitute the sodic soils with pH above 8.5, whereas chloride or sulphate ion dominates the saline soils with pH below 8.5. Some plants grow well in salt affected coastal areas, shores of backwaters lakes and marshy lands. The plants that thrive well in high salt concentrations are called halophytes. However, some plants that cannot withstand even 10% of seawater are called glycophytes or non-halophytes (Gorham 1995; Cherian et al. 1999; Parida and Das 2005; Yadav et al. 2011; Mane et al. 2011).

1.4 Reclamation and Management Strategies

Reclamation is the process of restoring disturbed land into a cultivable soil. While as management is the sum total of all procedures to protect soil and increase its performance. This reclamation process of saline soil includes the following methods:

1.4.1 Physical Method

Physical method of reclamation of saline soil includes the following processes:

1.4.1.1 Scarping

Scraping is the temporary method for soil reclamation in which salt layer of soil surface is scrapped off by mechanical means and the lower layer with less salt content is used for cultivation. Since, with the lowering of ground level in relation to water table salt accumulates again, this method has resulted in limited success. Thus, it again intensifies the problem. This method is rarely used because it involves high cost (Gupta and Gupta 1987).

1.4.1.2 Flushing

Another method of desalinization is to flush the soil with water. In this method accumulated salt on the surface is washed away by water which is associated with very low permeability and high salinity content in their surface layer. This method has little practical significance and shows high efficiency at the beginning but gradually decreases as saline concentration begins to fall.

1.4.1.3 Leaching

In this method, the excess amount of salt can be removed by applying water onto the soil surface. The soluble salts dissolved and transported through the soil are consequently removed from the root zone through drainage. Nevertheless, the amount of liquid required for this process varies and depends on the various characteristics of the soil (Gupta and Gupta 1987). The quantity of soluble salt leached per unit volume of water applied describes the efficiency of leaching (Tanji 1990) and this depends on the (a) uptake of water (b) uniformity of distribution of water on the soil surface, and (c) sufficiency of drainage. The major factors like primary salt content, soil salinity requirement after leaching and depth of root zone becomes imperative to provide a reliable estimate of the quantity of water required for leaching. Leaching of the salts from the soil is usually done by two methods (a) continuous leaching (b) intermittent leaching.

Continuous ponding is the traditional method for leaching of salts from surface irrigated lands (Tanji 1990). In this method of leaching, the water flows in the macropores with salts in the micropores to diffuse to the mobile water. Being faster, this method is also extensively used when time becomes the limiting factor.

Intermittent ponding is most effective method for leaching when water is the limiting factor. Less than 30–35% of water is needed for intermitted ponding than continuous ponding, but the main disadvantage of this method is that it is much slower than continuous ponding method (Gupta and Gupta 1987).

1.4.2 Chemical Method

Gypsum, sulfuric acid and farm yard manure amendments are applied especially in sodic or saline sodic soils. Sodic and saline sodic soil reclamation needs a different approach than saline soils which might be more costly. Furthermore an increase in the infiltration rates is to be required in sodic soils for reclamation that can be achieved by mechanical and chemical measures. Gypsum is considered as one of the most helpful soil amendment to leach out cations (Na^+) from the soil. Gypsum is a slight soluble salt of calcium and sulphate and hence it will react slowly with the soil. The gypsum amount required will vary widely depending on the percentage of exchangeable sodium and soil texture.

It has been investigated that chemical amendment-based technology has been established to reclaim the alkali/sodic soils for proper regulation of salts in root zone (Singh 2009). This includes field leveling, bunding, soil sampling to know the sodicity status for working out amendment dose application of gypsum/pyrite as per the need of the soil followed by rice-wheat rotation for 3–4 years including sesbania as a green manure crop after wheat harvest in April.

1.4.3 Biological Method

Leaching and water quality appears to be limited for the knowledge of soil amendments. By biological reclamation the use of saline wastelands can be made possible through effective management strategies by using salt tolerant plants. To reclaim salt affected soils of unproductive agricultural lands by biological means may be a feasible choice. Agricultural land fed with rain water, lack of irrigation water, shallow and brackish groundwater suggests that the salt affected lands may better be cultivated with crops tolerant to low to moderate saline conditions. Several varieties of salt tolerant crops like rice, wheat and mustard have been developed with a good economic yield both in high pH alkali soils and in saline soils (Singh and Sharma 2006). Among these rice varieties like CSR10, CSR13, CSR19, CSR23, CSR27, and CSR30 can be cultivated in soils with pH ranging from 9.4 to 9.8 and EC values of 6–11 dSm^{-1}. Salt tolerant varieties with their level of tolerance to soil salinity and alkalinity has been shown in Table 1.3.

Table 1.3 Suggested salt tolerant varieties

		Adaptability	
Crop	Tolerant varieties	Sodic with pH	Saline with EC (dSm^{-1})
Rice	CSR 10[a], CSR 11, CSR 12, CSR 13[a]	9.8–10.2	6–11
	CSR19, CSR23[a], CSR27[a], CSR30[a], CSR1,	9.4–9.8	6–11
	CSR2, CSR3, CSR4[a], CST7-1[a],	9.4–9.8	6–9
Wheat	KRL 1–4[a], WH157	<9.3	6–10
	Raj3077, KRL19[a]	<9.3	6–10
Barley	DL200, Ratna, BH97, DL348	8.8–9.3	
Indian Mustard	Pusa Bold, Varuna	8.8–9.2	6–8
	Kranti, CS52[a], CSTR330-1,	8.8–9.3	6–9
	CST609-B 10, CS54[a]	8.8–9.3	6–9
Gram	Karnal Chana 1	<9.0	<6.0
Sugar beet	Ramonskaaya 06, Maribo Resistapoly	9.5–10	<6.5
Sugarcane	Co453, Co1341	<9.0	ECe–10

[a]Institute varieties released by central varietal released committee.
Source: (Singh 2009)

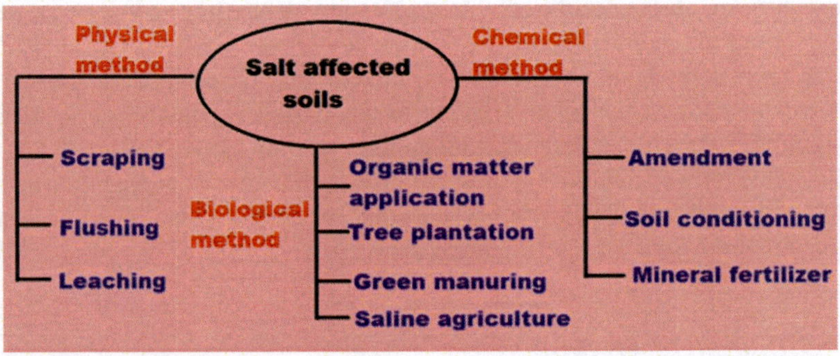

Fig. 1.1 Integrated management practices for the reclamation of salt affected soils

The biological reclamation techniques provide the earnings to the farmers without the need of costly drainage and reclamation work. Salt-tolerant trees and shrub species especially *Atriplex, Eucalyptus*, *Chenopodium*, *Suaeda*, *Salicornia*, *Kochia*, *Sesbania*, *Salsola, Juncus* and others can be grown efficiently in those areas of the land that cannot be restored (Aronson 1985; Le-Houerou 1986). For land reclamation purpose several investigators have studied the cultivation of salinity and sodocity-tolerant plants like grasses (Qadir et al. 1996), agronomic and horticultural crops (Ahmad et al. 1992, 2006, 2008a, b, 2010a, 2012a; Azooz et al. 2011), forest species. Furthermore this process can be made more effective by combining various ameliorative methods. As a result the outcome of this interaction becomes more impressive rather applied singly (Fig. 1.1).

1.5 Plant Adaptations to Salt Stress

Under osmotic stress, plants accumulate osmotically active compounds called osmolytes in order to lower the osmotic potential. These are referred to as compatible metabolites because they do not apparently interfere with the normal cellular metabolism of the cell (Ahmad and Sharma 2008; Ahmad and Prasad 2012a, b). The primary function of compatible solutes is to maintain cell turgor and thus providing the driving gradient for water uptake. Recent studies indicate that compatible solutes can also act as free-radical scavengers or chemical chaperones by directly stabilizing membranes and/or proteins (Mc Neil et al. 1999; Diamant et al. 2001). Glycerol and sucrose helps by were discovered by empirical methods to protecting the biological macromolecules against the damaging effects of salinity. Later, a systematic examination of the molecules, which accumulate in halophytes and halo-tolerant organisms, led to the identification of a variety of molecules also able to provide protection (Arabawa and Timasheff 1985; Wiggins 1990). These molecules are not highly charged, but are polar, highly soluble and have a larger hydration shell. Such molecules will be preferentially solubilized in the bulk water of the cell where they could interact directly with the macromolecules.

1.5.1 Glycine-Betaine

Plants synthesise glycine betaine-a major osmolyte that have been widely observed in various plant species conferring salt-tolerance (Rhodes and Hanson 1993; Hanson et al. 1994; Ahmad and Sharma 2008; Chen and Murata 2011; Koyro et al. 2012). Highly tolerant species, i.e., *Spartina* and *Distichlis* show highest accumulation, low tolerant accumulate average levels while as sensitive one's show low levels of accumulation of glycine-betaine (Rhodes et al. 1989). Glycinebetaine is synthesized from choline in two steps, the first being catalyzed by choline mono-oxygenase leading to synthesis of betaine aldehyde, which is further oxidized by betaine-aldehyde dehydrogenase (Ahmad and Sharma 2008; Chen and Murata 2011; Koyro et al. 2012). Genetic evidence showed that glycine-betaine has been obtained to develop the salinity tolerance for barley and maize (Rhodes et al. 1989; Grumet and Hanson 1986). Among these two plants, isogenic barley lines show different abilities to adjust osmotically. Transgenic rice plants expressing betaine-aldehyde dehydrogenase converted high levels of exogenously applied betaine aldehyde to glycine-betaine than did wild-type plants. The elevated level of glycinebetaine in transgenic plants conferred significant tolerance against salt, cold and heat stress (Chen and Murata 2011). Transgenic plants expressing bacterial gene for the synthesis of GB is given in table 1.4.

Table 1.4 Transgenic plants expressing bacterial gene for the synthesis of GB and their tolerance to salt stress

Gene	Plant species	Tolerance to	Reference
codA	*Arabidopsis*	Salt, Chilling stress	Hayashi et al. 1997
cox	*Arabidopsis*	Salt, drought, freezing	Huang et al. 2000
codA	*Arabidopsis*	Salt	Sulpice et al. 2003
codA	*Brassica juncea*	Salt	Prasad et al. 2000
codA	*Diospyras kaki*	Salt	Gao et al. 2000
codA	*Lycopersicon esculantum*	Chilling, salt	Park et al. 2004
cox	*Nicotiana tabacum*	Salt	Huang et al. 2000
codA	*Oryza sativa*	Salt	Mohanty et al. 2002
cox	*Oryza sativa*	Salt	Su et al. 2006
codA	*Soalnum tuberosum*	Salt, drought	Ahmad et al. 2008
BADH	*Nicotiana tabacum*	Salt	Zhou et al. 2008
BADH	*Nicotiana tabacum*	Salt	Yang et al. 2008
CMO	*Nicotiana tabacum*	Salt, drought	Zhang et al. 2008
BADH	*Lycopersicon esculantum*	Salt	Zhou et al. 2007
CMO	*Oryza sativa*	Salt	Shirasawa et al. 2006
OsCMO	*Nicotiana tabacum*	Salt	Luo et al. 2012
SoBADH	*Ipomoea batatas*	Salt	Fan et al. 2012

1.5.2 Polyamines

Any stress factor like osmotic stress, low pH, potassium deficiency, nutrient deficiency or light leads to accumulation of polyamines. Among polyamines, putrescine accumulation is correlated with increased argenine decarboxylase (ADC) activity in oats. Similar studies reports that transgenic carrot cells over expressing ornithine decarboxylase (ODC) cDNA considerably show that these cells were significantly more tolerant to both salt stress as well as water stress (Bohnert et al. 1995). It has been demonstrated that increased ethylene synthesis and seed germination leads to the suppression of polyamine biosynthesis (Gallardo et al. 1995). Lin and Kao (1995) observed that spermidine increases and putrescine decreases in the shoot and roots of rice seedlings and their accumulation including spermine with ADC activity signifies a specific role in salt-tolerance (Ahmad et al. 2012b). Polyamines such as spermine and spermidine are derived from methionine and ornithine while as putriscine from arginine. The first step involves the decarboxylation of ornithine catalysed by ODC (Ahmad et al. 2012b). Furthermore polyamines and their corresponding enzyme activities are substantially enhanced under salt and drought stress (Lefevre and Lutts 2000). Nuclear DNA is found to be stabilized by histones in eukaryotic organisms while as putrescine and polyamines take over the role of histones in bacteria besides the regulation of DNA in plant mitochondria and chloroplasts. Moreover polyamines stimulate several protein biosynthesis mostly through nucleic acid interaction. They also play a key role in stabilization of bio-membranes.

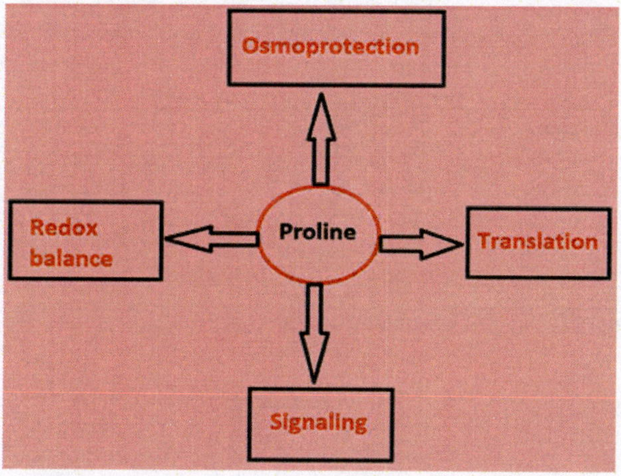

Fig. 1.2 Multiple function of proline

1.5.3 Proline

Proline plays a key role in osmoregulation in plants subjected to hyperosmotic stresses, primarily drought and salinity stress (Ahmad and Jhon 2005; Ahmad et al. 2006, 2010a, 2012a; Ahmad and Sharma 2010). Accumulation of proline is a means of adaptation to abiotic stress and has been observed in rye grass for the first time by Kemble and MacPherson (1954). There are various other compatible solutes including glycine betaine (Mc Cue and Hanson 1990), and polyols that have been shown to accumulate in plants subjected to osmotic stress conditions (Adams et al. 1992). Proline accumulation influences stress tolerance in different ways (Fig. 1.2). Proline acts as a molecular chaperone protecting protein integrity and thereby increase the activities of many enzymes. Besides plants, proline accumulation has been found most widely distributed osmolyte in eubacteria, protozoa, marine invertebrates and algae (Mc Cue and Hanson 1990). Investigations shows that enhancement of proline is due to the stimulation of biosynthetic pathway in plants. Glutamate or ornithine is utilized for the synthesis of proline, glutamate being the primary precursor in osmotically stressed cells (Ahmad and Sharma 2008; Koyro et al. 2012). Transcripts corresponding to both cDNAs that accumulate against NaCl stress are found to play a regulatory key to build up strategies for overproduction of proline in a selected plant species. Furthermore the intermediates of proline biosynthetic pathways also enhance the expression of numerous genes that are regulated osmotically in rice (Iyer and Caplan 1998). Evidence also shows that proline degradation in the mitochondria is linked to respiratory electron transport system as well as ATP production. Expression of P5CS transgenic rice from moth-bean led to stress-induced overproduction of the P5CS enzyme and proline accumulation in transgenic rice plants under the control of an inducible promoter. While as second generation (RI) transgenic plants observed an enhancement of biomass in response to salt and water stress (Zhu et al. 1998.). Table 1.5 shows response of different plants towards proline accumulation under salt stress.

Table 1.5 Responses of biochemical attributes under salt stress

	Species	Response to salinity	References
Soluble protein	*Pisum sativum*	Increase	Ahmad and Jhon 2005
	Oryza sativa	Decrease	Alamgir and Ali 1999
	Vicia faba	Decrease	Gadallah 1999
	Amaranthus tricolor	Decrease	Wang and Nil 2000
	Bruguiera parviflora	Decrease	Parida et al. 2002
	Pancratium maritimum	Increases at low salinity Decrease at high salinity	Khedr et al. 2003
	Arabidopsis thaliana	Increase	Quintero et al. 1996
	Morus alba	Increase	Ahmad and Sharma 2010
	Brassica juncea var. Bio902	Increase	Mittal et al. 2012
	Brassica juncea var. Urvashi	Decrease	Mittal et al. 2012
	Beta vulgarus	Decrease	Jamil et al. 2012a
	Oryza sativa	Decrease	Jamil et al. 2012b
	Portulaca oleraceae	Decrease	Rahdari et al. 2012
	Setaria italica	Increase	Hendawy et al. 2012
	Portulaca oleraceae	Increases	Rahdari et al. 2012
	Prunus species	Decrease	Sorkheh et al. 2012
	Borago officinalis	Decrease	Enteshari et al. 2011
Proline	*Pisum sativum*	Increase	Ahmad and Jhon 2005
	Portulaca oleraceae	Increases	Rahdari et al. 2012
	Matricaria chamomilla	Increases	Heidari and Sarani 2012
	Borago officinalis	Increases	Enteshari et al. 2011
	Brassica juncea var. Bio902	Increase	Mittal et al. 2012
	Brassica juncea var. Urvashi	Decrease	Mittal et al. 2012
	Suaeda maritima	Increase	Rajaravindran and Natarajan 2012
	Lycopersicon esculantum	Increase	Babu et al. 2012
	Morus alba	Increase	Ahmad and Sharma 2010
Carbohydrates	*Portulaca oleraceae*	Increases	Rahdari et al. 2012
	Matricaria chamomilla	Increases	Heidari and Sarani 2012
	Beta vulgaris L.	Increases	Dadkhah 2010
	Borago officinalis	Decrease	Enteshari et al. 2011
	Prosopis alba	Increase in soluble carbohydrate	Meloni et al. 2004
	Morus alba	Increase	Ahmad and Sharma 2010

1.5.4 Carbohydrates

Osmotic potential accounts more than 50% of sugar in glycophytes subjected to saline conditions (Cram 1976). Despite a significant decrease in net CO_2 assimilation rate its accumulation in plants have been widely reported (Murakeozy et al. 2003).

Accumulation of carbohydrates plays a central role in osmoprotection, osmotic adjustment, carbon storage and radical scavenging under salt stress (Ahmad and Sharma 2008; Koyro et al. 2012). Trehalose, have shown to accumulate and protects membranes and proteins in cells against water deficit and reduced aggregation of denatured proteins (Singer and Lindquist 1998). At the same time suppressive effect has been observed by trehalose on apoptotic cell death (Yamada et al. 2003) suggesting the presence of trace amounts in vascular plants, including major crops, though the definite role of this osmolyte in metabolism is still unclear.

Role of sugars in adaptation of plants to salinity have been concluded to be universally associated with salt tolerance. However, this does not signify the role of indicator for salt tolerance in breeding programs for some species. Table 1.5 shows response of different plants towards carbohydrate accumulation under salt stress.

1.5.5 Proteins

Salt-induced proteins in plants have been classified into two major groups (Mansour 2000), i.e., salt stress proteins, which accumulate only due to salt stress, and stress associated proteins, that accumulates in response to various abiotic stress like heat, cold, drought, water logging, and high and low mineral nutrients. Protein accumulation also provides a storage pool of nitrogen to be re-utilized later (Singh et al. 1987) and also a key role in osmotic adjustment. Large number of cytoplasmic proteins cause alterations in cytoplasmic viscosity of the cells stimulated by salinity (Hasegawa et al. 2000). Proteins have shown to increase on exposure to salt stress and can be synthesized *de novo* in response to salt stress or may be present constitutively at low concentration (Pareek et al. 1997). Protein with 26 kDa named as osmotin have been detected in tobacco in response to salt stress (Singh et al. 1987). In salt stressed *Mesembryanthemum crystallinum* another osmotin-like protein was also observed to increase as compared to non-stressed plants (Thomas and Bohnert 1993). In barley, two 26 kDa polypeptides, identified as germin which is not immunologically related to osmotin, have been found to increase in response to salt stress (Hurkman et al. 1991). Similar reports have been found in radish with 22 kDa protein (Lopez et al. 1994). Salt tolerant shows higher soluble proteins than salt sensitive species of barley (Hurkman et al. 1989), sunflower, finger millet (Uma et al. 1995), and rice (Pareek et al. 1997). Table 1.5 shows response of different plants towards protein accumulation under salt stress.

1.6 Reactive Oxygen Species and Antioxidants

Adaptation of the plant cell to high salinity involves osmotic adjustment and the compartmentation of toxic ions, whereas an increasing body of evidence suggests that high salinity also induces the generation of reactive oxygen species (ROS) and oxidative stress (Savouré et al. 1999; Ahmad et al. 2008a, b, 2010a, b, c, 2011, 2012a). The oxygen in the atmosphere enabled respiratory mechanism and electron transport system which use molecular oxygen (O) as final electron acceptor, which led to the formation of ROS in cells (Temple et al. 2005; Ahmad et al. 2008a, b, 2010a, b, c, 2011, 2012a). Although, atmospheric oxygen is relatively non-reactive, it can give rise to reactive oxygen intermediates which include superoxide anion (O_2^-), hydrogen peroxide (H_2O_2), hydroxyl radical (OH^-) and singlet oxygen ($^1O_2$) (Scandalios 2005).

ROS are produced continuously by photosynthesis, photorespiration and CO_2 assimilation in plants. ROS will act as damaging, protective or signaling factor depends on the delicate equilibrium between ROS production and scavenging at the proper site and time (Ahmad et al. 2008a, b, 2010a). Plant cells have developed a comprehensive array of antioxidant defense to prevent the formation of ROS or to limit their damaging effects.

1.6.1 Antioxidants

Salt stress is complex and imposes a water deficit because of osmotic effects on a wide variety of metabolic activities (Greenway and Munns 1980; Cheeseman 1988). Although a wide range of genetic adaptations to saline conditions has been observed and a number of significant physiological responses have been associated with tolerance, underlying mechanisms of salt tolerance in plants are still poorly understood. The effects of various environmental stresses in plants are known to be mediated, at least in part, by an enhanced generation of reactive oxygen species (ROS) including $\cdot O_2$, H_2O_2, and $\cdot OH$ (Hernandez et al. 2000; Benavides et al. 2000; Ahmad et al. 2008a, b, 2010b, c, 2011). These ROS are highly reactive and can alter normal cellular metabolism through oxidative damage to membranes, proteins, and nucleic acids; they also cause lipid peroxidation, protein denaturation, and DNA mutation (Imlay 2003; Ahmad et al. 2008a, b, 2010b, c, 2011). To prevent damage to cellular components by ROS, plants have developed a complex antioxidant system. The primary components of this system include carotenoids, ascorbate, glutathione, and tocopherols, in addition to enzymes such as superoxide dismutase (SOD), catalase (CAT), glutathione peroxidase (GPX), peroxidases, and the enzymes involved in ascorbate–glutathione cycle (Foyer and Halliwell 1976), such as ascorbate peroxidase (APX) and glutathione reductase (GR) (Ahmad et al. 2008a, b, 2010b, c, 2011). Many components of this antioxidant defense system can be found in various subcellular compartments (Hernandez et al. 2000; Ahmad et al. 2011). The scavenging of ROS by increased activation of antioxidant enzymes can improve salt tolerance (Alscher et al. 2002). A relationship between salt tolerance and increased activation of antioxidant enzymes has been demonstrated in Plantago (Sekmen et al. 2007), pea (Hernandez et al. 2000; Ahmad et al. 2008a, b), Arabidopsis, rice (Dionisio-Sese and Tobita 2007), tomato, soybean, maize (Azevedo Neto et al. 2006), broad bean (Azooz et al. 2011), mustard (Ahmad 2010; Ahmad et al. 2010a, 2012a).

Under salt stress, increase in activity of SOD, APX, GR, DHAR, CAT and POX as well as higher antioxidant activity in tolerant species/varieties have been reported by various workers (Ahmad et al. 2010a, 2012a; Azooz et al. 2011; Koyro et al. 2012).

Superoxide dismutase (SOD) is the first defense agent against ROS being the major scavenger of $\cdot O_2^-$ (Almoguera et al. 1995). High SOD activity protects the plant against the superoxide radical, it cannot be considered solely responsible for membrane protection against peroxidation because it converts $O^{2\bullet-}$ to H_2O_2, which is also a ROS. Current studies have shown that over-expression of mitochondrial Mn-SOD in transgenic *Arabidopsis thaliana* (Wang et al. 2004) and chloroplastic Cu/Zn-SOD in transgenic *Nicotiana tabacum* (Badawi et al. 2004) can provide enhanced tolerance to salt stress. Similar results have been found in *Morus alba* (Sudhakar et al. 2001; Ahmad et al. 2010a), *Triticum aestivum* (Sairam et al. 2002), *Lycopersicon* sp (Mittova et al. 2002), *Pisum sativum* (Ahmad et al. 2008a, b), *Vicia faba* (Azooz et al. 2011) and *Brassica juncea* (Ahmad 2010; Ahmad et al. 2010a, 2012a). Earlier studies suggested that the increased SOD activity enables the plant to resist the potential oxidative damage caused by NaCl salinity exposure (Khan et al. 2002; Panda and Khan 2003; Ahmad and Umar 2011).

Ascorbate peroxidase is a hydrogen peroxide-scavenging enzyme found in higher plants, algae, and some cyanobacteria (Asada 1992). Ascorbate peroxidase (APX) is a multigenic family with various isoforms in which cytosolic APX plays a fundamental role in non-photosynthetic tissues, by preventing H_2O_2 dependent inhibition of cytosolic enzymes (Verniquet et al. 1991). APX in the mechanisms of salt tolerance has been substantiated at protein level (Elkahouia et al. 2005; Masood et al. 2006; Koca et al. 2007). APX activity had a key role in response to salt stress in the comparison of the activities of antioxidant enzymes in salt-sensitive and salt-tolerant cultivars (Gueta-Dahan et al. 1997; Ahmad et al. 2008a, b, 2010a, 2012a; Ahmad and Umar 2011).

GR is one of the three enzymes, which catalyze reactions that maintain large pool of GSH and ascorbate in the H_2O_2 scavenging path way in chloroplasts (Yousuf et al. 2012). There are reports showing that GR activity increased in NaCl-tolerant pea variety as compared to NaCl-sensitive pea (Hernandez et al. 2000). The salt treatment had little effect on the activity of glutathione reductase (Lee et al. 2001), and it was suggested that its lower activity in the stressed roots could be due to some acclimation or an inability to maintain a high GSH/GSSG ratio (Mittova et al. 2000; Khan and Panda 2008).

Catalase is the main scavenger of H_2O_2 in peroxisomes, converting it to water and molecular oxygen (Willekens et al. 1995). CAT activity has been found to increase under salt stress in soybean (Comba et al. 1998), tobacco (Bueno et al. 1998), cucumber (Lechno et al. 1997), mulberry (Sudhakar et al. 2001; Ahmad et al. 2010a) and mustard (Ahmad et al. 2012a). Azevedo Neto et al. (2006) also found higher CAT activity in two maize cultivars differing in salt tolerance.

In plant cells, the most important reducing substrate for H_2O_2 detoxification is ASC. Ascorbic acid is an important antioxidant, which reacts not only with H_2O_2 but also with O^{2-}, OH and lipid hydroperoxidases (Reddy et al. 2004; Ahmad et al. 2008a, b, 2010a, b, c, 2011). Ascorbic acid can act as the "terminal antioxidant" because the redox potential of the AA/monodehydro ascorbate (MDA) pair is lower than that of most of the bioradicals (Scandalios et al. 1997). Several studies have revealed that ascorbic acid plays an important role in improving plant tolerance to abiotic stress (Shalata and Neumann 2001; Athara et al. 2008; Ahmad et al. 2008a, b, 2009, 2010b, c, 2011; Ahmad and Umar 2011).

Glutathione plays an important role in the protection against oxidative stress (Ahmad et al. 2008a, b, 2009, 2010b, c, 2011). It is involved in the ascorbate/glutathione cycle and in the regulation of protein thiol-disulphide redox status of plants in response to abiotic and biotic stress (Mullineaux and Rausch 2005; Yousuf et al. 2012). Ruiz and Blumwald (2002) and Mullineaux and Rausch (2005) reported that glutathione content in wild canola plants increased under salt stress, this suggests a possible protective mechanism against salt induced oxidative damage.

1.7 Conclusion and Future Perspective

Plants always experience the fluctuations of environment that causes stress and leads to crop loss worldwide. Salt stress is one of the most damaging abiotic stresses caused due various factors including human activities. In arid and semi-arid regions the salinity is intensified due to fertilizers and irrigation with saline ground water. Salinity in soils affect water availability due to limitation of water uptake of the plants. The ions Na^+ and Cl^- also hampers the assimilation, transport and distribution of essential mineral nutrients within the plant. It has been observed that nutrient assimilation especially K^+ and Ca^{2+} is reduced in the rooting medium under high levels of the NaCl, which ultimately leads to ion imbalances of K^+, Ca^{2+} and Mg^{2+} compared to Na^+. It has been reported that salinity is responsible for the inhibition of cell division and cell enlargement in plants. Overall growth of the plant is also affected with salinity that is why plants showed stunted growth under saline environment. Osmotic damage due to osmotic stress could occur as a result of high concentrations of Na^+ in the leaf apoplast, since Na^+ enters leaves in the xylem stream and is left behind as water evaporates. During stress conditions plants need to maintain internal water potential below that of soil and maintain turgor and water uptake for growth. This requires an increase in osmotica, either by uptake of soil solutes or by synthesis of metabolic (compatible) solutes.

Accumulation of different compatible solutes have been reported during salt stress in different plants. The compatible solutes protect the plants from stress through different courses, including contribution to cellular osmotic adjustment, detoxification of reactive oxygen species, protection of membrane integrity, and stabilization of enzymes/proteins. Some solutes perform an extra function of protection of cellular components from dehydration injury and are called as osmoprotectants. These solutes include proline, sucrose, polyols, trehalose and quaternary ammonium compounds (QACs) such as glycine betaine, alaninebetaine, proline, etc.

Apart from osmotic stress salt is responsible for the generation of ROS in cells that leads to oxidative stress. The generation of these ROSs is due to the imbalance between the production and scavenging machinery of ROS. The unquenched ROS react spontaneously with organic molecules and cause membrane lipid peroxidation, protein oxidation, enzyme inhibition and DNA and RNA damage.

Under severe stress conditions this ROS ultimately leads to cell death. Plants have evolved mechanisms that allow them to adapt and survive under abiotic stress. The production of ROS is however kept under tight control by a versatile and cooperative antioxidant system that modulates intracellular ROS concentration and sets the redox-status of the cell. Plants overexpressing antioxidant enzymes have been engineered with the aim of increasing stress tolerance by directly modifying the expression of these ROS scavenging enzymes. Many workers have reported the positive effects of SOD, CAT, APX, GR, MDHAR, AsA, glutathione, etc. in combating oxidative damage to the cell. There can be no doubt that transgenic plants will be invaluable in assessing the precise role that main antioxidants and ROS play in the functional network that controls stress tolerance.

One of the most important problems before the plant biologists is to develop stress tolerant plants with maximum yield. Since the development of modern biotechnology, a vast research has been carried out to understand the various approaches that plants have adopted to overcome the environmental stresses. Transgenic research proved to be invaluable tool for the development of stress tolerant crops. The advancement in omics is being used in elucidating important plant processes in response to various abiotic stresses. The road to engineering such tolerance into sensitive species is still far from us. Much effort is still required to uncover in detail each product of genes induced by salt stress and signal transduction pathways. Plant biologists should look forward for defined set of markers to predict tolerance towards a particular type of stress with a definite degree of assurance.

References

Adams P, Thomas JC, Vernon DM, Bohnert HJ, Jensen RG (1992) Distinct cellular and organismic responses to salt stress. Plant Cell Physiol 33(8):1215–1223

Ahmad P (2010) Growth and antioxidant responses in mustard (*Brassica juncea* L.) plants subjected to combined effect of gibberellic acid and salinity. Arch Agro Soil Sci 56(5):575–588

Ahmad P, Jhon R (2005) Effect of Salt stress on growth and biochemical parameters of *Pisum sativum* L. Arch Agro Soil Sci 51(6):665–672

Ahmad P, Prasad MNV (2012a) Environmental adaptations and stress tolerance in plants in the era of climate change. Springer Science + Business Media, New York

Ahmad P, Prasad MNV (2012b) Abiotic stress responses in plants: metabolism, productivity and sustainability. Springer Science + Business Media, New York

Ahmad P, Sharma S (2008) Salt stress and phyto-biochemical responses of plants. Plant Soil Environ 54(3):89–99

Ahmad P, Sharma S (2010) Physio-biochemical attributes in two cultivars of mulberry (*M. alba*) under $NaHCO_3$ stress. Int J Plant Produc 4(2):79–86

Ahmad P, Umar S (2011) Oxidative stress: role of antioxidants in plants. Studium Press, New Delhi

Ahmad P, Sharma S, Srivastava PS (2006) Differential physio-biochemical responses of high yielding varieties of Mulberry (*Morus alba*) under alkalinity (Na_2CO_3) stress *in vitro*. Physiol Mol Biol Plants 12(1):59–66

Ahmad P, Jhon R, Sarwat M, Umar S (2008a) Responses of proline, lipid peroxidation and antioxidative enzymes in two varieties of *Pisum sativum* L. under salt stress. Int J Plant Produc 2(4):353–366

Ahmad P, Sarwat M, Sharma S (2008b) Reactive oxygen species, antioxidants and signaling in plants. J Plant Biol 51(3):167–173

Ahmad P, Jeleel CA, Azooz MM, Nabi G (2009) Generation of ROS and non-enzymatic antioxidants during abiotic stress in Plants. Bot Res Intern 2(1):11–20

Ahmad P, Jaleel CA, Salem MA, Nabi G, Sharma S (2010a) Roles of Enzymatic and non-enzymatic antioxidants in plants during abiotic stress. Crit Rev Biotechnol 30(3):161–175

Ahmad P, Jaleel CA, Sharma S (2010b) Antioxidative defence system, lipid peroxidation, proline metabolizing enzymes and Biochemical activity in two genotypes of *Morus alba* L. subjected to NaCl stress. Russ J Plant Physiol 57(4):509–517

Ahmad P, Umar S, Sharma S (2010c) Mechanism of free radical scavenging and role of phytohormones during abiotic stress in plants. In: Ashraf M, Ozturk M, Ahmad MSA (eds) Plant adaptation and phytoremediation. Springer, Dordrecht/Heidelberg/London/New York, pp 99–108

Ahmad P, Nabi G, Jeleel CA, Umar S (2011) Free radical production, oxidative damage and antioxidant defense mechanisms in plants under abiotic stress. In: Ahmad P, Umar S (eds) Oxidative stress: role of antioxidants in plants. Studium Press, New Delhi, pp 19–53

Ahmad P, Hakeem KR, Kumar A, Ashraf M, Akram NA (2012a) Salt-induced changes in photosynthetic activity and oxidative defense system of three cultivars of mustard (*Brassica juncea* L.). Afr J Biotechnol 11(11):2694–2703

Ahmad P, Kumar A, Gupta A, Hu X, Hakeem KR, Azooz MM, Sharma S (2012b) Polyamines: Role in Plants Under Abiotic Stress. In: Ashraf M, Ozturk M, Ahmad MSA, Aksoy A (eds) Crop production for agricultural improvement. Springer, Dordrecht/Heidelberg/London/New York, pp 490–512

Ahmad S, Keatinge JDH, Ali A, Khan BR (1992) Selection of barley lines suitable for spring sowing in the arid highlands of Baluchistan Sarhad. J Agric 8:49–56

Alamgir ANM, Ali MY (1999) Effect of salinity on leaf pigments, sugar and protein concentrations and chloroplast ATPAase activity of rice (Oryza sativa L.). Bangladesh J Bot 28:145–149

Almoguera C, Coca MA, Jouanin L (1995) Differential accumulation of sunflower tetra-ubiquitin mRNA during zygotic embryogenesis and developmental regulation of their heat shock response. Plant Physiol 107:765–773

Alscher RG, Erturk N, Heath LS (2002) Role of superoxide dismutases (SODs) in controlling oxidative stress in plants. J Exp Bot 53:1331–1341

Apel K, Hirt H (2004) Reactive oxygen species: metabolism, oxidative stress, and signal transduction. Annu Rev Plant Biol 55:373–399

Arabawa T, Timasheff SN (1985) The stabilization of proteins by osmolytes. J Biophys 47:411–414

Aronson JA (1985) Economic halophytes, A Global Review. In: Wickens GE, Gooding JR, Fields DV (eds) Plant for arid lands. George Allen and Unwin, London, pp 177–188

Asada K (1992) Ascorbate peroxidase a hydrogen peroxide scavenging enzyme in plants. Plant Physiol 85:235–241

Ashraf M (1994) Organic substances responsible for salt tolerance in *Eruca sativa*. Biol Plant 36:255–259

Ashraf M (2009) Biotechnological approach of improving plant salt tolerance using antioxidants as markers. Biotechnol Adv 27:84–93

Athara H, Khanb A, Ashraf M (2008) Exogenously applied ascorbic acid alleviates salt-induced oxidative stress in wheat. Env Exp Bot 63:224–231

Azevedo Neto AD, Prisco JT, Enéas-Filho J, Abreu CEB, Gomes-Filho E (2006) Effect of salt stress on antioxidative enzymes and lipid peroxidation in leaves and roots of salt-tolerant and salt-sensitive maize genotypes. J Environ Exp Bot 56:87–94

Azooz MM, Youssef AM, Ahmad P (2011) Evaluation of salicylic acid (SA) application on growth, osmotic solutes and antioxidant enzyme activities on broad bean seedlings grown under diluted seawater. Inter J Plant Physiol Biochem 3:253–264

Babu MA, Singh D, Gothandam KM (2012) The effect of salinity on growth, hormones and mineral elements in leaf and fruit of tomato cultivar PKM1. J. Anim Plant Sci 22(1):159–164

Badawi GH, Yamauchi Y, Shimada E, Sasaki R, Kawano N, Tanaka K (2004) Enhanced tolerance to salt stress and water deficit by over expressing superoxide dismutase in tobacco (*Nicotiana tabacum*) chloroplasts. Plant Sci 166:919–928

Benavides MP, Marconi PL, Gallego SM, Comba ME, Tomaro ML (2000) Relationship between antioxidant defence systems and salt tolerance in *Solanum tuberosum*. Aust J Plant Physiol 27:273–278

Bhargava GP (2005) Overview of problem of salt affected soils in India. In: Singh KN, Sharma PC (eds) Crop improvement for management of salt affected soils. Central Soil Salinity Research Institute, Karnal, pp 1–10

Bohnert HJ, Nebson DW, Jensen RG (1995) Adaptation to environmental stress. Plant Cell 7:1099–1111

Bueno P, Piqueras A, Kurepa J, Savoure A, Verbruggen N, Montagu VM, Inze D (1998) Expression of antioxidant enzymes in responses to abscisic acid and high osmoticum in tobacco BY-2 cell cultures. Plant Sci 138:27–34

Cheeseman JM (1988) Mechanism of salinity tolerance in plants. Plant Physiol 87:547–550

Chen THH, Murata N (2011) Glycinebetaine protects plants against abiotic stress: mechanisms and biotechnological applications. Plant Cell Environ 34:1–20

Cherian S, Reddy MP, Pandya JB (1999) Studies on salt tolerance in *Avicennia marina* (Forstk.) Vierh: effect of NaCl salinity on growth, ion accumulation and enzyme activity. Indian J Plant Physiol 4:266–270

Comba ME, Benavides MP, Tomaro ML (1998) Effect of salt stress on antioxidant defence system in soybean root nodules. Aust J Plant Physiol 25:665–671

Cram WJ (1976) Negative feedback regulation of transport in cells. The maintenance of turgor volume and nutrient supply. In: Encyclopedia of plant physiology, Transport in plants 2, Part A. Cells 2: 284–316

Dadkhah AR (2010) Effect of Long Term Salt Stress on Gas Exchange and Leaf Carbohydrate Contents in Two Sugar Beet (Beta vulgaris L.) Cultivars. Res J Biol Sci 5(8):512–516

Diamant S, Eliahu N, Rosenthal D, Goloubino P (2001) Chemical chaperones regulate molecular chaperones in vitro and in cells under combined salt and heat stresses. J Biol Chem 276:39586–39591

Dionisio-Sese ML, Tobita S (2007) Antioxidant responses of rice seedlings to salinity stress. Plant Sci 135:1–9

Elkahouia S, Hernández JA, Abdellyc C, Ghrira R, Limama F (2005) Effects of salt on lipid peroxidation and antioxidant enzyme activities of *Catharanthus roseus* suspension cells. Plant Sci 168:607–613

Enteshari S, Alishavandi R, Delavar K (2011) Interactive effects of silicon and NaCl on some physiological and biochemical parameters in Borago officinalis L. Iranian J Plant Physiol 2(1):315–320

FAO (2007) Extent and causes of salt-affected soils in participating countries. Global network on integrated soil management for sustainable use of salt-affected soils, AGL

Fan W, Zhang M, Zhang H, Zhang P (2012) Improved Tolerance to Various Abiotic Stresses in Transgenic Sweet Potato (Ipomoea batatas) Expressing Spinach Betaine Aldehyde Dehydrogenase. PLoS ONE 7(5): e37344

Flowers TJ, Yeo AR (1995) Breeding for salinity tolerance in crop plants: where next? Aust J Plant Physiol 22:875–884

Flowers T, Troke PF, Yeo AR (1977) The mechanism of salt tolerance in halophytes. Ann Rev Plant Physiol 28:89–121

Foyer CH, Halliwell B (1976) Presence of glutathione and glutathione reductase in chloroplasts: a proposed role in ascorbic acid metabolism. Planta 133:21–25

Gadallah MAA (1999) Effects of proline and glycinebetaine on Vicia faba responses to salt stress. Biol Plant 42:249–257

Gallardo M, Gallardo ME, Matilla AJ, de Ruedo PM, Sanchez-Calle IM (1995) Inhibition of polyamine synthesis by cyclohexylamine stimulates the ethylene pathway and accelerates the germination of *Cicer arietinum* seeds. Physiol Plant 91:9–16

Gao M, Sakamoto A, Miura K, Murata N, Sugiura A, Tao R (2000) Transformation of Japanese persimmon (Diospyros kaki Thunb.) with a bacterial gene for choline oxidase. Mol Breed 6:501–510

Gorham J (1995) Mechanism of salt tolerance of halophytes. In: Choukr Allah CR, Malcolm CV, Handy A (eds) Halophytes and biosaline agriculture. Marcel Dekker, New York, pp 207–223

Greenway H, Munns R (1980) Mechanisms of salt tolerance in non-halophytes. Annu Rev Plant Physiol 31:149–190

Grumet R, Hanson AD (1986) Glycinebetaine accumulation in barley. Aust J Plant Physiol 13:353–364

Gueta-Dahan Y, Yaniv Z, Zilinskas BA, Ben-Hayyim G (1997) Salt and oxidative stress: similar and specific responses and their relation to salt tolerance in citrus. Planta 203:460–469

Gupta SK, Gupta IC (1987) Land development and leaching in management of saline soils and waters. Mohan Primlani, New Dehli, pp 136–152

Hanson AD, Rathinasabapathi B, Rivoal J, Burnet M, Dillon MO, Gage DA (1994) Osmoprotective compound in the Plumbaginaceae: a natural experiment in metabolic engineering of stress tolerance. Proc Natl Acad Sci USA 91:306–310

Hasegawa PM, Bressan RA, Zhu JK, Bohnert HJ (2000) Plant cellular and molecular responses to high salinity. Annu Rev Plant Physiol/Plant Mol Biol 51:463–499

Hayashi H, Mustardy L, Deshnium P, Ida M, Murata N (1997) Transformation of Arabidopsis thaliana with the codA gene for choline oxidase; accumulation of glycinebetaine and enhanced tolerance to salt and cold stress. Plant J 12:133–142

Heidari M, Sarani S (2012) Growth, biochemical components and ion content of Chamomile (Matricariachamomilla L.) under salinity stress and iron deficiency. J Saudi Soc Agricul Sci 11(1):37–42

Hendawy SF, El-Sherbeny SE, Hussein MS, Youssef AA (2012) Evaluation of some cultivars of foxtail plants under salinity conditions. J Applied Sci Res 8(2): 620–627

Hernandez JA, Jimenez A, Mullineaux P, Sevilla F (2000) Tolerance of pea (*Pisum sativum* L.) to long-term salt stress is associated with induction of antioxidant defences. Plant Cell Environ 23:853–862

Huang J, Hirji R, Adam L, Rozwadowski KL, Hammerlindl JK, Keller WA, Selvaraj G (2000) Genetic Engineering of Glycinebetaine Production toward Enhancing Stress Tolerance in Plants: Metabolic Limitations. Plant Physiol 122:747–756

Hurkman WJ, Fornari CS, Tanaka CK (1989) A comparison of the effect of salt on polypeptide and translatable mRNA in roots of a salt tolerant and salt sensitive cultivar of barley. Plant Physiol 90:1444–1456

Hurkman WJ, Rao HP, Tanaka CK (1991) Germin like polypeptides increase in barley roots during salt stress. Plant Physiol 97:366–374

Imlay JA (2003) Pathways of oxidative damage. Annu Rev Microbiol 57:395–418

Iyer S, Caplan A (1998) Products of proline catabolism can induce osmotically regulated genes in rice. Plant Physiol 116:203–211

Jaleel CA, Gopi R, Manivannan P, Panneerselvam R (2007a) Responses of antioxidant defense system of *Catharanthus roseus* (L.) G. Don. to paclobutrazol treatment under salinity. Acta Physiol Planta 29:205–209

Jaleel CA, Manivannan P, Sankar B, Kishorekumar A, Gopi R, Somasundaram R, Panneerselvam R (2007b) Water deficit stress mitigation by calcium chloride in *Catharanthus roseus*: effects on oxidative stress, proline metabolism and indole alkaloid accumulation. Coll Surf B: Biointerf 60:110–116

Jaleel CA, Manivannan P, Sankar B, Kishorekumar A, Gopi R, Somasundaram R, Panneerselvam R (2007c) *Pseudomonas fluorescens* enhances biomass yield and ajmalicine production in *Catharanthus roseus* under water deficit stress. Coll Surf B: Biointerf 60:7–11

Jaleel CA, Lakshmanan GMA, Gomathinayagam M, Panneerselvam R (2008) Triadimefon induced salt stress tolerance in *Withania somnifera* and its relationship to antioxidant defense system. South Africa J Botany 74(1):126–132

Jamil M, Ashraf M, Rehman S, Ahmad M, Rha ES (2012a) Salinity induced changes in cell membrane stability, protein and RNA contents. Afr J Biotechnol 11(24):6476–6483

Jamil M, Bashir S, Anwar S, Bibi S, Bangash A, Ullah F, Rha ES (2012b) Effect of salinity on physiological and biochemical characteristics of different varieties of rice. Pak J Bot 44:7–13
Kemble AR, Macpherson HT (1954) Sorbitol and proline as intracellular osmotic solutes in the green alga *Stichococcus bacillaris*. Can J Bot 56:676–679
Khan MH, Panda SK (2008) Alterations in root lipid peroxidation and antioxidative responses in two rice cultivars under NaCl-salinity stress. Acta Physiol Plant 30:91–89
Khan MH, Singha LB, Panda SK (2002) Changes in antioxidant levels in *Oriza sativa* L. roots subjected to NaCl-salinity stress. Acta Physiol Plant 24:145–148
Khan MA, Shirazi MU, Ali M, Mumtaz S, Sherin A, Ashraf MY (2006) Comparative performance of some wheat genotypes growing under saline water. Pak J Bot 38:1633–1639
Khedr AHA, Abbas MA, Wahid AAA, Quick WP, Abogadallah GM (2003): Proline induces the expression of salt stress responsive proteins and may improve the adaptation of Pancratium maritimum L. to salt stress. J Exp Bot 54:2553–2562
Koca H, Bor M, Ozdemir F, Turkan I (2007) The effect of salt stress on lipid peroxidation, antioxidative enzymes and proline content of sesame cultivars. Env Exp Bot 60:344–351
Koyro HW, Ahmad P, Geissler N (2012) Abiotic stress responses in plants: an overview. In: Ahmad P, Prasad MNV (eds) Environmental adaptations and stress tolerance of plants in the era of climate change. Springer Science + Business Media, New York, pp 1–28
Lechno S, Zamski E, Telor E (1997) Salt stress-induced responses in cucumber plants. J Plant Pysiol 150(1–2):206–211
Lee DH, Kim YS, Lee CB (2001) The inductive responses of the antioxidant enzymes by salt stress in the rice (*Oryza sativa*). J Plant Physiol 158:737–745
Lefevre I, Lutts S (2000) Effects of salt and osmotic stress on free polyamine accumulation in moderately salt resistant rice cultivar AIW 4. Int Rice Res Note 25:36–37
Le-Houerou HN (1986) Salt-tolerant plants of economic value in the Mediterranean basin. Reclam Reveg Res 5:319
Lin CC, Kao CH (1995) Levels of endogenous polyamines and NaCl inhibited growth of rice seedlings. Plant Growth Regul 17:15–20
Lopez F, Vansuyt G, Fourcroy P, Case-Delbart F (1994) Accumulation of a 22-kDa protein and its mRNA in the leaves of *Raphanus sativus* in response to salt stress or water stress. Physiol Plant 91:605–614
Luo D, Niu X, Yu J, Yan J, Gou X, Lu BR, Liu Y (2012) Rice choline monooxygenase (OsCMO) protein functions in enhancing glycine betaine biosynthesis in transgenic tobacco but does not accumulate in rice (Oryza sativa L. ssp. japonica). Plant Cell Rep DOI: 10.1007/s00299-012-1276-2
Mane AV, Deshpande TV, Wagh VB, Karadge BA, Samant JS (2011) A critical review on physiological changes associated with reference to salinity. Inter J Environ Sci 1:1192–1216
Mansour MMF (2000) Nitrogen containing compounds and adaptation of plants to salinity stress. Biol Plant 43:491–500
Masood A, Shah NA, Zeeshan M, Abraham G (2006) Differential response of antioxidant enzymes to salinity stress in two varieties of Azolla (*Azolla pinnata* and *Azolla tilieuloides*). Env Exper Bot 58:216–222
Mc Cue KF, Hanson AD (1990) Salt-inducible betaine aldehyde dehydrogenase from sugar beet: cDNA cloning and expression. Trends Biotechnol 8:358–362
Mc Neil SD, Nuccio ML, Hanson AD (1999) Betaines and related osmoprotectants: targets for metabolic engineering of stress resistance. Plant Physiol 120:945–949
Meloni DA, Gulotta MR, Martinez CA, Oliva M (2004) The effects of salt stress on growth, nitrite reduction and proline and glycinebetaine accumulation in Prosopis alba. Braz J Plant Physiol 16:39–46
Miflin B (2000) Crop improvement in the 21st century. J Exp Bot 51:1–8
Mittal S, Kumari N, Sharma V (2012) Differential response of salt stress on Brassica juncea: photosynthetic performance, pigment, proline, D1 and antioxidant enzymes. Plant Physiol Biochem 54:17–26
Mittova V, Volokita M, Guy M, Tal M (2000) Activities of SOD and the ascorbate-glutathione cycle enzymes in sub-cellular compartments in leaves and roots of the cultivated tomato and its wild salt tolerant relative *Lycopersicon pennellii*. Physiol Plant 110:42–51

Mittova V, Tal M, Volokita M, Guy M (2002) Salt stress induces up-regulation of an efficient chloroplast antioxidant system in the salt-tolerant wild tomato species *Lycopersicon pennel* but not in the cultivated species. Plant Physiol 115:393–400

Mohanty A, Kathuria H, Ferjani A, Sakamoto A, Mohanty P, Murata N, Tyagi AK (2002). Transgenics of an elite indica rice variety Pusa Basmati 1 harbouring the codA gene are highly tolerant to salt stress. Theor. Appl. Genet. 106:51–57

Mullineaux PM, Rausch T (2005) Glutathione, Photosynthesis and the redox regulation of stress responsive gene expression. Photosynth research 47:459–474

Murakeozy EP, Nagy Z, Duhaze C, Bouchereau A, Tuba Z (2003) Seasonal changes in the levels of compatible osmolytes in three halophytic species of inland saline vegetation in Hungary. J Plant Physiol 160:395–401

Panda SK, Khan MH (2003) Salt stress influences lipid peroxidation and antioxidants in the leaf of an indica rice (Oryza sativa L). Physiol Mol Biol Plants 9:273–278

Pareek A, Singla SL, Grover A (1997) Salt responsive proteins/genes in crop plants. In: Jaiwal PK, Singh RP, Gulati A (eds) Strategies for improving salt tolerance in higher plants. Oxford and IBH Publication, New Delhi, pp 365–391

Park EJ, Jeknic Z, Sakamoto A, DeNoma J, Yuwansiri R, Murata N, Chen THH (2004) Genetic engineering of glycinebetaine synthesis in tomato protects seeds, plants, and flowers from chilling damage. Plant J 40:474–487

Parida AK, Das AB (2005) Salt tolerance and salinity effect on plants: a review. Ecotoxicol Environ Saf 60:324–349

Parida AK, Das AB, Das P (2002): NaCl stress causes changes in photosynthetic pigments, proteins and other metabolic components in the leaves of a true mangrove, Bruguiera parviflora, in hydroponic cultures. J Plant Biol 45:28–36

Postel S (1999) Redesigning irrigated agriculture. In: Brown LR, Flavin C, French H (eds) State of the world 2000. WW Norton, New York/London, pp 39–58

Prasad KVSK, Sharmila P, Kumar PA, Pardha Saradhi P (2000) Transformation of Brassica juncea (L.) Czern with bacterial codA gene enhances its tolerance to salt stress. Mol Breed 6:489–499

Qadir M, Qureshi RH, Ahmad N, Ilyas M (1996) Salt-tolerant forage cultivation on a saline sodic field for biomass production and soil reclamation. Land Degrad Devel 7(2):11–18

Quintero FJ, Garciadeblas B, Rodriguez-Navarro A (1996): The SAL1 gene of Arabidopsis, encoding an enzyme with 3'(2'),5'-bisphosphate nucleotidase and inositol polyphosphate 1-phosphatase activities, increases salt tolerance in yeast. Plant Cell 8:529–537

Rahdari P, Tavakoli S, Hosseini SM (2012) Studying of salinity stress effect on germination, proline, sugar, protein, lipid and chlorophyll content in Purslane (Portulaca oleracea L.) Leaves. J Stress Physiol Biochem 8(1):182–193

Rajaravindran M, Natarajan S (2012) Effect of NaCl Stress on Biochemical and Enzymes Changes of the Halophyte Suaeda maritima Dum. Inter J Res Plant Sci 2(1):1–7

Reddy AR, Chaitanya KV, Vivekanandan M (2004) Drought-induced response of photosynthesis and antioxidant metabolism in higher plants. J Plant Physiol 161:1189–1202

Rhoades G (1998) Managed professionals: unionized faculty and restructuring academic labor. State University of New York Press, Albany

Rhodes D, Hanson AD (1993) Quaternary ammonium and tertiary sulphonium compounds in higher plants. Annu Rev Plant Physiol Mol Biol 44:357–384

Rhodes D, Rich PJ, Brunk DG, Ju GC, Rhodes JC, Pauly MH, Hansen LA (1989) Development of two isogenic sweet corn hybrids differing for glycinebetaine content. Plant Physiol 91:1112–1121

Richards LA (1954) Diagnosis and improvement of saline and alkali soils. USDA Agriculture Handbook 60, Washington D.C.

Ruiz JM, Blumwald E (2002) Salinity induced glutathione synthesis in *Brassica napus*. Planta 214:965–969

Sairam RK, Rao KV, Srivastava GC (2002) Differential response of wheat genotypes to long term salinity stress in relation to oxidative stress, antioxidant activity and osmolyte concentration. Plant Sci 163:1037–1046

Savouré A, Thorin D, Davey M, Hua XJ, Mauro S, Van Montagu M, Inzé D, Verbruggen N (1999) NaCl and $CuSO_4$ treatments trigger distinct oxidative defense mechanism in *Nicotiana plumbaginifolia* L. Plant Cell Environ 22:387–396

Scandalios JG (2005) Oxidative stress: molecular perception and transduction of signals triggering antioxidant gene defenses. Braz J Med Biol Res 38:995–1014

Scandalios JG, Guan L, Polidoros AN (1997) Catalases in plants: gene structure, properties, regulation and expression. In: Scandalios JG (ed) Oxidative stress and the molecular biology of antioxidant defenses. Cold Spring Harbor Laboratory, Cold Spring Harbor, New York, pp 343–406

Sekmen AH, Turkan I, Takio S (2007) Difference in responses of antioxidative enzymes and lipid peroxidation to salt stress in salt tolerant *Plantago martima* and salt sensitive *Plantago media*. Physiol Plant 131:399–411

Shalata A, Neumann PM (2001) Exogenous ascorbic acid (vitamin c) increases resistance to salt tolerance and reduced lipid peroxidation. J Exp Bot 364:2207–2211

Sharma SS, Dietz KJ (2009) The relationship between metal toxicity and cellular redox imbalance. Trends Plant Sci 14:43–50

Shi QH, Zhu ZJ (2008) Effects of exogenous salicylic acid on manganese toxicity, element contents and antioxidative system in cucumber. Environ Exp Bot 63:317–326

Shirasawa K, Takabe T, Takabe T, Kishitani S (2006) Accumulation of Glycinebetaine in Rice Plants that Overexpress Choline Monooxygenase from Spinach and Evaluation of their Tolerance to Abiotic Stress. Ann Bot 98 (3):565–571

Singer MA, Lindquist S (1998) Multiple effects of trehalose on protein folding in vitro and in vivo. Mol Cell 1:639–648

Singh G (2009) Salinity-related desertification and management strategies: Indian experience. Land Degrad Develop (Special Issue: Desertification) 20(4):367–385

Singh KN, Sharma PC (2006) Salt tolerant varieties released for saline and alkali soils. Central Soil Salinity Research Institute, Karnal

Singh NK, Bracken CA, Hasegawa PM, Handa AK, Buckel S, Hermodson MA, Pfankoch F, Regnier FE, Bressan RA (1987) Characterization of osmotin. A thaumatin-like protein associated with osmotic adjustment in plant cells. Plant Physiol 85:529–536

Smirnoff N (2000) The role of active oxygen in the response of plants to water deficit and desiccation. New Phytol 125:27–58

Sorkheh K, Shiran B, Rouhi V, Khodambashi M, Sofo A (2012) Salt stress induction of some key antioxidant enzymes and metabolites in eight Iranian wild almond species. Acta Physiol Plant 34:203–213

Srivalli B, Chinnusamy V, Khanna-Chopra R (2003) Antioxidant defense in response to abiotic stresses in plants. J Plant Biol 30:121–139

Stockle CO (2001) Environmental impact of irrigation. In: IV International Congress of agricultural engineering, Chillan, Chile, 9–11 May 2001

Sudhakar C, Lakshmi A, Giridara Kumar S (2001) Changes in the antioxidant enzyme efficacy in two high yielding genotypes of mulberry (*Morus alba* L.) under NaCl salinity. Plant Sci 161:613–619

Su J, Hirji R, Zhang L, He C, Selvaraj G, Wu R (2006) Evaluation of the stress-inducible production of choline oxidase in transgenic rice as a strategy for producing the stress-protectant glycine betaine. J. Exp. Bot. (March 2006) 57 (5):1129–1135

Sulpice R, Tsukaya H, Nonaka H, Mustardy L, Chen THH, Murata N (2003) Enhanced formation of flowers in salt-stressed Arabidopsis after genetic engineering of the synthesis of glycine betaine. Plant J 36:165–176

Szabolcs I (1994) Soil salinization. In: Pessarakli M (ed) Handbook of plant crop stress. Marcel Dekker, New York, pp 3–11

Tanji KK (1990) Salinity assessment and management. Manual and Reports on Engineering Practices 71, American Society of Civil Engineers, New York

Temple MD, Perrone GG, Dawes IW (2005) Complex cellular responses to reactive oxygen species. Trends Cell Biol 15:319–326

Thomas JC, Bohnert HJ (1993) Salt stress perception and plant growth regulators in the halophyte *Mesembryanthemum crystallinum*. Plant Physiol 103:1299–1304

Triantaphylidès C, Krischke M, Hoeberichts FA, Ksas B, Gresser G, Havaux M, Van Breusegem F, Mueller MJ (2008) Singlet oxygen is the major reactive oxygen species involved in photooxidative damage to plants. Plant Physiol 148:960–968

Tuteja N, Ahmad P, Panda BB, Tuteja R (2009) Genotoxic stress in plants: shedding light on DNA damage, repair and DNA repair helicases. Mutat Res 681:134–149

Uma S, Prasad TG, Kumar MU (1995) Genetic variability in recovery growth and synthesis of stress proteins in response to polyethylene glycol and salt stress in finger millet. Ann Bot 76:43–49

Verniquet F, Gaillard J, Neuburger M, Douce R (1991) Rapid inactivation of plant aconitase by hydrogen peroxide. Biochem J 276:643–648

Wang SL, Lin CY, Guo YL, Lin LY, Chou WL, Chang LW (2004) Infant exposure to polychlorinated dibenzo-*p*-dioxins, dibenzofurans and biphenyls (PCDD/Fs, PCBs)-correlation between prenatal and postnatal exposure. Chemosphere 54:1459–1473

Wang Y, Nil N (2000): Changes in chlorophyll, ribulose biphosphate carboxylase-oxygenase, glycine betaine content, photosynthesis and transpiration in Amaranthus tricolor leaves during salt stress. J Hortic Sci Biotechnol 75:623–627

Wiggins PM (1990) Role of water in some biological processes. Microbiol Rev 54:432–449

Willekens H, Inze D, Van Montagu M, Van Camp W (1995) Catalases in plants. Mol Breed 1: 207–228

Yadav S, Irfan M, Ahmad A, Hayat S (2011) Causes of salinity and plant manifestations to salt stress: a review. J Environ Biol 32:667–685

Yamada T, Takatsu Y, Manabe T, Kasumi M, Marubashi W (2003) Suppressive effect of trehalose on apoptotic cell death leading to petal senescence in ethylene-insensitive flowers of gladiolus. Plant Sci 164:213–221

Yang X, Liang Z, Wen X, Lu C (2008) Genetic engineering of the biosynthesis of glycinebetaine leads to increased tolerance of photosynthesis to salt stress in transgenic tobacco plants. Plant Mol Biol 66:73–86

Yeo AR (1999) Predicting the interaction between the effects of salinity and climate change on crop plants. Sci Hortic (Amsterdam) 78:159–174

Yousuf PY, Hakeem KR, Chandna R, Ahmad P (2012) Role of Glutathione Reductase in Plant Abiotic Stress. In: Ahmad P, Prasad MNV (eds) Abiotic stress responses in plants: metabolism, productivity and sustainability. Springer Science + Business Media, New York, pp 149–158

Zhang J, Tan W, Yang XH, Zhang HX (2008) Plastid-expressed choline monooxygenase gene improves salt and drought tolerance through accumulation of glycine betaine in tobacco. Plant Cell Rep 27:1113–1124

Zhou S, Chen X, Zhang X, Yinxin Li (2008) Improved salt tolerance in tobacco plants by co-transformation of a betaine synthesis gene BADH and a vacuolar Na+/H+ antiporter gene SeNHX1. Biotechnol Lett 30:369–376

Zhou SF, Chen XY, Xue XN, Zhang XG, Li YX (2007) Physiological and Growth Responses of Tomato Progenies Harboring the Betaine Alhyde Dehydrogenase Gene to Salt Stress. J Integ Plant Biol 49:628–637

Zhu BC, Su J, Chang MC, Verma DPS, Fan YL, Wu R (1998) Over expression of a delta-pyrroline-5-carboxylate synthetase gene and analysis of tolerance to water and salt stress in transgenic rice. Plant Sci 139:41–48

Chapter 2
Plant Response to Salt Stress and Role of Exogenous Protectants to Mitigate Salt-Induced Damages

Mirza Hasanuzzaman, Kamrun Nahar, and Masayuki Fujita

2.1 Introduction

World agriculture is facing a lot of challenges like producing 70% more food for an additional 2.3 billion people by 2050 while at the same time fighting with poverty and hunger, consuming scarce natural resources more efficiently and adapting to climate change (FAO 2009). However, the productivity of crops is not increasing in parallel with the food demand. The lower productivity in most of the cases is attributed to various abiotic stresses. Curtailing crop losses due to various environmental stressors is a major area of concern to cope with the increasing food requirements (Shanker and Venkateswarlu 2011).

As a sessile organism, plants often experience abiotic stress like salinity, drought, high or low temperature, flooding, metal toxicity, ozone, UV-radiations, herbicides, etc., which pose serious threat to the crop production (Bhatnagar-Mathur et al. 2008; Ahmad and Prasad 2012a, b). The complex nature of the environment along with its

M. Hasanuzzaman
Laboratory of Plant Stress Responses, Department of Applied Biological Science, Faculty of Agriculture, Kagawa University, Miki-cho, Kita-gun, Kagawa 761-0795, Japan

Department of Agronomy, Faculty of Agriculture, Sher-e-Bangla Agricultural University, Sher-e-Bangla Nagar, Dhaka 1207, Bangladesh
e-mail: mhzsauag@yahoo.com

K. Nahar
Laboratory of Plant Stress Responses, Department of Applied Biological Science, Faculty of Agriculture, Kagawa University, Miki-cho, Kita-gun, Kagawa 761-0795, Japan

Department of Agricultural Botany, Faculty of Agriculture, Sher-e-Bangla Agricultural University, Sher-e-Bangla Nagar, Dhaka 1207, Bangladesh

M. Fujita (✉)
Laboratory of Plant Stress Responses, Department of Applied Biological Science, Faculty of Agriculture, Kagawa University, Miki-cho, Kita-gun, Kagawa 761-0795, Japan
e-mail: fujita@ag.kagawa-u.ac.jp

P. Ahmad et al. (eds.), *Ecophysiology and Responses of Plants under Salt Stress*,
DOI 10.1007/978-1-4614-4747-4_2,

unpredictable conditions and global climate change are increasing gradually which is creating the situation more adverse (Mittler and Blumwald 2010). Abiotic stresses remain the greatest constraint to crop production worldwide. It has been projected that more than 50% of yield reduction is the direct result of abiotic stresses (Rodríguez et al. 2005; Acquaah 2007). The major abiotic stresses like drought, high salinity, cold, and heat negatively influence the survival, biomass production and yield of staple food crops up to 70% (Vorasoot et al. 2003; Kaur et al. 2008; Ahmad et al. 2010a; Thakur et al. 2010; Mantri et al. 2012; Ahmad et al. 2012); hence, threaten the food security worldwide.

Salinity is one of the most brutal environmental factors limiting the productivity of crop plants because most of the crop plants are sensitive to salinity caused by high concentrations of salts in the soil. A considerable amount of land in the world is affected by salinity which is increasing day by day. More than 45 million hectares (M ha) of irrigated land which account to 20% of total land have been damaged by salt worldwide and 1.5 M ha are taken out of production each year due to high salinity levels in the soil (Pitman and Läuchli 2002; Munns and Tester 2008). On the other hand, increased salinity of agricultural land is expected to have destructive global effects, resulting in up to 50% loss of cultivable lands by the middle of the twenty-first century (Mahajan and Tuteja 2005). In most of the cases, the negative effects of salinity have been attributed to increase in Na^+ and Cl^- ions in different plants hence these ions produce the critical conditions for plant survival by intercepting different plant mechanisms. Although both Na^+ and Cl^- are the major ions which produce many physiological disorders in plants, Cl^- is the most dangerous (Tavakkoli et al. 2010). Salinity at higher levels causes both hyperionic and hyperosmotic stress and can lead to plant demise. The outcome of these effects may cause membrane damage, nutrient imbalance, altered levels of growth regulators, enzymatic inhibition and metabolic dysfunction, including photosynthesis which ultimately leads to plant death (Mahajan and Tuteja 2005; Hasanuzzaman et al. 2012a). High salt concentration in the soil or in the irrigation water can also have a devastating effect on plant metabolism, disrupting cellular homeostasis and uncoupling major physiological and biochemical processes. Biochemical and molecular studies of salt stress responses in plants have revealed significant increases of reactive oxygen species (ROS), including singlet oxygen ($^1O_2$), superoxide (O_2^-), hydroxyl radical ($OH^\bullet$) and hydrogen peroxide (H_2O_2) (Tanou et al. 2009; Ahmad et al. 2010a, 2012; Ahmad and Umar 2011). However, the effect of salt stress on plants depends on the concentration and time of exposure of salt, plant genotypes and environmental factors.

Mechanisms of salt tolerance, not yet completely clear, can be explained to some extent by stress adaptation effectors that mediate ion homeostasis, osmolyte biosynthesis, toxic radical scavenging, water transport and long distance response co-ordination (Hasegawa et al. 2000). However, attempts to improve yield under stress conditions by plant improvement have been largely unsuccessful, primarily due to the multigenic origin of the adaptive responses. Therefore, a well-focused approach combining the molecular, physiological, biochemical and metabolic aspects of salt tolerance is essential to develop salt-tolerant crop varieties. Exploring suitable ameliorants or stress alleviant is one of the tasks of plant biologists. In recent decades exogenous protectant such as osmoprotectants (proline, glycinebetaine, trehalose, etc.), plant hormone (gibberellic acids, jasmonic acids, brassinosterioids, salicylic acid, etc.), antioxidants (ascorbic acid, glutathione, tocopherol, etc.), signaling molecules (nitric oxide, hydrogen peroxide, etc.), polyamines (spermidine, spermine, putrescine), trace elements (selenium, silicon, etc.) have been found effective in mitigating the salt induced damage in plant (Hoque et al. 2007; Ahmad et al. 2010a, 2012; Azzedine et al. 2011; Hasanuzzaman et al. 2011a, b; Hayat and Ahmad 2011; Hossain et al. 2011; Poór et al. 2011; Ioannidis et al. 2012; Nounjan et al. 2012; Rawia et al. 2011; Iqbal et al. 2012; Tahir et al. 2012; Yusuf et al. 2012). These protectants showed the capacity to enhance the plant's growth, yield as well as stress tolerance under salinity.

This chapter provides a comprehensive review of the major responses of plants to saline environments and the mechanisms by which growth and development and physiology of plants are affected by salinity. We also discuss the nature and types of salinity and the possible mechanism of salt stress in plants. Finally, we focus the issue of using exogenous protectants to mitigate the salt-induced damages in plants.

2.2 Causes and Types of Salinity

Among the abiotic stresses, salinity is the most destructive factor which limits the crop productivity considerably. A large area of land in the world is affected by salinity which is increasing day by day. Salinity is more prominent problem in irrigated crop lands. Worldwide, around 17% of the cultivated land is under irrigation and irrigated agriculture contributes more than 30% of the total agricultural production (Hillel 2000). It is estimated that at least 20% of total irrigated lands in the world is salt-affected (Pitman and Läuchli 2002). However, the statistics varies depending on sources. According to the FAO Land and Nutrition Management Service (2008), 6.5% of the total land in the world is affected by salt (either salinity or sodicity) which accounts for 831 Mha of land (Table 2.1).

There are different causes of the development of soil salinity. The major forms are viz. (i) natural or primary salinity and (ii) secondary or human-induced salinity. Primary salinity is occurred due to the long-term natural accumulation of salts in the soil or surface water. This is a natural process which is caused mainly by weathering of parent materials containing soluble salts through break down of rocks containing Cl^- of Na^+, Ca^{2+} and Mg^{2+} and sometimes SO_4^{2-} and CO_3^{2-}. In addition, deposition of sea salt carried by wind and rain is also a reason, which varies with the types of soil. Secondary salinity occurs due to anthropogenic activities that disrupt the hydrologic balance of the soil between water applied (irrigation or rainfall) and water used by crops (transpiration) (Munns 2005; Garg and Manchanda 2008). In many irrigated areas, the water table has raised due to excessive amounts of applied water together with insufficient drainage. Most of the irrigation systems of the world have caused secondary salinity, sodicity or waterlogging (Garg and

Table 2.1 Variation in salt-affected areas in the world, in million hectares (M ha)

Region	Total area (M ha)	Saline soils		Sodic soils	
		M ha	%	M ha	%
Africa	1,899	39	2.0	34	1.8
Asia, the Pacific and Australia	3,107	195	6.3	249	8.0
Europe	2,011	7	0.3	73	3.6
Latin America	2,039	61	3.0	51	2.5
Near East	1,802	92	5.1	14	0.8
North America	1,924	5	0.2	15	0.8
Total	12,781	397	3.1	434	3.4

Source: FAO land and plant nutrition service (2008)

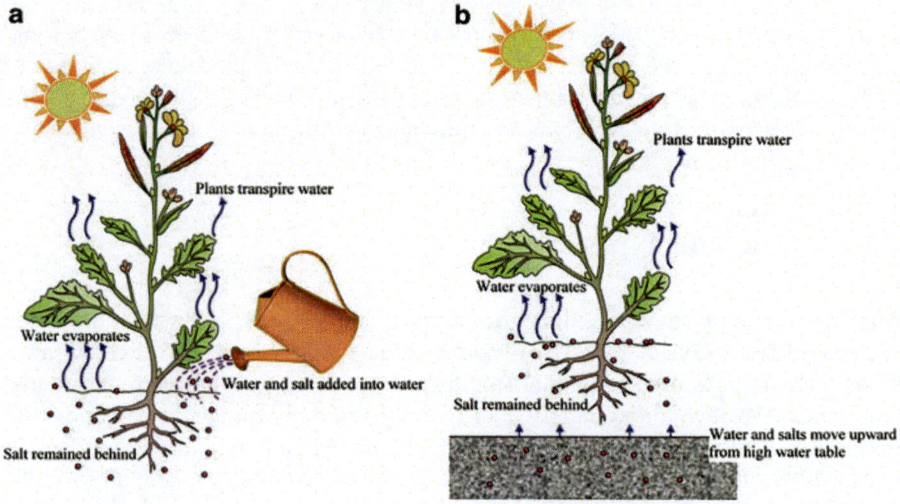

Fig. 2.1 Salinization in crop lands caused by salty irrigation water or rise of water table with saline water

Manchanda 2008). In irrigated lands, after irrigation, the water applied to the soil is consumed by the crop or evaporates directly from the moist soil. The excess salt is remained and accumulated in the soil which is called salinization (Fig. 2.1a). It is sometimes recognizable by a whitish layer of dry salt on the soil surface. In addition, salted groundwater may also contribute to salinization. Due to excessive irrigation and improper drainage the water table rises which allow the salty groundwater to reach in the upper soil layers and rhizosphere (Fig. 2.1b).

Based on the nature, characteristics and plant growth relationships in salt affected soils, two main types of soils have been coined by Szabolcs (1974). These are:

A) Saline soils-The soluble salts are chiefly NaCl and Na_2SO_4 and sometimes also contain appreciable quantities of Cl^- and SO_4^- of Ca^{2+} and Mg^{2+}. These soils contain sufficient neutral soluble salts to pose negative effect on growth of most crop plants.

Table 2.2 Classification of water quality based on total salt concentration (Pitman and Läuchli 2002)

Water designation	Total dissolved salts (mg L^{-1})	EC (dS m^{-1})
Fresh water	<500	<0.6
Slightly brackish	500–1,000	0.6–1.5
Brackish	1,000–2,000	1.5–3.0
Moderately saline	2,000–5,000	3.0–8.0
Saline	5,000–10,000	8.0–15.0
Highly saline	10,000–35,000	15.0–45.0

Table 2.3 Different measures of soil salinity

Measurement and units	Application
Conductivity (dS m^{-1})	Soils
Conductivity (μS cm^{-1})	Irrigation and river water
Total dissolved salts (mg L^{-1})	Irrigation and river water
Molarity of NaCl (mM)	Laboratory

B) Sodic soils – These soils contain Na^+ salts capable of alkaline hydrolysis, mainly Na_2CO_3. Previously these soils have also been termed as 'Alkali'.

Further categories of salt-affected soils which, though less extensive, are commonly found in different parts of the world are:

C) Acid-sulfate soils: These soils have pH below 3.5 to 4.0 and found within a 50 cm depth that is directly or indirectly caused by H_2SO_4 formed by the oxidation of pyrite (FeS_2) or other reduced S compounds which is accelerated by brackish and saline mangrove swamps. Apart from high salinity, this soil also responsible for iron (Fe) and aluminium (Al) toxicities and deficiency of phosphorus (P) (Pons 1973; Abrol et al. 1988).

D) Degraded sodic soils: These soils are an advanced stage of soil development coming from the washing out of salts. In this process there is a affinity for the dispersed clay and organic matter to move down the profile resulting in the formation of a dark, extremely compact layer having a sharply defined upper surface and merging gradually into the subsoil with increasing depth. These soils originally had enough exchangeable Na^+ but that most of this Na^+ has been lost through leaching (Abrol et al. 1988).

Based on the salt concentration, the saline water is classified into different types as presented in Table 2.2.

Soil salinity is measured by electrical conductivity (EC). The international system (SI) unit of EC is dS m^{-1}. Salinity is also measured as mM which is vastly used in laboratory experiment (Table 2.3). In the field, the salinity of soil water or irrigation water is measured in terms of its electrical conductivity or in terms of osmotic potential. Pure water is a very poor conductor of electric current; the conductivity of a water sample is due to the ions dissolved in it. Generally, the higher the salt concentration in water, the greater it's electrical conductivity and the lower its osmotic potential/pressure (Taiz and Zeiger 2006).

2.3 Nature and Mechanisms of Salt Stress

Most crops do not grow well on soils that contain salts. One reason is that salt causes a reduction in rate and amount of water that plant roots can take up from the soil. Also, some salts are toxic to plants when present in high concentration. The highly tolerant crops can withstand a salt concentration of the saturation extract up to 10 $g L^{-1}$. The moderately tolerant crops can withstand salt concentration up to 5 $g L^{-1}$. The limit of the sensitive group is about 2.5 $g L^{-1}$ (Brouwer et al. 1985). Some plants are more tolerant to a high salt concentration than others. Some examples are given in the Table 2.4.

Some of the negative effects of salinity have been caused mainly by Na^+ and Cl^- ions in plants and these ions produce the decisive conditions for plant survival by intercepting different plant mechanisms. Plant roots are generally affected due to Na^+ and Cl^- along with other cations present in the soils in different concentration (1–150 mM for glycophytes; more for halophytes). However, the uptake of these ions depends on the plant growth stage, genetic characters and environmental factors like temperature, relative humidity and light intensity. Excessive amount of salt in cultivated soils retards the growth, limits economic yield and even lead plants to death. There are some points at which salt transport is regulated. These are: (i) selective uptake from the soil solution, (ii) loading of xylem, (iii) removal of salt from the xylem in the upper part of the plant, (iv) loading of the phloem and (v) excretion through salt glands or bladders (Munns et al. 2002a, b; Fig. 2.2). For a salt tolerant plant growing for some time in a soil solution of 100 mM NaCl, the root concentrations of Na^+ and Cl^- are typically about 50 mM, the xylem concentration

Table 2.4 Major crops showing different salt-tolerance levels (Brouwer et al. 1985)

Highly tolerant	Moderately tolerant	Sensitive
Hordeum vulgare (Barley)	*Triticum aestivum* (Wheat)	*Pisum sativum* (Pea)
Beta vulgaris (Sugarbeet)	*Lycopersicon esculentum* (Tomato)	*Phaseolus* spp. (Beans)
Gossypium spp. (Cotton)	*Avena sativa* (Oat)	*Saccharum officinarum* (Sugarcane)
Asparagus spp.	*Medicago sativa* (Alfalfa)	*Trifolium pratense* (Red clover)
Spinacia oleracea (Spinach)	*Oryza sativa* (Rice)	*Pyrus communis* (Pear)
Phoenix dactylifera (Date palm)	*Zea mays* (Maize)	*Malus domestica* (Apple)
	Linum usitatissimum (Flax)	*Citrus aurantium* (Orange)
	Solanum tuberosum (Potato)	*Prunus* spp.
	Daucus carota (Carrot)	
	Allium cepa (Onion)	
	Cucumis sativus (Cucumber)	
	Punica granatum (Pomegranate)	
	Ficus carica (Fig)	
	Olea europaea (Olive)	
	Vitis vinifera (Grape)	

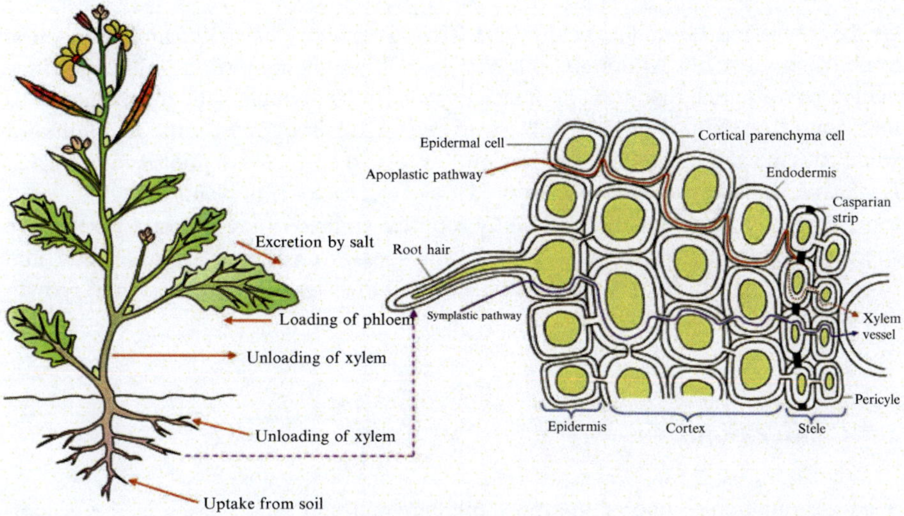

Fig. 2.2 Transport and regulation of salt in soil-plant system

about 5 mM, and the concentration in the oldest leaf as high as 500 mM (Munns 2002a). The toxic ions move into the plant with the water flow. The ions move from soil to the vascular system of the root by symplastic and apoplastic pathways. In symplastic pathway, water enters into the roots through plasma membranes of epidermis and further cell-to-cell movement occurs through plasmodesmata until the xylem becomes saturated. In apoplastic pathway, water enters through intracellular spaces to unload the salt in xylem (Fig. 2.2). Differential osmotic potential is the dynamic force of energy driven pathways, i.e. symplastic, while apoplastic is a non-energy driven pathway. Hence, based on osmotic potential, plant can control the toxic ions like Na^+ to enter into the cell through energy driven pathway (Garciadeblas et al. 2003).

2.4 Plant Responses to Salt Stress

High salinity causes both hyperionic and hyperosmotic stresses and can lead to plant death (Hasegawa et al. 2000). It is reported that plants growing under saline conditions are affected in three ways: reduced water potential in root zone causing water deficit, phytotoxicity of ions such as Na^+ and Cl^- and nutrient imbalance depressing uptake and transport of nutrients. Na^+ competes with K^+ for binding sites essential for cellular functions (Munns 2002a). Excess salt concentration also enhances the osmotic potential of soil matrix which restricts the water uptake by plants. Sodium is the primary toxic ion, because it interferes with K^+ uptake as well as and disturbs stomatal regulation which ultimately causes water loss and necrosis.

On the other hand, Cl^--induces chlorotic toxicity symptoms due to impaired production of chlorophyll (Chl). Although both Na^+ and Cl^- are the major ions which produce many physiological disorders in plants, especially Cl^-, which is the most dangerous than Na^+ (Tavakkoli et al. 2010). In plant cells, Cl^- is required for the regulation of some enzyme activities in the cytoplasm. It is also a co-factor in photosynthesis and is involved in turgor and pH regulation. However, it is toxic to plants at high concentrations, with critical levels for toxicity reported to be 4–7 mg g^{-1} for Cl^--sensitive species and 15–50 mg g^{-1} for Cl^--tolerant species (Xu et al. 2000; White and Broadley 2001). Higher accumulation of Cl^- led to a significant reduction in growth and water use efficiency in plants.

2.4.1 Germination

Seed germination is one of the most fundamental and vital phases in the growth cycle of plants that determine plant establishment and the yield of the crops. The available literature revealed the effects of salinity on the seed germination of various crops like *Oryza sativa* (Xu et al. 2011), *Triticum aestivum* (Akbarimoghaddam et al. 2011), *Zea mays* (Carpıcı et al. 2009; Khodarahmpour et al. 2012), *Brassica* spp. (Ibrar et al. 2003; Ulfat et al. 2007), *Glycine max* (Essa 2002), *Vigna* spp., (Jabeen et al. 2003) and *Helianthus annuus* (Mutlu and Buzcuk 2007). It is well established that salt stress has negative correlation with seed germination and vigor (Rehman et al. 2000). Higher level of salt stress inhibits the germination of seeds while lower level of salinity induces a state of dormancy (Khan and Weber 2008). Salinity have many-fold effects on the germination process: it alters the imbibition of water by seeds due to lower osmotic potential of germination media (Khan and Weber 2008), causes toxicity which changes the activity of enzymes of nucleic acid metabolism (Gomes-Filho et al. 2008), alters protein metabolism (Yupsanis et al. 1994; Dantas et al. 2007), disturbs hormonal balance (Khan and Rizvi 1994), and reduces the utilization of seed reserves (Promila and Kumar 2000; Othman et al. 2006). It may also negatively affect the ultrastructure of cell, tissue and organs (Koyro 2002; Rasheed 2009). However, there are various internal (plant) and external (environmental) factors that affect seed germination under saline conditions which includes nature of seed coat, seed dormancy, seed age, seed polymorphism, seedling vigor, temperature, light, water and gasses (Wahid et al. 2011). The germination rates and percentage of germinated seeds at a particular time varies considerably among species and cultivars. Läuchli and Grattan (2007) proposed a generalized relationship between percent germination and time after adding water at different salt levels (Fig. 2.3).

In *Solanum lycopersicum*, high concentrations of salt (150 mM NaCl) in the germination media significantly delays onset and reduced the rate of germination (Foolad and Lin 1997, 1998). Further investigation in *S. lycopersicum*, Kaveh et al. (2011) found a significantly negative correlation between salinity and the rate and

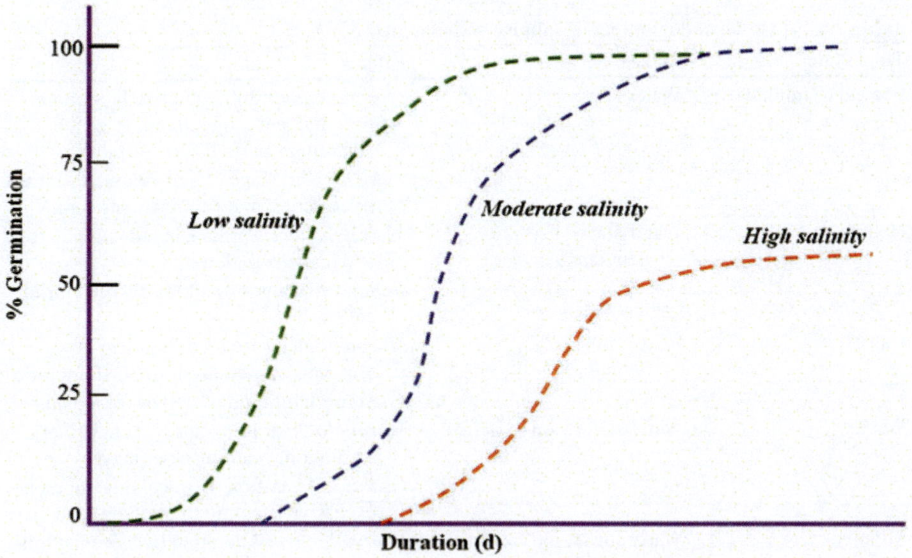

Fig. 2.3 Relationship between rate of germination and time after sowing at different salinity levels

percentage of germination which resulted in delayed germination and reduced germination percentage. Cuartero and Fernandez-Munoz (1999) reported that seeds need 50% more days to germinate at 80 mM NaCl and about 100% more days at 190 mM NaCl than control. Neamatollahi et al. (2009) reported that increasing of NaCl concentration in priming treatments reduced germination percentage due to higher osmotic pressures. Lombardi and Lupi (2006) reported that an increase in NaCl concentration progressively retarded and decreased germination of *Hordeum secalinum*, where 10-day treatment with 400 and 500 mM NaCl caused 40% and 38% reductions in germination rate, respectively. Bordi (2010) reported that the germination percentage in *B. napus* significantly reduced at 150 and 200 mM NaCl. Germination rate also decreased on increasing concentration of salinity levels. Compared with control, germination percentage and germination speed were decreased by 38% and 33%, respectively at 200 mM NaCl. This was caused due to ionic imbalance, osmotic regulation disorders and finally decreased water absorption by seeds. While studying with four rice cultivars, we observed a significant reduction in germination rate when exposed to various concentration of salt (30–150 mM). However, the sensitive cultivars were more prone to germination reduction under salt stress (Hasanuzzaman et al. 2009). In *Vigna radiata*, germination percentage decreased up to 55% when irrigated with 250 mM NaCl (Nahar and Hasanuzzaman 2009). In a recent study, Khodarahmpour et al. (2012) observed drastic reduction in germination rate (32%), length of radicle (80%) and plumule (78%), seedling length (78%) and seed vigor (95%) in *Zea mays* seeds exposed to 240 mM NaCl.

Table 2.5 Time-dependent effect of salinity on plant growth

Time scale	Causes	Effects
Second to minutes	Water stress	*Morphological*: Immediate reduction in root and leaf elongation rate which is sometimes partially recoverable. *Cellular*: Shrinkage of cell volume followed by restoration due to regaining turgor
Hours	Water stress, Ca^{2+} deficiency	*Morphological*: Permanent reduction in root and leaf elongation Cellular: Changes rheological behavior of cell wall
Days	Water stress, Ca^{2+} deficiency	*Morphological*: Reduction in leaf emergence, increase in root: shoot ratio *Cellular:* Inhibition of cell development
Weeks	Water stress, ion toxicity	*Morphological*: Reduced branches/tiller formation, death of older leaves *Cellular*: Alteration of apical development, excessive accumulation of Na^+ and Cl^-
Months	Water stress, ion toxicity	*Morphological*: Alteration in flowering time and reduced seed production. Immature death of plants *Cellular*: Alteration in the development of reproductive organs, Reduction of assimilate production

2.4.2 Growth

One of the initial effects of salt stress on plant is the reduction of growth rate. Salinity can affect growth of plant in various ways. First, the presence of salt in the soil reduces the water uptake capacity of the plant, and this causes quick reduction in the growth rate. This first phase of the growth response is due to the osmotic effect of the soil solution containing salt, and produces a package of effects similar to water stress (Munns 2002b). The mechanisms by which salinity affects growth of a plant depend on the time scale over which the plant is exposed to salt (Table 2.5). Munns (2002b) summarized the sequential events in a plant grown in saline environment. He stated that "In the first few seconds or minutes, water is lost from cells and shrinked. Over hours, cells recover their original volume but the elongation rates are still reduced which led to lower growth rates of leaf and root. Over days, cell division rates are also affected, and contribute to lower rates of leaf and root growth. Over weeks, changes in vegetative development and over months changes in reproductive development can be seen". Later on, Munns (2005) developed the 'two-phase growth response to salinity' for better understanding the temporal differences in the responses of plants to salinity (Fig. 2.4). The first phase of growth reduction is a quicker process which is due to osmotic effect. The second phase, on the other hand, is much slower process which is due to the salt accumulation in leaves, leading to salt toxicity in the plants. The later one may results in death of

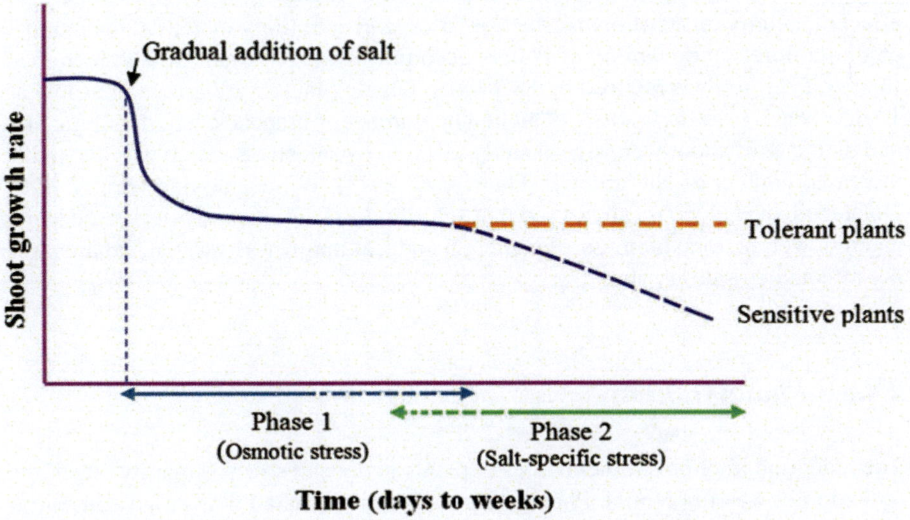

Fig. 2.4 Overview of the two-phase growth response to salinity for plant differing in salt sensitivity

leaves and reduce the total photosynthetic leaf area which reduce the supply of photosynthate in plants and ultimately affect the yield. With annual species, the timescale is day or week, depending on species and salinity level. With perennial species, the timescale is months or year. During phase 1, growth of both genotypes is reduced due to the osmotic effect of the saline solution adjacent to roots. During phase 2, leaves of more sensitive genotype are died and the photosynthetic capacity of the plant is greatly reduced which imposes an additional effect on growth. Upon addition of salt at one step, the growth rate plummets to zero or below and takes 1–24 h to regain the new steady rate, depending on the extent of the osmotic shock (Munns 2002a).

In plants, where Na^+ and Cl^- build up in the transpiring leaves over a long period of time, resulting in high salt concentration and leaf death. Leaf injury and death are attributed to the high salt load in the leaf that exceeds the capacity of salt compartmentation in the vacuoles, causing salt to build up in the cytoplasm to toxic levels (Munns 2002a, 2005; Munns et al. 2006). There are abundant literature indicating that plants are particularly susceptible to salinity during the seedling and early vegetative growth stage. In our study, we observed a remarkable reduction in plant height and tiller number and leaf area index in *O. sativa* plants grown in saline soil (Hasanuzzaman et al. 2009). Under saline condition, some crops are most sensitive during vegetative and early reproductive stages, less sensitive during flowering and least sensitive during the seed filling stage. In all these studies, seed weight is the yield component of interest but similar conclusions regarding growth stage sensitivity were obtained with both determinate crops (the grain crops) and indeterminate (cowpea) crops (Läuchli and Grattan 2007). Khatun and Flowers (1995) studied the effect of salinity on sterility and seed set in *O. sativa*. Salinity increased the number of sterile florets and viability of pollen, becoming more pronounced with increased salinity. Seed set was reduced by 38% when female plants were grown in as low as 10 mM NaCl. In *Suaeda salsa*, plant height, number of branches, length of branches and diameter of shoot were significantly affected by salt stress which was due to the increased content of Na^+ and Cl^- (Guan et al. 2011). While studying with *G. max*, Dolatabadian et al. (2011) observed that salinity stress significantly decreased shoot and root weight, total biomass, plant height and leaf number. However, leaf area was not affected by salinity stress.

2.4.3 Photosynthesis

The reduction in photosynthetic rates in plants under salt stress is mainly due to the reduction in water potential. Photosynthesis is also inhibited when high concentrations of Na^+ and/or Cl^- are accumulated in chloroplasts. As photosynthetic electron transport is relatively insensitive to salts, either carbon metabolism or photophosphorylation may be affected due to salt stress (Sudhir and Murthy 2004). A positive correlation between salt stress induced photosynthetic rate and yield has been obtained in different crops (Pettigrew and Meredith 1994; Sudhir and Murthy 2004). Fisarakis et al. (2001) reported a positive growth inhibition caused by salinity associated with a marked inhibition of photosynthesis. However, there are many reports showing no or little relationship between growth and photosynthetic capacity (Rogers and Noble 1992; Hawkins and Lewis 1993). In fact, the effect of salinity on photosynthetic rate depends on salt concentration as well as plant species or genotypes. There is evidence that at low salt concentration salinity sometimes stimulate photosynthesis. For instance, in *Bruguiera parviflora*, Parida et al. (2004) observed that rate of photosynthesis increased at low salinity while decreased at high salinity, whereas stomatal conductance remained unchanged at low salinity and decreased at high salinity. There are some other factors that reduced photosynthetic rates under salt stress are: enhanced senescence, changes in enzyme activity, induced by alterations in cytoplasmic structure and negative feedback by reduced sink activity (Iyengar and Reddy 1996). The reduction in stomatal conductance which results in restricting the availability of CO_2 for carboxylation reactions is also a factor that reduces photosynthesis under stress (Brugnoli and Björkman 1992). It was reported that stomatal closure minimizes loss of water through transpiration and this affects light-harvesting and energy-conversion systems thus leading to alteration in chloroplast activity (Iyengar and Reddy 1996). Higher stomatal conductance in plants is known to increase CO_2 diffusion into the leaves and thereby favor higher photosynthetic rates. One of the most notable effects of salt stress is the alteration of photosynthetic pigment biosynthesis (Maxwell and Johnson 2000). The decrease in Chl content under salt stress is a commonly reported phenomenon and in various studies and the Chl concentration were used as a sensitive indicator of the cellular metabolic state (Chutipaijit et al. 2011). In *Oryza sativa* leaves, the reduction of

Fig. 2.5 Salt-induced chlorosis in wheat and rapeseed leaves. Hydropinically grown wheat (6-day-old) and rapeseed (12-day-old) seedlings were exposed to salt stress for 4 days and 2 days, respectively

Chl *a* and *b* contents of leaves was observed after NaCl treatment (200 mM NaCl, 14 days) where reduction of the Chl *b* content of leaves (41%) was affected more than the Chl *a* content (33%) (Amirjani 2011). In another study, *Oryza sativa* exposed to 100 mM NaCl showed 30%, 45% and 36% reduction in Chl *a*, Chl *b* and carotenoids (Car) contents as compared to control (Chutipaijit et al. 2011). Saha et al. (2010) observed a linear decrease in the levels of total Chl, Chl *a*, Chl *b*, Car and xanthophylls as well as the intensity of Chl fluorescence in *Vigna radiata* under increasing concentrations of NaCl treatments. Compared to control, the pigment contents decreased on an average, by 31% for total Chl, 22% for Chl *a*, 45% for Chl *b*, 14% for carotene and 19% for xanthophylls (Saha et al. 2010). Associated with the decline in pigment levels, there was an average 16% loss of the intensity of Chl fluorescence as well. In our recent study, we observed a higher chlorosis in wheat and rapeseed leaves when subjected to salt stress (Fig. 2.5).

2.4.4 Water Relation

According to Romero-Aranda et al. (2001) increase of salt in the root medium can lead to a decrease in leaf water potential and, hence, may affect many plant processes. Osmotic effects of salt on plants are the result of lowering of the soil water potential due to increase in solute concentration in the root zone. At very low soil water potentials, this condition interferes with plant's ability to extract water from the soil and maintain turgor. However, at low or moderate salt concentration (higher soil water potential), plants adjust osmotically (accumulate solutes) and maintain a potential gradient for the influx of water. Salt treatment caused a significant decrease in relative water content (RWC) in sugar beet varieties (Ghoulam et al. 2002). According to Katerji et al. (1997), a decrease in RWC indicates a loss of turgor that results in limited water availability for cell extension processes. Steudle (2000) reported that in transpiring plants, water is thought to come from the soil to the root xylem through apoplastic pathway due to hydrostatic pressure gradient. However, under salt stressed condition, this situation changes because of the restricted transpiration. Under these situations, more of water follows cell-to-cell path, flowing across membranes of living cells (Vysotskaya et al. 2010).

2.4.5 Nutrient Imbalance

It is well-established that crop performance may be adversely affected by salinity-induced nutritional disorders. However, the relations between salinity and mineral nutrition of crops are very complex (Grattan and Grieve 1999). The nutritional disorders may result from the effect of salinity on nutrient availability, competitive uptake, transport or distribution within the plant. Numerous reports indicated that salinity reduces nutrient uptake and accumulation of nutrients into the plants (Rogers et al. 2003; Hu and Schmidhalter 2005). However, very few evidences exist that addition of nutrients at levels above those considered optimal in non-saline environments, improves crop yield (Grattan and Grieve 1999). In fact, these processes may occur simultaneously and whether they affect the crop yield or quality depends on the toxic level, composition of salts, the crop species and surrounding environment (Grattan and Grieve 1999). Numerous plant studies have demonstrated that salinity could reduce N accumulation in plants. Decreased N uptake under saline conditions occurs due to interaction between Na^+ and NH_4^+ and/or between Cl^- and NO_3^- that ultimately reduce the growth and yield of the crop (Rozeff 1995). This reduction in NO_3^- uptake is associated with Cl^- antagonism (Bar et al. 1997) or reduced water uptake under saline conditions (Lea-Cox and Syvertsen 1993). The availability of P was reduced in saline soils due to (a) ionic strength effects that reduced the activity of PO_4^{3-}, (b) phosphate concentrations in soil solution was tightly controlled by sorption processes and (c) low solubility of Ca-P minerals. Hence, it is noteworthy that phosphate concentration in field grown agronomic crops decreased as salinity increased (Qadir and Schubert 2002). Different plant studies indicated that high level of external Na^+ caused a decrease in both K^+ and Ca^{2+} concentrations in plant tissues of many plant species (Hu and Schmidhalter 1997, 2005; Asch et al. 2000). This reduction in K^+ concentration in plant tissue might be due to the antagonism of Na^+ and K^+ at uptake sites in the roots, the influence of Na^+ on the K^+ transport into xylem or the inhibition of uptake processes (Suhayda et al. 1990). In another study, Hu and Schmidhalter (1997) also stated that Mg^{2+} concentration decreased due to salinity in *T. aestivum* leaves.

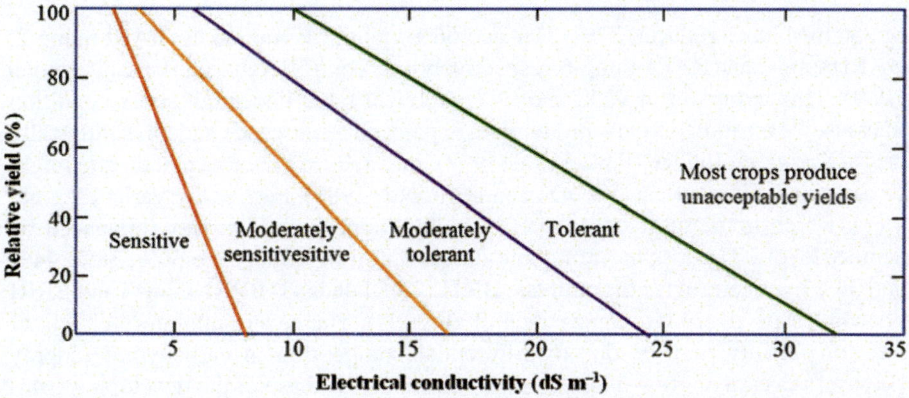

Fig. 2.6 Relative yield in response to different salinity levels and varying degree of salt tolerance (Mass 1986)

The availability of micronutrients in saline soils is dependent on the solubility of micronutrients, the pH of soil solution, redox potential of the soil solution and the nature of binding sites on the organic and inorganic particle surfaces. In addition, salinity can differently affect the micronutrient concentrations in plants depending upon crop species and salinity levels (Oertli 1991). Micronutrient deficiencies are very common under salt stress because of high pH (Zhu et al. 2004).

2.4.6 Yield

The above mentioned effects of salt stress on plants ultimately lead to reduction of yield of crop which is most countable effect of salt stress in agriculture. Except some halophytes, yield of most of the crops reduced greatly due to salt stress. Tolerance and yield stability are multigenic traits that are complicated to establish in crops since salt stress may be imposed continuously or intermittently, or become gradually more severe and at any stage during development (Yokoi et al. 2002). Crop species have exhibited substantial differences in salt tolerance based on their relative yields. Relative yield often exhibits a linear decrease after a threshold salinity has been reached (Fig. 2.6), and salt tolerance has been defined in terms of two parameters: the threshold electrical conductivity and the percent decrease in relative yield per unit of electrical conductivity in dS m^{-1} above the threshold. It was observed that relative yield varied greatly depending on the salinity levels and the degree of tolerance (Mass 1986). Different yield components of *V. radiata* were significantly affected by salinity stress as reported by Nahar and Hasanuzzaman (2009). Number of pods per plant, seeds per pod and seed weight were negatively correlated with salinity levels. The reproductive growth of *V. radiata* was also affected by salinity as the number of pods per plant substantially decreased with increasing salinity levels. An application of 250 mM NaCl reduced 77%, 73% and 66% yield in *V. radiata* cv. BARI mung-2, BARI mung-5 and BARI mung-6, respectively over control (Nahar and Hasanuzzaman 2009). This reduction of yield and its component rated under salt stress condition may also be attributed to low production, expansion, senescence and physiologically less active green foliage (Wahid et al. 1997), thus reduced photosynthetic rate might be a supplementary effect (Seemann and Critchley 1985). In *O. sativa* varieties, grain yield, which is the ultimate product of yield components is greatly influenced by salinity levels. The loss of grain yield due to 150 mM salinity are 50%, 38%, 44% and 36% over control for the cultivars BR11, BRRI dhan41, BRRI dhan44 and BRRI dhan46, respectively (Hasanuzzaman et al. 2009). The severe inhibitory effects of salts on fertility may be due to differential competition in carbohydrate supply between vegetative growth and constrained supply of these to the developing panicles (Murty and Murty 1982). Also reduced viability of pollen under stress condition could result in failure of seed set (Abdullah et al. 2001). Grain yield reduction of rice varieties due to salt stress is also reported earlier by Linghe and Shannon (2000) and Gain et al. (2004). As reported by Greenway and Munns (1980), after some time in 200 mM NaCl, a salt-tolerant species such as sugar beet might have a reduction of only 20% in dry weight, a moderately tolerant species such as cotton might have a 60% reduction, and a sensitive species such as soybean might be dead. On the other hand, a halophyte such as *Suaeda maritima* might be growing at its optimum rate (Flowers et al. 1986).

2.4.7 Salinity Induced Oxidative Stress

Salt stress can lead to stomatal closure, which reduces CO_2 availability in the leaves and inhibits carbon fixation, exposing chloroplasts to excessive excitation energy which in turn increase the generation of reactive oxygen species (ROS) such as superoxide ($O_2^{\bullet -}$), hydrogen peroxide (H_2O_2), hydroxyl radical (OH•) and singlet oxygen ($^1O_2$) (Parida and Das 2005; Ahmad and Sharma 2008; Ahmad et al. 2010a, 2011). On the other hand, as salt stress is complex and imposes a water deficit because of osmotic effects on a wide variety of metabolic activities (Greenway and Munns 1980; Cheeseman 1988). This water deficit leads to the formation of ROS (Halliwell and Gutteridge 1985; Elstner 1987). ROS are highly reactive and may cause cellular damage through oxidation of lipids, proteins and nucleic acids (Pastori and Foyer 2002; Apel and Hirt 2004; Ahmad et al. 2010a, b). In many plant studies, it was observed that production of ROS is increased under saline conditions (Hasegawa et al. 2000) and ROS-mediated membrane damage has been demonstrated to be a major cause of the cellular toxicity by salinity in different crop plants such as rice, tomato, citrus, pea and mustard (Gueta-Dahan et al. 1997; Dionisio-Sese and Tobita 1998; Mittova et al. 2004; Ahmad et al. 2009, 2010b). Long-term salinity treatments (EC 5.4 and 10.6 dS m^{-1}, 60 days) caused significant increase in H_2O_2 and lipid peroxidation in wheat seedlings, which were higher in salt-sensitive cultivar than salt-tolerant cultivar (Sairam et al. 2002). In recent study, increased lipid peroxidation and levels of H_2O_2 was observed with increased salinity in *B. napus* (Hasanuzzaman et al. 2011a) and *T. aestivum* (Hasanuzzaman et al. (2011b).

2.5 Role of Exogenous Protectants to Mitigate Salt-Induced Damages

Numerous research results have indicated that exogenous application of osmoprotectants, plant hormones, antioxidants, signaling molecules, polyamines and trace elements provided significant protection against salt-induced damages in plants (Table 2.6). These protectants enhanced salt stress tolerance by enhancing their germination, growth, development, photosynthesis, antioxidative capacities and yield.

2.5.1 Osmoprotectants

2.5.1.1 Proline

The accumulation of osmolytes such as proline (Pro) is a well-known adaptive mechanism in plants against salt stress conditions. It has also been suggested that Pro accumulation can serve as a selection criterion for the tolerance of most species to stressed conditions (Parida and Das 2005; Ashraf and Foolad 2007; Ahmad et al. 2009). Since the first report on Pro accumulation in wilting perennial rye grass (Kemble and MacPherson 1954), a number of research works has been carried out concerning the role of Pro as a compatible osmolyte and osmoprotectant and its roles in salt stress tolerance. Several studies have attributed an antioxidant feature to Pro, suggesting ROS scavenging activity and Pro acting as a $^1O_2$ quencher (Smirnoff and Cumbes 1989; Matysik et al. 2002). Working with *Arabidopsis* mutants, Werner and Finkelstein (1995) found that a Pro-deficient mutant, selected for its ability to germinate on saline media, was unable to continue growth on that media because it could not accumulate Pro to the equivalent level of the wild type. Proline also induces the expression of salt-stress-responsive proteins and may improve the plant adaptation to salt-stress (Khedr et al. 2003). They reported that severe salt stress inhibited the activities of antioxidant enzymes catalase (CAT) and peroxidase (POD) in *Pancratium maritimum* plants, but the activities of these enzymes were significantly higher in the presence of Pro than in its absence. It was expected that up-regulation of antioxidant system offered by Pro protect plants against NaCl-induced oxidative damage. Hoque et al. (2008) showed that Pro improves salt tolerance in *Nicotiana tabacum* plants by increasing the activity of enzymes involved in the antioxidant defense system. Earlier, it has been reported that Pro protects higher plants against osmotic stresses not only by adjusting osmotic pressure but also by stabilizing many

Table 2.6 Summary of the protective effects of different exogenous protectants under salt stress

Name of the crop	Salinity dose and duration	Dose of protectant	Protective effects	References
Olea europaea L. cv. Chemlali	100 and 200 mM NaCl, 6 months	25 and 50 mM Pro	Modulated antioxidative enzyme activities Increased photosynthetic activity and plant growth Maintained suitable plant water status	Ahmed et al. (2010)
Nicotiana tabacum BY-2 cells	200 mM, NaCl, 7 days	20 mM Pro	Increased fresh weight Enhanced the activities of POD and CAT	Hoque et al. (2007)
Oryza sativa L. sub sp. *indica*	150 mM NaCl, 14 days	50 mM GB, 14 days	Increased Chl and Car content Increased WUE Increased seed weight and yield	Cha-Um and Kirdmanee (2010)
Nicotiana tabacum BY-2 cells	200 mM, NaCl, 7 days	20 mM GB	Increased fresh weight Increased the activity of POD	Hoque et al. (2007)
Oryza sativa L. cv. Nipponbare	25 mM NaCl, 12 h	1 and 5 mM Pro and GB	Suppressed Na^+-enhanced apoplastic flow to reduce Na^+ uptake in rice plants Increased K^+/Na^+ ratio	Sobahan et al. (2009)
Cucumis melo L. cv. Yuhuang and cv. Xuemei	100 mM NaCl, 5 days	0.2 mM Pro	Increased fresh and dry weights Increased *Pn*, *F*v/*F*m, ΦPSII and Chl content Reduced the $O_2^{\bullet -}$ level and the H_2O_2 content Enhanced activities of SOD, POD, APX, CAT and DHAR	Yan et al. (2011)
Oryza sativa L. cv. KDML105	100 mM NaCl, 6 days	10 mM Pro	Increased fresh and dry weight Reduced the Na^+/K^+ ratio Increased endogenous Pro and transcript levels of P5CS and P5CR Decreased the activity of the antioxidant enzymes and upregulated the transcription of genes encoding several antioxidant enzymes	Nounjan et al. (2012)
Oryza sativa L. cv. Nipponbare	150 mM NaCl, 5 days	5 mM GB	Prevented the salt-induced swelling of thylakoids, disintegration of grana staking and intergranal lamellae and disruption of mitochondria	Rahman et al. (2012)
Oryza sativa L. cv. KDML105	100 mM NaCl, 6 days	10 mM Tre	Increased fresh and dry weight Reduced the Na^+/K^+ ratio and strongly decreased endogenous Pro	Nounjan et al. (2012)

Zea mays, cv. Giza 2	−0.2 MPa, 14 days	10 mM Tre, presoaking for 8 h	Increased plant height, root and shoot dry weight and leaf relative water content Increased photosynthetic pigments and nucleic acids content Increased soluble sugars and soluble protein content	Zeid (2009)
Medicago sativa	NaCl 15 dS m^{-1}, 7 days	0.5 mM SA, 6 h (Pretreatment)	Increased germination percentage, seed vigor index and growth parameters Decreased electrolyte leakage	Torabian (2010)
Brassica juncea	NaCl 150 mM, 3 days	10 μM SA spray	Improved plant growth Decreased electrolyte leakage Improved photosynthesis and transpiration rate Enhanced activities of CAT, POX and SOD	Yusuf et al. (2012)
Viola odorata L.	NaCl 50 mM, 120 days	SA at 30 mg L^{-1}, 120 days	Increased plant height, root length and biomass Decreased Na^+ and Cl^- content	Hussain et al. (2011a)
Pisum sativum cv. Lincoln	70 mM NaCl, 25 days	SA 25–100 μM, 7 days × 2 spray	Reduced plant growth Imbalanced antioxidant metabolism	Barba-Espín et al. (2011)
Capsicum annuum cv. Beldi	4 gL^{-1} NaCl, 28 days	0.5 mg L^{-1} EBR, 28 days	Increase plant growth, RWC and photosynthetic pigments Decreased electrolyte leakage	Houimli et al. (2010)
Cucumis sativus L. cv Zhongnong 8	250 mM NaCl, 48 h	5 μM EBR, 48 h	Improved seed germination	Wang et al. (2011a)
Zea mays L. cv. Partap-1	25, 50 and 75 mM NaCl, 30 days	10^{-8}, 10^{-6} and 10^{-4} mM HBR, 12 h seed soaking	Enhanced antioxidant enzymes' activities and increased protein content Decreased lipid peroxidation	Arora et al. (2008)
Vigna sinensis	25, 50, 100, and 150 mM, 45 days	0.05 ppm brassino-lide (2 sprays)	Increased fresh weight, dry weight and length of root and shoot Increased antioxidant enzymes; activities Decreased lipid peroxidation	El-Mashad and Mohamed (2012)
Capsicum annuum L. cv. Beldi	70 mM NaCl, 21 days	10 μM EBR, 21 days	Increased relative growth rate (RGR), net assimilation rate (NAR) and leaf area ratio (LAR) Increased photosynthesis, stomatal conductance Increased WUE	Samira et al. (2012)

(continued)

Table 2.6 (continued)

Name of the crop	Salinity dose and duration	Dose of protectant	Protective effects	References
Oryza sativa L. cv. IR29 and IR651	100 mM NaCl, 4, 12, 24, 48, 96 and 168 h	20 μM ABA (spray)	Increased root and shoot dry weight Decreased Na^+ content	Saedipour (2011)
Oryza sativa L. cv. IR6	50 and 75 mM NaCl, 25 days	10 μM ABA, 24 h	Decreased Na^+ and Cl^- concentrations and Na^+/K^+ ratio Increased K^+ and Ca^{2+} concentrations Increased Pro accumulation and soluble sugar content Increased yield	Gurmani et al. (2011)
Zea mays L. cv. DK 647	100 mM NaCl, 100 days	1 or 2 mM IAA (spray)	Reduced Na^+ concentration Increased those of Ca^{2+} and K^+ Increase Chl *a* and Chl *b* content Decreased electrolyte leakage and Pro content	Kaya et al. (2009)
Brassica juncea	100 and 150 mM NaCl, 45 days	75 mg l^{-1} GA_3 (spray)	Increased fresh and dry weight Increased Pro and GB biosynthesis Decreased lipid peroxidation and electrolyte leakage Increased Chl content	Ahmad et al. (2009)
Saccharum officinarum L.	9 dS m^{-1}, 50 days	100 ppm GA (spray)	Increased sugar content Increased dry weight and protein content	Shomeili et al. (2011)
Beta vulgaris L. cv. Tianjin qing pielan	4.7, 9.4 and 14.1 dS m^{-1}, 8 and 15 days	100, 150 and 200 mg L^{-1} GA_3 (Presoaking)	Improved seed germination Increased fresh and dry weight	Jamil and Rha (2007)
Brassica juncea	25 or 50 mM NaCl, 20 days	10 μM GA_3 (spray)	Increased Chl content, stomatal conductance and net photosynthetic rate	Shah (2007)
Glycine max cv. Hwangkeum	100 mM NaCl, 13 days	0.5, 1.0 and 5.0 μM GA3, 13 days	Increased plant height and fresh weight Regulated the phytohormonal balance	Hamayun et al. (2010)
Pisum sativum L. cv. Ran 1	30 mM NaCl, 4 and 7 days	10 μM JA, 3 days	Increased the photosynthesis, RWC and protein content Decrease of Na+and Cl- accumulation in the shoot	Fedina and Tsonev (1997)

Glycine max	60 mM NaCl, 2 weeks	20 and 30 μM MeJA, 24 h	Increased plant growth Increased Chl content, leaf photosynthetic rate, leaf transpiration rate and Pro content Increased ABA levels	Yoon et al. (2009)
Lycopersicon esculentum Mill. cv. M82	300 mM NaCl, 9 h	0.5 mM AsA, 24 h	Increased survival rate of seedlings Decreased lipid peroxidation	Shalata and Neumann (2001)
Cicer arietinum L.	20 and 40 mM NaCl, 6 weeks	4 mM AsA	Increased the contents of Chl *a* and Chl stability index (CSI %)	Beltagi (2008)
Triticum durum Desf. var. Waha	150 mM NaCl, 2 weeks	0.7 mM AsA	Increased leaf area Improved Chl and Car contents and enhanced Pro accumulation Decreased H_2O_2 content	Azzedine et al. (2011)
Triticum aestivum *L.* cv. S-24 and MH-97	100 mM NaCl, 4 weeks	50, 100 mg L^{-1} AsA (spray)	Protected the photosynthetic machinery	Khan and Weber (2008)
Glycine max L. cv. SHAR and DPX	12.5 and 50 mM NaCl, 10 days	400 mg L^{-1}, 4 h	Increased growth Enhanced antioxidant activities	Dehghan et al. (2011)
Saccharum sp. cv. HSF 240	100, 120 and 140 mM NaCl, 30 days	0.5 mM AsA, 24 h	Increased root growth Increased activities of antioxidant enzymes Increased soluble protein contents	Munir and Aftab (2011)
Brassica napus L. cv. Serw and cv. Pactol	100 mM and 200 mM NaCl, 3 weeks	100 mg L^{-1} GSH, 24 h	Increased growth Increased photosynthetic pigments Enhanced antioxidant enzymes' activities	Kattab (2007)
Tagetes erecta L.	1,500 ppm NaCl	100 and 200 ppm GSH	Increased plant height, no. of branches, fresh and dry weight of herb and flowers, no. of flowers Increased total carbohydrates(%), total phenols and zathaphylls pigment content Increased mineral ions percentage	Rawia et al. (2011)

(continued)

Table 2.6 (continued)

Name of the crop	Salinity dose and duration	Dose of protectant	Protective effects	References
Allium cepa L., Giza 6	150 mM NaCl, 3 h	0.5 mM GSH, 2 h	Ameliorated NaCl-induced plasma membrane changes and maintained its permeability and cell viability	Salama and Al-Mutawa (2009)
Triticum aestivum	6 and 9 dS m^{-1}, 30, 60 and 90 days	100 mg L^{-1} GSH, 6 h	Increased growth and yield Improved endogenous antioxidant content	Sakr and El-Metwally (2009)
Helianthus annuus L. cv. Hysun 336 and Euroflor	1.56, 4.68 and 7.83 dS m^{-1}	25 and 50 mg L^{-1} α-Tocopherol, 12 h	Enhanced antioxidant enzymes' activities Improved mineral nutrient uptake	Rady et al. (2011)
Triticum aestivum L. cv. Giza 168	0.12, 0.35 and 0.70% NaCl, 65 days	α-Tocopherol 100 mg L^{-1} (spray)	Decreased the Na^+ and Cl^- content Increased the K, Ca and Mg content. Increased antioxidant enzymes' activities Decreased the levels of H_2O_2 and lipid peroxidation	Farouk (2011)
Triticum aestivum L. cv. Huaimai 17	300 mM NaCl, 1–5 days	100 µM SNP, 20 h	Increased seed germination Enhanced seed respiration rate and ATP synthesis	Zheng et al. (2009)
Cucumis sativus L. cv. Jinchun 2	50 mM NaCl, 8 days	100 µM SNP, 8 days	Increased seedling growth, photosynthetic pigment content, Pro accumulation, net photosynthetic rate, *gs* and Tr	Fan et al (2007)
Cucumis sativus L. cv. Jinchun 2	50 mM NaCl, 8 days	100 µM SNP, 8 days	Increased growth and dry matter partitioning Increased PAs biosynthesis	Fan et al. (2010)
Cicer arietinm L. cv HC-3	25 mM NaCl, 2, 4 and 6 days	0.2 and 1 mM SNP, 2, 4 and 6 days	Increased RWC Decreased relative membrane injury	Sheokand et al. (2010)
Lycopersicom esculentum Mill. cv. Hufan1480 and Hufan2496	100 mM, 8 days	100 µM SNP, 8 days	Increased plant growth and biomass accumulation	Wu et al. (2011)

Oryza sativa L.	80 mM NaCl, 5 days	100 and 200 µM SNP, 16 h	Increased seed germination	Habib et al. (2010)
Tricitum aestivum L. cv. Pradip	150, and 300 mM NaCl, 4 days	1 mM SNP, 1 days	Increased the content of non-enzymatic antioxidant Enhanced the activities of antioxidant enzymes Decreased lipid peroxidation	Hasanuzzaman et al. (2011b)
Triticum aestivum L. cv.MH-97	150 mM NaCl, variable time	1, 40, 80 and 120 mM H_2O_2, 8 h	Improved photosynthetic capacity Increased antioxidant defense Greater tissue K^+, Ca^{2+}, NO_3^- $PO4_3^-$ levels and improved K^+:Na^+ ratio	Wahid et al. (2007)
Hordeum vulgare L. cv. Alfa	150 mM NaCl, 4 and 7 days	1 and 5 µM H_2O_2, 2 days	Improved CO_2 fixation Enhanced antioxidant defense	Fedina et al. (2009)
Capsicum annuum L. cv. California Wonder	NaCl, 200 mM, 10 days	1.5 mM H_2O_2, 24 h	Increased seed germination Accelerated flowering and increased fruit yield	Yadav et al. (2011)
Oryza sativa L. cv. IKP and Pokkali	50 and 100 mM, 5 days	1 mM Put, 5 days	Reduces Na^+ accumulation Increased PA content	Quinet et al. (2010)
Punica granatum L. cv. Rabbab	40, 80 and 120 mM NaCl, 72 h	1 and 2 mM Put and Spd	Decreased Na and Cl content Increased Pro content	Amri et al. (2011)
Citrus karna Raf.	3.0 dS m^{-1}, 90 days	50 mg L^{-1} Put	Reduced the membrane injury index Increased RWC, photosynthetic rate, and pigments content Improved the activities of SOD and POD Increased Pro content	Sharma et al. (2011)
Oryza sativa L. cv. Pokkali and KDML105	150 mM NaCl, 7 days	1 MM Spd, 24 h	Improved growth Increased membrane stabilization Efficient scavenging of free radicals and decreased MDA Maintained K^+/Na^+ status	Saleethong et al. (2011)
Cucumis sativus L. cv. Changchun mici and Jinchun No. 2	50 mM NaCl, 7 days	0.1 mM Spd, 7 days	Reduced induced membrane damage Increased growth and photosynthesis Increase in PA and Pro contents Up-regulated antioxidant enzyme activities	Duan et al. (2008)

(continued)

Table 2.6 (continued)

Name of the crop	Salinity dose and duration	Dose of protectant	Protective effects	References
Sorghum bicolor (L.) Moench	180 mM NaCl, 7 days	0.25 mM Spm	Increased growth Increased antioxidant capacity	Chai et al. (2010)
Oryza sativa L. cvs. M-1-48, Nonabokra and Gobindobhog	200 mM NaCl, 15 days	1 mM Spd or 1 mM Spm, 15 days	Increased Chl content Decreased Na^+ content Decreased MDA and H_2O_2 levels Increased antioxidant metabolism	Roychoudhury et al. (2011)
Cucumis sativus L. cv. Polan	50 mM NaCl, 14 days	5, 10, or 20 μM Na_2SeO_4, 14 days	Improved the growth rate, photosynthetic pigments and Pro contents Decreased Cl^- ions Enhanced antioxidative capacity	Hawrylak-Nowak (2009)
Cucumis sativus	2,000 ppm NaCl, 7 days	1 ppm Se, 14 days	Increased activities of POD, CAT, SOD, APX and PAL Reduction in electrolyte leakage and MDA content	Walaa et al. (2010)
Brassica napus L. cv. BINA Sharisha 3	100 and 200 mM NaCl	25 μM Na2SeO4, 48 h	Improved antioxidative capacity Decreased MDA and H_2O_2 levels Decreased chlorosis	Hasanuzzaman et al. (2011a)
Brassica napus L. cv. BINA Sharisha 3	100 and 200 mM NaCl, 48 h	1 mM SiO_2, 48 h	Enhanced antioxidative defense Decreased MDA and H_2O_2 levels	Hasanuzzaman and Fujita (2011b)
Brassica napus L. cv. Hayola	150 mN NaCl, 25 days	2 mM Na2SiO3, 25 days	Increased the ROS scavenging capacity Decreased tissue Na^+ contents Maintained the membrane integrity of root cells	Hashemi et al. (2010)

Zea mays	120 mM NaCl, 28 days	0.4 -3.2 mM $Si(OH)_4$, 28 days	Increased growth Increased CO_2 assimilation rate (*A*), *gs*, Tr, and leaf sub-stomatal CO_2 concentration	Parveen and Ashraf (2010)
Medicago sativa L. cv. Zhongmu No. 1 and Defor	120 mM NaCl, 15 days	1 mM K_2SiO_3, 15 days	Increased activities of antioxidant enzymes	Wang et al. (2011b)
Glycine max L. cv. Taekwangkong	80 mM NaCl, 14 days	2.5 mM Na_2SiO_3, 14 days	Increased plant growth Improved Chl content	Lee et al. (2010)
Triticum aestivum cv. SARC-3) and Auqab 2000	150 mM NaCl, 10 days	2 mM Si, 12 days	Improved growth Decreased Na^+ and Na^+:K^+ ratio	Tahir et al. (2012)

functional units such as complex II electron transport, membranes, and proteins and enzymes such as RuBisCo (Hamilton and Heckathorn 2001). Proline perform these functions by protecting the photosynthetic apparatus (Ashraf et al. 2008), by functioning as an oxygen radical scavenger (Heuer 2003), and by displaying an antioxidant activity (Okuma et al. 2004). While studying with olive trees, Ahmed et al. (2010) observed that Pro supplements seemed to improve salt tolerance in olive tree by modulating some antioxidative enzyme activities, photosynthetic activity, and thus maintained better plant growth and water status. Moreover, the decrease of soluble sugar content in Pro treated-plants revealed the important osmoprotective effect played by added Pro. The Pro application mitigated the reduction of growth and photosynthetic activity under salt stress in olive trees. The increment rate of leaf RWC in the presence of 25 and 50 mM Pro was 4.45% and 6.67%, respectively, in comparison to values recorded in 100 mM NaCl-treated plants. In 200 mM NaCl plus Pro-treated plants, this increase was 1.14 times for 25 mM Pro and 1.19 times for 50 mM Pro higher than those recorded in severe salt stress treatment (200 mM NaCl). Deivanai et al. (2011) demonstrated that rice seeds pretreated with Pro (1, 5 and 10 mM) and grown at different NaCl concentrations counteracted the adverse effect of salt. Pretreatment of Pro with a concentration of 1 mM was found to be effective and stimulated cellular activities, whereas 10 mM Pro was ineffective in improving plant growth under high level of salt (300 and 400 mM NaCl).

Hoque et al. (2007) examined the growth and activities of antioxidant enzymes in tobacco Bright Yellow-2 (BY-2) culture cells in suspension under salt stress and found that both Pro and betaine mitigated the inhibition of growth of BY-2 cells under salt stress. Salt stress significantly decreased the activities of superoxide dismutase (SOD), catalase (CAT) and peroxidase (POD) in BY-2 cells. However, exogenous application of Pro or betaine alleviated the reduction in CAT and POD activities but not SOD activity under salt stress. Neither Pro nor betaine directly scavenged $O_2^{\bullet-}$ or H_2O_2. They also concluded that the mitigating effect of Pro was more than that of betaine because of its superior ability to increase the activities of antioxidant enzymes. Sobahan et al. (2009) further reported that exogenous Pro and betaine suppressed Na^+-enhanced apoplastic flow to reduce Na^+ uptake in rice plants. In their study, addition of Pro or betaine to the saline medium suppressed Na-induced trisodium-8-hydroxy-1,3,6-pyrenetrisulphonic acid (an apoplastic tracer) uptake and Na^+ accumulation, while the K^+ content was slightly increased, which led to a high K^+/Na^+ ratio under saline conditions. Lima-Costa et al. (2008) cultured a salt-sensitive *Citrus sinensis* 'Valencia late' cell line which had a lower growth rate and accumulates Pro when exposed to salt (>200 mM NaCl). However, the addition of exogenous Pro to this cell line was evaluated in terms of cell metabolism. Thus, a positive influence on the relieve of salt stress symptoms due to the presence of exogenous Pro 5 mM and 100 mM NaCl was obtained, with increased growth of this salt sensitive citrus cell line. Yan et al. (2011) found that application of exogenous 0.2 mM Pro to salinized nutrient solution alleviated the decrease in fresh and dry weights of *Cucumis melo* seedlings. Exogenous Pro significantly alleviated the decrease of *Pn*, *F*v/*F*m, ΦPSII and Chl content under saline conditions. Compared with NaCl alone, exogenous Pro also reduced the $O_2^{\bullet-}$ level and the H_2O_2

content which was accompanied by the enhanced activities of SOD, POD, APX, CAT and DHAR. In a recent study, Nounjan et al. (2012) observed that salt stress resulted in growth reduction, increase in the Na^+/K^+ ratio, increase in Pro level and up-regulation of Pro synthesis genes (pyrroline-5-carboxylatesynthetase, P5CS; pyrroline-5-carboxylate reductase, P5CR) as well as accumulation of H_2O_2, increased activity of antioxidative enzymes (SOD, POX, APX, CAT) and transcript up-regulation of genes encoding antioxidant enzymes (Cu/ZnSOD, MnSOD, CytAPX, CatC) of *O. sativa* seedlings. On the other hand, exogenous Pro supplementation under salt stress condition reduced the Na^+/K^+ ratio, further increased the endogenous Pro and transcript levels of P5CS and P5CR, but decreased the activity of the antioxidant enzymes. In addition, the transcription of genes encoding several antioxidant enzymes was upregulated.

2.5.1.2 Glycinebetaine

Glycinebetaine (GB) is a small organic metabolite soluble in water and non-toxic at high concentrations which can potentially play a protective role against salt stress (Ashraf and Foolad 2007; Chen and Murata 2008). The major role of GB in plants exposed to salt is probably protecting cells by osmotic adjustment (Gadallah 1999), protein stabilization (RuBisCo) (Mäkelä et al. 2000), photosynthetic apparatus protection (Allakhverdiev et al. 2003; Cha-Um and Kirdmanee 2010), and reduction of ROS (Ashraf and Foolad 2007).

Lutts (2000) suggested that GB may have a positive impact on both absorption and translocation of monovalent cations in salt-stressed rice plants and that its synthesis by transferring gene coding for choline monooxygenase (CMO) may constitute an interesting goal for genetic engineering in this species. From their study, it was observed that the presence of GB in nutrient solution had no deleterious effect on unstressed plants, but it clearly improved surviving percentages and growing abilities of salt-treated plants. The positive effect of exogenous GB was associated with reduced Na^+ accumulation and with the maintenance of K^+ concentration in all parts of salinized plants. Rahman et al. (2002) reported the beneficial effect of GB on the ultrastructure of salt-stressed *O. sativa* seedlings. While exposed to 150 mM NaCl, the seedlings ultrastructural damages such as swelling of thylakoids, disintegration of grana staking and intergranal lamellae and disruption of mitochondria have been reported. However, these damages were largely prevented by pretreatment of plants with GB. These effects might be due to the production of many vacuoles in the root cells which acted as store of Na^+ and prevented its accumulation in the shoots. In their experiment, Cha-Um and Kirdmanee (2010) applied GB as foliar spray in salt-sensitive *O. sativa* plants exposed to 150 mM of NaCl stress. The results showed that GB treated plants maintained water use efficiency (WUE) and pigment stabilization, leading to high photosynthetic performance in Chl *a* fluorescence and CO_2 assimilation, and increasing plant height under salt stress condition which provided a notion that exogenous application of GB in optimum doses should be used as a short-term technique for the improvement of salt tolerance in *O. sativa*.

2.5.1.3 Trehalose

Trehalose (Tre) functions as compatible solute and is upregulated in plants under abiotic stress (Zeid 2009; López-Gómez and Lluch 2012). It plays an osmoprotective role in physiological responses, enhancing the plant's tolerance to abiotic stress. Nounjan et al. (2012) reported that exogenous Tre treatment under salt stress condition reduced the Na^+/K^+ ratio and strongly decreased endogenous Pro in *O. sativa* seedlings. Transcription of P5CS and P5CR was enhanced while the activities of SOD and POX were decreased and the activity of APX increased and the transcription of all antioxidant enzyme genes upregulated. However, exogenous Tre did not alleviate growth inhibition during salt stress. Pre-soaking maize seeds with Tre (10 mM) showed better performance under salinity stress condition as reported by Zeid (2009). Trehalose pretreatment alleviated the adverse effects of salinity stress in maize seedlings. Hill-reaction activity, photosynthetic pigments and nucleic acids content increased in response to Tre application. Trehalose treatment also ameliorated salinity stress through stabilization of plasma membranes by decreasing the rate of ion leakage, and increasing the ratio of K^+/Na^+ in the leaves of *Z. mays* seedlings.

However, some results indicated that Tre genetically engineered plants exhibit altered morphology, possibly caused by toxicity of high trehalose concentrations, indicating that Tre is a noncompatible solute (Schluepmann et al. 2003; Cortina and Culianez-Maciá 2005). Thus the role of Tre as compatible solute in plants under abiotic stress is still under discussion. Although transgenic plants with microbial Tre biosynthesis often lead to developmental aberrations, diverse studies have shown that Tre accumulation is involved in protecting plants from stress like salinity (López-Gómez and Lluch 2012). These results are promising for the generation of crops resistant to stress. Further studies are required for better understanding the role of Tre in plant protection.

2.5.2 *Plant Hormones*

2.5.2.1 Abscisic Acid

Abscisic acid (ABA) is an important phytohormone that plays an important role in response to various abiotic stresses and stress signaling. ABA also play important roles in many physiological processes like seed dormancy and delays in germination, development of seeds, acceleration of stomatal closure, synthesis of storage proteins and lipids, leaf senescence, etc. (Tuteja 2007). One of the major functions of ABA seems to be the regulation of plant water balance and osmotic stress tolerance. Although the direct relation between stress tolerance and increased levels of ABA does not always exist during the last two decades, it has been well established that ABA is a vital cellular signal that mediates the expression of a number of salt and water deficit-responsive genes. Koornneef et al. (1998) reported several ABA deficient mutants viz. *aba1*, *aba2* and *aba3* in *Arabidopsis*. ABA deficient mutants for *N. tabacum*, *L. esculentum* and *Z. mays* have also been reported (Swamy and Smith 1999). It was observed that without any stress treatment the growth of these mutants is comparable to wild type plants. However, under salt stress ABA deficient mutants showed poor growth (Xiong et al. 2001).

Upon exposure to salinity, plant shows a proportional increase in ABA concentration which is mostly correlated with water potential of leaf or soil. This suggests that salt-induced endogenous ABA is due to water deficit rather than ionic toxicity (Zhang et al. 2006a). This may not be similar to the prolonged increasing of endogenous ABA levels that can occur in association with slowly increasing salinity stresses in nature or field situations (Etehadnia et al. 2008). Increase of endogenous ABA concentration in leaf tissue for salt stressed crop were reported in many plant studies (Cramer and Quarrie 2002; Kang et al. 2005; Cabot et al. 2009; Atkinson and Urwin 2012; Babu et al. 2012). Jeschke et al. (1997) reported that the increase of ABA concentration in the xylem is correlated with reduced leaf conductance and general inhibition of leaf growth. Salt stress stimulated ABA synthesis in roots and its xylem transport are well correlated to the stomatal reactions. This may be explained by the fact, when roots are directly exposed to salt, ABA in roots stimulates ion accumulation in vacuoles which may be necessary for adaptation to saline conditions (Jeschke et al. 1997). Later, Fricke et al. (2004) observed that ABA induced the increase of xylem water potential as well as water uptake to the plant under saline condition.

While studying with salt sensitive (IR29) and tolerant (IR651) varieties of indica rice (*O. sativa*) to a range of salinity (0 and 100 mM NaCl), Saeedipour (2011) observed that tolerance of IR29 to saline stress was generally improved by ABA treatment and leaf Na^+ content reduced to their respective control treatment. This ABA effect was evident in IR29 (sensitive) the ability to recover from stress increased up to 7-fold. Independent of the saline treatment, the absolute endogenous leaf ABA content in sensitive variety was significantly more than tolerant one. However, upon stress, the increase in endogenous ABA synthesis was higher in tolerant than in sensitive varieties. On the other hand, using inhibitor of ABA synthesis, the opposite effect was observed in most of the cases. In another experiment, Gurmani et al. (2011) found that the addition of ABA to *O. sativa* cv. IR-6 has a significant role in reducing salinity stress. ABA was found to be effective in reducing Na^+ and Cl^- concentrations and Na^+/K^+ ratio, increasing K^+ and Ca^{2+} concentrations, Pro accumulation, soluble sugar content. Compared to NaCl alone, ABA treatment increased grain yield by 21%. Keskin et al. (2010) reported that the *MAPK4*-like, *TIP1* and *GLP1* genes were induced more rapidly in response to ABA treatment in *T. aestivum*.

2.5.2.2 Indole Acetic Acid

Although there are very few reports regarding the relationship between auxin level and salt stress in plants as well as the role of auxin in alleviating salt-induced damages. Some research reports indicated that indole acetic acid (IAA) responded to salinity in crop plants. The differences in IAA content under stress conditions appeared to be similar to those of ABA (Ribaut and Pilet 1991).

A significant decline in the level of IAA in the root system of *T. aestivum* plants under salinity was reported by Sakhabutdinova et al. (2003). Similar reduction of IAA level after NaCl treatment was reported in *O. sativa* (Nilsen and Orcutt 1996) and *L. esculentum* (Dunlap and Binzel 1996). Pre-soaking *T. aestivum* seeds with IAA alleviated the growth inhibiting effect of salt stress (Sastry and Shekhawa 2001; Afzal et al. 2005). There is a report that an increase in IAA contents promotes the formation of an attraction signal in the leaf growth zone in response to salt stress (Akhiyarova et al. 2005). It was reported that germination of wheat seed declined with higher salinity level, while this adverse effect was reversed by treatment of seeds with IAA (Gulnaz et al. 1999). Later, Akbari et al. (2007) showed that application of auxin increased hypocotyls length, seedling fresh and dry weight and hypocotyls dry weight of wheat plants under salinity. In *Z. mays* plants, foliar application of IAA, especially at 2 mM, counteracted some of the salt induced adverse effects by enhancing essential inorganic nutrients as well as by maintaining membrane permeability (Kaya et al. 2009). Exogenous IAA significantly reduced Na^+ concentration and increased those of Ca^{2+} and K^+. Application of IAA also increased Chl content, RWC and grain yield while electrolyte leakage and Pro content decreased (Kaya et al. 2009). In contrary, it was observed that when IAA was added with NaCl, the root growth was inhibited (Jemâa et al. 2011).

2.5.2.3 Gibberellic Acid

Gibberellic acids (also called Gibberellin A_3, GA, and GA_3) are generally involved in growth and development; they control seed germination, leaf expansion, stem elongation and flowering (Magome et al. 2004; Kim and Park 2008). Additionally, GAs interact with other hormones to regulate various metabolic processes in the plants. However, many conflicting theories have been put forward concerning their interactions (Yang et al. 1996; Van Huizen et al. 1997). In order to alleviate deleterious effects of salinity, different types of phytohormones have been used. Among them, GA_3 have been the main focus of some plant scientists. Innumerable works have confirmed the potential of GA_3 to synergistically improve crop performance under normal conditions. In recent decades, light has been thrown on the influence of GA_3 during salt stress (Kaya et al. 2009). Maggio et al. (2010) reported that GA_3 treatment in *L. esculentum* reduced stomatal resistance and enhanced plant water use at low salinity. GA_3-priming-induced increase in *T. aestivum* grain yield was attributed to the GA_3-priming-induced modulation of ions uptake and partitioning (within shoots and roots) and hormones homeostasis under saline conditions (Iqbal and Ashraf 2010). Under saline conditions, seed germination has been improved by application of GA_3 and in this experiment, growth and grain yield of wheat were decreased with increasing salinity levels, but increased relatively by seed treatment with GA_3 (Kumar and Singh 1996). In addition, GAs interacts with other hormones to regulate various metabolic processes in the plants. In *B. juncea*, the application of

10 μM GA_3 appeared to mitigate the adverse effects of salinity stress on the overall performance and productivity. Application of GA_3 significantly increased leaf area, dry mass, leaf Chl content, stomatal conductance and photosynthesis rate compared to salt alone (Shah 2007). In another study, application of GA_3 counteracted the adverse effects of NaCl salinity on relative water content, electrolyte leakage and Chl content (Ahmad et al. 2009). GA_3 was sufficient to attenuate partially the stimulatory effect of NaCl supply on Pro and GB biosynthesis in *B. juncea* (Ahmad 2010). Application of GA_3 also reduced lipid peroxidation in the leaves, which was increased during salt stress and thus indicated that application of GA_3 reduced the harmful effects of salinity and increased resistance to salinity (Ahmad et al. 2009). In sugarcane plantlets, foliar application of GA_3 (100 ppm) play an important role on imparting salt tolerance in terms of enhancing nutrient uptake, as well as the morphological and physiological aspects. The inhibition of the growth of sugarcane plantlets by salt stress was removed by GA_3. Exogenous GA also increased sugar and soluble protein content, while Chl content remained unchanged (Shomeili et al. 2011). The application of GA_3 reduced the inhibitory effect of NaCl on growth attributes and photosynthetic pigments in *Hibiscus sabdariffa* by inducing the enzyme activity and enhancing RWC and thus GA_3 helped in the tolerance of plants to salt stress (Ali et al. 2011). Priming of *Beta vulgaris* seeds with GA_3 increased the final germination percentage and germination rate under saline condition. Priming is also responsible for the alleviation of adverse effect of salt stress on sugar beet in terms of root and shoot length and root and shoot fresh weights of plants. (Jamil and Rha 2007). Hamayun et al. (2010) reported that exogenous GA_3 also mitigated the adverse effects of salt stress in *Glycine max* by regulating the level of phytohormones, thus aids the plant in resuming its normal growth and development. Phytohormonal analysis of soybean showed that the level of bioactive gibberellins (GA_1 and GA_4) and jasmonic acid increased in GA_3 treated plants, while the endogenous ABA and salicylic acid (SA) contents declined under the same treatment (Hamayun et al. 2010). Recently, Iqbal and Ashraf (2010) reported that increased grain yield in *Triticum aestivum* was attributed to the GA_3-priming-induced modulation of ions uptake and partitioning (within shoots and roots) and hormones homeostasis under saline conditions. However, the mechanisms by which GA_3-priming induce salt tolerance in plants are not yet clear. Salinity perturbs the hormonal balance in plants. The hormonal homeostasis under salt stress therefore might be the possible mechanism of GA_3-induced plant salt tolerance. Iqbal and Ashraf (2010) hypothesized that pre-sowing treatment with GA_3 could modulate growth by interacting with other endogenous plant hormones.

2.5.2.4 Jasmonic Acid

Jasmonic acid (JA) and its methyl esters are ubiquitous in plants and have hormone properties. These are important cellular regulators involved in diverse developmental processes, such as seed germination, root growth, fertility, fruit ripening, senescence and stomatal closure (Wasternack and Hause 2002; Cheong and Choi 2003; Hossain et al. 2011). Like other phytohormones JA have both synergistic and antagonistic effects. Jasmonate derivatives induce the accumulation of so-called JA-induced-proteins that were found in all plant species tested. However, the role of most of the derivatives of JA is still unclear.

Jasmonates are involved in plant responses to various abiotic stresses and elicit unique responses (Rohwer and Erwin 2008). There are few reports on the role of exogenous JA in plant response to salt stress. It has been reported that jasmonate treatments (or endogenous of these compounds) is accompanied by the synthesis of abundant proteins in response to abiotic stress, called JIPs (Sembdner and Parthier 1993). Pretreatment of *Pisum sativum* seedlings with 10 μM JA counteracted the effect of salt stress by increasing the photosynthesis, RWC and protein content (Fedina and Tsonev 1997). Exogenously supplied JA itself plays the role of a stressor that causes typical stress responses like accumulation of free Pro, high photorespiration, etc. Pretreatment with JA also leads to a decrease of Na^+ and Cl^- accumulation in the shoot. This protection established the involvement of MeJA in osmoregulation or osmoprotection based on increased Pro accumulation and decreased ion accumulation (Fedina and Tsonev 1997). Pedranzani et al. (2003) reported that JA levels in *L. esculentum* cultivars changed in response to salt-stress and JA increase was observed in salt tolerant cultivar from the beginning of salinization, while in salt sensitive cultivar JA level decreased after 24 h of salt treatment. Exogenous JA application after salt treatment may change the balance of endogenous hormones, such as ABA, which provides an important clue for understanding the protection mechanisms against salt stress (Kang et al. 2005). Seo et al. (2005) reported that treatment with JA in the presence of salt stress increased the GAs content. However, endogenous content of bioactive GA_1 was higher in post-treatment by JA than in pre-treatment by JA. In *G. max*, treatment with exogenous JA mitigated the harmful effect of NaCl (50 mM NaCl). The greatest yield (157% of control) was obtained from soybean plants sprayed with JA (Sheteawi 2007). JA also reduced the salt effects on seed carbohydrates, lipids, proteins, N, P and K. Yoon et al. (2009) observed that pretreatment with MeJA (20 and 30 μM) counteracted the negative effects of NaCl stress on plant growth, Chl content, leaf photosynthetic rate, leaf transpiration rate, and Pro content of hydroponically gown *G. max* seedlings. Pretreatment with MeJA also significantly increased ABA levels. Kang et al. (2005) reported that post-application with exogenous JA can ameliorate salt-stressed rice seedlings, especially the salt-sensitive rather than the salt-tolerant cultivar. However, there seems to be little information about how salinity affects endogenous JA levels in plants.

2.5.2.5 Salicylic Acid

Salicylic acid (SA) is a common plant-produced phenolic compound and a potential endogenous plant hormone that plays an important role in plant growth and development. The role of SA is intensively studied in plant responses to biotic stress. In recent years the involvement of SA in the response to abiotic stresses has widely been studied (El Tayeb 2005; Ahmad et al. 2011). However, the actual role of SA in abiotic stresses remains unresolved. Several methods of application (soaking the seeds prior to sowing, adding to the hydroponic solution, irrigating, or spraying with SA solution) have been shown to protect various plant species against abiotic stress by inducing a wide range of processes involved in stress tolerance mechanisms (Horvath et al. 2007).

El Tayeb (2005) found that SA application to barley induced a pre-adaptive response to salt stress, enhanced the synthesis of Chl *a*, Chl *b* and Car, and maintained membrane integrity, leading to improvement of plant growth. SA-pretreated plants exhibited less Ca^{2+} and more accumulation of K^+, and soluble sugars in roots under saline condition (El Tayeb 2005). *Zea mays* treated with SA exhibited increased growth, decreased lipid peroxidation and membrane permeability, which were increased by salt stress (Gunes et al. 2007). In mungbean plants SA alleviates salt-induced decrease in photosynthesis and minimizes the leaf Na^+, Cl^-, and H_2O_2 content (Nazar et al. 2011). This was accompanied by increased N and S assimilation through inducing the activity of NR and ATPs. Exogenous SA also improves grain yield under salt stress in *T. aestivum* (Arfan et al. 2007). The application of SA via root drenching protected *Lens esculentum* against NaCl stress and increased photosynthetic rates under salt stress (Stevens et al. 2006; Poór et al. 2011).

It was found that SA treatment caused accumulation of both ABA and IAA in *T. aestivum* seedlings under salinity. However, the SA treatment did not influence on cytokinin content. Thus, protective SA action includes the development of antistress programs and acceleration of normalization of growth processes after removal of stress factors (Sakhabutdinova et al. 2003). Gémes et al. (2011) suggested that, the cross-talk of signaling pathways induced by SA and high salinity may occur at the level of ROS and NO production. They observed that SA-induced generation of H_2O_2 and NO are considered to be functional links of cross-tolerance to various stressors. SA-stimulated pre-adaptation state was beneficial in the acclimation to subsequent salt stress in *Solanum lycopersicum* (Gémes et al. 2011). At the whole-plant level, SA-induced massive H_2O_2 accumulation only at high concentrations (1–10 mM), which later leads to death of the plant. Torabian (2011) reported that pre-treatment with SA induced adaptive responses in *Medicago sativa* plant under salinity stress and consequently, encouraged protective reactions in biotic membranes which improved the growth of seedlings. SA pre-treatment improved growth and resulted in higher resistance of plants to salinity, so that it increased germination percentage, seed vigor index and growth parameters of the seedlings. Also, salinity intensified electrolyte leakage, while SA decreased it and this decrease was stronger at SA concentration (Torabian 2011).

Erdal et al. (2011) investigated the effects of foliar-application of SA on salt sensitivity of *T. aestivum*. They observed that salt-induced deleterious effect in wheat seedlings were significantly alleviated by the SA treatment. SA can be used as a signal molecule to investigate plant defense to abiotic stress. After the application of SA, increasing tolerance of wheat seedlings to salt stress may be related to increases in antioxidative enzyme activity. Exogenous SA treatment significantly increased the fresh and dry weights in both root and shoots of wheat plants under salt stress. In parallel to increasing antioxidant activity, SA treatment decreased H_2O_2 content when compared to plants growing under salt stress without SA. In *Brassica juncea*, Yusuf et al. (2012) reported that SA enhanced the level of antioxidant system (SOD, CAT and POX) both under stress and stress-free conditions. However, the influence of SA on antioxidant system was more pronounced under stressful condition, therefore, suggesting that the elevated level of antioxidant system might be responsible for increased tolerance of *B. juncea* plants to NaCl stress.

However, some studies demonstrate that application of SA (0.5 mM) may promote the formation of ROS in the photosynthetic tissues and increase oxidative damage during salt and osmotic stresses. For instance, Barba-Espín et al. (2011) studied the effect of SA treatment on the response of *P. sativum* plants to salinity. NaCl-induced damage to leaves was increased by SA, which was correlated with a reduction in plant growth. The content of AsA and GSH in leaves of salt-treated plants increased in response to SA, although accumulation of the respective DHA and GSSG occurred. An increase in H_2O_2 also occurred in leaves of salt-exposed plants treated with SA. Negative effect of SA in the *P. sativum* plants exposed to NaCl was also correlated with an imbalance in antioxidant metabolism. Generally, deficiency of SA or a very high level of SA increases plant susceptibility to abiotic stresses. The optimal concentration (0.1–0.5 mM for most plants) enhances abiotic stress tolerance.

2.5.2.6 Brassinosteroids

Brassinosteroids (BRs) is the most recent group of phytohormones and is a class of over 40 polyhydroxylated sterol derivatives, ubiquitously distributed in all kinds of plant. Their strong growth-inducing capacity, recognized as early as prior to their identification in 1979, tempted the scientists to visualize the practical importance of this group of phytohormones (Hayat and Ahmad 2011). Recently, the physiological, cellular, and molecular mechanisms by which BRs regulate various aspects of plant development are being discovered (Yang et al. 2011). Although there are different kinds of BR, 24-epibrassinolide (EBR) is one of the most widely used BR. The BRs have a potential application in agriculture to increase yield and to stimulate crop growth under stress (Houimli et al. 2010; El-Mashad and Mohamed 2012; Hayat and Ahmad 2011).

The effect of EBR and 28-homobrassinolide (HBR) on the inhibition of germination and seedling growth of rice (*O. sativa*) induced by salinity stress was studied by Anuradha and Rao (2001). They reported that application of BRs reverse the inhibitory effect on germination and seedling growth. The activation of seedling growth by BRs under salinity stress was associated with enhanced levels of nucleic acids and soluble proteins (Anuradha and Rao 2001). The effect of BR on *H. vulgare* leaf cell ultrastructure was examined under salt stress. Leaf segments were pre-incubated in either BR solution or water and then incubated in 0.5 M NaCl solution in presence or absence of BR. BR had no effect on leaf cell ultrastructure under normal conditions. However, damages imposed by salt stress on nuclei and chloroplasts were significantly reduced by BR treatment (Krishna 2003). When *Capsicum annuum* seedlings were sprayed with EBR in the presence of NaCl, it significantly ameliorated the adverse effects of salinity by increasing the RWC, photosynthetic pigments and decreasing the electrolyte leakage. Exogenous BR also increased the fresh and dry weight of plant parts under salt stressed condition (Houimli et al. 2010). Another recent study reported the ameliorative effect of EBR and ethylene on germination of *Cucumis sativus* seeds in presence of NaCl (250 mM). The reduction in ethylene evolution from imbibed seeds by salt stress was attenuated by EBR. In maize seedlings, pre-sowing treatments of HBR enhanced the activities of antioxidative enzymes (SOD, GPX, CAT, GR, APX) and minimized the lipid peroxidation thus helps the plants to withstand the oxidative stress induced by salt stress. In *T. aestivum*, exogenous application of EBR increased plant biomass under saline condition, but it had no prominent effect on accumulation of different mineral nutrients (Shahbaz and Ashraf 2007). Foliar application of BRs also increased yield attributes of salt treated *T. aestivum* and significantly overcome the negative effect of salinity on crop productivity and photosynthetic pigments (Eleiwa et al. 2011). Salinity reduced sugar (reducing and non-reducing), total carbohydrate and protein percentage of grains while foliar application with BRs significantly increased gradually all the chemical constituents. Foliar application of BRs also significantly increased the concentration and total uptake of macro and micronutrients (N, P, K, Fe, Mn, Zn and Cu) in straw and grains (Eleiwa and Ibrahim 2011). El-Mashad and Mohamed (2012) also reported that brassinolide enhanced tolerance of *V. sinensis* plants to NaCl. They observed that foliar spray of BR (0.05 ppm) mitigated salt stress by inducing activities of enzymatic and non-enzymatic antioxidants, e.g., SOD, POX, polyphenol oxidase, AsA, tocopherol, and GSH. Recently, Samira et al. (2012) showed that the EBR treated plants had greater relative growth rate compared to untreated plants when exposed to salt stress. Application of EBL increased photosynthesis by increasing stomatal conductance in salt stressed plants and may have contributed to the enhanced growth. The water use efficiency was also improved by the application of EBL.

2.5.3 *Antioxidants*

2.5.3.1 Ascorbic Acid

Ascorbate (AsA) or ascorbic acid (Vitamin C) is an important antioxidant in plant tissue which is synthesized in cytosol of higher plants primarily from conversion of D-glucose to AsA. AsA has been shown to have an essential role in several physiological processes in plants, including growth, differentiation, and metabolism. It functions as a reductant for many free radicals, thereby minimizing the damage caused by oxidative stress. Plant with higher amount of AsA content showed better protection against oxidative stress. Ascorbate influences many enzyme activities, minimizing the oxidative damage through synergic function with other antioxidants

(Foyer and Noctor 2005a, b). Ascorbic acid reacts with a range of ROS such as $^1O_2$, $O_2^{\bullet -}$, HO• and H_2O_2, which is the basis of its antioxidant action (Shigeoka et al. 2002; Foyer 2004). The role of AsA as a cofactor for a range of oxygenase and hydroxylase enzymes is also dependent on its reducing activity (De Tullio 2004). Upon oxidation, by loss of one electron, the monodehydroascorbate (MDHA) radical is formed and this is usually the initial product of AsA oxidation in biological systems (Smirnoff and Pallanca 1996; Noctor and Foyer 1998). AsA can also directly scavenge and regenerate tocopherol from tocopheroxyl radicals, thus providing membrane protection (Li and Jin 2007). AsA also plays a role as a co-factor of violaxanthin de-epoxidase, thus sustaining dissipation of excess excitation energy (Pourcel et al. 2007).

AsA plays an important role in plant stress tolerance. Under stressed condition plants showed different capacity of AsA metabolism which is due to the variation of AsA synthesis and regeneration. Different studies showed that AsA content in leaves of stressed plants tends to increase with increasing levels of salt stress (Mohamed et al. 2010). Agarwal and Shaheen (2007) reported that AsA concentration in leaves of *Momordica charantia* increased under NaCl stress as compared to control. Increase in AsA concentration due to salinity was reported by other researchers (Panda and Upadhyay 2004; Parida et al. 2004).

Exogenous application of AsA influences many enzyme activities and minimizes the damage caused by oxidative processes through a synergic function with other antioxidants (Shalata and Neumann 2001; Athar et al. 2008). This multiplicity of functions has led some researchers to suggest that, in addition to being a powerful antioxidant and redox buffer, AsA may be a signaling molecule involved in the regulation of complex processes such as the senescence of plants and their response to O_3, photo-oxidative conditions, or pathogen attack (Pastori et al. 2003). Exogenous application of AsA helps the *L. esculentum* seedling to recover from salt stress (Shalata and Neumann 2001). They observed that the addition of exogenous AsA to the root medium remarkably increased seedling survival and decreases lipid peroxidation. Hamada and Al-Hakimi (2009) found that exogenously applied AsA were generally effective partially or completely countering the inhibitory effects of salt stress on net photosynthetic rate, pigments biosynthesis and membrane integrity by exerting a stimulatory action on these parameters, especially in plants subjected to moderate and low salinity levels. The leakage of K^+ was also reduced by the application of AsA. Khan et al. (2006) applied AsA as foliar spray (0, 50, 100 mg L^{-1}) on *T. aestivum* grown in hydroponics. They observed that foliar spray with AsA improved the growth of non-stressed plants of both cultivars, but did not alleviate the adverse effects of salt stress on plants. However, salt-induced reduction in leaf Chl *a* was improved with AsA application. AsA application enhanced the Na^+ accumulation in the leaves of salt stressed plants of both cultivars, but it did not change the K^+ accumulation in the leaves and roots of the salt stressed plants. In vitro experiments were performed by Zeid et al. (2008) in *H. annuus* seedling, to determine responses of *T. aestivum* calli to AsA concentrations (0, 250, 500, 1,000 and 2,000 ppm) under different levels of sea water (0%, 15%, 30% and 45%) and to determine suitable concentrations of AsA to enhance tolerance to salinity. Beltagi

(2008) observed significant synergistic effect between NaCl (40 mM) and AsA treatment, where AsA increased the contents of Chl *a* and Chl stability index (CSI %) in *Cicer arietinum*. Khafagy et al. (2009) observed that pre-soaking of *C. annuum* seeds in AsA partially counteracted the harmful effect of NaCl salinity. Chl *a* and *b* concentrations significantly increased in AsA pre-soaked salt-stressed seedlings compared to the seedlings subjected to salt only.

In a recent study, Azzedine et al. (2011) reported that the application of vitamin C was effective to mitigate the adverse effect of salt stress on plant growth due to increased leaf area, improved Chl and Car contents, enhanced Pro accumulation and decreased H_2O_2 content. Dehghan et al. (2011) reported that exogenously applied AsA counteracts the adverse effects of salt stress on growth of *Glycine max* seedlings which was cultivar specific. AsA induced enhancement in growth of salt-stressed plants coupled with an increase in CAT, POD and SOD activities. Ascorbic acid pretreatment to in vitro-grown sugarcane plants enhance their salt tolerance by enhancing CAT and POD activities soluble protein contents as well as better root length (Munir and Aftab 2011).

2.5.3.2 Glutathione

Glutathione (GSH) is a strong antioxidant which prevents damage to important cellular components caused by ROS (Pompella et al. 2003). It also plays an indirect role in protecting membranes by maintaining α-tocopherol and zeaxanthin in the reduced state. It can also function directly as a free radical scavenger by reacting with $^1O_2$, $O_2^{\bullet -}$ and $HO^{\bullet}$. GSH protects proteins from denaturation caused by oxidation of protein thiol groups under stress. In addition, GSH is a substrate for glutathione peroxidase (GPX) and glutathione-*S*-transferases (GST), which are also involved in the removal of ROS (Noctor et al. 2002).

Glutathione accumulates to high concentrations, especially in stress situations. Increase in GSH concentrations during stress offsets stress-initiated oxidation of GSH and causes changes in gene expression directly or through interaction with regulatory proteins and/or transcription factors. This increase is equally important in signal transduction and defense against ROS and is through a multilevel control mechanism, which includes coordinate activation of genes encoding GSH biosynthetic enzymes and GR (Srivalli and Khanna-Chopra 2008). Thus, GSH acts as a redox sensor of environmental cues, and increase in GSH helps plants to tolerate oxidative stress. Likewise, GSH also plays a protective role in salt tolerance by maintaining the redox state. Investigation on the enzymatic pathways leading to GSH synthesis in wild type and salt-tolerant *B. napus* plants showed assimilation of sulfur and the biosynthesis of cysteine and GSH in order to mitigate salt-induced oxidative stress (Ruiz and Blumwald 2002; Hussain et al. 2008). Sumithra et al. (2006) found that GSH concentration in the salt-stressed mungbean leaves of cv. Pusa Bold was higher than cv. CO 4, whereas GSSG concentration was higher in the leaves of CO 4 than in those of Pusa Bold, indicating that Pusa Bold was more tolerant than CO 4 as the levels of lipid peroxidation and H_2O_2 concentration in Pusa Bold was lower than in CO 4 under salt stress. In addition, maintaining a high ratio of GSH/GSSG plays an important role in salt tolerance (Hasanuzzaman et al. 2011a, b). Salt-tolerant cultivars of cotton had a higher GSH/GSSG ratio than salt-sensitive lines under saline conditions (Gossett et al. 1996).

Using buthionine sulfoximine (BSO) and exogenous GSH, Gossett et al. (1996) investigated the importance of maintaining sufficient GSH pools. BSO reduced the growth of the control cell line by 94%, whereas the NaCl-tolerant cell line showed significantly less growth reduction (Gossett et al. 1996). When medium containing BSO was supplemented with exogenous GSH, growth was restored in both cell lines; however, when GSSG was added to medium with BSO, growth was almost completely restored in only the NaCl-tolerant cell line. This was most likely due to the elevation of GR activity and an increased ability to convert GSSG to GSH. Kattab (2007) reported that *B. napus* seed priming with GSH improved seedling resistance probably by enhancing the activities of antioxidant enzymes (SOD, GPX, POX and APX. Exogenous GSH (0.5 mM) maintained plasma membrane permeability under NaCl stress and cell viability in *Allium cepa* (Salama and Al-Mutawa 2009). However, this effect could partially alleviate the harmful effect of salinity stress which reflected on growth and yield of *T. aestivum* plant. In *Tagetes erecta*, application of GSH (100 or 200 ppm) was found to be effective in increasing plant height, no. of branches, fresh and dry weight of herb and flowers, no. of flowers, total carbohydrates (%), total phenols, xanthophyll pigment content and mineral ion percentage under saline (1,500 ppm NaCl) conditions (Rawia et al. 2011).

2.5.3.3 Tocopherol

Tocopherols belong to Vitamin E family of amphiphilic antioxidants, with the subfamily of tocotrienols. Tocopherols and tocotrienols are synthesized by higher plant plastids and by cyanobacteria. There are four tocopherol and tocotrienol isomers (α, β, γ and δ). Relative antioxidant activity of the tocopherol isomers *in vivo* is $\alpha>\beta>\gamma>\delta$ and hence α-tocopherol has the highest antioxidant activity (Garg and Manchanda 2009). Tocopherols contribute to reduce ROS levels (mainly $^1O_2$ and OH•) in photosynthetic membranes and limits the extent of lipid peroxidation by reducing lipid peroxyl radicals (LOO•) to their corresponding hydroperoxides (Maeda et al. 2005). In addition, tocopherols are part of an intricate signaling network controlled by ROS, antioxidants, and phytohormones, and are therefore good candidates to influence cellular signaling in plants (Munné-Bosch 2007). However, the ability to critically assess the physiological roles of tocopherol has only recently become available with characterization of plant and cyanobacterial mutants affected in its biosynthesis and transgenic plants with increased tocopherol and tocotrienol content.

Several lines of evidence indicate that α-tocopherol plays a major role in plant stress tolerance, keeping an adequate redox state in chloroplasts (Munné-Bosch 2005). However, studies on tocopherol-deficient plants have recently revealed that α-tocopherol is not essential for plant survival under optimal conditions, and that α-tocopherol deficiency leads to only a slightly increased susceptibility to photooxidative stress (Kanwischer et al. 2005). Rady et al. (2011) presoaked *H. annuus* seeds with exogenous α-tochopherol and exposed to saline soil with different salinity levels (EC 1.56, 4.68 and 7.83 dS m^{-1}). They observed that salinity induced decreased in total soluble sugars content and the activities of CAT, POX, PPO and PAL were significantly altered by exogenous α-tochopherol. Treatment with α-tochopherol also improved the mineral nutrient content in the plant with concomitant increase in Pro, free amino acids and total phenol contents in both cultivars. Farouk (2011) reported that α-tocopherol could minimize salt-induced leaf senescence in *T. aestivum*. Exogenous α-tocopherol also enhanced the antioxidant enzyme activities under salt stress which rendered the lower level of H_2O_2 and lipid peroxidation. Salt stressed plants supplemented with α-tocopherol decreased the Na^+ and Cl^- content but increased the K^+, Ca^{2+} and Mg^{2+} contents (Farouk 2011).

2.5.4 Signaling Molecules

2.5.4.1 Nitric Oxide

Nitric oxide (NO) is a gaseous biological molecule which is involved in physiological responses to various abiotic stresses. Recently, NO has emerged as an important signaling molecule and antioxidant. NO triggers many kinds of redox-regulated (defense-related) gene expressions, directly or indirectly, to establish plant stress tolerance (Sung and Hong 2010). Several reports indicated that the application of exogenous NO donors confers tolerance to various abiotic stresses including salinity (Hossain et al. 2010; Xiong et al. 2010; Hasanuzzaman et al. 2011a; Bai et al 2011; Liu et al. 2011). NO exerts a protective function against oxidative stress mediated by reaction with lipid radicals, which stops the propagation of lipid oxidation; scavenge the $O_2^{\bullet-}$ and formation of peroxynitrite ($ONOO^-$) that can be neutralized by other cellular processes. It also helps in the activation of antioxidant enzymes (SOD, CAT, APX, GPX, GR, POX, etc.) and functions as a signaling molecule in the cascade of events leading to gene expression. These mechanism together confer enhance protection against oxidative stress (Hasanuzzaman et al. 2010a; Misra et al. 2011). However, whether or not endogenous NO has an antioxidant function is debatable.

Uchida et al. (2002) reported enhanced tolerance to salt stress (100 mM NaCl, 8 days) in rice seedlings pre-treated with NO (1 μM SNP, 2 days). This pre-treatment induced the activity of antioxidant enzymes (SOD, CAT and APX). Enhanced seed germination and root growth of *Lupinus luteus* seedlings (Kopyra and Gwóźdź 2003) and increased growth and dry weight of *Z. mays* seedlings (Zhang et al. 2006b) were also observed with the treatment of NO donor under salt stressed condition. Fan et al. (2007) showed that exogenous NO (100 μM SNP) significantly alleviated the salt injury to cucumber seedlings and increased seedling growth. In addition, photosynthetic pigment content, Pro as well as the activity of SOD, POD, CAT and APX were also increased. Similarly, net photosynthetic rate, stomatal conductance, and transpiration rate also increased significantly. However, exogenous NO donor markedly decreased membrane permeability, rate of $O_2^{\cdot-}$ production, the contents of MDA and H_2O_2, and intercellular CO_2 concentration. Treating *H. vulgare* leaves with exogenous NO (50 μM SNP), Li et al. (2008) observed that NO could alleviate the damage of salt stress (50 mM NaCl) which was reflected by decreased ion leakage, MDA and H_2O_2 content. Additionally, the presence of the NO donor enhanced the activities of SOD, APX and CAT. David et al. (2010) reported that NO enhanced biochemical adaptation during the seedling growth of *H. annuus* under salinity conditions (40–120 mM NaCl). They found an increased Na^+/K^+ ratio (4-fold) in roots, and Na^+ was rapidly transported to the cotyledons, which registered a concomitant increase in this ratio. They also concluded that the origin of this endogenous generation of NO appears to be mediated by NO synthase (NOS) activity (David et al. 2010). Exogenous NO supplementation as SNP has significant ameliorating effect against NaCl induced oxidative damage in *Cicer arietinum* leaves as observed by Sheokand et al. (2010) who exposed 5-day-old plants to NaCl treatment (250 mM) alone and in combination with two concentrations of SNP (0.2 and 1 mM) for 2, 4 and 6 days. Both the SNP treatments had a positive effect on antioxidant enzymes SOD, CAT, APX, GR and DHAR under salt stress. NaCl treatment resulted in a decline in the GSH/GSSG and AsA/DHA ratio; however, SNP treatments increased the reduced form of both the metabolites thus elevating the ratio of GSH/GSSG and AsA/DHA. Exogenous NO partially decreased MDA and H_2O_2 content. When exposed to NO donors, NO-associated salt priming action was evident in halophytes in tolerating high salinity during germination and early growth stages (Molassiotis et al. 2010) which was due to the better induction of antioxidant enzyme activity in response to high salinity conditions. Recently, Corpas et al. (2011) reported that under salt stress the osmotic stress-activated protein kinase (NtOSAK) is activated by NO and confer stress signals. While studying with *L. esculentum* cv. Hufan1480 and Hufan2496, Wu et al. (2011) observed notable improvement of growth and enhanced antioxidant defense in salt-stressed (100 mM NaCl) plants when treated with exogenous NO (100 μM SNP). They observed that in presence of 100 μM SNP under salt stress, the reduction in shoot and root dry mass declined to 16% and 3%, respectively in Hufan1480, and to 21% and 6%, respectively in Hufan2496. The MDA content of Hufan1480 and Hufan2496 decreased significantly by 22% and 12% over the salt treatment, respectively. The rate of $O_2^{\bullet-}$ production in Hufan1480 and Hufan2496 decreased significantly by 20% and 17%, respectively by application of 100 μM SNP under salt stress. A remarkable increase in the activities of SOD, POD, CAT, APX, AsA and GSH were recorded by NO treatments under stress condition. In our recent study, we observed that exogenous NO modulated the ROS detoxification systems in *T. aestivum* seedlings (Hasanuzzaman et al. 2011a). The seedlings pretreated with NO donor (1 mM SNP, 24 h) when exposed to salt (150 and 300 mM NaCl, 4 days) showed an increase in the AsA and GSH contents and the GSH/GSSG ratio as well as the activities of monodehydroascorbate reductase (MDHAR), dehydroascorbate reductase (DHAR), GR, GST and GPX as compared to the seedlings without NO pretreatment, which ultimately decreased the contents of MDA and H_2O_2.

2.5.4.2 Hydrogen Peroxide

It is generally thought that H_2O_2 is a ROS and for many years and it was viewed as the inevitable but unwanted by-product of an aerobic respiration. But recent studies have shown that it has important role in redox signaling in regulating normal processes, including oxidative stress and thus it has been established as a 'necessary evil for cell signaling' (Rhee 2006). The function of H_2O_2 as a signaling molecule in transduction of stress signals to the alteration of expression profiles of target genes was also studied in plants (Hung et al. 2005; Hernandez et al. 2010). The connection between H_2O_2 and signaling networks has been extensively documented for a number of stress responses (Larkindale and Knight 2002; Apel and Hirt 2004; Cheeseman 2007).

Recent studies have demonstrated that pretreatment of plants with exogenous H_2O_2 confers abiotic stress tolerance including salinity. Azevedo Neto et al. (2005) reported that addition of H_2O_2 to the nutrient solution induces salt tolerance by enhanced activities of antioxidants and reduced peroxidation of membrane lipids in leaves and roots of maize as an acclimation response. Wahid et al. (2007) reported that *T. aestivum* seeds soaked in 1, 40, 80 and 120 mM H_2O_2 and subsequent growing in saline condition (150 mM NaCl) showed that the level of H_2O_2 in seedlings arising from H_2O_2-treated seeds grown under salinity was markedly lower than the salinized controls, suggesting the operation of antioxidant system in them. These seedlings also exhibited better photosynthetic capacity, particularly the stomatal conductance (gs), thus improving the leaf gas exchange due to stomatal component of photosynthesis. Moreover, H_2O_2 treatment improved leaf water relations and maintained turgor. Although Na^+ and Cl^- content increased due to salinity, H_2O_2-treated seedlings displayed greater tissue K^+, Ca^{2+}, NO_3^- PO_4^{3-} levels and improved K^+:Na^+ ratio. Exogenous H_2O_2 treatment enhanced the membrane properties, as revealed from greatly reduced relative membrane permeability (RMP) and less altered ion leakage pattern (comparable to water controls). Fedina et al. (2009) reported that pretreated *Hordeum vulgare* seedlings with 1 and 5 μM H_2O_2 for 2 days followed by exposure to 150 mM NaCl for 4 and 7 days showed higher rate of $^{14}CO_2$ fixation with lower MDA, H_2O_2 and Pro contents in comparison to the seedlings subjected NaCl stress only. In addition, Cl^- content in the leaves of NaCl treated plants considerably less in pre-treated plants. These results clearly indicated that H_2O_2 metabolism is involved as a signal in the processes of salt tolerance. Recently, Yadav et al. (2011) observed that seeds of *C. annuum* primed with H_2O_2 (1.5 mM) showed enhance tolerance to salt stress (NaCl, 200 mM, 10 days). The plants grown from primed seeds flowered earlier and also produced more number of fruits.

2.5.5 *Polyamines*

Polyamines (PAs) are ubiquitous low-molecular-weight aliphatic amines that are involved in regulation of plant growth and development. PAs are also implicated in a wide range of environmental stress tolerance in plants. New roles are being discovered every day for these interesting molecules in the plant world. In higher plants, the most common PAs are spermidine (Spd), spermine (Spm) and their diamine obligate precursor putrescine (Put). Like PAs displaying high biological activity are involved in a wide array of fundamental processes in plants, such as replication and gene expression, growth and development, senescence, membrane stabilization, enzyme activity modulation and adaptation to abiotic stresses (Kuznetsov and Shevyakova 2007; Gill and Tuteja 2010; Hussain et al. 2011b; Shu et al 2012; Alet et al. 2012). However, the precise physiological function and mechanism of action of PAs still remain unclear. In contrast to the reliable works on the role of PAs in plants defense against biotic and abiotic stresses, few reports recently indicated that PAs may act as cellular signals in intrinsic talk with hormonal pathways including ABA (Alcazar et al. 2010a, b; Gill and Tuteja 2010). Additionally, PAs like Spm and Spd are regarded as potent inducers of NO in plants which is another potent signaling molecule (Tun et al. 2006).

Changes in plant PA metabolism occurs in response to a variety of abiotic stresses (Alcazar et al. 2006; Gill and Tuteja 2010). These changes in cellular PA under stress only provide clues on its possible implication in stress response, but they do not provide evidence of its role in counteracting stress. Hence, to understand whether PA actually protect cells from stress-induced damages, exogenous application of PA, which is expected to increase endogenous PA, has been investigated before or during stress (Velikova et al. 2000; Navakouidis et al. 2003; Wang et al. 2007). It has been reported that exogenous application of PAs could alleviate salt-induced reduction in photosynthetic efficiency, but this effect is strongly depended both on PAs concentration or types and stress levels (Duan et al. 2008). The efficiency of PSII (Fv/Fm) measured in leaves of salt-stressed *Cucumis sativus* seedlings was not much influenced by 1 mM Spd application, although Spd could ameliorate plant growth and increase net photosynthetic rate (P_N), *Gs*, intercellular CO_2 concentration (Ci), actual efficiency of photosystem II (ΦPSII) and the coefficient of photochemical quenching (qP) of *C. sativus* seedlings subjected to salinity (Li et al. 2007). In another study, 10 mM Put alleviated the reduction of salt stress on P_N. However, Put had no effect on gas exchange and transpiration rate (Tr), and aggravated the reduction of salt stress on Ci. The result obtained by Zhang et al. (2009) suggested that Put strongly affects photosynthetic apparatus involving in enhancement of photochemical quenching rather than regulation of stomatal closure or opening. Several publications have reported that changes of endogenous PA level and forms are involved in regulating the photochemical efficiency of salt-stressed plants, and PAs metabolism-related enzymes are closely correlated with photosynthesis. Exogenous PAs increased bound Spd contents in chloroplasts to enhance the photosynthetic capacities of *Z. mays* exposed to salt stress (Liu et al. 2006). Yamaguchi et al. (2006) reported specific role of Spm during high salt stress using an *Arabidopsis* double knockout mutant plant (*acl5/spms*) which cannot produce Spm. The mutant showed higher sensitivity to high salt than wild type plants. This phenotype was cured by exogenous Spm but not by other PAs, i.e. Put and Spd, suggesting a strong link between Spm-deficiency and NaCl-hypersensitivity. Duan et al. (2008) applied exogenous Spd to salinized nutrient solution which resulted in alleviation of the salinity-induced membrane damage, growth and photosynthesis inhibition, together with an increase in PA and Pro contents as well as antioxidant enzyme activities in the roots of *C. sativus*. Seedlings of *Sorghum bicolor* were subjected to salt stress (180 mM NaCl, 7 days) supplemented with 0.25 mM Spm, showed improved growth and partial increase in activities of POX and GR with concomitant decreased in MDA content. However, Spm had no effects on soluble protein and Pro content in response to salt stress (Chai et al. 2010).

Quinet et al. (2010) found that Put differently influences the effect of salt stress on PA metabolism and ethylene synthesis in *O. sativa* cultivars differing in salt resistance. Exogenous Put reduced Na^+ accumulation in shoots and roots of salt-treated plants of susceptible cultivar while no change was obtained in tolerant one. Amri et al. (2011) showed that the use of different degrees of exogenous PA can reduce the effects of salt stress on growth of *Punica granatum*. PA treated seedlings under salt stressed condition did not show clear differences in plant growth, however it significantly reduced the Na^+ and Cl^- content and increased Pro content. Application of Put reduced the membrane injury index and increased RWC, photosynthetic rate, and pigments content of *Citrus karna* under saline conditions compared to plants exposed to NaCl in the absence of Put (Sharma et al. 2011). Application of Put alone or in combination also improved the activities of SOD, POD and Pro content under saline conditions. More importantly, application of Put increased K^+ and reduced Na^+ and Cl^- concentrations in leaf tissues which indicated that Put could improve the tolerance of salt-susceptible Karna khatta by regulating absorption and accumulation of ions as well as improving antioxidant enzyme activities (Sharma et al. 2011). In *C. sativus* seedlings, exogenous Put regulated ion distribution in salt-stressed plants, especially, by preventing the accumulation of Na^+ and Cl^- in the leaves which were associated with an improvement of the actual PS II efficiency which rendered the plants more tolerant to salt stress (Shu et al. 2010). The photosynthesis of salt-stressed *C. sativus* was enhanced by exogenous Spd that mitigates the decreased stomatal conductance under salt stress as reported by Li et al. (2007). Spd had better effects on the growth and photosynthesis in leaves of salt-stressed seedlings than roots. Recently, Zhang et al. (2011) observed that exogenous Put concentrations significantly increased growth, photosynthesis and decreased lipid peroxidation of *C. sativus* seedlings under salt stress. Improved photosynthesis in plants due to Put application was due to its modulation capacity of photosynthetic proton circuit as recently reported by Ioannidis et al. (2012). Gupta et al. (2012) found a Ca^{2+} independent auto regulatory cytoplasmic protein which is phosphorylated in root cytosolic fraction during NaCl/ABA/Spd treatment indicating its importance in salinity mediated signal transduction. Anjum (2011) reported that application of Spd (0.1 or 0.5 mM) to the saline nutrient solution and its weekly sprays (1 or 5 mM) on NaCl-stressed plants improved leaf number, Chl content, Fv/Fm, net photosynthetic rate, and N content; increased total Spd and Spm contents; and reduced Na^+ contents in Troyer citrange (*Poncirus trifoliata* × *Citrus sinensis*). In *O. sativa* seedlings, exogenous PAs (Spd and Spm) reversed the inhibitory effect of salinity which was conferred by preventing growth inhibition or various forms of cellular damages, maintaining proper K^+/Na^+ balance or triggering the level of osmolytes and activity of antioxidant enzymes and thus led the plants to tolerate the salt stress (Roychoudhury et al. 2011). Saleethong et al. (2011) investigated the effects of exogenously supplied Spd in two rice cultivars differing in salt tolerance. The major effect of exogenous Spd offered protective roles on salinity-stressed plants by stabilizing membrane, scavenging free radicals and maintaining K^+/Na^+ status. These results indicated that exogenous PAs can be applied as short-term pretreatment prior to introduction of salt stress to increase salt tolerance.

2.5.6 Trace Elements

Selenium (Se), and silicon (Si) are considered as beneficial elements for plants: they are not required by all plants but can promote plant growth and may be essential for particular taxa. These beneficial elements have been reported to enhance resistance to abiotic stresses such as drought, salinity, and nutrient toxicity or deficiency (Hasanuzzaman et al. 2010a, b; Hasanuzzaman and Fujita 2011b; Tahir et al. 2012). In case of Se, the beneficial effects of low doses of Se have received little attention compared to toxic effects that typically occur at higher concentrations. Better understanding of the effects of beneficial elements is important to improve crop productivity and enhance plant nutritional value for a growing world population.

2.5.6.1 Selenium

During last two decades the physiological roles of Se in plants have been studied by many researchers although Se has not been confirmed to be an essential micronutrient in higher plants. There are several evidences on its positive effect on plant growth and productivity at low concentrations (Turakainen et al. 2004; Hasanuzzaman et al. 2010a, b; Hasanuzzaman and Fujita 2012; Hasanuzzaman et al. 2012b). However, the specific physiological mechanisms underlying the beneficial role of Se in plants have not been clearly elucidated. It is already established that the plants supplemented with Se have shown enhanced resistance to certain abiotic stresses including salinity (Djanaguiraman et al. 2005; Filek et al. 2008; Hawrylak-Nowak 2009; Cartes et al. 2010; Chu et al. 2010; Djanaguiraman et al. 2010; Hasanuzzaman and Fujita 2010; Hasanuzzaman et al. 2010b; Yao et al. 2010a, b; Hasanuzzaman and Fujita 2011b; Hasanuzzaman et al. 2011b). One of the major effects of Se on abiotic stress tolerance is associated with its antioxidative capacity (Djanaguiraman et al. 2005; Hasanuzzaman et al. 2011b; Hasanuzzaman and Fujita 2011a).

A plenty of research results have shown the ability of Se to protect plants from salt stress-induced damages when applied at low concentration. The interaction of Se with soil salinity has been studied earlier by Terry et al. (2000). Kong et al. (2005) reported that at low concentrations (1–5 μM), Se tends to stimulate the growth, the activities of SOD and POD, as well as the accumulation of water-soluble sugar in leaves of sorrel (*R. patientia* × *R. tianshanicus*) seedlings. However, at higher concentrations (10–30 μM), Se exerted diminished beneficial effects on growth and enzyme activities. Results revealed that SOD and POD activity of salt-stressed seedlings increased when exposed to concentrations ranging 1–5 μM Se. At concentrations between 10 and 30 μM, there were adverse effects on both enzymes compared with that at 5 μM Se. In *C. sativus* leaves, Se treatments at 5 and 10 μM significantly improved the growth rate and increased the photosynthetic pigments and Pro contents when subjected to salt stress (Hawrylak-Nowak 2009). Additionally, Se enhanced the salt tolerance of *G. max* seedlings by protecting the cell membrane against lipid peroxidation (Djanaguiraman et al. 2005). Se-treated plants have also increase Pro content (Djanaguiraman et al. 2005). However, the mechanisms and the reasons for Pro accumulation in Se-supplied plants have not been fully investigated. Walaa et al. (2010) observed that NaCl-induced lipid peroxidation which led to increase the percentage of electrolyte leakage, were effectively minimized when the seedlings were pretreated with Se. Se-supplemented seedlings also showed enhanced antioxidant activities and Pro content. In our recent study, we investigated the regulatory role of exogenous Se in the antioxidant defense systems in *B. napus* seedlings exposed to salt stress (Hasanuzzaman et al. 2011b). Twelve-day-old seedlings, grown in Petri dishes, were supplemented with Se (25 μM Na_2SeO_4) and salt (100 and 200 mM NaCl) separately and in combination, and grown for 48 h. The AsA content of the seedlings decreased significantly with increased salt stress, while the amount of GSH and GSSG increased. In addition, the APX and GST activity increased significantly with increased salt concentration (both at 100 and 200 mM NaCl), while GPX activity increased only at moderate salt stress (100 mM NaCl). Glutathione reductase activity remained unchanged at 100 mM NaCl, while it decreased under severe (200 mM NaCl) salt stress. The CAT, MDHAR and DHAR activities decreased with increasing concentration of salt stress, whereas a sharp decrease of these activities was observed under severe salt stress (200 mM NaCl). Concomitant increases in the levels of H_2O_2 and MDA were also measured. However, further investigation revealed that Se treatment had a synergistic effect: in salt-stressed seedlings, it increased the AsA and GSH contents, and the activities of CAT, APX, MDHAR, DHAR, GR, GST and GPX. As a result, addition of Se in salt-stressed seedlings led to a reduction in the levels of H_2O_2 and MDA as compared to salt stress alone.

2.5.6.2 Silicon

Silicon (Si) is the second most abundant element on the earth crust after oxygen and it is accumulated in plants at a rate comparable to those of macronutrient elements like Ca, Mg and P (Epstein 1999). Although Si is a major constituent of plants, to date its essentiality has not been completely established. It may be considered a 'quasi essential' element for plants because its deficiency can cause various dysfunction in regards to plant growth, development and reproduction. In addition, supplementation with Si exerts a number of beneficial effects on growth and yield of several plant species (Richmond and Sussman 2003; Pilon-Smits et al. 2009). Over last two decades, numerous studies have been performed to understand the possible mechanism(s) for Si-enhanced resistance and/or tolerance of higher plants to abiotic stresses as well as the mechanism of Si uptake and transport in higher plants (Liang et al. 2007). It was found to stimulate enzymatic and non-enzymatic antioxidant under stressful condition (Liang et al. 2007). The possible mechanisms of Si-mediated protective effects under salt stress may include increased plant water status (Romero-Aranda et al. 2006), enhanced photosynthetic activity and maintenance of ultra structure of leaf organelles (Shu and Liu. 2001), Dismutation of ROS (Zhu et al. 2004), immobilization of toxic Na^+ (Liang et al. 2003), reduced Na^+ uptake in plants and enhanced K^+ uptake (Liang et al. 2005; Tahir et al. 2006) and higher K^+: Na^+ selectivity (Hasegawa et al. 2000). In particular, Si was recently proven to mitigate salinity stress by enhancing Na^+ exclusion and decreasing lipid membrane peroxidation through stimulation of enzymatic and non-enzymatic antioxidants (Saqib et al. 2008; Hasanuzzaman and Fujita 2011b). The protective effect of Si on salinity has been examined in *O. sativa* (Lekklar and Chaidee 2011), *T. aestivum* (Tuna et al. 2008; Tahir et al. 2012), *Z. mays* (Moussa 2006), *H. vulgare* (Liang et al. 2005), *B. napus* (Hashemi et al. 2010), *L. esculentum* (Romero-Aranda et al. 2006), and *C. sativus* (Zhu et al. 2004).

In *B. napus*, exogenous Si ameliorated the deleterious effects of salinity on the growth through lowering tissue Na^+ contents, maintaining the membrane integrity of root cells as evidenced by reduced lipid peroxidation and lignifications; and increased ROS scavenging capacity (Hashemi et al. 2010). While studying with *Z. mays*, Parveen and Ashraf (2010) reported that under saline condition exogenously applied Si significantly increased growth of plant. Exogenously applied Si also improved some key plant gas exchange characteristics such as net CO_2 assimilation rate (*A*), stomatal conductance (*gs*), transpiration (*E*), and leaf sub-stomatal CO_2 concentration (*Ci*) and thus acted as beneficial elements for improving salt tolerance of *Z. mays* plants. Applying Si to *Medicago sativa* could alter the activity of antioxidative enzyme of one or several organs of plants to improve the salt tolerance (Wang et al. 2011b). The plants under NaCl stress supplemented with Si significantly increased APX activity in root, shoot and leaves, and CAT activity in leaves, and POD activity in shoots, but decreased the SOD activity in roots under salt stress (Wang et al. 2011b). Ali et al. (2011) found that Si supplementation into the root medium significantly improved the K^+ and K^+:Na^+ ratio, leaf water potential and stomatal conductance, but reduced the Na^+ in *T. aestivum*. These plants also showed a concomitant increase in number of tillers, number of grains per spike, grain and straw yield with Si application both under optimal and stressful conditions which suggested that Si application in soil medium is beneficial in profoundly affecting physiological phenomena and improving wheat growth under salt stress. Lima et al. (2011) reported that Si application in the nutrient solution significantly increased growth parameters and decreased ion leakage in *Z. mays* seedlings, whereas this response was not observed in *Vigna unguiculata*. In *Glycine max*, an addition of Si to salt stressed plants substantially alleviated the adverse effects of NaCl on growth, as it enhanced endogenous GA_3, while reducing the levels of ABA and Pro (Lee et al. 2010). In *Saccharum officinarum*, Si-enhanced tolerance to salt which was ascribed to decreased Na^+ concentration and increased K^+ with a resultant improvement in K^+/Na^+ ratio (Ashraf et al. 2010). Recently, we investigated the beneficial role of exogenous Si (1 mM SiO_2) in the antioxidant defense and MG detoxification systems of rapeseed seedlings exposed to salt stress (100 and 200 mM NaCl). We observed that Si treatment had a synergistic effect on salt-stressed seedlings by increasing the AsA and GSH contents, the GSH/GSSG ratio, and the activities of APX, MDHAR, DHAR, GR, GST, GPX, CAT, Gly I, and Gly II. The addition of Si also showed reduced levels of H_2O_2 and MDA in salt-stressed seedlings compared to salt stress alone. Our results suggested that the exogenous application of Si rendered the plants more tolerant to short-term salt stress-induced oxidative damage by enhancing their antioxidant defense and MG detoxification systems (Hasanuzzaman and Fujita 2011b). Tahir et al. (2012) reported that application of Si increased shoot and root dry weight and plant water contents in both normal and saline conditions. Shoot Na^+ and Na^+:K^+ ratio also decreased with Si application under stress conditions. Improved growth of salt-stressed wheat by Si application was mainly attributed to improved plant water contents in shoots, chlorophyll content, decreased Na^+ and increased K^+ concentrations in shoots as well as maintained membrane permeability (Tahir et al. 2012).

2.6 Conclusion and Future Perspectives

Based on a plenty of research findings, it is clear that salt stress has devastating effect on the growth, development, physiology and yield of plants. However, the response to salinity differs greatly among various plant species, the levels of stress as well as the environmental condition. In recent years, the biochemical responses of plants to salt stress have been studied intensively. Information on the tolerance mechanism is useful for developing new cultivars that are adaptable in salinity environments although defining salt tolerance is quite difficult because of the complex nature of salt stress and the wide range of plant responses. The use of exogenous protectants under salt stress condition has been found to be very much effective to alleviate salt-induced damages. Phytohormones are thought to be the most important endogenous substances involved in the mechanisms of tolerance or susceptibility of plants. However, the exact mechanism of protection and signal transduction pathways are still unclear. The appropriate dose and duration of treatment of the exogenous protectants and the proper methods of application should be studied more precisely. In addition, further investigations considering molecular approaches are needed to reveal the underlying mechanisms of protection under stressful condition.

Acknowledgements We express our sincere thanks to Md. Iqbal Hosen, Key Laboratory of Biodiversity and Biogeography, Kunming Institute of Botany, Chinese Academy of Sciences, Kunming 650201, Yunnan, China for providing several supporting document regarding salt stress. We also thank Md. Mahabub Alam, Faculty of Agriculture, Kagawa University, Japan for his continuous support during manuscript preparation. We apologize to all researchers for those parts of their work that were not cited in the manuscripts because of the page limitation.

References

Abdullah Z, Khan MA, Flowers TJ (2001) Causes of sterility in seed set of rice under salinity stress. J Agron Crop Sci 167:25–32

Abrol IP, Yadav JSP, Massoud FI (1988) Salt-affected soils and their management. FAO Soils Bulletin 39, Food and Agriculture Organization of the United Nations, Rome, Italy

Acquaah G (2007) Principles of plant genetics and breeding. Blackwell, Oxford, p 385

Afzal I, Basra S, Iqbal A (2005) The effect of seed soaking with plant growth regulators on seedling vigor of wheat under salinity stress. J Stress Physiol Biochem 1:6–14

Agarwal S, Shaheen R (2007) Stimulation of antioxidant system and lipid peroxidation by abiotic stresses in leaves of *Momordica charantia*. Braz J Plant Physiol 19:149–161

Ahmad P (2010) Growth and antioxidant responses in mustard (*Brassica juncea* L.) plants subjected to combined effect of gibberellic acid and salinity. Arch Agron Soil Sci 56:575–588

Ahmad P, Prasad MNV (2012a) Abiotic stress responses in plants: metabolism, productivity and sustainability. Springer, New York

Ahmad P, Prasad MNV (2012b) Environmental adaptations and stress tolerance in plants in the era of climate change. Springer Science + Business Media, New York

Ahmad P, Sharma S (2008) Salt stress and phyto-biochemical responses of plants. Plant Soil Environ 54:89–99

Ahmad P, Umar S (2011) Oxidative stress: role of antioxidants in plants. Studium Press, New Delhi

Ahmad P, Jeleel CA, Azooz MM, Nabi G (2009) Generation of ROS and non-enzymatic antioxidants during abiotic stress in Plants. Bot Res Intern 2:11–20

Ahmad P, Jaleel CA, Salem MA, Nabi G, Sharma S (2010a) Roles of Enzymatic and non-enzymatic antioxidants in plants during abiotic stress. Crit Rev Biotechnol 30(3):161–175

Ahmad P, Jaleel CA, Sharma S (2010b) Antioxidative defence system, lipid peroxidation, proline metabolizing enzymes and Biochemical activity in two genotypes of *Morus alba* L. subjected to NaCl stress. Russ J Plant Physiol 57:509–517

Ahmad P, Nabi G, Ashraf M (2011) Cadmium-induced oxidative damage in mustard [*Brassica juncea* (L.) Czern. & Coss.] plants can be alleviated by salicylic acid. South Afr J Bot 77:36–44

Ahmad P, Hakeem KR, Kumar A, Ashraf M, Akram NA (2012) Salt-induced changes in photosynthetic activity and oxidative defense system of three cultivars of mustard (*Brassica juncea* L.). Afr J Biotechnol 11:2694–2703

Ahmed CB, Rouina BB, Sensoy S, Boukhriss M, Abdullah FB (2010) Exogenous proline effects on photosynthetic performance and antioxidant defense system of young olive tree. J Agric Food Chem 58:4216–4222

Akbari G, Sanavy SA, Yousefzadeh S (2007) Effect of auxin and salt stress (NaCl) on seed germination of wheat cultivars (*Triticum aestivum* L.). Pakistan J Biol Sci 10:2557–2561

Akbarimoghaddam H, Galavi M, Ghanbari A, Panjehkeh N (2011) salinity effects on seed germination and seedling growth of bread wheat cultivars. Trakia J Sci 9:43–50

Akhiyarova GR, Sabirzhanova IB, Veselov DS, Frike V (2005) Participation of plant hormones in growth resumption of wheat shoots following short-term NaCl treatment. Rus J Plant Physiol 52:788–792

Alcazar R, Marco F, Cuevas JC, Patron M, Ferrando A, Carrasco P, Tiburcio AF, Altabella T (2006) Involvement of polyamines in plant response to abiotic stress. Biotechnol Lett 28:1867–1876

Alcazar R, Altabella T, Marco F, Bortolotti C, Reymond M, Knocz C, Carrasco P, Tiburcio AF (2010a) Polyamines: molecules with regulatory functions in plant abiotic stress tolerance. Planta 231:1237–49

Alcazar R, Planas J, Saxena T, Zarza X, Bortolotti C, Cuevas J, Bitrian M, Tiburcio AF, Altabella T (2010b) Putrescine accumulation confers drought tolerance in transgenic *Arabidopsis* plants overexpressing the homologous Arginine decarboxylase2 gene. Plant Physiol Biochem 48:547–52

Alet AI, Sánchez DH, Cuevas JC, Marina M, Carrasco P, Altabella T, Tiburcio AF, Ruiz OA (2012) New insights into the role of spermine in *Arabidopsis thaliana* under long-term salt stress. Plant Sci 182:94–100

Ali HM, Siddiqui MH, Basalah MO, Al-Whaibi MH, Sakran AM, Al-Amri A (2011) Effects of gibberellic acid on growth and photosynthetic pigments of *Hibiscus sabdariffa* L. under salt stress. Afr J Biotechnol 11:800–804

Allakhverdiev SI, Hayashi H, Nishiyama Y, Ivanov AG, Aliev JA, Klimov VV, Murata N, Carpemtier R (2003) Glycinebetaine protects the D1/D2/*Cyt b 559* complex of photosystem II against photo-induced and heat-induced inactivation. J Plant Physiol 160:41–49

Amirjani MR (2011) Effect of salinity stress on growth, sugar content, pigments and enzyme activity of rice. Int J Bot 7:73–81

Amri E, Mirzaei M, Moradi M, Zare K (2011) The effects of spermidine and putrescine polyamines on growth of pomegranate (*Punica granatum* L. cv 'Rabbab') in salinity circumstance. Int J Plant Physiol Biochem 3:43–49

Anjum MA (2011) Effect of exogenously applied spermidine on growth and physiology of citrus rootstock Troyer citrange under saline conditions. Turk J Agric For 35:43–53

Anuradha S, Rao SSR (2001) Effect of brassinosteroids on salinity stress induced inhibition of seed germination and seedling growth of rice (*Oryza sativa* L.). Plant Growth Regul 33:151–153

Apel K, Hirt H (2004) Reactive oxygen species: metabolism, oxidative stress and signal transduction. Annu Rev Plant Biol 55:373–399

Arfan M, Athar HR, Ashraf M (2007) Does exogenous application of salicylic acid through the rooting medium modulate growth and photosynthetic capacity in differently adapted spring wheat cultivated under salt stress? J Plant Physiol 6:685–694

Arora N, Bhardwaj R, Sharma P, Arora HK (2008) 28-Homobrassinolide alleviates oxidative stress in salt-treated maize (*Zea mays* L.) plants. Braz J Plant Physiol 20:153–157

Asch F, Dingkuhn M, Miezan K, Doerffling K (2000) Leaf K/Na ratio predicts salinity induced yield loss in irrigated rice. Euphytica 113:109–118

Ashraf M, Foolad MR (2007) Roles of glycine betaine and proline in improving plant abiotic stress resistance. Environ Exp Bot 59:206–216

Ashraf M, Athar HR, Harris PJC, Kwon TR (2008) Some prospective strategies for improving crop salt tolerance. Adv Agron 97:45–109

Ashraf R, Ahmad R, Bhatti AS, Afzal M, Sarwar A, Maqsood MA, Kanwal S (2010) Amelioration of salt stress in sugarcane (*Saccharum officinarum* L.) by supplying potassium and silicon in hydroponics. Pedosphere 20:153–162

Athar HR, Khan A, Ashraf M (2008) Exogenously applied ascorbic acid alleviates salt-induced oxidative stress in wheat. Environ Exp Bot 63:224–231

Atkinson NJ, Urwin PE (2012) The interaction of plant biotic and abiotic stresses: from genes to the field. J Exp Bot. doi:10.1093/jxb/ers100

Azevedo Neto AD, Prisco JT, Eneas-Filho J, Medeiros J-VR, Gomes-Filho E (2005) Hydrogen peroxide pre-treatment induces stress acclimation in maize plants. J Plant Physiol 162:1114–1122

Azzedine F, Gherroucha H, Baka M (2011) Improvement of salt tolerance in durum wheat by ascorbic acid application. J Stress Physiol Biochem 7:27–37

Babu MA, Singh D, Gothandam KM (2012) The effect of salinity on growth, hormone and mineral elements in leaf and fruit of tomato cultivar PKM1. J Anim Plant Sci 22:159–164

Bai X, Yang L, Tian M, Chen J, Shi J, Yang Y, Hu X (2011) Nitric oxide enhances desiccation tolerance of recalcitrant *Antiaris toxicaria* seeds via protein *S*-nitrosylation and carbonylation. PLoS ONE 6:e20714

Bar Y, Apelbaum A, Kafkafi U, Goren R (1997) Relationship between chloride and nitrate and its effect on growth and mineral composition of avocado and citrus plants. J Plant Nutr 20:715–731

Barba-Espín G, Clemente-Moreno MJ, Álvarez S, García-Legaz MF, Hernández JA, Díaz-Vivancos P (2011) Salicylic acid negatively affects the response to salt stress in pea plants. Plant Biol 13:909–917

Beltagi MS (2008) Exogenous ascorbic acid (vitamin C) induced anabolic changes for salt tolerance in chick pea (*Cicer arietinum* L.) plants. Afr J Plant Sci 2:118–123

Bhatnagar-Mathur P, Vadez V, Sharma KK (2008) Transgenic approaches for abiotic stress tolerance in plants: retrospect and prospects. Plant Cell Rep 27:411–424
Bordi A (2010) The influence of salt stress on seed germination, growth and yield of canola cultivars. Not Bot Hort Agrobot Cluj 38:128–133
Brouwer C, Goffeau A, Heibloem M (1985) Irrigation water management: training manual No. 1 – Introduction to irrigation. Food and Agriculture Organization of the United Nations, Rome, Italy
Brugnoli E, Björkman O (1992) Growth of cotton under continuous salinity stress: influence on allocation pattern, stomatal and non-stomatal components of photosynthesis and dissipation of excess light energy. Planta 187:335–347
Cabot C, Sibole JV, Barcelo J, Poschenrieder C (2009) Abscisic acid decreases leaf Na^+ exclusion in salt-treated *Phaseolus vulgaris* L. J Plant Growth Regul 28:187–192
Carpıcı EB, Celık N, Bayram G (2009) Effects of salt stress on germination of some maize (*Zea mays* L.) cultivars. Afr J Biotechnol 8:4918–4922
Cartes P, Jara AA, Pinilla L, Rosas A, Mora ML (2010) Selenium improves the antioxidant ability against aluminium-induced oxidative stress in ryegrass roots. Ann Appl Biol 156:297–307
Chai YY, Jiang CD, Shi L, Shi TS, Gu WB (2010) Effects of exogenous spermine on sweet sorghum during germination under salinity. Biol Plant 54:145–148
Cha-Um S, Kirdmanee C (2010) Effect of glycinebetaine on proline, water use, and photosynthetic efficiencies, and growth of rice seedlings under salt stress. Turk J Agric For 34:517–527
Cheeseman JM (1988) Mechanism of salinity tolerance in plants. Plant Physiol 87:547–550
Cheeseman JM (2007) Hydrogen peroxide and plant stress: a challenging relationship. Plant Stress 1:4–15
Chen THH, Murata N (2008) Glycinebetaine: an effective protectant against abiotic stress in plants. Trend Plant Sci 13:499–505
Cheong JJ, Choi YD (2003) Methyl jasmonate as a vital substance in plants. Trends Genet 19:409–413
Chu J, Yao X, Zhang Z (2010) Responses of wheat seedlings to exogenous selenium supply under cold stress. Biol Trace Elem Res 136:355–363
Chutipaijit S, Cha-um S, Sompornpailin K (2011) High contents of proline and anthocyanin increase protective response to salinity in *Oryza sativa* L. spp. indica. Aust J Crop Sci 5:1191–1198
Corpas FJ, Leterrier M, Valderrama R, Airaki M, Chaki M, Palma JM, Barroso JB (2011) Nitric oxide imbalance provokes a nitrosative response in plants under abiotic stress. Plant Sci 181:604–611
Cortina C, Culianez-Maciá FA (2005) Tomato abiotic stress enhanced tolerance by trehalose biosynthesis. Plant Sci 169:75–82
Cramer GR, Quarrie SA (2002) Abscisic acid is correlated with the leaf growth inhibition of four genotypes of maize differing in their response to salinity. Funct Plant Biol 29:111–115
Cuartero J, Fernandez-Munoz R (1999) Tomato and salinity. Sci Hort 78:83–125
Dantas BF, De Sá Ribeiro L, Aragão CA (2007) Germination, initial growth and cotyledon protein content of bean cultivars under salinity stress. Rev Bras de Sementes 29:106–110
David A, Yadav S, Bhatla SC (2010) Sodium chloride stress induces nitric oxide accumulation in root tips and oil body surface accompanying slower oleosin degradation in sunflower seedlings. Physiol Plant 140:342–354
De Tullio MC (2004) How does ascorbic acid prevent scurvy? A survey of the nonantioxidant functions of vitamin C. In: Asard H (ed) Vitamin C: its function and biochemistry in animals and plants. Garland Science/BIOS Scientific Publishers, London/New York, pp 176–190
Dehghan G, Rezazadeh L, Habibi G (2011) Exogenous ascorbate improves antioxidant defense system and induces salinity tolerance in soybean seedlings. Acta Biol Szeged 55:261–264
Deivanai S, Xavier R, Vinod V, Timalata K, Lim OF (2011) Role of exogenous proline in ameliorating salt stress at early stage in two rice cultivars. J Stress Physiol Biochem 7:157–174
Dionisio-Sese ML, Tobita S (1998) Antioxidant responses of rice seedlings to salinity stress. Plant Sci 135:1–9

Djanaguiraman M, Devi DD, Shanker AK, Sheeba A, Bangarusamy U (2005) Selenium – an antioxidative protectant in soybean during senescence. Plant Soil 272:77–86

Djanaguiraman M, Prasad PVV, Seppanen M (2010) Selenium protects sorghum leaves from oxidative damage under high temperature stress by enhancing antioxidant defense system. Plant Physiol Biochem 48:999–1007

Dolatabadian A, Modarressanavy SAM, Ghanati F (2011) Effect of salinity on growth, xylem structure and anatomical characteristics of soybean. Not Sci Biol 3:41–45

Duan J, Li J, Guo S, Kang Y (2008) Exogenous spermidine affects polyamine metabolism in salinity-stressed Cucumis sativus roots and enhances short-term salinity tolerance. J Plant Physiol 165:1620–1635

Dunlap JR, Binzel ML (1996) NaCl reduces indole-3-acetic acid levels in the roots of tomato plants independent of stress-induced abscisic acid. Plant Physiol 112:379–384

El Tayeb MA (2005) Response of barley grains to the interactive effect of salinity and salicylic acid. Plant Growth Regul 45:215–224

Eleiwa ME, Ibrahim SA (2011) Influence of brassinosteroids on wheat plant (*Triticum aestivum* L.) production under salinity stress conditions. II. Chemical constituent and nutritional status. Aust J Basic Appl Sci 5:49–57

Eleiwa ME, Bafeel SO, Ibrahim SA (2011) Influence of brassinosteroids on wheat plant (*Triticum aestivum* L.) production under salinity stress conditions. I. Growth parameters and photosynthetic pigments. Aust J Basic Appl Sci 5:58–651

El-Mashad AAA, Mohamed HI (2011) Brassinolide alleviates salt stress and increases antioxidant activity of cowpea plants (*Vigna sinensis*). Protoplasma 249:625–635

Elstner EF (1987) Metabolism of activated oxygen species. In: Davies DD (ed) The biochemistry of plants, vol II, Biochemistry of metabolism. Academic, San Diego, pp 252–315

Epstein E (1999) Silicon: its manifold roles in plants. Ann Appl Biol 155:155–16

Erdal S, Aydın M, Genisel M, Taspınar MS, Dumlupinar R, Kaya O, Gorcek Z (2011) Effects of salicylic acid on wheat salt sensitivity. Afr J Biotechnol 10:5713–5718

Essa TA (2002) Effect of salinity stress on growth and nutrient composition of three soybean (*Glycine max* L. Merrill) cultivars. J Agron Crop Sci 88:86–93

Etehadnia M, Waterer DR, Tanino KK (2008) The method of ABA application affects salt stress responses in resistant and sensitive potato lines. J Plant Growth Regul 27:331–341

Fan H, Guo S, Jiao Y, Zhang R, Li J (2007) Effects of exogenous nitric oxide on growth, active oxygen species metabolism, and photosynthetic characteristics in cucumber seedlings under NaCl stress. Front Agric China 1:308–314

Fan H-F, Du C-X, Guo S-R (2010) Nitric oxide enhances salt tolerance in cucumber seedlings by regulating free polyamine content. Environ Exp Bot. doi:10.1016/j.envexpbot.2010.09.007

FAO (2008) Land and plant nutrition management service. htpp://www.fao.org/ag/agl/agll/spush

FAO (2009) High level expert forum – how to feed the world in 2050. Economic and Social Development Department, Food and Agricultural Organization of the United Nations, Rome

Farouk S (2011) Ascorbic acid and α-Tocopherol minimize salt-induced wheat leaf senescence. J Stress Physiol Biochem 7:58–79

Fedina IS, Tsonev TD (1997) Effect of pretreatment with methyl jasmonate on the response of *Pisum sativum* to salt stress. J Plant Physiol 151:735–740

Fedina IS, Nedeva D, Çiçek N (2009) Pre-treatment with H_2O_2 induces salt tolerance in barley seedlings. BIiol Plant 53:321–324

Filek M, Keskinen R, Hartikainen H, Szarejko I, Janiak A, Miszalski Z, Golda A (2008) The protective role of selenium in rape seedlings subjected to cadmium stress. J Plant Physiol 165:833–844

Fisarakis I, Chartzoulakis K, Stavrakas D (2001) Response of Sultana vines (*V. vinifera* L.) on six rootstocks to NaCl salinity exposure and recovery. Agric Water Manage 51:13–27

Flowers TJ, Hajibagheri MA, Clipson NJW (1986) Halophytes. Quart Rev Biol 61:313–337

Foolad MR, Lin GY (1997) Genetic potential for salt tolerance during germination in *Lycopersicon* species. HortScience 32:296–300

Foolad MR, Lin GY (1998) Genetic analysis of low temperature tolerance during germination in tomato, *Solanum lycopersicum* Mill. Plant Breed 117:171–176

Foyer (2004) The role of ascorbic acid in defense networks and signaling in plants. In: Asard H (ed) Vitamin C: its function and biochemistry in animals and plants. Garland Science/BIOS Scientific Publishers, London/New York, pp 73–91

Foyer CH, Noctor G (2005a) Redox homeostasis and antioxidant signaling: a metabolic interface between stress perception and physiological responses. Plant Cell 17:1866–1875

Foyer CH, Noctor G (2005b) Oxidant and antioxidant signaling in plants: a re-evaluation of the concept of oxidative stress in a physiological context. Plant Cell Environ 28:1056–1071

Fricke W, Akhiyarova G, Veselov D, Kudoyarova G (2004) Rapid and tissue-specific changes in ABA and in growth rate in response to salinity in barley leaves. J Exp Bot 55:1115–1123

Gadallah MAA (1999) Effect of proline and glycinebetaine on *Vicia faba* responses to salt stress. Biol Plant 42:249–257

Gain P, Mannan MA, Pal PS, Hossain MM, Parvin S (2004) Effect of salinity on some yield attributes of rice. Pak J Biol Sci 7:760–762

Garciadeblas B, Senn ME, Bañuelos MA, Rodriquez-Navarro A (2003) Sodium transport and HKT transporters: the rice model. Plant J 34:788–801

Garg N, Manchanda G (2008) Salinity and its effects on the functional biology of legumes. Acta Physiol Plant 30:595–618

Garg N, Manchanda G (2009) ROS generation in plants: boon or bane? Plant Biosys 143:81–96

Gémes K, Poór Horváth E, Kolbert Z, Szopkó D, Szepesi Á, Tari I (2011) Cross-talk between salicylic acid and NaCl-generated reactive oxygen species and nitric oxide in tomato during acclimation to high salinity. Physiol Plant 142:179–192

Ghoulam C, Foursy A, Fares K (2002) Effects of salt stress on growth, inorganic ions and proline accumulation in relation to osmotic adjustment in five sugar beet cultivars. Environ Exp Bot 47:39–50

Gill SS, Tuteja N (2010) Polyamines and abiotic stress tolerance in plants. Plant Signal Behav 51:26–33

Gomes-Filho E, Machado Lima CRF, Costa JH, da Silva AC, da Guia Silva Lima M, de Lacerda CF, Prisco JT (2008) Cowpea ribonuclease: properties and effect of NaCl-salinity on its activation during seed germination and seedling establishment. Plant Cell Rep 27:147–157

Gossett DR, Banks SW, Millhollon EP, Lucas MC (1996) Antioxidant response to NaCl stress in a control and a NaCl-tolerant cotton cell line grown in the presence of paraquat, buthionine sulfoximine and exogenous glutathione. Plant Physiol 112:803–809

Grattan SR, Grieve CM (1999) Salinity-mineral nutrient relations in horticultural crops. Sci Hortic 78:127–157

Greenway H, Munns R (1980) Mechanisms of salt tolerance in nonhalophytes. Annu Rev Plant Physiol 31:149–190

Guan B, Yu J, Chen X, Xie W, Lu Z (2011) Effects of salt stress and nitrogen application on growth and ion accumulation of Suaeda salsa plants. Intl Conf Remote Sens Environ Transport Engin, 24–26 June 2011. pp. 8268–8272

Gueta-Dahan Y, Yaniv Z, Zilinskas BA, Ben-Hayyim G (1997) Salt and oxidative stress: similar and specific responses and their relation to salt tolerance in citrus. Planta 204:460–469

Gulnaz AJ, Iqbal J, Azam F (1999) Seed treatment with growth regulators and crop productivity. II. Response of critical growth stages of wheat (*Triticum aestivum* L.) under salinity stress. Cereal Res 27:419–426

Gunes A, Inal A, Alpaslan M, Eraslan F, Bagci EG, Cicek N (2007) Salicylic acid induced changes on some physiological parameters symptomatic for oxidative stress and mineral nutrition in maize (*Zea mays* L.) grown under salinity. J Plant Physiol 164:728–736

Gupta K, Gupta B, Ghosh B, Sengupta DN (2012) Spermidine and abscisic acid-mediated phosphorylation of a cytoplasmic protein from rice root in response to salinity stress. Acta Physiol Plant 34:29–40

Gurmani AR, Bano A, Khan SU, Din J, Zhang JL (2011) Alleviation of salt stress by seed treatment with abscisic acid (ABA), 6-benzylaminopurine (BA) and chlormequat chloride (CCC) optimizes ion and organic matter accumulation and increases yield of rice (*Oryza sativa* L.). Aust J Crop Sci 5:1278–1285

Habib N, Ashraf M, Ahmad MSA (2010) Enhancement in seed germinability of rice (*Oryza sativa* L.) by pre-sowing seed treatment with nitric oxide (NO) under salt stress. Pak J Bot 42:4071–4078
Halliwell B, Gutteridge JMC (1985) Free radicals in biology and medicine. Clarendon, Oxford
Hamada AM, Al-Hakimi AM (2009) Exogenous ascorbic acid or thiamine increases the resistance of sunflower and maize plants to salt stress. Acta Agron Hung 57:335–347
Hamayun M, Khan SA, Khan AL, Shin JH, Ahmad B, Shin D-H, Lee I-J (2010) Exogenous gibberellic acid reprograms soybean to higher growth and salt stress tolerance. J Agric Food Chem 58:7226–7232
Hamilton EW III, Heckathorn SA (2001) Mitochondrial adaptation to NaCl. Complex I is protected by antioxidants and small heat shock proteins, whereas complex II is protected by proline and betaine. Plant Physiol 126:1266–1274
Hasanuzzaman M, Fujita M (2011a) Selenium pretreatment upregulates the antioxidant defense and methylglyoxal detoxification system and confers enhanced tolerance to drought stress in rapeseed seedlings. Biol Trace Elem Res 143:1758–1776
Hasanuzzaman M, Fujita M (2011b) Exogenous silicon treatment alleviates salinity-induced damage in *Brassica napus* L. seedlings by up-regulating the antioxidant defense and methylglyoxal detoxification system. Abstract of Plant Biology 2011, American Society of Plant Biology. http://abstracts.aspb.org/pb2011/public/P10/P10001.html
Hasanuzzaman M, Fujita M (2012) Selenium and plant health: the physiological role of selenium. In: Aomori C, Hokkaido M (eds) Selenium: sources, functions and health effects. Nova Publishers, New York. ISBN 978-1-61942-062-5
Hasanuzzaman M, Fujita M, Islam MN, Ahamed KU, Nahar K (2009) Performance of four irrigated rice varieties under different levels of salinity stress. Int J Integ Biol 6:85–90
Hasanuzzaman M, Hossain MA, Fujita M (2010a) Physiological and biochemical mechanisms of nitric oxide induced abiotic stress tolerance in plants. Am J Plant Physiol 5:295–324
Hasanuzzaman M, Hossain MA, Fujita M (2010b) Selenium in higher plants: physiological role, antioxidant metabolism and abiotic stress tolerance. J Plant Sci 5:354–375
Hasanuzzaman M, Hossain MA, Fujita M (2011a) Nitric oxide modulates antioxidant defense and the methylglyoxal detoxification system and reduces salinity-induced damage of wheat seedlings. Plant Biotechnol Rep 5:353–365
Hasanuzzaman M, Hossain MA, Fujita M (2011b) Selenium-induced up-regulation of the antioxidant defense and methylglyoxal detoxification system reduces salinity-induced damage in rapeseed seedlings. Biol Trace Elem Res 143:1704–1721
Hasanuzzaman M, Hossain MA, da Silva JAT, Fujita M (2012a) Plant responses and tolerance to abiotic oxidative stress: antioxidant defenses is a key factors. In: Bandi V, Shanker AK, Shanker C, Mandapaka M (eds) Crop stress and its management: perspectives and strategies. Springer, Berlin, pp 261–316
Hasanuzzaman M, Hossain MA, Fujita M (2012b) Exogenous selenium pretreatment protects rapeseed seedlings from cadmium-induced oxidative stress by up regulating the antioxidant defense and methylglyoxal detoxification systems. Biol Trace Elem Res. doi:10.1007/s12011-012-9419-4
Hasegawa P, Bressan RA, Zhu JK, Bohnert HJ (2000) Plant cellular and molecular responses to high salinity. Annu Rev Plant Physiol Plant Mol Biol 51:463–499
Hashemi A, Abdolzadeh A, Sadeghipour HR (2010) Beneficial effects of silicon nutrition in alleviating salinity stress in hydroponically grown canola, *Brassica napus* L., plants. Soil Sci Plant Nutr 56:244–253
Hawkins HJ, Lewis OAM (1993) Combination effect of NaCl salinity, nitrogen form and calcium concentration on the growth and ionic content and gaseous properties of *Triticum aestivum* L. cv. Gamtoos. New Phytol 124:161–170
Hawrylak-Nowak B (2009) Beneficial effects of exogenous selenium in cucumber seedlings subjected to salt stress. Biol Trace Elem Res 132:259–269
Hayat S, Ahmad A (2011) Brassinosteroids: a class of plant hormone. Springer, Dordrecht
Hernandez M, Fernandez-Garcia N, Diaz-Vivancos P, Olmos E (2010) A different role for hydrogen peroxide and the antioxidative system under short and long salt stress in *Brassica oleracea* roots. J Exp Bot 61:521–535

Heuer B (2003) Influence of exogenous application of proline and glycinebetaine on growth of salt stressed tomato plants. Plant Sci 165:693–699

Hillel D (2000) Salinity management for sustainable irrigation. The World Bank, Washington, DC

Hoque MA, Banu MNA, Okuma E, Amako K, Nakamura K, Shimoishi Y, Murata Y (2007) Exogenous proline and glycinebetaine increase NaCl-induced ascorbate-glutathione cycle enzyme activities, and proline improves salt tolerance more than glycinebetaine in tobacco Bright Yellow-2 suspension-cultured cells. J Plant Physiol 164:1457–1468

Hoque MA, Okuma E, Nakamara Y, Shimoishi Y, Murata Y (2008) Proline and glycinebetaine enhance antioxidant defense and methylglyoxal detoxification systems and reduce NaCl-induced damage in cultured tobacco cells. J Plant Physiol 165:813–824

Horvath E, Szalai G, Janda T (2007) Induction of abiotic stress tolerance by salicylic acid signaling. J Plant Growth Regul 26:290–300

Hossain KK, Itoh RD, Yoshimura G, Tokuda G, Oku H, Cohen MF, Yamasaki H (2010) Effects of nitric oxide scavengers on thermoinhibition of seed germination in *Arabidopsis thaliana*. Russ J Plant Physiol 57:222–232

Hossain MA, Munemasa S, Uraji M, Nakamura Y, Mori IC, Murata Y (2011) Involvement of endogenous abscisic acid in methyl jasmonate-induced stomatal closure in *Arabidopsis*. Plant Physiol 156:430–438

Houimli SIM, Denden M, Mouhandes BD (2010) Effects of 24-epibrassinolide on growth, chlorophyll, electrolyte leakage and proline by pepper plants under NaCl-stress. EurAsia J Biosci 4:96–104

Hu Y, Schmidhalter U (1997) Interactive effects of salinity and macronutrient level on wheat. J Plant Nutr 20:1169–1182

Hu Y, Schmidhalter U (2005) Drought and salinity: a comparison of their effects on mineral nutrition of Plants. J Plant Nutr Soil Sci 168:541–549

Hung S-H, Yu C-W, Lin CH (2005) Hydrogen peroxide functions as a stress signal in plants. Bot Bull Acad Sin 46:1–10

Hussain TM, Chandrasekhar T, Hazara M, Sultan Z, Saleh BZ, Gopal GR (2008) Recent advances in salt stress biology – a review. Biotechnol Mol Biol Rev 3:8–13

Hussain K, Nawaz K, Majeed A, Ilyas U, Lin F, Ali K, Nisar MF (2011a) Role of exogenous salicylic acid applications for salt tolerance in violet (*Viola Odorata* L.). Sarhad J Agric 27:171–175

Hussain SS, Ali M, Ahmad M, Siddique KHM (2011b) Polyamines: natural and engineered abiotic and biotic stress tolerance in plants. Biotechnol Adv 29:300–311

Ibrar M, Jabeen M, Tabassum J, Hussain F, Ilahi I (2003) Salt tolerance potential of *Brassica juncea* Linn. J Sci Tech Univ Peshawar 27:79–84

Ioannidis NE, Cruz JA, Kotzabasis K, Krame DM (2012) Evidence that putrescine modulates the higher plant photosynthetic proton circuit. PLoS One 7:e29864

Iqbal M, Ashraf M (2010) Gibberellic acid mediated induction of salt tolerance in wheat plants: growth, ionic partitioning, photosynthesis, yield and hormonal homeostasis. Environ Exp Bot. doi:10.1016/j.envexpbot.2010.06.002

Iqbal N, Masood A, Khan NA (2012) Phytohormones in salinity tolerance: ethylene and gibberellins cross talk. In: Khan NA, Nazar R, Iqbal N, Anjum NA (eds) Phytohormones and abiotic stress tolerance in plants. Springer, Berlin, pp 77–98

Iyengar ERR, Reddy MP (1996) Photosynthesis in highly salt-tolerant plants. In: Pessaraki M (ed) Handbook of photosynthesis. Marcel Dekker, New York, pp 897–909

Jabeen M, Ibrar M, Azim F, Hussain F, Ilahi I (2003) The effect of sodium chloride salinity on germination and productivity of Mung bean (*Vigna mungo* Linn.). J Sci Tech Univ Peshawar 27:1–5

Jamil M, Rha ES (2007) Gibberellic acid (GA_3) enhance seed water uptake, germination and early seedling growth in sugar beet under salt stress. Pak J Biol Sci 10:654–658

Jemâa E, Saïda A, Sadok B (2011) Impact of indole-3-butyric acid and indole-3-acetic acid on the lateral roots growth of *Arabidopsis* under salt stress conditions. Aust J Agric Eng 2:18–24

Jeschke WD, Peuke AD, Pate JS, Hartung W (1997) Transport, synthesis and catabolism of abscisic acid (ABA) in intact plants of castor bean (*Ricinus communis* L.) under phosphate deficiency and moderate salinity. J Exp Bot 48:1737–1747

Kang DJ, Seo YJ, Lee JD, Ishii R, Kim KU, Shin DH, Park SK, Jang SW, Lee IJ (2005) Jasmonic acid differentially affects growth, ion uptake and abscisic acid concentration in salt-tolerant and salt-sensitive rice cultivars. J Agron Crop Sci 191:273–282

Kanwischer M, Porfirova S, Bergmüller E, Dörmann P (2005) Alterations in tocopherol cyclase activity in transgenic and mutant plants of *Arabidopsis* affect tocopherol content, tocopherol composition, and oxidative stress. Plant Physiol 137:713–723

Katerji N, Van Hoorn JW, Hamdy A, Mastrorilli M, Moukarzel E (1997) Osmotic adjustment of sugar beets in response to soil salinity and its influence on stomatal conductance, growth and yield. Agric Water Manage 34:57–69

Kattab H (2007) Role of glutathione and polyadenylic acid on the oxidative defense systems of two different cultivars of canola seedlings grown under saline condition. Aust J Basic Appl Sci 1:323–334

Kaur G, Kumar S, Nayyar H, Upadhyaya HD (2008) Cold stress injury during the pod-filling phase in chickpea (*Cicer arietinum* L.): effects on quantitative and qualitative components of seeds. J Agron Crop Sci 194:457–464

Kaveh H, Nemati H, Farsi M, Jartoodeh SV (2011) How salinity affect germination and emergence of tomato lines. J Biol Environ Sci 5:159–163

Kaya C, Tuna AL, Yokas I (2009) The role of plant hormones in plants under salinity stress. In: Ashraf M, Ozturk M, Athar HR (eds) Salinity and water stress: improving crop efficiency. Springer, Berlin, pp 45–50

Kemble AR, MacPherson HT (1954) Liberation of amino acids in perennial ray grass during wilting. Biochem J 58:46–59

Keskin BC, Sarikaya AT, Yuksel B, Memon AR (2010) Abscisic acid regulated gene expression in bread wheat. Aust J Crop Sci 4:617–625

Khafagy MA, Arafa AA, El-Banna MF (2009) Glycinebetaine and ascorbic acid can alleviate the harmful effects of NaCl salinity in sweet pepper. Aust J Crop Sci 3:257–267

Khan MA, Rizvi Y (1994) Effect of salinity, temperature and growth regulators on the germination and early seedling growth of *Atriplex griffithii* var. Stocksii Can J Bot 72:475–479

Khan MA, Weber DJ (2008) Ecophysiology of high salinity tolerant plants (tasks for vegetation science), 1st edn. Springer, Amsterdam

Khan A, Ahmad MSA, Athar HUR, Ashraf M (2006) Interactive effect of foliarly applied ascorbic acid and salt stress on wheat (*Triticum aestivum* L.) at the seedling stage. Pak J Bot 38:1407–1414

Khatun S, Flowers TJ (1995) Effects of salinity on seed set in rice. Plant Cell Environ 18:61–67

Khedr AHA, Abbas MA, Wahid AAA, Quick WP, Abogadallah GM (2003) Proline induces the expression of salt-stress-responsive proteins and may improve the adaptation of *Pancratium maritimum* L. to salt-stress. J Exp Bot 54:2553–2562

Khodarahmpour Z, Ifar M, Motamedi M (2012) Effects of NaCl salinity on maize (*Zea mays* L.) at germination and early seedling stage. Afr J Biotechnol 11:298–304

Kim S-G, Park C-M (2008) Gibberellic acid-mediated salt signaling in seed germination. Plant Signal Behav 3:877–879

Kong L, Wang M, Bi D (2005) Selenium modulates the activities of antioxidant enzymes, osmotic homeostasis and promotes the growth of sorrel seedlings under salt stress. Plant Growth Regul 45:155–163

Koornneef M, Leon-Kloosterziel KM, Schwartz S, Zeevaart JAD (1998) The genetic and molecular dissection of abscisic acid biosynthesis and signal transduction in *Arabidopsis*. Plant Physiol Biochem 36:83–89

Kopyra M, Gwóźdź EA (2003) Nitric oxide stimulates seed germination and counteracts the inhibitory effect of heavy metals and salinity on root growth of *Lupinus luteus*. Plant Physiol Biochem 41:1011–1017

Koyro H-W (2002) Ultrastructural effects of salinity in higher plants. In: Läuchli A, Lüttge U (eds) Salinity: environment – plants – molecules. Kluwer, Amsterdam, pp 139–157

Krishna P (2003) Brassinosteroid-mediated stress responses. J Plant Growth Regul 22:289–297
Kumar B, Singh B (1996) Effect of plant hormones on growth and yield of wheat irrigated with saline water. Ann Agric Res 17:209–212
Kuznetsov VV, Shevyakova NI (2007) Polyamines and stress tolerance of plants. Plant Stress 1:50–71
Larkindale J, Knight M (2002) Protection against heat stress-induced oxidative damage in *Arabidopsis* involves calcium, abscisic acid, ethylene, and salicylic acid. Plant Physiol 128:682–695
Läuchli A, Grattan SR (2007) Plant growth and development under salinity stress. In: Jenks MA, Hasegawa PM, Mohan JS (eds) Advances in molecular breeding towards drought and salt tolerant crops. Springer, Berlin, pp 1–32
Lea-Cox JD, Syvertsen JP (1993) Salinity reduces water use and nitrate-N-use efficiency of citrus. Ann Bot 72:47–54
Lee SK, Sohn EY, Hamayun M, Yoon JY, Lee IJ (2010) Effect of silicon on growth and salinity stress of soybean plant grown under hydroponic system. Agroforest Syst 80:333–340
Lekklar C, Chaidee A (2011) Roles of silicon on growth of Thai Jasmine rice (*Oryza sativa* L. cv. KDML105) under salt stress. Agric Sci J 42:45–48
Li JM, Jin H (2007) Regulation of brassinosteroid signaling. Trends Plant Sci 12:37–41
Li J, Gao X-H, Guo S-R, Zhang R-H, Wang X (2007) Effects of exogenous spermidine on photosynthesis of salt-stressed *Cuellmis sativus* seedlings. Chin J Ecol 26:1595–1599
Li Q-Y, Niud H-B, Yind J, Wanga M-B, Shao HB, Deng DZ, Chen X-X, Ren J-P, Li Y-C (2008) Protective role of exogenous nitric oxide against oxidative-stress induced by salt stress in barley (*Hordeum vulgare*). Colloids Surf B: Biointer 65:220–225
Liang YC, Chen Q, Liu Q, Zhang W, Ding R (2003) Effects of silicon on salinity tolerance of two barley cultivars. J Plant Physiol 160:1157–1164
Liang YC, Zhang WO, Chen J, Ding R (2005) Effect of silicon on H^+-ATPase and H^+-PPase activity, fatty acid composition and fluidity of tonoplast vesicles from roots of salt-stressed barley (*Hordeum vulgare* L.). Environ Exp Bot 53:29–37
Liang Y, Sun W, Zhu YG (2007) Christie P: mechanisms of silicon mediated alleviation of abiotic stresses in higher plants: a review. Environ Poll 147:422–428
Lima MDA, de Castro VF, Vidal JB, Enéas-Filho J (2011) Silicon application on plants of maize and cowpea under salt stress. Rev Ciênc Agron 42:398–403
Lima-Costa ME, Ferreira S, Duarte A, Ferreira AL (2008) Alleviation of salt stress using exogenous proline on a citrus cell line. Acta Hort 868:109–112
Linghe Z, Shannon MC (2000) Salinity effects on seedling growth and yield components of rice. Crop Sci 40:996–1003
Liu J, Zhou YF, Zhang WH, Liu YL (2006) Effects of exogenous polyamines on chloroplast-bound polymine content and photosynthesis of corn suffering salt stress. Acta Bot Boreal-Occident Sin 26:254–258
Liu X, Wang L, Liu L, Guo Y, Ren H (2011) Alleviating effect of exogenous nitric oxide in cucumber seedling against chilling stress. Afr J Biotechnol 10:4380–4386
Lombardi T, Lupi B (2006) Effect of salinity on the germination and growth of *Hordeum secalinum* Schreber (Poaceae) in relation to the seeds after-ripening time. Atti Soc tosc Sci nat Mem Serie B 113:37–42
López-Gómez M, Lluch C (2012) Trehalose and abiotic stress tolerance. In: Ahmad P, Prasad MNV (eds) Abiotic stress responses in plants: metabolism, productivity and sustainability. Springer, New York, pp 253–265
Lutts S (2000) Exogenous glycinebetaine reduces sodium accumulation in salt-stressed rice plants. IRRN 25:39–40
Maeda H, Sakuragi Y, Bryant DA, DellaPenna D (2005) Tocopherols protect *Synechocystis* sp. Strain PCC 6803 from lipid peroxidation. Plant Physiol 138:1422–1435
Maggio A, Barbieri G, Raimondi G, De Pascale S (2010) Contrasting effects of GA_3 treatments on tomato plants exposed to increasing salinity. J Plant Growth Regul 29:63–72

Magome H, Yamaguchi S, Hanada A, Kamiya Y, Odadoi K (2004) Dwarf and delayed-flowering 1, a novel Arabidopsis mutant deficient in gibberellins biosynthesis because of overexpression of a putative AP2 transcription factor. Plant J 37:720–729

Mahajan S, Tuteja N (2005) Cold, salinity and drought stresses: an overview. Arch Biochem Biophys 444:139–158

Mäkelä P, Kärkkäinen J, Somersalo S (2000) Effect of glycine betaine on chloroplast ultrastructure, chlorophyll and protein content, and RuBPCO activity in tomato grown under drought or salinity. Biol Plant 43:471–475

Mantri N, Patade V, Penna S, Ford R, Pang E (2012) Abiotic stress responses in plants: present and future. In: Ahmad P, Prasad MNV (eds) Abiotic stress responses in plants: metabolism, productivity and sustainability. Springer, New York, pp 1–19

Mass EV (1986) Salt tolerance of plants. Appl Agric Res 1:12–26

Matysik J, Alia, Bhalu B, Mohanty P (2002) Molecular mechanisms of quenching of reactive oxygen species by proline under stress in plants. Curr Sci 82:525–532

Maxwell K, Johnson GN (2000) Chlorophyll fluorescence – A practical guide. J Exp Bot 51:659–668

Misra AN, Misra M, Singh R (2011) Nitric oxide: A ubiquitous signaling molecule with diverse role in plants. Afr J Plant Sci 5:57–74

Mittler R, Blumwald E (2010) Genetic engineering for modern agriculture: challenges and perspectives. Annu Rev Plant Biol 61:443–462

Mittova V, Guy M, Tal M, Volokita M (2004) Salinity upregulates the antioxidative system in root mitochondria and peroxisomes of the wild salt-tolerant tomato species *Lycopersicon pennellii*. J Exp Bot 55:1105–1113

Mohamed MA, Matter MA, Saker MM (2010) Effect of salt stress on some defense mechanisms of transgenic and wild potato clones (*Solanum tuberosum* L.) grown *in vitro*. Nat Sci 12:181–193

Molassiotis A, Tanou G, Diamantidis G (2010) NO says more than 'YES' to salt tolerance: salt priming and systemic nitric oxide signaling in plants. Plant Signal Behav 5:209–2012

Moussa HR (2006) Influence of exogenous application of silicon on physiological response of salt-stressed maize (*Zea mays* L.). Int J Agri Biol 8:293–297

Munir N, Aftab F (2011) Enhancement of salt tolerance in sugarcane by ascorbic acid pretreatment. Afr J Biotechnol 10:18362–18370

Munné-Bosch S (2005) The role of α-tocopherol in plant stress tolerance. J Plant Physiol 162:743–748

Munné-Bosch S (2007) α-Tocopherol: a multifaceted molecule in plants. Vitam Horm 76:375–392

Munns R (2002a) Comparative physiology of salt and water stress. Plant Cell Environ 25:239–250

Munns R (2002b) Salinity, growth and phytohormones. In: Läuchli A, Lüttge U (eds) Salinity: environment – plants – molecules. Kluwer, The Netherlands, pp 271–290

Munns R (2005) Genes and salt tolerance: bringing them together. New Phytol 167:645–663

Munns R, Tester M (2008) Mechanisms of salinity tolerance. Annu Rev Plant Biol 59:651–681

Munns R, James RA, Lauchli A (2006) Approaches to increasing the salt tolerance of wheat and other cereals. J Exp Bot 57:1025–1043

Murty PSS, Murty KS (1982) Spikelet sterility in relation to nitrogen and carbohydrate contents in rice. Ind J Plant Physiol 25:40–48

Mutlu F, Buzcuk S (2007) Salinity induced changes of free and bound polyamine levels in Sunflower (*Helianthus annuus* L.) roots differing in salt tolerance. Pak J Bot 39:1097–1102

Nahar K, Hasanuzzaman M (2009) Germination, growth, nodulation and yield performance of three mungbean varieties under different levels of salinity stress. Green Farming 2:825–829

Navakouidis E, Lutz C, Langebartels C, Lutz-Meindl U, Kotzabasis K (2003) Ozone impact on the photosynthetic apparatus and the protective role of polyamines. Bioch Biophys Acta 1621:160–169

Nazar R, Iqbal N, Syeed S, Khan NA (2011) Salicylic acid alleviates decreases in photosynthesis under salt stress by enhancing nitrogen and sulfur assimilation and antioxidant metabolism differentially in two mungbean cultivars. J Plant Physiol 168:807–815

Neamatollahi E, Bannayan M, Ghanbari A, Haydari M, Ahmadian A (2009) Does hydro and osmo-priming improve fennel (*Foeniculum vulgare*) seeds germination and seedlings growth? Not Bot Hort Agrobot Cluj 37:190–194

Nilsen E, Orcutt DM (1996) The physiology of plants under stress – abiotic factors. Wiley, New York, pp 118–130

Noctor G, Foyer CH (1998) Ascorbate and glutathione: keeping active oxygen under control. Annu Rev Plant Physiol Plant Mol Biol 49:249–279

Noctor G, Gomez L, Vanacker H, Foyer CH (2002) Interactions between biosynthesis, compartmentation and transport in the control of glutathione homeostasis and signalling. J Exp Bot 53:1283–304

Nounjan N, Nghia PT, Theerakulpisut P (2012) Exogenous proline and trehalose promote recovery of rice seedlings from salt-stress and differentially modulate antioxidant enzymes and expression of related genes. J Plant Physiol 169:596–604

Oertli JJ (1991) Nutrient management under water and salinity stress. In: Proceeding of the symposium on nutrient management for sustained productivity. Dept. Soils Punjab Agric. Unver. Ludhiana, India. P. 138–165

Okuma E, Murakami Y, Shimoishi Y, Tada M, Murata Y (2004) Effects of exogenous application of proline and betaine on the growth of tobacco cultured cells under saline solutions. Soil Sci Plant Nutr 50:1301–1305

Othman Y, Al-Karaki G, Al-Tawaha AR, Al-Horani A (2006) Variation in germination and ion uptake in barley genotypes under salinity conditions. World J Agric Sci 2:11–15

Panda SK, Upadhyay RK (2004) Salt stress injury induces oxidative alterations and antioxidative defence in the roots of *Lemna minor*. Biol Plant 48:249–253

Parida AK, Das AB (2005) Salt tolerance and salinity effect on plants: a review. Ecotoxicol Environ Saf 60:324–349

Parida AK, Das AB, Mohanty P (2004) Investigations on the antioxidative defense responses to NaCl stress in a mangrove, *Bruguiera parviflora*: differential regulations of isoforms of some antioxidative enzymes. Plant Growth Regul 42:213–226

Parveen N, Ashraf M (2010) Role of silicon in mitigating the adverse effects of salt stress on growth and photosynthetic attributes of two maize (*Zea mays* L.) cultivars grown hydroponically. Pak J Bot 42:1675–1684

Pastori GM, Foyer CH (2002) Common components, networks, and pathways of cross-tolerance to stress. The central role of "Redox" and abscisic acid-mediated controls. Plant Physiol 129:460–468

Pastori G, Kiddle G, Antoniw J, Bernard S, Veljovic-Jovanovic S, Verrier PJ, Noctor G, Foyer CH (2003) Leaf vitamin C contents modulate plant defense transcripts and regulate genes that control development through hormone signaling. Plant Cell 15:939–951

Pedranzani H, Racagni G, Alemano S, Miersch O, Ramirez I, Pena-Cortes H, Taleisnik E, Machado-Domenech E, Abdala G (2003) Salt tolerant tomato plants show increased levels of jasmonic acid. Plant Growth Regul 41:149–158

Pettigrew WT, Meredith WR (1994) Leaf gas exchange parameters vary among cotton genotypes. Crop Sci 34:700–705

Pilon-Smits EAH, Quinn CF, Tapken W, Malagoli M, Schiavon M (2009) Physiological functions of beneficial elements. Curr Opin Plant Biol 12:267–274

Pitman MG, Läuchli A (2002) Global impact of salinity and agricultural ecosystems. In: Läuchli A, Lüttge U (eds) Salinity: environment – plants – molecules. Kluwer, Dordrecht, pp 3–20

Pompella A, Visvikis A, Paolicchi A, De Tata V, Casini AF (2003) The changing faces of glutathione, a cellular protagonist. Biochem Pharm 66:1499–1503

Pons LJ (1973) Outline of genesis, characteristics, classification and improvement of acid sulfate soils. In: Dost H (ed) Acid sulfate soils. Proc Int Symp ILRI Publ 18:3–23

Poór P, Gémes K, Horváth F, Szepesi A, Simon ML, Tari I (2011) Salicylic acid treatment via the rooting medium interferes with stomatal response, CO_2 fixation rate and carbohydrate metabolism in tomato, and decreases harmful effects of subsequent salt stress. Plant Biol 13:105–114

Pourcel L, Routaboul JM, Cheynier V (2007) Flavonoid oxidation in plants: from biochemical properties to physiological functions. Trends Plant Sci 12:29–36

Promila K, Kumar S (2000) *Vigna radiata* seed germination under salinity. Biol Plant 43:423–426

Qadir M, Schubert S (2002) Degradation processes and nutrient constraints in sodic soils. Land Degrad Dev 13:275–294

Quinet M1, Ndayiragije A, Lefèvre I, Lambillotte B, Dupont-Gillain CC, Lutts S (2010) Putrescine differently influences the effect of salt stress on polyamine metabolism and ethylene synthesis in rice cultivars differing in salt resistance. J Exp Bot 61:2719–2733

Rady MM, Sadak MS, El-Bassiouny HMS, Abd El-Monem AA (2011) Alleviation the adverse effects of salinity stress in sunflower cultivars using nicotinamide and α-Tocopherol. Aust J Basic Appl Sci 5:342–355

Rahman MS, Miyake H, Takeoka Y (2002) Effect of exogenous glycinebetaine on growth and ultrastructure of salt-stressed rice seedlings (*Oryza sativa* L.). Plant Prod Sci 5:33–44

Rasheed R (2009) Salinity and extreme temperature effects on sprouting buds of sugarcane (*Saccharum officinarum* L.): Some histological and biochemical studies. Ph.D. thesis, Dept. of Botany, Univ. of Agriculture, Faisalabad, Pakistan

Rawia Eid A, Taha LS, Ibrahiem SMM (2011) Alleviation of adverse effects of salinity on growth, and chemical constituents of marigold plants by using glutathione and ascorbate. J Appl Sci Res 7:714–721

Rehman S, Harris PJC, Bourne WF, Wilkin J (2000) The relationship between ions, vigour and salinity tolerance of Acacia seeds. Plant Soil 220:229–233

Rhee SG (2006) H_2O_2, a necessary evil for cell signaling. Science 312:1882–1883

Ribaut JM, Pilet PE (1991) Effect of water stress on growth, osmotic potential and abscisic acid content of maize roots. Physiol Plant 81:156–162

Richmond KE, Sussman M (2003) Got silicon? The non-essential beneficial plant nutrient. Curr Opin Plant Biol 6:268–272

Rodríguez M, Canales E, Borrás-Hidalgo O (2005) Molecular aspects of abiotic stress in plants. Biotecnol Aplic 22:1–10

Rogers ME, Noble CL (1992) Variation in growth and ion accumulation between two selected populations of *Trifolium repens* L. differing in salt tolerance. Plant Soil 146:131–136

Rogers ME, Grieve CM, Shannon MC (2003) Plant growth and ion relations in lucerne (*Medicago sativa* L.) in response to the combined effects of NaCl and P. Plant Soil 253:187–194

Rohwer CL, Erwin JE (2008) Horticultural applications of jasmonates: a review. J Hortic Sci Biotechnol 83:283–304

Romero-Aranda R, Soria T, Cuartero S (2001) Tomato plant-water uptake and plant-water relationships under saline growth conditions. Plant Sci 160:265–272

Romero-Aranda MR, Jurado O, Cuartero J (2006) Silicon alleviates the deleterious salt effect on tomato plant growth by improving plant water status. J Plant Physiol 163:847–855

Roychoudhury A, Basu S, Sengupta DN (2011) Amelioration of salinity stress by exogenously applied spermidine or spermine in three varieties of indica rice differing in their level of salt tolerance. J Plant Physiol 168:317–328

Rozeff N (1995) Sugarcane and salinity – A review paper. Sugarcane 5:8–19

Ruiz JM, Blumwald E (2002) Salinity-induced glutathione synthesis in *Brassica napus*. Planta 214:965–969

Saeedipour S (2011) Salinity tolerance of rice lines related to endogenous abscisic acid (ABA) level synthesis under stress. Afr J Plant Sci 5:628–633

Saha P, Chatterjee P, Biswas AK (2010) NaCl pretreatment alleviates salt stress by enhancement of antioxidant defense system and osmolyte accumulation in mungbean (*Vigna radiata* L. Wilczek). Indian J Exp Biol 48:593–600

Sairam RK, Roa KV, Srivastava GC (2002) Differential response of wheat genotypes to long term salinity stress in relation to oxidative stress, antioxidant activity and osmolyte concentration. Plant Sci 163:1037–1046

Sakhabutdinova AR, Fatkhutdinova DR, Bezrukova MV, Shakirova FM (2003) Salicylic acid prevents the damaging action of stress factors on wheat plants. Bulg J Plant Physiol 314–319

Salama KHA, Al-Mutawa MM (2009) Glutathione-triggered mitigation in salt-induced alterations in plasmalemma of onion epidermal cells. Int J Agric Biol 11:639–642

Saleethong P, Sanitchon J, Kong-ngern K, Theerakulpisut P (2011) Pretreatment with spermidine reverses inhibitory effects of salt stress in two rice (*Oryza sativa* L.) cultivars differing in salinity tolerance. Asian J Plant Sci 10:245–254

Samira IM-H, Dridi-Mouhandes B, ben Mansour-Gueddes S, Denden M (2012) 24-Epibrassinolide ameliorates the adverse effect of salt stress (NaCl) on pepper (*Capsicum annuum* L.). J Stress Physiol Biochem 8:232–240

Saqib M, Zorb C, Schubert S (2008) Silicon-mediated improvement in the salt resistance of wheat (*Triticum aestivum*) results from increased sodium exclusion and resistance to oxidative stress. Funct Plant Biol 35:633–639

Sastry EVD, Shekhawa KS (2001) Alleviatory effect of GA_3 on the effect of salt at seedling stage in wheat (*Triticum aestivum*). Indian J Agric Res 35:226–231

Schluepmann H, Pellny T, Van Dijken A, Smeekens S, Paul MJ (2003) Trehalose 6-phosphate is indispensable for carbohydrate utilisation and growth in *Arabidopsis thaliana*. Proc Natl Acad Sci USA 100:6849–6854

Seemann JR, Critchley C (1985) Effects of salt stress on the growth, ion contents, stomatal behaviour and photosynthetic capacity of a salt sensitive species, *Phaseolus vulgaris* L. Planta 164:66–69

Sembdner G, Parthier B (1993) The biochemistry and physiology and molecular actions of jasmonates. Ann Rev Plant Physiol Plant Mol Biol 44:569–586

Seo HS, Kim SK, Jang SW, Choo YS, Sohn EY, Lee IJ (2005) Effect of jasmonic acid on endogenous gibberellins and abscisic acid in rice under NaCl stress. Biol Plant 49:447–450

Shah SH (2007) Stress on mustard as affected by gibberellic acid application. Gen Appl Plant Physiol 33:97–106

Shahbaz M, Ashraf M (2007) Influence of exogenous application of brassinosteroid on growth and mineral nutrients of wheat (*Triticum aestivum* L.) under saline conditions. Pak J Bot 39:513–522

Shalata A, Neumann PM (2001) Exogenous ascorbic acid (Vitamin C) increases resistance to salt stress and reduces lipid peroxidation. J Exp Bot 52:2207–2211

Shanker AK, Venkateswarlu B (2011) Abiotic stress in plants – mechanisms and adaptations. InTech, Rijeka, p ix

Sharma DK, Dubey AK, Srivastav M, Singh AK, Sairam RK, Pandey RN, Dahuja A, Kaur C (2011) Effect of putrescine and paclobutrazol on growth, physiochemical parameters, and nutrient acquisition of salt-sensitive citrus rootstock Karna khatta (*Citrus karna* Raf.) under NaCl Stress. J Plant Growth Regul 30:301–311

Sheokand S, Bhankar V, Sawhney V (2010) Ameliorative effect of exogenous nitric oxide on oxidative metabolism in NaCl treated chickpea plants. Braz J Plant Physiol 22:81–90

Sheteawi S (2007) Improving growth and yield of salt-stressed soybean by exogenous application of jasmonic acid and ascobin. Int J Agri Biol 9:473–478

Shigeoka S, Ishikawa T, Tamoi M, Miyagawa Y, Takeda T, Yabuta Y, Yoshimura K (2002) Regulation and function of ascorbate peroxidase isoenzymes. J Exp Bot 53:1305–1319

Shomeili M, Nabipour M, Meskarbashee M, Memari HR (2011) Effects of gibberellic acid on sugarcane plants exposed to salinity under a hydroponic system. Afr J Plant Sci 5:609–616

Shu LZ, Liu YH (2001) Effects of silicon on growth of maize seedlings under salt stress. Agroenviron Prot 20:38–40

Shu S, Sun J, Guo S-R, Li J, Liu C-J, Wang C-Yi Du C-X (2010) Effects of exogenous Putrescine on PS II photochemistry and ion distribution of cucumber seedlings under salt stress. Acta Hort Sin 37:1065–1072

Shu S, Guo S-R, Yuan L-Y (2012) A review: polyamines and photosynthesis. In: Najafpour MM (ed) Advances in photosynthesis – fundamental aspects. InTech, Rijeka, pp 439–464

Smirnoff N, Cumbes QJ (1989) Hydroxyl radical scavenging activity of compatible solutes. Phytochemistry 28:1057–1060

Smirnoff N, Pallanca JE (1996) Ascorbate metabolism in relation to oxidative stress. Biochem Soc Trans 24:472–478

Sobahan MA, Arias CR, Okuma E, Shimoishi Y, Nakamura Y, Hirai Y, Mori IC, Murata Y (2009) Exogenous proline and glycinebetaine suppress apoplastic flow to reduce Na^+ uptake in rice seedlings. Biosci Biotechnol Biochem 73:2037–2042

Srivalli S, Khanna-Chopra R (2008) Role of glutathione in abiotic stress tolerance. In: Khan NA, Singh S, Umar S (eds) Sulfur assimilation and abiotic stress in plants. Springer, Berlin/Heidelberg, pp 207–225

Steudle E (2000) Water uptake by roots: effects of water deficit. J Exp Bot 51:1531–1542

Stevens J, Senaratna T, Sivasithamparam K (2006) Salicylic acid induces salinity tolerance in tomato (*Lycopersicon esculentum* cv. Roma): associated changes in gas exchange, water relations and membrane stabilisation. Plant Growth Regul 49:77–83

Sudhir P, Murthy SDS (2004) Effects of salt stress on basic processes of photosynthesis. Photosynthetica 42:481–486

Suhayda CG, Giannini JL, Briskin DP, Shannon MC (1990) Electrostatic changes in *Lycopersicon esculentum* root plasma membrane resulting from salt stress. Plant Physiol 93:471–478

Sumithra K, Jutur PP, Carmel BD, Reddy AR (2006) Salinity-induced changes in two cultivars of *Vigna radiata*: responses of antioxidative and proline metabolism. Plant Growth Regul 50:11–22

Sung CH, Hong JK (2010) Sodium nitroprusside mediates seedling development and attenuation of oxidative stresses in Chinese cabbage. Plant Biotechnol Rep 4:243–251

Swamy PM, Smith B (1999) Role of abscisic acid in plant stress tolerance. Curr Sci 76: 1220–1227

Szabolcs I (1974) Salt affected soils in Europe. Martinus Nijhoff, The Hague, 63 p

Tahir MA, Rahmatullah Aziz T, Ashraf M, Kanwal S, Muhammad A (2006) Beneficial effects of silicon in wheat under salinity stress-pot culture. Pak J Bot 38:1715–1722

Tahir MA, Aziz T, Farooq M, Sarwar G (2012) Silicon-induced changes in growth, ionic composition, water relations, chlorophyll contents and membrane permeability in two salt-stressed wheat genotypes. Arch Agron Soil Sci 58:247–256

Taiz L, Zeiger E (2006) Plant physiology, 4th edn. Sinaur Associates, Sunderland

Tanou G, Molassiotis A, Diamantidis G (2009) Induction of reactive oxygen species and necrotic death-like destruction in strawberry leaves by salinity. Environ Exp Bot 65:270–281

Tavakkoli E, Rengasamy P, McDonald GK (2010) High concentrations of Na^+ and Cl^- ions in soil solution have simultaneous detrimental effects on growth of faba bean under salinity stress. J Exp Bot 61:4449–4459

Terry N, Zayed AM, de Souza MP, Tarun AS (2000) Selenium in higher plants. Annu Rev Plant Physiol Plant Mol Biol 51:401–432

Thakur P, Kumar S, Malik JA, Berger JD, Nayyar H (2010) Cold stress effects on reproductive development in grain crops: an overview. Environ Exp Bot 67:429–443

Torabian AR (2010) Effect of salicylic acid on germination and growth of alfalfa (*Medicago sativa* L.) seedlings under water potential loss at salinity stress. Plant Ecophysiol 2:151–155

Torabian AR (2011) Effect of salicylic acid on germination and growth of alfalfa (*Medicago sativa* L.) seedlings under water potential loss at salinity stress. Plant Ecophysiol 2:151–155

Tun NN, Santa-Catarina C, Begum T, Silveria V, Handro W, Floh EIS, Scherer GFE (2006) Polyamines induce rapid biosynthesis of nitric oxide (NO) in *Arabidopsis thaliana* seedlings. Plant Cell Physiol 47:346–54

Tuna AL, Kaya C, Higgs D, Murillo-Amador B, Aydemir S, Girgin AR (2008) Silicon improves salinity tolerance in wheat plants. Environ Exp Bot 62:10–16

Turakainen M, Hartikainen H, Seppänen MM (2004) Effects of selenium treatments on potato (*Solanum tuberosum* L.) growth and concentrations of soluble sugars and starch. J Agric Food Chem 52:5378–5382

Tuteja N (2007) Abscisic acid and abiotic stress signaling. Plant Signal Behav 2:135–138

Uchida A, Jagendorf AT, Hibino T, Takabe T (2002) Effects of hydrogen peroxide and nitric oxide on both salt and heat stress tolerance in rice. Plant Sci 163:515–523

Ulfat M, Athar H, Ashraf M, Akram NA, Jamil A (2007) Appraisal of physiological and biochemical selection criteria for evaluation of salt tolerance in canola (*Brassica napus* L.). Pak J Bot 39:1593–1608

Van Huizen R, Ozga JA, Reinecke DM (1997) Seed and hormonal regulation of gibberellins 20-oxidase in pea pericarp. Plant Physiol 115:123–128

Velikova V, Yordanov I, Edreva A (2000) Oxidative stress and some antioxidant systems in acid rain-treated bean plants: protective role of exogenous polyamines. Plant Sci 151:59–66

Vorasoot N, Songsri P, Akkasaeng C, Jogloy S, Patanothai A (2003) Effect of water stress on yield and agronomic characters of peanut. Songklanakarin J Sci Technol 25:283–288

Vysotskaya L, Hedley PE, Sharipova G, Veselov D, Kudoyarova G, Morris J, Jones HG (2010) Effect of salinity on water relations of wild barley plants differing in salt tolerance. AoB Plant. doi:10.1093/aobpla/plq006

Wahid A, Rao R, Rasul E (1997) Identification of salt tolerance traits in sugarcane lines. Field Crop Res 54:9–17

Wahid A, Perveena M, Gelania S, Basra SMA (2007) Pretreatment of seed with H2O2 improves salt tolerance of wheat seedlings by alleviation of oxidative damage and expression of stress proteins. J Plant Physiol 164:283–294

Wahid A, Farooq M, Basra SMA, Rasul E, Siddique KHM (2011) Germination of seeds and propagules under salt stress. In: Pessarakli M (ed) Handbook of plant and crop stress, 3rd edn. CRC Press, Boca Raton, pp 321–337

Walaa AE, Shatlah MA, Atteia MH, Sror HAM (2010) Selenium induces antioxidant defensive enzymes and promotes tolerance against salinity stress in cucumber seedlings (*Cucumis sativus*). Arab Univ J Agric Sci 18:65–76

Wang X, Shi G, Xu Q, Hu J (2007) Exogenous polyamines enhance copper tolerance of *Nymphoides peltatum*. J Plant Physiol 164:1062–1070

Wang B, Zhang J, Xia X, Zhang W-H (2011a) Ameliorative effect of brassinosteroid and ethylene on germination of cucumber seeds in the presence of sodium chloride. Plant Growth Regul. doi:10.1007/s10725-011-9595-9

Wang X, Wei Z, Liu D, Zhao G (2011b) Effects of NaCl and silicon on activities of antioxidative enzymes in roots, shoots and leaves of alfalfa. Afr J Biotechnol 10:545–549

Wasternack C, Hause B (2002) Jasmonates and octadecanoids – signals in plant stress response and development. In: Moldav K (ed) Progress in nucleic acid research and molecular biology, vol 72. Academic, New York, pp 165–221

Werner JE, Finkelstein RR (1995) *Arabidopsis* mutants with reduced response to NaCl and osmotic stress. Physiol Plant 93:659–666

White PJ, Broadley MR (2001) Chloride in soils and its uptake and movement within the plant: a review. Ann Bot 88:967–988

Wu X, Zhu W, Zhang H, Ding H, Zhang HJ (2011) Exogenous nitric oxide protects against salt-induced oxidative stress in the leaves from two genotypes of tomato (*Lycopersicom esculentum* Mill.). Acta Physiol Plant 33:1199–1209

Xiong L, Ishitini M, Lee H, Zhu JK (2001) The *Arabidopsis LOS5/ABA3* locus encodes a molybdenum cofactor sulfurase and modulates cold stress and osmotic stress responsive gene expression. Plant Cell 13:2063–2083

Xiong J, Fu G, Tao L, Zhu C (2010) Roles of nitric oxide in alleviating heavy metal toxicity in plants. Arch Biochem Biophys 497:13–20

Xu G, Magen H, Tarchitzky J, Kafkafi U (2000) Advances in chloride nutrition of plants. Adv Agron 68:97–150

Xu S, Hu B, He Z, Ma F, Feng J, Shen W, Yan J (2011) Enhancement of salinity tolerance during rice seed germination by presoaking with hemoglobin. Int J Mol Sci 12:2488–2501

Yadav PV, Kumari M, Ahmed Z (2011) Seed priming mediated germination improvement and tolerance to subsequent exposure to cold and salt stress in *Capsicum*. Res J Seed Sci 4:125–136

Yamaguchi K, Takahashi Y, Berberich T, Imai A, Miyazaki A, Takahashi T, Michael A, Kusano T (2006) The polyamine spermine protects against high salt stress in *Arabidopsis thaliana*. FEBS Lett 580:6783–6788

Yan Z, Guo S, Shu S, Sun J, Tezuka T (2011) Effects of proline on photosynthesis, root reactive oxygen species (ROS) metabolism in two melon cultivars (*Cucumis melo* L.) under NaCl stress. Afr J Biotechnol 10:18381–18390

Yang T, Davies PJ, Reid JB (1996) Genetic dissection of the relative roles of auxin and gibberellin in the regulation of stem elongation in intact light-grown peas. Plant Physiol 110:1029–1034

Yang C-J, Zhang C, Lu Y-N, Jin J-O, Wang X-L (2011) The mechanisms of Brassinosteroids' action: From signal transduction to plant development. Mol Plant 4:588–600

Yao X, Chu J, Ba C (2010a) Responses of wheat roots to exogenous selenium supply under enhanced ultraviolet-B. Biol Trace Elem Res 137:244–252

Yao XQ, Chu JZ, Ba CJ (2010b) Antioxidant responses of wheat seedlings to exogenous selenium supply under enhanced ultraviolet-B. Biol Trace Elem Res 136:96–105

Yokoi S, Bressan RA, Hasegawa PM (2002) Salt stress tolerance of plants. JIRCAS Working Report 2002. pp. 25–33

Yoon JY, Hamayun M, Lee S-K, Lee I-J (2009) Methyl jasmonate alleviated salinity stress in soybean. J Crop Sci Biotechnol 12:63–68

Yupsanis T, Moustakas M, Domiandou K (1994) Protein phosphorylation-dephosphorylation in alfalfa seeds germinating under salt stress. J Plant Physiol 143:234–240

Yusuf M, Fariduddin Q, Varshney P, Ahmad A (2012) Salicylic acid minimizes nickel and/or salinity-induced toxicity in Indian mustard (*Brassica juncea*) through an improved antioxidant system. Environ Sci Pollut Res 19:8–18

Zeid IM (2009) Trehalose as osmoprotectant for maize under salinity-induced stress. Res J Agric Biol Sci 5:613–622

Zeid FA, El Shihy OM, Ghallab AERM, Ibrahim FEZA (2008) Effect of exogenous ascorbic acid on wheat tolerance to salinity stress conditions. Arab J Biotech 12:149–174

Zhang J, Jia W, Yang J, Ismail AM (2006a) Role of ABA in integrating plant responses to drought and salt stresses. Field Crop Res 97:111–119

Zhang YY, Wang LL, Liu YL, Zhang Q, Wei QP, Zhang WH (2006b) Nitric oxide enhances salt tolerance in maize seedlings through increasing activities of proton-pump and Na^+/H^+ antiport in the tonoplast. Planta 224:545–555

Zhang RH, Li J, Guo SR, Tezuka T (2009) Effects of exogenous putrescine on gas exchange characteristics and chlorophyll fluorescence of NaCl-stressed cucumber seedlings. Photosynth Res 100:155–162

Zhang R-H, Guo S-R, Duan Z-G (2011) Effects of exogenous putrescine concentrations on growth, photosynthesis and lipid peroxidation of cucumber seedlings under salt stress. Jiangsu J Agr Sci 27:836–841

Zheng C, Jiang D, Liu F, Dai T, Liu W, Jing Q, Cao W (2009) Exogenous nitric oxide improves seed germination in wheat against mitochondrial oxidative damage induced by high salinity. Environ Exp Bot 67:222–227

Zhu ZJ, Wei GQ, Li J, Qian QQ, Yu JQ (2004) Silicon alleviates salt stress and increases antioxidant enzymes activity in leaves of salt-stressed cucumber (*Cucumis sativus* L.). Plant Sci 167:527–533

Chapter 3
Effects of Salinity on Ion Transport, Water Relations and Oxidative Damage

Maduraimuthu Djanaguiraman and P.V. Vara Prasad

3.1 Introduction

Soils with high salt concentration can limit productivity of agricultural and other economically important crops (Ahmad and Prasad 2012a, b). Salt affected soils are generally categorized into three types: saline soils (containing excess water-soluble salts), sodic soils (containing excess exchangeable sodium), and saline-sodic soils (containing excess salts and exchangeable sodium). In saline soils, the dominant exchangeable cations are calcium and magnesium. Saline soils have electrical conductivity (EC) above 4 dSm^{-1}, sodium adsorption ratio (SAR) below 13, and exchangeable sodium percentage (ESP) below 15, and soil structure and soil water movement is not a serious problem. Sodic soils have EC below 4 dSm^{-1}, have pH greater than 8.5, SAR above 13, and ESP above 15, and generally have poor soil structure that prevents water movement into and through the soils. Sodic soils have high exchangeable sodium, compared to calcium and magnesium. Saline-sodic soils are high in both sodium and other salts, with EC above 4 dSm^{-1}, SAR above 13, and ESP above 15, soil pH can be below or above 8.5, and generally have good soil structure and adequate water movement through soil profile.

Dramatic increase in world population in recent times challenges the present agricultural production system, since more food is needed for growing populations. Increased food production can be achieved by increasing cultivated land, increasing crop productivity or increasing cropping intensity and minimizing pre- and post-harvest yield losses. Increasing pressure on land area brought agriculture to marginal and salt-affected regions. Similarly, increased crop intensification, fertilizer use and hot and dry environments and depletion of underground water have increased problems associated with salts. It is estimated that at least 20 % of all irrigated lands

M. Djanaguiraman • P.V.V. Prasad (✉)
2004 Throckmorton Plant Sciences Center, Department of Agronomy,
Kansas State University, Manhattan, KS 66506, USA
e-mail: vara@ksu.edu

P. Ahmad et al. (eds.), *Ecophysiology and Responses of Plants under Salt Stress*,
DOI 10.1007/978-1-4614-4747-4_3, © Springer Science+Business Media, LLC 2013

are salt-affected (Pitman and Lauchli 2002). The total global area of salt-affected soils has recently been estimated to be approximately 830 million hectares (Martinez-Beltran and Manzur 2005) and most of it associated with saline or associated sodic soils. Among the various sources of soil salinity, irrigation combined with poor drainage is the most serious because water will evaporate but salts remain and accumulate in the soil. In addition, salts in irrigation water are serious problem in certain regions.

In saline condition the soil solution has many dissolved toxic ions, which can cause external and internal stress and damage to the plants. Sodium chloride constitutes majority of the salts. Higher concentration of sodium ion is toxic to most plants, and in some plants growth is also inhibited by chloride ions. High salt content in the soil solution causes decreased water potential representing osmotic stress in the root zone, further entry of ions in the cell creates ionic stress. Hence, salinity stress is a combination of osmotic and ionic effects. Many plants have evolved several mechanisms either to exclude salt from their cells or to tolerate its presence within the cells. However, the physiological, biochemical and molecular mechanisms of salt tolerance in plants are not yet sufficiently understood, which slows development of salt tolerant crops. This chapter provides a brief overview of effects of soil salinity on plant growth and tolerance mechanisms.

3.2 Effects of Salinity on Growth, Development and Yield

Plant responds to salinity in two phases: in a rapid phase, plants exhibit osmotic effects and inhibits growth of young leaves, and in a slower ionic phase, plants accelerates senescence of mature leaves (Munns and Tester 2008). Plants undergo characteristic morphological changes from the time of imposition of stress until maturity. Immediately after salinization, cells will dehydrate and shrink, but after some time it absorbs water and maintains its volume (i.e. recovery) (Ahmad and Prasad 2012a, b). Even though, recovery favors cell elongation, cell division is reduced, leading to lower rates of leaf and root growth. If the stress intensity is severe the plants will show visual injury due to excessive ions absorbed. During salinity stress the growth reduction occurs immediately after exposure due to osmotic changes outside the root zone (physiological drought). Later slower effects are created by salt accumulation in leaves leading to toxicity (Munns et al. 2006). Most of the annuals are tolerant to salinity at germination stage but sensitive during emergence and early vegetative stage (Maas and Grattan 1999). Tolerance towards salinity increases with plant age as the plants develops adaptation window and acclimates. Tolerance during seed germination and emergence is based on percent survival and during later stages it is based on relative growth reductions and rate of senescence. Salinity decreased seed germination percent and it was associated with synthesis of antioxidant enzymes; in addition elevated synthesis of antioxidant enzymes were observed in tolerant genotypes of rice (Djanaguiraman et al. 2003). Salinity also delays seedling emergence and if the stress is severe enough, crop production can be decreased severely (Maas and Grattan 1999). Salinity stress affects both vegetative and reproductive development; salinity often reduces shoot growth more than root growth (Djanaguiraman et al. 2006) and can reduce the number of florets per panicles, increase sterility and affect the time of flowering and maturity in rice (Djanaguiraman et al. 2003). In rice the total shoot biomass was much reduced in salt stressed plants relative to non-stressed plants (Djanaguiraman et al. 2003). The decrease in biomass is mainly attributed to ion imbalance. In experiments with wheat, sorghum and cowpea (Maas and Poss 1989a, b; Maas et al. 1986), vegetative and early reproductive stages were found to be more sensitive than seed filling stages. In wheat, salt stress is imposed when the shoot apex is in vegetative stage, can adversely affect spike development and decrease yields of wheat (Maas and Grieve 1990). Similarly, stress during spike or panicle differentiation, causes earlier reproductive stage development but with reduced number of spikelets per panicle. Anthesis also occurred earlier in salt-stressed plants compared to non-stressed plant. In addition, salinity stress during reproductive stages of development decrease pollen viability and fertilization, resulting in decreased grain numbers in several cereal and legume crops. In terms of crop management, avoiding salinity stress just prior to and during flower (spike) development will help to set initial yield potential, actual grain numbers and minimizes yield losses.

3.3 Effects of Salinity on Ion Transport

Light and nutrients are the main inputs for plants to grow and complete its life cycle. The essential elements of plant growth and development significantly contribute to the metabolic processes. The essential mineral nutrients are of importance in numerous biological functions and each had its own function in the cell (Hajiboland 2012; Nieves-Cordones et al. 2012). Energy storage, structural integrity and roles in redox reactions were some of the biological functions of these nutrients. Although these essential mineral nutrients were imperative for plant survival, excessive soluble salts in the soil had some deleterious effects on most plants. According to their response to high salt concentrations, plants could be divided into two groups. Glycophytes, that are sensitive to high salinity, and halophytes that are tolerant to saline soils. Experimentally, halophytes are of importance since they have the ability to cope with salinity stress and studies on them might lead to the discovery of salinity tolerance mechanisms.

The effect of salinity stress in plants could be generally classified into primary and secondary effects. Primary salt injuries include direct and specific toxic effects as well as indirect effects, such as metabolic disturbance and inhibition of growth and development. Secondary salt effects include nutrient deficiency and osmotic dehydration. Sodium chloride stress decreased growth and yield by (i) nutritional deficiencies shown by decrease in nitrate, potassium, water content, and increase in root to shoot ratio, (ii) lowered nitrogen compound synthesis via inhibition of nitrate reduction and ammonia assimilation and (iii) protein catabolism at high salt concentration, since sodium chloride enhanced the proteolytic activity and ammonia production.

3.3.1 Ion Homeostasis

Homeostasis can be defined as the tendency of a cell or organism to maintain its internal steady state, even in response to any environmental stimulus tending to disturb normality, because of the coordinate responses of its constituent components. Homeostasis of ion concentration plays an important role in the physiology of all living cells. Regulation of ion fluxes is important to regulate such that the concentration of essential ions is greater and the toxic ions is below the range that can create ion imbalance (Hajiboland 2012; Nieves-Cordones et al. 2012). Cells use mainly two transport systems for ions, primary active transport that uses energy, and secondary transports that occurs through channel proteins and co-transporters. The two main ions that play important role in cell are K^+ and Na^+. Cells generally maintain high concentrations of K^+ and low concentrations of Na^+ (Hajiboland 2012; Nieves-Cordones et al. 2012). Intracellular K^+ and Na^+ homeostasis is important for active function of enzymes, maintenance of membrane potential and osmotic potential of the cell for cell volume regulation and cell function (Hajiboland 2012; Nieves-Cordones et al. 2012). Homeostasis of K^+ and Na^+ is even more important under salinity. Plants survival under saline condition mainly depends on the ion homeostasis. In salinity stress, when Na^+ moves into plant cells and reaches toxic levels it disrupts enzymes and cell function (Ahmad et al. 2010a, b, c, 2011, 2012a). To prevent negative effects on plant cells, excessive Na^+ have to be moved out of cytoplasm and compartmentalized in vacuole. Unlike animal cells, plant cells do not have special Na^+ or K^+ driven ATPase, and are dependent upon H^+ driven ATPase and pyrophosphatases. In recent years many transports of K^+ and Na^+ have been identified (Zhu 2003). The increase in salt concentration inside the plant reduces the cellular osmotic potential, which confronts ion toxicity and oxidative stress (Ahmad et al. 2010a, b, c, 2011). Under salinity conditions, variations in proton gradients that are generated by membrane bound H^+ pumps are crucial for maintaining cytoplasm homeostasis (Pons et al. 2011). Rapid activation of tonoplast-bound pumps was observed in lines that were tolerant to salinity stress. The Na^+/H^+ antiport activity was generally limited to tolerant lines (Pons et al. 2011). Overexpression of Sodium proton antiport (NHX) genes are shown to improve salt tolerance (Pons et al. 2011; Li et al. 2011).

3.3.2 Ion Transport

Salinity stress imposes two stresses in plants. First is water deficit, which is due to high solute concentration, and the second is ion stress, which alters the Na^+ and Cl^- ion concentration mainly by reduction in K^+ ion concentration and increase in Na^+ concentration. This increase in Na^+ is mainly due to similarity between hydrated Na^+ and K^+ radii, and is difficult to discriminate between them. In vitro protein synthesis requires protein concentration of 100–150 mM whereas these K^+ concentrations can be inhibited by Na^+ concentration of 100 mM mainly through competition by Na^+ on K^+ binding site.

3.3.3 Sodium Transport

It is important to know the sodium uptake, transport and compartmentation under saline environmental condition. In order to ameliorate Na^+ toxicity three major strategies have been observed: (a) inhibition of Na^+ influx; (b) improving Na^+ compartmentation in vacuoles via tonoplast Na^+/H^+ antiporters; and (c) increasing Na^+ extrusion via plasma membrane Na^+/H^+ antiporters (Blumwald et al. 2000; Zhu 2001; Munns and Tester 2008).

Na^+ flux upon NaCl shock showed two phases of responses with high Na^+ influx, with drastic reduction in net influx and within several minutes there exhibited a drastic Na^+ efflux in the plant (Shabala 2000). Further Babourina et al. (2000) concluded that Na^+ efflux was mainly due to Na-extrusion system. The NaCl induced Na^+/H^+ exchange in root tissues and cells was inhibited by amiloride (a Na^+/H^+ antiporter inhibitor), indicating that the Na^+ extrusion in stressed roots is the result of an active Na^+/H^+ antiport across the plasma membrane. In case of salt sensitive plants, salt induced Na^+ efflux was correspondingly inhibited by the application of amiloride, the inhibitor of Na^+/H^+ antiporter (Sun et al. 2009).

Sodium entry into cells of roots is mainly by enormous negative membrane potential across the plasma membrane with passive transport system of Na^+. Sodium enters into plants mainly by high affinity K^+ transporter HKT1 (Maser et al. 2002) and non-selective cation channels (Amtmann and Sanders 1999). Na^+ uptake through the apoplast also plays a major role in entry of Na^+ into plants especially by transpiration stream (Yeo et al. 1999). This apoplast pathway is affected by root development and silica deposition in the cell wall. Na^+ uptake through non-selective cation channels is partially sensitive to calcium. This shows that calcium ion plays very important role in preventing Na^+ uptake by the plants (Tester and Davenport 2003).

The *Arabidopsis AtHKT1* protein mediation of Na^+ influx was observed by Uozumi et al. (2000). A screen for suppressor mutations of the salt-hypersensitive *Arabidopsis* mutant *sos3* leads to identification of mutant alleles of *AtHKT1* (Laurie et al. 2002). *athkt1* suppression of salt overly sensitive, (*sos3*) is due to reduced Na^+ accumulation. In wheat, *HKT1* (K^+- Na^+ transporter) also helps in Na^+ influx under salt stress. *AtHKT1* shows negative regulation to salt overly sensitive 3 (*SOS3*) and its compartments whereas there is no regulation from *SOS3* on *AtHKT1* but suppresses *athkt1* due to Na^+ entry. Even though, *At*HKT1 (Na^+ influx carrier) has intracellular Na^+ toxic effect, it has an important role in plant salt tolerance mainly by acting as a transporter in the long-distance transport of Na^+ from root to shoot (Maser et al. 2002).

Calcium sensors (Calcium Binding Proteins) *CBL10* belongs to the family of calcineurin-B-like-proteins (*CBLs*) which specifically interact with a family of serine-threonine protein kinases designated as CBL-interacting protein kinases (*CIPKs*) (Ahmad et al. 2012a; Sarwat et al. 2012). *cbl10* mutants are involved in mediating salt stress tolerance which could be revealed by T-DNA insertion showing a reduced Na^+ and increased K^+ content. Localization studies of green fluorescent protein fusion proteins suggest that *CBL10* is localized to moving punctate structures (endosomes) and at the tonoplast. Yeast-two hybrid and BiFC analysis identified the salt tolerance factor *CIPK24* (SOS2) as predominant interaction partner and revealed

CBL10/CIPK24 complex formation at the tonoplast. *CBL10* and *CIPK24* constitute a novel Ca^{2+}-regulated salt tolerance pathway that regulates the sequestration/compartmentalization of Na^+ into vacuoles of green tissues.

3.3.4 Sodium Compartmentalization

Vacuolar sequestration of Na^+ not only lowers the Na^+ concentration on the cytoplasm but also contributes to osmotic adjustment to maintain water uptake from saline solutions. Other organelles like plastids and mitochondria are also involved in the Na^+ compartmentation at the subcellular level. Cytosolic concentration of Na^+ remains non-toxic both in halophytes and glycophytes in spite of Na^+ influx. The compartmentalization of Na^+ into vacuoles has altered deleterious effect on Na^+ in the cytosol. This thus allows NaCl to act as osmoticum maintaining osmotic potential, thereby driving water inside the plant.

In the current model for intracellular Na^+ sequestration, yeast cells rely on an endosomal H^+-ATPase to establish a H^+ gradient that can drive Na^+ and Cl^- influx via Na^+/H^+ antiporters and chloride channels, respectively (Gaxiola et al. 1999). According to this model, increased H^+ influx into an endosomal compartment will enhance cation sequestration via the Nhx1 Na^+/H^+ antiporter. In *Arabidopsis*, the AtNHX family (Na^+/H^+ antiporters) is involved in Na^+ compartmentation. *AtNHX1* and *AtNHX2* are localized in the tonoplast and they are upregulated by abscisic acid (ABA) or osmotic stress. The vacuolar H^+- ATPase components also increased in response to salt stress. Thus, over expression of AtNHX1 in various plants or of the vacuolar H^+-pyrophosphatase in *Arabidopsis* was reported to enhance plant salt tolerance substantially (Gaxiola et al. 2001).

Salt stress regulation of *AtNHX1* expression is not impaired in the *Arabidopsis sos1*, *sos2* or *sos3* mutants. However, mutations that causes ABA deficiency or the ABA-*insensitive1 (abi1)* mutation partially disrupts *AtNHX1* up-regulation by salt stress (Yokoi et al. 2002). This suggests that SOS-independent, ABA-dependent pathway regulates the expression of the vacuolar antiporter in response to salt stress. However, the SOS pathway appears to regulate the activity of vacuolar Na^+/H^+ antiporters (Qiu et al. 2002).

Interestingly, loss-of-function mutations in the plasma membrane H^+-ATPase (*PMA1*) confer Na^+ tolerance. The mutant cells (*pma1-4*) had a number of phenotypes, which includes reduced Na^+ influx, increased Na^+ tolerance in response to low extracellular pH and increased intracellular Na^+ levels. These results suggested that in addition to the limited Na^+ influx, a mechanism for intracellular sequestration of Na^+ might exist (Nass et al. 1997).

3.3.5 Sodium Efflux

Na^+ efflux from one cell to another cause problem, therefore, it is must to consider the efflux on specific tissues in whole plants. In *Arabidopsis*, Na^+/K^+ antiporter present in the plasma membrane encoded by SOS1 gene catalyzes Na^+ efflux. SOS1 family is present in salt stress condition and involved in transport of Na^+ and not K^+ or Li^+ ion. Its activity is mainly in root epidermal cells especially at the root tip and in cells bordering the vascular tissue throughout the plant. Thus maximum efflux of Na^+ was observed via roots rather than leaves. SOS1 has various roles like (a) Na^+ efflux into root medium; (b) providing time for Na^+ storage in the vacuole by slowing down Na^+ accumulation in the cytoplasm; and (c) controlling long-distance Na^+ transport between roots and leaves by loading Na^+ into and unloading Na^+ from the xylem and phloem. Its role in long distance transport plays an important part in coordination between transpirational Na^+ flow and the vacuolar sequestration of Na^+ in leaves. H^+-ATPase in plasma membrane generates the driving force for Na^+ transport by SOS1. Disruption of AHA4 (root endodermis specific plasma membrane H^+-ATPase) in *Arabidopsis* plants increase sensitivity to salt stress, whereas overexpression of H^+-pyrophosphatase, AVP1 has shown to increase salt as well as drought tolerance.

SOS3 is a myristolated calcium binding protein that is capable of sensing the calcium signal that is elicited by salt stress. SOS2 is a serine/threonine protein kinase. In the presence of calcium, SOS3 activates SOS2 recombinant protein to plasma membrane which ultimately activates SOS1 gene, which on activation provides salt tolerance to mutant plants which was defective in all endogenous Na^+ transporters (*sos1*, *sos2 and sos3*) in yeast (Quintero et al. 2002).

The research has shown that a better understanding of mechanism of Na^+ transport in the plants will help improve salinity tolerance in plants (Plett and Moller 2010).

3.3.6 Potassium Transport

Under salt stress condition, high K^+/Na^+ ratio is important for maintaining cellular metabolism. Plasma membrane is depolarized under this condition and reduces the electrochemical driving force of Na^+ uptake and cause drastic K^+ efflux from both roots and mesophyll cells and reduces K^+ pools and comprises the metabolic competence of the cell. Increased uptake of K^+ is difficult to attain during saline condition due to direct competition of Na^+ under K^+ binding sites on transport system and also by reduced electrochemical potential difference for passive K^+ uptake. Hence it is too difficult to maintain the cytosolic K^+ ion concentration. These can be done mainly by either up-regulation or down-regulation of K^+ transporter genes (*AtKC1*), which will reflect the tolerance or sensitivity to saline condition (Pilot et al. 2003). At the active level of these genes, K^+ channels are regulated by protein kinases and phosphatases. The *Arabidopsis sos* mutants have a growth defect under K^+-limiting conditions. *Athkt1* mutations suppress not only the salt-hypersensitivity but also the K^+ acquisition defect of the *sos3* mutant. The involvement of the SOS pathway could be indirect. A defect in Na^+ efflux in the *sos* mutant may lead to excessive cytoplasmic Na^+ that is inhibitory to K^+ uptake transporters such as AKT1. Under K^+ limiting conditions, inhibitory levels of cytoplasmic Na^+ may arise in the *sos* mutants, even when grown in media that is not supplemented with extra NaCl (Zhu et al. 1998). Salinity tolerant genotypes are capable of maintaining higher xylem K^+/Na^+ ratios and efficiently sequester the accumulated Na^+ in the leaves, which is achieved by efficient loading of K^+ into xylem (Shabala et al. 2010). They also showed that K^+ – permeable voltage-sensitive channels are involved in xylem loading and operate in a feedback manner to maintain a constant K^+/Na^+ ratio in the xylem sap.

3.3.7 *Chlorine Transport*

Cl^- toxicity plays a major role in case of woody plants in response to salt stress condition. However, these are less investigated comparing with the Na^+, which has intensive studies (Teakle and Tyerman 2010). Cl^- influx is coupled with Na^+ thermodynamically in the mesophyll cells of leaves (Shabala 2000). An increase in Cl^- influx was observed through protoplast (Teakle and Tyerman 2010). In case of *P. euphratica* resistant plant, Cl^- efflux was more comparing with the salt sensitive plants.

3.3.8 *Calcium*

Calcium ion (Ca^{2+}) can regulate and also alleviate negative influence of salinity stress on plant growth (Ahmad et al. 2012a; Sarwat et al. 2012). Application of Ca^{2+} can minimize the leaves of K^+ and also protect cell membranes from damage caused by Na^+ through minimizing its uptake and prevent Na^+ to the cell walls. Wu and Wang (2012) showed that at low salinity, Ca^{2+} decreased root Na^+ accumulation, increased shoot K^+ accumulation and enhanced the selective absorption and transport capacity of K^+ over Na^+ in rice. However, at high salinity Ca^{2+} did not have any effect on Na^+ and K^+ accumulation. Na^+ efflux and Na^+ influx were remarkably reduced Ca^{2+} under both low and high salinity, their ratio was lowered only under low salt stress (Wu and Wang 2012). They suggest that Ca^{2+} could regulate K^+/Na^+ homeostasis in rice at low salinity by enhancing the selectivity for K^+ over Na^+, reducing the Na^+ influx and efflux, and lowering the futile cycling of Na^+.

3.4 Effects of Salinity on Water Relations and Osmotic Adjustment

Salinity causes negative effects on the plant water status (Ahmad and Sharma 2010; Ahmad et al. 2010a, 2012b; Azooz et al. 2011). The mechanism controlling the internal plant water status may involve water uptake or water conservation by the plant; and also internal plant water conductance during drought. The plant water status differed significantly among cultivars exposed to the same period of water exclusion. Leaf water content is a better indicator of the water status of a plant. Plants are able to maintain the turgor potential at the same value for lower values of leaf water potential owing to their osmotic adjustment. Salinity decreases leaf relative water content, stomatal conductance, leaf water potential, solute potential and turgor potential (Ahmad et al. 2012b; Ahmad and Prasad 2012a, b). Water use efficiency of plants; calculated, as the ratio between total plant dry weight and total plant water uptake did not change under salinity stress conditions. Under salinity stress conditions, plants with high water content were more tolerant than plants with lower values. Increasing the salinity of irrigation water caused a reduction in the water potential, osmotic and pressure potentials of both leaf and root in plants. The detrimental effects of salinity on plant water status and metabolism of plants were more pronounced at flowering rather than the vegetative stage.

Salinity imposes turgor loss in plant tissue, is due to decrease in water potential. In salinity stress, plants escape from dehydration, by reducing their osmotic potential and by adjusting with osmolytes, which helps in transport, accumulation and compartmentation of inorganic ions and organic solutes. Leaf osmotic potential and leaf ion accumulation are highly correlated with total dry matter production of a plant. External high osmotic potential reduces the plant water retention, which in turn affects the gas exchange, photosynthetic rate and protein synthesis.

The role of decreased turgor potential on growth is unclear. In some plants species it was observed that salinity causes decrease in turgor potential that leads to growth cessation, while in other species it was observed that salinity increased turgor potential. This cellular response may be due to osmotic adjustment and also due to differences in sensitivity of these plant species to salinity stress. Cultivars within species are also known to differ in water relations in response to salinity stress. In general, salt sensitive cultivars have higher leaf turgor than salt tolerant cultivars, this may be related to reason that salt sensitive plants may not be able to efficiently exclude salts from their roots when compared to salt tolerant cultivars (Ashraf 2004; Ahmad et al. 2006, 2010a; Ahmad 2010; Ahmad and Sharma 2010).

The relation between turgor, tissue elongation and growth is contradictory due to the reasons for what controls cell wall expansion. In addition to turgor there are several other factors which influences growth, such as phytohormones (Ahmad et al. 2010c). Ashraf (2004) suggested although salinity influences water relations in plants, the response and measurements of water potential and turgor potential alone may not help in discriminating differences among salt tolerant and salt susceptible genotypes. A better and improved understanding of mechanisms that confer tolerance to salinity are necessary. Aquaporins are water channel proteins that are present in most organisms from bacteria to higher plants; they are known to be the major facilitators of the movement across membranes (reviewed in chapter 3). The role of these water channels in maintaining water balance and movement across membranes under osmotic stress is not clear. Water channels could facilitate water uptake in roots if their presence is associated with accumulation of metabolites that help in osmotic adjustment. In contrast, removal of water channels on the membranes can restrict water movement or water loss. In addition, the amount of water channels in the membranes of the vacuole may help in separation or concentrations of ions. As the knowledge about role of aquaporins is increasing, it is important to understand the influence of salinity on the aquaporins and water movements and water relations in plants.

Osmotic adjustment refers to the lowering of the osmotic potential due to the net accumulation of solutes in response to water deficit or salinity stress (Ahmad and Sharma 2008; Ahmad et al. 2012b). Osmoregulation could be defined as the regulation of osmotic potential within a cell by the addition or removal of solutes from solution until the intracellular osmotic potential is approximately equal to the potential of the medium surrounding the cell. Osmotic adjustment is an important mechanism in salinity tolerance, because it enables (a) continuation of cell expansion; (b) stomatal and photosynthetic adjustments; (c) better plant growth; and (d) yield production. The compounds mainly involved in osmotic adjustment are the soluble sugars, organic acids, free amino acids and potassium ion. The degree of the osmoregulatory processes is affected by the rate of stress, stress preconditioning, the organ type and age, and the genetic variation between and within species. The non-toxic compatible organic solutes are accumulated in the cytoplasmic compartment of cells; and inorganic ions toxic to metabolic processes were restricted to the vacuolar compartment. The leaf osmotic adjustment of plants was linearly proportional to salinity stress tolerance. Water flow through the soil-plant-atmosphere continuum is mainly due to gradient of decreasing water potential. Water potential of pure water is 0, and increasing salinity or concentrations of salts in the water decreases water potential and as such sharp increases in salt will hold water osmotically away from plants, and to limit restrictions on water uptake, plants must generate increasingly lower water potentials to allow continued water flux. This is often achieved through salt tolerance, where solute concentrations can be changed through production of compatible solutes. Osmotic adjustment in plant cells can also be mediated through accumulation of Na^+ and Cl^- to generate sufficient turgor for survival and growth at higher salinity stress. Such salt tolerant plants generally decrease leaf water potential below the soil water potentials. Research also showed that genotypes tolerant to salinity stress had mechanisms to prevent high Na^+ and Cl^- in leaves (Nemati et al. 2011).

Photosynthesis is an important biochemical pathway for survival and growth of plants. Photosynthetic rate are reduced due to ultra-structural changes in chloroplast under salinity condition. Photosystem II, electron transport chain and CO_2 assimilation were the photosynthesis components mainly affected by salinity. The stomatal conductance was decreased by salinity stress. Leaf photosynthetic rates were maintained in salt tolerant varieties, but declined in the sensitive varieties at higher salinity levels. This reduction in photosynthetic rate may be due to the increased stomatal resistance and thereby lower intercellular carbon dioxide concentration. Under salinity treatment, plants expressed two phases of photosynthetic inhibition: in the first phase, decrease in photosynthesis was gradual, whereas in the second phase it was rapid and accompanied by a decline of the energy conversion efficiency of photosystem II strongly related to adverse effects of salinity. The decrease was due to reduced carbon dioxide assimilation associated with a decline in stomatal conductance, water use efficiency and Rubisco activity, as well as slower electron transport of photosystem II under severe salinity stress. Photosynthesis is decreased due to two main reasons, (a) reduction in chlorophyll content; and (b) reduction in leaf area. Chlorophyll pigments play a major role in plant productivity, as they are one of the principal pigments responsible for photosynthesis. Decrease in the total chlorophyll and chlorophyll fractions may be attributed to the increased chlorophyllase activity and partly due to the interference of ions with the *de novo* synthesis of proteins; the structural components of chloroplast. Compartmentalization of ions into vacuoles, cytoplasm and chloroplast may decrease the damage to photosynthetic system under salinity stress.

3.5 Effects of Salinity on Compatible Solutes

Some plants have the potential to increase cellular concentrations of osmotically active compounds called as compatible solutes (Ahmad and Jhon 2005; Ahmad et al. 2006, 2010a,2012b; Ahmad and Sharma 2008). This happens when cells are stressed (particularly drought or increased salinity). They are termed compatible due to the fact that they allow normal cellular metabolism at high concentrations. Compatible solutes help in maintenance of ion homeostasis and water relations (Ahmad 2010; Ahmad et al. 2010a; 2012b; Azooz et al. 2011; Katare et al. 2012). Compatible solutes do not interfere with function or structure of the protein, but only help in alleviating the negative effects of high ion concentrations on the enzymes, stabilizing proteins, protein complexes, membranes and cell function under stress conditions. To accumulate the ionic balance in the vacuoles, cytoplasm accumulates compatible solutes as they are low-molecular-mass compounds and do not interfere with normal biochemical reactions (Parida and Das 2005; Zhifang and Loescher 2003; Ahmad and Sharma 2008).

Some examples of compatible solutes include proline, mannitol, glycine betaine, sugars, and some nitrogen containing compounds. Their function is mostly an osmoregulatory function; in addition some compatible solutes are also capable of scavenging free radicals. Salinity caused an increase in the proline content in leaves, stem and roots of most plants (Ahmad and Sharma 2008; Ahmad 2010; Azooz et al. 2011; Katare et al. 2012). Proline is an important compatible solute in halophytic plants as well as some glycophytes. Intermediates of proline biosynthesis and catabolism induced the expression of several osmotically regulated genes. Proline accumulation may be due to the increased activity of pyroline-5-carboxylase reductase and reduced level of proline oxidase observed during episodes of salinity (Nounjan et al. 2012). Accumulation of proline in rice was a symptom of salt stress injury, which was due to increase in the levels of Ornithine-Δ-Aminotransferase (OAT) and its precursor glutamate. Mannitol, a sugar alcohol produced to cope with salt stress, is synthesized via the action of mannose-6-phosphate reductase (M6PR) (Zhifang and Loescher 2003). Arabidopsis plants were transformed with M6PR gene, the plants were phenotypically similar and mature plants showed high levels of salt tolerance compared to wild type, and completed their normal life cycle and produced seeds in soils with high salt (300 mM NaCl). These results demonstrates role of developing tolerance by means of increased biosynthesis of mannitol (Zhifang and Loescher 2003).

Transgenic cotton plants that accumulated glycinebetaine in larger quantities generally more tolerant to salt stress than wild-type (Zhang et al. 2011). Levels of endogenous glycinebetaine were positively correlated with relative water content and osmotic adjustment. These results indicate that glycinebetaine in transgenic plants not only maintains the integrity of the cell membranes but also alleviated osmotic stress caused by high salinity (Zhang et al. 2011). Other solutes such as polyols, carbohydrates (glucose, fructose, sucrose, fructons, starch), nitrogen containing compounds (amino acids, proteins, quaternary ammonium compounds), and polyamines are accumulated under salt stress (Parida et al. 2002; Ahmad and Sharma 2008.

In some cases, although exogenous application of osmoprotectants did not alleviate growth inhibition during salt stress, they exhibited pronounced beneficial effect during recovery period showing higher percentage of recovery in plants treated with salt stress and protein or trehalose (Nounjan et al. 2012). Recent study showed that osmotic adjustment provided by various compound can vary based on the intensity of the salinity stress. Under low and moderate salinity, inorganic cations were the major contributors of the osmotic adjustment in roots, followed by the sugars, while the relative contribution of proline and free amino acids was less (Misic et al. 2012). While, osmotic adjustment under severe salinity stress appeared to be mediated by accumulation of organic compounds. In addition inorganic ions can also play significant role for osmotic adjustment and plants ability to maintain K^+ levels and the involvement of compounds such as putrescine efflux in maintaining ionic balance under high salinity conditions (Orsini et al. 2011.

3.6 Effects of Salinity on Oxidative Damage

Oxygen molecule, a free radical was introduced in the earth's atmosphere by the O_2 evolving photosynthetic organisms. In chloroplast, oxygen generated as a product of photosynthesis is capable of accepting electrons passing through the photosystem thereby producing reactive oxygen species (ROS) (Ahmad et al. 2010b, c, 2011). Under normal conditions of plant growth, there is a dynamic equilibrium between ROS and antioxidants; this balance is interrupted by salinity stress. The change in equilibrium cause sudden increase in intracellular level of ROS favoring oxidative reactions and promotes oxidative stress in plants (Ahmad et al. 2008, 2010b). Salinity, accompanied by high concentrations of salts primarily induces ionic imbalance and hyper-osmotic stress in plants. Due to disruption of homeostasis and water balance, cellular membranes, enzyme activity and photosynthetic apparatus are affected with an over production of ROS in salt stressed plants. The increase in ROS is responsible for causing oxidative damage to cells known as oxidative burst. Membranes are primary targets of salinity injury to cell and cellular organelles due to the fact that ROS has high potential to react with unsaturated fatty acids resulting in peroxidation of essential membrane lipids in the membranes of cell and intercellular organelles (Ahmad et al. 2010b, c, 2011). Peroxidation of membranes leads to leaky membranes leading to loss of electrochemical gradient, loss of homeostasis, loss of cellular contents, rapid desiccation and cell death. The damage associated with salt stress increases the lipid peroxidation, electrolyte leakage and hydrogen peroxide to a greater extent in salt sensitive genotypes or species (Hu et al. 2012; Ahmad et al. 2008, 2010a).

3.6.1 Reactive Oxygen Species

Oxygen is essential for plant life. It is generated in the electron transport during light reaction of photosynthetic pathway. The reduction of O_2 to H_2O is a mixed blessing that provides energy for higher organisms. Due to incomplete reduction of O_2 favored by stress factors, excess production of ROS takes place, which are highly reactive and capable of oxidizing biological molecules. Under salt stress, stomatal closure limits the availability of CO_2 but the light driven electron transport continues at higher rates. The demand for photosynthesis is thereby limited at the expense of light absorption by leaves and hence the excess excitation energy causes decrease in electron transport. The transport of electrons to alternate acceptors was created due to imbalance between the consumption of NADPH during carbon fixation and regeneration of NADP at PS I (Tanaka et al. 1999; Foyer and Noctor 2000; Tausz et al. 2004). In plants, ROS such as H_2O_2 (hydrogen peroxide), $^1O_2$ (singlet oxygen), O_2^- (superoxide radical) and OH (hydroxyl radical, it is a neutral form of the hydroxide ion OH^-) are produced in different cellular components as byproducts of various metabolic pathways (Ahmad et al. 2008, 2010b, 2011; Ahmad and Umar 2011). Under steady state, these ROS are scavenged by the respective antioxidants. Abiotic stress factors cause rapid increase in accumulation of ROS that cannot be balanced by the antioxidative defense system (Ahmad et al. 2008, 2010b, c, 2011). The formation of ROS is initiated by the reduction of molecular oxygen. Reduction of molecular oxygen continues in a series of steps leading to formation of several O_2 radical species. The steps involve the production of O_2^-, H_2O_2 and OH^- by using one or two or three electrons respectively in reduction of oxygen. Decreased photosynthesis under salinity stress increases the formation of ROS molecules and also changes the dynamics of enzymes that detoxify ROS species (Ahmad et al. 2008, 2010b, c, 2011).

3.6.2 Superoxide Radical (O_2^-)

Superoxide radical is generated by univalent reduction of molecular oxygen during electron transfer. Here, only one electron is involved for reduction. O_2^- is moderately reactive and first generated ROS with approximately 2–4 μs of half-life. It is generated in the thylakoid membrane bound primary electron of photosystem I. O_2^- induces the formation of other ROS like $^1O_2$ and OH^- (Halliwell 2006) and these highly reactive molecules cause lipid peroxidation in membranes. $^1O_2$ is generated from superoxide radical by the reduction of Fe^{3+} and furthermore, O_2^- generates

H_2O_2 through dismutation catalysed by superoxide dismutase and OH^- is produced by the production of H_2O_2 radicals. The generation of OH^- by the reactions involving O_2^-, H_2O_2 and iron is called Haber-Weiss reaction and Fenton's reaction is the oxidation of Fe^{2+} by H_2O_2. Salinity results in increased production of superoxide radicals (Ahmad et al. 2010b, 2011; Ahmad and Umar 2011).

3.6.3 *Hydrogen Peroxide (H_2O_2)*

H_2O_2 is moderately reactive ROS generated by the univalent reduction of O_2^- catalyzed by superoxide dismutase and has long lifetime of 1 ms. It plays an important role in physiological processes such as photorespiration, photosynthesis, senescence, stomatal movement and growth and development. At high concentrations, H_2O_2 causes cell death in plants.

3.6.4 *Singlet Oxygen ($^1O_2$)*

$^1O_2$ is generated by transfer of excess excited energy of O_2 and not electron transfer. The primary source of $^1O_2$ is the chlorophyll pigments associated with the electron transport system and its lifetime is approximately 3 μs (Hatz et al. 2007). During photosynthesis, chlorophyll triplet state reacts with $^3O_2$ to form $^1O_2$. This $^1O_2$ is responsible to cause detrimental effect on whole photosynthetic machinery. $^1O_2$ act as an oxidizing agent of biological molecules causing cell death in plants (Krieger-Liszkay et al. 2008) and also induces the production of anti-microbial secondary metabolites (Flors et al. 2006). Salinity stress produces higher quantities of singlet oxygen species.

3.6.5 *Hydroxyl Radical (OH^-)*

Hydroxyl radical is the most reactive oxidant in cells and is generated from O_2^- and H_2O_2 in the presence of Fe through Fenton's reaction. OH^- plays a key role in lipid peroxidation and also damages DNA, protein and other cell constituents leading to cell death (Vranova et al. 2002). As other oxidants, higher salt concentrations in cells results in accumulation of hydroxyl radical.

3.7 Effects of Salinity on Antioxidants

Plants have the ability to protect themselves from oxidants (e.g. ROS) by employing specific mechanisms in the organelles such as chloroplast, mitochondria and peroxisomes. The mechanism includes the activation of enzymatic antioxidants and non-enzymatic antioxidants. The enzymatic antioxidants include superoxide dismutase (SOD), catalase (CAT), ascorbate peroxidase (APX), monodehydro ascorbate reductase (MDHAR), dehydroascorbate reductase (DHAR), glutathione reductase (GR) (Noctor and Foyer 1998; Smirnoff 2002; Ahmad et al. 2009, 2010a,b; 2012a, b; Ahmad and Umar 2011), glutathione-S-transferases (GST) and glutathione peroxidases (GPX), and non enzymatic antioxidants are ascorbic acid (AA), flavones, carotenoids, anthocyanins and tocopherols (Mittler 2002; Ahmad et al. 2009, 2010b, c, 2011). These antioxidants are generally increased under salinity stress conditions and often provide protection to membranes and cell function. However the response of antioxidant can be stage specific, at some stages the activities can be decreased due to high salinity stress, and these changes the activities can be increased once the plants are recovered from the stress (Sekmen et al. 2012). Salinity tolerance among different genotypes can be related to induction and sustained expression and production of highly regulated antioxidant mechanisms (Mhadhbi et al. 2011). Furthermore, tolerance behavior of certain genotypes could be related to induction of antioxidant genes in plant roots, leading to more efficient enzyme stimulation and protection.

3.7.1 Superoxide Dismutase (SOD)

SOD belongs to the family of metaloenzymes and is ubiquitous in every cellular component of the plant. It catalyses the disproporationation of O_2^- to molecular oxygen and H_2O_2 (Giannopolitis and Ries 1977). SOD exists in different isoforms such as manganese containing SOD (Mn-SOD), copper-zinc containing SOD (Cu/Zn-SOD), iron containing SOD (Fe-SOD) and nickel containing SOD (Ni-SOD), which are localized in different cellular components. Mn-SOD is found in mitochondria and peroxisomes, Cu/Zn-SOD is predominant in cytosolic fractions and chloroplasts and Fe-SOD in chloroplast compartments. Several environment stresses induce the generation of SOD in plants cells, which is the first line of defense against the toxic oxygen radicals. Salinity increases the activity of SOD in different plant species and a different mechanism was reported to operate in oxidative stress injury (Yu and Rengel 1999a, b; Ahmad et al. 2008, 2010b, c, 2011).

3.7.2 Catalases (CATs)

CATs present in peroxisomes are tetrameric heme containing enzymes involved in dismutation of H_2O_2 into H_2O and O_2. They play a significant role in scavenging H_2O_2 generated during β-oxidation of fatty acids in seed germination, photorespiration and purine catabolism. CAT exhibit high turnover rate i.e., one molecule of CAT can detoxify nearly 6 million molecules of H_2O_2 to H_2O and O_2 per minute. Though CAT mainly reacts with H_2O_2, it also can react with some organic hydroperoxides such as methyl hydrogen peroxide (MeOOH) (Ali and Alqurainy 2006). Like SOD, CAT also exists in different isoforms in higher plants (Polidoros and Scandalios 1999). A significant increase in CAT activity was noted in *C. arietinum* leaves (Eyidogan and Oz 2005) and roots (Kukreja et al. 2005) under salt stress treatment.

Catalase activity assays were performed before and after salt stress treatments in *B. gymnorrhiza* and it was found that the enzyme activity increased 4.9 fold under salt stress compared to non-stressed plants (Takemura et al. 2000).

3.7.3 Ascorbate-Glutathione Cycle Enzymes

The enzymes of ascorbate-glutathione cycle include APX, MDHAR, DHAR and GR. Ascorbate peroxidase is involved in oxidation of ascorbate to MDA (monodehydroascorbate). It is essential for regeneration of ascorbate from MDA to maintain the H_2O_2 scavenging system in chloroplasts (Ahmad et al. 2008, 2010b, c, 2011; Ahmad and Umar 2011; Yousuf et al. 2012). MDHAR catalyses reduction of MDA to ascorbate. Dehydroascorbate is generated if MDA is not converted to ascorbate. Dehydroascorbate is converted into ascorbate by the thiol enzyme, DHAR, but the contribution of MDHAR is greater than that of DHAR (Asada 1994; Minkov et al. 1999). Another potential enzyme belonging to the ascorbate-glutathione cycle is glutathione reductase (GR). Glutathione is involved in conversion of Dehydroascorbate to ascorbate through DHAR. Glutamyl cysteine glycine (GSH), a metabolic regulatory molecule is oxidized to glutathione di sulphide (GSSG) and GSH is being regenerated by GR. Many studies indicated that ascorbate-glutathione pathway enzymes were up regulated under salinity stress. The increases in the activity of the above enzymes are higher in tolerant variety than susceptible varieties. Lee et al. (2001) showed that salinity stress increased the activity of APX in rice leaves but not in rice roots. There was an early increase in GR activity in NaCl-exposed shoot cultures of rice (Fadzilla et al. 1997). Salinity stress enhanced the expression of APX and GR genes (Kawasaki et al. 2001). However, Lopez et al. (1996) demonstrated that APX activity and not the mRNA level, was enhanced under salinity stress in *Raphanus sativus* plants. Transgenic tobacco seedling that overexpress a cDNA encoding an enzyme with both GST and GPX stimulated and improved seed germination and seedling growth under salt stress (Roxas et al. 1997).

Sulfur is part of several organic molecules in plants and present it thiol (–SH) groups in proteins such as GSH or non protein thiols such as glutathione. Cysteine is the precursor or S-donor for most of the other organic S compounds in plants. Under salinity stress there is a high demand for cysteine which emphasizes the role of sulfuric compounds as osmolytes or antioxidants. Nazar et al. (2011) provided a detailed review on aspects related effects of salinity on physiology and metabolism of plants and the importance of S in salinity tolerance or salinity induced responses.

3.7.4 Non-enzymatic Antioxidants

Apart from the enzymatic antioxidants, plants also produce different non-enzymatic antioxidants that are capable of scavenging the toxic ROS and reduce oxidative stress (Ahmad et al. 2008, 2009, 2010b, c, 2011). The most important non-enzymatic antioxidants are ascorbic acid (AA) and glutathione. Ascorbic acid/vitamin C is the most powerful, abundant and water-soluble antioxidant, which reduces the harmful effects of ROS. It is found in many plant cell types, organelles especially in photosynthetic cells, meristems and in some fruits and apoplast. Higher concentration of AA is found in mature leaves. In leaves and chloroplasts, AA remains available in reduced form under normal physiological conditions. In mitochondria, AA is synthesized and transported to other cell components by proton-electron chemical gradient or by facilitated diffusion AA is the most powerful ROS detoxifying compound because it has the ability to donate electrons in a number of enzymatic reactions. AA can directly quench H_2O_2, O_2, and OH. H_2O_2 is reduced by AA to H_2O through ascorbate peroxidase reaction (Foyer et al. 1997). It imparts membrane protection by regenerating tocopherols from tocopheroxyl radical. Therefore, a higher level of AA is important to minimize oxidative stress and regulate plant metabolic processes (Athar et al. 2008). Increase in ascorbate levels and the maintenance of the redox state in the cell is very critical for root growth and development in salt stress (Hernandez et al. 2010). Several findings indicated that the salinity induced wilting was associated with increases in the cellular activity of damaging ROS and the elevating effect of AA on seedling survival was associated with the partial inhibition of ROS production. The decrease in level of lipid peroxidation by ROS in root, stem and leaf tissues during salt stress was decreased by exogenous ascorbic acid application (Shalata and Neumann 2001). Zhang and Kirkham (1996) reported similar inhibitory effects of exogenous AA on lipid peroxidation in sunflower seedlings exposed to osmotically induced water-stress. Tripeptide glutathione is a major source of non-protein thiols and a crucial metabolite in plants. It has been detected as reduced form in all cell compartments such as cytosol, chloroplasts, endoplasmic reticulum, vacuole, mitochondria, chloroplasts, peroxisomes and apoplast. It is essential to regulate the normal reduced state of cells which thereby reduces the injurious effects of stress induced ROS (Wang et al. 2008). Reduced glutathione (GSH) plays an important role in regeneration of ascorbic acid, water-soluble antioxidant through the ascorbate-glutathione cycle and marks its role in antioxidative defense system. It is also involved in the detoxification of dehydroascorbate reductase. Glutathione maintains the redox equilibrium of cellular compartments in combination with oxidized form of glutathione (GSSG). Ruiz and Blumwald (2002) observed a threefold increase in glutathione content in wild-type *Brassica napus* L. plants exposed to salinity stress. The induction of glutathione synthesis during salinity stress suggests a possible protective mechanism against salinity induced oxidative damage. Some other non enzymatic antioxidants such as carotenoids, flavonoids, tocopherols and polyamine have some protection against oxidative stress by acting as scavengers of ROS molecules. Salinity also induce changes in phenolic compounds (Petridis et al. 2012) where biosynthesis of phenols and oleuropein was stimulated in leaves. There was strong correlation between total phenol content and antioxidant activity in both leaves and roots (Petridis et al. 2012). In addition to antioxidants, plant hormones especially brassinosteriods and salicylic acid can influence tolerance of plants to salinity stress. The role of exogenous application of brassinosteriods and salicylic acid in regulation of various biochemical and physiological processes that can lead to improved salt tolerance (Ashraf et al. 2010). It was observed that although salicylic acid is not essential for germination under normal growth conditions, but is plays a promotive role in seed germination under high salinity by reducing oxidative damage (Lee et al. 2010).

3.8 Possible Strategies to Improve Salt Tolerance

Identification of reliable morphological, physiological or biochemical traits under salinity stress are an integral part of any successful breeding program. Ashraf and Harris (2004) conducted review of various biochemical indicators of salinity tolerance in plants. They argued that at present there are no well-defined indicators for salinity tolerance available to assist plant breeders in the improvement of salinity tolerance in important agricultural crops. Thus, there is a need to determine distinct traits and underlying biochemical mechanisms of salinity tolerance. Ashraf and Harris (2004) provided about 11 different biomarkers which include soluble sugars, proteins, amino acids, ammonium compounds, polyamines, polyols, antioxidants, ATPase. As there is a large variability in responses to salinity stress by various plant species and genotypes within species, it is more valuable if biochemical indicators are specified for individual species rather than generalized for all species (Ashraf and Harris 2004).

Salinity stress is complex in nature; hence there is no single definite morphological marker available for identification of tolerant or sensitive lines. Therefore, several parameters may be used in combination for the effective and reproducible screening. The tolerance of plants to salinity stress mainly depends on the low Na^+ and Cl^- as well as maintenance of high nutrient concentration especially K^+ homeostasis. There are extensive genetic diversities in plants varying from halophytes (native to saline environment) to glycophytes (salt sensitive) for preferential K^+ accumulation. Halophytes tolerate extreme salinity because of special anatomical and morphological adaptations or mechanisms (Flowers et al. 1986). The cytotoxic ions like Na^+ and Cl^- are compartmentalized in vacuoles and used as osmotic solutes both in halophytes and glycophytes (Blumwald et al. 2000; Niu et al. 1995). It follows then that many of the molecular entities that mediate ion homeostasis and salt stress signaling are similar in all plants (Hasegawa et al. 2000). Plant genetic model *Arabidopsis* a glycophyte, will be required to delineate if salt tolerance is affected most by formation or function of genes or more by differences in the expression of common genes either due to transcriptional or post-transcriptional control (Zhu 2001).

NaCl is the principal stress-causing agent; research focus has been the transport systems that are involved in utilization of Na^+ as an osmotic solute (Blumwald et al. 2000). Intracellular Na^+ homeostasis and salt tolerance are modulated by Ca^{2+} and high $[Na^+]_{ext}$ negatively affects K^+ acquisition. Na^+ competes with K^+ for uptake through common transport systems and does this effectively since the $[Na^+]_{ext}$ in saline environments is usually considerably greater than $[K^+]_{ext}$. Ca^{2+} enhances K^+/Na^+ selective intracellular accumulation (Maathuis et al. 1996). Research of the last decade has defined many of the molecular entities that mediate Na^+ and K^+ homeostasis and given insight into the function of Ca^{2+} in the regulation of these transport systems. The salt overly sensitive (SOS) stress-signaling pathway was identified to be a pivotal regulator of plant ion homeostasis and salt tolerance (Hasegawa et al. 2000; Sanders 2000). This signaling pathway functionally resembles the yeast calcineurin cascade that controls Na^+ influx and efflux across the plasma membrane (Bressan et al. 1998). Expression of an activated form of calcineurin in plants enhances salt tolerance further implicating the functional similarity between the calcineurin and the SOS pathways (Pardo et al. 1998). Little is known about the mechanistic entities that are responsible for Cl^- transport or the regulation of Cl^- homeostasis (Hedrich 1994).

Role of cytosolic N^+ exclusion in roots as a means of salinity tolerance and method of functional assessment was developed (Cuin et al. 2011). They showed that active efflux of N^+ from root epidermal cells was mediated by a SOS1-like homolog, energized by the plasma membrane H^+-ATPase. SOS1-like efflux activity was highest in salt tolerant genotypes. The salt tolerant plants had an enhanced ability to sequester large quantities of Na^+ in to the vacuoles of roots cells, as revealed by confocal microscopy using sodium green stain, while the sensitive lines had greater proportion of Na^+ in the root cell cytosol (Cuin et al. 2011).

Zhang et al. (2001) found that transgenic *Brassica napus* plants overexpressing *AtNHX1*, a vacuolar Na^+/H^+ antiport from *Arabidopsis thaliana*, were able to grow, flower, and produce seeds in the presence of high sodium chloride. Although the transgenic plants grown in high salinity accumulated higher sodium; the dry weight and growth of these plants were only marginally affected by the high salt concentration. Moreover, seed yields and the seed oil quality were not affected by the high salinity of the soil. Strategies to modify ion transport and understanding the mechanisms of transport will help improve our knowledge and help develop strategies to improve salinity tolerance (Plett and Moller 2010). These results demonstrate the potential use of these transgenic plants for agricultural use in saline soils. They concluded, that the modification of a single trait significantly improved the salinity tolerance of this crop plant, suggest that with a combination of breeding and transgenic plants it could be possible to produce salt-tolerant crops with far fewer target traits than had been anticipated.

Munns and Tester (2008) provided a comprehensive review on mechanisms of salinity tolerance with particular emphasis on food grains crops. They summarized that plant adaptation to salinity are of three distinct types (a) osmotic stress tolerance – in this case a reduced response to osmotic stress would result in greater leaf growth and stomatal conductance, but the resulting increased leaf would benefit only plants that have sufficient soil waters; (b) Na^+ exclusion from leaf blades – in this case Na^+ by roots ensures that Na does not accumulate to toxic concentrations within leaves, a failure to exclude Na^+ manifests its toxic effects and causes premature death of older leaves; and (c) Tissue tolerance to Na^+ and Cl^- – this focuses on compartmentalization of Na^+ and Cl^- at the cellular or intercellular level to avoid concentrations within cytoplasm, especially in the mesophyll cells of the leaf. In addition to these increased production of osmoprotectants and antioxidant productions systems also have some potential and needs further exploration.

3.9 Conclusion and Future Perspective

Salinity is an important abiotic stress that limits growth and yield of many crop species. Salinity stress causes changes in ion homeostasis and ion transport (Na, K, Cl), water relations (changes water, osmotic and turgor potential of plants), and associated osmotic adjustments. Salinity stress results in production of several ROS or free radicals resulting in oxidative damage of membranes leading to loss of cell function. Salinity stress can also cause accumulation of compatible solutes and stimulate production of antioxidants which can protect membranes and cell organelles. Some of the adaptations to salinity stress in plants include (i) accumulation of osmoprotectants; (ii) sodium (Na^+) or chloride (Cl^-) exclusion; (iii) tissue tolerance to accumulated Na^+ or Cl^-; and (iv) detoxification of ROS. Plant species and genotypes within species show differential responses to salinity stress. Various traits that govern salinity tolerance can be identified by studying physiological and biochemical mechanisms, which can lead to identification of new genetic sources of salinity tolerance. Salt tolerance in crop plants can be improved by identifying new genetic materials through screening for individual and combinations of above mentioned adaptation traits. Plant species and genotypes within species are known to differ in their tolerance to salinity. In addition, several transgenic plants with increased expression of specific transports, sequesters, compatible solutes have shown increased tolerance to salinity. Therefore, opportunities exist for selection, traditional breeding method and molecular approaches. Although our understanding of physiological basis for tolerance, susceptibility or genetic variability has improved, there are still several issues related to screening large populations, efficient phenotypic techniques, relevance of field and controlled environment studies, performance of transgenic under field conditions, identification of key genes associated with tolerance and their function at whole crop level and on crop productivity. Recent advances in molecular genetics, functional genomics, proteomics, and metabolomic analyses along with enhanced understanding of biology of tolerance, certainly provides hope for improving salinity tolerance of food crops.

Acknowledgements Authors would like to thank Kansas State University and United States Aid for International Development Agency for supporting Crop Physiology and Production Program. Contribution no 12-385-B from the Kansas Agricultural Experiment Station.

References

Ahmad P (2010) Growth and antioxidant responses in mustard (*Brassica juncea* L.) plants subjected to combined effect of gibberellic acid and salinity. Arch Agro Soil Sci 56(5):575–588

Ahmad P, Jhon R (2005) Effect of salt stress on growth and biochemical parameters of *Pisum sativum* L. Arch Agro Soil Sci 51(6):665–672

Ahmad P, Prasad MNV (2012a) Environmental adaptations and stress tolerance in plants in the era of climate change. Science+Business Media, New York, http://f6mail.rediff.com/prism/@mail/http:/help.yahoo.com/l/in/yahoo/mail/classic/context/context-07.htmlSpringer

Ahmad P, Prasad MNV (2012b) Abiotic stress responses in plants: metabolism, productivity and sustainability. Science+Business Media, New York, http://f6mail.rediff.com/prism/@mail/http:/help.yahoo.com/l/in/yahoo/mail/classic/context/context-07.htmlSpringer

Ahmad P, Sharma S (2008) Salt stress and phyto-biochemical responses of plants. Plant Soil Environ 54(3):89–99

Ahmad P, Sharma S (2010) Physio-biochemical attributes in two cultivars of mulberry (*M. alba*) under $NaHCO_3$ stress. Int J Plant Produc 4(2):79–86

Ahmad P, Umar S (2011) Oxidative stress: Role of antioxidants in plants. Studium Press, New Delhi

Ahmad P, Sharma S, Srivastava PS (2006) Differential physio-biochemical responses of high yielding varieties of Mulberry (*Morus alba*) under alkalinity (Na_2CO_3) stress *in vitro*. Physiol Mol Biol Plants 12(1):59–66

Ahmad P, Sarwat M, Sharma S (2008) Reactive oxygen species, antioxidants and signaling in plants. J Plant Biol 51(3):167–173

Ahmad P, Jeleel CA, Azooz MM, Nabi G (2009) Generation of ROS and non-enzymatic antioxidants during abiotic stress in Plants. Bot Res Intern 2(1):11–20

Ahmad P, Jaleel CA, Salem MA, Nabi G, Sharma S (2010a) Roles of enzymatic and non-enzymatic antioxidants in plants during abiotic stress. Crit Rev Biotechnol 30(3):161–175

Ahmad P, Jaleel CA, Sharma S (2010b) Antioxidative defence system, lipid peroxidation, proline metabolizing enzymes and Biochemical activity in two genotypes of *Morus alba* L. subjected to NaCl stress. Russ J Plant Physiol 57(4):509–517

Ahmad P, Umar S, Sharma S (2010c) Mechanism of free radical scavenging and role of phytohormones during abiotic stress in plants. In: Ashraf M, Ozturk M, Ahmad MSA (eds) Plant adaptation and phytoremediation. Springer, Dordrecht/Heidelberg/London/New York, pp 99–108

Ahmad P, Nabi G, Jeleel CA, Umar S (2011) Free radical production, oxidative damage and antioxidant defense mechanisms in plants under abiotic stress. In: Ahmad P, Umar S (eds) Oxidative stress: role of antioxidants in Plants. Studium Press, New Delhi, pp 19–53

Ahmad P, Bhardwaj R, Tuteja N (2012a) Plant signaling under abiotic stress environment. In: Ahmad P, Prasad MNV (eds) Environmental adaptations and stress tolerance of plants in the era of climate change. Springer, Science+Business Media, New York, pp 297–323

Ahmad P, Hakeem KR, Kumar A, Ashraf M, Akram NA (2012b) Salt-induced changes in photosynthetic activity and oxidative defense system of three cultivars of mustard (*Brassica juncea* L.). Afr J Biotechnol 11(11):2694–2703

Ali AA, Alqurainy F (2006) Activities of antioxidants in plants under environmental stress. In: Motohashi N (ed) The Lutein-prevention and treatment for diseases. Transworld Research Network, India, pp 187–256

Amtmann A, Sanders D (1999) Mechanisms of NaCl uptake by plant cells. Adv Bot Res 29:75–112

Asada K (1994) Production and action of active oxygen species in photosynthetic tissues. In: Foyer CH, Mullineaux PM (eds) Causes of photooxidative stress and amelioration of defense systems in plants. CRC Press, Boca Raton, pp 77–104

Ashraf M (2004) Some important physiological selection criteria for salt tolerance in plants. Flora 199:361–375

Ashraf M, Harris PJC (2004) Potential biochemical indicators of salinity tolerance in plants. Plant Sci 166:3–16

Ashraf M, Akram NA, Arteca RN, Foolad MR (2010) The physiological, biochemical, and molecular roles of brassinosteriods and salicylic acid in plant processes and salt tolerance. Critical Rev Plant Sci 29:162–190

Athar HR, Khan A, Ashraf M (2008) Exogenously applied ascorbic acid alleviates salt-induced oxidative stress in wheat. Env Exp Bot 63:224–231

Azooz MM, Youssef AM, Ahmad P (2011) Evaluation of salicylic acid (SA) application on growth, osmotic solutes and antioxidant enzyme activities on broad bean seedlings grown under diluted seawater. Inter J Plant Physiol Biochem 3:253–264

Babourina O, Leonova T, Shabala S, Newman I (2000) Effect of sudden salt stress on ion fluxes in intact wheat suspension cells. Ann Bot 85:759–767

Blumwald E, Aharon GS, Apse MP (2000) Sodium transport in plant cells. Biochim Biophys Acta 1465:140–151

Bressan RA, Hasegawa PM, Pardo JM (1998) Plants use calcium to resolve salt stress. Trends Plant Sci 3:411–412

Cuin TA, Bose J, Stefano G, Jha D, Tester M, Mancuso S, Shabala S (2011) Assessing the role of root plasma membrane and tonoplast Na^+/H^+ exchangers in salinity tolerance in wheat: in planta quantification methods. Plant Cell Environ 34:947–961

Djanaguiraman M, Ramadass R, Durga Devi D (2003) Effect of salt stress on germination and seedling growth in rice genotypes. Madras Agric J 90:50–53

Djanaguiraman M, Annie Sheeba J, Durga Devi D, Bangarusamy U (2006) Rice can acclimate to lethal level of salinity by pretreatment with sub-lethal level of salinity through osmotic adjustment. Plant Soil 284:363–373

Eyidogan F, Oz MT (2005) Effect of salinity on antioxidant responses of chickpea seedlings. Acta Physiol Plant 29:485–493

Fadzilla NM, Finch RP, Burdon RH (1997) Salinity, oxidative stress and antioxidant responses in shoot cultures of rice. J Exp Bot 48:325–331

Flors C, Fryer MJ, Waring J, Reeder B, Bechtold U, Mullineaux PM, Nonell S, Wilson MT, Baker NR (2006) Imaging the production of singlet oxygen *in vivo* using a new fluorescent sensor, Singlet Oxygen Sensor Green. J Exp Bot 57:1725–1734

Flowers TJ, Hajibagheri MA, Clipson NJW (1986) Halophytes. The Quart Rev Biol 61:313–337

Foyer HC, Noctor G (2000) Oxygen processing in photosynthesis: regulation and signaling. New Phytol 146:359–388

Foyer CH, Lopez-Delgardo H, Dat JF, Scott IM (1997) Hydrogen peroxide and glutathione associated mechanisms of acclimatory stress tolerance and signaling. Physiol Plant 100:241–254

Gaxiola RA, Rao R, Sherman A, Grisa P, Alper SL, Fink GR (1999) The *Arabidopsis thaliana* proton transporters, *AtNhx1* and *Avp1*, can function in cation detoxification in yeast. Proc Nat Acad Sci USA 96:1480–1485

Gaxiola RA, Li J, Undurraga S, Dang LM, Allen GJ, Alper SL, Fink GR (2001) Drought and salt tolerant plants result from overexpression of the AVP1 HR-pump. Proc Nat Acad Sci USA 98:11444–11449

Giannopolitis CN, Ries SK (1977) Superoxide dismutase I. Occurrence in higher plants. Plant Physiol 59:309–314

Hajiboland R (2012) Effect of micronutrient deficiencies on plants stress responses. In: Ahmad P, Prasad MNV (eds) Abiotic stress responses in plants: metabolism, productivity and sustainability. Springer, Science + Business Media, New York, pp 283–329

Halliwell B (2006) Reactive species and antioxidants. Redox biology is a fundamental theme of aerobic life. Plant Physiol 141:312–322

Hasegawa PM, Bressan RA, Zhu JK, Bohnert HJ (2000) Plant cellular and molecular responses to high salinity. Ann Rev Plant Physiol Plant Mol Biol 51:463–499

Hatz S, Lambert JDC, Ogilby PR (2007) Measuring the lifetime of singlet oxygen in a single cell: addressing the issue of cell viability. Photochem Photobiol Sci 6:1106–1116

Hedrich R (1994) Voltage-dependent chloride channels in plant cells: identification, characterization, and regulation of guard cell anion channel. Curr Topics Membr 42:1–33

Hernandez M, Fernandez-Garcia N, Diaz-Vivancos P, Olmos E (2010) A different role for hydrogen peroxide and the antioxidative system under short and long salt stress in *Brassica oleracea* roots. J Exp Bot 61:521–535

Hu LX, Li HY, Pang HC, Fu JM (2012) Responses of antioxidant gene, protein and enzymes to salinity stress in two genotypes of perennial ryegrass (*Lolium perenne*) differing in salt tolerance. J Plant Physiol 169:146–156

Katare DP, Nabi G, Azooz MM, Aeri V, Ahmad P (2012) Biochemical modifications and enhancement of psoralen content in salt-stressed seedlings of *Psoralea corylifolia* Linn. J Funct Environ Bot 2(1):x–xx

Kawasaki S, Borchert C, Deyholos M, Wang H, Brazille S, Kawai K, Galbraith D, Bohnert HJ (2001) Gene expression profiles during the initial phase of salt stress in rice. Plant Cell 13:889–905

Krieger-Liszkay A, Fufezan C, Trebst A (2008) Singlet oxygen production in photosystem II and related protection mechanism. Photosyn Res 98:551–564

Kukreja S, Nandval AS, Kumar N, Sharma SK, Sharma SK, Unvi V, Sharma PK (2005) Plant water status, H_2O_2 scavenging enzymes, ethylene evolution and membrane integrity of *Cicer arietinum* roots as affected by salinity. Biol Plant 49:305–308

Laurie S, Feeney KA, Maathuis FJM, Heard PJ, Brown SJ, Leigh RA (2002) A role for HKT1 in sodium uptake by wheat roots. Plant J 32:139–149

Lee DH, Kim YS, Lee CB (2001) The inductive responses of the antioxidant enzymes by salt stress in the rice (*Oryza sativa* L.). J Plant Physiol 158:737–748

Lee S, Klim SG, Park CM (2010) Salicylic acid promotes seed germination under high salinity by modulating antioxidant activity in Arabidopsis. New Phytol 188:626–637

Li MR, Lin XJ, Li HQ, Pan XP, Wu GJ (2011) Overexpression of AtNHX5 improved tolerance to both salt and water stress in rice (*Oryza sativa* L.). Plant Cell Tissue Culture 107:283–293

Lopez F, Vansuyt G, Casse-Delbart F, Fourcroy P (1996) Ascorbate peroxidase activity, not the mRNA level, is enhanced in salt-stressed *Raphanus sativus* plants. Physiol Plant 97:13–20

Maas EV, Grattan SR (1999) Crop yields as affected by salinity. In: Skaggs RW, Van Schilfgaarde J (eds.) Agricultural drainage. Agron. Monograph 38. ASA, CSSA, SSA, Madison, pp. 55–108

Maas EV, Grieve CM (1990) Spike and leaf development in salt-stressed wheat. Crop Sci 30:1309–1313

Maas EV, Poss JA (1989a) Salt sensitivity of wheat at different growth stages. Irrig Sci 10:29–40

Maas EV, Poss JA (1989b) Salt sensitivity of cowpea at various growth stages. Irrig Sci 10: 313–320

Maas EV, Poss JA, Hoffman GJ (1986) Salinity sensitivity of sorghum at three growth stages. Irrig Sci 7:1–11

Maathuis FJM, Verlin D, Smith FA, Sanders D, Ferneandez JA, Walker NA (1996) The physiological relevance of Na^+ coupled K^+ transport. Plant Physiol 112:1609–1616

Martinez-Beltran J, Manzur CL (2005) Overview of salinity problems in the world and FAO strategies to address the problem. In: Proceedings of the International Salinity Forum, Riverside, April 2005, pp. 311–313

Maser P, Eckelman B, Vaidyanathan R, Horie T, Fairbairn DJ, Kubo M, Yamagami M, Yamaguchi K, Nishimura M, Uozumi N, Robertson W, Sussman MR, Schroeder JI (2002) Altered shoot/root NaCl distribution and bifurcating salt sensitivity in Arabidopsis by genetic disruption of the NaCl transporter AtHKT1. FEBS Lett 531:157–161

Mhadhbi H, Fotopoulos V, Mylona PV, Jebara M, Aouani ME, Polidoros AN (2011) Antioxidant gene-enzyme responses in *Medicago truncatula* genotypes with different degree of sensitivity to salinity. Physiol Plant 141:201–214

Minkov IN, Jahoubjan GT, Denov ID, Toneva AT (1999) Photooxidative stress in higher plants. In: Pessarakli M, Dekker M (ed.) Plant and crop stress. (2nd ed.). New York, pp. 499–525

Misic D, Siler B, Zivkovic JN, Simonovic A, Maksimovic V, Budimir S, Janosevic D, Durickovic M, Nicolic M (2012) Contribution of inorganic cations and organic compounds to osmotic adjustment in root cultures of two *Centaurium* species differing in tolerance to salt stress. Plant Cell Tissue Organ Culture 108:389–400

Mittler R (2002) Oxidative stress, antioxidants and stress tolerance. Trends Plant Sci 7:405–410

Munns R, Tester M (2008) Mechanisms of salinity tolerance. Ann Rev Plant Biol 59:651–681

Munns R, James RA, Lauchli A (2006) Approaches to increasing the salt tolerance of wheat and other cereals. J Exp Bot 57:1025–1043

Nass R, Cunningham KW, Rao R (1997) Intracellular sequestration of sodium by a novel Na^+/H^+ exchanger in yeast is enhanced by mutations in the plasma membrane H^+-ATPase. J Biol Chem 272:26145–26152

Nazar R, Masood A, Syeed S, Khan NA (2011) Understanding the significance of sulfur in improving salinity tolerance in plants. Environ Exp Bot 70:80–87

Nemati I, Moradi F, Gholizadeh S, Esmaeili MA, Bihamta MR (2011) The effect of salinity stress on ions and soluble sugars distribution in leaves, leaf sheaths and roots of rice (*Oryza sativa* L.) seedlings. Plant Soil Environ 57:26–33

Nieves-Cordones M, Alemán F, Fon M, Martínez V, Rubio F (2012) K^+ nutrition, uptake, and its role in environmental stress in plants. In: Ahmad P, Prasad MNV (eds) Environmental adaptations and stress tolerance of plants in the era of climate change. Springer, Science + Business Media, New York, pp 85–112

Niu X, Bressan RA, Hasegawa PM, Pardo JM (1995) Ion homeostasis in NaCl stress environments. Plant Physiol 109:735–742

Noctor G, Foyer CH (1998) Ascorbate and glutathione: keeping active oxygen under control. Ann Rev Plant Physiol Plant Mol Biol 49:249–279

Nounjan N, Nghia PT, Theerakulpisut P (2012) Exogenous proline and trehalose promote recovery of rice seedling from salt-stress and differentially modulate antioxidants enzymes and expression of related genes. J Plant Physiol 169:596–604

Orsini F, Accorsi M, Gianquinto G, Dinelli G, Antognoni F, Carrosco KBR, Martinez EA, Alnayef M, Marotti I, Bosi S, Biondi S (2011) Beyond the ionic and osmotic response to salinity in *Chenopodium quinoa*: functional elements of successful halophytism. Func Plant Biol 38:818–831

Pardo JM, Reddy MP, Yang S, Maggio A, Huh GH, Matsumoto T, Coca MA, Paino D, Urazo M, Koiwa H, Yun DJ, Watad AA, Bressan RA, Hasegawa PM (1998) Stress signaling through Ca^{2+}/calmodulin-dependent protein phosphatase calcineurin mediates salt adaptation in plants. Proc Nat Acad Sci USA 95:9681–9686

Parida AK, Das AB, Das P (2002) NaCl stress causes changes in photosynthetic pigments, proteins and other metabolic components in the leaves of a true mangrove, *Bruguiera parviflora*, in hydroponic cultures. J Plant Biol 45:28–36

Paridia AK, Das AB (2005) Salt tolerance and salinity effects on plants: a review. Ecotoxicol Environ Safety 60:324–349

Petridis A, Therios I, Samouris G, Tananaki C (2012) Salinity-induced changes in phenolic compounds in leaves and roots of four olive cultivars (*Olea europaea* L.) and their relationship to antioxidant activity. Environ Exp Bot 76:37–43

Pilot G, Gaymard F, Mouline K, Cherel I, Sentenac H (2003) Regulated expression of Arabidopsis shaker KR channel genes involved in K^+ uptake and distribution in the plant. Plant Mol Biol 51:773–787

Pitman MG, Lauchli A (2002) Global impact of salinity and agricultural ecosystems. In: Lauchli A, Luttge U (eds) Salinity: Environment – Plants – Molecules. Kluwer Academic, Dordrecht, pp 3–20

Plett DC, Moller IS (2010) Na plus transport in glycophytic plants: what we know and would like to know. Plant Cell Environ 33:612–626

Polidoros NA, Scandalios JG (1999) Role of hydrogen peroxide and different classes of antioxidants in the regulation of catalase and glutathione S-transferase gene expression in maize (*Zea mays* L.). Physiol Plant 106:112–120

Pons R, Cornejo MJ, Sanz A (2011) Differential salinity-induced variations in the activity of H^+ pumps and Na^+/H^+ antiporters that are involved in cytoplasm ion homeostasis as a function of genotype and tolerance level in rice cell lines. Plant Physiol Biochem 49:1399–1409

Qiu Q, Schumaker KS, Zhu JK (2002) Regulation of vacuolar membrane Na^+/H^+ exchange activity in *Arabidopsis thaliana* by the SOS pathway. Abstract 200, Plant Biology 2002, 3–7 August 2002, Denver

Quintero FJ, Ohta M, Shi H, Zhu JK, Pardo JM (2002) Reconstitution in yeast of the Arabidopsis SOS signaling pathway for NaCl homeostasis. Proc Nat Acad Sci USA 99:9061–9066

Roxa VP, Smith RK, AllenER Allen RD (1997) Overexpression of glutathione S-transferase/glutathione peroxidase enhances the growth of transgenic seedling during stress. Nat Biotechnol 15:988–991

Ruiz JM, Blumwald E (2002) Salinity-induced glutathione synthesis in *Brassica napus*. Planta 214:965–969

Sanders D (2000) Plant biology: the salty tale of *Arabidopsis*. Curr Biol 10:486–488

Sarwat M, Ahmad P, Nabi G, Hu X (2012) Ca^{2+} signals: the versatile decoders of environmental cues. Crit Rev Biotechnol doi: 10.3109/07388551.2012.672398.

Sekmen AH, Turkan I, Tanyolac ZO, Ozfidan C, Dinc A (2012) Different antioxidant defense responses to salt stress during germination and vegetative stages of endemic halophyte *Glypsophila oblanceolata* Bark. Environ Exp Bot 77:63–76

Shabala S (2000) Ionic and osmotic components of salt stress specifically modulate net ion fluxes from bean leaf mesophyll. Plant Cell Environ 23:825–837

Shabala S, Cuin TA, Pang JY, Percey W, Chen ZH, Conn S, Eing C, Wegner LH (2010) Xylem ionic relations and salinity tolerance in barley. Plant J 61:829–853

Shalata A, Neumann PM (2001) Exogenous ascorbic acid (vitamin C) increases resistance to salt stress and reduces lipid peroxidation. J Exp Bot 52:2207–2211

Smirnoff N (2002) Antioxidants and reactive oxygen species in plants. Wiley-Blackwell, New Jersey

Sun J, Chen S, Dai S, Wang R, Li N, Shen X (2009) NaCl induced alternations of cellular and tissue ion fluxes in roots of salt-resistant and salt-sensitive *Popular* species. Plant Physiol 149:1141–1153

Takemura T, Hanagata N, Sugihara K, Baba S, Karube I, Dubinsky Z (2000) Physiological and biochemical responses to salt stress in the mangrove, *Bruguiera gymnorrhiza*. Aqua Bot 68:15–28

Tanaka Y, Hibino T, Hayashi Y, Tanaka A, Kishitani S, Takabe T (1999) Salt tolerance of transgenic rice overexpressing yeast mitochondrial Mn-SOD in chloroplasts. Plant Sci 148:131–138

Tausz M, Ircelj H, Grill D (2004) The glutathione system as a stress marker in plant ecophysiology: is a stress-response concept valid? J Exp Bot 55:1955–1962

Teakle NL, Tyerman SD (2010) Mechanism of Cl^- transport contributing to salt tolerance. Plant Cell Environ 33:566–589

Tester M, Davenport R (2003) NaCl tolerance and NaCl transport in higher plants. Ann Bot 91:503–527

Uozumi N, Kim EJ, Rubio F, Yamaguchi T, Muto S, Tsuboi A, Bakker EP, Nakamura T, Schroeder JI (2000) The *Arabidopsis* HKT1 gene homolog mediates inward NaCl currents in *Xenopus laevis* oocytes and NaCl uptake in *Saccharomyces cerevisiae*. Plant Physiol 122:1249–1259

Vranova E, Inze D, Van Brensegem F (2002) Signal transduction during oxidative stress. J Exp Bot 53:1227–1236

Wang S, Assmann SM, Fedoroff NV (2008) Characterization of the *Arabidopsis* heterotrimeric G protein. J Biol Chem 283:13913–13922

Wu GQ, Wang SM (2012) Calcium regulates K^+/Na^+ homeostasis in rice (*Oryza sativa* L.) under saline conditions. Plant Soil Environ 58:121–127

Yeo AR, Flowers SA, Rao G, Welfare K, Senanayake N, Flowers TJ (1999) Silicon reduces sodium uptake in rice (*Oryza sativa* L.) in saline conditions and this is accounted for by a reduction in the transpirational bypass flow. Plant Cell Environ 22:559–565

Yokoi S, Quintero FJ, Cubero B, Ruiz MT, Bressan RA, Hasegawa PM, Pardo JM (2002) Differential expression and function of *Arabidopsis thaliana NHX* Na^+/H^+ antiporters in the salt stress response. Plant J 30:529–539

Yousuf PY, Hakeem KR, Chandna R, Ahmad P (2012) Role of glutathione reductase in plant abiotic stress. In: Ahmad P, Prasad MNV (eds) Abiotic stress responses in plants: metabolism, productivity and sustainability. Springer, Science + Business Media, New York, pp 149–158

Yu Q, Rengel Z (1999a) Drought and salinity differentially influence activities of superoxide dismutases in narrow-leafed lupins. Plant Sci 142:1–11

Yu Q, Rengel Z (1999b) Micronutrient deficiency influences plant growth and activities of superoxide dismutases in narrow-leafed lupins. Ann Bot 83:175–182

Zhang JX, Kirkham MB (1996) Lipid peroxidation in sorghum and sunflower seedlings as affected by ascorbic acid, benzoic acid and propyl gallate. J Plant Physiol 149:489–493

Zhang HX, Hodson JN, Williams JP, Blumwald E (2001) Engineering salt-tolerant *Brassica* plants: characterization of yield and seed oil quality in transgenic plants with increased vacuolar sodium accumulation. Proc Nat Aca Sci USA 98:12832–12836

Zhang KW, Guo N, Lian LJ, Wang J, Lv S, Zhang JR (2011) Improved salt tolerance and seed cotton yield in cotton (*Gossypium hirsutum* L.) by transformation with betA gene for glycinebetaine synthesis. Euphytica 181:1–16

Zhifang G, Loescher WH (2003) Expression of a celery mannose 6-phosphate reductase in *Arabidopsis thaliana* enhances salt tolerance and induces biosynthesis of both mannitol and glucosyl-mannitol dimmer. Plant Cell Environ 26:275–283

Zhu JK (2001) Plant salt tolerance. Trends Plant Sci 6:66–71

Zhu JK (2003) Regulation of ion homeostasis under salt stress. Curr Opinion Plant Biol 6:441–445

Zhu JK, Liu J, Xiong L (1998) Genetic analysis of salt tolerance in *Arabidopsis thaliana:* evidence of a critical role for potassium nutrition. Plant Cell 10:1181–1192

Chapter 4
Halotolerance in Lichens: Symbiotic Coalition Against Salt Stress

David Delmail, Martin Grube, Delphine Parrot, Jeanne Cook-Moreau, Joël Boustie, Pascal Labrousse, and Sophie Tomasi

4.1 Introduction

Lichens cover about 8% of land surface, they occur in almost all terrestrial habitats of the world, including the most inclement environments such as continental areas of Antarctica to sand deserts (Green et al. 2012; Maphangwa et al. 2012). Among these extreme habitats, coasts and rocky sea shores are colonized by adapted halotolerant species (Ahmadjian 1995; Gilbert 2000; Grube 2010; Grube and Blaha 2005). Certain lineages of lichens are abundant in the littoral to supralittoral belts, where they are exposed to both salt and dry stress and towards the poles, to cold stress as well. On rocky seashores, they form an ecological zonation with each level subjected to different exposures of salt and immersion (Sibaja-Cordero and Troncoso 2011). Maritime influences are not limited to seashores, but may also extend many kilometers inland (Nash III and Lange 1988). For example, the prehistoric monument of Stonehenge in England is colonized by maritime specimens (Gilbert 2000).

In intertidal environments, salinity is the main factor influencing lichen zonation, but there are still few available data concerning the different levels of halotolerance. These lichens tolerate fluctuating conditions of humidity and salt that are unsuitable

D. Delmail (✉) • D. Parrot • J. Boustie • S. Tomasi
Laboratory of Pharmacognosy and Mycology – UMR CNRS 6226 ISCR PNSCM,
University of Rennes 1 (European University of Brittany), 35043 Rennes, France
e-mail: david.delmail@wanadoo.fr

M. Grube
Institute of Plant Sciences, Karl-Franzens-Universität Graz, 8010 Graz, Austria

J. Cook-Moreau
Department of Biochemistry and Molecular Genetics,
University of Limoges (School of Medicine), 87025 Limoges, France

P. Labrousse
Laboratory of Botany and Cryptogamy – GRESE EA 4330,
University of Limoges (Faculty of Pharmacy), 87025 Limoges, France

P. Ahmad et al. (eds.), *Ecophysiology and Responses of Plants under Salt Stress*,
DOI 10.1007/978-1-4614-4747-4_4,

for most other organisms, and adapt at various levels of organization. In this chapter, ecology and mechanisms of adaptation to complex stressful environmental conditions at cellular and biochemical levels are described in relation to symbiont diversity as well as implications of phenotypic characteristics and genetics.

4.2 Ecology, Diversity and Morpho-anatomical Adaptations

As salt-rich habitats of lichens are mostly represented by marine and maritime environments, our focus will be on lichens encountered in coastal areas, although some lichens also can occur inland (e.g. arid steppes) with elevated salt conditions. The upper zones along coasts are influenced by prolonged sun exposure, wind-dependent fluctuations in salinity and temperature, whereas lower zones are subjected to almost constant submersion and wave action. Organisms and particularly lichens have developed diverse means of adapting to survive such environmental extremes.

According to Fletcher (1973a, b) sublittoral, littoral, supralittoral, and terrestrial zones can be differentiated. In these zones, most of lichens are weathered on rocks and distributed according to three levels of immersion: (i) permanently immersed in sea water, corresponding to marine lichens, (ii) partially immersed and (iii) above the upper water line but exposed to saline particles suspended in the atmosphere, corresponding to maritime lichens.

The zone below the littoral belt hosts about a dozen of described lichen species but this number is regarded as low in comparison to the number of marine fungi (Kohlmeyer and Kohlmeyer 1979), some of which might be involved in other associations with photoautotrophs. The crustose lichen genus *Verrucaria* is particularly well represented along the littoral belt. Only one submersed lichen, *Halographis runica* has been described recently (Kohlmeyer and Volkmann-Kohlmeyer 1988).

The littoral zone is easily recognized by the black belt on rocks corresponding to lichens with dark pigments (melanins) visible in sea-tide environments. *Hydropunctaria maura* is the most ubiquitous species forming distinct dark band. Additional pyrenocarpous lichens dominate this area: the hardly distinguishable *Collemopsidium halodytes* and *Lichina pygmea* form dark spots on rocks (and on barnacles and limpets for *C. halodytes*) appearing during low-tide periods (Fig. 4.1). This intertidal zone is inhabited by a limited number of lichen species (16 species in Great Britain according to Gilbert (2000)). Although not really proven to be in strict correlation with salinity, the genus *Lichina* and some members of *Pyrenocollema* and *Verrucaria,* are usually found in marine habitats (Hawksworth 2000).

Unusual forms are also found on macroalgae and are described as mycophycobiosis or parasitism. These cases include a symbiosis with brown algae: *Mycophyscias ascophylli* on *Ascophyllum nodosum* or *Pelvetia caniculata*, *Collemopsidium pelvetiae* on the latter, and *Wahlenbergiella tavaresiae* recently described on *Petroderma maculiforme* (Gueidan et al. 2011; Moe 1997; Sanders et al. 2004).

In the supralittoral zone, most of species are saxicolous lichens which form three to four colour belts (dark, orange, grey-green and white). Exposure to waves and saline foam decreases from the mesic-supralittoral zone to the aerohaline zone. To some

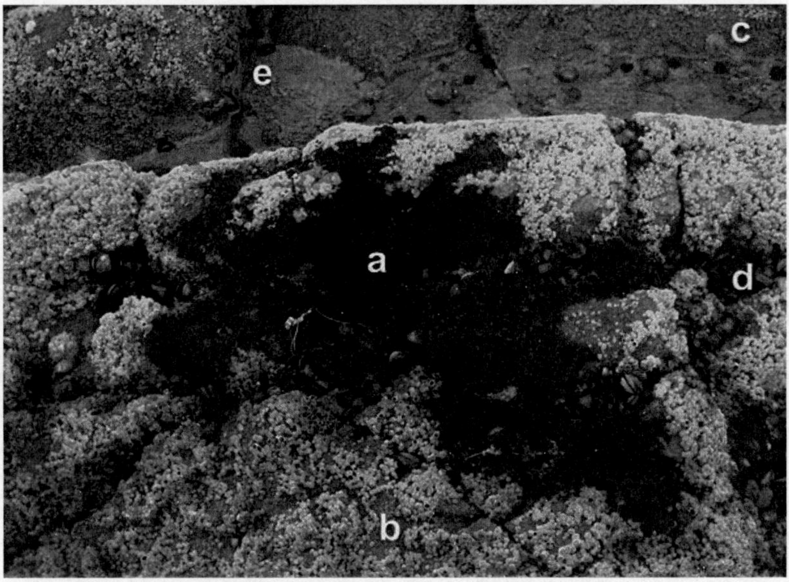

Fig. 4.1 A patch of brown halophilic *Lichina pygmaea* (*a*) among barnacles (*b*), limpets (*c*), mussels (*d*) and sea anemones (*e*) (Plouzané, France). Total immersion of this patch during ~12 h per day at ~34 g.l^{-1} NaCl

extent, the intertidal zone can overlap the mesic supralittoral level. There, a second fructicose brown-dark Lichinomycetes, *Lichina confinis* can be encountered in association with various orange *Caloplaca* species (e.g. *Caloplaca thallincola*). Such placoid lichens form the characteristic "orange belt" due to anthraquinones as typical secondary metabolites (Arup 1995). In the upper part of this zone, the foliose and ubiquitous *Xanthoria* lichens may develop (e.g. *X. aureola* along coasts of Brittany).

In the xeric supralittoral zone (or terrestrial "white" zone), a variety of crustose (e.g. *Buellia, Lecanora, Ochrolechia*), foliose (e.g. *Parmelia, Xanthoria*) or fructicose (e.g. *Roccella, Ramalina*) lichens are present.

Most of the 700 species recognized in tidal zones can be found in this upper part (aerohalin zone) but the optimal habitats could be under salt-free conditions for some of them (e.g. *Cladonia* species). A number of these species are halophilic or halotolerant but not strictly dependent on salt levels as most of them are also found at a distance from the coast.

In the aerohaline zone, fructicose lichen species are restricted from coastal to inland sites influenced by maritime conditions of fog and high humidity. The Californian species *Ramalina menziesii*, with thalli resembling hair-nets, maintains low levels of photosynthesis for only 2–3 h per day. Their morphological adaptation includes surface increase by reticulate structures that enhance water holding capacity of the thallus when tresses are ventilated off the ocean (Rundel 1974). This adaptation is apparently unrelated to salinity but considerable variations in salt sensitivity exist among other maritime and marine species. Nash III and Lange (1988) reported that species growing near the ocean, such as *Caloplaca coralloides*,

Dendrographa minor and *Niebla cephalota* were tolerant to sea salt while the species restricted to inland sites were very sensitive. Fletcher (1976) reported that marine and maritime lichens are relatively tolerant to long-term salinity fluctuations. Nevertheless, no intraspecific variation could be observed for the effect of NaCl on growth inhibition of maritime and non-maritime *Ramalina* species (Takahagi et al. 2002). The ability to develop salt tolerance seems to fluctuate according to the species. No consistent trends were thus noted such as morpho-anatomic adaptations of haloterant lichen species. As for algae, morphologically analogous lichen species have similar nutrient-uptake characteristics, independently of their taxonomic classification. Species distribution on the shore depends on their ability to grow according to salt concentrations, light exposure and nitrogen load. As reported above, a distribution of the thallus shapes and organization is observed. While lichens with anatomically poorly stratified thallus such as the black *Verrucaria* appear in the littoral fringe, followed by stratified crustose lichens with apothecia in the lower supralittoral zone (e.g. *Caloplaca*). The stratified foliose lichens appear in the middle and upper supralittoral zones (e.g. *Xanthoria*). Moreover, the presence in the lower supralittoral zone of microfructicose lichens (such as *Lichina* or *Spilonema* species) attached loosely to rocks could be due to their inability to physically withstand the pounding of waves on upper zones (Brodo and Sloan 2004). Constant anatomical differences are observed among closely related crustose and foliose Lecanorales species from coastal (supralittoral) and inland zones, and suggest a possible role of hyphal walls for osmotic buffering (Grube and Blaha 2005). However, the morphological differences in species of the littoral zones (i.e. high content of melanised pigments strongly conglutinated and densely arranged hyphae) cannot be strictly related to the salt-rich habitat. Crustose lichens are known to grow more slowly than foliose lichens, possibly explains their ability to survive in the presence of various stress factors (polyextremotolerance).

4.3 Salt Stress and Halotolerance

High salt concentrations in the environment lead to a decrease in hydric potential affecting water availability in poikilohydric organisms such as lichens (Hasegawa et al. 2000). Indeed, poikilohydry means (i) that lichens rely directly on the environment for water and (ii) that their hydric potential tends to equilibrium with the ecosystem water status. When sufficiently hydrated, lichens are active but when their water content falls, they become inactive and/or dormant.

In addition to the hyperosmotic shock due to high salinity and the generated subsequent oxidative stress (Bohnert et al. 1995), deleterious consequences of high NaCl concentration also include ion toxicity and nutrient imbalance (Serrano et al. 1999; Hasegawa et al. 2000; Rodríguez-Navarro 2000). In consequence, lichens have developed various biochemical and physiological mechanisms to respond and adapt to these stresses and thus acquire tolerance. Like in plants, adaptation to stress may be mediated by both pre-existing and induced defenses (Jakab et al. 2005).

4.3.1 *Desiccation*

One of the main consequences of salt stress is the loss of intracellular water. Desiccation tolerance could be summarized as the ability to equilibrate the internal water potential with that of the environment, and then resume normal function when rehydrated. Indeed, below a water-content threshold of 10% there is not enough water to form a monolayer around macromolecules thus stopping enzyme reactions and metabolism (Billi and Potts 2002; Green et al. 2011). Moreover, water loss is a dynamic phenomenon and seasonal acclimation can occur in lichens (Lange and Green 2005). Therefore, physiological processes involved in desiccation tolerance are directly linked to the autoecological conditions and recent history of the considered organisms. The association of water-loss-sensitive lichens with autoecological changes influence desiccation suggests that there are some forms of metabolic cost to maintain tolerance (Green et al. 2011).

4.3.1.1 General Physiological Effects

In salt-resistant species, cells present several acclimation mechanisms that allow their survival. They have the capability to mediate reversibility of disturbances and to maintain the integrity of all cell organelles and components. To prevent desiccation and protect all cell elements, lichens react with antioxidant systems. During the steps of water loss and recovery, reactive oxygen species (ROS) may accumulate in lichen tissues, especially when received-light intensity is excessive as photosynthesis is disrupted and declines with loss of water content (Green et al. 2011). ROS are involved in many damages and disturbances as, for example, they could affect enzyme structure and induce membrane-lipid peroxidation (Kranner et al. 2005).

Moreover, lichens produce and accumulate many metabolites also known as compatible solutes which do not disrupt the normal metabolic pathways. Frequently observed metabolites with an osmolyte function are sugars (e.g. fructose, sucrose), sugar alcohols and complex sugars (e.g. fructans). In addition, charged metabolites such as glycine-betaine, amino acids (e.g. proline (Sekmen et al. 2012)) and ectoin, are also produced. The accumulation of these osmolytes facilitates the osmotic adjustment in cells. Water moves from high water potential to low water potential and accumulation of these osmoprotectants reduces the water potential preventing intracellular water loss (Mahajan and Tuteja 2005).

4.3.1.2 Impact on Photosynthesis

To live well, you must live hidden (*bene vixit, bene qui latuit*). This famous dictum from the French philosopher René Descartes well defines the necessity of protecting photobionts in halotolerant lichens to ensure efficient photosynthesis. As intertidal lichens live in sunny environments without any shade, they have developed a photoprotection system including the synthesis of photoprotective pigments

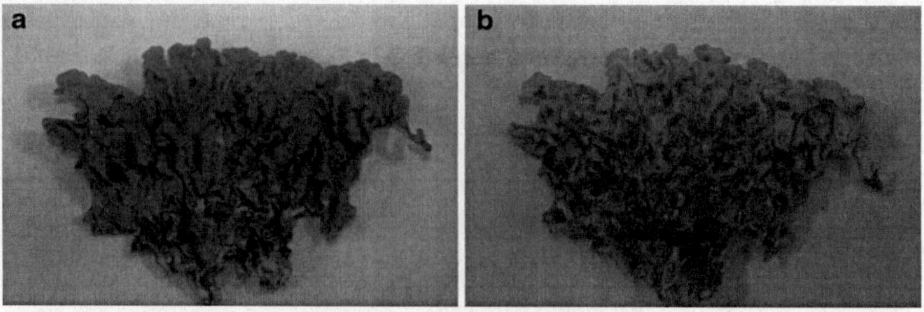

Fig. 4.2 *Upper side* (**a**) and *lower side* (**b**) of halotolerant *Xanthoria aureola*. (**a**) parietin derivatives confer dark pigmentation to the cortex; (**b**) despite rock fragments, scarce pigmentation confers a lighter shade. Scale bar: 1 cm

(e.g. atranorin, calycin, 1'-chloropannarin, epiphorellic acid (I and II), pannarin, parietin, rhizocarpic acid, usnic acid) (Fernández et al. 1998; Hidalgo et al. 2002; Øvstedal et al. 2009; Raikou et al. 2011; Rancan et al. 2002). For instance, the coastal *Xanthoria aureola* like most species from the Teloschistales order, has a yellow (or orange) pigmentation due to the presence of parietin derivatives (anthraquinones) in its cortex. The intensity of pigmentation is variable and linked to the environment. In shady conditions, the thallus is drab (grey and green) without any hint of yellow (Smith et al. 2009). Parietin synthesis is correlated with light intensity as this compound is approximately five times more concentrated in specimens exposed to sea cliffs than in those living in underbrush or growing on trees (Gauslaa and Ustvedt 2003). Parietin efficiently absorbs solar radiation both in the 400–500 nm photosynthetically-active-radiation (PAR) band and in the ultraviolet (UV)-B range (Solhaug and Gauslaa 1996). These pigments are usually located as tiny extracellular crystals in the top layer of the upper cortex next to microalgal and cyanobacterial photobionts. Depending on their absorption spectrum, these pigments protect the photosynthetic partner, and therefore its physiological processes from solar radiation (Gauslaa and Solhaug 2004; Solhaug and Gauslaa 1996). A concentration gradient can also be observed in thallus of *Xanthoria* with its upper side colored by photoprotective compounds and its lower one less- or un-colored, suggesting involvement of the mycobiont in protection of the photobiont (Fig. 4.2). These pigments protect against UV radiation especially within lichen genus or species that synthesize brown melanin pigment which has an excellent absorption (from 240 to 320 nm) (Gauslaa and Solhaug 2001) (e.g. *Lichina confinis*, *Lichina pygmaea*). However, Solhaug et al. (2003) highlight that, during desiccation, the parietin and melanin biosynthesis pathways are disrupted leading to ineffective protection of the photobiont photosystems. In the absence of these photoprotectors, free mycobionts show severe light damage which causes a permanent reduction in quantum efficiency (Fv/Fm) and in quantum yield for photosynthetic O_2 production and photosynthetic capacity (Solhaug and Gauslaa 1996). Because metabolism is lower or absent during desiccation, photooxidative damage especially due to ROS and their oxidative consequences, accumulates with length of exposure to light radiation while metabolic repair of disruptions is not possible (Gauslaa and Solhaug 1999; Gray et al. 2007). Other UV-absorbing compounds such as mycosporine-like amino acids are known to be produced by algae, cyanobacteria and marine lichens such as *Lichina* (Roullier et al. 2011) along with other amino-acid derivatives (Roullier et al. 2010) and sugar-derivative compounds.

A few studies have addressed the impact of salinity stress on lichen photosynthesis (Matos et al. 2011; Matthes-Sears et al. 1987; Nash III and Lange 1988; Smith and Gremmen 2001). Parameters other than CO_2 exchanges and chlorophyll fluorescence, will be considered as impact of salt stress on photosynthesis of halotolerant species are obvious when considering the correlation between water content, external level of salinity and photosynthesis response. When exposed during 4.5 days to 70 g L^1 NaCl (natural conditions 34 g L^{-1} NaCl), four intertidal species from France give different responses depending on their location on the seashore and therefore their halotolerance potential (Fig. 4.3). One less adapted species seems to be *Xanthoria aureola* as its respiration rate, raw photosynthesis, net photosynthesis and water content decrease by 79%, 85%, 88% and 24%, respectively. All these losses are correlated together ($0.97 > |\rho| > 0.76$) at a significant level of $p = 0.05$. This species is located higher up on the seashore and is never immersed but only subject to sparse saline ocean spray. This could explain its poor halotolerance and low capability to rapidly acclimate during salt stress. *Ramalina cuspidata* and *Ramalina siliquosa*, which often co-occur on rocks, present analogous variations in their physiological parameters, respectively: respiration rate (–78% and–70%), raw photosynthesis (–72% and –71%), net photosynthesis (–66% and –71%) and water content (–28% and –29%). However, *R. cuspidata* shows more significant correlations between parameter variations. Respiration is correlated with raw photosynthesis (–0.76) and water content (–0.76); the last two also being correlated (0.95 and 0.87 in *R. siliquosa*). Located closer to the shore line than *X. aureola*, some *Ramalina* patches may be immersed for a few minutes when the tide comes in, or when sprayed with saline droplets due to wind or waves. Despite drastic effects, the impact of NaCl on physiological parameters reveals better salt-stress acclimation in *Ramalina* rather than *X. aureola*, especially as net photosynthesis is less affected. As seen further, photosynthesis remains one of the most important processes in lichens as it provides the carbon backbone necessary for osmoprotective synthesis and the pool of reduced nicotinamide adenine dinucleotide phosphate (NADPH) involved in antioxidant-system functioning. *Lichina pygmaea* has a negative correlation (–0.71) between reduced rates of respiration (–42%) and raw photosynthesis (–18%). In contrast, net photosynthesis and water content increase by 14% and 5%, respectively, and are positively correlated (0.84). These results are consistent with the extreme adaptation capabilities of *L. pygmaea* to salt stress as this species is located near exclusive marine animals on the foreshore (Fig. 4.1) and remains totally immersed during several hours per day. The positive correlation between net photosynthesis and water content seems to be consistent with the involvement of this physiological process in the regulation of water loss through the production of osmolytes. Moreover, respiration and photorespiration losses may be linked to a reduced catabolism to maintain the intracellular osmoprotector rates as high as possible.

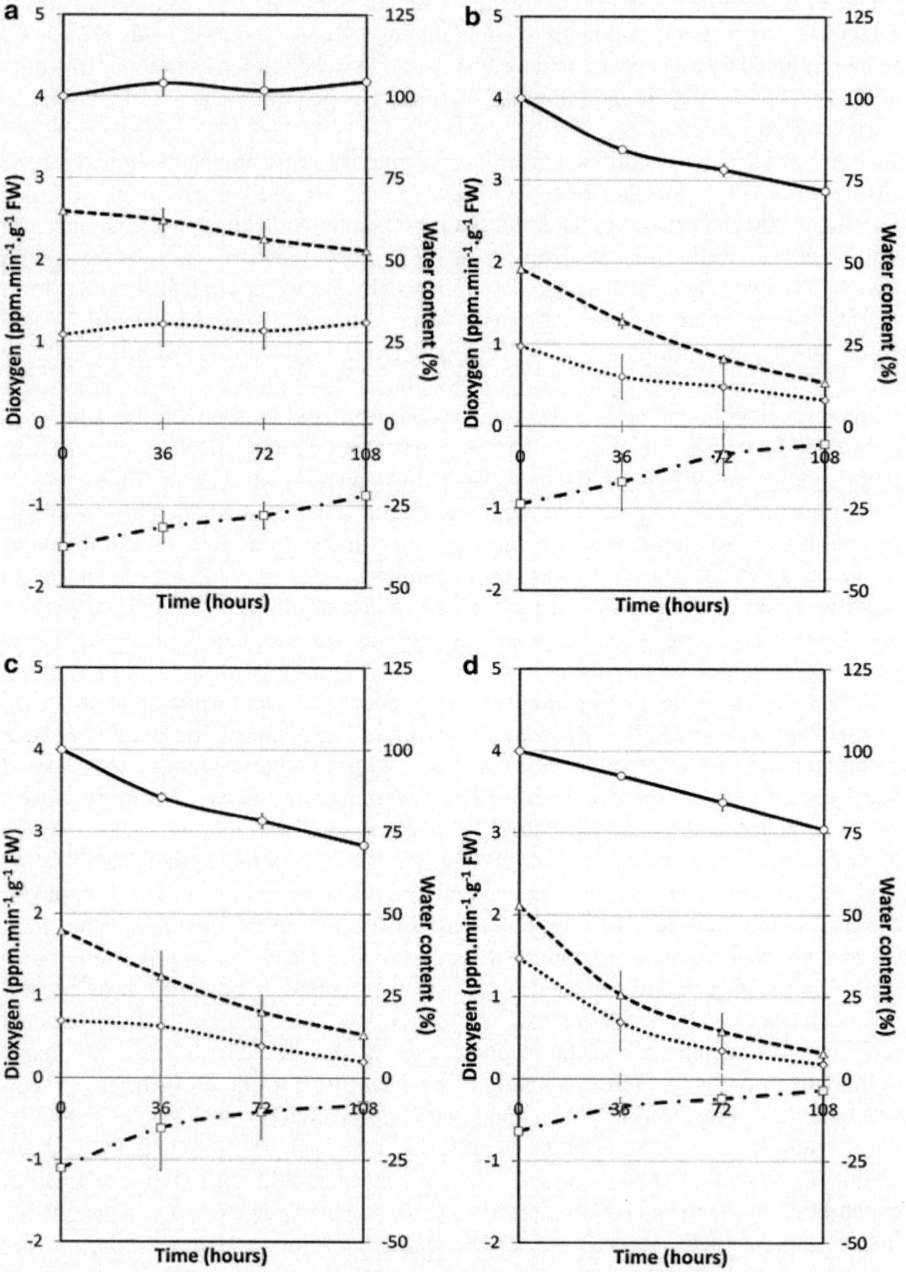

Fig. 4.3 Evolution of respiration (– • –), raw-photosynthesis (– –), net-photosynthesis (• • •) and water-content (—) rates of four halotolerant lichen species during 108 h of exposure to 70 g.l^{-1} NaCl (photoperiod: 14 h:10 h, ~500 lux; dioxygen measured with dissolved oxygen meter HI 9146, HANNA Instruments) (Delmail et al. unpublished data). (**a**) *Lichina pygmaea*; (**b**) *Ramalina cuspidata*; (**c**) *Ramalina siliquosa*; (**d**) *Xanthoria aureola*

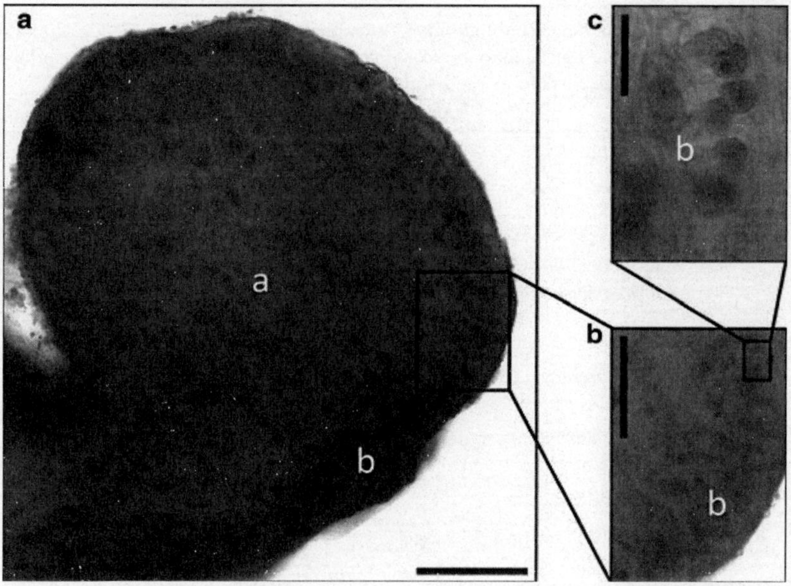

Fig. 4.4 Perithecium of *Lichina pygmaea* (longitudinal section) with fungus (*a*) (**a**) and cyanobacteria *Calothrix* (*b*) (**a, b, c**). Scale bars: (**a**), 10 μm; (**b**), 5 μm; (**c**), 1 μm

4.3.1.3 Osmotic Stress Tolerance

Osmotic Metabolites Associated with Photobiont Biodiversity

Although the common genera of lichen photobionts are cyanobacteria *Nostoc*, as well as the green algae *Trebouxia* and *Trentepohlia*, it should be noted that on the basis of Tschermak-Woess (1988) data, nearly 40 genera of photobionts may exist (e.g. *Cryptococcus*, *Calothrix* (Fig. 4.4)) (Friedl and Büdel 2008; Grube and Muggia 2010; Keddy 2007).

According to Ahmadjian (1993a) there were several misidentifications and taxonomic changes, and the number of photobionts is approximately 25. The photobiont genera commonly recognized are listed in Table 4.1. These classifications are in urgent need of revision, as it has been noticed that some genera (e.g. *Chlorella*) are polyphyletic. Recent molecular work also suggests that the species concepts within the photobiont genera are poorly defined (e.g. Muggia et al. 2010), and there are yet undescribed lineages in the *Trebouxia* genus.

As mentioned above, poikilohydric organisms do not control their water relations with the environment. During the water loss induced by salt stress, lichens and especially their associated photobiont(s) must survive severe drought stress. During this disturbance, lichen cells (from mycobionts and photobionts) display similar features: deformation of the cell membrane system and strong condensation of the protoplast (Honegger 1998). However, cyanobionts and phycobionts show some peculiarities at the physiological level during desiccation.

Table 4.1 List of the main lichen photobionts (Ahmadjian 1993b; Friedl and Büdel 2008; Keddy 2007; Ozenda and Clauzade 1970). Taxonomic synonyms are not listed here. Cyanobacterial genera including anhydrobiotic species are marked with an asterisk (*)

Photobiont	Genus	Main representative lichen taxa	Main osmolytes
Cyanobacteria	*Anabaena*	*Stereocaulon*	Sugars: sucrose trehalose Sugars derivatives: 2-O-glucopyranosyl-glycerol glucosyl glycerol glucosyl glycerate Betaines: glycine-betaine glutamate-betaine Amino acids: arginine proline
	Anacystis		
	Aphanocapsa		
	*Calothrix**	*Lichina*	
	*Chroococcidiopsis**		
	*Chroococcus**	*Cora*, *Phylliscum*, *Pyrenopsidium*, (*Lichinella*?)	
	Cyanosarcina		
	*Dichothrix**	*Placynthium nigrum*	
	Entophysalis		
	Fischerella		
	*Gloeocapsa**	*Anema*, *Gloeoheppia*, *Peccania*, *Psorotichia*, *Pyrenopsis*, *Synalissa*, *Thyrea*	
	Hyella	*Thelidium litorale*, *Paraphysothele halodytes*	
	*Hyphomorpha**		
	*Myxosarcina**		
	*Nostoc**	Collemaceae, Cyanophyllineae, Nephromaceae, Pannariaceae, Peltigeraceae, Pyrenidiaceae, *Stereocaulon*, Stictaceae	
	*Rivularia**		
	*Scytonema**	Ephebaceae, Heppiaceae, Pannariaceae, Pyrenothricaceae, *Stereocaulon*, Thelotremaceae	
	*Stigonema**	*Ephebe*, *Spilonema*, *Stereocaulon*	

(continued)

Table 4.1 (continued)

Photobiont	Genus	Main representative lichen taxa	Main osmolytes
Green algae	*Asterochloris*		Sugars :
	Cephaleuros	*Strigula*	sucrose
	Chlorella	Caliciales, Lecanorales, *Lecidea*, (*Cladonia*?)	Sugars derivatives: glycerol
	Chlorococcum		ribitol
	Chlorosarcina	*Lecidea lapicida* and *L. plana*	sorbitol
	Cladophora	*Blodgettia*	erythritol
	Coccobotrys	Verrucariales, *Lecidea humosa*	isofloridoside
	Coccomyxa	Baeomycetaceae, *Icmadophila*, Peltigeraceae	mannitol Amino acids: proline
	Desmococcus	Verrucariales	glycine
	Dictyochloropsis	Peltigeraceae, Stictaceae	alanine
	Dilabifilium	Verrucariales	glutamate
	Elliptochloris		
	Friedmannia		
	Gloeocystis	*Catillaria*, *Gloeolecta bryophaga*, *Lecidea*	
	Hyalococcus	*Dermatocarpon*	
	Leptosira	*Thrombium epigeum*	
	Myrmecia	*Bacidia*, *Catillaria*, *Dermatocarpon*, *Lecidea*, *Psomora hypnorum*, *Sarcogyne simplex*, *Verrucaria*	
	Phycopeltis	Arthoniaceae, Opegraphaceae, Strigulaceae, Thelotremaceae	
	Physolinum		
	Pleurococcus	*Dermatocarpon miniatum*, *Endocarpon pallidum*, *Lecidea coarcta*, *Staurothele umbrina*, *S. catalepta*, *Thelidium*, *Verrucaria*	
	Prasolia sp.	*Mastodia tessellate*	
	Pseudochlorella	*Lecidea*	
	Pseudochlorococcus		
	Stichococcus	Caliciaceae, *Staurothele*	
	Trebouxia	*Buellia*, *Caloplaca*, *Cladonia*, *Lecanora*, *Lecidea*, *Parmelia*, *Physcia*, *Pilophoron*, *Stereocaulon*, *Verrucaria*, *Xanthoria*	
	Trentepohlia	Arthoniales, *Cystocoleus ebeneus*, *Gyalecta jenensis*, Ostropales, Sphaeriales	
	Trochiscia	*Polyblastia*	
Xanthophyceae	*Botrydiopsis*		
	Heterococcus	Verrucariales, *Verrucaria elaeomelaena* and *V. Laevata*	
Phaeophyceae	*Petroderma*	Verrucariales	

In cyanobacteria, one mechanism of water loss tolerance is known as "anhydrobiosis". This phenomenon consists in their capability to survive to nearly total dehydration (Büdel 2011). Before establishment of any symbiosis relationships, only 10–28% of *Chroococcidiopsis* cells in aggregates avoid subcellular damages (e.g. DNA fragmentation, loss of phycobiliprotein autofluorescence, plasmalemma disruption) whereas in the lichenized state, all cells survive desiccation. This observation could be generalized to all lichenized cyanobacteria and Büdel (2011) indicates that the mycobiont provides an optimal environment for a damage-free drying of the cyanobiont. Numerous cyanobionts belong to genera including anhydrobiotic species (Table 4.1). To cope with this stress, the cyanobacteria produce compatible solutes also designed as osmolytes or osmoprotective compounds, including sugars (sucrose and trehalose particularly), sugar derivatives (e.g. 2-O-glucopyranosyl-glycerol) and betaines (e.g. glycine-betaine, glutamate-betaine) (Klähn and Hagemann 2011; Lüttge 2011; Mao et al. 2010). During salt stress in *Synechocystis,* Ferjani et al. (2003) demonstrated the importance of the compatible solute glucosylglycerol to avoid deleterious effects of NaCl on cell division and size. In addition, a close relationship exists between the synthesis of this compound and salinity levels as glucosylglycerol-phosphate synthase, the key enzyme of glucosylglycerol biosynthesis in salt-stressed *Synechocystis* cells, activated by NaCl addition (Hagemann et al. 2001). Heat Shock Proteins (HSP), a class of molecular chaperones (Al-Waibi 2011), and a 36 kDa Acidic Water Stress Protein (WspA; a class of polypeptides with a structural role in cell stability (Dadheech 2010)) protecting cell structure are also secreted during the dehydration/rehydration phase in cyanobacteria like *Tolypothrix byssoidea* and *Nostoc commune* (Adhikary 2003; Potts et al. 2005). During saline adaptation a set of salt-stress proteins such as flavodoxin in *Synechocystis* is expressed (Fulda and Hagemann 1995). Flavodoxin is a very potent counterstress molecule now used in transgenic plants to enhance their salt stress resistance (Coba de la Peña et al. 2010; Matias et al. 2008). A major threat to cyanobacteria during desiccation is photoinhibition. To cope with this stress the desiccation tolerant cyanobacteria maintain their chlorophyll and are defined as homoiochlorophyllous. To dissipate energy particularly during drought stress, cyanobacteria develop several defense systems: (i) carotenoid pigments (zeaxanthin and cathaxanthin) prevent oxidative stress, (ii) inactivation of PSII and (iii) rapid turnover of protein D1 from the PSII reaction centre. Each of them leads to a rapid recovery of photosynthetic activity during rehydration. Thus homoiochlorophylly allows a rapid restoration of photosynthetic activity during rehydration (Lüttge 2011).

In phycobionts, green algae are able to produce polyols that can be allocated to the mycobionts. Depending on the genus of green alga, the polyol composition differs in the lichen thallus. *Coccomyxa*, *Myrmecia* or *Trebouxia* containing lichen produce ribitol whereas sorbitol and erythritol are synthesized by *Hyalococcus* and *Trentepohlia* lichens (Hill and Ahmadjian 1972). Polyols such as sugars certainly enhance the desiccation tolerance of lichens as they significantly increase the intracellular osmotic potential and limit water loss. A possible mechanism of protection is the formation of a glassy-state in the cytoplasm, known as vitrification, limiting harmful chemical reactions and preventing cellular collapse and modification of pH and ionic strength (Beckett et al. 2008). Specific proteins named dehydrins are synthesized during dehydration and rehydration and are certainly involved in desiccation tolerance of *Trebouxia* (Gasulla et al. 2009). Due to their random coil structure, dehydrins protect cells against dehydration through water binding and protection of protein structure (Rorat 2006). As in cyanolichens, photoprotection mechanisms exist in chlorolichen like *Lobaria pulmonaria* but the molecular mechanisms of energy dissipation are still unclear leading Heber et al. (2010) to suggest the existence of desiccation-responsive chlorophyll protein. Similarly, Kosugi et al. (2009) suggested that a fungal product enhanced the desiccation tolerance of the photobiont through increased resistance to photoinhibition.

To conclude, an important feature should be pointed out: photobiont stress tolerance, if existing in the sole photosynthetic organism, is generally enhanced by symbiosis with the fungal partner. This quite expected feature needs many more studies to understand the underlying mechanisms of stress tolerance in lichens (Kranner et al. 2008).

Osmotic Metabolites in Mycobionts

The majority of recent studies on mechanisms of mycobiont desiccation-tolerance have focused on antioxidant activities. Recently developed tools of molecular biology, particularly from the -omics disciplines, have just started to be applied to lichens and especially mycobionts. Thus, many unsolved questions remain including how mycobionts survive desiccation and what is the relationship between mycobiont taxon and osmolyte diversity. Only information generalized to fungal cells exists concerning the fundamental mechanism of osmoprotection during salt stress and water loss (e.g. metabolite synthesis and storage). As the main strategy for halotolerance, fungi use compatible solutes. These metabolites are found in halotolerant species and could constitute amount of weight of the mycobiont. The dry weight of the intertidal *Xanthoria elegans* contains approximately 5.4% arabitol, 4.3% mannitol and 1.7% ribitol (Aubert et al. 2007). The suggested protection mechanism of these species during dehydration is preferential exclusion: sugars and polyols are all preferentially excluded from the surface of proteins keeping the proteins preferentially hydrated (Green et al. 2011). These taxa synthesize glycerol as one of the main osmolytes and its accumulation correlates with environmental NaCl concentration. Polyols like mannitol are also naturally produced but their synthesis is reduced when salt stress is intense. Their exclusive implication in osmoprotection and osmoregulation is discussed. As in plants, they may also act as ROS scavenger to prevent oxidative stress (da Silva Graça 2004).

Uptake of compatible solutes from the immediate environment has often been implicated in osmoregulation. Acyclic polyols such as D-arabinitol, erythritol, galactitol, mannitol, ribitol, sorbitol and xylitol, are a group of polyols that can be used as storage compounds or osmoprotective compounds. Fungi poorly metabolize these compounds and it has been reported that they are transported through the fungal plasmalemma by a non saturable process (Canh et al. 1975). It has also been reported that most of these polyols could be transported through a proton symporter which seems to be active only starting at 1 M NaCl in the environment (Lages et al. 1999). Glycerol may also be absorbed from the exterior cell environment. Previous studies suggested the existence of active glycerol transport in halotolerant fungi through co-transport of this polyol with Na^+ or K^+, resulting in Na-detoxification and water-loss regulation (Lucas et al. 1990; Van Zyl et al. 1991). Later, a second active-transport system still involved in glycerol absorbance, was also described: a H^+-symporter (da Silva Graça 2004; Lages and Lucas 1995, 1997).

In lichens, there is no experiment which determines the origin of these environmental osmolytes. However, as some new symbiotic partners (epilichenic bacteria) are recently considered in the lichen association, these organisms may be implied in osmolyte production in the external environment of the mycobiont. They can release metabolites during stress regulation or after death during cell-wall destruction. As a consequence, they may play one of the main roles in water-loss inhibition and therefore in lichen halotolerance.

In mycobionts, betaine lipids are also represented but to a lesser extent than photobionts (~40–50% of total extrachloroplastic polar lipids, whereas it does not exceed 5–10%) (Kotlova and Sinyutina 2005). These compounds remain more stable than phospholipids during dehydration, certainly due to their differences in chemical structure and less sensitivity to hydrolysis. As betain lipids can substitute for phospholipids as a plasmalemma constituent (Künzler and Eichenberger 1997), Kotlova and Sinyutina (2005) implies that lichen plasmalemma is enriched with diacylglyceroltrimethylhomoserine (DGTS) during dehydration. Long-term conservation of DGTS rates is important for the maintenance of intact membrane structures, but also for water-potential regulation. Acidic lipids such as betaine lipids are actively bound by dehydrins (proteins enhancing the water-retaining capacity of the cytosol) involved in the regulation of intracellular water potential (Koag et al. 2003; Kotlova and Sinyutina 2005).

Osmotic Metabolites Associated with Bacteriobiont Biodiversity

Lichens present a wide variety of microbial communities, commonly known as microbionts or bacteriobionts (Grube and Berg 2009; Grube et al. 2009; Bates et al. 2011) (Fig. 4.5). Studies reveal that the lichen-associated bacterial community constitutes an integrative part of the lichen thallus. Thus, lichens may be considered as a mini-ecosystem where bacteria contribute to nutrient acquisition, recycling and antagonism (Cardinale et al. 2012b; Schneider et al. 2011). Lichen symbiosis can adapt to a wide range of habitat conditions (e.g. osmotic stress conditions) by development of adaptation mechanisms (Grube 2010).

Bacterial communities are abundant and millions of bacteria cells can be present per gram of lichen thallus (Cardinale et al. 2008; Grube et al. 2009). Their abundance and diversity are dependent on many factors (lichen species, age and part of thallus, sun exposure, substrate, geographic region and patterns of fungal secondary metabolites) (Bjelland et al. 2011; Cardinale et al. 2012b).

Fig. 4.5 Scanning electron micrography of the upper cortex of *Stereocaulon montagneanum* showing epilichenic bacterial colonies. Scale bar: 10 μm

Before the rise of genomics, studies based on culture-dependent approaches detected the presence of various bacterial taxa such as *Azotobacter, Bacillus, Beijerinckia, Clostridium* and *Pseudomonas* (Grube and Berg 2009). Approaches based on 16 S rRNA gene sequencing have revealed a greater diversity of lichen-associated bacterial communities than previously supposed (Cardinale et al. 2006; González et al. 2005; Liba et al. 2006). Several recent studies reported that Alphaproteobacteria are dominant in these communities (Bates et al. 2011; Cardinale et al. 2008; 2012a, b; Grube et al. 2009; Schneider et al. 2011). This taxon can represent more than 60% of all lichenized Proteobacteria in the studied lichens (Cardinale et al. 2008; Schneider et al. 2011). However, recent reports demonstrated that other bacterial lineages could also be abundant, such as Acidobacteria (Hodkinson et al. 2011). Other taxa are also described like abundant Firmicutes, Actinobacteria and other Proteobacteria (Cardinale et al. 2006, 2008, 2011, 2012a) and less-represented *Chloroflexi, Deinococcus* or Verrucomicrobia (Bates et al. 2011; Cardinale et al. 2006, 2008; Grube et al. 2009) (Table 4.2).

Studies focused on these communities show new species. An et al. (2008, 2009) and Li et al. (2007) isolated and identified three new strains of Actinobacteria from lichens collected in Japan: *Schumannella luteola* sp. nov., *Leifsonia lichenia* sp. nov. and *Nocardioides exalbidus* sp. nov. New strains of *Streptomyces* were also isolated from *Cladonia* (*C. gracilis* (strain L-4-4) and *C. uncialis* (*Streptomyces uncialis*)) and from Japanese lichens (strains RI104-LiC106 and RI104-LiB101) (Cheenpracha et al. 2010; Davies et al. 2005; Motohashi et al. 2010; Williams et al. 2008). However, little information on the diversity and abundance of these bacterial communities is currently available, as most of them are uncultivable yet. It could be noted that Bjelland et al. (2011) have highlighted for the first time Archaea (abundant and highly diversified community) to be associated with rock-inhabiting lichens like intertidal *Hydropunctaria maura*.

During osmotic stress caused by salt stress, bacteria accumulate cytoplasmic solutes (Csonka 1989; Kempf and Bremer 1998) (Table 4.2). Like Eukaryotes, they increase their intracellular solute pool by storing large amounts of compatible solutes,

Table 4.2 Example of lichen-associated bacterial communities and their potential osmolytes

Host lichen	Photobiont	Bacteriobiont	Potential osmolytes	References
Canoparmelia crozalsiana	*Trebouxia*	Proteobacteria, Gammaproteobacteria (Stenotrophomonas) (Serratia)	Glycine-betaine, trehalose	Csonka 1989 Liba et al. 2006
Cladonia rangiferina and *Cladonia coccifera*	*Trebouxia*	Proteobacteria, Betaproteobacteria (Burkholderia), Firmicutes (Paenibacillus)	γ-aminobutyrate, glutamate, glutamine, proline	Cardinale et al. 2006 Csonka 1989 Killham and Firestone 1984 Morbach and Krämer 2002
Cladonia gracilis and *Cladonia uncialis*	*Trebouxia*	Actinobacteria (Streptomyces)	Alanine, γ-aminobutyrate, glutamate, glutamine, proline	Cheenpracha et al. 2010 Davies et al. 2005 Killham and Firestone 1984 Morbach and Krämer 2002 Williams et al. 2008
Hypogymnia physodes	*Trebouxia*	Firmicutes (Paenibacillus)	γ-aminobutyrate, glutamate, proline	Cardinale et al. 2006 Killham and Firestone 1984 Morbach and Krämer 2002

either by in vivo synthesis or by direct uptake from the close environment to restore the internal-external balance (Kempf and Bremer 1998). The spectrum of osmoprotectant compounds used by microorganisms is generally limited: K^+ ions, sugars (threhalose), polyols (glucosylglycerol and glycerol), free amino acids (alanine, γ-aminobutyrate, glutamate, glutamine and proline), derivatives thereof (ectoin and proline-betaine), quaternary amines and their sulfonium analogues (carnitine, dimethylsulfoniopropionate and glycine-betaine), sulfate esters (choline-O-sulfate), N-acetylated diamino acids and small peptides (N-acetylglutaminylglutamine amide and Nδ-acetylornithine) (Csonka 1989; Kempf and Bremer 1998). However ectoin, glycine-betaine, proline and threhalose are probably the most used compatible solutes among bacteria (Morbach and Krämer 2002).

Various responses against osmotic stress occur within the different bacteria phyla. Gram-positive (Firmicutes and Actinobacteria) and Gram-negative bacteria do not use the same mechanisms. Killham and Firestone (1984) and Morbach and Krämer (2002) reported that Gram-positive bacteria belonging to Actinobacteria (e.g. *Streptomyces*, *Corynebacterium*) or Firmicutes (*Bacillus*, *Paenibacillus*), accumulate different osmolytes depending on the intensity of osmotic stress. During salt stress, these bacteria preferentially accumulate proline and γ-aminobutyrate while Gram-positive bacteria stock glutamate, particularly during hyperosmotic stress (Killham and Firestone 1984; Morbach and Krämer 2002). *Bacillus subtilis* and two *Streptomyces* (*S. griseus* and *S. californicus*) answer to osmotic stress by increased proline synthesis (Morbach and Krämer 2002) and free amino-acid pools (alanine, glutamine and proline) (Killham and Firestone 1984). In contrast, Gram-negative bacteria accumulate glutamate (in lesser amount glutamine) and free amino acids such as proline and/or alanine in response to salt stress (Csonka 1989). Gammaproteobacteria, particularly Enterobacteriales like *Escherichia coli* or potential epilichenic bacteria (*Pantoea sp.* and *Serratia sp.*) accumulate glycine-betaine and threhalose (Csonka 1989). Proline synthesis increases in Firmicutes (e.g. *Bacillus sp.*, *Paenibacillus sp.*) with severity of osmotic stress (Csonka 1989).

Exceptions occur among halotolerant Eubacteria and halophilic Archaea whose entire physiology has been geared to life in saline environments. These bacteria store very high intracellular concentrations of ions to limit any nutrient imbalance (Kempf and Bremer 1998).

Lichen-associated bacterial communities appear to be abundant and diverse, but their functional involvement in the lichen symbiosis still remains largely unexplored. However, Schneider et al. (2011) demonstrated that proteins implied in the transport of amino acids, nucleotides, coenzymes and lipids, exist in bacteria and fungi. Accordingly, the reallocation of resources in lichens could also be mediated by bacteria (Schneider et al. 2011).

4.3.2 Induction of Oxidative Stress and Antioxidant Activity

Salt toxicity causes damage to the cytoskeleton, makes leaky membranes, and changes protein structures so their activity is disturbed and/or reduced (Beckett et al. 2008).

All these consequences could lead to oxidative stress. Lichens use dioxygen as an energy source for their development through microalgae and/or cyanobacteria. However, due to ionic imbalance and photooxidation, this aerobic process could lead to the production of ROS which are chemically diversified reactive molecules containing oxygen (Kranner et al. 2003). ROS are natural byproducts of metabolism and play important roles in homeostasis, cell signaling and apoptosis. However, under salt stress, their levels can increase dramatically leading to disruption and damage in cell compartments (Kranner et al. 2005).

Incomplete dioxygen reduction through cytochromes from the respiratory chain implies ROS production as singlet oxygen ($^1O_2$) and superoxide radical ($O_2^{\bullet-}$) that leads to the synthesis of hydroxyl radical ($^{\bullet}OH$), hydroperoxyl radical ($^{\bullet}O_2H$) and hydrogen peroxide (H_2O_2) (Fig. 4.6). The alkoxyl ($RO^{\bullet}$) and peroxyl ($RO_2^{\bullet}$) radicals are the consequences of the membrane phospholipid peroxidation or lipoperoxidation, by ROS (Edreva 2005; Delmail and Labrousse 2012; Lagadic et al. 1997; Li et al. 1994; Thompson et al. 1987).

At the same time, the photosynthetic electron transport chains could produce high concentrations of ROS. Electrons tetravalently reduce intracellular oxygen to water. But some electrons could leak from many sites along the electron transport chain, resulting in a univalent reduction of dioxygen to form the extremely reactive superoxide radical which can dismutate to form hydrogen peroxide (Alscher et al. 2002). This last reaction is spontaneous or catalyzed by one of the superoxide dismutases (Fig. 4.6) depending on the cell compartment where the reaction occurs: manganese-superoxide dismutase (peroxisome, microalga mitochondria), iron-superoxide dismutase (microalga chloroplast or cyanobacteria cytosol) or copper/zinc-superoxide dismutase (cytosol, microalga chloroplast) (Delmail and Labrousse 2012; Fornazier et al. 2002; Gill and Tuteja 2010; Pereira et al. 2002).

Hydrogen peroxide is not a free radical due to all its matched electrons. However, it has a strong toxic potential and diffuses far from its synthesis site. It can pass through biological membranes via aquaporins as its chemical structure is close to water (Bienert et al. 2006, 2007; Parent et al. 2008). The concentration of this oxidative compound is regulated by antioxidant enzymes such as the ascorbate peroxidase, catalase or glutathione peroxidase (Fig. 4.6) (Chou et al. 2012). These proteins use NADPH produced during the photosynthesis for their functioning (Fig. 4.6). However, ROS could disrupt the photosynthetic electron transport chains in thylakoid membranes and some electrons are deflected. Without NADPH normal synthesis, lichens use a cytosolic secondary catabolic pathway to produce it, the pentose phosphate pathway (Delmail and Labrousse 2012; Kranner 2002; Kruger and von Schaewen 2003).

Hydrogen peroxide could be also produced through the bivalent reduction of oxygen in the presence of oxidases like peroxisomal glycolate oxidase or amine oxidase (Parent et al. 2008). Hydrogen-peroxide toxicity is also linked to its implication in the synthesis of hydroxyl and hydroperoxyl radicals through the Haber-Weiss and Fenton reactions (Fig. 4.6). Like their ROS parent, these short-lifespan radicals diffuse easily through biological membranes and could affect and disturb all organelles and cell compartments. They are also mainly implied in lipoperoxidation

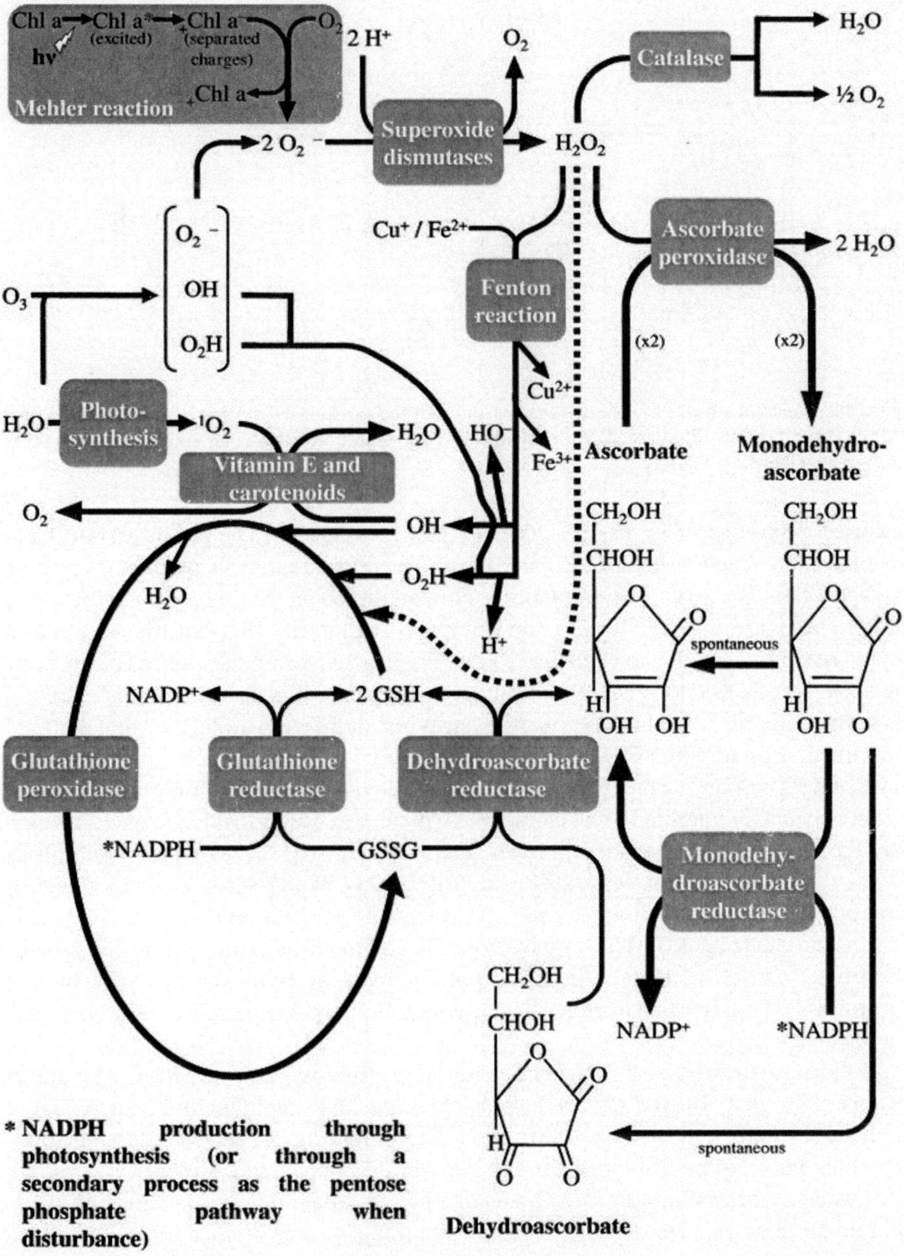

Fig. 4.6 Main antioxidant pathways including enzymes and scavengers (based on Delmail and Labrousse 2012). For easier comprehension, certain reactions are not equilibrated due to the substrate diversity. Chl a, chlorophyll a; GSH, glutathione; GSSG, glutathione disulfide; $NADP^+$, oxidized nicotinamide adenine dinucleotide phosphate; NADPH, reduced nicotinamide adenine dinucleotide phosphate

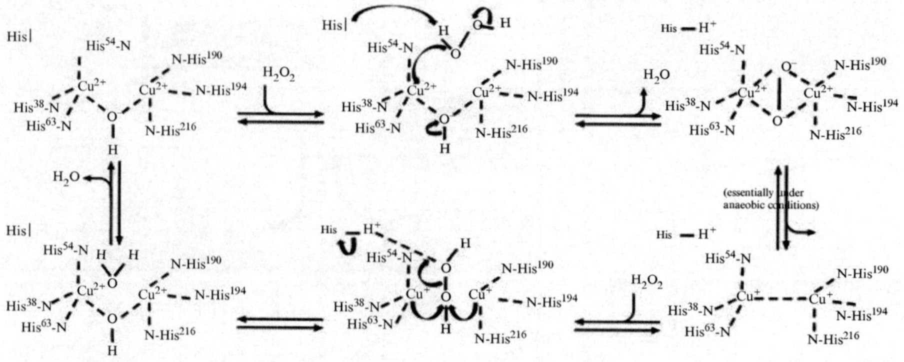

Fig. 4.7 Mechanism of H_2O_2 detoxification through the tyrosinase catalase cycle. His, histidine. Modified and reprinted (adapted) with permission from García-Molina et al. 2005. Copyright 2005 American Chemical Society

(Cortés-Rojo et al. 2011; Edreva 2005; Lagadic et al. 1997). The produced fatty-acid radical then reacts with molecular oxygen, thereby creating a peroxyl fatty acid radical. This latter reacts with another phospholipid, producing a new radical and a lipid peroxide, or a cyclic peroxide if it reacts with itself. This cycle continues as a chain reaction mechanism (Schaich 2005). The process terminates when two radicals react and produce a non-radical compound. It happens when the radical concentration is sufficiently high. Living organisms have evolved different molecules that increase termination by trapping ROS (Paramesha et al. 2011). Among such compounds, the most important are scavengers mainly constituted with lipid-soluble antioxidants: α-tocopherol (or vitamin E) and carotenoids (β-caroten, xanthophylls) (Fig. 4.6) which are the key antioxidants in membranes (Delmail et al. 2011; Delmail and Labrousse 2012; Kranner and Lutzoni 1999; Smirnoff 1993). Water-soluble antioxidants of low molecular weight also exist such as glutathione (γ-glutamyl-cysteinyl-glycine) and ascorbate (Fig. 4.6). They are hydrophilic and their main function is cell protection from oxidative damage in liquid phases, particularly in the cytosol (Beckett et al. 2008). They act as scavengers and produce radicals that are reduced by NADPH-dependent enzymes.

In a particular case, ROS do not only affect internal cell structures. Oxidants such as H_2O_2 may diffuse across the plasmalemma and stay in the mixed apoplast of lichens (Henzler and Steudle 2000). This last compartment corresponds to the contact area between photobiont cells and mycobiont hyphae and is constituted by the cell walls of these symbionts. H_2O_2 can damage plasmalemma-bound and cell-wall-bound proteins and therefore their respective integrity and permeability. As lichens are mainly constituted with fungi, they lack extracellular peroxidases and catalases despite their strong capacity to detoxify external ROS (Beckett and Minibayeva 2007). However in the sub-order Peltigerineae, significant extracellular tyrosinase activity is correlated with reduction of H_2O_2. This copper-containing enzyme catalyzing the oxidation of phenols (e.g. tyrosine) may function here through a catalase-like mechanism (García-Molina et al. 2005) (Fig. 4.7). As the Peltigerinean lichens are more sensitive to water loss than other species (Beckett et al. 2003), they produce more ROS during dehydration process and need better protection against oxidative stress. In this case, the tyrosinase catalase cycle may play a major role in maintaining the integrity of cell-wall and plasmalemma components (Beckett et al. 2008). However, as this detoxification process among lichens was discovered early, one point remains unclear. H_2O_2 has always been considered as an inactivator of several copper containing enzymes, such as fungal tyrosinase (Andrawis and Kahn 1985). High H_2O_2 concentrations oxidize a methionine residue in position 374 of the fungal tyrosinase active site to methionine sulfoxide, and this event inactivates the protein (Schweikardt et al. 2007). Moreover, all methionine sulfoxide reductases (msrA and msrB) which keep check on methionine sulfoxide in fungi are very sensitive to ROS (Soriani et al. 2009). As a significant correlation was previously established between the mixed-apoplast H_2O_2 content among Peltigerineae and their tyrosinase activity, it is essential to elucidate in further studies about the mechanisms involved to prevent inactivation of tyrosinase in lichens.

Considering all these elements, ROS are considered as lichenotoxic compounds. However, it is currently admitted that their synthesis in relation to respiratory and photosynthetic metabolism plays an essential role in life and death of cells. They could play an alternative role and act as cell signalization molecules to establish some defense mechanisms towards xenobiotic stress (Delmail and Labrousse 2012; Parent et al. 2008).

4.3.3 Nutrient Imbalance and Toxicity

Symbiotic association in lichens is an opportunity for all symbionts to benefit from nutrient mobilization and compound synthesis by each organism. They influence the mineral-nutrient input composition by enhancing or selecting uptake of certain nutrients. As each symbiotic partner has its own sensitivity to salt stress, symbiotic dependency increases with NaCl concentration and remains a necessary step to colonize and live in extreme environments.

4.3.3.1 Nitrogen

Fungi are unable to fix dinitrogen as it is a characteristic of prokaryotes. They obtain nitrogen through the absorption of amino acids, nucleotides and other organic compounds. In the case of lichens, photobiont cyanobacteria provide in addition to carbon, fixed nitrogen to other symbionts (fungi and/or microalgae). The importance of molecular nitrogen fixation is reflected in the phenotypic and physiological adaptations of lichen-associated cyanobacteria (e.g. numerous nitrogen-fixing heterocysts, or specialized nitrogen-fixing cells, in symbiotic *Nostoc* compared to free-living filaments) (e.g. *Lichina* species). Further adaptation is found in tripartite symbiosis where cyanobacteria are concentrated in specific organs called cephalodias, where they fix nitrogen and are protected from high oxygen concentrations. In tripartite symbiosis, photosynthesis is mainly due to micro-alga photobionts which supply fungi and cyanobacteria with photosynthates (Honegger 2001; Kneip et al. 2007). In case of salt stress, NaCl interferes with nitrogen acquisition and utilization by influencing different stages of nitrogen metabolism such as NO_3^- uptake and reduction, and protein synthesis (Frechilla et al. 2001). Indeed, NaCl induces a sharp decline in nitrogenase activity as it is reported to inhibit the assembly of nitrogenase components by binding to the iron-containing protein. In addition, inhibition of nitrogenase activity is due to (i) disturbed electron transport, (ii) protection reduction of heterocyst nitrogenase from oxygen due to disturbed plasmalemma permeability and (iii) inadequate supply of photosynthetically generated ATP and reductants from the photosynthetic microorganisms. The heterocyst number is positively correlated with salt concentration. This could be a response to nitrogen stress induced by the inhibitory effect of NaCl on nitrogenase. However, such an increase in heterocyst frequency does not induce an increase in dinitrogen-fixation and heterocysts are supposed to be physiologically defective (Rai et al. 2001).

Alternative mechanisms for obtaining nitrogen in lichens have been poorly investigated but the presence of nitrogen-fixing bacteria may help in better assimilation of nitrogen by the host lichen during salt stress as observed among halotolerant plant species (dos Santos et al. 2010). These microorganisms have been traditionally reported in symbiotic associations with plants (e.g. *Rhizobia* with Fabaceae, *Frankia* with non-legume plants), and as endophytes (Cocking 2003), but sparse information is available for interactions with lichens (Grube and Berg 2009). Free-living and endophytic chemo-organotrophic diazotrophic bacteria are able to release nitrogenated compounds including amino acids, vitamins (Miyasaka et al. 2003), and phytohormones (Thuler et al. 2003) to nearby organisms. Liba et al. (2006) reported that some nitrogen-fixing bacteria from five genera (*Acinetobacter*, *Pantoea*, *Pseudomonas*, *Serratia* and *Stenotrophomonas*), show the potential for establishing relationships with lichens, although their assessment is based on enrichment cultures and does not adequately reflect the abundance of these strains in situ. The ecological role of these microorganisms could be linked to the nutritional needs of lichens, and it could be of great advantage to live in limited-nutrient-availability environments. Bacterial excreted amino acids (mainly glutamate, isoleucine, leucine, methionine and tyrosine) may be absorbed by mycobionts and photobionts to be incorporated into their carbon bone, sparing the lichen the expense of synthesizing its own amino acids (Liba et al. 2006). The influence of phytohormones from nitrogen-fixing bacteria (e.g. soluble ethylene derivatives like ethylene glycol) on lichen symbionts is currently unknown but they may play an important role in regulating the development of surrounding plants when transferred from the lichen surface to soil by rain or waves. To acquire more available resources (such as nutrients, water or light) from the environment, allelopathy is a biochemical strategy that controls competitors in terms of survival, growth and reproduction.

4.3.3.2 Phosphorus

The limited availability of soluble phosphates is also one of the great obstacles to colonization of a limited-nutrient-availability environment. Environmental salinity may disrupt phosphorous absorption as orthophosphate ions (PO_4^{3-}) could form precipitates with calcium (Ca^{2+}), magnesium (Mg^{2+}) and zinc (Zn^{2+}) (Evelin et al. 2009). Because these complexes are more difficult to be uptaken by fungi, minerals become less available. In aquatic ecosystems, as well as in isolated lichen symbionts, the growth of free-living algal and fungal cells has been shown to be phosphorous-limited (Makkonen et al. 2007). In phosphorous-depleted cyanobacteria, chlorophyll and phycocyanin contents decrease, associated with a reduction in nitrogen uptake resulting in a loss of carbon fixation (Lewitus and Caron 1990). In *Anabaena*, a loss of heterocysts, a decline in chlorophyll a, protein, RNA, and cellular phosphorous is specifically noted (Healey 1973). Moreover, during phosphorous starvation in microalgae, the total phospholipid content decreases and phospholipids are substituted by non-phosphorous lipids like glycolipids to support membrane functionality and maintained photosynthesis. An alternative source of phosphorous could be phospholipid recycling which releases free inorganic phosphate (Okanenko et al. 2011).

As nitrogen-fixing bacteria can solubilize phosphates, they could also play a key role to overcome this hurdle. This type of benefit can be attributed to bacteria isolated from lichens, similarly to what genera *Pantoea* and *Pseudomonas* can do for vascular plants when they act as endophytes or are located in the rhizosphere. (Liba et al. 2006). Several free *Pantoea* and *Pseudomonas* strains are halotolerant and live at high salt levels (>2.5 M NaCl) (Gopalakrishnan et al. 2006; Margesin and Schinner 2001) which increases the possibility of involvement in saline tolerance of lichens.

Enhanced uptake of phosphorous by lichens under salt conditions may reduce the negative effects of Na^+ and Cl^- ions by maintaining tonoplast integrity. This facilitates the compartmentalization within vacuoles and selective ion intake, thereby preventing these elements from interfering in metabolic pathways (Evelin et al. 2009).

4.3.3.3 Potassium and Sodium

Na^+ ions are competitors with K^+ for binding sites essential for several cell functions in fungi, microalgae, cyanobacteria and bacteria (Becker 1994; da Silva Graça 2004; Rodríguez-Navarro 2000). Under normal growth conditions, sodium is continuously excreted to maintain low intracellular levels. Intracellular sodium content may vary greatly both within populations and among different species (da Silva Graça 2004). Potassium is an absolute requirement for many cell functions, such as intracellular pH and osmotic regulation, protein synthesis, and enzyme activation (Becker 1994; Evelin et al. 2009; Rodríguez-Navarro 2000). These functions cannot be replaced by sodium as competition under very high salinity levels disturbs the cytosol ionic balance and disrupts metabolic pathways (Evelin et al. 2009). However, there is a case of functional Na^+ substitution for K^+ in fungal cells, where sodium restores the cell volume, cellular pH, and enhances growth. The most beneficial effects of Na^+ might be due to a complex regulation that allows small, but not excessive Na^+ uptake. Such a system mediates both potassium and sodium influxes. When K^+ reaches the mM range and Na^+ is concentrated in the environment, the Michaelis-Menten constant (K_m) decreases to μM values for the internal K^+ concentration while the $K_{m\,Na+}$ remains in the mM range. As a consequence, the ratio between $K_{m\,K+}$ and $K_{m\,Na+}$ may be as low as 1:700. When Na^+/K^+ concentration ratio is lower than 700:1, Na^+ influx is less than that of K^+, and Na^+ concentration in cells is at non-toxic levels. But when the ratio is higher than 700:1, Na^+ influx is larger than K^+ and Na^+ may substitute for a significant part of the K^+ content, becoming toxic (Rodríguez-Navarro 2000). Thus, the toxicity limit in saline conditions depends on the considered lichen species and its capability to regulate the Na^+ influx. In general, sodium content increases with increasing external NaCl concentration. For example, an environment with a 3 M Na^+ concentration may implicate intracellular values from 0.5 M to >3 M in moderately halophilic bacteria, depending on species and growth conditions (da Silva Graça 2004).

4.3.3.4 Chloride

There is much less to be said about the function of this anion in biological systems. Chloride provides a negative charge in the formation of membrane potentials and is responsible for the regulation of intracellular osmotic pressures. A number of proteins function as Cl^- channels in biological membranes (Li and Weinman 2002). A few enzymes (e.g. α-amylase which catalyze starch hydrolysis in fungi) require Cl^- for functioning (Feller et al. 1996), and the haloperoxidases use Cl^-, Br^- and I^- as substrates along with H_2O_2 to halogenate aromatic amino acids or to produce bactericidal compounds (Morrison and Schonbaum 1976). Chloride has also been found to function as a bridging ligand between heme a and Cu_B in the oxidized form of cytochrome oxidase (Fabian et al. 2001). Manganese oxidation, catalyzed by chlorophyll photochemistry in photosystem II, is the key step in redox reactions leading to formation of O_2 from H_2O. Chloride is also required for Mn oxidation by photosystem II in thylakoid membranes and it has been demonstrated that the anion is required for Mn redox reactions immediately preceding oxidation of H_2O to O_2 (Yocum 2008).

Chloride toxicity has not been previously studied in lichens. Considering each symbiont, no toxicity has yet been reported at fungal cell levels in excess conditions. However, no concrete information is available concerning this absence of effects. It is supposed that a limited uptake of the anion, very active extrusion (due to the negative-inside membrane potential) or some chloride-resistant enzymes are involved in this process (da Silva Graça 2004). In cyanobacteria, Cl^- inhibits ribulose-1,5-bisphosphate-carboxylase activity from *Aphanothece halophytica*, thus disrupting photosynthesis and decreasing carbon fixation. The intracellular chloride concentration increased from 35 mM to 150 mM, when NaCl concentration in the culture medium increased from 500 mM to 2 M (Incharoensakdi and Takabe 1988).

Excess Cl^- anions may be compartmentalized in vacuole, thereby preventing them from interfering with metabolic pathways (Evelin et al. 2009). Some reports on halophilic bacteria exist but results are highly variable, from relatively low values (55 and 140 mM Cl^- in *Halomonas halodenitrificans* and *Salinivibrio costicola*, respectively, in the presence of 1 M Cl^-) to values as high as 700–980 mM Cl^- in *Pseudomonas halosaccharolytica* grown under 1–3 M Cl^- (Ventosa et al. 1998) showing the high tolerance capabilities of bacteria such as Gamma-proteobacteria known to be present on lichen thalli as mentioned above.

4.3.4 Genetics Studies

Grube and Blaha (2005) studied the phylogenetic relationships of certain genes which are involved in conferring halotolerance or halophily. They considered genera containing marine and non-marine taxa to detect the supposed involvement of these genes in halotolerance through their evolution and expression. They focused on the synthesis of a brown pigment, dihydroxynaphthalene (DHN) melanin. Earlier studies claimed that fungal melanins produced in brackish and marine environments, confer protection against hyperosmotic shock as they act as osmoprotective compounds (Butler and Day 1998). Polyketide synthase (PKS) genes are implied in the synthesis of 1,8-DHN, precursor of DHN-melanin. Those PKS first use acetate as a precursor to produce 1,3,6,8-tetrahydroxynaphthalene. Then, a hydroxynaphthalene reductase converts this last compound to scytalone. Dehydration of scytalone forms 1,3,8-trihydroxynaphthalene, which is converted to 1,8-DHN after an additional reduction and dehydration step. Finally, oxidative polymerization of 1,8-DHN by a laccase, yields the DHN-melanin (Tsai et al. 1999). A genetic analysis performed by Grube and Blaha (2005) revealed the presence of scytalone dehydratase (SCD) in six Verrucariales species, suggesting potential synthesis of DHN-melanin. This enzyme plays a key role both in dehydration of both scytalone and vermelone. In addition, the strong analogy between phylogenetic trees of SCD and PKS tends to underline their common involvement in synthesis.

Alternatively, some fungi are able to synthesize melanin via L-3,4-dihydroxyphenylalanine by a pathway that resembles mammalian melanin biosynthesis (Eisenman and Casadevall 2012). No information is available on trihydroxynaphthalene reductase in the study by Grube and Blaha (2005), despite the fact that this enzyme is essential for the next step of vermelone dehydration (Kogej et al. 2004; Liao et al. 2001). So, if alternative melanin production is suggested in the studied species, it could be supposed that only 2-hydroxyjuglone synthesis occurs from the exclusive oxidation of 1,3,8-trihydroxynaphthalene (which could eventually justify a transcription/translation of PKS and SCD and a shunt in DHN-melanin synthesis). These authors opened new perspectives by considering the role of genetics in saline adaptation. However, further data on enzyme expression and activity are needed to establish the melanin synthesis pathways in halotolerant species, including the pathways involved in brown-pigment production.

As also suggested by these authors, many more genes involved in salt tolerance need to be detected from all lichen symbionts in the future. Using reverse transcription and amplified fragment-length polymorphism, Grube and Blaha (2005) focused on the specific transcriptional responses during salt stress (1 M NaCl) in several specimens of *Parmelia subrudecta*. In salt-treated thalli, a 400-bp cDNA fragment missing in untreated thalli, was identified. This differential transcription may result from gene (over-) expression induced by salt stress. These experiments only gave a first glimpse into differential transcription under salt stress in a lichen that normally lives in non-saline environments. Sequencing of transcripts would provide further insights (by generating and analyses of expressed sequence tag (EST)-profiles), especially when different species of lichens, salt-tolerant and -intolerant will be compared.

4.4 Conclusion and Future Perspectives

The lichen symbiotic association, comprising fungi, algae and/or cyanobacteria, and associated bacterial communities was among the first recognized symbiosis (De Bary 1879). However, today, we still have little insight into the versatile adaptations to diverse environmental conditions of lichen symbiosis including halotolerance as a fascinating aspect. Lichens seem to have particular pre-adaptations to the saline habitat, but we still need to explore the role of genes and their regulation by specific signals. Halotolerance in lichens implies a succession of molecular and physiological events that leads to the regulation of hydric potential, photosynthesis, osmolarity and oxidative processes. Reactions to salt stress at symbiotic stage is clearly different from those of isolated partners. Each symbiont develops both specific arrays of metabolites for osmotic adjustments and well adapted ionic mechanisms to limit nutrient-imbalance or it activates certain antioxidant pathways involved in free-radical and oxidative-compound regulation. Genetics also plays a major role as NaCl can stimulate the expression of genes involved in salt-stress acclimation.

This chapter mainly focuses on lichens from intertidal habitats or those exposed to marine salt spray. However inland lichens that are exposed as well to saline conditions are poorly known. They live in uncommon habitats such as manmade lagoons (Gilbert 2001), salt lakes (Johnson 2007) and salt deserts (D'Antonio and Haubensak 2005). It would be especially interesting to study the extreme mechanisms of lichen drought adaptation in the latter environment as all conditions for halophily and xerophily occur. Ecophysiological properties, metabolomics and genetics of these fascinating holobionts merit increased attention by lichenologists in the near future.

Acknowledgments This research was supported by the Université de Rennes 1, the Université Européenne de Bretagne, the Ministère de l'Enseignement supérieur et de la Recherche, the Institut National des Sciences Appliquées de Rennes and the UMR CNRS 6226 SCR. The authors sincerely thank Dr. Frédéric Barrière and Dr. Marie-Laurence Abasq-Paofai for technical support, Dr. Françoise Lohézic-Le Dévéhat and Friardi Ismed for providing SEM photography, and Dr. Béatrice Legouin-Gargadennec for valuable discussions.

References

Adhikary SP (2003) Heat shock proteins in the terrestrial epilithic cyanobacterium *Tolypothrix byssoidea*. Biol Plant 47:125–128

Ahmadjian V (1993a) The lichen symbiosis. Wiley, New York

Ahmadjian V (1993b) The lichen photobiont – what can it tell us about lichen systematic. Bryologist 96:310–313

Ahmadjian V (1995) Lichens are more important than you think. BioScience 45:123–124

Alscher RG, Erturk N, Heath LS (2002) Role of superoxide dismutases (SODs) in controlling oxidative stress in plants. J Exp Bot 53:1331–1341

Al-Waibi MH (2011) Plant heat-shock proteins: a mini review. J King Saud Univ 23:139–150

An SY, Xiao T, Yokota A (2008) *Schumannella luteola* gen. nov., sp. nov., a novel genus of the family Microbacteriaceae. J Gen Appl Microbiol 54:253–258

An SY, Xiao T, Yokota A (2009) *Leifsonia lichenia* sp. nov., isolated from lichen in Japan. J Gen Appl Microbiol 55:339–343

Andrawis A, Kahn V (1985) Inactivation of mushroom tyrosinase by hydrogen peroxide. Phytochemistry 24:397–405

Arup U (1995) Littoral species of *Caloplaca* in North America: a summary and key. Bryologist 98:129–140

Aubert S, Juge C, Boisson AM, Gout E, Bligny R (2007) Metabolic processes sustaining the reviviscence of lichen *Xanthoria elegans* in high mountain environments. Planta 226: 1287–1297

Bates ST, Cropsey GW, Caporaso JG, Knight R, Fierer N (2011) Bacterial communities associated with the lichen symbiosis. Appl Environ Microb 77:1309–1314

Becker EW (1994) Microalgae: biotechnology and microbiology. Cambridge University Press, Irthlingborough

Beckett RP, Minibayeva FV (2007) Rapid breakdown of exogenous extracellular hydrogen peroxide by lichens. Physiol Plant 129:588–596

Beckett RP, Minibayeva FV, Vylegzhanina NV, Tolpysheva T (2003) High rates of extracellular superoxide reduction by lichens in the Suborder Peltigerineae correlate with indices of high metabolic activity. Plant Cell Environ 26:1827–1837

Beckett RP, Kranner I, Minibayeva FV (2008) Stress physiology and the symbiosis. In: Nash TH III (ed) Lichen biology. Cambridge University Press, Cambridge, pp 134–151

Bienert GP, Schjoerring JK, Jahn TP (2006) Membrane transport of hydrogen peroxide. Bioch Biophys Acta – Biomembranes 1758:994–1003

Bienert GP, Møller ALB, Kristiansen KA, Schulz A, Møller IM, Schjoerring JK, Jahn TP (2007) Specific aquaporins facilitate the diffusion of hydrogen peroxide across membranes. J Biol Chem 282:1183–1192

Billi D, Potts M (2002) Life and death of dried prokaryotes. Res Microbiol 153:7–12

Bjelland T, Grube M, Hoem S, Jorgensen SL, Daae FL, Thorseth IH, Øvreås L (2011) Microbial metacommunities in the lichen-rock habitat. Environ Microbiol Rep 3:434–442

Bohnert HJ, Nelson DE, Jensen RG (1995) Adaptations to environmental stresses. Plant Cell 7:1099–1111

Brodo IM, Sloan NA (2004) Lichen zonation on coastal rocks in Gwaii Haanas National Park Reserve, Haida Gwaii (Queen Charlotte Islands), British Columbia. Can Field-Naturalist 117: 405–424

Büdel B (2011) Cyanobacteria: habitats and species. In: Lüttge U, Beck E, Bartels D (eds) Plant desiccation tolerance, ecological studies 215. Springer, Berlin, pp 11–21

Butler MJ, Day AW (1998) Fungal melanins: a review. Can J Microbiol 44:1115–1136

Canh DS, Horák J, Kotyk A, Říhová L (1975) Transport of acyclic polyols in *Saccharomyces cerevisiae*. Folia Microbiol 20:320–325

Cardinale M, Puglia AM, Grube M (2006) Molecular analysis of lichen-associated bacterial communities. FEMS Microbiol Ecol 57:484–495

Cardinale M, Vieira de Castro J Jr, Müller H, Berg G, Grube M (2008) *In situ* analysis of the bacterial community associated with the reindeer lichen *Cladonia* arbuscula reveals predominance of alphaproteobacteria. FEMS Microbiol Ecol 66:63–71

Cardinale M, Grube M, Berg G (2011) *Frondihabitans cladoniiphilus* sp. nov., an actinobacterium of the family microbacteriaceae isolated from lichen, and emended description of the genus *Frondihabitans*. Int J Sys Evol Microbiol 61:3033–3038

Cardinale M, Grube M, Castro JV Jr, Müller H, Berg G (2012a) Bacterial taxa associated with the lung lichen *Lobaria pulmonaria* are differentially shaped by geography and habitat. FEMS Microbiol Lett 329:111–115

Cardinale M, Steinova J, Rabensteiner J, Berg G, Grube M (2012b) Age, sun and substrate: triggers of bacterial communities in lichens. Environ Microbiol Rep 4:23–28

Cheenpracha S, Vidor NB, Yoshida WY, Davies J, Chang LC (2010) Coumabiocins A-F, aminocoumarins from an organic extract of *Streptomyces sp.* L-4-4. J Nat Prod 73:880–884

Chou TS, Chao YY, Kao CH (2012) Involvement of hydrogen peroxide in heat shock- and cadmium-induced expression of ascorbate peroxidase and glutathione reductase in leaves of rice seedlings. J Plant Physiol 169:478–486

Coba de la Peña T, Redondo FJ, Manrique E, Lucas MM, Pueyo JJ (2010) Nitrogen fixation persists under conditions of salt stress in transgenic *Medicago trunculata* plants expressing a cyanobacterial flavodoxin. Plant Biotechnol J 8:954–965

Cocking EC (2003) Endophytic colonization of plant roots by nitrogen-fixing bacteria. Plant Soil 252:169–175

Cortés-Rojo C, Estrada-Villagómez M, Calderón-Cortés E, Clemente-Guerrero M, Mejía-Zepeda R, Boldogh I, Saavedra-Molina A (2011) Electron transport chain dysfunction by H_2O_2 is linked to increased reactive oxygen species production and iron mobilization by lipoperoxidation: Studies using *Saccharomyces cerevisiae* mitochondria. J Bioenerg Biomemb:135–147

Csonka LN (1989) Physiological and genetic responses of Bacteria to osmotic stress. Microbiol Rev 53:121–147

D'Antonio C, Haubensak K (2005) Variation in biotic crusts, shrubs, rainfall and resistance to cheatgrass invasion in the Great Basin. Ecological Society of America 2005, Montréal

da Silva Graça MMC (2004) Salt stress response of the extremely halotolerant yeast *Candida halophila* (*syn versatilis*) CBS 4019 – Biochemical and physiological studies. Universidade do Minho – Escola de Ciências, Braga

Dadheech N (2010) Desiccation tolerance in cyanobacteria. Afr J Microbiol Res 4:1584–1593

Davies J, Wang H, Taylor T, Warabi K, Huang XH, Andersen RJ (2005) Uncialamycin, a new enediyne antibiotic. Org Lett 7:5233–5236

De Bary A (1879) Die Erscheinung der Symbiose. Karl J Trübner, Strasbourg

Delmail D, Labrousse P (2012) Plant ageing, a counteracting agent to xenobiotic stress. In: Nagata T (ed) Senescence. InTech Publishers, Rijeka, pp 89–106

Delmail D, Buzier R, Simon S, Hourdin P, Botineau M, Labrousse P (2011) HPLC method for the analysis of α-tocopherol from *Myriophyllum alterniflorum*. Chem Nat Compd 47:679–680

dos Santos Freire MBG, Lira-Cadete L, de Farias ARB, Kuklinsky-Sobral J, dos Santos KCF (2010) Isolation of *Atriplex nummularia*-associated halotolerant bacteria and bioprospecting by nitrogen fixing bacteria in saline-sodic soil. 19th World Congress of Soil Science – Soil Solutions for a Changing World, Brisbane

Edreva A (2005) Generation and scavenging of reactive oxygen species in chloroplasts: a submolecular approach. Agric Ecosyst Environ 106:119–133

Eisenman HC, Casadevall A (2012) Synthesis and assembly of fungal melanin. Appl Microbiol Biotechnol 93:931–940

Evelin H, Kapoor R, Giri B (2009) Arbuscular mycorrhizal fungi in alleviation of salt stress: a review. Ann Bot 104:1263–1280

Fabian M, Skultety L, Brunel C, Palmer G (2001) The cyanide stimulated dissociation of chloride from the catalytic center of oxidized oxidase. Biochemistry 40:6061–6069

Feller G, Bussy O, Houssier C, Gerday C (1996) Structural and functional aspects of chloride binding to *Alteromonas haloplanctis* alpha-amylase. J Biol Chem 271:23836–23841

Ferjani A, Mustardy L, Sulpice R, Marin K, Suzuki I, Hagemann M, Murata N (2003) Glucosylglycerol, a compatible solute, sustains cell division under salt stress. Plant Physiol 131:1628–1637

Fernández E, Reyes A, Hidalgo ME, Quilhot W (1998) Photoprotector capacity of lichen metabolites assessed through the inhibition of the 8-methoxypsoralen photobinding to protein. J Photochem Photobiol B 42:195–201

Fletcher A (1973a) The ecology of marine (littoral) lichens on some rocky shores of Anglesey. Lichenologist 5:368–400

Fletcher A (1973b) The ecology of marine (supralittoral) lichens on some rocky shores of Anglesey. Lichenologist 5:401–422

Fletcher A (1976) Nutritional aspects of marine and maritime lichen ecology. In: Brown DH, Hawksworth DL, Bailey RH (eds) Lichenology: progress and problems. Academic, New York, pp 359–384

Fornazier RF, Ferreira RR, Vitória AP, Molina SMG, Lea PJ, Azevedo RA (2002) Effects of cadmium on antioxidant enzyme activities in sugar cane. Biol Plant 45:91–97

Frechilla S, Lasa B, Ibarretxe L, Lamsfus C, Aparicio Trejo P (2001) Pea response to saline stress is affected by the source of nitrogen nutrition (ammonium or nitrate). Plant Growth Regul 35:171–179

Friedl T, Büdel B (2008) Photobionts. In: Nash TH III (ed) Lichen biology. Cambridge University Press, Cambridge, pp 9–26

Fulda S, Hagemann M (1995) Salt treatment induces accumulation of flavodoxin in the cyanobacterium *Synechocystis* sp. PCC6803. J Plant Physiol 146:520–536

García-Molina F, Hiner ANP, Fenoll LG, Rodríguez-Lopez JN, García-Ruiz PA, García-Cánovas F, Tudela J (2005) Mushroom tyrosinase: catalase activity, inhibition, and suicide inactivation. J Agric Food Chem 53:3702–3709

Gasulla F, Gomez De Nova P, Esteban-Carrasco A, Zapat JM, Barreno E, Guéra A (2009) Dehydration rate and time of desiccation affect recovery of the lichenic algae *Trebouxia erici*: alternative and classical protective mechanisms. Planta 231:195–208

Gauslaa Y, Solhaug KA (1999) High-light damage in air-dry thalli of the old forest lichen *Lobaria pulmonaria* – interactions of irradiance, exposure duration and high temperature. J Exp Bot 50:697–705

Gauslaa Y, Solhaug KA (2001) Fungal melanins as a sun screen for symbiotic green algae in the lichen *Lobaria pulmonaria*. Oecologia 126:462–471

Gauslaa Y, Solhaug KA (2004) Photoinhibition in lichens depends on cortical characteristics and hydration. Lichenologist 36:133–143

Gauslaa Y, Ustvedt EM (2003) Is parietin a UV-B or a blue-light screening pigment in the lichen *Xanthoria parietina*? Photoch Photobio Sci 2:424–432

Gilbert O (2000) Lichens. Harper Collins, London

Gilbert O (2001) The lichen flora of coastal saline lagoons in England. Lichenologist 33:409–417

Gill SS, Tuteja N (2010) Reactive oxygen species and antioxidant machinery in abiotic stress tolerance in crop plants. Plant Physiol Bioch 48:909–930

González I, Ayuso-Sacido A, Anderson A, Genilloud O (2005) Actinomycetes isolated from lichens: evaluation of their diversity and detection of biosynthetic gene sequences. FEMS Microbiol Ecol 54:401–415

Gopalakrishnan M, Kumar AS, Mody K, Jha B (2006) Biosurfactant production by marine bacteria. Nat Acad Sci Lett 29:95–101

Gray DW, Lewis LA, Cardon ZG (2007) Photosynthetic recovery following desiccation of desert green algae (Chlorophyta) and their aquatic relatives. Plant Cell Environ 30:1240–1255

Green TGA, Sancho LG, Pintado A (2011) Ecophysiology of Desiccation/Rehydration Cycles in Mosses and Lichens. In: Lüttge U, Beck E, Bartels D (eds) Plant desiccation tolerance, ecological studies 215. Springer, Berlin, pp 89–120

Green TGA, Brabyn L, Beard C, Sancho LG (2012) Extremely low lichen growth rates in Taylor Valley, Dry Valleys, continental Antarctica. Polar Biol 35:535–541

Grube M (2010) Die hard: lichens. In: Seckbach J, Grube M (eds) Symbioses and stress: joint ventures in biology. Springer, Dordrecht, pp 509–523

Grube M, Berg G (2009) Microbial consortia of bacteria and fungi with focus on the lichen symbiosis. Fungal Biol Rev 23:72–85

Grube M, Blaha J (2005) Halotolerance and lichen symbioses. In: Gunde-Cimerman N, Oren A, Plemenitas A (eds) Adaptation to life at high salt concentrations in Archaea, Bacteria, and Eukarya. Springer, Dordrecht, pp 471–488

Grube M, Muggia L (2010) Identifying algal symbionts in lichen symbioses. In: Nimis PL, Lebbe RV (eds) Tools for identifying biodiversity: progress and problems. Proceedings of the International Congress (Museum National d'Histoire Naturelle), Paris, pp 295–299

Grube M, Cardinale M, Vieira de Castro J Jr, Müller H, Berg G (2009) Species-specific structural and functional diversity of bacterial communities in lichen symbioses. Intern Soc Microbial Ecol 3:1105–1115

Gueidan C, Thüs H, Pérez-Ortega S (2011) Phylogenetic position of the brown algae-associated lichenized fungus *Verrucaria tavaresiae* (Verrucariaceae). Bryologist 114:563–569

Hagemann M, Effmert U, Kerstan T, Schoor A, Erdmann N (2001) Biochemical characterization of glucosylglycerol-phosphate synthase of *Synechocystis sp.* strain PCC 6803: comparison of crude, purified, and recombinant enzymes. Curr Microbiol 43:278–283

Hasegawa PM, Bressan RA, Zhu JK, Bohnert HJ (2000) Plant cellular and molecular responses to high salinity. Ann Rev Plant Physiol Plant Mol Biol 51:463–499

Hawksworth DL (2000) Freshwater and marine lichen-forming fungi. Fungal Div 5:1–7

Healey FP (1973) Characteristics of phosphorus deficiency in *Anabaena*. J Phycol 9:383–394

Heber U, Bilger W, Turk R, Lange OL (2010) Photoprotection of reaction centres in photosynthetic organisms: mechanisms of thermal energy dissipation in dessicated thalli of the lichen *Lobaria pulmonaria*. New Phytol 185:459–470

Henzler T, Steudle E (2000) Transport and metabolic degradation of hydrogen peroxide in *Chara coralline*: model calculations and measurements with the pressure probe suggest transport of H_2O_2 across water channels. J Exp Bot 51:2053–2066

Hidalgo ME, Fernández E, Ponce M, Rubio C, Quilhot W (2002) Photophysical, photochemical, and thermodynamic properties of shikimic acid derivatives: calycin and rhizocarpic acid (lichens). J Photochem Photobiol B 66:213–217

Hill DJ, Ahmadjian V (1972) Relationship between carbohydrate movement and the symbiosis in lichens with green algae. Planta 103:267–277

Hodkinson BP, Gottel NR, Schadt CW, Lutzoni F (2011) Photoautotrophic symbiont and geography are major factors affecting highly structured and diverse bacterial communities in the lichen microbiome. Environ Microbiol 14:147–161

Honegger R (1998) The lichen symbiosis – what is so spectacular about it? Lichenologist 30: 193–212

Honegger R (2001) The symbiotic phenotype of lichen-forming ascomyces. In: Hock B (ed) The mycota; IX fungal associations. Springer, Berlin, pp 165–188

Incharoensakdi A, Takabe T (1988) Determination of intracellular chloride ion concentration in a halotolerant cyanobacterium *Aphanothece halophytica*. Plant Cell Physiol 29:1073–1075

Jakab G, Ton J, Flors V, Zimmerli L, Métraux JP, Mauch-Mani B (2005) Enhancing *Arabidopsis* salt and drought stress tolerance by chemical priming for its abscisic acid responses. Plant Physiol 139:267–274

Johnson AM (2007) Food abundance and energetic carrying capacity for wintering waterfowl on the Great Salt Lake wetlands. Oregon State University, Corvallis

Keddy PA (2007) Positive interactions: mutualism, commensalism, and symbiosis. In: Keddy PA (ed) Plant and vegetation: origins, processes, consequences. Cambridge University Press, Cambridge, pp 336–402

Kempf B, Bremer E (1998) Uptake and synthesis of compatible solutes as microbial stress responses to high-osmolality environments. Arch Microbiol 170:319–330

Killham K, Firestone MK (1984) Salt stress control of intracellular solutes in *Streptomycetes* indigenous to saline soils. Appl Environ Microb 47:301–306

Klähn S, Hagemann M (2011) Compatible solute biosynthesis in cyanobacteria. Environ Microbiol 13:551–562

Kneip C, Lockhart P, Voß C, Maier UG (2007) Nitrogen fixation in eukaryotes – New models for symbiosis. BMC Evol Biol 7:55

Koag MC, Fenton RD, Wilkens S, Close TJ (2003) The binding of maize DHN1 to lipid vesicles. Gain of structure and lipid specificity. Plant Physiol 131:309–316

Kogej T, Wheeler MH, Lanišnik Rižner T, Gunde-Cimerman N (2004) Evidence for 1,8-dihydroxynaphthalene melanin in three halophilic black yeasts grown under saline and non-saline conditions. FEMS Microbiol Lett 232:203–209

Kohlmeyer J, Kohlmeyer E (1979) Marine mycology. The higher fungi. Academic Press, New York

Kohlmeyer J, Volkmann-Kohlmeyer B (1988) *Halographis* (Opegraphales), a new endolithic lichenoid from corals and snails. Can J Bot 66:1138–1141

Kosugi M, Arita M, Shizuma R, Moriyama Y, Kashino Y, Koike H, Satoh K (2009) Responses to desiccation stress in lichens are different from those in their photobionts. Plant Cell Physiol 50: 879–888

Kotlova ER, Sinyutina NF (2005) Changes in the content of individual lipid classes of a lichen *Peltigera aphthosa* during dehydration and subsequent rehydration. Russ J Plant Physiol 52: 35–42

Kranner I (2002) Glutathione status correlates with different degrees of desiccation tolerance in three lichens. New Phytol 154:451–460

Kranner I, Lutzoni F (1999) Evolutionary consequences of transition to a lichen symbiotic state and physiological adaptation to oxidative damage associated with poikilohydry. In: Lerner HR (ed) Plant response to environmental stress: from phytohormones to genome reorganization. Dekker, New York, pp 591–628

Kranner I, Zorn M, Turk B, Wornik S, Beckett RP, Batic F (2003) Biochemical traits of lichens differing in relative desiccation tolerance. New Phytologist 160:167–176

Kranner I, Cram WJ, Zorn M, Wornik S, Yoshimura I, Stabentheiner E, Pfeifhofer HW (2005) Antioxidants and photoprotection in a lichen as compared with its isolated symbiotic partners. P Natl Acad Sci USA 102:3141–3146

Kranner I, Beckett R, Hochman A, Nash TH III (2008) Desiccation tolerance in lichen: a review. Bryologist 111:576–593

Kruger NJ, von Schaewen A (2003) The oxidative pentose phosphate pathway: structure and organisation. Curr Opin Plant Biol 6:236–246

Künzler K, Eichenberger W (1997) Betaine lipids and zwitterionic phospholipids in plants and fungi. Phytochemistry 46:883–892

Lagadic L, Caquet T, Amiard JC (1997) Biomarqueurs en écotoxicologie : principes et définitions. In: Lagadic L, Caquet T, Amiard JC (eds) Biomarqueurs en écotoxicologie. Masson, Paris, pp. 1–10

Lages F, Lucas C (1995) Characterization of a glycerol/H^+ symport in the halotolerant yeast *Pichia sorbitophila*. Yeast 11:111–119

Lages F, Lucas C (1997) Contribution to the characterization of glycerol active uptake in *Saccharomyces cerevisiae*. Bioch Biophys Acta – Bioenergetics 1322:8–18

Lages F, Silva-Graça M, Lucas C (1999) Active glycerol uptake is a mechanism underlying halotolerance in yeasts: a study of 42 species. Microbiology 145:2577–2585

Lange OL, Green TGA (2005) Lichens show that fungi can acclimate their respiration to seasonal changes in temperature. Oecologia 142:11–19

Lewitus AJ, Caron DA (1990) Relative effects of nitrogen or phosphorus depletion and light intensity on the pigmentation, chemical composition, and volume of *Pyrenomonas salina* (Cryptophyceae). Mar Ecol Prog Ser 61:171–181

Li X, Weinman SA (2002) Chloride channels and hepatocellular function: prospects for molecular identification. Annu Rev Physiol 64:609–633

Li Y, Trush MA, Yager JD (1994) DNA damage caused by reactive oxygen species originating from a copper-dependent oxidation of the 2-hydroxy catechol of estradiol. Carcinogenesis 15:1421–1427

Li B, Xie CH, Yokota A (2007) *Nocardioides exalbidus* sp. nov., a novel actinomycete isolated from lichen in Izu-Oshima Island, Japan. Actinomycetologica 21:22–26

Liao DI, Basarab GS, Gatenby AA, Valent B, Jordan DB (2001) Structures of trihydroxynaphthalene reductase-fungicide complexes: implications for structure-based design and catalysis. Structure 9:19–27

Liba CM, Ferrara FIS, Manfio GP, Fantinatti-Garboggini F, Albuquerque RC, Pavan C, Ramos PL, Moreira-Filho CA, Barbosa HR (2006) Nitrogen-fixing chemo-organotrophic bacteria isolated from cyanobacteria-deprived lichens and their ability to solubilize phosphate and to release amino acids and phytohormones. J Appl Microbiol 101:1076–1086

Lucas C, da Costa M, van Uden N (1990) Osmoregulatory active sodium-glycerol co-transport in the halotolerant yeast *Debaryomyces hansenii*. Yeast 6:187–191

Lüttge U (2011) Cyanobacteria: multiple stresses, desiccation-tolerant photosynthesis and dinitrogen fixation. In: Lüttge U, Beck E, Bartels D (eds) Plant desiccation tolerance, ecological studies 215. Springer, Berlin, pp 23–43

Mahajan S, Tuteja N (2005) Cold, salinity and drought stresses: an overview. Arch Biochem Biophys 444:139–158

Makkonen S, Hurri RSK, Hyvärinen M (2007) Differential responses of lichen symbionts to enhanced nitrogen and phosphorus availability: an experiment with *Cladina stellaris*. Ann Bot 99:877–884

Mao X, Olman V, Stuart R, Paulsen IT, Palenik B, Xu Y (2010) Computational prediction of the osmoregulation network in *Synechococcus sp*. WH8102. BMC Genomics 11:291

Maphangwa KW, Musil CF, Raitt L, Zedda L (2012) Differential interception and evaporation of fog, dew and water vapour and elemental accumulation by lichens explain their relative abundance in a coastal desert. J Arid Environ 82:71–80

Margesin R, Schinner F (2001) Potential of halotolerant and halophilic microorganisms for biotechnology. Extremophiles 5:73–83

Matias D, Zurbriggen MD, Tognetti VB, Fillat MF, Hajirezae MR, Valle EM, Carrillo N (2008) Combating stress with flavodoxin: a promising route for crop improvement. Trends Biotechnol 26:531–537

Matos P, Cardoso-Vilhena J, Figueira R, Sousa AJ (2011) Effects of salinity stress on cellular location of elements and photosynthesis in *Ramalina canariensis* Steiner. Lichenologist 43:155–164

Matthes-Sears U, Nash TH III, Larson DW (1987) Salt loading does not control CO_2 exchange in *Ramalina menziesii* Tayl. New Phytol 106:59–69

Miyasaka NRS, Thuler DS, Floh EIS, Handro W, Toledo MBD, Gagioti SM, Barbosa HR (2003) During stationary phase, *Beijerinckia derxii* shows nitrogenase activity concomitant with the release and accumulation of nitrogenated substances. Microbiol Res 158:309–315

Moe RL (1997) *Verrucaria tavaresiae* sp. nov., a marine lichen with a brown algal photobiont. Bull Californ Lichen Soc 4:7–11

Morbach S, Krämer R (2002) Body shaping under water stress: osmosensing and osmoregulation of solute transport in Bacteria. ChemBioChem 3:384–397

Morrison M, Schonbaum GR (1976) Peroxidase-catalyzed halogenation. Annu Rev Biochem 45:861–888

Motohashi K, Takagi M, Yamamura H, Hayakawa M, Shin-Ya K (2010) A new angucycline and a new butenolide isolated from lichen-derived *Streptomyces spp*. J Antibiot 63:545–548

Muggia L, Zellnig G, Rabensteiner J, Grube M (2010) Morphological and phylogenetic study of algal partners associated with the lichen-forming fungus *Tephromela atra* from the Mediterranean region. Symbiosis 51:149–160

Nash TH III, Lange OL (1988) Responses of lichens to salinity: concentration and time-course relationships and variability among Californian species. New Phytol 109:361–367

Okanenko A, Tarand N, Kosyk O (2011) Sulfurcontaining plant lipids. ABEГA, Kiev

Øvstedal DO, Tønsberg T, Elvebakk A (2009) The lichen flora of Svalbard. Sommerfeltia 33:1–393

Ozenda P, Clauzade G (1970) Les lichens: étude biologique et flore illustrée. Masson, Paris
Paramesha S, Vijay R, Bekal M, Kumari S, Pushpalatha KC (2011) A study on lipid peroxidation and total antioxidant status in diabetes with and without hypertension. Res J Pharm Biol Chem Sci 2:329–334
Parent C, Capelli N, Dat J (2008) Formes réactives de l'oxygène, stress et mort cellulaire chez les plantes. CR Biol 331:255–261
Pereira GJG, Molina SMG, Lea PJ, Azevedo RA (2002) Activity of antioxidant enzymes in response to cadmium in *Crotalaria juncea*. Plant Soil 239:123–132
Potts M, Slaughter SM, Hunneke FU, Garst JF, Helm RF (2005) Desiccation tolerance of Prokaryotes: application of principles to human cells. Integr Comp Biol 45:800–809
Rai V, Tiwari SP, Rai AK (2001) Effect of NaCl on nitrogen fixation of unadapted and NaCl-adapted *Azolla pinnata–Anabaena azollae*. Aquat Bot 71:109–117
Raikou V, Protopapa E, Kefala V (2011) Photo-protection from marine organisms. Rev Clin Pharmacol Pharmacokinet 25:131–136
Rancan F, Rosan S, Boehm K, Fernández E, Hidalgo ME, Quihot W, Rubio C, Boehm F, Piazena H, Oltmanns U (2002) Protection against UVB irradiation by natural filters extracted from lichens. J Photochem Photobiol B 68:133–139
Rodríguez-Navarro A (2000) Potassium transport in fungi and plants. Bioch Biophys Acta 1469: 1–30
Rorat T (2006) Plant dehydrins – tissue location, structure and function. Cell Mol Biol Lett 11: 536–556
Roullier C, Chollet-Krugler M, Van de Weghe P, Lohezic-Le Devehat F, Boustie J (2010) A novel aryl-hydrazide from the marine lichen *Lichina pygmaea*: isolation, synthesis of derivatives, and cytotoxicity assays. Bioorg Med Chem Lett 20:4582–4586
Roullier C, Chollet-Krugler M, Pferschy-Wenzig EM, Maillard A, Rechberger GN, Legouin-Gargadennec B, Bauer R, Boustie J (2011) Characterization and identification of mycosporine-like compounds in cyanolichens. Isolation of mycosporine hydroxyglutamicol from Nephroma laevigatum Ach. Phytochemistry 72:1348–1357
Rundel PW (1974) Water relations and morphological variation in *Ramalina menziesii* Tayl. Bryologist 77:23–32
Sanders WB, Moe RL, Ascaso C (2004) The intertidal marine lichen formed by the pyrenomycete fungus *Verrucaria tavaresiae* (Ascomycotina) and the brown alga *Petroderma maculiforme* (Phaeophyceae): thallus organization and symbiont interaction. Am J Bot 91:511–522
Schaich KM (2005) Lipid oxidation: theoretical aspects. In: Shahidi F (ed) Bailey's industrial oil and fat products. Wiley, Hoboken, pp 269–355
Schneider T, Schmid E, De Castro Jr JV, Cardinale M, Eberl L, Grube M, Berg G, Riedel K (2011) Structure and function of the symbiosis partners of the lung lichen (*Lobaria pulmonaria* L. Hoffm.) analyzed by metaproteomics. Proteomics 11:2752–2756
Schweikardt T, Olivares C, Solano F, Jaenicke E, García-Borrón JC, Decker H (2007) A three-dimensional model of mammalian tyrosinase active site accounting for loss of function mutations. Pigment Cell Res 20:394–401
Sekmen AH, Turkan I, Tanyolac ZO, Ozfidan C, Dinc A (2012) Different antioxidant defense responses to salt stress during germination and vegetative stages of endemic halophyte *Gypsophila oblanceolata* Bark. Environ Exp Bot 77:63–76
Serrano R, Mulet JM, Rios G, Marquez JA, de Larriona IF, Leube MP, Mendizabal I, Pascual-Ahuir A, Proft MRR, Montesinos C (1999) A glimpse of the mechanism of ion homeostasis during salt stress. J Exp Bot 50:1023–1036
Sibaja-Cordero JA, Troncoso JS (2011) Upper and lower limits of rocky shore organisms at different spatial scales and wave exposure (Islas CÍES, NW Spain). Thalassas 27:81–100
Smirnoff N (1993) The role of active oxygen in the response of plants to water deficit and desiccation. New Phytol 127:27–58
Smith VR, Gremmen NJM (2001) Photosynthesis in a sub-Antarctic shore-zone lichen. New Phytol 149:291–299

Smith CW, Aptroot A, Coppins BJ, Fletcher A, Gilbert OL, James PW, Wolseley PA (2009) The lichens of Great Britain and Ireland. The British Lichen Society, London

Solhaug KA, Gauslaa Y (1996) Parietin, a photoprotective secondary product of the lichen *Xanthoria parietina*. Oecologia 108:412–418

Solhaug KA, Gauslaa Y, Nybakken L, Bilger W (2003) UV-Induction of sun-screening pigments in lichens. New Phytol 158:91–100

Soriani FM, Kress MR, Fagundes de Gouvêa P, Malavazi I, Savoldi M, Gallmetzer A, Strauss J, Goldman MH, Goldman GH (2009) Functional characterization of the *Aspergillus nidulans* methionine sulfoxide reductases (msrA and msrB). Fungal Genet Biol 46:410–417

Takahagi T, Yamamoto Y, Kinoshita Y, Takeshita S, Yamada T (2002) Inhibitory effects of sodium chloride on induction of tissue cultures of lichens of *Ramalina* species. Plant Biotechnol 19: 53–55

Thompson JE, Legge RL, Barber RF (1987) The role of free radicals in senescence and wounding. New Phytol 105:317–344

Thuler DS, Floh EIS, Handro W, Barbosa HR (2003) *Beijerinckia derxii* releases plant growth regulators and amino acids in synthetic media independent of nitrogenase activity. J Appl Microbiol 95:799–806

Tsai HF, Wheeler MH, Chang YC, Kwon-Chung KJ (1999) A developmentally regulated gene cluster involved in conidial pigment biosynthesis in *Aspergillus fumigatus*. J Bacteriol 181: 6469–6477

Tschermak-Woess E (1988) The algal partner. In: Galun M (ed) CRC handbook of lichenology, vol 1. CRC Press, Boca Raton, pp 39–92

Van Zyl PJ, Prior BA, Kilian SG (1991) Regulation of glycerol metabolism in *Zygosaccharomyces rouxii* in response to osmotic stress. Appl Microbiol Biot 36:369–374

Ventosa A, Nieto JJ, Oren A (1998) Biology of moderately halophilic aerobic bacteria. Microbiol Mol Biol Rev 62:504–544

Williams DE, Davies J, Patrick BO, Bottriell H, Tarling T, Roberge M, Andersen RJ (2008) Cladoniamides A-G, tryptophan-derived alkaloids produced in culture by *Streptomyces uncialis*. Org Lett 10:3501–3504

Yocum CF (2008) The calcium and chloride requirements of the O_2 evolving complex. Coord Chem Rev 252:296–305

Chapter 5
Changes in Photosystem II in Response to Salt Stress

Anjana Jajoo

5.1 Introduction

The population of world is expected to increase from about 6 billion people in year 2000 to more than 10 billion in 2050. To feed this increasing population, the average world cereal yield will need to reach 5 $t\,ha^{-1}$ from its present 3 $t\,ha^{-1}$. At the same time, global climatic changes are posing more stressed conditions for the crops to grow and environmental stresses represent the most limiting factors for agricultural productivity. In general, abiotic stress often causes a series of changes at whole plant as well as molecular level that unfavorably affect growth, development and productivity of the plant. Abiotic stresses, singly or in combination, result in both general and specific detrimental effects on plant growth and development. A variety of distinct abiotic stresses exist, such as water (drought, flooding), high and low light, extreme temperature (chilling, freezing, heat), salinity, heavy metals (Allakhverdiev et al. 2008).

Looking to the severity of the decrease in crop yield caused by salinity (high salt stress) it is essential to understand physiology of the salt stressed plants. Knowledge of the type of damage that occurs to the plant during salt stress will direct research to design genetic modifications which may provide the plants with more salt tolerant machinery. This chapter particularly deals with the effects of salt stress on the process of photosynthesis the efficiency of which is responsible for the overall crop yield. If we can improve photosynthesis under salt stress conditions, definitely it will have a positive impact on the crop yield and production. Since in the process of photosynthesis, photosystem II is one of the most stress-susceptible components, we have focused more on it. We have also focused on the utility of chlorophyll a fluorescence induction kinetics to get quick information about the efficiency of photochemical reactions taking place in the leaf in the field conditions. This minireview presents a brief depiction

A. Jajoo (✉)
School of Life Science, Devi Ahilya University, Indore 452 017, MP, India
e-mail: anjanajajoo@hotmail.com

P. Ahmad et al. (eds.), *Ecophysiology and Responses of Plants under Salt Stress*,
DOI 10.1007/978-1-4614-4747-4_5,

and recent advances related to the effects of salt stress on photosynthesis, particularly, Photosystem II and shall be helpful to gain better understanding of the phenomenon of salinity stress and hence designing strategies to cope up with it. At the same time, Understanding the adaptive mechanisms exercised by the plants in natural stress conditions will help to improve salt tolerance in crop plants.

5.1.1 Environmental Stresses and Their Impact on Plant Growth

Plants are able to "sense" environmental changes and subsequently "respond" to stress. Most plants grow in environmental conditions that are, to a considerable degree, unfavorable to their growth.

Each environmental factor usually has a minimum and maximum level, beyond which plants cannot survive. Due to their stationary status, plants can only make metabolic and structural adjustment to cope with biological and non-biological trauma (Allahverdiyev et al. 2011).

Drought, salinity, extreme temperatures (cold and heat) and oxidative stress are often interrelated; these conditions singularly or in combination induce cellular damage. Due to the complex nature of stresses, not a single sensor, but multiple sensors may be responsible for perception of stress stimuli.

5.1.2 High Salt Stress

More than 800 million ha of land throughout the world are salt-affected (http://www.fao.org/ag/agl/agll/spush/). Growing concentration of salts in the rhizosphere have been because of natural causes such as salty raining waters near around the coasts, contamination from the parental rocks and oceanic salts and cultivation practices (Mahajan and Tuteja 2005). Increased salinisation of arable lands accompanied with water stress can result in 50% loss of arable lands by the year 2050. Sustainable salinization of arable land is getting more widespread and thus decreasing the yield from formerly productive soil everywhere in the world. Therefore, it requires increased water-use efficiency and salt tolerance for agricultural production in an ever-decreasing area of arable land (Tiburcio et al. 2012). Basically the term salinity implies high concentration of salts in soil, it is NaCl that constitutes the most part in soil salinity and that is why all plants have evolved several mechanisms to regulate NaCl accumulation or exclusion. High salinity causes both hyperionic and hyperosmotic stresses and finally affects the growth of the plant.

There is widespread occurrence of salt-affected soils and some plants are adapted (halophytes) to grow on such soils. However, most of our crops are salt-sensitive. As a consequence, salinity is a major threat to agriculture, especially in areas where secondary salinisation has developed through irrigation or deforestation. Attempts to improve the salt tolerance of crops have met with very limited success, due to the complexity of the genetic and physiological traits (Flowers and Flowers 2005).

5.2 General Effects of Salt Stress on Plants

Plant salt stress is a condition where excessive salts in soil solution cause inhibition of plant growth or plant death. No toxic substance has been found to restrict plant growth more as compared to salt, so salt stress presents an increasing threat to plant agriculture (Zhu 2007). Saline soil is characterized by toxic levels of chlorides and sulfates of sodium. The electrical conductivity of saturation extracts of saline soil is more than 4.0 dS/m (40 mM NaCl). The problem of soil salinity is increasing because of several reasons including the use of sea water for irrigation, improper drainage, salt accumulation in the root zone in arid and semi-arid regions due to high evaporative demand and insufficient leaching of ions as the rainfall is inadequate (Chinnusamy and Zhu 2003).

5.2.1 *Effects of High Salt Stress on Plant Growth*

Plants which have capacity to grow on highly saline environments are traditionally classified as glycophytes or halophytes (Flowers et al. 1977). Being the natural inhabitants of highly saline soils, halophytes efficiently excludes salts from their roots and leaves and some can tolerate salts that are more than twice the concentration of seawater. Salinity includes ionic stress (mainly due to Na^+, Cl^-, and SO_4^{2-}), osmotic stress, and secondary stresses such as nutritional imbalances and oxidative stress for glycophytes (Zhu 2002). Besides Na^+, some plant species are also sensitive to chloride, the major anion found in saline soils. High concentrations of Na^+ disturb osmotic balance and results in "physiological drought", preventing plant water uptake. Halophytic plants that are tolerant of sodium toxicity are probably inhibited by osmotic stress.

Salt stress affects several important processes such as growth, photosynthesis, protein synthesis, and energy and lipid metabolism (Parida and Das 2005; Abbaspour et al. 2012). Some of the major effects of salt stress on plant growth and the mechanism of tolerance of salt stress by plants have been summarized in Fig. 5.1. Salinity causes increases in epidermal thickness, mesophyll thickness, palisade cell length, palisade diameter, and spongy cell diameter in leaves of bean, cotton, etc. Salt stress has various effects on plant physiological processes such as increased respiration rate and ion toxicity, changes in plant growth, mineral distribution, and membrane instability resulting from calcium displacement by sodium, membrane permeability, and decreased efficiency of photosynthesis (Sudhir and Murthy 2004). Salt stress can lead to stomatal closure, which reduces CO_2 availability in the leaves and inhibits its carbon fixation, exposing chloroplasts to excessive excitation energy, which in turn result in generation of reactive oxygen species (ROSs) (Parida and Das 2005; Ahmad and Sharma 2008). To cope with the detrimental effects of salt stress, plants have evolved many biochemical and molecular mechanisms. Some of the biochemical strategies are (i) selective buildup or exclusion of salt ions, (ii) control of ion uptake by roots and transport into leaves, (iii) ion compartmentalization, (iv) synthesis

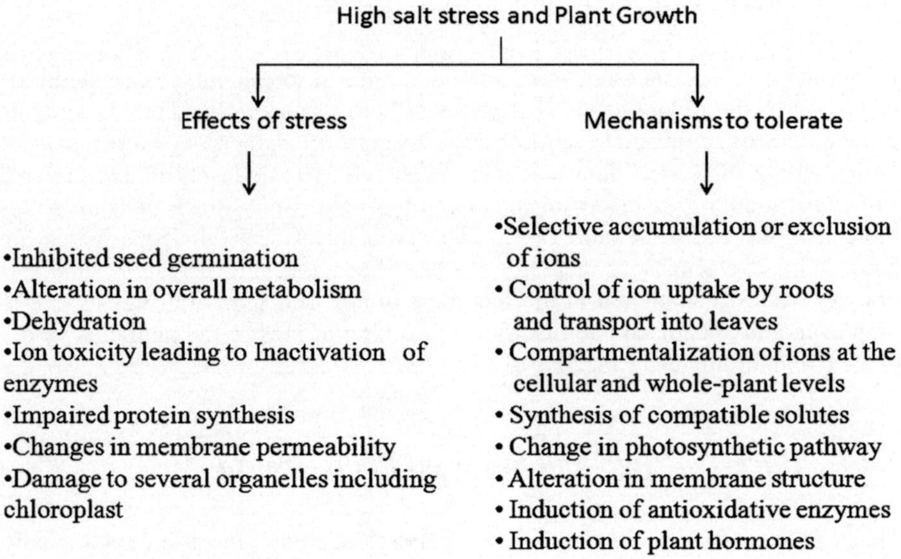

Fig. 5.1 Effects of salt stress on plant growth

of compatible osmolytes, (v) alteration in photosynthetic pathway, (vi) changes in membrane structure, (vii) induction of antioxidative enzymes and (viii) stimulation of phytohormones (Parida and Das 2005). Salinity is detrimental to plant growth as it causes nutritional constraints by decreasing uptake of phosphorus, potassium, nitrate and calcium, ion cytotoxicity and osmotic stress. Under salinity, ions like Na^+ and Cl^- penetrate the hydration shells of proteins and interfere with the function of these proteins. Ionic toxicity, osmotic stress, and nutritional defects under salinity lead to metabolic imbalances and oxidative stress (Chinnusamy and Zhu 2003).

Understanding the mechanisms of plant salt tolerance will lead to effective means to breed or genetically engineer salt-tolerant crops.

5.2.2 *Effects of High Salt Stress on Photosynthesis*

Several physiological processes contribute to limitation of plant growth by environmental factors. However, the dominant physiological process is photosynthesis. Plant growth as biomass production is a measure of net photosynthesis and, therefore, environmental stresses affecting photosynthesis also affect the growth and ultimately crop yield.

Photosynthesis is one of the most important metabolic processes in plants and its study provides information about the general "health" of plants. Photosynthesis serves as a global stress sensor in plants, algae and cyanobacteria. The modifications

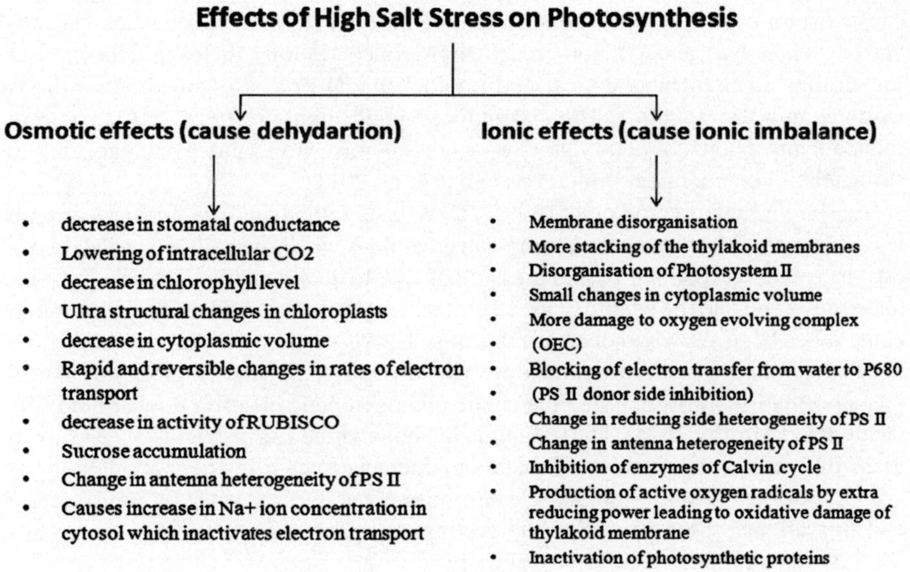

Fig. 5.2 Effects of salt stress including osmotic effects and ionic effects on the basic process of photosynthesis

of the chloroplast in response to various environmental stresses has been a topic of interest always. Photosynthesis occurs in two stages: the light reactions and the dark reactions. The chloroplast is the organelle where photosynthesis occurs in photosynthetic eukaryotes. Light reactions occur in the thylakoid stacks of the grana in the chloroplasts. Dark reactions occur in the stroma. Photosynthesis converts light energy into chemical energy via electron transport through pigment-protein complexes, Photosystem II (PSII) and Photosystem I (PSI). The light energy absorbed by pigments initiate the primary photochemical events followed by reactions that ultimately form the stable organic compounds. Since the two photosystems regulate photosynthetic efficiency and hence net productivity considerable attention has been paid to the effects of environmental stress on these photosystems.

Response of photosynthesis to drought and salinity stress is highly complex. It involves the interplay of limitations taking place at different sites of the cell/leaf and at different time scales in relation to plant development (Chaves et al. 2009). The stress is sensed at the levels of pigment composition, structural organization, primary photochemistry and the CO_2 fixation (Biswal et al. 2011; Shu et al. 2012; Mittal et al. 2012). Salt stress causes either short or long-term effects on photosynthesis. The short-term effect occurs after a few hours or within 1 or 2 days of the onset of exposure while long term effect arises after exposure to salt stress for several days.

Salt stress is exhibited in two ways: by causing a change in the osmoticum of the surroundings (osmotic stress) and by causing change in the ionic composition of the medium (ionic stress). As shown in Fig. 5.2, osmotic and ionic effects on photosynthesis are manifested in different ways. The accumulation of salt ions in plants can cause osmotic stress, ionic toxicity and induce nutritional deficiencies (Munns 2002). When Na^+ and Cl^- ions reach high concentrations in leaves, they cause impairment in both biochemical and photochemical processes of photosynthesis (Munns and Tester 2008). The salt-induced ionic toxicity effects are capable to induce acute photosynthetic damages (photochemistry and gas exchange) due to stomatal and biochemical limitations (Silva et al. 2011).

As shown by Downton et al. (1985), there is development of a thicker leaf with less chlorophyll per unit area during salt treatment which permitted stomatal conductance and intercellular partial pressure of CO_2 to decline without restricting photosynthesis and had the benefit of greatly increasing water use efficiency. Chlorophyll content has been shown to decrease in salt sensitive/susceptible plants as compared to salt-tolerant plants. At the same time salt stress increases the efficiency of photophosphorylation by stimulating the cyclic photosynthetic electron flow around PS1 (Sudhir and Murthy 2004). The reduction in photosynthetic rate has also been shown to be due to the reduction in stomatal conductance which restricts availability of CO_2 for carboxylation. Stomatal closure minimizes loss of water by transpiration and this affects chloroplast light-harvesting and energy-conversion systems thus leading to alteration in chloroplast activity.

Photosynthetic rate is lower in salt-treated plants, but the photosynthetic potential is not greatly affected when rates are expressed with regard to chlorophyll or leaf area. Decreases in photosynthetic rate are due to several factors: (1) dehydration of cell membranes which reduce their permeability to CO_2, (2) salt toxicity, (3) reduction of CO_2 supply because of hydroactive closure of stomata, (4) enhanced senescence induced by salinity, (5) changes of enzyme activity induced by changes in cytoplasmic structure (Parida and Das 2005). Electron microscopy has shown that the thylakoidal structure of the chloroplasts becomes disorganized, the number and size of plastoglobuli increases, and their starch content decreases in plants treated with NaCl (Hernandez et al. 1999). In the mesophyll of sweet potato leaves, thylakoid membranes of chloroplast are swollen and most are lost under severe salt stress (Mitsuya et al. 2000).

5.3 Effects of Salt Stress on Photosystem II

5.3.1 General Effects of High Salt Stress on PS II

The four major protein components of the photosynthetic electron transport chain are Photosystem II (PSII), Photosystem I (PSI), the cytochrome ($Cytb_6f$) complex, and ATP synthase. The majority of PS II reaction centers (RC) with their main light harvesting complex (LHC) II are located in the grana while Photosystem I (PSI) is localized in stroma-exposed thylakoid membranes. PS II is a multi-subunit complex whose function is to organize the chlorophylls for light harvesting and harbor the electron transport cofactors needed for the oxidation of water.

Effects of salt stress in cyanobacterium *Spirulina platensis* (Sudhir et al. 2005) showed a decrease in PS II mediated activity and an increase in PS I activity. It was ascribed to changes in the thylakoid membrane protein profile which led to the decreased energy transfer from light harvesting antenna to PS II. Salt adapted cells can maintain a high conversion efficiency of excitation energy through the down regulation of PS II RCs (Lu and Vonshak 2002). In the cyanobacterium, salt stress inhibits the apparent quantum efficiency of photosynthesis and photosystem II (PSII) activity while stimulating photosystem I (PSI) activity and dark respiration significantly. Salt stress also results in a decrease in overall activity of the electron transport chain (Lu and Vonshak 1999). Experimental evidence shows that at low salinity (100 mM) PSII mediated electron transport activity increases while a decrease in PS II activity is observed at high salinity in *B. parviflora* (Parida et al. 2003). High salt stress has a negative influence on PS II activity and the effect carried with the duration of stress application and on the cultivar used. In cynobacteria, under salt stress there is loss in chlorophyll protein (47 kDa) and a core membrane linker protein 94 kDa that can attach phycobilisome to thylakoid (Garnier et al. 1994). The 23 kDa protein which is extrinsically bound to PS II is also dissociated under salt stress (Sudhir et al. 2005). The short term stress leads to acclamatory changes in the functional aspects of PS II.

5.3.2 Effects of Salt Stress on Chl a Fluorescence Induction Kinetics

Chl *a* fluorescence kinetics is an informative tool for studying effects of different environmental stresses on photosynthesis (Kalaji et al. 2011). Due to its intricate connection with several processes taking place during the conversion of light due to its intricate connection with the numerous processes taking place during the energy conversion of light into a stable chemical form, Chl *a* fluorescence has proven to be an open window in the heart of the photosynthesis process (Papageorgiou and Govindjee 2004; Stirbet and Govindjee 2011). Chl *a* fluorescence is becoming a popular tool for plant management especially photosynthesis research. Chl *a* fluorescence originates mainly from PS II. Most interesting aspect of this technique is that chlorophyll fluorescence parameters start to fluctuate before other visual symptoms due to stress appear. Various parameters arising from fluorescence measurements can be exploited to gain information about status and efficiency of different components of PSII in Early stages of development and stress. When a dark-adapted photosynthetic sample is illuminated, Chl *a* fluorescence emission exhibits a fast rise to a maximum followed by a decline to a steady state over some minutes (Stirbet and Govindjee 2011). Analysis of the intermediate data points of the fast fluorescence rise forms the basis of the so-called "OJIP curve", whose shape is universal for all photosystems containing Chl *a*. The OJIP transient has the potential to be used for the characterization of the photochemical quantum yield of PS II photochemistry, and the electron transport activity. A representative OJIP curve has

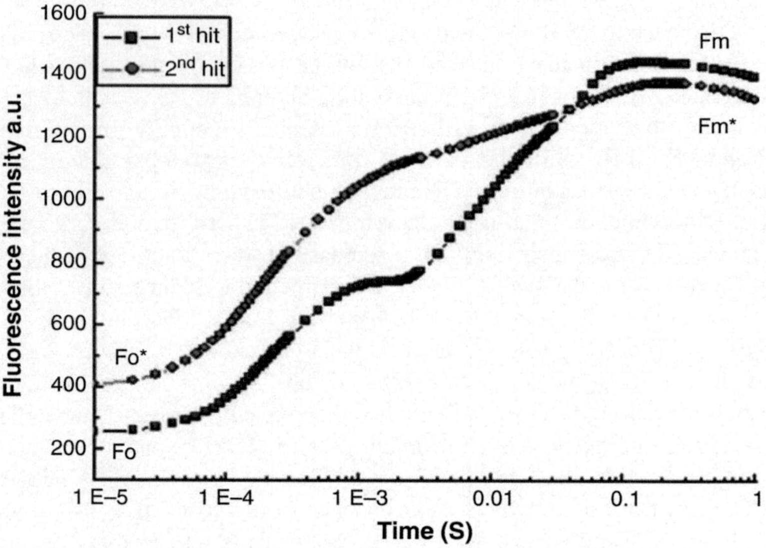

Fig. 5.3 A representative fluorescence induction curve (first hit) and a curve obtained after second hit. The graphs have time axes in logarithmic scale

been shown in Fig. 5.3 illustrating various intermediate phases. The OJIP transient is the fast Chl *a* fluorescence rise, measured at high light intensities [generally at 3000 μmol photons m-2 s-1]. The OJIP transient starts at O or Fo [minimum fluorescence, all Q_A is oxidized] and reaches a maximum called P or Fm (all Q_A is reduced) in ~200 ms. The intermediate steps are called J and I and they are situated at ~2 ms and ~30 ms. The O to J phase is due to the net photochemical reduction of Q_A to Q_A-. The intermediate I step and the final P step have been proposed to be due to existence of fast and slow reducing plastoquinone (PQ) pool, as well as due to different redox states of the reaction centers (RC) of PS II which reduces the PQ pool (Govindjee 1995; Haldimann and Strasser 1999). The popular JIP-test is a tool to analyse the polyphasic rise of the Chl *a* fluorescence transient and has been developed to investigate in vivo the "vitality" of plants and the adaptive behavior of the photosynthetic apparatus to different stresses (Christen et al. 2007) like high temperature, salinity stress (Mehta et al. 2010b). By measuring the yield of chlorophyll fluorescence, information about changes in the efficiency of photochemistry can be obtained (Maxwell and Johnson 2000). At a given moment, the physiological state of the sample determines the shape of the fluorescence transient of any sample. Diagnosis and early detection of various stresses using this non-invasive method is highly useful even more because field investigation can be conducted with high laboratory precision.

Polyphasic chlorophyll *a* fluorescence transient was measured to evaluate the effects of high salt stress on the photochemical efficiency of PS II. The OJIP transient represents the successive reduction of electron transport pool of PS II (Govindjee 1995). As evident in Fig. 5.4, the intensity of fluorescence in the induction curve

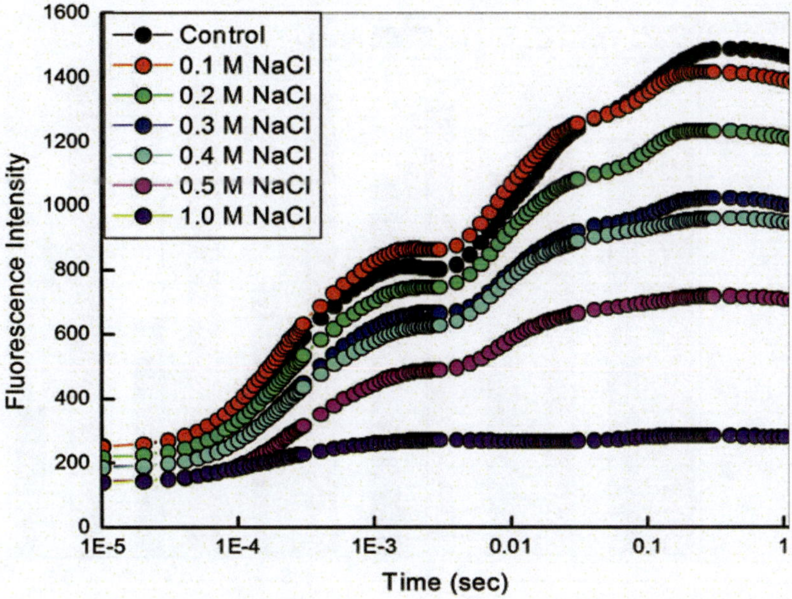

Fig. 5.4 The OJIP Chl *a* fluorescence transient curves (log time scale) in wheat leaves gradually exposed to various concentration of NaCl

decreased with increase in NaCl concentration. A significant decline in the minimal fluorescence (Fo), variable fluorescence (Fv) and maximal fluorescence (Fm) was observed with an increase in salt concentration. This decline in the fluorescence yield of leaves can be attributed to an inhibition of electron flow at oxidizing site of PS II (Lu and Vonshak 2002). The decrease in Fm and fluorescence at J, I, P has been suggested to be due to two reasons, first by inhibition of electron transport at the donor side of the PS II which results in the accumulation of P_{680}^{+} and second due to a decrease in the pool size of Q_A^-. Area over the fluorescence induction curve between Fo and Fm is proportional to the pool size of the electron acceptor Q_A on the reducing side of PS II. The area is dramatically reduced in case the electron transfer from reaction center to quinone pool is blocked. A decrease in area over the fluorescence curve with increase in NaCl concentration has been observed which suggests that high salt stress inhibits the electron transfer rates at the donor side of PS II. Fv/Fm ratio was not affected significantly in high salt treatment. ABS/RC i.e. effective antenna size of an active reaction centers, is influenced by ratio of active/inactive RCs and with increase in NaCl concentration the value of ABS/RC increased.

The kinetics of relative variable fluorescence (Vj) can give information about effects of high salt stress in electron transport chain on acceptor side of PS II. Vj is equivalent to (Fj – Fo/Fm – Fo) where Fj is the fluorescence at J step i.e. at 2 ms. Efficiency with which a trapped exciton can move an electron in to the electron transport chain further than Q_A^- (Ψo, which is calculated as ETo/TRo) was also measured. Increase in the value of Vj by 29% and a decrease in the value of Ψo by

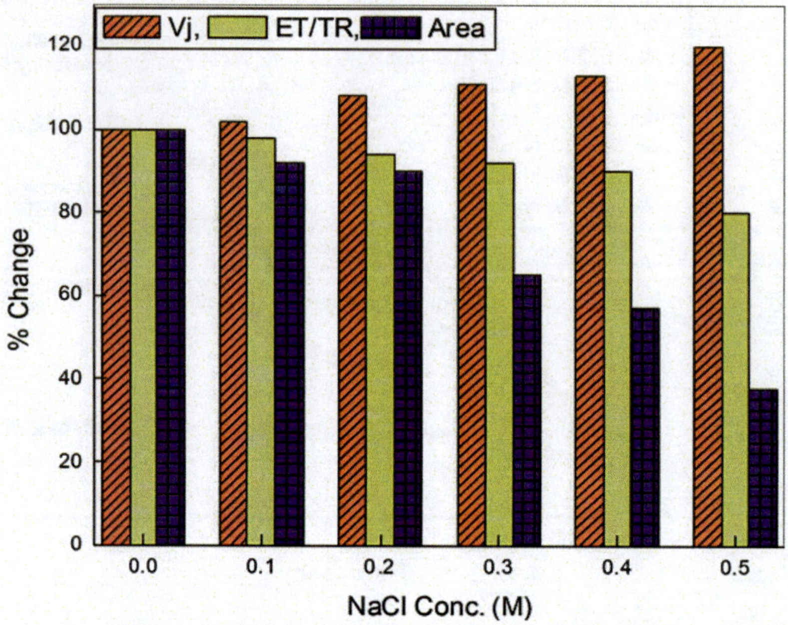

Fig. 5.5 Change (in %) variable fluorescence (Vj), efficiency of forward electron transfer (ETo/TRo) beyond Q_A and area over the curve in response to salt stress in wheat leaves

26% (Fig. 5.5) in 0.5 M NaCl treatment suggested a loss in reoxidation capacity of Q_A^- and an inhibition of electron transport at the acceptor side of PS II (Lu and Vonshak 1999) and also beyond Q_A^-. In recovery studies it was observed that the damage at acceptor side of PS II was recovered completely while damage at the donor side of PS II was recovered more than 80%. The rapid decline in photosynthesis under NaCl stress is reversible and specific to osmotic stress, where as the slow decline is irreversible and specific to ionic stress (Zhang and Xing 2008).

Another important parameter of JIP test is the performance index (PI) which is an indicator of sample vitality. It is the combined measurement of three functions: amount of photosynthetic reaction centers (RC/ABS), the maximal energy flux which reaches to the PS II reaction centers and the rate of electron transport at the onset of illumination. PI can be calculated as

$$PI_{ABS} = RC / ABS.\Phi_{Po} / (1 - \Phi_{Po}).\Psi_{Eo} / (1 - \Psi_{Eo})$$

Where Φ_{Po}, is the exciton trapped per photon absorbed and Ψ_{Eo}, is the probability that an electron can move further than Q_A^-. With increase in NaCl concentration a significant decrease in the value of performance index was observed and its value became half of the control in 0.5 M NaCl treatment (Mehta et al. 2010a). Since Fv/Fm ratio was not decreased significantly with high salt stress it can be said that high salt stress did not influence the number of quanta absorbed per unit time. The ratio $\Psi_{Eo}/(1 - \Psi_{Eo})$ decreased with increase in NaCl concentration and became 57% of the

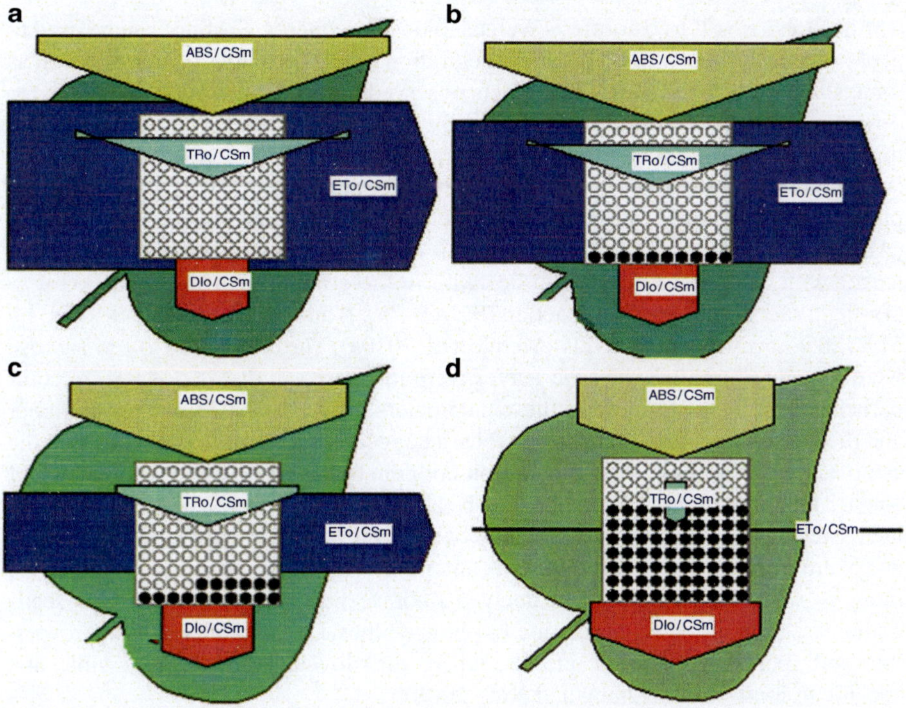

Fig. 5.6 Energy pipeline leaf model of phenomenological fluxes (per cross section, CS) in gradually stressed wheat leaves exposed to various concentrations of NaCl gradually. (**a**) Control (**b**) 0.3 M NaCl (**c**) 0.5 M NaCl and (**d**) 1 M NaCl. The value of each parameter can be seen in relative changes in width of each *arrow*. Active RCs are shown as *open circles* and inactive RCs are *closed circles*

control, suggesting that the efficiency of the forward electron transport rates were decreased. These results are in accordance with the data of ET/TR (Fig. 5.5).

It is possible to visualize the derived parameters by means of dynamic energy pipeline model of the photosynthetic apparatus (Krüger et al. 1997). The leaf model deals with the phenomenon-logical energy fluxes (per cross-section). Electron transport in a PS II cross section (ETo/CS) deals with the reoxidation of reduced Q_A via electron transport over a cross section of active and inactive RCs (Force et al. 2003). At various NaCl concentrations, a decrease in the electron transport per excited cross section (ETo/CS) due to inactivation of reaction center complex was observed. Density of the active reaction centers (RC/CS) reflects the number of active RCs in PS II cross section (indicated as open circles) which decreased with increase in salt concentration (Fig. 5.6). A decrease in RC/CS ratio suggests that the active RCs are converted into inactive RCs. ABS/CS is the number of photons absorbed by an excited PS II cross section (Force et al. 2003). At high salt concentration, a decrease in the energy absorbed per excited cross section (ABS/CS) was observed indicating that the energy absorption efficiency of PS II was decreased with increase in salt concentration.

Salt stress involves osmotic as well as ionic components. Sorbitol can be used to give osmotic stress while high salt concentration (NaCl) provides osmotic as well as ionic stress. A comparison of the effects observed in these two cases will help us to differentiate between the effects caused by osmotic and ionic components of the high salt stress. Osmotically, 0.5 M NaCl should behave as 1 M sucrose. Hyperosmotic conditions cause an efflux of water through water channel thereby decreasing cytoplasmic volume and reversibly inactivating the photosynthetic machinery (Allakhverdiev and Murata 2008). It was observed that the treatment of wheat leaves with 1 M sucrose caused a decrease in the efficiency of light reaction $\Phi_{Po}/(1-\Phi_{Po})$, rate of biochemical reaction ($\Psi_{Eo}/(1-\Psi_{Eo})$) and performance index (PI) by 11%, 20% and 30% respectively (Mehta et al. 2010a). The effects due to 1 M sucrose were recovered totally when the leaves were immersed in distilled water. In comparison, the effects observed in these parameters in 0.5 M NaCl were much higher and the effects were not totally reversible. It suggests that the effects observed in the samples treated with 0.5 M NaCl exhibit both the osmotic and ionic components of NaCl. The initial reversible effects may be ascribed to the osmotic aspects while the later, irreversible effects may be because of the ionic aspects of NaCl (Allakhverdiev and Murata 2008). These results are in contention with earlier studies in *Arabidopsis thaliana* where measurement of delayed fluorescence in high salt stressed seeds demonstrated that the rapid decline in photosynthesis under NaCl stress is reversible and specific to osmotic effects, where the slow decline is irreversible and specific to ionic stress (Zhang and Xing 2008).

Thus according to Mehta et al. (2010a, b) high salt stress inhibits the electron transport rates by ~75% at the donor and by ~25% at the acceptor side of PS II. As compared to acceptor side, the donor side of PS II is significantly affected by high salt stress. Inactive PS II centers increased with increasing salt concentration. Complete recovery of the damage caused at the acceptor side was observed, while damage to donor side could be recovered by more than 80%.

5.3.3 *Effects of High Salt Stress on Heterogeneity of PS II*

It is well established that PSII of higher plants is not homogenous in nature (Lavergne and Briantais 1996). The PSII varies in its structure and function both and this diverse nature of PSII is known as Photosystem heterogeneity. The concept of PSII heterogeneity originated in order to explain the biphasic nature of the kinetics of primary PSII activity. Two main types of PSII heterogeneity have been studied widely i.e., PSII antenna heterogeneity and PSII reducing side heterogeneity. On the basis of the differences in the antenna size the concept of α, β and γ centers has been introduced while on the basis of acceptor/reducing side function, Q_B-reducing and Q_B-non-reducing centers have been defined. Extent and nature of PSII heterogeneity may vary under different physiological conditions (Tongra et al. 2011; Lavergne and Briantais 1996) i.e. salinity stress, temperature stress (Mathur et al. 2010a, b), etc.

On the basis of the biphasic data obtained from fluorescence kinetics, presence of two distinct populations of PSII centers (termed as PSIIα and $PSII_\beta$) in the chloroplast was suggested (Melis and Homann 1976; Melis and Duysens 1979; Black et al. 1986). In a step-wise process in the development of PSII units, the addition of about 80 Chl to LHC II-inner portion of the antenna of $PSII_\gamma$ (contains ~ 50 Chl) yields $PSII_\beta$ (~ 130 Chl). In a second step, the addition of another 80chl to LHC II-peripheral part increases the antenna size to yield $PSII_\alpha$ [20]. The dominant form, PSII α, is localized in the grana partition regions [Andersson and Melis 1983] and is responsible for the majority of the water oxidation activity and plastoquinone reduction. These centers possess a Chl a core complex, an accessory Chl *a-b* light harvesting inner antenna (LHC II-inner), and a peripheral antenna (LHC II-peripheral) containing a combined total of about 210–250 Chl a and Chl b molecules (Morrissey et al. 1989). These have a higher absorption cross-section area due to association with the peripheral Chla/b LHCs. $PSII_\alpha$ are characterized by a large light harvesting antenna and excited states transfer between PSII units is possible in them as exhibited by the sigmoidal fluorescence rise when measured with DCMU. In contrast, $PSII_\beta$ are mainly located in stromal region of thylakoid membranes and are characterized by smaller light harvesting antenna of $PSII_\alpha$ and there is no possibility of the excited states transfer between PS IIs as reflected in an exponential fluorescence rise in the presence of DCMU. Smaller antenna size has been ascribed to the absence of peripheral LHC II in PSII. The α and β centers are similar in terms of their intrinsic trapping and fluorescence properties (Melis 1991). However they show difference in the recombination rate (reopening) in the α and β centers (Melis and Homann 1976). In addition to differences in their lateral location, the two types of PSII differ in terms of their kinetic properties, apparent mid-point potential of their primary electron acceptors, connectivity to the plastoquinone pool and their DCMU sensitivity as well (Sundby et al. 1986). The $PSII_\gamma$ is localized in stroma lamellae region, has the smallest antenna size among the three components and has the longest lifetime. $PSII_\alpha$ is believed to be the major 'normal' PSII centers whereas $PSII_\beta$ and $PSII_\gamma$ represent the two minor groups of 'abnormal' PSII centers with low quantum efficiencies due to their slow electron donation systems. The slow rate of $PSII_\beta$ and PSII γ as compared to the α center might be due to slow electron donation to their reaction center, which might undergo many turnovers via back reaction under continuous excitation, until their reduced primary acceptors were stabilized by the electron donation into the system. Some important characteristics of PS IIα, β and γ centres are shown in Fig. 5.7.

In addition to heterogeneity on the antenna size, PSII centers also display heterogeneity related to the reducing side of Q_A^- in relation to electron flow to the plastoquinone pool. It has been shown that a number of PS II centers, though photochemically competent, are unable to transfer electrons efficiently from electron acceptor Q_A^- to secondary electron acceptor Q_B (Lavergne 1982; Graan and Ort 1986; Guenther et al. 1988). These centers are termed as PS II Q_B-non-reducing using Lavergne's nomenclature (Lavergne 1982). In such centers Q_A^- can be reoxidized only by a back reaction with the donor side of PSII (Schanker and Strasser 2005). Q_B-non-reducing differs from Q_B reducing center in being incapable of

PSII α :
▪dominant form
▪is localized in the grana partition regions.
▪characterized by a large light harvesting antenna (210-250 Chl *a* and *b*)
▪exhibit possibility of excited states transfer between PSII units that is reflected in a sigmoidal fluorescence rise when measured with DCMU.

PSIIβ :
▪are mainly located in stromal region of thylakoid membranes
▪are characterized by ~2.5 times smaller light harvesting antenna of PSIIα.
▪impossibility of the excited states transfer between PS IIs that is reflected in an exponential fluorescence rise when measured with DCMU.

PSIIγ:
▪antenna size of center is supposed to be very very small center. And they are photochemically inactive.

Fig. 5.7 Important characteristics of PS II alpha (α) and beta (β) and gamma (γ) centers

reducing the PQ pool. Q_B-non-reducing is normally either equal to $PSII_\beta$ or a subset of it (Guenther et al. 1988). It is found that the electron transfer from the primary acceptor Q_A to PQ is more than 1,000 times slower in the inactive centers as compared to the active centers (Hsu 1992).

The biological processes involved in adaptation to stress in plants are complex. Changes in several biological macromolecules including proteins and lipids are involved in these adaptive mechanisms. Rearrangement in PS II seems to be an adaptive mechanism of plants to tolerate stress conditions.

To evaluate the effect of high salt stress on reducing side and antenna size heterogeneity of Photosystem II, the chlorophyll *a* fluorescence transient curves were measured in the absence and presence of DCMU.

5.3.3.1 Effect of High Salt Stress on Reducing Side Heterogeneity

Reducing side heterogeneity of PSII, relative amount of Q_B reducing and Q_B-non-reducing centers was measured by the double-hit method as described in Strasser and Tsimilli 1998. According to this method, fluorescence measurement is induced by two subsequent pulses (each of 1 s). The first pulse (denoted as first hit) is given after a dark period which is long enough to ensure the reopening of all reaction centers. It is followed by a second pulse (denoted as second hit) and the duration of the dark interval between two hits was 500 ms. As shown in Table 5.1, the amount of Q_B non-reducing centers increase in salt stressed leaves. In control leaves, the Q_B non-reducing centers were found to be 13% which became 31% in 0.5 M salt treatment.

Table 5.1 Amounts of Q_B non reducing and Q_B reducing centers in response to high salt stress in wheat leaves

NaCl Concentration (M)	% of Q_B non-reducing centers	% of Q_B reducing centers
Control	13 ± 1	87 ± 2
0.1	14 ± 1	86 ± 2
0.2	20 ± 1	80 ± 2
0.3	22 ± 1	78 ± 1
0.4	24 ± 1	76 ± 2
0.5	31 ± 1	69 ± 3
Recovery	31 ± 1	69 ± 2
1.0	34 ± 1	66 ± 3
Recovery	33 ± 1	67 ± 3

In leaves treated with 0.5 M NaCl no recovery was observed and the number of non Q_B reducing centers were almost the same as in the 0.5 M NaCl treated leaves. This result suggests that the damage at the reducing side of PS II was permanent.

5.3.3.2 Effects of High Salt Stress on Antenna Heterogeneity

The kinetics of complementary area of DCMU treated fluorescence induction curve was calculated by the equation $[B = \int(Fm - Ft)dt]$, where B is the double normalized (between 0 and 1) kinetics of complementary area (Strasser et al. 2000). The B kinetics of the first light pulse was fitted with three exponentials which correspond to three different types of PS II centres (PS II α, β and γ centers) differentiated on the basis of their lifetimes. As shown in Fig. 5.8, in wheat leaves grown in control conditions, the lifetime of the fastest α component was found to be 0.41 ms, contributing to 71% of the total amplitude. The β component was about 3.8 fold slower (life time ~1.34 ms) and contributed to about 27% of the total amplitude. The γ component had longest lifetime (8.79 ms) and contributed only to 2% of the total amplitude in control leaves. With increase in salt concentration the percentage of α centers decrease while that of β and γ centers increase. The relative ratio of α: β: γ centers in control leaves was 71:27:2 while it becomes 33:40:25 in salt stressed leaves (1 M NaCl). It was observed in recovery studies that changes caused by 0.5 M NaCl was almost recoverable but those caused by 1 M NaCl were partially recoverable. These results indicate that the damage caused due to high salt stress in antenna size heterogeneity were not permanent but temporary and largely reversible, suggesting that the α, β and γ centers were interconvertible to most extent.

It is known that the relative variable fluorescence $[(Ft - Fo)/(Fm - Fo)]$ is directly proportional to the number of closed RCs as well as linearly related to the rate at which centers close. This has been explained by the connectivity of PS II units. $PSII_\alpha$ showed a non-exponential (sigmoid) rise while $PSII_\beta$ were characterized by an exponential rise of the time course of complementary area (CA) whereas (Melis and Homann 1976). The non-exponential fluorescence rise of $PSII_\alpha$ is generally reflects energetic connectivity between these PSIIs while the exponential rise for

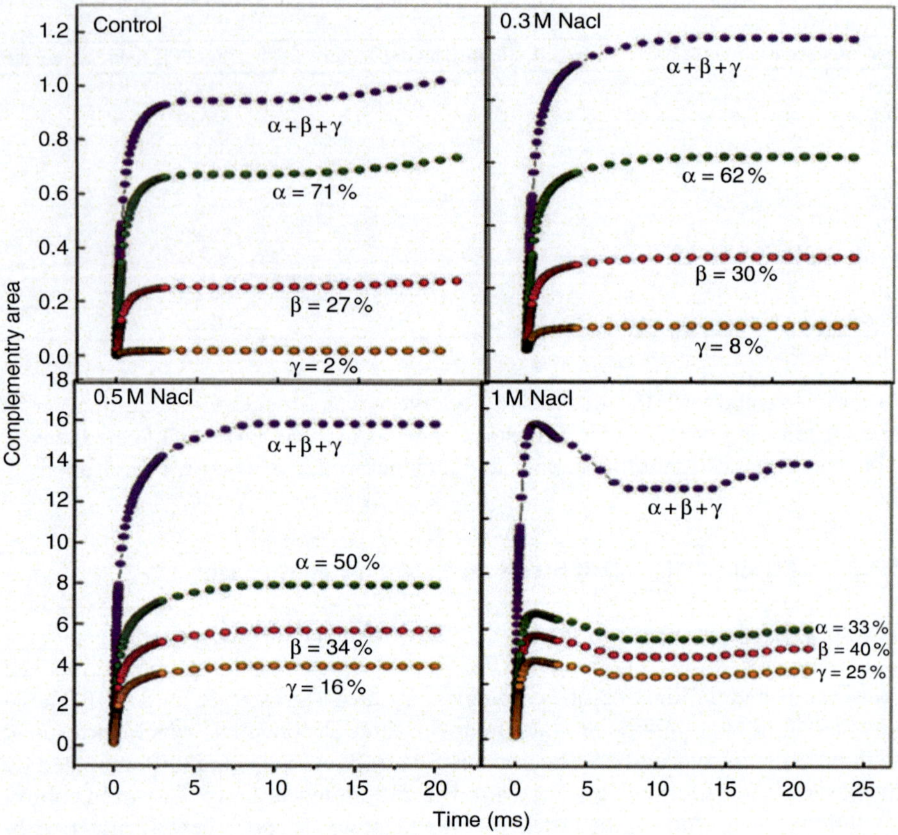

Fig. 5.8 Complementary area curves (linear time scale) showing percentage of alpha (α) and beta (β) and gamma (γ) centers in control and salt-treated wheat leaves

$PSII_{\beta}$ reflected the mutual energetic separation of these PSII. In salt stressed leaves, the α component (sigmoidal phase) of chlorophyll *a* fluorescence induction curve decreased while the β component (exponential phase) increased (Mehta et al. 2010b). A loss in connectivity also reflects that the fraction of closed RCs i.e. Q_B-non-reducing centers has increased (Strasser and Tsimilli 1998).

The energy pipeline models of the photosynthetic apparatus also help to study the antenna and reducing side heterogeneity of PS II (Krüger et al. 1997; Strasser 1987; Strasser et al. 2000) which help to calculate specific energy fluxes. The parameter ABS/RC demonstrates average antenna size and expresses the total absorption of PSII antenna chlorophylls divided by the number of active (in the sense of Q_A reducing) reaction centers. The parameter TRo/RC refers only to the active (Q_A to Q_A^-) centers (Force et al. 2003). Under high salt stress, the flux ratios ABS/RC, TRo/RC and DIo/RC increased (Mehta et al. 2010b). The ratio of ABS/RC seems to have increased due to inactivation of some active RCs. TRo/RC which represents the maximal rate by which an exciton is trapped by the RC resulting in the reduction of Q_A, increased indicating that all the Q_A has been reduced but it is not able to oxidize back due to stress. It also means that under high salt stress, the reoxidation of Q_A^- is inhibited so that Q_A cannot transfer electrons efficiently to Q_B. The parameter DIo/RC reflects the ratio of the dissipation of untrapped excitation energy from all RCs with respect to the number of active RCs. Dissipation may occur in various ways such as heat, fluorescence and energy transfer to other systems. It is also influenced by the ratios of active/inactive RCs. The ratio of total dissipation to the amount of active RCs (DIo/RC) increased due to the high dissipation of energy from the active RCs. All these energy flux ratios conclusively suggest that the number of inactive centers have increased due to high salt stress in wheat leaves.

Thus it is concluded that increase in salt concentration caused an alteration in antenna and reducing side heterogeneity of PS II. An increase in the relative amounts of Q_B non-reducing centers and a change in the relative amounts of α, β and γ centers were caused by an increase in the salt concentration. Salt stress led to the conversion of the active α or center into inactive β and γ centers. Recovery studies suggested that the change in antenna size was recovered while the changes in reducing side heterogeneity could not be recovered.

5.4 Conclusion and Future Perspectives

Light and dark reaction of photosynthesis are inhibited by salt stress. High salt stress inhibited the electron transport rates at the donor side of PS II by ~75% while the acceptor side was inhibited by ~25%. Thus under high salt stress, the donor side of PS II is more significantly affected as compared to acceptor side. Inactive PS II centers increase with increasing salt concentration. Most of the damage caused by high salt stress is recovered when the normal conditions were restored. Increase in salt concentration cause an alteration in PS II heterogeneity as well. Increase in salt concentration cause a change in the relative amounts of α, β and γ centers and an increase in the relative amounts of Q_B non-reducing centers. High salt stress leads to the conversion of the active α centers into inactive β and γ centers. The changes in antenna heterogeneity are recovered while changes in reducing side heterogeneity are not recovered while that of. Alteration in heterogeneity of PS II seems to be an adaptive mechanism of plants to face harsh environmental conditions. The structure and function of PS II is manipulated temporarily under high salt stress in the form of change in heterogeneity. Following are the unresolved areas which need to be addressed in the future:

1. Most of the stresses are inter-related and their relationships are complex. Synergistic effects of various stresses are known, for ex. salt stress inhibits the repair of PSII from light induced damage. Such relationships need to be examined in detail and with various stress combinations.
2. Normally, individual stress is monitored in lab conditions that obviously do not simulate the conditions in the field where plant experiences multiple stresses

simultaneously. However in field studies it is difficult to distinguish between effects of individual stress. Study of individual stress in lab and in field conditions should be carried out, interpreted and differentiated very cautiously.
3. Further research to differentiate between the osmotic and ionic components of salt stress is required.

Acknowledgement Department of Science and Technology (DST) is thankfully acknowledged for the Project (INT/RFBR/P-98). Author thanks Ms. Pooja Mehta for her help with the figures.

References

Abbaspour H, Afshari H, Abdel-Wahhab A (2012) Influence of salt stress on growth, pigments, soluble sugars and ion accumulation in three pistachio cultivars. J Medicinal Plants Res V 6(12):2468–2473

Ahmad P, Sharma S (2008) Salt stress and phytobiochemical responses of plants – a review. Plant Soil Environ 54:89–99

Allahverdiyev S, Atilla A, Ismail BS, Sahmurova A (2011) Response of photosystem II and photosynthetic pigments to salt and Baikal EM1 in tree seedlings. A J Biotechnol 10(4):535–538

Allakhverdiev SI, Murata N (2008) Salt stress inhibits photosystem II and I in cynobacteria. Photosynth Res 98:529–539

Allakhverdiev SI, Kreslavski VD, Klimov VV, Los DA, Carpentier R, Mohanty P (2008) Heat stress: an overview of molecular responses in photosynthesis. Photosynth Res 98:541–450

Andersson JM, Melis A (1983) Localization of different photosystems in separate regions of chloroplast membranes. Proc Natl Acad Sci USA 80:745–749

Biswal B, Joshi PN, Raval MK, Biswal UC (2011) Photosynthesis, a global sensor of environmental stress in green plants: stress signalling and adaptation. Current Sci 101:47–56

Black MT, Brearley TH, Horton P (1986) Heterogeneity in chloroplast photosystem II. Photosynth Res 89:193–207

Chaves MM, Flaxes J, Pinheiro C (2009) Photosynthesis under drought and salt stress regulation mechanism from whole plant to cell. Ann Bot 103:551–556

Chinnusamy V, Zhu JK (2003) Plant salt tolerance. Topics in current genetics. In: Hirt K, Shinozaki (eds) Plant responses to abiotic stress, vol 4 H. Springer, Berlin/Heidelberg, pp 241–270

Christen D, Schőnmanna S, Jermini M, Strasser RJ, Defago G (2007) Characterization and early detection of grapevine [*Vitis Vinifera*] stress responses to esca disease by *in situ* chlorophyll fluorescence and comparison with drought stress. Environ Exp Bot 60:504–514

Downton WJS, Grant WJR, Robinson SP (1985) Photosynthetic and stomatal responses of spinach leaves to salt stress. Plant Physiol 77:85–88

Flowers TJ, Flowers SA (2005) Why does salinity pose such a difficult problem for plant breeders? Agri Water Management 78:15–24

Flowers TJ, Troke PF, Yeo AR (1977) The mechanism of salt tolerance in halophytes. Annu Rev Plant Physiol 28:89–121

Force L, Critchley C, van Rensen JS (2003) New fluorescence parameters for monitoring photosynthesis in plants. Photosynth Res 78:17–33

Garnier F, Dubacq JP, Thomas JC (1994) Evidence for a transient association of new proteins with the *Spirulina* maxima phycobilisomes in relation to light intensity. Plant Physiol 106:747–754

Govindjee (1995) Sixty-three years since Kautsky: chlorophyll *a* fluorescence. Aust J Plant Physiol 22:131–160

Graan T, Ort DR (1986) Detection of oxygen-evolving photosystem II centers inactive in plastoquinone reduction. Biochim Biophys Acta 852:320–330

Guenther JE, Nemson JA, Melis A (1988) Photosystem stoichiometry and chlorophyll antenna size in *Dunaliella salina* (green algae). Biochim Biophys Acta 934:108–117

Haldimann P, Strasser RJ (1999) Effects of anaerobiosis as probed by the polyphasic Chl *a* fluorescence rise kinetics in pea. Photosynth Res 62:67–83

Hernandez JA, Campillo A, Jimenez A, Alacon JJ, Sevilla F (1999) Response of antioxidant systems and leaf water relations to NaCl stress in pea plants. New Phytol 141:241–251

Hsu BD (1992) The active photosystem II centers can make a significant contribution to the initial fluorescence rise from Fo to Fi. Plant Sci 81:169–174

Kalaji HM, Govindjee, Bosa K, Kos'cielniak J, Gołaszewska KZ (2011) Effects of salt stress on photosystem II efficiency and CO_2 assimilation of two Syrian barley landraces. Env Exp Bot 73:64–72

Krüger GHJ, Tsimilli MM, Strasser RJ (1997) Light stress provokes plastic and elastic modification in structures and function of photosystem II in *camellia* leaves. Physiol Plant 101:265–277

Lavergne J (1982) Two types of primary acceptors in chloroplasts photosystem II. Photobiochem Photobiophys 3:257–285

Lavergne J, Briantais JM (1996) Photosystem II heterogeneity. In: Ort RD, Yocum CF (eds) Oxygeneic photosynthesis: the light reactions. Kluwer, Dordrecht, pp 265–287

Lu C, Vonshak A (1999) Characterization of PS II photochemistry in salt-adapted cells of cyanobacterium *Spirulina platensis*. New Phytol 141:231–239

Lu C, Vonshak A (2002) Effects of salinity stress on photosystem II function in cyanobacterial *Spirulina platensis* cells. Physiol Plant 114:405–413

Mahajan S, Tuteja N (2005) Cold, salinity and drought stresses: an overview. Arch Biochem Biophys 444:139–158

Mathur S, Jajoo A, Mehta P, Bharti S (2010a) Analysis of elevated temperature-induced inhibition of photosystem II by using chlorophyll *a* fluorescence induction kinetics in wheat leaves (*Triticum aestivum*). Plant Biol 13:1–6

Mathur S, Allakhverdiev SI, Jajoo A (2010b) Analysis of high temperature stress on the dynamics of antenna size and reducing side heterogeneity of photosystem II in wheat (*Triticum aestivum*). Biochim Biophys Acta 1807:22–29

Maxwell K, Johnson GN (2000) Chlorophyll fluorescence – a practical guide. J Exp Bot 51: 659–668

Mehta P, Allakhverdiev SI, Jajoo A (2010a) Characterization of photosystem II heterogeneity in response to high salt stress in wheat leaves (*Triticum aestivum*). Photosynth Res 105:249–255

Mehta P, Jajoo A, Mathur S, Bharti S (2010b) Chlorophyll *a* fluorescence study revealing effects of high salt stress on photosystem II in wheat leaves. Plant Physiol Biochem 48:16–20

Melis A (1991) Dynamics of photosynthetic membrane composition and function. Biochim Biophys Acta 1058:87–106

Melis A, Duysens LMN (1979) Biphasic energy conversion kinetics and absorbance difference spectra of photosystem II of chloroplasts. Evidence for two different photosystem II reaction centers. Photochem Photobiol 29:373–382

Melis A, Homann PH (1976) Heterogeneity of the photochemical centers in system II of chloroplasts. Photochem Photobiol 23:343–350

Mitsuya S, Takeoka Y, Miyake H (2000) Effects of sodium chloride on foliar ultrastructure of sweet potato (*Ipomoea batatas Lam.*) plantlets grown under light and dark conditions *in vitro*. J Plant Physiol 157:661–667

Mittal S, Kumari N, Sharma V (2012) Differential response of salt stress on Brassica juncea: photosynthetic performance, pigment, proline, D1 and antioxidant enzymes. Plant Physiol Biochem 54:17–26

Morrissey PJ, Glick RE, Melis A (1989) Supramolecular assembly and function of subunits associated with the chlorophyll a/b light-harvesting complex II (LHC II) in soybean chloroplasts. Plant Cell Physiol 30:335–344

Munns R (2002) Comparative physiology of salt and water stress. Plant Cell Environ 25:239–250

Munns R, Tester M (2008) Mechanisms of salinity tolerance. Ann Rev Plant Biol 59:651–681
Papageorgiou GC, Govindjee (2004) Chlorophyll *a* Fluorescence: a signature of photosynthesis. In: Advances in photosynthesis and respiration. Springer, Dordrecht
Parida AK, Das AB (2005) Salt tolerance and salinity effects on plants: a review. Ecotoxicol Environ Saf 60:324–349
Parida AK, Das AB, Mittra B (2003) Effects of NaCl stress on the structure, pigment complex composition and photosynthetic activity of mangrove *Bruguiera parviflora* chloroplasts. Photosynthetica 41:191–200
Schanker G, Strasser RJ (2005) Quantification of non-QB-reducing centers in leaves using a far-red pre-illumination. Photosynth Res 84:145–151
Shu S, Guo SR, Sun J, Yuan LY (2012) Effects of salt stress on the structure and function of the photosynthetic apparatus in *Cucumis sativus* and its protection by exogenous putrescine. Physiol Plant. doi:10.1111/j.1399-3054.2012.01623.x
Silva EN, Ribeiro RV, Ferreira-Silva SL, Viégas RA, Silveira JAG (2011) Salt stress induced damages on the photosynthesis of physic nut young plants. Sci Agric (Piracicaba Braz) 68:62–68
Stirbet A, Govindjee (2011) On the relation between the Kautsky effect [chlorophyll *a* fluorescence induction] and photosystem II: basics and applications of the OJIP fluorescence transient. J Photochem Photobiol B Biol 104:236–257
Strasser RJ (1987) Energy pipeline model of the photosynthetic apparatus. In: Biggins J (ed) Progress in photosynthesis research. Matinus Nijhoff, Dordrecht, pp 717–720
Strasser RJ, Tsimilli MM (1998) Activity and heterogeneity of PS II probed in vivo by the Chlorophyll *a* fluorescence rise O-[K]-J-I-P. In: Garab G (ed) Photosynthesis: mechanisms and effects. Kluwer Academic, Dordrecht, pp 4321–4324
Strasser RJ, Srivastava A, Tsimilli MM (2000) The fluorescence transient as a tool to characterize and screen photosynthetic samples. In: Yunus M, Pathre U, Mohanty P (eds) Probing photosynthesis: mechanisms. Regulation and adaptation. Taylor & Francis, London, pp 443–483
Sudhir P, Murthy SDS (2004) Effects of salt stress on basic processes of photosynthesis. Photosynthetica 42(4):481–486
Sudhir PR, Pogoryelov D, Kovacs L, Garab G, Murthy SD (2005) The effects of salt stress on photosynthetic electron transport and thylakoid membrane proteins in the cyanobacterium *Spirulina platensis*. J Biochem Mol Biol 38:481–485
Sundby C, Melis A, Mäenpää P, Andersson B (1986) Temperature-dependent changes in the antenna size of photosystem II. Reversible conversion of Photosystem II α to Photosystem IIβ. Biochim Biophys Acta 851:475–483
Tiburcio AF, Wollenweber B, Zilberstein A, Koncz C (2012) Abiotic stress tolerance. Plant Sci 182:1–2
Tongra T, Mehta P, Mathur S, Agrawal D, Bharti S, Los D, Allakhverdiev SI, Jajoo A (2011) Computational analysis of pH induced changes in Chlorophyll *a* fluorescence transients. Biosystems 103:158–163
Zhang L, Xing D (2008) Rapid determination of damage to photosynthesis caused by salt and osmotic stress using delayed fluorescence of chloroplast. Photochem Photobiol Sci 7:352–360
Zhu JK (2002) Salt and drought stress signal transduction in plants. Ann Rev Plant Biol 53: 247–273
Zhu JK (2007) Plant salt stress. In: Encyclopedia of life sciences. Wiley, Chichester, doi: 10.1002/9780470015902.a0001300.pub2

Chapter 6
Adaptive Plasticity of Salt-Stressed Root Systems

Boris Rewald, Oren Shelef, Jhonathan E. Ephrath, and Shimon Rachmilevitch

6.1 Introduction

With food production having to meet the demands of a growing world population, and global climate warming and salinization protruding, understanding plant species' reaction to abiotic stresses is of vital importance. Salinity is one of the most severe environmental factors inhibiting the growth and yields of crop plants; >7% of the world's land area and up to 50% of irrigated agricultural areas are salt-affected (Pitman and Läuchli 2004; Munns 2005; Rengasamy 2006).

Salinity can cause several challenges for plants, including water stress, mal-nutrition and accumulation of excess ions to potentially toxic levels. Salt tolerance differs dramatically between species/varieties and ontogenetic stages and is usually assessed by a reduction in biomass production, yield or survival rates. While salt exclusion, compartmentation and osmoregulation are the mechanisms particularly considered to increase the salt tolerance of plants, adaptation to salinity is determined by the integrating effects of several mechanisms at the cell, tissue and organ level (Zekri and Parsons 1992; Maas 1993; Kozlowski 1997; Parida and Das 2005; Munns and Tester 2008). Root systems can exhibit enormous plasticity on the level of biomass, morphology and/or physiology in response to different environmental parameters, like water and nutrient availability (Wang et al. 2009; Gruber et al. 2011) or excess ions (Deak and Malamy 2005; Rewald et al. 2011b, d). However, despite the likelihood

B. Rewald (✉)
Institute of Forest Ecology, University of Natural Resources and Life Sciences (BOKU), Peter-Jordan-Straße 82, 1190 Vienna, Austria
e-mail: rewald@rootecology.de

O. Shelef • J.E. Ephrath • S. Rachmilevitch
French Associates Institute for Agriculture and Biotechnology of Drylands, Ben-Gurion University of the Negev, Sede Boqer Campus, 84990 Midreshet Ben-Gurion, Israel

P. Ahmad et al. (eds.), *Ecophysiology and Responses of Plants under Salt Stress*,
DOI 10.1007/978-1-4614-4747-4_6, © Springer Science+Business Media, LLC 2013

that differences among root systems may (partially) underlie distinct salt tolerances, information on the physiological and phenotypical plasticity of root systems under salt stress is scant (Neumann 1995; Munns 2002; Vadez et al. 2007).

This chapter reviews modifications among root system size and architecture, morphological and anatomical root traits, root longevity and respiration, and root functions, such as water and nutrient uptake under salinity. It explores the question of whether changes in roots are caused indirectly, e.g. by osmotic stress or ion toxicity, or whether they could be an active response of plants, which may be of potential adaptive significance. In addition, a short overview on the properties of saline soils is given.

6.2 Saline Soils

Saline soils are both naturally occurring and the consequence of anthropogenic activities ("secondary salinization"; Ghassemi et al. 1995). Salts can be deposited in the soil from wind and rain, as well as through the weathering of rocks. The types of salinity, based on soil processes and water availability, are (i) groundwater-associated salinity, (ii) transient salinity and (iii) irrigation-caused salinity (Munns 2002; Rengasamy 2006). Three factors especially contribute to secondary salinization in agricultural areas: (i) poor irrigation management and lack of suitable drainage; (ii) irrigation with saline water; and (iii) rising groundwater tables due to vegetation changes. Salt-affected soils are more extensive in arid and semi-arid regions compared with humid regions because low precipitation and high potential evapotranspiration facilitate salt accumulation (Abrol et al. 1988; World Bank 2007) and the used water sources often contain high amounts of salts (Mostafazadeh-Fard et al. 2009).

Root system characteristics are directly influenced by the properties of saline soils. In general, salt-affected soils are characterized by high concentrations of soluble salts and low organic matter and nitrogen contents (Ashraf and Rehman 1999; Green et al. 2008). The exact relationship between soil solution salinity and the chemical and physical soil properties depends on the dissolved salt. The most common salts present in irrigation or ground water are either chlorides, sulfates, and carbonates (CO_3^-) or bicarbonates of calcium, magnesium, sodium, and potassium; in this chapter, we will concentrate mainly on NaCl salinity effects on root systems because it is one of the most dominant salts in saline soils (Rengasamy 2006).

Salt anions affect soil properties directly by increasing salinity and indirectly by affecting the exchangeable sodium, calcium, and magnesium ratios. Changes in the concentrations of cations on exchange sites and in soil solutions lead to changes in pH and ion concentrations and, ultimately, to the disruption of biogeochemical cycles. When salts are dissolved in a solution, they often disassociate, resulting in high pH values and the formation of sodium-dominated (sodic) soils when other salt ions are leached from the soil profile (Miller and Donahue 1995). For example, (bi-) carbonates are most common in arid and semi-arid areas of the western United States.

When soil moisture is reduced, either by evaporation, plant uptake, or drainage, $Ca(HCO_3)_2$ decomposes into solid calcium carbonate (lime), CO_2 and water. Through this process, calcium is removed from clay particles while sodium is left behind, creating a sodic soil from a calcium-dominated soil. Carbonates, when present, are usually found at pH > 8 and will cause calcium and magnesium to precipitate in drying soils. In brief, salinity disrupts mineral nutrient acquisition by plants in two ways. First, the ionic strength of the substrate can influence nutrient uptake and translocation. The second and more common mechanism by which salinity disrupts the mineral relations of plants is through reduction of nutrient availability by competition with major ions (e.g. Na^+ and Cl^-) in the substrate. These interactions often lead to Na^+-induced Ca^{2+} and/or K^+ deficiencies and Ca^{2+}-induced Mg^{2+} deficiencies (Grattan and Grieve 1992; Chhabra 1996). Soil salinity significantly reduces the absorption of mineral nutrients, especially phosphorus (P), because phosphate ions precipitate with Ca^{2+}, Mg^{2+} and Zn^{2+} ions in saline soils and become unavailable to plants.

Salt anions, such as chloride, can affect soil physical properties by causing fine particles to bind together into aggregates, a process known as flocculation, which is potentially beneficial in terms of soil aeration, root penetration, and root growth. However, Na^+ has the opposite effect of salinity on soils; sodium saturation may cause soil dispersion because of its relatively large size, single electrical charge and hydration status (Tanji 2002; Warrence et al. 2003). Soil dispersion hardens soil and blocks water infiltration, influencing resource availability and root foraging negatively. The major implication associated with decreased infiltration due to sodium-induced dispersion is reduced plant available water due to "surface crusting" and increased runoff. The effects of dispersion on soil hydraulic conductivity include reduced oxygen availability through waterlogged surface soil layers with implications for the soil rhizosphere activity and directly limited nutrient transport to roots within the transpirational mass flow (Norrström and Bergstedt 2001). In general, excess soluble salts in the root zone reduce the plant available water (Munns 2002). Basically, water potentials of saline soils are more negative and water is thus less available for plant uptake due to osmotic forces, even if volumetric soil water contents are as high as field capacity (Oron et al. 1999). However, the effects of salinity on soil physical properties depend strongly on the soil type, the salinity concentrations of soils and precipitation/irrigation water, and the relationship between salinity and sodicity. In general, sandy soils can withstand higher salinity in irrigation water because more dissolved salts will be removed from the root zone by artificial or precipitation-induced leaching; clayey soils are at a greater risk than coarse textured soils for excess sodium to bind to them and cause effects, such as dispersion.

Soils can be highly variable in salt concentration, soil moisture and nutrient availability in both space and time (Bazihizina et al. 2009 and references within; Rewald et al. 2011b, d). The factors contributing to this variability are the complex interactions of climate, topography, soil properties (e.g. texture, mulches), irrigation practices (e.g. drip irrigation, leaching events), and fluctuating groundwater levels. Thus, plants with distinct sources of water and nutrients are able to take up these resources mainly from the least or optimal saline source (Yakir and Yechieli 1995; Bazihizina et al. 2009). Furthermore, the physical and chemical properties of saline soils can be improved by reciprocal processes. Plants generally influence the physical properties of soils, such as porosity, aggregate stability and water retention, and can facilitate salt removal or subsoil enrichment with favourable cations ("phytoremediation"; McKee 1993; Ashraf et al. 2010 and references within).

In summary, salinity-affected soils change in a range of physical, chemical and biological properties besides possessing excessive amounts of salt ions. See recent reviews for further details on soil chemical and physical properties under excess salinity (Tanji 2002; Roberts et al. 2009).

6.3 Root Biomass and Root to Shoot Ratios

Salinity lowers the total photosynthetic capacity of the plant through decreased leaf growth and inhibited photosynthesis, limiting its ability to grow (Yeo 2007). Furthermore, the limitation of photo-assimilates under salinity results in competition among different physiological processes and organs (Munns and Termaat 1986).

Root biomass has been reported to be generally less affected by excess salinity than aboveground organs (Munns and Tester 2008). In accordance, the root: shoot ratio of *Citrus* spp. has been found to increase significantly, while the rootstock biomass was marginally reduced under salinity (Zekri and Parsons 1989; Rewald et al. 2012), and the shoot growth of olive seedlings was found to be more sensitive to salinity than their root growth (Perica et al. 2008; Cimato et al. 2010 and references therein). Increased root: shoot ratios are thought to improve the 'source: sink ratio' for water and nutrients under salinity (Albacete et al. 2008). In contrast, Rewald et al. (2011d) found that the fine root biomasses of two olive varieties were significantly reduced under salinity in absolute terms and standardized for trunk size; the decrease in root biomass was higher in the salt sensitive as compared with the salt tolerant olive variety. Similar differences in root: leaf biomass allocation under different drought regimes have been reported in beech seedlings (i.e. constant root: shoot ratios) and mature trees (i.e. reduced root: leaf biomass ratios under drought), indicating ontogenetic and time-dependent differences in the stress response of woody species (Rewald et al. 2011d and references within). The root biomass of *Capsicum annumm* and *Chloris gayana* plants were also found to be more inhibited by salt stress than shoot biomass, leading to reduced root: shoot ratios (Céccoli et al. 2011; Siddikee et al. 2011). Finally, root: shoot ratios were found unaltered in two *Prosopis* species under salinity (Villagra and Cavagnaro 2005). Thus, the biomass percentage allocated to roots, stems and leaves was found to differ widely under salinity and to be related to time of exposure, species/varieties and ontogenetic stages. Future studies are urgently needed to reassess the conclusion that "root biomass is [generally] less affected by excess salinity than aboveground organs" (Munns and Tester 2008).

How are root to shoot ratios regulated under salinity? In most cases, environmental signals, such as excess salt concentration, trigger the response of the plant to its growing conditions through changes in phytohormone concentrations, controlling the assimilate partitioning between different sink tissues (Sachs 2005; Hartig and Beck 2006). For example, Wolf et al. (1990) reported that salinity increased plant ABA concentration. Furthermore, it has been demonstrated that a decrease in cytokine (CK) supply from the root to the shoot could inhibit leaf growth and thus influence the root: shoot ratio (Van der Werf and Nagel 1996; Rahayu et al. 2005). There are also reports that salinity modifies the auxin indoleacetic acid (IAA) levels differently in the roots and shoots of tomato plants, however, these reports are contradictory (Dunlap and Binzel 1996; Albacete et al. 2008), and further research is needed to determine the mechanisms underlying root: shoot ratio regulation under salinity. See Ghanem et al. (2011) for a recent review on root hormones and root to shoot signalling.

6.4 Root Elongation, Morphology and Architecture

The shape and size of root systems (see above) is determined by the elongation of individual root tips, by the rate and location of lateral root development and by root longevity (see below). Water and nutrient uptake efficiency and foraging are largely determined by root system expansion, root architecture and root morphology; however, excess salinity affects these root processes differently (Bernstein and Kafkafi 2002).

6.4.1 Root Elongation Rates and Direction

An early plant response to excess salts is the inhibition of expansion growth in leaves and roots. Restrictions on root growth reduce the soil volume which can be explored and, thus, the availability and uptake of water and nutrients. Root elongation rates under salinity were found to be reduced in a wide range of species, such as *Arabidopsis thaliana*, *Citrus* spp., *Oryza* spp., *Triticum* spp., and *Zea mays* (Jones 1985; Zhu et al. 1998; Jbir et al. 2001) and are often used to determine the salt tolerance of species and varieties. However, the effect of NaCl upon the root elongation of *Agrostis stolonifera* populations was not correlated with the effect upon yield, nor with the Na^+ level of their native habitat (Tiku and Snaydon 1971) thus, root elongation might not be a universal indicator of salinity tolerance.

The rapid elongation of plant roots is primarily a consequence of the expansion of cells produced by a meristem division in the root tip. The early onset of growth inhibition can be based either on the toxic consequences of salt accumulation in the growing tissues, a reduction in water availability for cell expansion or, alternatively, an actively regulated plant stress response. Excessive salinity has been reported to inhibit both root cell production and cell expansion (Neumann 1995 and references within). Experiments by Cramer et al. (1988), using an iso-osmotic concentration of mannitol, suggested that the root growth inhibition in *Zea mays* is not due to osmotic factors on cell expansion but rather to the toxic effects of salt on the metabolism. This is confirmed by measurements of K^+ concentration in root tips, the main cation for turgor homeostasis; K^+ deposition in root tips is often maintained, even under high NaCl levels and when K^+/Na^+ ratios remain high (Jeschke and Stelter 1976; Zhong and Läuchli 1994; Wyn Jones and Gorham 2002). However, in melon plants, root elongation was suppressed under increased external, iso-osmotic concentrations of both NaCl and mannitol (Yermiyahu et al. 1997). Neumann et al. (1994) concluded that cell wall hardening in the elongating root tips is an important component of root growth inhibition induced by long-term salinization. In accordance, Neves et al. (2010) found that enhanced lignification (see below) solidifies the cell wall and restricts soybean root growth under salinity.

Root elongation in halophytes, such as mangroves and *Tamarix* spp., is found to be lower under salinity too, although growth reductions set in at higher salt concentrations. Regardless, the magnitude of reduced root elongation under salinity is highly dependent upon the mineral composition of the rooting medium (Tiku and Snaydon 1971; Cramer et al. 1988). The fact that increasing Ca^{2+} levels can increase root (and leaf) growth is especially well established (Neumann 1995). For example, Cramer et al. (1988) found that treatment with Ca^{2+} after the addition of NaCl partially restored *Zea mays* root growth; pre-treatment with Ca^{2+} completely prevented the inhibition of growth by salt stress. Similar, Yermiyahu et al. (1997) observed nearly optimal root growth for two melon varieties when 40% and 51% of the total plasma membrane charged sites were bound by Ca^{2+}, in a salt-resistant and a salt-sensitive melon variety, respectively.

Very little is known concerning the influence of excess salts on root growth direction and the underlying mechanisms by which plant roots need to overcome the gravity signal. Recently, curvature measurements by Sun et al. (2008) showed that the gravitropic response of *Arabidopsis* roots is greatly reduced upon exposure to salt stimuli in a dose-dependent manner. Using an ion gradient, these authors showed that the roots of seedlings on a salt-free medium grew downward first but then curved and grew upward toward the lower level of salt once a region with a higher salt concentration was reached (agravitropism or "negative halotropism"). In brief, exposure to salt stress caused a rapid degradation of amyloplasts in root columella cells. The altered root growth direction in response to salt was found to be correlated with root gravitropism control protein (*PIN2*) mRNA abundance, expression and localization of this protein and ion disequilibrium (Sun et al. 2008). In contrast, Shelef et al. (2010) observed that the halophyte *Bassia indica* developed horizontal roots, originating from the taproot, growing toward more saline soil regions. This phenomenon was termed "positive halotropism" and was suggested to be associated with optimal osmolarity along the growth path and with increased water and nutrient availability. However, future studies need to examine if other halophytes show similar growth pattern under salt gradients.

6.4.2 Root Morphology

Similar to reduced root elongation, salt stress is known to induce changes in root diameter. Reduced root diameters under salinity have been found in a range of plant types, such as desert halophytes, the grasses *Cnyodon dactylos*, *Oryza sativa*, *Triticum* spp., and seedlings of the tree species *Prosopis tamarugo* (Valenti et al. 1991; Samarajeewa et al. 1999; Yi et al. 2007; Hameed et al. 2010; Shafi et al. 2010). In *Populus* spp., the root diameter increment was significantly reduced under salinity (Ehlting et al. 2007). In contrast, studies on plants, such as *Hordeum* spp., *Gossypium hirsutum*, *Citrus volkameriana*, and *Tessaria absinthioides*, reported increased root diameter under salinity (Huang and Redmann 1995; Reinhardt and Rost 1995; Degano 1999; Rewald et al. 2012). While reduced root diameter can be explained by reduced cell expansion or cell division under salinity, mainly in the vascular tissues, increased root diameter might be caused by succulence of the cortex. Succulence is an anatomical adaptation, which, by increasing the volume of vacuoles, permits the accumulation of larger amounts of water and dissolved ions in leaves, shoots and roots (Munns 2002). Under the same NaCl concentration, the root succulence of two *Elaeagnus* species was found to be even slightly higher than leaf succulence (Wang et al. 2010). Degano (1999) rates root succulence as an adaptation to saline conditions; indeed, root succulence peaked in the halophyte *Atriplex halimus* under $CaCl_2$ salinity concentrations above optimal (Nedjimi et al. 2006). Besides increasing storage capacities, thicker roots are favourable to overcome increased soil strength, as prevails in some saline soils, and might have reduced maintenance respiration and turnover rates, resulting in reduced carbon costs below ground (see below). Because plant nutrient foraging is strongly affected by root diameter (via specific root length and areas), a likely cost of thicker roots is a reduced rate of nutrient uptake. Similar to findings on root elongation, some studies have indicated a strong influence of nutritional parameters on salt stress effects on root diameter. For example, root diameters in *Populus* spp. were more strongly affected by salt in ammonium-fed plants than in nitrate-fed plants (Ehlting et al. 2007), and changes in the root diameter of *Zea mays* under salinity were found to be modified or absent by the presence of Ca^{2+} (Evlagon et al. 1990). Future studies are needed to determine the (species-specific) cellular mechanisms and environmental conditions underlying the changes in root diameter under salinity.

Under constant tissue density, the root diameter is directly related to the specific root length (SRL) and the specific root area (SRA). Higher specific root lengths and areas have been linked to higher root hydraulic conductivity and higher rates of root proliferation, and there is some evidence to suggest that the "fineness" of the absorptive roots might be linked to a suite of traits influencing the rate of resource acquisition over the lifetime of the root tissue (Eissenstat and Achor 1999 and references within). Under salinity, the SRL increased in the grass *Chloris gayana* (Céccoli et al. 2010), and SRAs were found to be increased in mature *Olea europaea* trees and the halophyte *Bassia indica* (Shelef et al. 2010; Rewald et al. 2011b, d). Increased SRL and SRA may partially compensate for a loss in root system biomass under salinity. This hypothesis is matched by a study of Echeverria et al. (2008) which found that salinity increased the SRL of a salt-sensitive *Lotus glaber* genotype but did not affect the SRL of salt-tolerant genotype which had a much smaller reduction in root system biomass under salinity. Interestingly, the sensitive *Lotus* plants adjusted their SRL under salt stress to levels close to those displayed by tolerant genotype roots. This suggests that, in this latter genotype, the measured traits would have a more suitable magnitude to minimise salt-stress impact and that adjusting root traits is a more important adaptive mechanism in salt-sensitive genotypes. However, SRL and SRA do not always increase under salinity; in hydroponically grown *Citrus volkameriana* rootstocks, the SRA was significantly reduced when growing under 90 mM NaCl (Rewald et al. 2012). Furthermore, analysed by root order (i.e. the branching hierarchy), the SRAs of *Citrus* root orders one (i.e. root tips) to four were significantly reduced under salinity, while the SRA of root orders five and six did not change significantly. Future studies need to determine the advantages and disadvantages of root diameter changes in terms of carbon costs, resource uptake and salt exclusion under different salinities and growth media. We hypothesize that a trade-off exists between optimizing the structural carbon investment for surface area/root length for resource foraging and potentially higher carbon costs due to increased root turnover and maintenance respiration in thinner roots. For woody root systems, different norms between root orders need to be taken into account.

6.4.3 Root Architecture

It has been suggested that root system architecture (RSA) is one important component defining nutrient uptake capacity (Fitter et al. 1991; Barber 1995). Furthermore, if axial resistance, i.e. the resistance for longitudinal water transport in conduits, is a significant component of total root resistance under salinity (Passioura 1977; Rewald et al. 2011c), then root branching could significantly influence root hydraulic conductivity (Joly 1989).

Under salt stress, lateral root (LR) formation has been found to be less affected by salinity than root elongation (Bernstein and Kafkafi 2002); however, increased, unaltered or reduced LR formation has been reported. If water availability decreases (or osmotic stress increases), lateral root emergence is often repressed although LR initiation is largely unaffected (Nibau et al. 2008 and references within). In *Arabidopsis*, salt stress can induce a seriously reduced meristematic zone and a strong reduction in the number of lateral root primordia (LRP), accompanied by the down-regulation of several cell cycle genes (Burssens et al. 2000). Other authors reported reduced branching of salt-stressed woody root systems (e.g., *Olea europaea* seedling, Gucci and Tattini 1997). In contrast, salt stress may also cause an increase in LR number, e.g. in *Arabidopsis, Cicer arietinum* and *Oryza sativa* cultivars (He et al. 2005; Shukla et al. 2006; Nibau et al. 2008; Krishnamurthy et al. 2011).

The *NAC2* transcription factor and other auxin-response genes are up-regulated by NaCl, and its over-expression causes increased LR formation, specifically without a change in root length (He et al. 2005; Shukla et al. 2006). Echeverria et al. (2008) reported unaltered root branching under salinity in *Lotus glaber*. Root branching is controlled by both cellular signalling pathways and environmental signals. Alongside ABA, the lateral root development 2 gene (*LRD2*) may be required to determine the percentage of LRP that becomes lateral under normal and stress conditions (Deak and Malamy 2005; Malamy 2009). The involvement of auxins in the development of lateral roots has also been studied extensively (Casimiro et al. 2003; Fukaki et al. 2007). It has been reported that a redistribution of auxins occurs at the root apex of *Arabidopsis* under salinity stress (Wang et al. 2009), the gene AUX1 is involved in auxin transport and affects the lateral root initiation in *Arabidopsis* (Chhun et al. 2007). However, the study of Krishnamurthy et al. (2011) indicates that AUX1-like genes are not responsible for the initiation of LR development under stress in rice, as levels were invariant or declined under stress. Instead, the transcript levels of *Arf8* were up-regulated in *Oryza sativa* upon salt stress, indicating a role in lateral root development (Krishnamurthy et al. 2011). Thus, the molecular mechanisms underlying root branching under salt stress are largely unknown; future studies must address how different signals work together to direct pericycle cell behaviour and LR developmental processes.

Due to the fact that traits often vary according to the position of individual root segments among the root branching hierarchy (i.e. "root order" or "lateral number"; Pagès and Kervella 1990; Pregitzer et al. 2002; Rewald et al. 2011a), analysis by branching hierarchy is a powerful approach to understand complex root systems. For example, in the halophyte *Plantago maritima*, primary and lateral roots reacted differently to NaCl salinity (Rubinigg et al. 2004). Primary root length increased, even under severe salinity (200 mM), while the total length of LR and the number of first, second and third LR was reduced under salinity; under severe NaCl stress, no third order laterals were formed. Rubinigg and colleagues (2004) concluded that the increase in total LR length in plants under moderate salinity was mainly caused by increased length growth, while the decrease in total LR length under severe salinity was the consequence of the inhibition of LRP and/or the activation of apical meristems. Recently, Rewald et al. (2012) provided the first evidence that NaCl can reduce the total number of root orders and changes the biomass and surface area frequency among root orders of severely salt-stressed *Citrus* plants. By comparing differently salinity-tolerant olive varieties, Tattini et al. (1994) concluded that root branching is negatively related to salt tolerance. In contrast, the increased salinity tolerance of transgenic tobacco expressing the *CAP2* gene was related to a large increase in LR number (Shukla et al. 2006). Thus, it remains an open question if increased or decreased root branching is positively correlated with plants' salt tolerance. We hypothesise that the answer to this question is related to the uptake efficiency (in terms of carbon invested per uptake rate) of different root orders under salinity and root age-related differences in tissue differentiation (Rewald et al. 2012).

6.5 Root Turnover and Respiration

6.5.1 Root Turnover

Fine root turnover, i.e. the inverse longevity of roots, has been estimated to account for as much as 33% of global net primary production (NPP; Jackson et al. 1997). In trees, NPP_{root} may be >50% of NPP_{total}, with fine roots comprising a substantial part of total belowground NPP (Nadelhoffer and Raich 1992). On the global scale, root turnover is primarily influenced by temperature, but on a local scale, it is highly related to precipitation and soil conditions, including many abiotic and biotic stresses (Gill and Jackson 2000; Norby and Jackson 2000). Rates of fine root turnover have great consequences for carbon allocation within the tree, nutrient cycling and C sequestration into soils. It has been widely reported that the shoot growth of non-halophyte plants is generally more sensitive than the root growth to salinity (Munns and Termaat 1986); however, the effects of salinity on root growth and turnover are complex and rely on the specific species or variety, the ion composition of salts, and the media in which plants grow (Valiela et al. 1976; Snapp and Shennan 1992). While direct studies on the influence of salinity on root turnover rates are scarce, salt stress increased the root senescence in *Solanum lycopersicum* var. UC82B, while the root senescence was not influenced by salinity in the var. CX8303 (Snapp and Shennan 1992, 1994). Rewald et al. (2011b, d) examined two differently salt-tolerant varieties of mature olive trees, Barnea (tolerant) and Proline (sensitive); the mature trees were irrigated using either fresh or moderately saline water (EC 1.2 and 4.2 dS m^{-1}) for 11 years. Both varieties exhibited a reduced fine root biomass under salinity; however, the salt-tolerant Barnea trees had a higher root biomass: necromass ratio under moderate salinity, indicating lower root turnover rates and, thus, a more efficient carbon use below ground. Thus, root turnover rates and the ability of the fine root system to resist the deleterious effects of salinity seemed to affect the salt resistance of plants by influencing water uptake and carbon allocation. In contrast, Ramoliya and Pandey (2006) suggested that increased root turnover could be beneficial by removing excess ions from the plant and preventing the transport of salt ions to leaves. Beyond that, Kacprzyk et al. (2011) suggested that induced programmed cell death (PCD) in severely salt-stressed roots can be advantageous for the plant if it eradicates the most salt susceptible roots, which are subsequently replaced by better adapted ones (under non-lethal conditions). The view that root cell death under salinity does not occur stochastically but might be well-regulated to minimize negative effects for the plant is supported by the findings of Liu et al. (2007). Liu and colleagues (2007) found that cell death in *Oryza sativa root* progressed successively, starting from the outer layer cells in the epidermis and exodermis and, subsequently, to the endodermis and stele, suggesting a possible function of the dead cells in preventing the influx of excess salt ions into the stele and into the shoots, leading to enhanced salt exclusion; in contrast, cell death induced by PEG-induced osmotic stress occurred randomly in roots, allowing a better ability to recover after stress.

How is the vitality of root cells and tissues influenced by excess salinity? For example, nuclear deformation of the meristematic cells of barley root occurred already within 12 h of severe salt stress and was followed by DNA degradation (Katsuhara and Kawasaki 1996). Similar, nuclear deformation of meristematic root cells, followed by degradation of the nuclei in the apical region, occurred within 24 h of 150 mM in *Glycine max* (Liu et al. 2000). The nuclear degradation is often accompanied by apoptosis-like DNA fragmentation and can be identified as salt-induced PCD (Katsuhara and Kawasaki 1996; Liu et al. 2007; Kacprzyk et al. 2011). Nuclear deformation and degradation might be caused by cellular dehydration, but excessive cellular salt concentrations, like other biotic and abiotic stresses, are also known to induce oxidative stress. Oxidative stress is indicated by increased levels of lipid peroxidation and H_2O_2 in salt-stressed *Lycopersicon esculentum* root mitochondria (Mittova et al. 2004); Kawasaki et al. (2001) identified the salt-induced up-regulation of several antioxidant genes in *Oryza sativa* roots. Several studies have demonstrated that salt-tolerant species increase their antioxidant enzyme activities and antioxidant contents in response to salt stress, while salt-sensitive species did so to a lesser extent or showed no response (Jbir et al. 2001; Mittova et al. 2004; He et al. 2007). For example, salt stress decreased the activities of superoxide dismutase (SOD) and guaiacol peroxidases (POD) and the contents of ascorbate (ASC) and glutathione (GSH) in *Lycopersicon esculentum* root mitochondria. In contrast, both H_2O_2 and lipid peroxidation levels decreased in the mitochondria of salt-resistant *Lycopersicon pennellii* roots under salt stress, and the levels of ASC and GSH and the activities of SOD, several isoforms of ascorbate peroxidase (APX), and POD increased. Similarly, accumulation of the non-enzymatic antioxidant mannitol in roots and shoots has been shown to increase the salinity tolerance in transgenic tobacco plants (Tarczynski et al. 1992). In view of the considerable variations in the protective mechanisms against reactive oxygen species and transcriptional responses between salt-stressed roots and aboveground organs (Ma et al. 2006; Brinker et al. 2010), and possible external influences such as mycorrhization on enzymatic antioxidant activity (He et al. 2007), further studies are required to assess salinity's consequences for (root) tissue vitality and plant organ viability. However, while assessing the enzymatic and non-enzymatic antioxidant responses of *Poncirus trifoliata* × *Citrus sinensis*, a salt-sensitive citrus rootstock, to different levels of salinity, Arbona et al. (2003) found that the adverse effect on the growth of citrus rootstock was mainly due to a cellular intoxication by chloride ions and not to the salt-induced oxidative stress. Recently, Rewald et al. (2012) found higher levels of Na^+ and Cl^- accumulated in lower root orders, such as root tips, compared with higher, older root orders. This higher susceptibility of lower root orders to salt accumulation has likely consequences for root-order-specific turnover rates. Understanding the patterns and regulations of root turnover is vital for the understanding of the whole plant reaction to salt stress; however, the complexity of the specific species, salt and nutrient composition, and the synergism with other stresses makes it difficult to predict the effects of salinity on root turnover on a global scale, and future research is needed in order to understand these processes.

6.5.2 Root Respiration

Aerobic respiration refers to the biological processes by which reduced organic compounds are mobilized and subsequently oxidized in a controlled manner. During respiration, energy is incorporated into forms that can be readily utilized for the maintenance and development of plants. On a whole plant basis, over 24 h, respiration dissipates about 25% of the carbon fixed by plants, and this figure can rise to 100% under stress conditions (Taiz and Zeiger 1998). For over 100 years, it has been observed that the rate of plant respiration increases under salinity, and the term "salt respiration" has been acknowledged (Steward 1935; Schwarz and Gale 1981). "Salt respiration" results from increased energetic costs associated with ion accumulation and toxicity. An increase in the respiratory cost for maintenance, for active ion transport and/or for growth processes in roots encountering salinity is usually consistent with the occurrence of a high rate of root respiration while growth rate is reduced.

Salt respiration is not equally significant in all plants. For example, respiration in *Zea mays*, a relatively salt-sensitive crop, showed little response to salt (Schwarz and Gale 1981), but the root respiration of moderately-sensitive *Brassica oleracea* was reduced significantly under moderate salt stress (del Amor and Cuadra-Crespo 2011). Burchett et al. (1984) studied the relationships between growth parameters and root respiration under salinity in seedlings of the grey mangrove *Avicennia marina*. The seedlings were grown for 6–8 weeks in 100%, 50%, 25% and 0% seawater. The root respiration of root segments was measured and showed stimulation in the presence of salt. The rates of respiration were highest in 25% seawater; since higher concentrations were associated with a decline in the rate of root respiration from the maximum root respiration is not stimulated by high salt concentrations salt per se. Similarly, Epron et al. (1999) reported that the root and shoot biomass of *Quercus* seedlings was reduced after 9 days of watering with a nutrient solution containing either 50 or 250 mM NaCl; however, the specific respiration of roots was unaffected by the moderate salinity treatment, while it was reduced by 62% under severe NaCl stress. Bloom and Epstein (1984) studied root respiration in two differently salt-tolerant *Hordeum vulgare* cultivars; under control conditions, both varieties had similar rates of root respiration. However, root respiration in the sensitive cultivar doubled when 10 mM NaCl or KCl was added to the medium, and only 6 h after the removal of the salt, root respiration returned to control levels; the addition of 5 mM K_2SO_4 increased root respiration by less than 20%. In the salt-tolerant *Hordeum* cultivar, 10 mM NaCl or KCl increased the root respiration by only approximately 50%, and the recovery took only 2 h. Thus, the level of salt-induced root respiration in a variety may indicate its level of salt-tolerance.

6.6 Root Anatomy and Cell Ultra-structure

Salinity changes the (ultra-)structure of plant organs above ground (Koyro 2004; Yamane et al. 2004; Mahmoodzadeh 2008). Since roots are the first organs to be exposed to salinity, there is no wonder that salinity causes anatomical alterations in roots – from the cellular level to ultra-structural changes. The drivers that determine the anatomical and ultra-structural responses of plants to salinity include the type of salinity exposure (e.g., the character of the growth medium, the length of exposure, the concentration and type of salinity), plant taxonomy (susceptible or tolerant, glycophyte or halophyte) and plant maturation.

6.6.1 Cell Number and Size

Contradictory reports exist regarding the influence of salt stress on root cell numbers and cell sizes. Bernstein and Kafkafi (2002) stated that it is still largely unknown how salt stress affects the numbers of cells along the root. Strogonov (1964) and Hajibagheri et al. (1985) found that cell numbers in root tips are not affected by salinity but that cell size is often increased. Root cell size increases mainly due to the development of vacuoles, as was shown in sorghum (Koyro 1997) and *Thellungiella halophila* (Mei-fang et al. 2005), a process known as cellular vacuolation (Hajibagheri et al. 1985). However, not only vacuoles are enlarged; Koyro (1997) showed an increased quantity of vesicles in root cells as a response to salinity. Both processes increase the membrane surface and, thus, the plants' ability to perform ion compartment and exclusion. The increase in cytosolic membranes may differ between cell types; for example, Koyro (1997) showed that the middle cortex cells of sorghum roots had larger central vacuoles and the endodermal and epidermal cells had larger vesicles under salinity. In contrast, Azaizeh et al. (1992) and West et al. (2004) found decreased cell sizes in soybean and *Arabidopsis* roots. Furthermore, West et al. (2004) showed that reduced cell production was caused by a smaller number of dividing cells, i.e. a meristem size reduction. It has previously been shown that impaired root growth, caused by osmotic stress conditions, is associated with a reduced cell division activity (Samarajeewa et al. 1999); meristematic root cells are the most sensitive to salt stress (Huang and Van Steveninck 1990).

6.6.2 Cell Walls

In general, cell walls are increasingly lignified under stress. Lignin deposition increased in the vascular root tissues of *Phaseolus vulgaris* in response to salinity (Cachorro et al. 1993). Cell-wall-bound peroxidase (POD) is often considered the enzyme most directly involved in lignification (Almagro et al. 2009). In agreement, Neves et al. (2010) showed that POD activities increased in the roots of *Glycine max* seedlings under NaCl stress, resulting in increased root lignin contents. In the central vascular cylinder of NaCl-treated wheat roots, the cell walls were found to be much thicker than those in the control seedlings, which is consistent with an increased lignification (Jbir et al. 2001). However, no response was observed in *Lycopersicon esculentum* roots (Peyrano et al. 1997), and increasing concentrations of NaCl even reduced root lignin levels in *Oryza sativa* seedlings (Lin and Kao 2001). Some plant species develop phi thickening in root tissues; phi thickenings are modifications of the middle of the radial cell walls and consist of nitro-cellulose wall deposits that are impregnated with lignin (Degenhardt and Gimmler 2000). They form on the walls of certain cell layers in the root cortex and were long thought to increase the structural stability of roots only (e.g. in response to compacted soils). However, López-Pérez et al. (2007) found that phi thickenings increased in salt-stressed *Brassica oleracea* plants. Both the modification of lignin levels and the development of phi thickenings are considered to be controversial modifications under salinity stress, and further research is needed to explain their adaptive significance.

6.6.3 Stele

Valenti et al. (1991) found delayed xylem differentiation of the stelar tissues in *Prosopis tamarugo*. Reinoso et al. (2004) showed that reduction in the size of *P. strombulifera* root vascular tissues was due to salt-affected vascular cambium activity that decreased secondary phloem and secondary xylem production. Similarly, salt stress retarded primary xylem differentiation and induced acceleration of the development of secondary xylem in soybean roots (Hilal et al. 1998). Salinity often promotes the diminution of root xylem vessel diameter, probably caused by a repression in the development of metaxylem vessels and altered cambial activity; examples are reduced xylem vessel diameter in the roots of cotton, tomatoes, and the grass species *Triticum aestivum* and *Chloris gayana* (Strogonov 1964; Huang and Redmann 1995; Akram et al. 2002; Céccoli et al. 2011). Reduction of xylem vessels' diameter is often accompanied by an increase in the number of xylem vessels (Kozlowski and Pallardy 1997). This may protect the vascular system from the cavitation that is more expected in stress conditions. The thresholds needed to induce a modified xylem vessel differentiation differ between species. While in the roots of a glycophyte shrub, more and thinner vessels developed under mild salinity stress, in halophytes, the same response occurred only at high salinities (400–800 mM NaCl; Boughalleb et al. 2009) or was virtually absent (Mei-fang et al. 2005). In a study on three differently salt-tolerant *Olea europaea* varieties, neither the maximum nor the hydraulically weighed conduit diameters in roots were found to be significantly different under moderate saline irrigation (Rewald et al. 2011c). Wider and fewer xylem vessels have been found in both stems and some coarse roots after exposure to salinity (Eckstein et al. 1978; Rewald et al. 2011c).

A recent study on *Citrus* root orders under severe salinity revealed different reaction norms in ephemeral roots and more persistent woody roots under salinity, with measurable changes in conduit diameter only in higher, woody root orders (Rewald et al. 2012). Thus, small conduit diameters may be a common adaptation to an increased tension of the water column in the conducting system (e.g. Baas et al. 1983) but cannot explain the differences in salt-resistance per se. Rather, additional adaptation, such as the differentiation of root xylem parenchyma into transfer cells, allowing for more effective ion transport, can be associated with salt tolerance (Kramer et al. 1997).

6.6.4 Cortex, and Exo- and Endodermis

The cortex, being the outermost cell layers after the epidermis/exodermis, is the first tissue to encounter salinity. Examples of increased root cortex thickness, as a response to salinity, are found in the literature, e.g. in *Suadea maritima* (Hajibagheri et al. 1985). Recently, Boughalleb et al. (2009) found that the root cortex thickness was promoted under 100–200 mM NaCl in the xero-halopyte shrubs *Nitraria retusa* and *Atriplex halimus*. However, in some cases, salinity imposed a reduction in cortex development. Akram et al. (2002) reported a decreased size of the cortex under salinity stress in three varieties of *Triticum aestivum;* the most salt-sensitive variety showed a more pronounced decrease. Similar, the cortical parenchyma in *Chloris gayana* decreased under saline conditions (Céccoli et al. 2011), and reductions of cortex thickness under high salinity were also reported in shrubs (Boughalleb et al. 2009) and tree roots (Reinoso et al. 2004). Kurth et al. (1986) and Akram et al. (2002) found no significant change in the number of cortical cell layers, but it was the size of the affected cells which induced modified cortex sizes in *Zea mays* and *Triticum aestivum*. The previous studies indicate that low to moderate salinity can have a stimulating effect on root cortex size, especially in halophytic species, while high salinity seems to reduce the cortex width by reduced cell sizes. The changes in cortex thickness often modify the stele to cortex ratios. For example, Céccoli et al. (2011) reported that the ratio of vascular cylinder/cortical parenchyma was decreased in *Chloris gayana*. Higher stele: root cross-section area ratios were reported in shrub roots (Boughalleb et al. 2009) and in *Prosopis strombulifera* (Reinoso et al. 2004). Akram et al. (2002) reported a decreased size of both metaxylem and the area of the cortex under salinity stress in three varieties of *Triticum aestivum*. Thus, it remains an open question if modified cortex sizes or stele to cortex ratios have regulatory functions for salt retention or flow regulation under saline conditions and if different reaction norms exist in halo- and glycophytes.

The roots of virtually all vascular plants have an endodermis, and the majority of angiosperm roots also have an exodermis, both with a Casparian band (Enstone et al. 2003). The endodermal Casparian band prevents the unregulated movement of substances from the apoplast into the stele and also prevents the backflow of ions that moved into the stele symplastically and were then released into its apoplast.

In roots with a mature exodermis, the barrier to the apoplastic in-flow of ions occurs near the root surface, although the exodermis usually maturates further away from the basal side of the root then the endodermis (Enstone et al. 2003). Reinhardt and Rost (1995) and Cheng et al. (2012) showed that salinity induces exodermis development in roots and suggested that the formation/lignification of the exodermis can contribute to protection against water or solute loss and unwanted uptake of metals. However, Taleisnik et al. (1999) found that, in *Chloris gayana*, there was no difference in exodermal development compared with control roots. In the halophyte *Thellungiella halophile*, the cortex consisted of only two layers of exodermis and endodermis cells (Mei-fang et al. 2005). Several authors found that salinity stimulates endodermal differentiation (Baumeister and Merten 1981; Hajibagheri et al. 1985; Reinhardt and Rost 1995). All roots possess a Casparian band, a strip of cell wall material composed of lignin or suberin, which is deposited on the radial walls of the endodermis. Those substances are a major barrier for ions and, possibly, water in the apoplastic pathway. This insures that ions on their radial pathway towards the stele come through a selection process while crossing a plasma membrane. Thus, an additional deposition of suberin lamellae in endodermal cells contributes another barrier to ion uptake through the apoplastic pathway (Glenn et al. 1999). Walker et al. (1984) observed an intensified suberization in the endo- and exodermis in citrus rootstocks grown under salinity stress; Kramer et al. (1978), Hajibagheri et al. (1985) and Shannon et al. (1994) showed the development of the Casparian strip closer to the root apex in *Atriplex hastate* and *Suaeda maritima* after the addition of NaCl to the nutrient solution. In glycophytes, Reinhardt and Rost (1995) found Casparian bands and suberin lamellae development close to the root tip in cotton seedlings with exposure to salinity. Thus, the endodermis and exodermis are most likely important for plant resistance to salinity stress by preventing/reducing the unregulated flow of salt ions into the cortex or the stele, respectively; however, further studies on their development under saline conditions are needed.

6.6.5 Root Hairs

Wang et al. (2008) found that root hair length and density decreased significantly in a dose-dependent manner in both primary roots and the junction sites between roots and shoots in salt-stressed *Arabidopsis thaliana*. Root hair growth and development were sensitive to inhibition by salt ions but not to osmotic stress. Their analyses of overly salt-sensitive mutants indicated that salt-induced root hair response is caused by ion disequilibrium and seems to be an early adaptive mechanism reducing excessive ion uptake. Importantly, root epidermis exhibits a rapid reversible cellular plasticity in *Arabidopsis*, where plants can switch their phenotypes back and forth in response to the changes of salt stress. The disrupted development of root epidermis can be restored during a prolonged treatment of low level of salt, further indicating that an adaptive response takes place (Wang and Li 2008). However, root hair growth and development and their physiological role in response to salt stress are still largely unknown and need further investigation (Grierson and Schiefelbein 2009).

6.7 Root Physiology

6.7.1 Osmotic Adjustment and Root Hydraulics

Although water is not a limiting factor in all saline ecosystems (e.g., mangroves and saline water-irrigated orchards), excess salinity creates a physiological drought (Clought and Sim 1989). Previous studies related lower leaf conductivities under salt stress to leaf dehydration and reduced leaf water potentials (Gucci et al. 1997). Reduced water uptake rates under salinity were frequently measured, e.g. in *Olea europaea* and *Phaseolus vulgaris* (O'Leary 1969; Therios and Misopolinos 1988; Rewald et al. 2011d). This decrease can be caused by both osmotic and toxic effects, depending on the salt concentration present.

Reduced water uptake in saline environments is thought to be partially caused by decreased soil osmotic potentials and may be counterbalanced by accumulating osmolytes (Chartzoulakis 2005 and references therein). Osmoregulation allows cells to maintain turgor and turgor-dependent processes while keeping a water potential gradient allowing for water uptake. Plant cells accumulate three kinds of osmotica: salts, small organic solutes such as soluble carbohydrates and nitrogen-containing compounds, and hydrophilic proteins. Most organic osmotic compounds are present in low concentrations when the plant is not under salt stress, but types of osmotica vary largely with species. Inorganic osmotica, such as Na^+ and Cl^-, are frequently accumulated because salts are "cheap" and available; however, they are toxic in high concentrations. In *Populus euphratica*, glucose, fructose, sucrose and galactose are major soluble carbohydrates acting as osmotica in leaves (Bogeat-Triboulot et al. 2007). Nitrogen-containing osmotica include amino acids such as proline, amide and proteins, as well as quaternary ammonium compounds (betaines) and polyamines (e.g., Rabie and Almadini 2005). While hydrophilic, glycine-rich proteins are very effective osmotica, they are energetically expensive and are not as commonly accumulated (Niu et al. 1997). In *Hordeum vulgare*, NaCl stress enhanced the asparagine and glutamine pool in both roots and leaves (Yamaya and Matsumoto 1989), in *Olea europaea* salinity stimulated the biosynthesis of phenols and oleuropein (Petridis et al. 2012); however, the types and accumulation rates of osmotica differ, often widely, between plant organs. For example, in the subtribe *Astragalinae*, some species have a higher proline content in the roots, while in others, proline concentrations are higher in the leaves (Niknam and Ebrahimzadeh 2002). Studies addressing organ-specific osmotica accumulation under salinity are scarce; the extremophile *Thellungiella halophila,* an important salt-tolerant model plant, was found to accumulate salt ions and proline in roots and leaves, but soluble carbohydrates, organic acids and amino acids were major osmotica in the roots only (Liu and Zhao 2005). Beside the quantity of osmotica accumulated, the accumulation rate is important to attenuate short-term osmotic shocks which can cause the plasmolysis of root cells in the extreme (Munns 2002). In accordance, Petrusa and Winicov (1997) found that salt-tolerant *Medicago sativa* plants rapidly doubled the proline contents in their roots, whereas in salt-sensitive plants, the increase was slow.

Beside osmotic effects, changes in the hydraulic properties of the root system under salinity can be induced by changes in root system size, architecture, morphology (see above), anatomy and physiology. Silva et al. (2008) found that pepper plants, treated with a low concentration of NaCl or with a nutrient solution with the same osmotic value, decreased their root water uptake rate and root conductivity values to the same extent. However, when the NaCl concentration was doubled and the osmotic pressure of the nutrient solution rose to the same value, only plants treated with NaCl decreased their root water uptake rate and Lp_r values further. Thus, the decrease under higher concentrations of NaCl was caused either by toxic effects due to the accumulation of Na^+ and Cl^- ions in root tissues or by the imbalance in the acquisition of other nutrients. *Glycine max* root system permeability to water was reduced to a marked extent as a result of growing in a saline solution; growth in NaCl for 14 days at -0.17 and -0.26 MPa resulted in a reduction in Lp_r by 27% and 72% compared to the control, respectively (Joly 1989). Salinity has been found to reduce the Lp_r of various species, such as *Lupinus* spp. and *Arabidopsis thaliana* var. (Munns and Passioura 1984; Sutka et al. 2011). However, root hydraulic conductivity was found to be unaffected in experiments on intact tomato and sunflower plants (Shalhevet et al. 1976) and barley (Munns and Passioura 1984). Fiscus and Markhart (1979) and Joly (1989) demonstrated that root hydraulic conductivity (Lp_r) is related to the relative root biomass and root to shoot ratios. Rewald et al. (2011d) found higher sap flow densities in the coarse roots of a salt-tolerant *Olea europaea* variety which possessed higher fine root biomasses than a salt-sensitive variety under moderate saline irrigation (4.2 dS m^{-1}).

Water transport within plants can be divided into several discrete steps, one of which is radial flow in roots. Unless the roots are very long, as in woody plants, or tracheary elements are largely embolized during water stress, the radial pathway is commonly believed to be the path of greatest resistance to water flow in the plant (Steudle and Peterson 1998). The current model of radial water flow identifies three major pathways: (i) water traveling through the apoplast of the cells in the root cortex, toward the endodermis and the xylem vessels; (ii) symplastic water transfer where water goes through cells and remains in the cytoplasm, travelling in the membrane continuum (endoplasmic reticulum and plasmodesmata); and (iii) a pathway through the vacuoles of cells (Steudle and Peterson 1998; Steudle 2000). It is considered that (ii) and (iii) represent the cell-to-cell pathway, as these components are difficult to separate and both use membrane transporters ("aquaporins", see below). Several reports have shown intra- and inter-specific differences in the relative proportion of water travelling through each of these pathways (Vadez et al. 2007 and references within). Because the predominance of one pathway could have a strong influence on the regulation of water uptake, with or without salt-induced water stress, future studies are needed to determine potential changes in radial water pathways under salinity.

More generally, many attempts have been made to correlate root hydraulic conductivity (Lp_r) with anatomical features (for an overview see Steudle and Peterson 1998). Many investigators have focused on the endo- and exodermal suberin deposits since this hydrophobic substance would be expected to reduce the Lp_r of the root

(O'Leary 1969; Zimmermann et al. 2000). Indeed, excess salinity is known to increase the suberization of the endo- and exodermis (see above). However, only in some cases did the results show the expected correlation (Zimmermann and Steudle 1998; North and Nobel 1998; Krishnamurthy et al. 2011), but in others, they did not (Hodges and Vaadia 1964; Clarkson et al. 1987), possibly because other factors, such as passage cells, aquaporin expression and nutrient deficiencies, strongly influence Lp_r (Clarkson et al. 2000). Recently, Krishnamurthy et al. (2011) demonstrated that the apoplastic barriers deposited in *Oryza sativa* under moderate salinity stress resist the flow of bulk water and dissolved solutes, resulting in a reduced uptake of Na^+ into shoots and, consequently, in better survival under subsequent acute stress but also in a large reduction in hydraulic conductivity. The salinity-induced barriers had smaller pore sizes compared with the pre-existing barriers, and the reduction in hydraulic conductivity was consistent with data on the deposition of additional suberin during salt stress. Similarly, Karahara et al. (2004) suggested that salinity regulates the width of the Casparian strip in maize. The effects of NaCl on the ultrastructural morphology of this strip revealed that the radial width of the lignified region and the tightly adhering region of the plasma membrane both increased under salt stress, likely causing a decrease in radial conductivity. However, Sutka et al. (2011) found no correlation between suberization patterns and root hydraulic properties in salt-stressed *Arabidopsis thaliana* roots, requiring future investigation of the topic.

Aquaporins belong to the major intrinsic protein (MIP) family of trans-membrane channels, which permit the selective membrane passage of water (and a few other compounds) but not of H^+ and other ions (Chrispeels et al. 1995; Hill et al. 2004) through the plasma lemma (by PIPs) and the tonoplast (by TIPs). It has been reported that salinity influences aquaporin expression, most probably via phytohormones like ABA and gibberellic acid (Mariaux et al. 1998; Siefritz et al. 2002). The enhanced expression of the PIP1 gene and the abundance of its protein could contribute to regulating root water permeability and, consequently, to better tolerance of the osmotic stress generated by salinity (Aroca et al. 2007; Jahromi et al. 2008). Recently, Sutka et al. (2011) revealed a positive overall correlation between Lp_r and certain highly expressed PIP transcripts in *Arabidopsis thaliana*. The often observed initial decrease of Lp_r upon salt exposure may be caused by an osmotic shock as a result of an aquaporin conformational change caused by negative pressures (Wan et al. 2004); applying a NaCl concentration of 50 mM to maize roots in two steps (25 mM each) reduced the conductivity of cortical cells to a lesser extent than when 50 mM NaCl was applied all at once. See Aroca et al. (2012) for a recent review on aquaporin regulation under salinity. Additionally, in salt-stressed *Brassica* plants, the apoplastic pathway was reduced by 60% with respect to control plants, which probably indicates that phi thickenings (see above) can affect the apoplastic in-flow of water to the stele (López-Pérez et al. 2007). It has been shown that the functionality of aquaporins was greatly reduced in NaCl-treated broccoli plants in the short term (7 days) but to a lesser extent in the longer term (14 days) (López-Berenguer et al. 2006), which is directly related to water flow through the symplastic pathway. The fact that phi thickening developed after 14 days, when aquaporins' functionality was partially restored, might point to a rather long-term acclimation mechanism by which plants can control water uptake under excess salinity (López-Pérez et al. 2007).

Because salinity can influence the root branching pattern (see above), the possibility that salinity could alter the relative proportions of highly radial conductive and less radial conductive root segments is likely (Joly 1989; Rewald et al. 2012). The decrease in water flux rates of root-branches under salinity was mainly caused by a significant reduction (>80%) of the water uptake by root tips (Rewald et al. 2012). The degree of flux reduction was in accordance with measurements on apical segments of corn roots, among which Lp_r was reduced by 80% under salinity (Evlagon et al. 1990). Furthermore, if salinity decreased branching and if axial resistance is a significant component of total root resistance (Passioura 1972), then Lp_r based on radial resistance would underestimate the true value if membrane permeability remained unchanged.

Salt stress modifies primary and secondary xylem differentiation (see above), and Sánchez-Aguayo et al. (2004) found that the average number of lignified cells in vascular bundles was significantly greater in tomato plants under salinity. Presumably, this increase may enhance the cell-to-cell pathway for water transport and greater ion selectivity and uptake. A high vessel density offers a double advantage with respect to conductive safety. First, when the same number of vessels is cavitated, a higher percentage of the transport system remains functional in high vessel density wood compared with low vessel density wood (Baas et al. 1983; Villar-Salvador et al. 1997). Second, a high proportion of vessels are in contact with each other via inter-vessel pits since vessels do not follow a straight line but twist along their path (Kitin et al. 2004). Therefore, embolised vessels can be circumvented by means of the high number of alternative routes for the water transport. According to the air-seeding hypothesis, small vessel diameters can be associated with small pit pore diameters within a species and, thus, to cavitation resistance (Tyree and Sperry 1989). Therefore, declining vessel dimensions with an increase in physiological water stress were expected. Future studies need to test the hypothesis by Céccoli et al. (2011) that "a reduction in vascular tissue dimensions more than any reduction in other tissue, would limit growth under salt stress conditions".

6.7.2 *Nutrient Uptake and Symbiotic Interactions*

In saline soils, the presence of, for example, NaCl alters the nutritional balance of plants, resulting in high ratios of Na^+/Ca^{2+}, Na^+/K^+, Na^+/Mg^{2+}, Cl^-/NO_3^-, and $Cl^-/H_2PO_4^-$, which may cause reductions in growth (e.g., *Citrus* spp., Ruiz et al. 1997). Salt can affect nutrient uptake through competitive interactions between ions or by affecting the ion selectivity of membranes (Cramer et al. 1986; Grattan and Grieve 1992). Examples of these effects include Na^+-induced Ca^{2+} or K^+-deficiencies, or both, and Ca^{2+}-induced Mg^{2+}-deficiencies (Manchanda and Garg 2008). However, while the nutritional status of salt-stressed plants is addressed regularly, studies on the physiological changes underlying the modified nutrition are scarce. Furthermore, the decrease of uptake and the impaired physiological mechanisms under salinity often differ between species and varieties.

Potassium is an essential factor in protein synthesis, glycolytic enzymes, and photosynthesis, osmotica mediating cell expansion and other turgor-driven processes, highlighting the important role of potassium for salt tolerance in plants (Alemán et al. 2011). Protein synthesis requires physiological K^+ concentrations of 100–150 mM and is inhibited by Na^+ concentrations above 100 mM through competition by Na^+ for K^+-binding sites (Blumwald 2000 and references within). Potassium acquisition from the soil solution takes place through epidermal and cortical root cells. Plants use low-affinity (i.e. channels) and high-affinity transporters (e.g. HAK1-type) for potassium uptake from the growth medium (Alemán et al. 2011). Because high affinity transporters allow for a higher K^+/Na^+ discrimination than low affinity systems, root cells were suggested to shift from low to high affinity K^+ uptake under salinity. Indeed, increased K^+/Na^+ discrimination of a high affinity potassium transporter (HKT1) from wheat has been shown to increase the salt tolerance of yeast strains deficient in K^+ uptake (Rubio et al. 1995). However, sodium can inhibit K^+ transporters and HAK1-type gene expression in salt-sensitive plants, even in small concentrations and when plants were starved of K^+ in the presence of Na^+ (Kim et al. 1998; Alemán et al. 2011 and references within); the high-affinity potassium transporter (HAK1) in barley allows sodium permeation (Santa-Maria et al. 1997). However, in the halophyte *Tellungiella halophila*, salinity decreased HAK5 transcription less than in its salt-sensitive relative *Arabidopsis* (Volkov and Amtmann 2006), and in the halophyte *Messembryanthemum crystallinum*, some HAK genes were even up-regulated in the presence of salinity (Su et al. 2002). Because K^+/Na^+ homeostasis is key to salinity tolerance, a difference between salt-sensitive and salt-tolerant species may reside in the differential quantity (amount of protein) or quality (activity under salt stress) of K^+ and/or Na^+ transporters and channels in roots and aboveground organs (Alemán et al. 2009; Rubio et al. 2010).

Beside impaired K^+/Na^+ homeostasis, there is evidence that salt stress inhibits the uptake and transport of macro-nutrients, such as N and P (Maas et al. 1979; Aslam et al. 1984). Nitrogen is the element that plants require in the largest amounts and is needed for most plant cell components, including amino and nucleic acids. Phosphorus is, for example, a constituent of nucleic acids, phospholipids and -proteins, and ATP. Therefore, N and P deficiency rapidly inhibits plant growth. While saline conditions may reduce soil-N mineralization under severe salinity, thus lowering the N availability (Laura 1977), a reduced N and P uptake may be also attributed to a decreased transpiration rate and, thus, to impaired transportation of elements from soil into roots to shoots via mass-flow (Hu and Schmidhalter 2005). However, some studies showed that salinity instead reduces the NO_3^- concentrations without affecting the total N content, and that the addition of NO_3^- results in a reduction in Cl^- uptake and accumulation due to NO_3^-/Cl^- antagonism (Hu and Schmidhalter 2005 and references within). While the N form might influence the sensitivity of the plant to salinity, for example, indirectly via NH_4^+-sensitive K^+

transporters (Figueira 2009; Voigt et al. 2009), future studies should focus on alternative strategies to increase plants' resistance to salinity, e.g. by genetically engineering a salt-tolerant, high-affinity transport system for NH_4^+ and/or NO_3^- (Kshatriya et al. 2009). In legumes, salinity may act on the *Rhizobium* symbiosis directly via affecting infection and nodule development (e.g. by reduced numbers of root hairs) and nitrogen fixation capacity (Delgado et al. 1994 and references within) and indirectly by reducing available carbon (Brugnoli and Lauteri 1991). Decreased N_2 fixation may be due, in part, to the salt-induced decrease in bacteroid O_2 uptake capacity, respiration and leghemoglobin content (Delgado et al. 1994); leghemoglobins are monomeric, soluble proteins that conduct O_2 to the bacteroids at a low constant concentration, protecting nitrogenase against O_2 damage while allowing oxidative phosphorylation. For improving symbiotic nitrogen fixation under excess salinity, salt-tolerant legume cultivars, as well as *Rhizobium*-symbionts, have to be selected (Ventorino et al. 2011). See Manchanda and Garg (2008) for a recent review on salinity and its effects on legumes and symbionts.

Champagnol (1979) has shown that there are three types of reaction to salinity, with respect to the P content of plant tissues; P nutrition was increased in some crops (e.g. *Sorghum bicolor*, *Zea mays* and sesame), reduced in others (tomatoes, onions and barley), and unchanged in millet, cabbage, *Brassica oleracea* var., carrots and wheat. The availability of P can be reduced in saline soils because of ionic-strength effects that reduce the activity of P and because P concentrations in the soil solution are tightly controlled by absorption processes and by the low solubility of Ca-P minerals (Grattan and Grieve 1999); thus, some plant species will respond positively to an additional P supply. However, the study of the effect of NaCl and P nutrition on alfalfa by Rogers et al. (2003) illustrated that high or non-limiting P levels do not affect the response of alfalfa to NaCl. Martinez and Läuchli (1994) found that high NaCl inhibited P uptake in the mature root zone but enhanced P uptake in the root tips of *Gossypium hirsutum*. They speculate that salinity may induce alkalinization of the cytoplasm in root tip cells, increasing the transmembrane pH gradient and, therefore, P uptake. Results showed an inhibition of ^{32}P transport within the roots and from the roots to the shoots in the salt treatment in *Gossypium hirsutum* and *Lactuca sativa* (Martinez and Läuchli 1994; Martinez et al. 1996). Although P deficiency under salinity is highly species-specific, salinity inhibited P uptake, in general, more severely at a low P concentration than at a high P concentration (Martinez and Läuchli 1994). Thus, mycorrhizal symbionts could play an important role in promoting salinity. Although the mechanism by which AMF improves salt resistance remains unclear, many studies have indicated that the enhanced salt tolerance mainly contributes to mycorrhizal-mediated enhancement of host mineral nutrient uptake, especially of immobile soil nutrients such as P, Cu, and Zn (He et al. 2007 and references within).

Because aquaporins do not only mediate the transport of water in roots but also some solutes (Maurel et al. 2008 and references within), changes in aquaporin expression and function due to excess salinity are likely to result in changed uptake or efflux of some micro-nutrients. For example, the aquaporin Lsi1 is expressed on the distal sites of exo- and endodermal root cells of *Oryza* spp. and may contribute, together with the efflux transporter Lsi2, to the transport of silicon acid from the soil solution into the root xylem (Ma et al. 2007). Lsi homologs from *Zea mays* and *Hordeum vulgare* are present in all parts of the plasma membrane in the cells of the root epidermal cell layers and cortex but not in the endodermis (Mitani et al. 2009). To date, no studies have addressed the potentially significant impact of salinity on Lsi transporters or other solute transporting aquaporins in roots (Mitani-Ueno et al. 2011). This is surprising because B concentrations have been found to be reduced in *Triticum aestivum* shoots and grains under salinity (Holloway and Alston 1992). Silicon is a major mineral nutrient in certain plants, such as Gramineae; furthermore, salt stress is mitigated by silicic acid, probably by a reduction of the apoplastic sodium absorption or due to a closure of openings and spaces (Yeo et al. 1999; Zhu et al. 2004; Gong et al. 2006).

6.8 Conclusion and Future Perspective

It is known that the salt tolerance of a plant depends on the combination of regulative and protective traits above and below ground. This chapter gave several examples of root anatomical, morphological and physiological changes under salt stress; information on salinity effects on root systems' plasticity, and its adaptive significance, is still scarce compared to studies on aboveground organs. Furthermore, results are sometimes contradicting and there is a high demand for studies relating changes on the cellular level to tissue and organ functions. To understand the different tolerance strategies adapted by glycophyte and halophyte species to overcome the osmotic effect and ion-specific toxicities in salt-stressed habitats, special attention has to be paid to the overall role of root systems as source of water and nutrients and as sink of carbohydrates. Once identified, root system traits of salt resistance plant species could be used for future breeding efforts.

Acknowledgements B.R. acknowledges partial support by a postdoctoral fellowship awarded by the Jacob Blaustein Center for Scientific Cooperation (BCSC), Israel.

References

Abrol IP, Yadav JSP, Massoud FI (1988) Salt-affected soils and their management. Food and Agriculture Organization of the United Nations, Rome

Akram M, Akhtar S, Javed IUH, Wahid A, Rasul E (2002) Anatomical attributes of different wheat (*Triticum aestivum*) accessions/varieties to NaCl salinity. Intern J Agricul Biol 4:166–168

Albacete A, Ghanem ME, Martinez-Andujar C, Acosta M, Sanchez-Bravo J, Martinez V, Lutts S, Dodd IC, Perez-Alfocea F (2008) Hormonal changes in relation to biomass partitioning and shoot growth impairment in salinized tomato (*Solanum lycopersicum* L.) plants. J Exp Bot 59:4119–4131

Alemán F, Nieves-Cordones M, Martínez V, & Rubio F (2009) Differential regulation of the HAK5 genes encoding the high-affinity K+ transporters of Thellungiella halophila and Arabidopsis thaliana. Environmental and Experimental Botany, 65(2–3), 263–269. doi: 10.1016/j.envexpbot.2008.09.011

Alemán F, Nieves-Cordones M, Martínez V, Rubio F (2011) Root K^+ acquisition in plants: the *Arabidopsis thaliana* model. Plant Cell Physiol 52:1603–1612

Almagro L, Gómez Ros LV, Belchi-Navarro S, Bru R, Ros Barceló A, Pedreño MA (2009) Class III peroxidases in plant defence reactions. J Exp Bot 60:377–390

Arbona V, Flors V, Jacas J, García-Agustín P, Gómez-Cadenas A (2003) Enzymatic and non-enzymatic antioxidant responses of Carrizo citrange, a salt-sensitive citrus rootstock, to different levels of salinity. Plant Cell Physiol 44:388–394

Aroca R, Porcel R, Ruiz-Lozano JM (2007) How does arbuscular mycorrhizal symbiosis regulate root hydraulic properties and plasma membrane aquaporins in Phaseolus vulgaris under drought, cold or salinity stresses? New Phytol 173(4):808–816

Aroca R, Porcel R, Ruiz-Lozano JM (2012) Regulation of root water uptake under abiotic stress conditions. J Exp Bot 63:43–57

Ashraf M (1999) Interactive effect of salt (NaCl) and nitrogen form on growth, water relations and photosynthetic capacity of sunflower (Helianthus annum L.). Ann Appl Biol 135(2):509–513

Ashraf M, Ozturk M, Ahmad MSA (2010) Plant adaptation and phytoremediation. Springer, Dordrecht

Aslam M, Huffaker RC, Rains DW (1984) Early effects of salinity on nitrate assimilation in barley seedlings. Plant Physiol 76:321–325

Azaizeh H, Gunsé B, Steudle E (1992) Effects of NaCl and $CaCl_2$ on water transport across root-cells of maize (*Zea mays* L.) seedlings. Plant Physiol 99:886–894

Baas P, Werker E, Fahn A (1983) Some ecological trends in vessel characters. Iawa Bull 4:141–159

Barber SA (1995) Soil nutrient bioavailability: a mechanistic approach. Wiley, New York

Baumeister W, Merten A (1981) Growth and root anatomy of two subspecies of *Festuca rubra* L. in response to NaCl salinization of the culture solution. Angew Bot 55:401–408

Bazihizina N, Colmer TD, Barrett-Lennard EG (2009) Response to non-uniform salinity in the root zone of the halophyte *Atriplex nummularia*: growth, photosynthesis, water relations and tissue ion concentrations. Ann Bot 104:737–745

Bernstein N, Kafkafi U (2002) Root growth under salinity stress. In: Waisel Y, Eshel Y, Kafkafi U (eds) Plant roots: the hidden half. Marcel Dekker, New York, pp 787–805

Bloom A, Epstein E (1984) Varietal differences in salt-induced respiration in barley. Plant Sci Lett 35:1–3

Blumwald E, Aharon GS, Apse MP (2000) Sodium transport in plant cells. Biochim Biophys Acta (BBA) – Biomembr 1465(1–2):140–151. doi:10.1016/s0005-2736(00)00135-8

Bogeat-Triboulot MB, Brosché M, Renaut J, Jouve L, Le Thiec D, Fayyaz P, Vinocur B, Witters E, Laukens K, Teichmann T, Altman A, Hausman JF, Polle A, Kangasjarvi J, Dreyer E (2007) Gradual soil water depletion results in reversible changes of gene expression, protein profiles, ecophysiology, and growth performance in *Populus euphratica*, a poplar growing in arid regions. Plant Physiol 143:876–892

Boughalleb F, Denden M, Tiba B (2009) Anatomical changes induced by increasing NaCl salinity in three fodder shrubs, *Nitraria retusa*, *Atriplex halimus* and *Medicago arborea*. Acta Physiol Plant 31:947–960

Brinker M, Brosché M, Vinocur B, Abo-Ogiala A, Fayyaz P, Janz D, Ottow EA, Cullmann AD, Saborowski J, Kangasjärvi J, Altman A, Polle A (2010) Linking the salt transcriptome with physiological responses of a salt-resistant *Populus* species as a strategy to identify genes important for stress acclimation. Plant Physiol 154:1697–1709

Brugnoli E, Lauteri M (1991) Effects of salinity on stomatal conductance, photosynthetic capacity, and carbon isotope discrimination of salt-tolerant (*Gossypium hirsutum* L.) and salt-sensitive (*Phaseolus vulgaris* L.) C_3 non-halophytes. Plant Physiol 95:628–635

Burchett MD, Field CD, Pulkownik A (1984) Salinity, growth and root respiration in the Grey mangrove, *Avicennia marina*. Physiol Plantarum 60:113–118

Burssens S, Himanen K, van de Cotte B, Beeckman T, Van Montagu M, Inzé D, Verbruggen N (2000) Expression of cell cycle regulatory genes and morphological alterations in response to salt stress in *Arabidopsis thaliana*. Planta 211:632–640

Cachorro P, Ortiz A, Barcelo AR, Cerdá A (1993) Lignin deposition in vascular tissues of *Phaseolus vulgaris* roots in response to salt stress. Phyton Ann Rei Bot 33:33–40

Casimiro I, Beeckman T, Graham N, Bhalerao R, Zhang H, Casero P, Sandberg G, Bennett MJ (2003) Dissecting *Arabidopsis* lateral root development. Trends Plant Sci 8:165–171

Céccoli G, Ramos JC, Ortega LI, Acosta JM, Perreta MG (2011) Salinity induced anatomical and morphological changes in *Chloris gayana* Kunth roots. Biocell 35:9–17

Champagnol F (1979) Relationships between phosphate nutrition of plants and salt toxicity. Phosphorus Agric 76:35–43

Chartzoulakis KS (2005) Salinity and olive: growth, salt tolerance, photosynthesis and yield. Agric Water Manage 78:108–121

Cheng H, Wang YS, Ye ZH, Chen DT, Wang YT, Peng YL, Wang LY (2012) Influence of N deficiency and salinity on metal (Pb, Zn and Cu) accumulation and tolerance by *Rhizophora stylosa* in relation to root anatomy and permeability. Environ Poll 164:110–117

Chhabra R (1996) Soil salinity and water quality. Taylor and Francis, New York

Chhun T, Uno Y, Taketa S, Azuma T, Ichii M, Okamoto T, Tsurumi S (2007) Saturated humidity accelerates lateral root development in rice (*Oryza sativa* L.) seedlings by increasing phloem-based auxin transport. J Exp Bot 58:1695–1704

Chrispeels MJ, Daniels MJ, Weig A (1995) Aquaporins facilitate water transport through the plasma-membrane and the tonoplast. Plant Physiol 108:6

Cimato A, Castelli S, Tattini M, Traversi ML (2010) An ecophysiological analysis of salinity tolerance in olive. Environ Exp Bot 68:214–221

Clarkson DT, Robards AW, Stephens JE, Stark M (1987) Suberin lamellae in the hypodermis of maize (*Zea mays*) roots - Development and factors affecting the permeability of hypodermal layers. Plant Cell Environ 10:83–93

Clarkson DT, Carvajal M, Henzler T, Waterhouse RN, Smyth AJ, Cooke DT, Steudle E (2000) Root hydraulic conductance: diurnal aquaporin expression and the effects of nutrient stress. J Exp Bot 51:61–70

Clought BF, Sim RG (1989) Changes in gas exchange characteristics and water use efficiency of mangroves in response to salinity and vapour pressure deficit. Oecologia 79:38–44

Cramer GR, Läuchli A, Epstein E (1986) Effects of NaCl and $CaCl_2$ on ion activities in complex nutrient solutions and root-growth of cotton. Plant Physiol 81:792–797

Cramer GR, Epstein E, Läuchli A (1988) Kinetics of root elongation of maize in response to short-term exposure to NaCl and elevated calcium-concentration. J Exp Bot 39:1513–1522

Deak KI, Malamy J (2005) Osmotic regulation of root system architecture. Plant J 43:17–28

Degano CAM (1999) Morphology and anatomy of *Tessaria absinthioides* (Hook. et Arn.) DC. under salinity conditions [In Portuguese, English abstract]. Rev Bras Bot 22:357–363

Degenhardt B, Gimmler H (2000) Cell wall adaptations to multiple environmental stresses in maize roots. J Exp Bot 51:595–603

del Amor F, Cuadra-Crespo P (2011) Alleviation of salinity stress in broccoli using foliar urea or methyl-jasmonate: analysis of growth, gas exchange, and isotope composition. Plant Growth Regul 63:55–62

Delgado MJ, Ligero F, Lluch C (1994) Effects of salt stress on growth and nitrogen fixation by pea, faba-bean, common bean and soybean plants. Soil Biol Biochem 26:371–376

Dunlap JR, Binzel ML (1996) NaCI reduces indole-3-acetic acid levels in the roots of tomato plants independent of stress-induced abscisic acid. Plant Physiol 112:379–384

Echeverria M, Scambato AA, Sannazzaro AI, Maiale S, Ruiz OA, Menéndez AB (2008) Phenotypic plasticity with respect to salt stress response by *Lotus glaber*: the role of its AM fungal and rhizobial symbionts. Mycorrhiza 18:317–329

Eckstein D, Liese W, Plossl P (1978) Histometrische Untersuchungen zur unterschiedlichen Streusalztoleranz von Weiden (Salix spp.). Forstwissenschaftliches Centralblatt 97(6):335–341

Ehlting B, Dluzniewska P, Dietrich H, Selle A, Teuber M, Hansch R, Nehls U, Polle A, Schnitzler JP, Rennenberg H, Gessler A (2007) Interaction of nitrogen nutrition and salinity in Grey poplar (*Populus tremula* x *alba*). Plant Cell Environ 30:796–811

Eissenstat DM, Achor DS (1999) Anatomical characteristics of roots of citrus rootstocks that vary in specific root length. New Phytol 141:309–321

Enstone DE, Peterson A, Ma FS (2003) Root endodermis and exodermis: structure, function, and responses to the environment. J Plant Growth Regul 21:335–351

Epron D, Toussaint ML, Badot PM (1999) Effects of sodium chloride salinity on root growth and respiration in oak seedlings. Ann For Sci 56:41–47

Evlagon D, Ravina I, Neumann P (1990) Interactive effects of salinity and calcium on hydraulic conductivity, osmotic adjustment, and growth in primary roots of maize seedlings. Israel J Bot 39:239–247

Figueira E (2009) Pea cultivation in saline soils: influence of nitrogen nutrition. In: Khan MS, Zaidi A, Musarrat J (eds) Microbial strategies for crop improvement. Springer, Heidelberg, pp 267–286

Fiscus EL, Markhart AH (1979) Relationships between root-system water transport properties and plant size in *Phaseolus*. Plant Physiol 64:770–773

Fitter AH, Stickland TR, Harvey ML, Wilson GW (1991) Architectural analysis of plant-root systems. 1. Architectural correlates of exploitation efficiency. New Phytol 118:375–382

Fukaki H, Okushima Y, Tasaka M (2007) Auxin-mediated lateral root formation in higher plants. Int Rev Cytol 256:111–137

Ghanem M, Hichri I, Smigocki A, Albacete A, Fauconnier ML, Diatloff E, Martinez-Andujar C, Lutts S, Dodd I, Pérez-Alfocea F (2011) Root-targeted biotechnology to mediate hormonal signalling and improve crop stress tolerance. Plant Cell Rep 30:807–823

Ghassemi F, Jakeman AJ, Nix HA (1995) Salinisation of land and water resources: human causes, extent, management and case studies. Cab International, Wallingford

Gill RA, Jackson RB (2000) Global patterns of root turnover for terrestrial ecosystems. New Phytol 147:13–31

Glenn EP, Brown JJ, Blumwald E (1999) Salt tolerance and crop potential of halophytes. Crit Rev Plant Sci 18:227–255

Gong HJ, Randall DP, Flowers TJ (2006) Silicon deposition in the root reduces sodium uptake in rice (*Oryza sativa* L.) seedlings by reducing bypass flow. Plant Cell Environ 29:1970–1979

Grattan SR, Grieve CM (1992) Mineral element acquisition and growth-response of plants grown in saline environments. Agr Ecosyst Environ 38:275–300

Grattan SR, Grieve CM (1999) Salinity mineral nutrient relations in horticultural crops. Sci Hortic 78(1–4):127–157

Green SM, Machin R, Cresser MS (2008) Effect of long-term changes in soil chemistry induced by road salt applications on N-transformations in roadside soils. Environ Poll 152:20–31

Grierson C, Schiefelbein J (2009) Genetics of root hair formation. In: Murphy AS, Peer W, Schulz B (eds) The plant plasma membrane. Springer, Berlin, pp 1–25

Gruber V, Zahaf O, Diet A, Zélicourt A, Lorenzo L, Crespi M (2011) Impact of the environment on root architecture in dicotyledoneous plants. In: de Oliveira AC, Varshney RK (eds) Root genomics. Springer, Heidelberg, pp 113–132

Gucci R, Tattini M (1997) Salinity tolerance in olive. Horticultural Rev 21:177–214

Gucci R, Lombardini L, Tattini M (1997) Analysis of leaf water relations in leaves of two olive (Olea europaea) cultivars differing in tolerance to salinity. Tree Physiol 17(1):13–21

Hajibagheri MA, Yeo AR, Flowers TJ (1985) Salt tolerance in *Suaeda maritima* (L.) Dum. Fine-structure and ion concentrations in the apical region of roots. New Phytol 99:331–343

Hameed M, Ashraf M, Naz N, Al-Qurainy F (2010) Anatomical adaptations of *Cynodon dactylon* (L.) PERS., from the salt range Pakistan, to salinity stress. I. Root and stem anatomy. Pakistan J Bot 42:279–289

Hartig K, Beck E (2006) Crosstalk between auxin, cytokinins, and sugars in the plant cell cycle. Plant Biol 8:389–396

He XJ, Mu RL, Cao WH, Zhang ZG, Zhang JS, Chen SY (2005) AtNAC2, a transcription factor downstream of ethylene and auxin signaling pathways, is involved in salt stress response and lateral root development. Plant J 44:903–916

He ZQ, He CX, Zhang ZB, Zou ZR, Wang HS (2007) Changes of antioxidative enzymes and cell membrane osmosis in tomato colonized by arbuscular mycorrhizae under NaCl stress. Colloid Surfaces B 59:128–133

Hilal M, Zenoff AM, Ponessa G, Moreno H, Massa EM (1998) Saline stress alters the temporal patterns of xylem differentiation and alternative oxidase expression in developing Soybean roots. Plant Physiol 117:695–701

Hill AE, Shachar-Hill B, Shachar-Hill Y (2004) What are aquaporins for? J Membrane Biol 197:1–32

Hodges TK, Vaadia Y (1964) Chloride uptake and transport in roots of different salt status. Plant Physiol 39:109–114

Holloway RE, Alston AM (1992) The effects of salt and boron on growth of wheat. Aust J Agr Res 43:987–1001

Hu YC, Schmidhalter U (2005) Drought and salinity: a comparison of their effects on mineral nutrition of plants. J Plant Nutr Soil Sci 168:541–549

Huang J, Redmann RE (1995) Salt tolerance of *Hordeum* and *Brassica* species during germination and early seedling growth. Can J Plant Sci 75:815–819

Huang CX, Van Steveninck RFM (1990) The role of particular pericycle cells in apoplastic transport in root-meristems of barley. J Plant Physiol 135:554–558

Jackson RB, Mooney HA, Schulze ED (1997) A global budget for fine root biomass, surface area, and nutrient contents. PNAS 94:7362–7366

Jahromi F, Aroca R, Porcel R, Ruiz-Lozano JM (2008) Influence of salinity on the In vitro development of Glomus intraradices and on the In vivo physiological and molecular responses of mycorrhizal lettuce plants. Microb Ecol 55(1):45–53

Jbir N, Chaibi W, Ammar S, Jemmali A, Ayadi A (2001) Root growth and lignification of two wheat species differing in their sensitivity to NaCl, in response to salt stress. CR Acad Sci III-Vie 324:863–868

Jeschke WD, Stelter W (1976) Measurement of longitudinal ion profiles in single roots of *Hordeum* and *Atriplex* by use of flameless atomic-absorption spectroscopy. Planta 128:107–112

Joly RJ (1989) Effects of sodium-chloride on the hydraulic conductivity of soybean root systems. Plant Physiol 91:1262–1265

Jones MP (1985) Genetic analysis of salt tolerance in mangrove swamp rice. Rice Genetics I:411–422

Kacprzyk J, Daly CT, McDabe PF (2011) The botanical dance of death: programmed cell death in plants. In: Kader JC, Delseny M (eds) Advances in botanical research, vol 60. Academic, San Diego, pp 170–263

Karahara I, Ikeda A, Kondo T, Uetake Y (2004) Development of the Casparian strip in primary roots of maize under salt stress. Planta 219:41–47

Katsuhara M, Kawasaki T (1996) Salt stress induced nuclear and DNA degradation in meristematic cells of barley roots. Plant Cell Physiol 37:169–173

Kawasaki S, Borchert C, Deyholos M, Wang H, Brazille S, Kawai K, Galbraith D, Bohnert HJ (2001) Gene expression profiles during the initial phase of salt stress in rice. Plant Cell 13:889–905

Kim EJ, Kwak JM, Uozumi N, Schroeder JI (1998) AtKUP1: an *Arabidopsis* gene encoding high-affinity potassium transport activity. Plant Cell 10:51–62

Kitin PB, Fujii T, Abe H, Funada R (2004) Anatomy of the vessel network within and between tree rings of *Fraxinus lanuginosa* (Oleaceae). Am J Bot 91:779–788

Koyro HW (1997) Ultrastructural and physiological changes in root cells of sorghum plants (*Sorghum bicolor* x *S. sudanensis* cv. Sweet Sioux) induced by NaCl. J Exp Bot 48:693–706

Koyro HW (2004) Ultrastructural effects of salinity in higher plants. In: Läuchli A, Lüttge U (eds) Salinity: environment – plants – molecules. Springer, Den Hague, pp 139–157

Kozlowski TT (1997) Responses of woody plants to flooding and salinity. Tree Physiol Monograph 1:1–29

Kozlowski TT, Pallardy SG (1997) Physiology of woody plants. Elsevier, London

Kramer D, Läuchli A, Yeo AR, Gullasch J (1977) Transfer cells in roots of phaseolus-coccineus – ultrastructure and possible function in exclusion of sodium from shoot. Ann Bot 41(175):1031–1040

Kramer D, Anderson WP, Preston J (1978) Transfer cells in the root epidermis of *Atriplex hastata* L. as response to salinity: a comparative cytological and X-ray microprobe investigation. Aust J Plant Physiol 5:739–747

Krishnamurthy P, Ranathunge K, Nayak S, Schreiber L, Mathew MK (2011) Root apoplastic barriers block Na+ transport to shoots in rice (*Oryza sativa* L.). J Exp Bot 62:4215–4228

Kshatriya K, Singh J, Singh D (2009) Salt tolerant mutant of *Anabaena doliolum* exhibiting efficient ammonium uptake and assimilation. Physiol Mol Biol Plants 15:377–381

Kurth E, Cramer GR, Läuchli A, Epstein E (1986) Effects of NaCl and $CaCl_2$ on cell enlargement and cell production in cotton roots. Plant Physiol 82:1102–1106

Laura RD (1977) Salinity and nitrogen mineralization in soil. Soil Biol Biochem 9:333–336

Lin CC, Kao CH (2001) Cell wall peroxidase activity, hydrogen peroxide level and NaCl-inhibited root growth of rice seedlings. Plant Soil 230:135–143

Liu AR, Zhao KF (2005) Osmotica accumulation and its role in osmotic adjustment in *Thellungiella halophila* under salt stress [In Chinese, English abstract]. Zhi Wu Sheng Li Yu Fen Zi Sheng Wu Xue Xue Bao 31:389–395

Liu T, Van Staden J, Cress WA (2000) Salinity induced nuclear and DNA degradation in meristematic cells of soybean (*Glycine max* (L.)) roots. Plant Growth Regul 30:49–54

Liu SH, Fu BY, Xu HX, Zhu LH, Zhai HQ, Li ZK (2007) Cell death in response to osmotic and salt stresses in two rice (*Oryza sativa* L.) ecotypes. Plant Sci 172:897–902

López-Berenguer C, García-Viguera C, Carvajal M (2006) Are root hydraulic conductivity responses to salinity controlled by aquaporins in Broccoli plants? Plant Soil 279:13–23

López-Pérez L, Fernández-Garcia N, Olmos E, Carvajal M (2007) The phi thickening in roots of broccoli plants: an acclimation mechanism to salinity? Int J Plant Sci 168:1141–1149

Ma SS, Gong QQ, Bohnert HJ (2006) Dissecting salt stress pathways. J Exp Bot 57:1097–1107

Ma JF, Yamaji N, Mitani N, Tamai K, Konishi S, Fujiwara T, Katsuhara M, Yano M (2007) An efflux transporter of silicon in rice. Nature 448:209–212

Maas EV (1993) Salinity and citriculture. Tree Physiol 12:195–216

Maas EV, Ogata G, Finkel MH (1979) Salt-induced inhibition of phosphate transport and release of membrane proteins from barley roots. Plant Physiol 64:139–143

Mahmoodzadeh H (2008) Ultrastructural changes in shoot apical meristem of canola (*Brassica napus* cv. Symbol) treated with sodium chloride. Pakistan J Biol Sci 11:1161–1164

Malamy JE (2009) Lateral root formation. Ann Plant Rev 37:83–126

Manchanda G, Garg N (2008) Salinity and its effects on the functional biology of legumes. Acta Physiol Plant 30:595–618

Mariaux JB, Bockel C, Salamini F, Bartels D (1998) Desiccation- and abscisic acid-responsive genes encoding major intrinsic proteins (MIPs) from the resurrection plant *Craterostigma plantagineum*. Plant Mol Biol 38:1089–1099

Martinez V, Läuchli A (1994) Salt-induced inhibition of phosphate-uptake in plants of cotton (*Gossypium hirsutum* L.). New Phytol 126:609–614

Martinez V, Bernstein N, Läuchli A (1996) Salt-induced inhibition of phosphorus transport in lettuce plants. Physiol Plant 97:118–122

Maurel C, Verdoucq L, Luu DT, Santoni V (2008) Plant aquaporins: membrane channels with multiple integrated functions. Ann Rev Plant Biol 59:595–624

Mckee KL (1993) Soil physicochemical patterns and mangrove species distribution – reciprocal effects? J Ecol 81:477–487

Mei-fang D, Wang-jun Y, Fu-de S (2005) Relations between anatomic structures of vegetative organs of *Thellungiella halophila* and its brackish environment. Xibei Zhiwu Xuebao 25:1077–1082

Miller RW, Donahue RL (1995) Soils in our environment. Prudence Hall, Englewood
Mitani N, Chiba Y, Yamaji N, Ma JF (2009) Identification and characterization of maize and barley Lsi2-like silicon efflux transporters reveals a distinct silicon uptake system from that in rice. Plant Cell Online 21:2133–2142
Mitani-Ueno N, Yamaji N, Zhao FJ, Ma JF (2011) The aromatic/arginine selectivity filter of NIP aquaporins plays a critical role in substrate selectivity for silicon, boron, and arsenic. J Exp Bot 62:4391–4398
Mittova V, Guy M, Tal M, Volokita M (2004) Salinity up – regulates the antioxidative system in root mitochondria and peroxisomes of the wild salt – tolerant tomato species *Lycopersicon pennellii*. J Exp Bot 55:1105–1113
Mostafazadeh-Fard B, Mansouri H, Mousavi SF, Feizi M (2009) Effects of different levels of irrigation water salinity and leaching on yield and yield components of wheat in an arid region. J Irrig Drain E-ASCE 135:32–38
Munns R (2002) Comparative physiology of salt and water stress. Plant Cell Environ 25(2):239–250
Munns R (2005) Genes and salt tolerance: bringing them together. New Phytol 167:645–663
Munns R, Passioura JB (1984) Hydraulic resistance of plants. 3. Effects of NaCl in barley and lupin. Aust J Plant Physiol 11:351–359
Munns R, Termaat A (1986) Whole-plant responses to salinity. Aust J Plant Physiol 13:143–160
Munns R, Tester M (2008) Mechanisms of salinity tolerance. Ann Rev Plant Biol 59:651–681
Nadelhoffer KJ, Raich JW (1992) Fine root production estimates and belowground carbon allocation in forest ecosystems. Ecology 73:1139–1147
Nedjimi B, Daoud Y, Touati M (2006) Growth, water relations, proline and ion content of in vitro cultured *Atriplex halimus* subsp. *schweinfurthii* as affected by $CaCl_2$. Commun Biometry Crop Sci 1:79–89
Neumann PM (1995) Inhibition of root growth by salinity stress: toxicity or an adaptive biophysical response? In: Baluska F, Ciamporova M, Gasparikova O, Barlow PW (eds) Structure and function of roots. Kluwer, Dordrecht, pp 299–304
Neumann PM, Azaizeh H, Leon D (1994) Hardening of root cell-walls – a growth-inhibitory response to salinity stress. Plant Cell Environ 17:303–309
Neves GYS, Marchiosi R, Ferrarese MLL, Siqueira-Soares RC, Ferrarese O (2010) Root growth inhibition and lignification induced by salt stress in soybean. J Agron Crop Sci 196:467–473
Nibau C, Gibbs DJ, Coates JC (2008) Branching out in new directions: the control of root architecture by lateral root formation. New Phytol 179:595–614
Niknam V, Ebrahimzadeh H (2002) Free proline content in *Astragalus* species. Pakistan J Bot 34:129–134
Niu DK, Wang MG, Wang YF (1997) Plant cellular osmotica. Acta Biotheor 45:161–169
Norby RJ, Jackson RB (2000) Root dynamics and global change: seeking an ecosystem perspective. New Phytol 147:3–12
Norrström AC, Bergstedt E (2001) The impact of road de-icing salts (NaCl) on colloid dispersion and base cation pools in roadside soils. Water Air Soil Poll 127:281–299
North GB, Nobel PS (1998) Water uptake and structural plasticity along roots of a desert succulent during prolonged drought. Plant Cell Environ 21:705–713
O'Leary PW (1969) The effect of salinity on permeability of roots to water. Israel J Bot 18:1–9
Oron G, DeMalach Y, Gillerman L, David I, Rao VP (1999) Improved saline-water use under subsurface drip irrigation. Agric Water Manage 39:19–33
Pagès L, Kervella J (1990) Growth and development of root systems: geometrical and structural aspects. Acta Biotheor 38:289–302
Parida AK, Das AB (2005) Salt tolerance and salinity effects on plants: a review. Ecotox Environ Safe 60:324–349
Passioura JB (1972) The effect of root geometry on yield of wheat growing on stored water. Aust J Agric Res 23:745–752
Passioura JB (1977) Determining soil-water diffusivities from 1-Step outflow experiments. Aust J Soil Res 15:1–8

Perica S, Goreta S, Selak GV (2008) Growth, biomass allocation and leaf ion concentration of seven olive (*Olea europaea* L.) cultivars under increased salinity. Sci Hortic 117:123–129
Petridis A, Therios I, Samouris G, Tananaki C (2012) Salinity-induced changes in phenolic compounds in leaves and roots of four olive cultivars (*Olea europaea* L.) and their relationship to antioxidant activity. Environ Exp Bot 79:37–43
Petrusa LM, Winicov I (1997) Proline status in salt-tolerant and salt-sensitive alfalfa cell lines and plants in response to NaCl. Plant Physiol Biochem 35:303–310
Peyrano G, Taleisnik E, Quiroga M, deForchetti SM, Tigier H (1997) Salinity effects on hydraulic conductance, lignin content and peroxidase activity in tomato roots. Plant Physiol Biochem 35:387–393
Pitman M, Läuchli A (2004) Global impact of salinity and agricultural ecosystems. In: Läuchli A, Lüttge U (eds) Salinity: environment – plants – molecules. Springer, Dordrecht, pp 3–20
Pregitzer KS, DeForest JL, Burton AJ, Allen MF, Ruess RW, Hendrick RL (2002) Fine root architecture of nine North American trees. Ecol Monogr 72:293–309
Rabie GH, Almadini AM (2005) Role of bioinoculants in development of salt-tolerance of Vicia faba plants under salinity stress. Afr J Biotech 4:210–222
Rahayu YS, Walch-Liu P, Neumann G, Römheld V, von Wirén N, Bangerth F (2005) Root-derived cytokinins as long-distance signals for NO_3–induced stimulation of leaf growth. J Exp Bot 56:1143–1152
Ramoliya P, Pandey A (2006) Effect of salinization of soil on emergence, growth and survival of *Albizzia lebbek* seedlings. Trop Ecol 47:27–36
Reinhardt DH, Rost TL (1995) Salinity accelerates endodermal development and induces an exodermis in cotton seedling roots. Environ Exp Bot 35:563–574
Reinoso H, Sosa L, Ramirez L, Luna V (2004) Salt-induced changes in the vegetative anatomy of *Prosopis strombulifera* (Leguminosae). Can J For Res 82:618–628
Rengasamy P (2006) World salinization with emphasis on Australia. J Exp Bot 57:1017–1023
Rewald B, Ephrath JE, Rachmilevitch S (2011a) A root is a root is a root? – Water uptake rates of *Citrus* root orders. Plant Cell Environ 34:33–42
Rewald B, Rachmilevitch S, Ephrath JE (2011b) Salt stress effects on root systems of two mature olive cultivars. Acta Hortic 888:109–128
Rewald B, Leuschner C, Wiesman Z, Ephrath JE (2011c) Influence of salinity on root hydraulic properties of three olive varieties. Plant Biosyst 145:12–22
Rewald B, Rachmilevitch S, McCue MD, Ephrath JE (2011d) Influence of saline drip-irrigation on fine root and sap-flow densities of two mature olive varieties. Environ Exp Bot 72:107–114
Rewald B, Raveh E, Gendler T, Ephrath JE, Rachmilevitch S (2012) Phenotypic plasticity and water flux rates of *Citrus* root orders under salinity. J Exp Bot. doi:10.1093/jxb/err457
Roberts T, Lazarovitch N, Warrick AW, Thompson TL (2009) Modeling salt accumulation with subsurface drip irrigation using HYDRUS-2D. SSSAJ 73:233–240
Rogers ME, Grieve CM, Shannon MC (2003) Plant growth and ion relations in lucerne (*Medicago sativa* L.) in response to the combined effects of NaCl and P. Plant Soil 253:187–194
Rubinigg M, Wenisch J, Elzenga JTM, Stulen I (2004) NaCl salinity affects lateral root development in *Plantago maritima*. Funct Plant Biol 31:775–780
Rubio F, Gassmann W, Schroeder JI (1995) Sodium-driven potassium uptake by the plant potassium transporter HKT1 and mutations conferring salt tolerance. Science 270:1660–1663
Rubio JS, Garcia-Sanchez F, Rubio F, Garcia AL, Martinez V (2010) The importance of K+ in ameliorating the negative effects of salt stress on the growth of pepper plants. Eur J Hortic Sci 75:33–41
Ruiz D, Martínez V, Cerdá A (1997) Citrus response to salinity: growth and nutrient uptake. Tree Physiol 17:141–150
Sachs T (2005) Auxin's role as an example of the mechanisms of shoot/root relations. Plant Soil 268:13–19
Samarajeewa PK, Barrero RA, Umeda-Hara C, Kawai M, Uchimiya H (1999) Cortical cell death, cell proliferation, macromolecular movements and *rTip1* expression pattern in roots of rice (*Oryza sativa* L.) under NaCl stress. Planta 207:354–361

Sánchez-Aguayo I, Rodriguez-Galan JM, Garcia R, Torreblanca J, Pardo JM (2004) Salt stress enhances xylem development and expression of S-adenosyl-L-methionine synthase in lignifying tissues of tomato plants. Planta 220:278–285

Santa-Maria GE, Rubio F, Dubcovsky J, Rodriguez-Navarro A (1997) The HAK1 gene of barley is a member of a large gene family and encodes a high-affinity potassium transporter. Plant Cell 9:2281–2289

Schwarz M, Gale J (1981) Maintenance respiration and carbon balance of plants at low levels of sodium chloride salinity. J Exp Bot 32:933–941

Shafi M, Guoping Z, Bakht J, Khan MA, Islam EU, Khan MD, Raziuddin (2010) Effect of cadmium and salinity stress on root morphology of wheat. Pakistan J Bot 42:2747–2754

Shalhevet J, Maas EV, Hoffman GJ, Ogata G (1976) Salinity and the hydraulic conductance of roots. Physiol Plant 38:224–232

Shannon MC, Grieve CM, Francoise LE (1994) Whole-plant response to salinity. In: Wilkinson RE (ed) Plant environment interactions. Dekker, New York, pp 199–244

Shelef O, Lazarovitch N, Rewald B, Golan-Goldhirsh A, Rachmilevitch S (2010) Root halotropism: salinity effects on *Bassia indica* root. Plant Biosyst 144:471–478

Shukla RK, Raha S, Tripathi V, Chattopadhyay D (2006) Expression of CAP2, an APETALA2-family transcription factor from chickpea, enhances Growth and tolerance to dehydration and salt stress in transgenic tobacco. Plant Physiol 142:113–123

Siddikee MA, Glick BR, Chauhan PS, Yim WJ, Sa T (2011) Enhancement of growth and salt tolerance of red pepper seedlings (*Capsicum annuum* L.) by regulating stress ethylene synthesis with halotolerant bacteria containing 1-aminocyclopropane-1-carboxylic acid deaminase activity. Plant Physiol Biochem 49:427–434

Siefritz F, Tyree MT, Lovisolo C, Schubert A, Kaldenhoff R (2002) PIP1 plasma membrane aquaporins in tobacco: from cellular effects to function in plants. Plant Cell Online 14:869–876

Silva C, Martínez V, Carvajal M (2008) Osmotic versus toxic effects of NaCl on pepper plants. Biol Plantarum 52:72–79

Snapp SS, Shennan C (1992) Effects of salinity on root-growth and death dynamics of tomato, *Lycopersicon esculentum* MILL. New Phytol 121:71–79

Snapp SS, Shennan C (1994) Salinity effects on root-growth and senescence in tomato and the consequences for severity of Phytophthora root-rot infection. J Am Soc Horticultural Sci 119(3):458–463

Steudle E (2000) Water uptake by roots: effects of water deficit. J Exp Bot 51:1531–1542

Steudle E, Peterson CA (1998) How does water get through roots? J Exp Bot 49:775–788

Steward FC (1935) Mineral nutrition of plants. Ann Rev Biochem 4:519–544

Strogonov BP (1964) Physiological basis of salt tolerance of plants (as affected by various types of salinity). Israel Progr Sci Trans, Jerusalem

Su H, Golldack D, Zhao C, Bohnert HJ (2002) The expression of HAK-Type K^+ transporters is regulated in response to salinity stress in common ice plant. Plant Physiol 129:1482–1493

Sun F, Zhang W, Hu H, Li B, Wang Y, Zhao Y, Li K, Liu M, Li X (2008) Salt modulates gravity signaling pathway to regulate growth direction of primary roots in *Arabidopsis*. Plant Physiol 146:178–188

Sutka M, Li G, Boudet J, Boursiac Y, Doumas P, Maurel C (2011) Natural variation of root hydraulics in *Arabidopsis* grown in normal and salt-stressed conditions. Plant Physiol 155:1264–1276

Taiz L, Zeiger E (1998) Water balance of the plant. In: Taiz L, Zeiger E (eds) Plant physiology. Sinauer Associates, Sunderland, pp 81–101

Taleisnik E, Peyrano G, Córdoba A, Arias C (1999) Water retention capacity in root segments differing in the degree of exodermis development. Ann Bot 83:19–27

Tanji KK (2002) Salinity in the soil environment. In: Läuchli A, Lüttge U (eds) Salinity: environments – plants – molecules. Kluwer, Dordrecht, pp 21–52

Tarczynski MC, Jensen RG, Bohnert HJ (1992) Expression of a bacterial mtlD gene in transgenic tobacco leads to production and accumulation of mannitol. PNAS 89:2600–2604

Tattini M, Ponzio C, Coradeschi MA, Tafani R, Traversi ML (1994) Mechanisms of salt tolerance in olive plants. Acta Hort 356:181–184

Therios IN, Misopolinos ND (1988) Genotypic response to sodium chloride salinity of four major olive cultivars (*Olea europea* L.). Plant Soil 106:105–111

Tiku B, Snaydon R (1971) Salinity tolerance within the grass species *Agrostis stolonifera* L. Plant Soil 35:421–431

Tyree MT, Sperry JS (1989) Vulnerability of xylem to cavitation and embolism. Ann Rev Plant Physiol Plant Mol Biol 40:19–38

Vadez V, Krishnamurthy L, Kashiwagi J, Kholova J, Devi JM, Sharma KK, Bhatnagar P, Hoisington DA, Hash CT, Bidinger FR, Keatinge JDH (2007) Exploiting the functionality of root systems for dry, saline, and nutrient deficient environments in a changing climate. SAT eJournal 4:1–61

Valenti GS, Ferro M, Ferraro D, Riveros F (1991) Anatomical changes in *Prosopis tamarugo* PHIL. seedlings growing at different levels of NaCl salinity. Ann Bot 68:47–53

Valiela I, Teal JM, Persson NY (1976) Production and dynamics of experimentally enriched salt-marsh vegetation – belowground biomass. Limnol Oceanogr 21:245–252

Van Der Werf A, Nagel OW (1996) Carbon allocation to shoots and roots in relation to nitrogen supply is mediated by cytokinins and sucrose: opinion. Plant Soil 185:21–32

Ventorino V, Caputo R, De Pascale S, Fagnano M, Pepe O, Moschetti G (2011) Response to salinity stress of *Rhizobium leguminosarum* bv. *viciae* strains in the presence of different legume host plants. Ann Microbiol 62:811–823

Villagra PE, Cavagnaro JB (2005) Effects of salinity on the establishment and early growth of *Prosopis argentina* and *Prosopis alpataco* seedlings in two contrasting soils: implications for their ecological success. Austr Ecol 30:325–335

Villar-Salvador P, Castro-Diez P, Perez-Rontome C, Montserrat-Marti G (1997) Stem xylem features in three *Quercus* (Fagaceae) species along a climatic gradient in NE Spain. Trees-Struct Funct 12:90–96

Voigt EL, Caitano RF, Maia JM, Ferreira-Silva SL, De Macêdo CEC, Silveira JAG (2009) Involvement of cation channels and $NH4^+$-sensitive K^+-transporters in Na^+ uptake by cowpea roots under salinity. Biol Plantarum 53:764–768

Volkov V, Amtmann A (2006) *Thellungiella halophila*, a salt-tolerant relative of *Arabidopsis thaliana*, has specific root ion-channel features supporting K^+/Na^+ homeostasis under salinity stress. Plant J 48:342–353

Walker RR, Sedgley M, Blesing MA, Douglas TJ (1984) Anatomy, ultrastructure and assimilate concentrations of roots of *Citrus* genotypes differing in ability for salt exclusion. J Exp Bot 35:1481–1494

Wan XC, Steudle E, Hartung W (2004) Gating of water channels (aquaporins) in cortical cells of young corn roots by mechanical stimuli (pressure pulses): effects of ABA and of $HgCl_2$. J Exp Bot 55:411–422

Wang Y, Li X (2008) Salt stress-induced cell reprogramming, cell fate switch and adaptive plasticity during root hair development in *Arabidopsis*. Plant Signaling Behav 3:436–438

Wang Y, Zhang W, Li K, Sun F, Han C, Wang Y, Li X (2008) Salt-induced plasticity of root hair development is caused by ion disequilibrium in *Arabidopsis thaliana*. J Plant Res 121:87–96

Wang YN, Li KX, Li X (2009) Auxin redistribution modulates plastic development of root system architecture under salt stress in *Arabidopsis thaliana*. J Plant Physiol 166:1637–1645

Wang Y, Zhang XM, Cao ZH, Liu W, Wang Q (2010) Effect of salt stress on seedlings biomass of two varieties of *Elaeagnus* spp. [In Chinese, English abstract]. China For Sci Technol 24:25–28

Warrence NJ, Bauder JW, Pearson KE (2003) Basics of salinity and sodicity effects on soil physical properties. Montana State University, Bozeman

West G, Inzé D, Beemster GTS (2004) Cell cycle modulation in the response of the primary root of *Arabidopsis* to salt stress. Plant Physiol 135:1050–1058

Wolf O, Jeschke WD, Hartung W (1990) Long-distance transport of abscisic-acid in NaCl-treated intact plants of *Lupinus albus*. J Exp Bot 41:593–600

World Bank (2007) Agriculture for development. World Bank, Washington, DC

Wyn Jones RG, Gorham J (2002) Intra- and inter-cellular compartmentation of ions – a study in specificity and plasticity. In: Läuchli A, Lüttge U (eds) Salinity: environments – plants – molecules. Kluwer, Dordrecht, pp 159–180

Yakir D, Yechieli Y (1995) Plant invasion of newly exposed hypersaline Dead Sea shores. Nature 374:803–805

Yamane K, Rahman MS, Kawasaki M, Taniguchi M, Miyake H (2004) Pretreatment with a low concentration of methyl viologen decreases the effects of salt stress on chloroplast ultrastructure in rice leaves (*Oryza sativa* L.). Plant Production Sci 7:435–441

Yamaya T, Matsumoto H (1989) Accumulation of asparagines in NaCl-stressed barley seedlings [abstract]. Berichte des Ohara Instituts für Landwirtschaftliche Biologie – Okayama Universität 19:181–188

Yeo AR (2007) Salinity. In: Yeo AR, Flowers TJ (eds) Plant solute transport. Blackwell, Oxford, pp 340–365

Yeo AR, Flowers SA, Rao G, Welfare K, Senanayake N, Flowers TJ (1999) Silicon reduces sodium uptake in rice (*Oryza sativa* L.) in saline conditions and this is accounted for by a reduction in the transpirational bypass flow. Plant Cell Environ 22:559–565

Yermiyahu U, Nir S, BenHayyim G, Kafkafi U, Kinraide TB (1997) Root elongation in saline solution related to calcium binding to root cell plasma membranes. Plant Soil 191:67–76

Yi LP, Ma J, Li Y (2007) Impact of salt stress on the features and activities of root system for three desert halophyte species in their seedling stage. Sci China Ser D 50:97–106

Zekri M, Parsons LR (1989) Growth and root hydraulic conductivity of several citrus rootstocks under salt and polyethylene glycol stresses. Physiol Plantarum 77:99–106

Zekri M, Parsons LR (1992) Salinity tolerance of *Citrus* rootstocks: effects of salt on root and leaf mineral concentrations. Plant Soil 147:171–181

Zhong HL, Läuchli A (1994) Spatial-distribution of solutes, K, Na, Ca and their deposition rates in the growth zone of primary cotton roots – effects of NaCl and $CaCl_2$. Planta 194:34–41

Zhu JK, Liu J, Xiong L (1998) Genetic analysis of salt tolerance in *Arabidopsis*: evidence for a critical role of potassium nutrition. Plant Cell Online 10:1181–1192

Zhu ZJ, Wei GQ, Li J, Qian QQ, Yu JQ (2004) Silicon alleviates salt stress and increases antioxidant enzymes activity in leaves of salt-stressed cucumber (*Cucumis sativus* L.). Plant Sci 167:527–533

Zimmermann HM, Steudle E (1998) Apoplastic transport across young maize roots: effect of the exodermis. Planta 206:7–19

Zimmermann HM, Hartmann K, Schreiber L, Steudle E (2000) Chemical composition of apoplastic transport barriers in relation to radial hydraulic conductivity of corn roots (*Zea mays* L.). Planta 210:302–311

Chapter 7
Salt Tolerance in Rice: Present Scenario and Future Prospects

Asiya Hameed, Tabasum N. Qadri, M.M. Azooz, and Parvaiz Ahmad

7.1 Introduction

Agriculture has always been recognized as the world's most important industry. As is obvious, the world has now experienced more than ever before. Environmental conditions like salinity and drought, etc., influence plant metabolism directly or indirectly, thus reduces the growth and productivity of number of plants (Naz et al. 2010; Ahmad and Prasad 2012a, b). Many rice growing regions like China, Bangladesh, India, Thailand, Pakistan and Egypt experience severe salt problems to such an extent which was never felt previously. As a result shortage of rice has become a burning question in most of the areas of the world (Baby Joseph et al. 2010). Recently Japan was hit by Tsunami which contaminated about 20,000 ha of rice fields with salinated water.

In agriculturally advanced countries, technological innovations have developed so rapidly that one cannot keep pace with the array of new technology into the agricultural industry.

Salinity is one of the major factors limiting the production of rice. Rice being the staple food, has an exceptional agricultural importance. India has to increase its rice productivity by 3% per annum in order to meet future food requirements and to

A. Hameed (✉) • T.N. Qadri
Department of Botany, Hamdard University, Hamdard Nagar, New Delhi, India
e-mail: hameed901@gmail.com

M.M. Azooz
Department of Biological Sciences, Faculty of Science,
King Faisal University, Al-Hassa, Saudi Arabia

Department of Botany, Faculty of Science, South Valley University, 83523 Qena, Egypt

P. Ahmad (✉)
Department of Botany, Amar Singh College, Srinagar, Jammu and Kashmir, India
e-mail: parvaizbot@yahoo.com

P. Ahmad et al. (eds.), *Ecophysiology and Responses of Plants under Salt Stress*,
DOI 10.1007/978-1-4614-4747-4_7,

maintain self-sufficiency (Thiyagarajan and Selvaraju 2001). Rice (*Oryza sativa* L.) a major crop is indigenous to tropical and subtropical Southeastern Asia and Africa and sustains nearly one-half of the world population (Wu et al. 2004).

Rice crop though sensitive to soil salinity is recommended in the process of soil reclamation due to the fact that it can be cultivated in flooded conditions and the standing water in fields facilitates to leach salts from the soil (Bhumbla and Abrol 1978). In spite of its sensitivity to excess salinity, a remarkable genetic variation exists to improve salt tolerance among and within rice varieties (Sabouri and Biabani 2009; Flowers and Yeo 1981) and this variability can be dominated to identify, select and develop salt tolerant rice genotypes. Many salt-tolerant genotypes have been developed in rice cultivars i.e. Pokkali and Nona-Bokra that not only withstand high levels of salt but also maintain optimum yield levels (Akbar et al. 1985; Gregorio and Senadhira 1993). Various studies have been conducted for the salinity tolerance at the seedling stage (Sabouri and Sabouri 2009; Mohammadi-Nejad et al. 2008) little attention has been devoted to the reproductive stage (Sabouri and Biabani 2009).

Numerous literature are available for plants that endure stresses and thereby show better tolerance (Singla-Pareek et al. 2001; Zhu 2001, 2002; Sairam and Tyagi 2004; Yamaguchi and Blumwald 2005; Rodríguez et al. 2005; Vinocur and Altman 2005; Bajaj and Mohanty 2005; Sahi et al. 2006; Gao et al. 2007; Vij and Tyagi 2007; Ahmad et al. 2009, 2010a, b, 2011a, 2012; Ahmad 2010; Azooz et al. 2011; Katare et al. 2012). Thus various genes may possibly contribute to the crop plants to abiotic stresses.

Besides leaf area index (LAI), tolerant groups of rice productivity have shown less significance for Na^+/K^+ ratio and plays a key role to confer tolerance against salt stress (Natarajan et al. 2005).

7.2 Yield Reduction by Salinity

Salinity is one of the severe environmental problems leading to reduction in the growth and yield of rice (Ashraf 2009). It is most widely grown in coastal areas swamped with seawater in high tidal period, somewhat vulnerable to salinity (Mori and Kinoshita 1987). In the beginning of the growing period, salts accumulated by capillary rise in the topsoil gets released into the soil solution and floodwater.

Significant variation have been observed in rice on the onset of germination followed by susceptility to saline stress during seedling stage and subsequently withstand a bit more in vegetative phase (Heenan et al. 1988), afterwards it again shows a sensitive response during maturity i.e., reproductive phase (Flowers and Yeo 1981; Abdullah et al. 2001). Whereas on the other hand, the vegetative shoot biomass of rice, is often affected much less than reproductive growth (except for young seedlings) (Munns et al. 2002). Zeng and Shannon (2000) observed that the seedling growth gets affected at salinity levels as low as 1.9 dS/m, nevertheless did not turn into reduction in grain yield. Grieve et al. (1993) demonstrated that crops usually decrease with increased salinity from planting to maturity, however rice shows the opposite effect (Lutts et al. 1995).

7.3 Mechanism of Salt Tolerance in Rice Plant

Ionic and osmotic stress are the consequences of salinity stress in plants. Root growth inhibition is a common feature and seems to affect the osmotic gradient (i.e., osmotic pressure) more as compared to Na^+ content of the plant itself (Munns et al. 2002; Munns 2002).

Ionic toxicity of Na^+ largely competes with K^+ for binding sites which are crucial to metabolic process. While as osmotic component relates to the buildup of Na^+ and Cl^- in the apoplastic space in leaf tissues, elevated levels of Na^+ in the apoplastic spaces cause dehydration and consequently shoots accumulate more Na^+ than roots. Therefore sensitivity to these stresses becomes evident more towards shoots rather than roots.

Owing to high level of Na^+ and/or Cl^- ions, depicts various physiological responses (Greenway and Munns 1980; Hasegawa et al. 2000; Zhu 2001). Na^+ toxicity mostly cause damage in Gramineaceous crop rice (Tester and Davenport 2003). While cytosolic Na^+ remains 1 to 10 mM in higher plants under non-saline conditions (Taiz and Zeiger 2002). For efficient metabolic activities, potassium ion is required in the cytosol to maintain the optimum level (Taiz and Zeiger 2002; Cuin et al. 2003). However K^+ activates more than 50 enzymes, vulnerable to high Na^+ and Na^+/K^+ ratios (Munns et al. 2006). Thus causes disruption in metabolic activities by a challenger between Na^+ and K^+ (Bhandal and Malik 1988; Tester and Davenport 2003). K^+ efflux brings a change in plasma membrane permeability by displacing Ca^{2+} with an increase in Na^+ level (Cramer et al. 1989). As a result, an imbalance of the Na^+/K^+ ratio occurs in the cytosol and eventually interrupting the enzymatic reactions.

Regulation of Na^+ influx and efflux into cell or apoplast/vacuole is important to maintain the low cytosolic Na^+ concentration and has been demonstrated that rice cultivar (Pokkali) plays a valuable role in Na^+ sequestration (Maathuis and Sanders 2001; Carden et al. 2003; Kader and Lindberg 2005; Anil et al. 2007). Besides a role in compartmentalization, over-expression of OsNHX simultaneously enhances the rice plants to endure the salt stress (Fukuda et al. 2004; Chen et al. 2007). Similar findings have been observed in roots of etiolated rice seedlings by OsGR gene expressions (Tsai et al. 2005).

Apoplastic sequestration in salt-tolerant rice though not an efficient strategy but most of Na^+ in leaves show apoplastic streaming (Yeo et al. 1999). Ratio of cytosolic Na^+/K^+ have been found to get altered on exposure to salt stress, since the concentration of Na^+ is much higher than at normal condition and when the K^+ concentration increases plants suffer from high Na^+/K^+ ratio disrupting the cytosolic Na^+/K^+ against salinity stress (Greenway and Osmond 1972). Stimulation of OsHKT2;2 (previous name OsHKT2) in shoots than in roots of salt-tolerant cv. Pokkali has been reported although, OsHKT2;2 (K^+-Na^+ coupled transporter) does not mediate K^+-influx from a high K^+ solution in the absence of Na^+, and confers tolerance to salinity under high Na^+, possibly by increased ability of K uptake, as shown in *S. cerevisiae* (Horie et al. 2001). At high salt stress, IR20 in the xylem sap of root soil shows an extensively higher Na concentration than

Pokkali (Faiyue et al. 2010; Kavitha et al. 2012). Similar reports have been observed by Krishnamurthy et al. 2011. Quantitative trait loci (QTLs) with wide range of adaptations in tolerant plants in response to osmotic and ionic stresses (for Na^+, Cl^-, etc.) in turn challenges farmers to build-up salt-tolerant rice cultivars with increased yield (Haq et al. 2010; Michael et al. 2010).

7.4 Genes Responsible for Salt Tolerance

LeeI et al. (2003) demonstrated that remarkable variation exists in *indica* and *japonica* rice groups to develop salinity tolerance. Similar results have also been observed in Pokkali and Nona-Bokra (Akbar et al. 1985; Gregorio and Senadhira 1993).

Gregorio and Senadhira (1993) investigated that certain group of genes with additive effects affects the sodium and potassium uptake in rice i.e., one group for sodium exclusion and the other for potassium absorption. Similar reports have been found in dry weight rice seedlings (Akbar et al. 1985). Furthermore salt stress responsive genes have been identified by cDNA array (Xinjian et al. 2002). In salt-sensitive rice plant extensive genetic variability have proved to play an effective role to develop the rice tolerant to salinity stress (Mohammadi-Nejad 2008; 2010).

7.5 Antioxidant System in Salt Tolerance

Increased production of reactive oxygen species (ROS) oxidize various biological molecules like DNA, proteins and lipids resulting in large degrees of environmental stresses within several sub-cellular compartments of the plant cell (Ahmad and Umar 2011). Energy required to regulate the metabolism of the organisms is provided by the reduction of oxygen, although its reduction is a mixed blessing (Breusegem et al. 2001). Therefore naturally occurring antioxidant enzymes present in plant tissues acts as a safe guard by preventing the cells from ROS and other stress conditions (Ahmad et al. 2008, 2009, 2010a). It has been illustrated that over-expression of genes encoding these enzymes leads to the development of stress tolerant transgenic plants. Besides this photorespiration increases by over-expression of glutamine synthetase thereby confers salt-tolerance in rice cultivars. SOD being the first line of defense in the ROS scavenging, converts superoxide radical to H_2O_2 (Scandalios 1993). It has been investigated that different over-expressions of SOD by diverse plant species shows a protective role against abiotic stresses (Gupta et al. 1993a, b; McKersie et al. 1993, 1999, 2000; Badawi et al. 2004; Wang et al. 2004, 2005; McKersie et al. 1996; Badawi et al. 2004; Van Camp et al. 1996; Ahmad et al. 2010a; 2011b; 2012). Also cDNA encoding Cu/Zn-SOD as well as yeast mitochondrial Mn-SOD have been reported to confer salt tolerance in rice (Prashanth et al. 2007; Tanaka et al. 1999) respectively Hence SOD provides an important assessment to plants for multiple stress tolerance.

Overexpressions as well as enhancement of ascorbate peroxidase has been reported in different plant species to accomplish tolerance against salt and oxidative stress suggesting a possible role to abolish H_2O_2 from cells (Wang et al. 1999, 2005). Lu et al. (2007) developed transgenic Arabidopsis plants over-expressing two rice cytosolic APXs (OsAPXa and OsAPXb) and investigated that transgenic plants exhibited increased tolerance to salt stress than wild type plants. Salt tolerance has shown to be associated with high level of glutathione and also GSH/GSSG ratio (Tausz et al. 2004; Yousuf et al. 2012). Furthermore transgenic rice over-expressing rice glyII gene have shown to endure the harmful stresses of salinity (Singla-Pareek et al. 2007). Association of various genes of antioxidant defense pathway could be a positive approach to build up plants to withstand various types of stress. It has been established that gene expressions of H_2O_2 in response to salt stress enhances the activities of APX and GR in the roots of rice seedlings (Hong et al. 2007, 2009). Also in agreement with Tsai et al. (2004) shows the relative importance of Na^+ and Cl^- in NaCl-induced antioxidant systems in roots of rice seedlings. Likewise expressions of OsAPX and OsGR have been found by Shankhdhar et al. (2000) in rice roots under NaCl stress.

7.6 Conclusion and Future Prospects

Transgenic rice plants against salt stress though widely studied, however the mechanism regarding the salt-tolerant and salt-sensitive plants is not clear so far. Therefore scientists further needs to study the salt tolerant rice in order to enhance the productivity and quality of the plant.

The future poses a challenge to many agricultural countries to accelerate the production to meet global needs. As rice is the largest cereal crop in the world, the enhancement drive should not only encourage growth and endure salt stress in the farming process of rice but promote the initiative for Research and Development in the rice growing region. Improving tolerance is the shared responsibility of all concerned in rice growing industry and everyone involved in the process should unite to evolutionize the production of rice in the world.

References

Abdullah Z, Khan MA, Flowers TJ (2001) Causes of sterility in seed set of rice under salinity stress. J Agron Crop Sci 187:25–32

Ahmad P (2010) Growth and antioxidant responses in mustard (*Brassica juncea* L.) plants subjected to combined effect of gibberellic acid and salinity. Arch Agro Soil Sci 56(5):575–588

Ahmad P, Prasad MNV (2012a) Environmental adaptations and stress tolerance in plants in the era of climate change. Springer Science+Business Media, New York

Ahmad P, Prasad MNV (2012b) Abiotic stress responses in plants: metabolism, productivity and sustainability. Springer Science+Business Media, New York

Ahmad P, Umar S (2011) Oxidative stress: role of antioxidants in plants. Studium Press, New Delhi

Ahmad P, Sarwat M, Sharma S (2008) Reactive oxygen species, antioxidants and signaling in plants. J Plant Biol 51(3):167–173

Ahmad P, Jeleel CA, Azooz MM, Nabi G (2009) Generation of ROS and non-enzymatic antioxidants during abiotic stress in plants. Bot Res Intern 2(1):11–20

Ahmad P, Jaleel CA, Salem MA, Nabi G, Sharma S (2010a) Roles of enzymatic and non-enzymatic antioxidants in plants during abiotic stress. Crit Rev Biotechnol 30(3):161–175

Ahmad P, Jaleel CA, Sharma S (2010b) Antioxidative defence system, lipid peroxidation, proline metabolizing enzymes and Biochemical activity in two genotypes of *Morus alba* L. subjected to NaCl stress. Russ J Plant Physiol 57(4):509–517

Ahmad P, Nabi G, Ashraf M (2011a) Cadmium-induced oxidative damage in mustard [*Brassica juncea* (L.) Czern. & Coss.] plants can be alleviated by salicylic acid. South Afr J Bot 77:36–44

Ahmad P, Nabi G, Jeleel CA, Umar S (2011b) Free radical production, oxidative damage and antioxidant defense mechanisms in plants under abiotic stress. In: Ahmad P, Umar S (eds) Oxidative stress: role of antioxidants in plants. Studium Press, New Delhi, pp 19–53

Ahmad P, Hakeem KR, Kumar A, Ashraf M, Akram NA (2012) Salt-induced changes in photosynthetic activity and oxidative defense system of three cultivars of mustard (*Brassica juncea* L.). Afr J Biotechnol 11(11):2694–2703

Akbar M, Khush GS, Hillerislambers D (1985) Genetics of salt tolerance in rice. IRRI. Rice genetics. In: Proceeding of international rice genetics symposium, IRRI, Manila, pp 399–409

Anil VS, Krishnamurthy H, Mathew M (2007) Limiting cytosolic Na confers salt tolerance to rice cells in culture: a two-photon microscopy study of SBFI-loaded cells. Physiol Plant 129:607–621

Ashraf M (2009) Biotechnological approach of improving plant salt tolerance using antioxidants as markers. Biotechnol Adv 27:84–93

Azooz MM, Youssef AM, Ahmad P (2011) Evaluation of salicylic acid (SA) application on growth, osmotic solutes and antioxidant enzyme activities on broad bean seedlings grown under diluted seawater. Inter J Plant Physiol Biochem 3:253–264

Badawi GH, Yamauchi Y, Shimada E, Sasaki R, Kawano N, Tanaka K, Tanaka K (2004) Enhanced tolerance to salt stress and water deficit by overexpressing superoxide dismutase in tobacco (*Nicotiana tabacum*) chloroplasts. Plant Sci 166:919–928

Bajaj S, Mohanty A (2005) Recent advances in rice biotechnology-towards genetically superior transgenic rice. Plant Biotech J 3:275–307

Bhandal IS, Malik CP (1988) Potassium estimation, uptake, and its role in the physiology and metabolism of flowering plants. Intl Rev Cyt 110:205–254

Bhumbla DR, Abrol IP (1978) Saline and sodic soils. In: Soils and rice. International Rice Research Institute, Los Baños, pp 719–738

Breusegem FV, Vranova E, Dat JF, Inze D (2001) The role of active oxygen species in plant signal transduction. Plant Sci 161:405–414

Carden DE, Walker DJ, Flowers TJ, Miller AJ (2003) Single-cell measurements of the contributions of cytosolic Na^- and K^- to salt tolerance. Plant Physiol 131:676–683

Chen H, An R, Tang JH, Cui XH, Hao FS, Chen J, Wang XC (2007) Over-expression of a vacuolar Na^-/H^- antiporter gene improves salt tolerance in an upland rice. Mol Breed 19:215–225

Cramer GR, Epstein E, Läuchli A (1989) Na^+-Ca^{2+} interactions in barley seedlings: relationship to ion transport and growth. Plant Cell Environ 12:551–558

Cuin TA, Miller AJ, Laurie SA, Leigh RA (2003) Potassium activities in cell compartments of salt-grown barley leaves. J Exp Bot 54:657–661

Faiyue B, Vijayalakshmi C, Nawaz S, Nagato Y, Taketa S, Ichii M, Al-Azzawi M, Flowers TJ (2010) Studies on sodium bypass flow in lateral rootless mutants lrt1 and lrt2, and crown rootless mutant crl1 of rice (*Oryza sativa* L.). Plant Cell Environ 33:687–701

Flowers TJ, Yeo AR (1981) Variability of sodium chloride resistance within rice (*Oryza sativa* L.) varieties. New Phytol 88:363–373

Fukuda A, Nakamura A, Tagiri A, Tanaka H, Miyao A, Hirochika H, Tanaka Y (2004) Function, intracellular localization and the importance in salt tolerance of a vacuolar Na^+/H^+ antiporter from rice. Plant Cell Physiol 45:146–159

Gao JP, Chao DY, Lin HX (2007) Understanding abiotic stress tolerance mechanisms: recent studies on stress response in rice. J Integr Plant Biol 49:742–750

Greenway H, Munns R (1980) Mechanisms of salt tolerance in non halophytes. Annu Rev Plant Physiol 31:149–190

Greenway H, Osmond CB (1972) Salt responses of enzymes from species differing in salt tolerance. Plant Physiol 49:256–259

Gregorio GB, Senadhira D (1993) Genetic analysis of salinity tolerance in rice (*O. sativa* L.). Theor Appl Genet 86:333–338

Grieve CM, Lesch SM, Maas EV, Francois LE (1993) Leaf and spikelet primordia initiation in salt-stressed wheat. Crop Sci 22:1286–1294

Gupta AS, Heinen JI, Holaday S, Burket JJ, Allen RD (1993a) Increased resistance to oxidative stress in transgenic plants that overexpress chloroplastic Cu/Zn superoxide dismutase. Proc Natl Acad Sci USA 90:1629–1633

Gupta AS, Robert P, Webb A, Holaday S, Allen RD (1993b) Overexpression of superoxide dismutase protects plants from oxidative stress. Plant Physiol 103:1067–1073

Haq TU, Gorham J, Akhtar J, Akhtar N, Steele KA (2010) Dynamic quantitative trait loci for salt stress components on chromosome 1of rice. Func Plant Biol 37:634–645

Hasegawa PM, Bressan RA, Pardo JM (2000) The dawn of plant salt tolerance genetics. Trends Plant Sci 5:317–319

Heenan DP, Lewin LG, McCaffery DW (1988) Salinity tolerance in rice varieties at different growth stages. Aust J Exp Agric 28:343–349

Hong CY, Hsu Y, Tsai YC, Kao CH (2007) Expression of ascorbate peroxidase8 in roots of rice (Oryza *sativa* L.) seedlings in response to NaCl. J Exp Bot 58:3273–3283

Hong CY, Chao Y, Yang M, Cheng S, Cho S, Kao CH (2009) NaCl-induced expression of glutathione reductase in roots of rice (*Oryza sativa* L.) seedlings is mediated through hydrogen peroxide but not abscisic acid. Plant Soil 320:103–115

Horie T, Yoshida K, Nakayama H, Yamada K, Oiki S, Shinmyo A (2001) Two types of HKT transporters with different properties of Na^+ and K^+ transport in *Oryza sativa*. Plant J 27:129–138

Joseph Baby, Jini D, Sujatha S (2010) Biological and physiological perspectives of specificity in abiotic salt stress response from various rice plants. Asian J Agricul Sci 2(3):99–105

Kader MA, Lindberg S (2005) Uptake of sodium in protoplasts of salt-sensitive and salt-tolerant cultivars of rice, Oryza sativa L. determined by the fluorescent dye SBFI. J Exp Bot 56: 3149–3158

Katare DP, Nabi G, Azooz MM, Aeri V, Ahmad P (2012) Biochemical modifications and enhancement of psoralen content in salt-stressed seedlings of *Psoralea corylifolia* Linn. J Funct Environ Bot 2(1):65–74

Kavitha PG, Miller AJ, Mathew MK, Maathuis FJM (2012) Rice cultivars with differing salt tolerance contain similar cation channels in their root cells. J Exp Bot 63(8):3289–3296

Krishnamurthy P, Ranathunge R, Nayak S, Schreiber L, Mathew MK (2011) Root barriers block Na^+ traffic to shoots in rice (*Oryza sativa* L.). J Exp Bot 62:4215–4228

LeeI IS, Kim DS, Lee SJ, Song HS, Lim YP, Lee YI (2003) Selection and characterizations of radiation-induced salinity-tolerant lines in rice. Breed Sci 53:313–318

Lu Z, Liu D, Liu S (2007) Two rice cytosolic ascorbate peroxidases differentially improve salt tolerance in transgenic Arabidopsis. Plant Cell Rep 26:1909–1917

Lutts S, Kinet JM, Bouharmont J (1995) Changes in plant response to NaCl during development of rice (*Oryza sativa* L.) varieties differing in salinity resistance. J Exp Bot 46:1843–1852

Maathuis FJM, Sanders D (2001) Sodium uptake in Arabidopsis roots is regulated by cyclic nucleotides. Plant Physiol 127:1617–1625

McKersie BD, Chen Y, deBeus M, Bowley SR, Bowler C, Inzé D, D'Halluin K, Botterman J (1993) Superoxide dismutase enhances tolerance of freezing stress in transgenic alfalfa (*Medicago sativa* L.). Plant Physiol 103:1155–1163

McKersie BD, Bowley SR, Harjanto E, Leprince O (1996) Water-deficit tolerance and field performance of transgenic alfalfa overexpressing superoxide dismutase. Plant Physiol 111:1177–1181

McKersie BD, Bowley SR, Jones KS (1999) Winter survival of transgenic alfalfa overexpressing superoxide dismutase. Plant Physiol 119:839–848

McKersie BD, Murnaghan J, Jones KS, Bowley SR (2000) Iron-superoxide dismutase expression in transgenic alfalfa increases winter survival without a detectable increase in photosynthetic oxidative stress tolerance. Plant Physiol 122:1427–1437

Michael JT, Marjorie de Ocampo, Egdane J, Akhlasur Rahman M, Andres GS, Dante LA, Ellen Tumimbang-Raiz E, Eduardo B, Zeba IS, Rakesh KS, Glenn BG Abdelbagi MI (2010) Characterizing the salt quantitative trait locus for salinity tolerance in rice. 3(2–3):148–160

Mohammadi-Nejad G, Arzani A, Rezai AM, Singh RK, Gregorio GB (2008) Assessment of rice genotypes for salt tolerance using microsatellite markers associated with the saltol QTL. Afr J Biotech 7(6):730–736

Mohammadi-Nejad G, Singh RK, Arzani A, Rezaie AM, Sabouri H, Gregorio GB (2010) Evaluation of salinity tolerance in rice genotypes. Inter J Plant Produc 4(3):199–208

Mori IK, Kinoshita T (1987) Salt tolerance of rice callus clones. Rice Genet Newslett 4:112–113

Munns R, Husain S, Rivelli AR, James RA, Condon AG, Lindsay MP, Lagudah ES, Schachtman DP, Hare RA (2002) Avenues for increasing salt tolerance of crops, and the role of physiologically-based selection traits. Plant Soil 247:93–105

Munns R (2002) Comparative physiology of salt and water stress. Plant Cell Environ 25:239–250

Munns R, James RA, Läuchli A (2006) Approaches to increasing the salt tolerance of wheat and other cereals. J Exp Bot 57:1025–1043

Natarajan SK, Ganapathy M, Krishnakumar S, Dhanalakshmi R, Saliha BB (2005) Grouping of rice genotypes for salinity tolerance based upon grain yield and Na: K ratio under coastal environment. Res J Agric Biol Sci 1:162–165

Naz N, Hameed M, Ashraf M (2010) Eco-morphic response to salt stress in two halophytic grasses from the Cholistan desert, Pakistan. Pak J Bot 42:1343–1351

Nejad GM, Arzani A, Rezai AM, Singh RK, Gregorio GB (2008) Assessment of rice genotypes for salt tolerance using microsatellite markers associated with the saltol QTL. Afr J Biotech 7:730–736

Prashanth SR, Sadhasivam V, Parida A (2007) Overexpression of cytosolic copper/zinc superoxide dismutase from a mangrove plant *Avicennia marina* in indica rice var pusa basmati-1 confers abiotic stress tolerance. Transgenic Res 17(2):281–291

Rodríguez M, Canales E, Borrás-Hidalgo O (2005) Molecular aspects of abiotic stress in plants. Biotechnol Applic 22:1–10

Sabouri H, Biabani A (2009) Toward the mapping of agronomic characters on a rice genetic map: quantitative trait loci analysis under saline condition. Biotech 8:144–149

Sabouri H, Sabouri A (2009) New evidence of QTLs attributed to salinity tolerance in rice. Afr J Biotech 7:4376–4383

Sahi C, Singh A, Kumar K, Blumwald E, Grover A (2006) Salt stress response in rice: genetics, molecular biology, and comparative genomics. Funct Integr Genomics 6:263–284

Sairam RK, Tyagi A (2004) Physiology and molecular biology of salinity stress tolerance in plants. Curr Sci 86:407–421

Scandalios JG (1993) Oxygen stress and superoxide dismutases. Plant Physiol 101:7–12

Shankhdhar D, Shankhdhar SC, Mani SC, Pant RC (2000) In vitro selection for salt tolerance in rice. Biol Plant 43(3):477–480

Singla-Pareek SL, Reddy MK, Sopory SK (2001) Transgenic approach towards developing abiotic stress tolerance in plants. Proc Ind Nat Sci Acad 67:265–284

Singla-Pareek SL, Yadav SK, Pareek A, Reddy MK, Sopory SK (2007) Enhancing salt tolerance in a crop plant by overexpression of glyoxalase II. Transgenic Res 17:171–180

Taiz L, Zeiger E (2002) Plant Physiol, 3rd edn. Sinauer, Sunderland, 612

Tanaka Y, Hibino T, Hayashi Y, Tanaka A, Kishitani S, Takabe T, Yokota S, Takabe T (1999) Salt tolerance of transgenic rice overexpressing yeast mitochondrial Mn-SOD in chloroplasts. Plant Sci 148:131–138

Tausz M, Sircelj H, Grill D (2004) The glutathione system as a stress marker in plant ecophysiology: is a stress-response concept valid? J Exp Bot 55:1955–1962

Tester M, Davenport R (2003) Na^+ tolerance and Na^+ transport in higher plants. Ann Bot 91:503–527

Thiyagarajan TM, Selvaraju R (2001) Water saving in rice cultivation in India. In: Proceedings of an international workshop on water saving rice production systems, Nanjing University, pp 15–45

Tsai YC, Hong CY, Liu LF, Kao CH (2004) Relative importance of Na^+ and Cl^- in NaCl-induced antioxidant systems in roots of rice seedlings. Physiol Plant 122:86–94

Tsai YC, Hong CY, Liu LF, Kao CH (2005) Expression of ascorbate peroxidase and glutathione reductase in roots of rice seedlings in response to NaCl and H_2O_2. J Plant Physiol 162:291–299

Van Camp W, Capiau K, Van Montagu M, Inze D, Slooten L (1996) Enhancement of oxidative stress tolerance in transgenic tobacco plants overproducing Fe-superoxide dismutase in chloroplasts. Plant Physiol 112:1703–1714

Vij S, Tyagi AK (2007) Emerging trends in the functional genomics of the abiotic stress response in crop plants. Plant Biotechnol J 5:361–380

Vinocur B, Altman A (2005) Recent advances in engineering plant tolerance to abiotic stress: achievements and limitations. Curr Opin Biotech 16:123–132

Wang J, Zhang H, Allen RD (1999) Overexpression of an Arabidopsis peroxisomal ascorbate peroxidase gene in tobacco increases protection against oxidative stress. Plant Cell Physiol 40:725–732

Wang B, Luttge U, Ratajczak R (2004) Specific regulation of SOD isoforms by NaCl and osmotic stress in leaves of the C3 halophyte Suaeda salsa L. J Plant Physiol 161:285–293

Wang FZ, Wang QB, Kwon SY, Kwak SS, Su WA (2005) Enhanced drought tolerance of transgenic rice plants expressing a pea manganese superoxide dismutase. J Plant Physiol 162:465–472

Wu CJ, Cheng ZQ, Huang XQ, Yin SH, Cao KM, Sun CR (2004) Genetic diversity among and within populations of Oryza granulate from Yunnan of China revealed by RAPD and ISSR markers: implications for conservation of the endangered species. Plant Sci 167:35–42

Xinjian H, Jianquan C, Zhigang Z, Jinsong Z, Shouyi C (2002) Identification of salt-stress responsive genes in rice (*Oryza sativa* L.) by cDNA array. Sci China (Series C) 45(5):477–484

Yamaguchi T, Blumwald E (2005) Developing salt tolerant crop plants: challenges and opportunities. Trends Plants Sci 10:615–620

Yeo AR, Flowers SA, Rao G, Welfare K, Senanayake N, Flowers TJ (1999) Silicon reduces sodium uptake in rice (*Oryza sativa* L.) in saline conditions and this is accounted for by a reduction in the transpirational bypass flow. Plant Cell Environ 22:559–565

Yousuf PY, Hakeem KR, Chandna R, Ahmad P (2012) Role of glutathione reductase in plant abiotic stress. In: Ahmad P, Prasad MNV (eds) Abiotic stress responses in plants: metabolism, productivity and sustainability. Springer Science+Business Media, New York, pp 149–158

Zeng L, Shannon MC (2000) Salinity effects on seedling growth and yield components of rice. Crop Sci 40:996–1003

Zhu JK (2001) Plant salt tolerance. Trends Plant Sci 6:66–71

Zhu JK (2002) Salt and drought stress signal transduction in plants. Annu Rev Plant Biol 53: 247–273

Chapter 8
Aquaporins: Role Under Salt Stress in Plants

Renu Bhardwaj, Indu Sharma, Mukesh Kanwar, Resham Sharma, Neha Handa, Harpreet Kaur, Dhriti Kapoor, and Poonam

8.1 Introduction

Aquaporins (AQPs), a class of the integral membrane proteins, belong to a large family of major intrinsic proteins (MIP) which are primarily associated with the flow of water in the membrane of biological cells (Agre 2006; Aroca et al. 2012; Vera-Estrella and Bohnert 2011). The Nobel Prize in Chemistry for year 2003 was jointly awarded to Peter Agre (Johns Hopkins University) and Roderick MacKinnon for the discovery of AQPs and work on the structure/mechanism of potassium channels respectively. According to Agre, the AQPs were discovered "by serendipity" (Knepper and Nielsen 2004). During their research on the Rh blood group antigens, they isolated the Rh molecule along with another unknown molecule of 28 kDa size (and therefore called 28 K). At first, this second molecule was assumed to be a piece of the Rh molecule, or a contaminant, but further investigation revealed that it was an *undiscovered molecule with unknown function*. This molecule was abundant in red blood cells and kidney tubules, and found to be related to proteins of diverse origins like the proteins from brains of fruit flies, bacteria, the lenses of eyes and plant tissues. As per the suggestion of Dr. John Parker (hematology professor at the University of North Carolina to Agre), it was later discovered that this molecule was the long-sought water channel. Thus, the first AQP described the mammalian AQP1'aquaporin-1' (originally known as CHIP 28).

AQP1 was found in erythrocytes and renal tubuli facilitating the osmotical driven permeation of water across membranes (Denker et al. 1988; Preston and Agre 1991). It was classified into a large superfamily of intrinsic membrane proteins named major intrinsic proteins (MIP) according to the prototype from bovine lens (Kaldenhoff and Fischer 2006). In 1999, together with other research teams, Agre

R. Bhardwaj (✉) • I. Sharma • M. Kanwar • R. Sharma • N. Handa • H. Kaur • D. Kapoor • Poonam
Department of Botanical and Environmental Sciences, Guru Nanak Dev University, Amritsar 143005, Punjab, India
e-mail: dr.renubhardwaj@yahoo.in; renu_bhardwaj@rediffmail.com

P. Ahmad et al. (eds.), *Ecophysiology and Responses of Plants under Salt Stress*, DOI 10.1007/978-1-4614-4747-4_8, © Springer Science+Business Media, LLC 2013 reported the first high-resolution images of the three-dimensional structure of AQP1. Further studies using supercomputer simulations have identified the pathway of water as it moves through the channel and demonstrated how a pore can allow water to pass without the passage of small solutes. However, the first report of protein mediated water transport through membranes was provided by Gheorghe Benga in 1986 (Kuchel 2006).

Aquaporins form tetramers in the cell membrane, with each monomer acting as a water channel (Gonen and Walz 2006). The different AQPs contain differences in their peptide sequence that allows for the size of the pore in the protein to differ between AQPs (Ayadi et al. 2011). The resultant size of the pore directly affects the molecules which are able to pass through the pore. These proteins are made up of six transmembrane α-helices arranged in a right-handed bundle, with the amino and the carboxyl termini located on the cytoplasmic surface of the membrane (Fu and Lu 2007; Gonen and Walz 2006). The amino and carboxyl halves of the sequence show similarity to each other. There are also five inter-helical loop regions (A–E) that form the extracellular and cytoplasmic vestibules. Loops B and E are hydrophobic loops that contain the highly, although not completely conserved, asparagines-proline–alanine (NPA) motif, which overlap the middle of the lipid bilayer of the membrane forming a 3-D ‘hourglass’ structure where the water flows through. This overlap forms one of the two well-known channel constriction sites in the peptide, the NPA motif and a second and usually narrower constriction known as ‘selectivity filter’ or ar/R selectivity filter (Gonen and Walz 2006; Fu and Lu 2007; Maurel and Plassard 2011). The two highly conserved NPA motifs are the most important structural domains that play a crucial role in water-selective permeation in aquaporin water channels (Guan et al. 2010). However, the functions of NPA motifs in aquaporin (AQP) biogenesis remain largely unknown. Few AQP members with variations in NPA motifs such as AQP11 and AQP12 do not express in the plasma membrane, suggesting an important role of NPA motifs in AQP plasma membrane targeting. The NPA motifs may interact with other structural domains in the regulation of membrane trafficking during aquaporin biogenesis (Guan et al. 2010; Sorieul et al. 2011).

There are 13 known types of aquaporins in mammals, and 6 of these are located in the kidney, but the existence of many more is suspected. However, in the plants water is taken up from the soil through the roots, where it passes from the cortex into the vascular tissues either *via* apoplastic or symplastic pathways. The presence of AQPs in the cell membranes facilitates the transcellular symplastic pathway for water transport and hence adjusts the overall hydraulic conductivity (Leitão et al. 2012). When plant roots are exposed to mercuric chloride ($HgCl_2$), which is known to inhibit AQPs, the flow of water is greatly reduced while the flow of ions is not, supporting the view that there exists a mechanism for water transport independent of the transport of ions. In plants, AQPs are separated into four main homologous subfamilies, or groups: Plasma membrane Intrinsic Protein (PIP), Tonoplast Intrinsic Protein (TIP), Nodulin-26 like Intrinsic Protein (NIP), Small basic Intrinsic Protein (SIP). These subfamilies have later been divided into smaller evolutionary subgroups based on their DNA sequence (Kaldenhoff and Fischer 2006) and amino acid sequence similarities (El-Mesbahi et al. 2012). PIPs cluster into two subgroups,

PIP1 and PIP2, whilst TIPs cluster into five subgroups, TIP1, TIP2, TIP3, TIP4 and TIP5. Each subgroup is again split up into isoforms e.g. PIP1; 1, PIP1;2 (Maeshima and Ishikawa 2008). Moreover, a novel category of major intrinsic proteins (MIP) is recently identified in land plants, and named X (for unrecognized) intrinsic proteins (XIPs). Further studies in *Populus*, revealed highest polymorphism of XIP isoforms in this genus with nine PtXIP sequences distributed within three XIP groups have been observed (Lopez et al. 2012). However, comprehensive PtXIP gene expression patterns further emphasized that only two isoforms (PtXIP2;1 and PtXIP3;2) were transcribed in vegetative tissues. The silencing of plant AQPs has been linked to poor plant growth and even death of the plant. Besides their major role in plant water relations, the plant AQPs also facilitates the transport of small solutes such as glycerol, silicon, ammonium, urea, boric acid, CO_2, arsenite, and hydrogen peroxide within and between cells (Aroca et al. 2012; Vera-Estrella and Bohnert 2011). Recently, the role of AQPs in stress protection has been recognized in durum wheat (Ayadi et al. 2011), barley (Katsuhara et al. 2011), broccoli (Muries et al. 2011) and tomato (Sade et al. 2010) subjected to salinity stress. This chapter reviews the structural and functional properties of the plant AQPs illustrating their importance in the water homeostasis, nutrition, and signaling processes during stress.

8.2 The Plant Aquaporin Family

Aquaporin proteins are expressed in multiple isoforms. In *Arabidopsis* about 35 and in rice 33 homologs have been identified (Johanson et al. 2001; Sakurai et al. 2005; Quigley et al. 2001). About 55 isoforms have been reported in poplar (Gupta and Sankararamakrishnan 2009) and 71 in cotton (Park et al. 2010). Also, 36 isoforms in maize have been identified (Chaumont et al. 2001). These homologs of aquaporins are divided into four groups based on their sequence homology.

8.2.1 Tonoplast Intrinsic Protein (TIP)

As the name indicates, these types of aquaporins are present in the membranes of the plant vacuoles, or tonoplast (Fig. 8.1). Vacuole in the plant cell is a compartment for cellular storage and also takes part in turgor regulation, cell signaling and degradation. Therefore, vacuoles have to regulate the flux of water and small solutes across the vacuolar membrane which suggests that AQPs participate in all these processes. Consequently, the first proteins with AQP function in plants were identified in vacuolar membranes from *Arabidopsis thaliana* (Johnson et al. 1990; Maurel et al. 1993). This tonoplast intrinsic protein (TIP1;1, initially named γ-TIP) when expressed in *Xenopus* oocytes, was found to be highly water selective and impermeable for glycerol. Also, isolated vacuoles or tonoplast vesicles exhibited a comparable 100-fold higher permeance as compared to purified plasma membranes

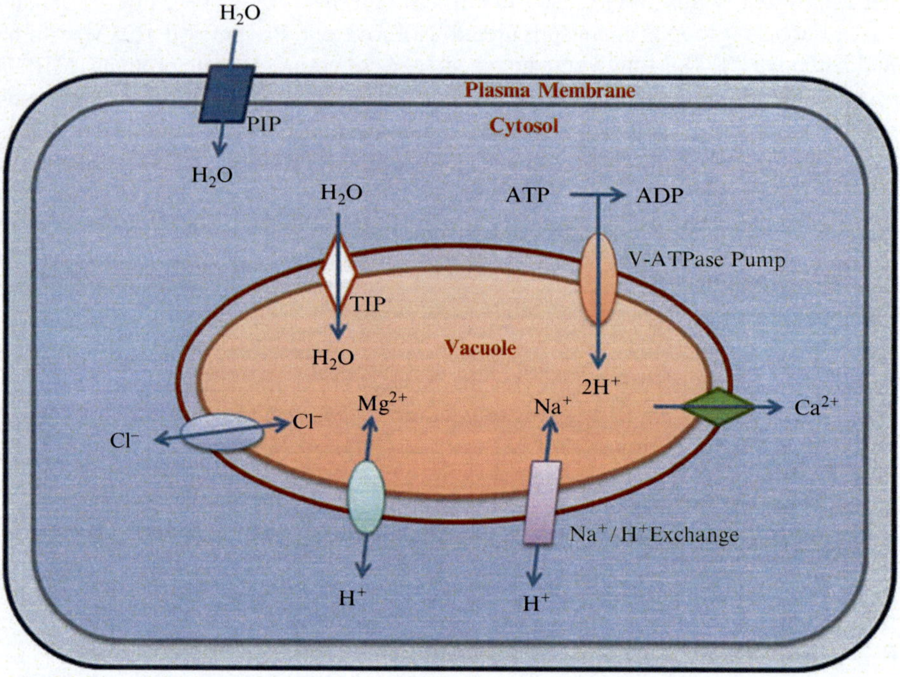

Fig. 8.1 Water transport in membranes of plant cell (PIP: plasma membrane intrinsic proteins; TIP: tonoplast intrinsic proteins)

when osmotic water permeability measurements were carried out (Maurel et al. 1997; Morillon and Lassalles 1999; Niemietz and Tyerman 1997). This shows that tonoplast has major participation in regulation of water flow in response to osmotic challenges like drought or salinity for a plant cell. Gerbeau et al. (1999) reported water flux which led to an increased permeability of solutes like urea and glycerol in purified tobacco tonoplast vesicles as compared to plasma membrane vesicles. Recent studies revealed the potential contribution of TIPs in transportation of NH_4^+/NH_3 from the cytoplasm into the vacuole. This had been shown in plants like *Arabidopsis*, wheat and also in oocytes of *Xenopus* (Holm et al. 2005; Jahn et al. 2004; Loque et al. 2005). Induction of osmotic stress led to TIP regulation and redistribution and this was investigated using ice plant (*Mesembryanthemum crystallinum*) (Vera-Estrella et al. 2004). Exposure to salt slightly decreased McTIP1;2 abundance while treatment with mannitol, sorbitol or abscisic acid caused an up-regulation (Kirch et al. 2000; Vera-Estrella et al. 1999). AQP expression in the presence of salt was also investigated in *Arabidopsis* roots (Boursiac et al. 2005). TIP and other AQP transcripts showed an extensive decrease in abundance after 4 h following exposure to salt. LiJuan et al. (2011) identified a novel TIP, AtSM34 in *Arabidopsis thaliana* and studied its role in osmotic stress in germinating seedlings. Overexpression of AtSM34 resulted in hypersensitivity in the presence of exogenous mannitol, sorbitol, and abscisic acid thereby causing significant delay in germination. Hence, TIPs are actively involved in osmoregulation and also take part in conductance of small solutes and gas. These functions indicate the link of TIPs to important metabolic pathways like urea cycle or amino acid synthesis.

8.2.2 Nodulin-26-Like Intrinsic Membrane Proteins (NIP)

These AQPs are close homologs of *GmNod26*, an abundant AQP in the peribacteroid membrane of symbiotic nitrogen-fixing nodules of soybean roots (Wallace et al. 2006). Soybean Nodulin 26 (Nod26) was described as a major integral protein and constitutes approximately 10% of total membrane protein (Fortin et al. 1985; Weaver et al. 1991). NIPs are present in non-leguminous plants and are usually localised in plasma and intracellular membranes (Ma et al. 2006; Mizutani et al. 2006; Takano et al. 2006). All proteins related to nodulin-26 have been included in NIP subfamily. Nod26 heterologously expressed in *Xenopus* oocytes showed a mercury-sensitive osmotic water permeability and conductance to glycerol (Rivers et al. 1997). NIPs in *Arabidopsis* and lotus formed functional glycerol permeases and exhibited partially low conductivity (Wallace et al. 2002; Weig and Jakob 2000; Weig et al. 1997). But NIP in zucchini enhanced the growth of urea transporter defective yeast mutant along with water permeability and lack glycerol conductivity (Klebl et al. 2003). Apart from the above functions, Hwang et al. (2010) hypothesized potential role of Nod26 from soybean in ammonia efflux. Therefore, Nod26 and other NIP proteins share the general multifunctional transport properties of water and uncharged solutes like glycerol and potentially ammonia. Compared to most water-selective AQPs, NIPs also have a lower rate of water transport.

8.2.3 Small Basic Intrinsic Protein (SIP)

It is the smallest subfamily covering 2–3 divergent aquaporin homologues. These aquaporins are small in size and are highly basic in nature. They possess a very short cytosolic N-terminal region which is the reason for their small size. In recent studies on *Arabidopsis*, the SIPs were mainly found in endoplasmic reticulum and scarcely present in plasma or vacuolar membranes (Ishikawa et al. 2005).

8.2.4 Plasma Membrane Intrinsic Proteins (PIP)

This is the largest subfamily of plant aquaporins having 13 members in *Arabidopsis*, 14 in maize, 11 in *Oryza sativa*, and 14 in *Populus trichocarpa* (Chaumont et al. 2000; Johanson et al. 2001). The most of the PIPs are localized in plasma membrane (Fig. 8.1) and are further divided into two phylogenetic groups *viz*. PIP1 and PIP2 (Zardoya 2005). The two subgroups differ in their length of N and C terminals. The amino acid residues at the selectivity filter are similar but PIP1 and PIP2 differ in their permeability properties and cellular functions (Wallace and Roberts 2004). Members of PIP2 subgroup are more efficient water channels as compared to PIP1 subgroup. In various studies using *Xenopus* oocytes or yeast membrane vesicles as heterologous expression system PIP2 aquaporins exhibit 5- to 20-fold increased water permeability compared to control values (Daniels et al. 1994; Tornroth-Horsefield et al. 2006; Weig et al. 1997). Some members in the PIP1 subfamily are capable of transporting glycerol (Biela et al. 1999; Moshelion et al. 2002), urea (Gaspar et al. 2003), and CO_2 (Uehlein et al. 2003). *Hordeum vulgare* PIP2;1 and tobacco PIP1 appear to transport CO_2 in addition to water (Uehlein et al. 2003; Flexas et al. 2006; Hanba et al. 2004). Hence, PIP aquaporins have a role in developmental and environmental response mechanisms which include general osmotic homeostasis, water uptake from the soil, transpiration, stomatal aperture control, cytosolic osmotic homeostasis, leaf movement, floral expansion, and postmeristematic elongation (Maurel et al. 2002; Tyerman et al. 2002).

Apart from the four major families of AQPs, few others like X intrinsic protein (XIP), hybrid intrinsic proteins (HIP) and GlpFlike intrinsic protein (GIP) have been identified by researchers. Members of these families are the least characterized and therefore studies have to be carried out to gain insight into solute transport, expression, post translational modifications and other properties.

8.3 Molecular and Cellular Properties

As discussed previously, AQPs are essentially a sub group of intrinsic water channel proteins belonging to the MIP family and are known for their role in facilitating bi-directional water flow and movement of low molecular weight molecules across cell membranes i.e. plasma and vacuolar membranes, after osmotic or hydrostatic pressure gradients get created (Chrispeels and Maurel 1994; Javot et al. 2003; Postaire et al. 2010). More than 150 AQPs have been identified in organisms ranging across the microbial world to the animal and plant kingdoms as well. Plants enclose a large number of AQPs with discrete cell type- and tissue defined expression patterns. Some of these are expressed as a result of constitutive gene regulation, whereas the expression of others is regulated in response to different environmental factors, such as drought and salinity. AQPs also have a strict ionic selectivity, and are in particular totally impermeable to protons. This is a crucial property in view of the important proton gradients that are present across plant membranes. Plant AQPs came into limelight with their discovery in intracellular membranes in soybean root cells (Sandal and Marcker 1988). Subsequently, their presence in *Arabidopsis* vacuolar membranes (Maurel et al. 1993) also further drew attention to their much elusive role in plant cells. Since the past two decades, around 13 AQP isoforms or homologues in the model plant *Arabidopsis*, are supposed to be predominantly intracellular, however despite all this, their function is intangible. The high number of plant intracellular AQPs has, of course, to be related to the high diversity of AQPs in plants compared with animals (Agre 1998; Maurel et al. 2008). The relevance of AQPs to plant cells needs to be understood at the molecular and cellular level to unravel their precise functions in homeostatic regulation and signal cascading. As this is a family with many structural homologues, their individual properties can be very specific and discrete, depending upon structural and phylogenetic rearrangements. Abundant studies have exposed an array of transport properties in between, as well as within plant AQP subclasses (Maurel et al. 2008). The most universal substances exchanged *via* these gated channels includes water, glycerol, urea, hydrogen peroxide (H_2O_2), metalloids, organic acids and gaseous compounds such as ammonia (NH_3) and carbon dioxide (CO_2). Subsequent studies on other plant AQPs apart from *Arabidopsis* cells revealed diverse functions, such as urea and glycerol transport in tobacco cells and rice Tonoplast Integral Proteins (Flexas et al. 2008; Gerbeau et al. 1999; Heckwolf et al. 2011; Becker 2011; Terashima and Ono 2002; Vera-Estrella and Bohnert 2011; Wang et al. 2011).

Regardless of their amino acid composition, AQPs can infuse small molecular sized neutral gases such as nitric oxide (NO) and CO_2 across membranes (Wu and Beitz 2007). However, the movement of larger solutes, such as salicylic acid, lactic acid or urea, appears to be challenging to most AQP isoforms, though some recent studies claim the contrary for the same. For uninhibited transport *via* these channels which are 'gated' two pore constrictions have been identified as important selectivity filters. One such constriction is formed by the NPA region. Water molecules form a single file in the aqueous pore and are reoriented upon interaction with the Asn residues of the NPA region (Tajkhorshid et al. 2002). This mechanism and a strong electrostatic field provide bases for blockade of proton conduction across the gated channels (de Groot et al. 2003). A second constriction, called Ar/R, is located on the extra-cytoplasmic mouth of the pore. It is formed by four residues including Aromatic residues and an Arg (R) residue, hence this name. This arrangement functions as a selectivity filter, due to stearic effects and the R residue acts as a site for electrostatic repulsion of protons (Fujiyoshi et al. 2002). One characteristic molecular feature that vitally point's out towards selective transport properties of AQPs is the R constriction, and dissimilarities in this site are thought to be the decisive reason for the broad permeant range exhibited by plant AQPs (Wallace and Roberts 2004; Bansal and Sankararamakrishnan 2007; Li et al. 2009a; Benga 2012). As a result of this differential sequence arrangements specific to the R constriction, categorization of AQPs depending on this structural motif offers a constructive tool to envisage substrate specificities. A simulative modelling approach, based on the presence of four residues in the pore region that form the R site, has resulted in the classification of plant AQPs into eight groups (Wallace and Roberts 2004). Pore diversity is mostly obvious for those AQPs whose expression is mainly intracellular. For example, members of the TIP family bunch in three groups, whereas members of the NIP and SIP families form two distinct subgroups each. On the contrary, all PIPs cluster in one group with strong structural and functional affinities to *Hs*AQP1, indicating their major role as water channels.

The detection of *At*TIP1; 1 which is the first water channel protein recorded in plants, in tonoplasts (Maurel et al. 1993), encouraged research on the water transport properties of the Tonoplast Proteins (TP). Maurel et al. (1997) analyzed vesicles from purified tobacco suspension cells for analyzing water transport abilities in TP vesicles. Further; experiments on isolated vacuoles from different plants confirmed the high osmotic water permeability across the tonoplasts (Morillon and Lassalles 1999). Progression or stress-induced changes in vacuolar membrane systems have been observed for different cell types. TIPs that have been primarily used as markers to follow changes in vacuole physiology, participate in these changes. Vera-Estrella et al. (2004) observed a rearrangement of vacuoles from suspension cell cultures of *M. crystallinum*, accompanied by re-distribution of *Mc*TIP1;2 into endosomal compartments, on account of mannitol treatment. This indicated that AQPs are capable of transporting water at a faster rate, and this resulted in restoration and maintenance of cell osmolarity under difficult osmotic conditions. Also, the rearrangement of the *Arabidopsis* TP is in accordance to formation of vacuolar macro domains, labelled by TIP1; 1 (Beebo et al. 2009). Interestingly, the vacuolar membranes of apposing TPs are characteristically rich in this AQP and their reshaping may lead to formation of membrane structures such as bulbs (Saito et al. 2002). These spherical structures have been cited in speedily growing cotyledon cells of *Arabidopsis.* This process gives rise to a controlled TP water permeability. Similar bulb-like structures are also reported in guard cells (Tanaka et al. 2007). Also, TIP activity in the TP has direct effects on vacuolar and cellular expansion and growth, in protoplasts from BY-2 cells. Sade et al. (2009) considered that *Sl*TIP2; 2 isoform of tomato (*Solanum lycopersicum*) which was contained in vacuoles enhanced water transport upon expression in *Arabidopsis* protoplasts. Over-expression of the protein in transgenic tomato improved yield parameters (plant biomass and fruit production) in field trials under normal conditions as well as under salt or drought stress conditions. On the other side, over-expression of *Sl*TIP2; 2 led to increased transpiration rates even under control conditions. Under drought and salt stress, the transgenic lines, compared with control plants, showed a smaller reduction in transpiration rate and a faster recovery after stress. These fluctuating changes indicated a deflection from isohydric behaviour to anisohydric behavior. In the case of the transgenic tomato plants, it was hence thought that vacuolar over-expression of *Sl*TIP2; 2 were able to evade stress-induced down regulation of endogenous TIPs and thus conferred anisohydric growth behavior. An *Arabidopsis* mutant lacking TIP1; 1 showed 40% reduction in length of primary root in comparison to wild type, when grown on glycerol containing media (Beebo et al. 2009). A proposition of TIPs in the transfer of urea was indicated in tobacco suspension cells, where, with the expression of *Nt*TIPa (a urea-transporting TIP), TP vesicles showed a 70-fold higher permeability to urea than the corresponding PM vesicles (Gerbeau et al. 1999).

In the recent years, urea transport has been reported in members of almost all TIP subclasses in *Arabidopsis*, including *At*TIP1;1, *At*TIP1;2, *At*TIP1;3, *At*TIP2;1, *At*TIP4;1 and *At*TIP5;1 (Liu et al. 2003; Soto et al. 2008). Plants also harbour a high-affinity urea transporter belonging to the sodium solute transporter protein family (Kojima et al. 2006). The transport of NH_3 was reported for members of the

TIP2 subfamily from wheat (*Ta*TIP2; 1) and *Arabidopsis* (*At*TIP2; 1, *At*TIP2; 3) (Jahn et al. 2004; Loque et al. 2005) and more recently for *At*TIP1; 2 (Dynowski et al. 2008). TIPs might rather play a role in subcellular partitioning of NH_3 and contribute to the detoxification of excess amounts of NH_3 in the cytosol. A correlation of AQPs with nitrogen transport, storage and/or assimilation has also been suggested by gene expression analyses. For instance, expression of *At*TIP1;2, *At*TIP2;1 and *At*TIP4;1 in roots and seedlings of *Arabidopsis* was up-regulated under conditions of nitrogen deficiency (Liu et al. 2003) while *At*TIP2;1 and *At*TIP2;3 showed an ammonium-induced up regulation (Loque et al. 2005). Similarly, AQPs for H_2O_2 include *At*TIP1; 1, *At*TIP1; 2 and *At*TIP2; 3 as (Bienert et al. 2007; Dynowski et al. 2008). In plants, H_2O_2 serves as a signalling molecule and it has been proposed that AQPs themselves could be subjected to H_2O_2-dependent regulation (Henzler et al. 2004; Boursiac et al. 2008).

Research in recent years on plant MIPs has revealed an expansion in the substrates found to be transported *via* these membrane channels. Hove and Bhave (2011) revealed that the molecular and cellular properties of AQPs depend upon the structural organization as well as on the nature of molecules being transported across the membranes. Many metabolic reactions and storage functions carried out by plant cells are required to be precise and localized and for this, higher plants have evolved with specifically high degree of sub cellular compartmentation. As a result of this diversification of functions, the cell organelles exhibit a huge facade of specific transport properties, each being supported by its own specialized protein equipment. AQPs have emerged as this class of central proteins that determine and organize organelle specialization. It has been realized that AQPs are not simple water channels but can exhibit varied transport-selectivity properties. Additionally, the large range of developmental, tissue-specific and sub cellular expression patterns that have been revealed can now largely be held accountable for the high multiplicity of AQP isoforms in plants. The intricate pattern of overlapping transport activities and subcellular localizations makes it difficult to understand the organelle- and/or isoform-specific functions of plant AQPs. Single molecular level studies are a recent approach to understand the molecular role of *Arabidopsis thaliana* PIP2; 1 *via* variable-angle evanescent wave microscopy and fluorescence correlation spectroscopy (Li et al. 2011). The molecular and functional characterization of protective proteins and aquaporins has revealed the importance of their regulation in response to various abiotic stresses (Hussain et al. 2011).

8.4 Mechanism of Transport

Although most plant AQPs show a high degree of homology to each other, despite this, they are contained in very specific sub cellular membranes. Very less is known about signal targeting of plant membrane proteins inclusive of AQPs. The early classification of plant AQPs in so-called PIPs and TIPs also indicates the possibility that the final membrane destination of a plant AQP is deduced from its primary sequence. Recent research has indicated that the intracellular localization of plant AQPs is influenced by a wide range of determinants, including targeting motifs, post-translational modifications and protein–protein interactions. Although the progress has mostly concerned the trafficking of PIPs to the plasma membrane (PM), it indicates a variety of mechanism that is expected to determine trafficking of intracellular AQPs. Usually, aquaporins are selective for water alone and aqua glyceroporins are known to be carriers to carrying water and small uncharged solutes including glycerol. The role of aquaporins as dual water and gated ion channels can be evaluated with various molecular and pharmacological tools. Finally, recent studies have shown that AQPs undergo modifications, such as phosphorylation (Prak et al. 2008), glycosylation (Vera-Estrella et al. 2004) or ubiquitination (Lee et al. 2009), which accompany a stress-induced or development-induced shift in their sub cellular localization. These aspects will have to be refined to understand in more detail the versatility and functional fine-tuning of intracellular AQPs. PIPs are shown to be primary channels mediating water uptake in plant cells. Water transport activity and mechanisms for the regulation of *Hordeum vulgare* PIP aquaporins, HvPIP2 have been successfully explored in *Xenopus* oocytes (Horie et al. 2011). Similarly, Major intrinsic proteins (MIPs) transport water and uncharged solutes across membranes. Recently, an uncharacterized MIP subfamily identified in the genomes of plants and fungi known as X Intrinsic Proteins (XIPs) have come to focus for their transport abilities. The genetic features, localization, expression, and transport functions of a group of *Solanaceae* XIPs have been recently (Bienert et al. 2011).

During the route of their production, synthesis, maturation followed by secretion, all AQP isoforms are transported through secretory pathways before reaching their objective compartments within the plant cell. Some AQPs, however, have been distinctively restricted at the input or at the start of this pathway, i.e. the endoplasmic reticulum (ER) (Ishikawa et al. 2005). AQPs that reside in the ER include SIPs (Johanson and Gustavsson 2002). Their specific localization in the rough ER and in other ER sub domains has been determined through expression of GFP–SIP fusion proteins in protoplasts of *Arabidopsis* suspension cultures, along with immune blots on membrane fractions purified from these cells. The ER compartment is the largest membrane surface area in plant cells. It is proposed that the complex sheet system of ER membranes is considered to act as giant vacuoles. By extending throughout the cytoplasm and surrounding the vacuoles, the ER might represent a barrier to intracellular water transport. Hence, the presence of AQPs may facilitate water fluxes across the double sheets of the ER membranes. In relation to their degenerated NPA motif, SIPs may also transport molecules other than water.

Plant endosomes are also intracellular organelles that denote variable biochemical composition and flexible structure. Endosomes are shown to enable the constitutive turnover of PM proteins, including the auxin-induced recycling of PIN proteins (Geldner et al. 2003) and the internalization of PIP AQPs (Paciorek et al. 2005). It is accepted, however, that the presence of PIPs in endosomes just reflects a dynamic regulation of their density at the PM and their sorting between storage or degradation paths. Chloroplasts are unique plant organelles that perform the essential yet basic process of photosynthesis. Because it requires the diffusion of atmospheric CO_2 into the stroma of the chloroplast, photosynthesis can be limited by the stomatal conductance and also by the mesophyll (or internal) conductance to CO_2. Terashima and Ono (2002) showed that mercurial compounds, which act as general and nonspecific inhibitors of AQPs, were able to decrease the internal conductance of *Vicia faba* and *Phaseolus vulgaris* without changing the assimilation rate of chloroplast CO_2. The implication of this transport property was eventually established by analysing transgenic tobacco plants that over expressed, or were deficient in, *Nt*AQP1. These materials revealed a strong relationship between the changing levels of *Nt*AQP1 and the corresponding *g*m values (Flexas et al. 2008; Uehlein et al. 2003). A mutant *Arabidopsis thaliana AtPIP1; 2* gene, has been characterized as a CO_2 transport-facilitating aquaporin in heterogonous expression systems with reduction in photosynthesis under the influence of varied atmospheric CO_2 concentrations (Heckwolf et al. 2011). Specific aquaporins as membrane intrinsic pore proteins are now known to have a function in the alteration of membrane CO_2 conductance (Kaldenhoff 2012).

Mitochondria are ubiquitous organelles of eukaryotic organisms. They play a crucial role in cell energy metabolism and harbour various anabolic and catabolic reactions. Mammalian AQP8 and AQP9 have been recently detected in inner mitochondrial membranes (Amiry-Moghaddam et al. 2003; Calamita et al. 2005). The membrane differentiated by the host plant cell to surround the bacterial endosymbiont is referred to as the peribacteroid or symbiosome membrane (Patriarca et al. 2004). Among the nodule-specific proteins (nodulins) of soybean (*Glycine max*), *Gm*NOD26 was identified as one of the most abundant proteins of the peribacteroid membrane (Fortin et al. 1987). This nodulin was actually the first MIP to be described in plants and further served as a founding member of the NIP subfamily. Recently, members of the Nodulin 26-like intrinsic protein (NIP) subfamily of plant aquaporins were shown to transport arsenite in *O. sativa* and *Arabidopsis*. Certain members of the rice PIP subfamily are also involved in Arsenic tolerance and transport (Mosa et al. 2012). Previously, it's known that the (*Oryza sativa* silicon transporter Lsi1 (OsNIP2; 1, an aquaporin channel) is the major entry for Aquaporins-6; plant Nodulin-26 route of arsenite into rice roots (Zhao et al. 2010). Also, gated ion channel activity has been shown). Hence, the entire movement and transportation of solutes and organic molecules with the cells happens *via* various sub families of gated ion channels called the AQPs present within the cellular and organelle membranes.

8.5 Molecular Mechanisms of Regulation of Plant Aquaporins

8.5.1 Transcriptional Regulation

Transcriptional regulation of AQPs is related to the regulation of water transport *via* variation in channel density in the target membranes. Expression of particular AQPs are specific for certain cell types and organs however, their expression vary diurnally and in response to environmental or developmental influences. It was suggested by Srivastava et al. (2010) that AQP genes PIP-1;4 regulate the loss of water under stress conditions. In the endodermis of *Arabidopsis* roots, expression of PIP1 AQP was greater than in its cortex (Schaffner 1998). Their expression levels are high in the regions of concentrated water flow (Hachez et al. 2006; Vandeleur et al. 2008). Cells in exodermis closely associated with the xylem vessels and the cells present in the phloem bundles showed strong antibody signals as compared to pre-immune serum controls. In the expression of MIPs, diurnal fluctuations have also been reported (Henzler et al. 1999; Lopez et al. 2003; Sakurai et al. 2005; Yamada et al. 1997) and they have also been correlated with water transport (Henzler et al. 1999; Moshelion et al. 2002; Vandeleur et al. 2005). Further, Janz and co-workers (2010) observed the low transcriptional responsiveness of *Populus euphratica* towards salt stress and thus possess the stimulation of stress protective genes. During the day, changes in expression in leaves have been correlated with the leaf water potential (Yamada et al. 1997). In case of maize ZmTIP2-3 expression started to increase prior to the light period and it was at its highest after 4 h of light (Lopez et al. 2003). According to Sakurai et al. (2005), diurnal fluctuations were observed in the roots of *Oryza sativa* PIP2 genes, OsPIP1;2 and OsPIP1;3 and they were at their peak after 3 h of onset of light and decreased to their minimum after 3 h of the onset of darkness. In response to many environmental factors like water stress, salinity, anoxia, nutrient depletion, hormones, low temperature and light, aquaporins have been found to be up and down-regulated (Bramley et al. 2007).

8.5.2 Post-translational Regulation

One method of the post-translational regulation of the activity of AQP is reversible phosphorylation. Water permeability is increased by phosphorylation of plant MIPs (Guenther et al. 2003; Johansson et al. 1998; Maurel et al. 1995). The process of de-phosphorylation occurs by a Ca^{2+} dependent protein kinase at two highly conserved serine residues, namely Ser 115 in cytosolic loop B and the other one is Ser 274, in the C-terminus. Residues present in the D-loop of SoPIP2;1 are involved in gating of the channel. According to the structure, loop-D has additional 4–7 amino acid residues, related to PIP subfamily, which caps the pore of the aquaporin occluding the pore from the cytosol (Tornroth-Horsefield et al. 2006). Water permeability of plasma membrane is regulated by cytosolic pCa (free calcium ion concentration) and pH. According to Alleva et al. (2006) in the storage roots of *Beta vulgaris*, measurements on plasma membrane vesicles have revealed very high water permeability, which was strongly regulated by pCa and pH. Gerbeau et al. (2002) reported that in the presence of magnesium and calcium ions, *Arabidopsis* Lpcell was reduced by 35% and 69% respectively. Two aquaporins namely, PIP1 and PIP2 have a histidine residue His 197, which is pH sensitive. When roots are subjected to anoxic stress due to a decrease in cytoplasmic pH, this helps in explaining the reduction in root hydraulic conductivity (Tournaire-Roux et al. 2003). It was observed that in different plants, the modification of PIP aquaporin expression might be one of the causes of salinity stress response (Marulanda et al. 2010).

Evidences for mechano-sensitive gating of AQPs have also been over-emphasized in the literature. For example, Wan and co-workers (2004) reported that Lpcell of maize root cortical cells decreased due to large pressure pulses. Activity of AQP based on Lpcell in *Chara* was also inhibited in the presence of high concentrations of osmotic solutes and with increasing the size of these solutes, decreased more strongly (Ye et al. 2005). Osmotic solutes also gate the water permeability of the symbiosome membrane containing NOD26 (Vandeleur et al. 2005). Interaction between different AQPs either in the membrane or *via* targeting to the membrane also regulate the AQP activity (Fetter et al. 2004). In *Xenopus* oocytes, osmotic water permeability was increased due to the co-expression of ZmPIP1;2, which has low activity with ZmPIP;4 or ZmPIP2;5 and it has been also reported for two grapevine AQPs namely VvPIP1;1 and VvPIP2;2 (Vandeleur et al. 2008). In living maize cell, physical interaction with PIP2 is needed to traffic PIP1 from the endoplasmic reticulum to the plasma membrane (Zelazny et al. 2007). Water/solute permeability is regulated by redistribution of AQPs *via* endomembrane vesicles and it occurs in response to osmotic stress (Vera-Estrella et al. 2004).

8.6 Aquaporin Activity/Functions

AQPs play essential role in the regulation of plant water status and expenses. Their gene expression depends upon the type of AQPs, plant organ and water stress level. Expression of AQP was responsive to water stress, related to homeostasis (Alexandersson et al. 2010) and it led in constant hydraulic conductivity and leaf water potential (Galmes et al. 2007). These proteins are responsible for water uptake and transport by roots, due to their involvement in root conductivity (Henzler et al. 1999; Tyerman et al. 2002). AQPs help in opening and closing the gate which regulates water movement in and out of cells. They trigger both reversible and irreversible circadian-regulated alterations in the cell volume of the leaves (Farre 2012). Involvement of AQPs in stem conductivity has been documented by Siefritz et al. (2002). According to Holbrook and Zwieniecki (1999) and Tyree et al. (1999), specific role of AQPs has been observed in the xylem conduit after recovering from drought induced embolism. Some AQPs have also been reported to perform certain specific activities like transport of ammonia and other substances (Tyerman et al. 2002; Luu and Maurel 2005), such as CO_2 in the leaf mesophyll during photosynthesis (Uehlein et al. 2003; Hanba et al. 2004; Flexas et al. 2006). AQPs help in passive transport of urea along a concentration gradient through a channel with the help of ZmNIP2;1, ZmNIP2;4 and ZmTIP4;4 genes (Gu et al. 2012; Witte 2011). According to Lovisolo et al. (2008), AQPs can account for well over 40% of the flow across roots under transpiring conditions. Some AQPs are constitutively expressed, while certain factors including different stimuli such as hormones, adverse environmental conditions like drought and salinity regulate the expression of other AQPs (Johansson et al. 1996; Vera-Estrella et al. 2004). Under water deficit conditions, their expression has been shown to increase, decrease or remain unaffected (Alexandersson et al. 2005). According to Kanai et al. (2011), in water uptake of roots, AQPs along with K-channel transporters regulate the stem diameter dynamics of green house *Lycopersicon esculentum* plants. During environmental changes, gene expression responses within hours and their expression pattern involves acclimation mechanisms with long term water shortage, remains to be elucidated. Expression of AQPs was studied in Richter-110 (*Vitis sp.*) by Galmes et al. (2007) during the water stress conditions. Their expression altered in leaves from no change to ten-fold. Due to the water stress imposition, stress maintenance and re-watering conditions, their expressions showed a similar pattern in leaves. During severe stressed conditions, these plants showed similar or increased expression of AQPs genes than control values. Rewatering led in maintenance or enhancement in the expression level of AQPs and their expression was also dependent on their specific type and intensity of stress.

Due to variation in AQPs abundance and activity, the plants could enhance rapidly water flux from the roots, in response to environmental changes (Luu and Maurel 2005). According to Henzler et al. (1999), AQP gene expression is stimulated at daytime, increasing water inflow as the transcription rate increases. Root hydraulic conductance could be changed during daytime according to gene expression (Henzler et al. 1999; Martinez-Ballesta et al. 2003a). It has been observed that during the day, leaf hydraulic conductance (K_{leaf}) increases in response to enhanced demand of water due to stomatal opening (Tsuda and Tyree 2000). K_{leaf} of walnut leaves (*Juglans regia*) was found low in dark conditions and due to exposure to sunlight, it was increased by 400% (Cochard et al. 2007). It was analyzed that low K_{leaf} in dark is associated with down-regulation of JrPIP2 AQP, whereas high K_{leaf} in light was due to up-regulation of JrPIP2. Similar results were analyzed by Tyree et al. (2005), where increase in K_{leaf} was observed by irradiance, which was induced by *de novo* expression of aquaporins. Enhancement in K_{leaf} is also due to activation of pre-existing AQPs and consequent increase in water transport.

There is also correlation between diurnal variation in root hydraulic conductivity and the expression of AQPs (Martinez-Ballesta et al. 2009). It has been reported that in *Lotus japonicus* (Henzler et al. 1999) and *Arabidopsis* (Martinez-Ballesta et al. 2003a), water uptake and AQP transcript levels were found to increase during the light cycle. Decrease in the amount or activity of AQPs in the root plasma membrane leads to large reduction in root hydraulic conductance under salinity (Carvajal et al. 1999; Martínez-Ballesta et al. 2000). It was reported that NaCl could influence negatively the function of AQPs, thus the water channels could be strongly reduced in number. AQPs help in mediating the mass flow of water within the plants. Protein abundance is the critical parameter for understanding their function at tissue, cell or subcellular level.

8.7 Aquaporins in Plant Growth and Development

8.7.1 Water Transport

Water is a necessary element of living organism. In plants water is transported from root to shoot system by hydrostatic and osmotic pressure gradient. The water evaporates form leaves of plant by transpiration and act as resistance to the water flow. But the roots can also play role as a barrier (Martre et al. 2001; Steudle 1994; Steudle and Peterson 1998). AQPs are major intrinsic proteins and assist in the flow of water across cellular membranes (plasma and vacuolar membranes), according to the osmotic or hydrostatic pressure gradients (Chrispeels and Maurel 1994). AQPs help in the water transport by opening and closing the gate which regulates the movement of water. AQPs which are present in the plasma membrane of plant help the plants to survive in the midst of the rapid changes in the water availability. It is investigated that in water deficit conditions tonoplast AQP gene is expressed for precise regulation of water in cauliflower can be associated with important cytological changes in the cells. It has been suggested that the decrease in root hydraulic conductance by salinity may have relation with the decrease in concentration or activity of AQP in the root plasma membrane (Carvajal et al. 1999; 2000). It is strongly supported that the aquaporins have more role in the water transport than previously thought, even in the conditions of transpiration (Knipfer and Fricke 2010, 2011; Fritz and Ehwald 2011). The experimental studies of Shelden (2007) show that there is a difference between the expression profile of AQPs to water stress in roots and petioles of two grapevine varieties. Galmes et al. (2007) has suggested that AQPs are responsive to the water stress as part of homeostasis leading to the constant leaf water potential and hydraulic conductivity. Kaldenhoff et al. (1998) has showed in *Arabidopsis* about the significance of PIPs in water transport using antisense technology. Morillon et al. (2001) has suggested that brassinolide has effect on the regulation of AQP activities in the *Arabidopsis thaliana* to be involved in the modification of water transport properties of cell membranes. Fetter et al. (2004) has illustrated that co-expression of ZmPIP1;2 and different maize PIP2isoforms led to increased water permeability, depending on the quantity of ZmPIP1;2 cRNA injected into *Xenopus* oocytes.

8.7.2 Nitrogen, Carbon, and Micronutrient Acquisition

Nitrogen is a basic constituent of many molecules and also used in the form of fertilizers. NO_3^- and NH_4^+ are the main forms to be applied as fertilizers. Gases like carbon dioxide and ammonia pass through the plasma membrane by dissolving in lipids and their transport is also reported to be facilitated by AQPs (Prasad et al. 1998; Terashima and Ono 2002). The less specific AQPs help in the transport of ammonia, urea, glycerol, carbon dioxide, boron, hydrogen peroxide (Biela et al. 1999;

Dordas et al. 2000; Gerbeau et al. 1999; Hanba et al. 2004; Henzler and Steudl 2000; Loque et al. 2005). It has been reported by Beebo et al. (2009) that *Arabidopsis* plants lacking TIP 1;1 when grown on glycerol has a reduced length of primary root as compared to the wild plants showing that AQPs has role in the transport of glycerol. In tobacco plants, the CO_2 assimilation rate increased in leaves over-expressing AtPIP1;2 (PIP1b) or NtAQP1(Aharon et al. 2003; Uehlein et al. 2003).

The first plant AQPs with glycerol permeability was found in soybean root nodules (Dean et al. 1999; Rivers et al. 1997). By using the cryo-electron microscopy and X-ray studies a mechanism of water and glycerol transport in *E. coli* by GlpF glycerol facilitator was found (de Groot et al. 2003; Fujiyoshi et al. 2002). Further Schuurmans et al. (2003) has suggested PsNIP-1 as the aquaglyceroporin of *Pisum*. Members of the NIP, PIP, and TIP subfamilies have been revealed to facilitate the transport of urea across membranes (Gerbeau et al. 1999; Klebl et al. 2003; Liu et al. 2003; Wallace and Roberts 2005). The aquaglyceroporin Fps1p from *Saccharomyces cerevisiae* has been exposed to mediate the transport of the undissociated form of acetic acid across yeast membranes (Mollapour and Piper 2007). In rice, Zhao et al. (2010) showed that OsNIP2:1 help in the efflux of $As(OH)_3$ approximately 15–20% of total Arsenic toxification and thus indicate that plant aquaporins help in the arsenic detoxification.

8.7.3 Plant Reproduction

Optimum temperature, moisture and oxygen are the basic needs of seed to germinate. Seed germination involves three phases' imbibition, synthesis of enzymes and cell elongation. Enzymes breakdown the stored material and produce energy needed for the growth of seedling. Different plants such as *Arabidopsis*, rice, canola, pea, spruce, and tobacco has been studied to find that AQPs help in the transport of water during seedling germination and establishment. Toole et al. in his studies has shown early this by his experiment. In his experiment mercury was used to inhibit the AQPs which result in the delayed seed germination and also delayed the seed coat rupture and radicle emergence (Toole et al. 1956). There are number of studies to show that AQP is involved in the imbibition of water during seed germination. Schuurmans et al. (2003) established that PIP1;1 are activated in mature seeds and during germination to help in the water absorption in *Pisum sativum*. There are almost 15 AQPs in the sexual reproducing organs of plants which show that water flow is important in this. In tobacco, NtPIP1; 1 and PIP2; 1 are expressed in reproductive organs, PIP1; 1 is highly expressed in the stigma, and both AQPs are expressed in the anther (Sakurai et al. 2005). There are many aquaproteins which are expressed in the zone of cell elongation and cell enlargement in plants like Q-TIP of *Arabidopsis* and maize and N-TIP of spinach. Several AQPs have been shown to be expressed in zones of cell enlargement and cell elongation e.g. Q-TIP of *Arabidopsis* (Ludevid et al. 1992) and maize (Chaumont and Barrieu 1998) and N-TIP of spinach (Karlsson et al. 2000).

8.7.4 Plant Movements and Rhythms

AQPs also play a specific role in the transport of NO which is a hydrophobic gas and it help the plants in many physiological processes (Herrera et al. 2006). Further Neill et al. (2008) has suggested that in plants, NO triggers signaling pathways which are involved in programmed cell death, pathogen defense, flowering, stomatal closure, and gravitropism. AQPs help in the transport of water which in turn helps in the movement of leaf and petal (Azad et al. 2004; Moshelion et al. 2002; Siefritz et al. 2004).

8.7.5 Photosynthesis

The rate of photosynthesis depends on the rate of CO_2 diffusion. As it has been found that aquaporins has role in CO_2 diffusion it shows increase in photosynthetic activity also (Aharon et al. 2003). There is increased net photosynthetic rate and leaf growth in tobacco plants which express NtAQP1 as compared to NtAQP1-silenced plants (Uehlein et al. 2003; 2008). Terashima *and* Ono has proposed that AQPs has implication in leaf internal CO_2 conductance. Vera-Estrella et al. (2012) has suggested a role for AQPs in maintaining water balance during CAM in *Mesembryanthemum crystallinum* and highlights the complexity of protein expression during the CAM cycle.

8.7.6 Symbioses: Mycorrhiza and Rhizobial Nodulation

The first plant NIP identified is *Gm*NOD26 is found to be symbiotically expressed to nitrogen-fixing nodules formed after infection of soybean by *Rhizobiaceae* bacteria. There is a 55% increase in Lp_r (root hydraulic conductivity) in poplar mycorrhized plants as compared to non-mycorrhiza plants which can be because of enhanced expression of aquaporins. The expression of AQP genes has been altered by the arbuscular mycorrhizal fungus *Glomus mosseae* in *Glycine max* and *Lactuca sativa* roots which apparently provided a mechanism that enhanced host plant tolerance to water deficit (Porcel et al. 2005). In bean (*Phaseolus vulgaris*), the expression of PIP1; 1 has been reduced by mycorrhizal formation but increased the expression of PIP1; 2. In *Medicago truncatula*, a number of PM (plasma membrane) proteins were differentially regulated by inoculation of *Glomus intraradices* (Valot et al. 2005). It has been reported that the expression of several PIP genes has been regulated by the presence of the ectomycorrhizal fungus *Amanita muscaria* in *Populus tremula -tremuloides* roots under optimal conditions (Marjanovic et al. 2005a). Mycorrhizal plants contain their own aquaporins in their fungal mycelia but there is very little information about their role in the water transport (Aroca et al. 2009; Lehto and Zwiazek 2011). Mycorrhizas may have an effect on the cell to cell water transport through effects on plant aquaporin expression and activity (Ruiz-Lozano and Aroca 2010).

8.7.7 *Plant Nutrition*

Like other living organisms plants also require nutrients for proper growth and development. These nutrients include nitrogen (N), phosphorus (P), potassium (K) calcium (Ca), sulphur (S), magnesium (Mg), Silicon (Si), boron (B), chlorine (Cl), manganese (Mn), iron (Fe), zinc (Zn), copper (Cu), molybdenum (Mo), nickel (Ni), selenium (Se), and sodium (Na). It has been found that *Mc*TIP1; 2, an aquaporin from *Mesembryanthemum*, to be permeable to K^+ (Vera-Estrella et al. 2004). Boron is important as it is a micronutrient. The boron permeability of purified plasma membrane vesicles which were obtained from squash (*Cucurbita pepo*) roots had been found that they are six times higher in the permeability as compared to the microsomal vesicles and boron permeation was partially inhibited by mercuric chloride or phloretin. Expression of a PIP1 aquaporin in oocytes has increased the boron permeability by about 30% (Dordas et al. 2000). It has been found that aquaporin AtNIP5; 1 is essential for the uptake of boron under boron limiting conditions (Takano et al. 2006). The first silicon (Si) transporter revealed in vascular plants was MIP OsNIP2; 1 from rice. The roles of different MIP isoforms have been found in rice, maize, and barley in Si uptake and distribution (Ma et al. 2006). Further studies by Hachez and Chaumont (2010) revealed that some AQPs are strict water channels but others can help in the transport of wide range of non polar solutes like urea or glycerol and also unusual permeants like gases carbon dioxide and nitric oxide which are non polar and polar gases like ammonia, reactive oxygen species hydrogen peroxide and metalloids antimonite, arsenite, boron and silicon.

8.8 Aquaporins in Defense Reaction

8.8.1 *Changes in Irradiance*

Light is a major environmental factor. Changes in light intensity and diurnal changes also affects the plant metabolism and so the water relations. Light influences the opening and closing of stomata and it affects the transpiration. The transcript abundance of SunTIP7 (a TIP homolog), in the Sunflower leave's guard cells is under diurnal control and it is maximum during sun setting, during the closure of stomata, it suggests the role of this aquaporin in efflux of water from guard cells (Sarda et al. 1997). Diurnal movement of leaves (a process where there is participation of AQPs) optimizes the light interception. Coordinate shrinking and swelling of leaves on the opposite sides of pulvinus (a motor organ), determines the movement of leaves in the Mimosaceae. Accompanying the leaf movement there is diurnal regulation of osmotic water permeability in the protoplasts isolated from *Samanea saman* pulvinus and it is maximum during morning and evening times (Moshelion et al. 2002). AQP gene expression and apoplastic water flow in bur oak

(*Quercus macrocarpa*) leaves in relation to the light response of leaf hydraulic conductance was studied by Voicu et al. (2009). It was revealed that putative AQP genes probably do not play a role in the light responses of hydraulic conductance at the transcript level, but they may function in regulating water homeostasis in leaves adapted to different light conditions. The hydraulic conductivity of the leaf vascular system (K_{leaf}) decreases under drought stress, possibly in response to Abscisic acid (ABA), which increases sharply in the xylem sap (ABA_{xyl}) during periods of drought (Shatil-Cohen et al. 2011).

Further, the study revealed that vascular bundle-sheath cells (BSCs) control K_{leaf} *via* the specific activity of BSC-AQPs. ABA fed to the leaf *via* the xylem (petiole) both decreased K_{leaf} and led to stomatal closure, replicating the effect of drought. In contrast, smearing ABA on the leaf blade, while also closing stomata, did not decrease K_{leaf} within 2–3 h of application, demonstrating that K_{leaf} does not depend entirely on stomatal closure. The BSCs showed decreased P_f in response to 'drought' and ABA treatment, and a reversible decrease with $HgCl_2$ (an AQP blocker). These P_f responses, absent in mesophyll cells, suggest stress-regulated AQP activity specific to BSCs, and imply a role for these cells in decreasing K_{leaf} *via* a reduction in P_f (Shatil-Cohen et al. 2011). The H_2O_2-regulated expression levels of all plasma membrane AQPs of *Arabidopsis thaliana* (AtPIPs) and the permeability of every AtPIP for H_2O_2 was determined in yeast by Hooijmaijers et al. (2012). It was observed that only certain isoforms of AtPIPs that were regulated by H_2O_2 treatment were permeable for H_2O_2 in yeast cells. It was further suggested the integrated regulation of AQP expression by H_2O_2 and emphasized the importance of capacity of individual AQP to transport H_2O_2.

8.8.2 Water, Salt, and Nutrient Stresses

There are several evidences of aquaporins role in various cellular processes in higher plants like solute and nutrient transport, water transport and stress responses (Johanson et al. 2001; Maurel et al. 2002; Maurel 2007; Tyerman et al. 1999, 2002). The importance of AQPs mediated symplastic water transport was suggested by lower water stress resistance and reduced root hydraulic conductivity due to downregulation of the expression of AQP gene Nt-AQP1 in Antisense transgenic *Nicotiana tabacum* plants (Siefritz et al. 2002). Reduced hydraulic conductivity of root cortex cells is exhibited in the Arabidopsis knockout mutant of PIP2;2, which expresses in roots (Javot et al. 2003). Plasma membrane aquaporin PIP1b overexpression in *Nicotiana tabacum* resulted in hypersensitivity to drought stress (Aharon et al. 2003). It suggested deleterious effect on the plants during water stress due to increased symplastic water transport *via* plasma membrane aquaporins. Also, it has been shown that the aquaporin gene family expression is downregulated by dehydration and other abiotic stresses in Arabidopsis (Alexandersson et al. 2005; Boursiac et al. 2005; Jang et al. 2004). A relationship in functioning of PIP2.1 (aquaporin) and Rma1H1 (a RING Membrane-Anchor E3 Ubiquitin Ligase Homolog), in transgenic *Arabidopsis* plants for drought tolerance was revealed (Lee et al. 2009). Postaire et al. (2010) investigated the significance of aquaporins for tissue water conductivity in *Arabidopsis thaliana* by using a combination of reverse genetics and pharmacological approaches. They found out that in aquaporin-mediated leaf water transport, AtPIP1;2 plays a major role and it is the main component of whole plant hydraulics.

8.8.3 Cold Stress

It is one of the abiotic stresses that affect the growth and development of crops and other plants and it results in huge losses in the crop produce. Jang et al. (2007) showed that transgenic Arabidopsis plants which overexpressed PIP1;4 or PIP2;5, showed increased water flow and so more germination under cold stress. Plant cells enhance their cold tolerance by expression of some cold related genes. In the experiment performed by Li et al. (2009b), GhTIP1; 1 expression was downregulated in roots but upregulated in cotyledons within a few hours after the cold treatment of cotton seedling. Lee et al. (2012) compared the effect of low root temperature on root cell water transport and growth between the *Arabidopsis thaliana* plants overexpressing PIP (plasma membrane intrinsic protein) 1;4 and PIP2;5 and the wild type *Arabidopsis thaliana* plants. They found out no significant differences in the relative growth rates of shoot and root between the different plant groups exposed to 23°C root temperature. There was a statistically significant and sharp reduction of the shoot and root relative growth rates in the plants overexpressing PIP1;4 and wild type plants when root zone temperature was decreased from 23°C to 10°C for 5 days. However, there was no significant differences in the relative root and shoot growth rates at both root zone temperatures in the PIP2;5 overexpressing plants.

8.8.4 Anoxia

Anoxia is the extreme form of hypoxia or "low oxygen" where there is complete deprivation of oxygen. In temperate latitudes, during winter or after irrigation flooding of soils occurs and it is a major agricultural problem because it results in acute deprivation of oxygen (anoxia) of plant roots. Plants respond early to anoxia by downregulation of water uptake by inhibiting the water permeability (hydraulic conductivity of roots). AQPs (water channel proteins) of the plasma membrane intrinsic protein (PIP) subgroup mediates water uptake by roots. During anoxic stress root water transport is regulated by cell pH through aquaporin gating (Tournaire-Roux et al. 2003). pH may be responsible for water channels gating or they could be mechanosensitive or osmosensitive, either directly or indirectly through phosphoregulation that is Ca^{2+} dependent (Johansson et al. 1998).

8.8.5 *Biotic Interactions*

Biotic interactions are sum total of all the effects of one organism on the other in a community. No organism can live in isolation, for survival it has to interact with the environment and the other organisms. Plant interactions with soil microorganisms are very important for mineral nutrition and metabolism of plants and it further plays a critical role in water relations and environmental stress tolerance in plants.

8.8.5.1 Nitrogen Fixing Bacteria

Association of nitrogen fixing bacteria (diazotrophs) with plant roots is a symbiotic relationship which is beneficial to both (it is called mutualism), in which nitrogen is converted into ammonia by enzyme nitrogenase. The reaction is:

$$N_2 + 8H^+ + 8e^- \rightarrow 2NH_3 + H_2$$

Ammonia produced in this way is used by the plant and the bacteria in turn get shelter and nutrients from plant roots. G*m*NOD26, the plant aquaporin to be identified first, is expressed in nitrogen fixing nodules formed in soybean roots due to the infection by *Rhizobiaceae* bacteria (Wallace et al. 2006). G*m*NOD26 is a peribacteroid membrane component (this is a plant origin membrane that covers the bacteroid and so helps in exchanges with the root cell). G*m*NOD26 is involved in solute transport, so it has been related to channel-mediated ammonia-import from the bacteroid but the evidence for this function is less (Niemietz and Tyerman 2000; Wallace et al. 2006).

8.8.5.2 Arbuscular Mycorrhizal (AM) Fungi

Arbuscular mycorrhiza (AM) belongs to the phylum *Glomeromycota* and it is the most common of all the mycorrhizal fungi. More than 80% species of vascular plants show symbiosis with this fungi. Interaction with this mycorrhiza results in large changes in the anatomy of root cells which involves the differentiation of periarbuscular membrane structures that are carbohydrates, mineral nutrient (phosphate) and water exchange site with the fungus. The differentiation of these structures results in large changes in TIP and PIP gene expression. Mycorrhized poplar plants show a 55% increase in Lpr relative to non-mycorrhized poplar plants. Also there are changes in the root anatomy; it is due to the enhanced expression of PIPs (Marjanovic et al. 2005b). Zawonik et al. (2011) assessed the plant performance and aquaporin expression of HvP1P2,1 gene in the roots of barley seedlings inoculated by Azospirillum growing under salt stress and non-stressed conditions.

8.8.5.3 Nematodes

When nematodes infect roots, there occur differentiation of giant cells that are the feeding sites for the parasite. The promotor sequence of a tobacco TIP gene includes the regulatory sequences that are involved in the response to nematode infection (Opperman et al. 1994). Increased expression of the AQP might be required for extensive delivery of solutes and water to the parasite, together with the giant cell's osmotic regulation. There is expression of a PIP1 homolog due to incompatible interaction with the parasite *Cuscuta reflexa.* It is probably due to the reason that the pathogen attachment induces the auxin-dependent hypodermal cell's elongation (Werner et al. 2001). Barzana et al. (2012) performed an experiment to see the cell to cell pathway versus apoplastic pathways relative contributions for water movement in the roots of arbuscular mycorrhizal and non-arbuscular mycorrhizal plants. In the results they found that in the roots of arbuscular mycorrhizal plants there was significant enhancement in the apoplastic water flow as compared to the non-arbuscular mycorrhizal plants and the increase was observed both in well-watered and drought conditions. Switching between apoplastic and cell to cell water transport (root aquaporins) and apoplastic pathways was modulated due to the presence of arbuscular mycorrhizal fungus in the roots of host plants. Arbuscular mycorrhizal plant's ability to switch between water transport pathways increases their ability to respond to water shortage conditions.

8.9 Plant Aquaporins: New Perspectives on Water and Nutrient Uptake in Saline Environment

8.9.1 Adaptation to Salinity Stress

Certain plant species like barley has been recognized as sodium chloride (NaCl) tolerant plant species (Katsuhara et al. 2011). A consistent hydraulic conductivity of barley roots were observed as compared to control plants, when in the soil solution concentration of NaCl was increased to 200 mmol L^{-1} over 10 days (Munns and Passioura 1984) and with increasing NaCl for a given transpiration rate, osmotic pressure of the xylem sap changed. According to Chen et al. (2005), relatively high concentration of 320 mmol L^{-1} of NaCl, caused 14-fold decrease in the potassium content in the NaCl sensitive varieties of barley as compared to only threefold reduction in salt tolerant varieties. Certain salt stress responsive proteins have been identified by Kamal et al. (2010) in China-108, Norin-61 and in Kantou-107 plants. Many reports suggested that aquaporins activity is also responsible for the reduction of hydraulic conductivity due to the NaCl stress (Boursiac et al. 2005; Carvajal et al. 1999; Martinez-Ballesta et al. 2003a). Thus, the expression signal of *Arabidopsis* plasma membrane intrinsic protein (PIP) and tonoplast intrinsic protein (TIP) aquaporins transcripts resulted reduction upto 60–75% in their abundance between 2 and 4 h after the salt treatment (Boursiac et al. 2005), which suggest that plants prepare themselves for an abrupt unexpected elevation due to salt concentration by virtue of the modulation of the water status of the cells at early stages of salt stress. According to Rodriguez-Gamir et al. (2012), PIP1 and PIP2 aquaporins mRNA transcripts of citrus roots do not show any effect, when these were subjected to salinity stress.

AQPs gating is considered as an adaptative mechanism for survival, when osmotic stress induces the water stress at the early stages of salt stress. Under salt stress the role of aquaporin has been analysed in pepper (Martinez-Ballesta et al. 2003b), *Arabidopsis* (Boursiac et al. 2005; Martinez-Ballesta et al. 2003a), broccoli (Lopez-Berenguer et al. 2006) and tomato plants (Sade et al. 2010). Mano (1996) observed the salt tolerance of 6,681 barley varieties, 368 isogenic lines and 353 wild *Hordeum* strains. When 4 days old seedlings of barley were treated with the salt concentration of 100 mmol L^{-1} for 5 days, it was observed that growth in salt- sensitive plants were reduced to 40% and in salt tolerant plants, it was only 15% (Ligaba and Katsuhara 2010). It was reported that under salt treatment, addition of 100 mmol L^{-1} NaCl in 1,743 variety of barley plants, which is salt sensitive reduced slightly the stomatal conductance, but K305 (salt tolerant) was not affected (Katsuhara et al. 2011), which indicated that the sensitivity of roots the salt could also be detected in the leaves. At 50 mmol L^{-1} NaCl, higher turgor pressure (P) was maintained by K305 as compared to 1,743 and these salt tolerant K305 found to be capable of modulating AQP gating. Salt tolerant variety rapidly inhibits water permeability values *via* enhancing hydrostatic half time of water exchange ($T_{1/2}{}^{w}$) and $_{\epsilon}$ values. It was suggested that reduction in water permeability value is attributable to the closing of AQP channels, which are determined by the application of mercury and ABA.

Down-regulation of several AQPs including HvPIP2;1 was observed in many species of plants under salt or osmotic stress of several hours to days (Jang et al. 2004; Katsuhara et al. 2002; Vandeleur et al. 2005; Zhu et al. 2005). Reduction in AQPs level can prevent cell death due to regular dehydration and gain time for intracellular osmotic adjustment. Water permeability of roots and salt sensitivity of transgenic rice plants were increased by continuous overexpression of HvPIP2;1 (Katsuhara et al. 2003). Lower down-regulation of cellular water permeability in such transgenic plants might induce relatively higher salts of water loss from roots or shoots and salt stress results in death with induced osmotic stress and dehydration.

8.9.2 Cellular Adaptations to Salinity

A very pivotal aspect out of the innumerable capabilities of AQPs is the versatility of membrane modulations and cellular adaptations to salt exposure and stress tolerance towards it. In particular, during exposure to salinity stress, the plant's water/sap condition stimulates specific pathways for osmotic regulation and water uptake and loss (Fricke and Peters 2002). One of the main responses of plants to salt stress is inhibition of their root water uptake capacity (i.e. root hydraulic conductivity, Lpr). In roots of most plant species investigated, drought or salt stresses also result in a marked decrease in Lpr (Maurel et al. 2002). The water uptake capacity of plant roots (i.e. their hydraulic conductivity, Lpr) is determined in large part by AQPs of the plasma membrane intrinsic protein (PIP) subfamily. AQPs are known to facilitate the uptake of soil water and mediate the regulation of root hydraulic conductivity (Lpr) in response to a large variety of environmental stresses including saline stress. A number of reports establish a decline of Lpr values under saline stress frequently (Navarro et al. 2003; Nedjimi 2009; Wan 2010; Muries et al. 2011; Sutka et al. 2011). The initial Lpr decrease upon salt exposure may be caused by an osmotic shock as a result of an AQP conformational change due to negative pressures (Wan et al. 2004).

Microarray experiments with gene specific tags were executed to investigate the expression of 35 genes of the *Arabidopsis* AQP family (Boursiac et al. 2005). Transcript AQP genes, mostly belonging to the PIP and TIP subfamilies were detected in un-treated roots and showed a sharp decline in signaling. It becomes indicative that exposure of roots to salt stress triggers changes in AQP expression at various levels. These changes include a synchronized transcriptional down-regulation and sub cellular reallocation of both PIPs and TIPs. Way back, Carvajal et al. (1999) reported *L*pr reduction by 70% as a result of NaCl exposure. The fact that residual *L*pr of salt-stressed *Arabidopsis* becomes opposed to mercury exposure was interpreted as the down regulation of AQP activity. In salt-stressed root cells of *Arabidopsis* and *Zea mays*, a coordinated down regulation of most AQP transcripts occurs. A study in *Hordeum vulgare* leaves suggested that there is an increased abundance of *HvPIP1; 6* transcripts in response to salt may reflect a role for this AQP in promoting remaining growth of the leaf under salt stress (Fricke et al. 2006). Also, accumulation of PIP proteins in roots of salt-treated plants for long periods (from 3 to15 days) has been found (Marulanda et al. 2010; Muries et al. 2011), indicating cell-to-cell pathway. Cellular and membrane modifications in AQP levels have been linked to salt levels as well as to some plant hormones much recently. The decrease in root water potential is caused by osmotic and toxic effects, depending on the salt concentration present. Silva et al. (2008) reported similar results in pepper plants treated with a low concentration (30 mM) of NaCl, or with a nutrient solution with the same osmotic value. Li et al. (2009c) investigated effect of phosphorus deficiency on ethylene production and Lpr in *Medicago falcata* L. It was concluded that ethylene induced by Phosphorous deficiency may play a crucial role in modulation of Lpr levels by affecting AQPs regulation in plants.

Prior to this, Boursiac et al. (2008) investigated two stimuli, salicylic acid (SA) and salt, as these are capable of inducing an accumulation of reactive oxygen species (ROS) and an inhibition of Lpr concurrently in the roots of *Arabidopsis* plants. The inhibition of Lpr by SA was partly checked by preventing the accumulation of hydrogen peroxide with application of exogenous catalase. It was proposed that ROS regulate *Arabidopsis* root AQPs through cell signalling mechanisms. Consequently, SA was identified as a new regulator of AQPs.

Martinez-Ballesta et al. (2008) studied two different effects of calcium, in plasma membrane vesicles and in protoplasts isolated from roots of *Capsicum annuum* L. Under saline conditions, PIP1 AQP abundance dwindled in protoplasts and plasma membrane vesicles, indicating inhibitory effects of NaCl on AQP functionality and protein abundance. Two different actions of Ca^{2+} that were observed included increase in free cytosolic calcium concentrations associated with stress perception leading to AQP closure and reduction in critical requirements of Ca^{2+} lead to an up regulation of AQPs, indicating that a positive role of calcium at whole plant level combined with an inhibitory mechanism at AQP level may work in the regulation of pepper root water transport under salt stress. The studies on genotypic differences related to root water transport and PIP aquaporin expulsion of *Arabidopsis thaliana* revealed the role of AQPs in plants exposed to salinity (Sutka et al. 2005). Role of TIP's has also been indicated in providing protection against saline stress. Peng et al. (2007) found that *Arabidopsis* plants over expressed a TIP aquaporin isolated and induced in seeds from the ginger plant were able to germinate, the seedlings grew even at toxic concentration of 150 mM NaCl. Later, Wang et al. (2011) found that *Arabidopsis* plants over expressed a TIP aquaporin from soybean in response to salt stress. Induction of salinity tolerant genes and the overall cellular responses generated in host plant cells need further scientific evaluation.

8.10 Conclusion and Future Perspective

The discovery of AQPs has emphasized their importance in maintaining plant homeostasis. Recent investigations highlighted that AQPs are indispensable proteins in trans-membrane water flow, thus directly or indirectly affect various plant processes like plant reproduction, photosynthesis, stomatal regulation, etc. Various studies revealed the important role of TIPs and PIPs but AQPs like SIPs and NIPs still need to be explored. However, the role of AQPs in assimilation of carbon or nitrogen and in uptake of micronutrient remains to be explored. Further, the role of XIPs, a new emerging family of MIPs needs an extensive investigation and its specific role in water and solute transport, interaction with other AQPs, expression and post translational modifications need to be highlighted. Novel classes of plant AQPs like HIPs and GIPs that have been discovered recently should be elucidated for their functions pertaining to their role in amelioration of various biotic and abiotic stresses in plants. At molecular level, these AQPs need to be studied for their contribution in transporting signaling molecules and thereby their role in signaling pathways needs to be explored. Cellular properties like concentrations of osmolytes, ROS, water contents, pH, etc., may change under the effect of environmental stresses. Such biochemical changes in the cell may modify AQP activity. Thus, there is critical demand to focus the investigations on functional and biological significance of AQPs in plants exposed to biotic and abiotic stresses. The future studies on AQPs mediated transport of heavy metals (As, Hg, Pb) and other metalloids might help in designing phytoremediation strategies, as well as for blocking the uptake of heavy metals in plants to limit food chain contamination (Mosa et al. 2012). Hence, there is a need to elucidate the molecular mechanisms of operation, interaction and regulation of these AQPs in normal as well as stressed conditions. In future, the detailed investigations into the transport specificity and fundamental mechanisms of subcellular trafficking of AQPs must be focused.

References

Agre P (1998) Aquaporin null phenotypes: the importance of classical physiology. Proc Natl Acad Sci USA 95:9061–9063

Agre P (2006) The aquaporin water channels. Proc Am Thoroc Soc 3:5–13

Aharon R, Shahak Y, Wininger S, Bendov R, Kapulnik Y, Galili G (2003) Overexpression of a plasma membrane aquaporin in transgenic tobacco improves plant vigor under favorable growth conditions but not under drought or salt stress. Plant Cell 15:439–447

Alexandersson E, Danielson JAH, Rade J, Moparthi VK, Fontes M, Kjellbom P, Johanson U (2010) Transcriptional regulation of aquaporins in accessions of *Arabidopsis* in response to drought stress. Plant J 61:650–660

Alexandersson E, Fraysse L, Sjövall-Larsen S, Gustavsson S, Fellert M, Karlsson M, Johanson U, Kjellbom P (2005) Whole gene family expression and drought stress regulation of aquaporins. Plant Mol Biol 59:469–484

Alleva K, Niemietz CM, Maurel C, Parisi M, Tyerman SD, Amodeo G (2006) Plasma membrane of *Beta vulgaris* storage root shows high water channel activity regulated by cytoplasmic pH and a dual range of calcium concentrations. J Exp Bot 57:609–621

Amiry-Moghaddam M, Williamson A, Palomba M, Eid T, de Lanerolle NC, Nagelhus EA (2003) Delayed K^+ clearance associated with aquaporin-4 mislocalization: phenotypic defects in brains of alpha-syntrophin-null mice. Proc Natl Acad Sci USA 100:13615–13620

Aroca R, Bago A, Sutka M, Paz JA, Cano C, Amodeo G, Ruiz-Lozano JM (2009) Expression analysis of the first arbuscular mycorrhizal fungi aquaporin described reveals concerted gene expression between salt-stressed and nonstressed mycelium. Mol Plant Microbe Interact 22: 1169–1178

Aroca R, Porcel R, Ruiz-Lozano JM (2012) Regulation of root water uptake under abiotic stress conditions. J Exp Bot 63(1):43–57

Ayadi M, Cavez D, Miled N, Chaumont F, Masmoudi K (2011) Identification and characterization of two plasma membrane aquaporins in durum wheat (*Triticum turgidum* L. subsp. durum) and their role in abiotic stress tolerance. Plant Physiol Biochem 49:1029–1039

Azad AK, Sawa Y, Ishikawa T, Shibata H (2004) Phosphorylation of plasma membrane aquaporin regulates temperature-dependent opening of tulip petals. Plant Cell Physiol 45:608–617

Bansal A, Sankararamakrishnan R (2007) Homology modeling of major intrinsic proteins in rice, maize and Arabidopsis: comparative analysis of transmembrane helix association and aromatic/arginine selectivity filters. BMC Struct Biol 7:27

Barzana G, Arora R, Paz JA, Chaumont F, Martinez-Ballesta MC, Carvajal M, Ruiz-Lozano JM (2012) Arbuscular mycorrhizal symbiosis increases relative apoplastic water flow in roots of the host plant under both well-watered and drought stress conditions. Ann Bot. doi:10.1093/aob/mcsoo7 pp 1-9

Becker A (2011) The involvement of aquaporins in ammonia/ammonium transport across root cell membranes of barley (*Hordeum vulgare* L.). Thesis Master of Science Ecology and Evolutionary Ecology, University of Toronto

Beebo A, Thomas D, Der C, Sanchez L, Leborgne-Castel N, Marty F, Schoefs B, Bouhidel K (2009) Life with and without AtTIP1;1, an *Arabidopsis* aquaporin preferentially localized in the apposing tonoplasts of adjacent vacuoles. Plant Mol Biol 70:193–209

Benga G (2012) On the definition, nomenclature and classification of water channel proteins (aquaporins and relatives). Mol Aspects Med, http://dx.doi.org/10.1016/j.mam.2012.04.003

Biela A, Grote K, Otto B, Hoth S, Hedrich R, Kaldenhoff R (1999) The *Nicotiana tabacum* plasma membrane aquaporin NtAQP1 is mercury-insensitive and permeable for glycerol. Plant J 18:565–570

Bienert GP, Bienert MD, Jahn TP, Boutry M, Chaumont F (2011) *Solanaceae* XIPs are plasma membrane aquaporins that facilitate the transport of many uncharged substrates. Plant J 6:306–317

Bienert GP, Muller AL, Kristiansen KA (2007) Specific aquaporins facilitate the diffusion of hydrogen peroxide across membranes. J Biol Chem 282:1183–1192

Boursiac Y, Boudet J, Postaire O, Luu DT, Tournaire-Roux C, Maurel C (2008) Stimulus-induced downregulation of root water transport involves reactive oxygen species-activated cell signalling and plasma membrane intrinsic protein internalization. Plant J 56:207–218

Boursiac Y, Chen S, Luu DT, Sorieul M, Van den Dries N, Maurel C (2005) Early effects of salinity on water transport in *Arabidopsis* roots, molecular and cellular features of aquaporin expression. Plant Physiol 139:790–805

Bramley H, Turner DW, Tyerman SD, Turner NC (2007) Water flow in the roots of crop species: the influence of root structure, aquaporin activity, and waterlogging. In: Sparks DL (ed) Advances in agronomy. Elsevier, Amsterdam

Calamita G, Ferri D, Gena P, Liquori GE, Cavalier A, Thomas D, Svelto M (2005) The inner mitochondrial membrane has aquaporin-8 water channels and is highly permeable to water. J Biol Chem 280:17149–17153

Carvajal M, Cerda A, Martínez V (2000) Does calcium ameliorate the negative effect of NaCl on melon root water transport by regulating aquaporin activity? New Phytol 145:439–447

Carvajal M, Martinez V, Alcaraz CF (1999) Physiological function of water channels as affected by salinity in roots of paprika pepper. Physiol Plant 105:95–101

Chaumont F, Barrieu F, Herman EM, Chrispeels MJ (1998) Characterization of a maize tonoplast aquaporin expressed in zones of cell division and elongation. Plant Physiol 117:1143–1152

Chaumont F, Barrieu F, Jung R, Chrispeels MJ (2000) Plasma membrane intrinsic proteins from maize cluster in two sequence subgroups with differential aquaporin activity. Plant Physiol 122:1025–1034

Chaumont F, Barrieu F, Wojcik E, Chrispeels MJ, Jung R (2001) Aquaporins constitute a large and highly divergent protein family in maize. Plant Physiol 125:1206–1215

Chen Z, Newman I, Zhou M, Mendham N, Zhang G, Shabala S (2005) Screening plants for salt tolerance by measuring Kþ flux: a case study for barley. Plant Cell Environ 28:1230–1246

Chrispeels MJ, Maurel C (1994) Aquaporins: the molecular basis of facilitated water movement through living plant cells? Plant Physiol 105:9–13

Cochard H, Venisse JS, Barigah TS, Brunel N, Herbette S, Guilliot A, Tyree MT, Sakr S (2007) Putative role of aquaporins in variable hydraulic conductance of leaves in response to light. Plant Physiol 143:122–133

Daniels MJ, Mirkov TE, Chrispeels MJ (1994) The plasma membrane of *Arabidopsis thaliana* contains a mercury-insensitive aquaporin that is a homolog of the tonoplast water channel protein TIP. Plant Physiol 106:1325–1333

de Groot BL, Frigato T, Helms V, Grubmüller H (2003) The mechanism of proton exclusion in the aquaporin-1 channel. J Mol Biol 333:279–293

Dean RM, Rivers RL, Zeidel ML, Roberts DM (1999) Purification and functional reconstitution of soybean nodulin-26, an aquaporin with water and glycerol transport properties. Biochemistry 38:347–353

Denker BM, Smith BL, Kuhajda FP, Agre P (1988) Identification, purification, and partial characterization of a novel Mr 28,000 integral membrane protein from erythrocytes and renal tubules. J Biol Chem 263:15634–15642

Dordas C, Chrispeels MJ, Brown PH (2000) Permeability and channel-mediated transport of boric acid across membrane vesicles isolated from squash roots. Plant Physiol 124:1349–1362

Dynowski M, Mayer M, Moran O, Ludewig U (2008) Molecular determinants of ammonia and urea conductance in plant aquaporin homologs. FEBS Lett 582:2458–2462

El-Mesbahi MN, Azcón R, Ruiz-Lozano JM, Aroca R (2012) Plant potassium content modifies the effects of arbuscular mycorrhizal symbiosis on root hydraulic properties in maize plants. Mycorrhiza. doi:10.1007/s00572-012-0433-3

Farre EM (2012) The regulation of plant growth by the circadian clock. Plant Biol 14:401–410

Fetter K, Van Wilder V, Moshelion M, Chaumont F (2004) Interactions between plasma membrane aquaporins modulate their water channel activity. Plant Cell 16:215–228

Flexas J, Bota J, Galme´s J, Medrano H, Ribas-Carbo M (2006) Keeping a positive carbon balance under adverse conditions: responses of photosynthesis and respiration to water stress. Physiol Plant 127:343–352

Flexas J, Ribas-Carbo M, Diaz-Espejo A, Galmés J (2008) Mesophyll conductance to CO_2: current knowledge and future prospects. Plant Cell Environ 31:602–661

Fortin M, Morrison N, Verma D (1987) Nodulin-26, a peribacteroid membrane nodulin is expressed independently of the development of the peribacteroid compartment. Nucleic Acids Res 15:813–824

Fortin MG, Zelechowska M, Verma DP (1985) Specific targeting of membrane nodulins to the bacteroid-enclosing compartment in soybean nodules. EMBO J 4:3041–3046

Fricke W, Peters WS (2002) The biophysics of leaf growth in salt-stressed barley – a study at the cell level. Plant Physiol 129:374–388

Fricke W, Akhiyarova G, Wei W, Alexandersson E, Miller A, Kjelldorn PO, Richardson A, Wojciechowski T, Scheriber L, Veselov D, Kudoyarova G, Volkov V (2006) The short term growth response to salt of the developing barley leaf. J Exp Bot 57:1079–1095

Fritz M, Ehwald R (2011) Mannitol permeation and radial flow of water in maize roots. New Phytol 189:210–217

Fu D, Lu M (2007) The structural basis of water permeation and proton exclusion in aquaporins. Mol Membr Biol 24(5–6):366–374

Fujiyoshi Y, Mitsuoka K, de Groot BL, Phillippsen A, Grubmuller H, Agre P, Engel A (2002) Structure and function of water channels. Curr Opin Struct Biol 12:509–515

Galmes J, Pou A, Alsina MM, Tomas M, Medrano H, Flexas J (2007) Aquaporin expression in response to different water stress intensities and recovery in Richter-110 (*Vitis sp.*): relationship with ecophysiological status. Planta 226:671–681

Gaspar M, Bousser A, Sissoëff I, Roche O, Hoarau J, Mahé A (2003) Cloning and characterization of ZmPIP1-5b, an aquaporin transporting water and urea. Plant Sci 165:21–31

Geldner N, Anders N, Wolters H, Keicher J, Kornberger W, Muller P, Delbarre A, Ueda T, Nakano A, Jürgens G (2003) The *Arabidopsis* GNOM ARF-GEF mediates endosomal recycling, auxin transport, and auxin-dependent plant growth. Cell 112:219–230

Gerbeau P, Amodeo G, Henzler T, Santoni V, Ripoche P, Maurel C (2002) The water permeability of *Arabidopsis* plasma membrane is regulated by divalent cations and pH. Plant J 30:71–81

Gerbeau P, Guclu J, Ripoche P, Maurel C (1999) Aquaporin Nt-TIPa can account for the high permeability of tobacco cell vacuolar membrane to small neutral solutes. Plant J 18:577–587

Gonen T, Walz T (2006) The structure of aquaporins. Q Rev Biophys 39(4):361–396

Gu R, Chen X, Zhou Y, Yuan L (2012) Isolation and characterization of three maize aquaporin genes, ZmNIP2;1, ZmNIP2;4 and ZmNIP4;4 involved in urea transport. BMB Rep 45(2): 96–101

Guan X-G, Su W-H, Yi F, Zhang D, Hao F, Zhang H-G, Liu Y-J, Feng X-C, Ma T-H (2010) NPA motifs play a key role in plasma membrane targeting of aquaporin-4. IUBMB Life 62(3): 222–226

Guenther JF, Chanmanivone N, Galetovic MP, Wallace IS, Cobb JA, Roberts DM (2003) Phosphorylation of soybean nodulin 26 on serine 262 enhances water permeability and is regulated developmentally and by osmotic signals. Plant Cell 15:981–991

Gupta AB, Sankararamakrishnan R (2009) Genome-wide analysis of major intrinsic proteins in the tree plant *Populus trichocarpa*: characterization of XIP subfamily of aquaporins from evolutionary perspective. BMC Plant Biol 9:134

Hachez C, Chaumont F (2010) Aquaporins: a family of highly regulated multifunctional channels. Adv Exp Med Biol 679:1–17

Hachez C, Moshelion M, Zelazny E, Cavez D, Chaumont F (2006) Localization and quantification of plasma membrane aquaporin expression in maize primary root: a clue to understanding their role as cellular plumbers. Plant Mol Biol 62:305–323

Hanba YT, Shibasaka M, Hayashi Y, Hayakawa T, Kasamo K, Terashima I, Katsuhara M (2004) Overexpression of the barley aquaporin HvPIP2;1 increases internal CO_2 conductance and CO_2 assimilation in the leaves of transgenic rice plants. Plant Cell Physiol 45:521–529

Heckwolf M, Pater D, Hanson DT, Kaldenhoff R (2011) The Arabidopsis thaliana aquaporin AtPIP1; 2 is a physiologically relevant CO_2 transport facilitator. Plant J 67:795–804

Henzler T, Waterhouse RN, Smyth AJ, Carvajal M, Cooke DT, Schäffner AR, Steudle E, Clarkson DT (1999) Diurnal variations in hydraulic conductivity and root pressure can be correlated with the expression of putative aquaporins in the roots of *Lotus japonicus*. Planta 210:50–60

Henzler T, Ye Q, Steudle E (2004) Oxidative gating of water channels (aquaporins) in *Chara* by hydroxyl radicals. Plant Cell Environ 27:1184–1195

Henzler T, Steudl E (2000) Transport and metabolic degradation of hydrogen peroxide in *Chara corallina*: model calculations and measurements with the pressure probe suggest transport of H_2O_2 across water channels. J Exp Bot 51:2053–2066

Herrera M, Hong NJ, Garvin JL (2006) Aquaporin-1 transports NO across cell membranes. Hypertension 48:157–164

Holbrook NM, Zwieniecki MA (1999) Embolism repair and xylem tension. Do we need a miracle? Plant Physiol 120:7–10

Holm LM, Jahn TP, Moller AL, Schjoerring JK, Ferri D, Klaerke DA, Zeuthen T (2005) NH_3 and NH^{4+} permeability in aquaporin expressing *Xenopus* oocytes. Pflugers Arch 450:415–428

Hooijmaijers C, Rhee JY, Kwak KJ, Chung GC, Horie T, Katsuhara M, Kang H (2012) Hydrogen peroxide permeability of plasma membrane aquaporins of *Arabidopsis thaliana*. J Plant Res 125:147–153

Horie T, Kaneko T, Sugimoto G, Sasano S, Panda SK, Shibasaka M, Katsuhara M (2011) Mechanisms of water transport mediated by PIP aquaporins and their regulation via phosphorylation events under salinity stress in barley roots. Plant Cell Physiol 52:663–675

Hove RM, Bhave M (2011) Plant aquaporins with non-aqua functions: deciphering the signature sequences. Plant Mol Biol 75:413–430

Hussain SS, Iqbal MT, Arif MA, Amjad M (2011) Beyond osmolytes and transcription factors: drought tolerance in plants *via* protective proteins and aquaporins. Biol Plant 55:401–413

Hwang JH, Ellingson SR, Roberts DM (2010) Ammonia permeability of the soybean nodulin 26 channel. FEBS Lett 584:4339–4343

Ishikawa F, Suga S, Uemura T, Sato MH, Maeshima M (2005) Novel type aquaporin SIPs are mainly localized to the ER membrane and show cell-specific expression in *Arabidopsis thaliana*. FEBS Lett 579:5814–5820

Jahn TP, Moller AL, Zeuthen T, Holm LM, Klaerke DA, Mohsin B, Kuhlbrandt W, Schjoerring JK (2004) Aquaporin homologues in plants and mammals transport ammonia. FEBS Lett 574:31–36

Jang JY, Kim DG, Kim YO, Kim JS, Kang H (2004) An expression analysis of a gene family encoding plasma membrane aquaporins in response to abiotic stresses in *Arabidopsis thaliana*. Plant Mol Biol 54:713–725

Jang JY, Lee SH, Rhee JY, Chung GC, Ann SJ, Kang H (2007) Transgenic *Arabidopsis* and Tobacco plants overexpressing an aquaporin respond differently to various abiotic stresses. Plant Mol Biol 64:621–632

Janz D, Behnke K, Schnitzler JP, Kanawati B, Schmitt-Kopplin P, Polle A (2010) Pathway analysis of the transcriptome and metabolome of salt sensitive and tolerant poplar species reveals evolutionary adaption of stress tolerance mechanisms. BMC Plant Biol 10:150

Javot H, Lauvergeat V, Santoni V (2003) Role of a single aquaporin isoform in root water uptake. Plant Cell 15:509–522

Johanson U and Gustavsson S (2002) A new subfamily of major intrinsic proteins in plants. Mol Biol Evol 19:456–461

Johanson U, Karlsson M, Johansson I, Gustavsson S, Sjovall S, Fraysse L, Weig AR, Kjellbom P (2001) The complete set of genes encoding major intrinsic proteins in *Arabidopsis* provides a framework for a new nomenclature for major intrinsic proteins in plants. Plant Physiol 126: 1358–1369

Johansson I, Karlsson M, Shukla VK, Chrispeels MJ, Larsson C, Kjellbom P (1998) Water transport activity of the plasma membrane aquaporin PM28A is regulated by phosphorylation. Plant Cell 10:451–460

Johansson I, Larsson C, Ek B, Kjellbom P (1996) The major integral proteins of spinach leaf plasma membranes are putative aquaporins and are phosphorylated in response to Ca^{2+} and apoplastic water potential. Plant Cell 8:1181–1191

Johnson KD, Hofte H, Chrispeels MJ (1990) An intrinsic tonoplast protein of protein storage vacuoles in seeds is structurally related to a bacterial solute transporter (GIpF). Plant Cell 2:525–532

Kaldenhoff R, Fischer M (2006) Functional aquaporin diversity in plants. Biochim Biophys Acta 1758(8):1134–1141

Kaldenhoff R, Grote K, Zhu JJ, Zimmermann U (1998) Significance of plasmalemma aquaporins for water-transport in *Arabidopsis thaliana*. Plant J 14:121–128

Kaldenhoff R (2012) Mechanisms underlying CO_2 diffusion in leaves. Curr Opin Plant Biol http://dx.doi.org/10.1016/j.pbi.2012.01.011

Kamal AHM, Kim KH, Shin KH, Choi JS, Baik BK, Tsujimoto H, Heo HY, Park CS, Woo SH (2010) Abiotic stress responsive proteins of wheat grain determined using proteomics technique. Austr J Crop Sci 4(3):196–208

Kanai S, Moghaieb RE, El-Shemy HA, Panigrahi R, Mohapatra PK, Ito P, Nguyen NT, Saneoka H, Fujita K (2011) Potassium deficiency affects water status and photosynthetic rate of the vegetative sink in green house tomato prior to its effects on source activity. Plant Sci 180:368–374

Karlsson M, Johansson I, Bush M, McCann MC, Maurel C, Larsson C, Kjellbom P (2000) An abundant TIP expressed in mature highly vacuolated cells. Plant J 21:83–90

Katsuhara M, Akiyama Y, Koshio K, Shibasaka M, Kasamo K (2002) Functional analysis of water channels in barley roots. Plant Cell Physiol 43:885–893

Katsuhara M, Koshio K, Shibasaka M, Hayashi Y, Hayakawa T, Kasamo K (2003) Over-expression of a barely aquaporin increased the shoot/root ratio and raised salt sensitivity in transgenic rice plants. Plant Cell Physiol 44:1378–1383

Katsuhara M, Rhee JY, Sugimoto G, Chung GC (2011) Early response in water relations influenced by NaCl reflects tolerance or sensitivity of barley plants to salinity stress via aquaporins. Soil Sci Plant Nutr 57(1):50–60

Kirch HH, Vera-Estrella R, Golldack D, Quigley F, Michalowski CB, Barkla BJ, Bohnert HJ (2000) Expression of water channel proteins in *Mesembryanthemum crystallinum*. Plant Physiol 123:111–124

Klebl F, Wolf M, Sauer N (2003) A defect in the yeast plasma membrane urea transporter Dur3p is complemented by CpNIP1, a Nod26-like protein from zucchini (*Cucurbita pepo* L.), and by *Arabidopsis thaliana* delta-TIP or gamma-TIP. FEBS Lett 547:69–74

Knepper MA, Nielsen S (2004) Peter Agre, 2003 Nobel Prize winner in chemistry. J Am Soc Nephrol 15(4):1093–1095

Knipfer T, Fricke W (2010) Root pressure and a solute reflection coefficient close to unity exclude a purely apoplastic pathway of radial water transport in barley (*Hordeum vulgare*). New Phytol 187:159–170

Knipfer T, Fricke W (2011) Water uptake by seminal and adventitious roots in relation to whole-plant water flow in barley (*Hordeum vulgare* L.). J Exp Bot 62:717–733

Kojima S, Bohner A, von Wiren N (2006) Molecular mechanisms of urea transport in plants. J Membr Biol 212:83–91

Kuchel PW (2006) The story of the discovery of aquaporins: convergent evolution of ideas-but who got there first? Cell Mol Biol 52(7):2–5

Lee HK, Cho SK, Son O, Xu Z, Hwang I, Kim WT (2009) Drought stress-induced Rma1H1, a ring membrane-anchor E3 ubiquitin ligase homolog, regulates aquaporin levels via ubiquitination in transgenic *Arabidopsis* plants. Plant Cell 21:622–641

Lee SH, Chung GC, Jang JY, Ahn SJ, Zwiazek JJ (2012) Overexpression of PIP2;5 aquaporin alleviates effects of low root temperature on cell hydraulic conductivity and growth in *Arabidopsis thaliana* Plant Physiol, doi:10.1104/pp.112.194506

Lehto T, Zwiazek JJ (2011) Ectomycorrhizas and water relations of trees: a review. Mycorrhiza 21: 71–90

Leitão L, Prista C, Moura TF, Loureiro-Dias MC, Soveral G (2012) Grapevine aquaporins: gating of a tonoplast intrinsic protein (TIP2;1) by cytosolic pH. PLoS One 7(3):e33219. doi:10.1371/journal.pone.0033219

Li RY, Ago Y, Liu WJ, Mitani N, Feldmann J, McGrath SP, Ma JF, Zhao FJ (2009a) The rice aquaporin Lsi1 mediates uptake of methylated arsenic species. Plant Physiol 150:2071–2080
Li DD, Tai FJ, Zhang ZT, Li Y, Zheng Y, Wu YF, Li XB (2009b) A cotton gene encodes a tonoplast aquaporin that is involved in cell tolerance to cold stress. Gene 438(1–2):26–32
Li X, Wang X, Yang Y, Li R, He Q, Fang X, Luu DT, Maurel C, Lin J (2011) Single-molecule analysis of PIP2; 1 dynamics and partitioning reveals multiple modes of *Arabidopsis* plasma membrane aquaporin regulation. Plant Cell 23:3780–3797
Li YS, Mao XT, Tian QY, Li LH, Zhang WH (2009c) Phosphorus deficiency-induced reduction in root hydraulic conductivity in *Medicago falcata* is associated with ethylene production. Environ Exp Bot 67:172–177
Ligaba A, Katsuhara M (2010) Insight into the salt tolerance mechanism in barley (*Hordeum vulgare*) from the comparisons of cultivars that differ in salt sensitivity. J Plant Res 123:105–118
LiJuan L, Fei R, PengCheng W, QiJun C, Jia C, XueChen W (2011) Identification of AtSM34, a novel tonoplast intrinsic protein-interacting polypeptide expressed in response to osmotic stress in germinating seedlings. Chinese Sci Bull 56:3518–3530
Liu LH, Ludewig U, Gassert B (2003) Urea transport by nitrogen-regulated tonoplast intrinsic proteins in *Arabidopsis*. Plant Physiol 133:1220–1228
Lopez D, Bronner G, Brunel N, Auguin D, Bourgerie S, Brignolas F, Carpin S, Tournaire-Roux C, Maurel C, Fumanal B, Martin F, Sakr S, Label P, Julien JL, Gousset-Dupont A, Venisse JS (2012) Insights into *Populus* XIP aquaporins: evolutionary expansion, protein functionality, and environmental regulation. J Exp Bot. doi:10.1093/jxb/err404
Lopez M, Bousser AS, Sissoeff I, Gaspar M, Lachaise B, Hoarau J, Mahe A (2003) Diurnal regulation of water transport and aquaporin gene expression in maize roots: contribution of PIP2 proteins. Plant Cell Physiol 44:1384–1395
Lopez-Berenguer C, Garcia-Viguera C, Carvajal M (2006) Are root hydraulic conductivity responses to salinity controlled by aquaporins in broccoli plants? Plant Soil 279:13–23
Loque D, Ludewig U, Yuan L, Von-Wiren N (2005) Tonoplast intrinsic proteins AtTIP2;1 and AtTIP2;3 facilitate NH3 transport into the vacuole. Plant Physiol 137:671–680
Lovisolo C, Tramontini S, Flexas J, Schubert A (2008) Mercurial inhibition of root hydraulic conductance in *Vitis* spp. rootstocks under water stress. Environ Exp Bot 3:178–182
Ludevid D, Hofte H, Himelblau E, Chrispeels MJ (1992) The expression pattern of the tonoplast intrinsic protein gamma-tip in *Arabidopsis-thaliana* is correlated with cell enlargement. Plant Physiol 100:1633–1639
Luu DT, Maurel C (2005) Aquaporins in a challenging environment: molecular gears for adjusting plant water status. Plant Cell Environ 28:85–96
Ma JF, Tamai K, Yamaji N, Mitani N, Konishi S, Katsuhara M, Ishiguro M, Murata Y, Yano M (2006) A silicon transporter in rice. Nature 440:688–691
Maeshima M, Ishikawa F (2008) ER membrane aquaporins in plants. Pflugers Arch 456(4): 709–716
Mano Y (1996) Studies on the breeding and evaluation of germplasm for salt tolerance in barley. Special report for Barley Germplasm Center, Okayama University, 2:1–89
Marjanovic Z, Uehlein N, Kaldenhoff R, Zwiazek JJ, Weiss M (2005a) Aquaporins in poplar: what a difference a symbiont makes! Planta 222:258–268
Marjanovic Z, Uwe N, Hampp R (2005b) Mycorrhiza formation enhances adaptive response of hybrid poplar to drought. Ann NY Acad Sci 1048:496–499
Martinez-Ballesta MC, Aparicio F, Pallas V, Martinez V, Carvajal M (2003a) Influence of saline stress on root hydraulic conductance and PIP expression in *Arabidopsis*. J Plant Physiol 160: 689–697
Martinez-Ballesta MC, Lopez-Perez L, Muries B, Munos-Azcarate O, Carvajal M (2009) Climate change and plant water balance: the role of aquaporins - a review. In: Climate Change, Intercropping, Pest Control and Beneficial Micro-organisms. Sustainable Agriculture Reviews 2. (Ed. E. Lichtfouse). Springer Science, New York
Martinez-Ballesta MC, Cabanero F, Olmos E, Periago PM, Maurel C, Carvajal M (2008) Two different effects of calcium on aquaporins in salinity stressed pepper plants. Planta 228:15–25

Martinez-Ballesta MC, Diaz R, Martinez V, Carvajal M (2003b) Different blocking effects of $HgCl_2$ and NaCl on aquaporins of pepper plants. J Plant Physiol 160:1487–1492

Martínez-Ballesta MC, Martínez V, Carvajal M (2000) Regulation of water channel activity in whole roots and in protoplasts from roots of melon plants grown under saline conditions. Australian J Plant Physiol 27:685–691

Martre P, North GB, Nobel PS (2001) Hydraulic conductance and mercury-sensitive water transport for roots of *Opuntia acanthocarpa* in relation to soil drying and rewetting. Plant Physiol 126:352–362

Marulanda A, Azcon R, Chaumont F, Ruiz-Lozano JM, Aroca R (2010) Regulation of plasma membrane aquaporins by inoculation with a *Bacillus megaterium* strain in maize (*Zea mays* L.) plants under unstressed and salt-stressed conditions. Planta 232:533–543

Maurel C, Javot H, Lauvergeat V, Gerbeau P, Tournaire C (2002) Molecular physiology of aquaporins in plants. Int Rev Cytol 215:105–148

Maurel C, Kado RT, Guern J, Chrispeels MJ (1995) Phosphorylation regulates the water channel activity of the seed-specific aquaporin alpha-tip. EMBO J 14:3028–3035

Maurel C, Plassard C (2011) Aquaporins: for more than water at the plant-fungus interface? New Phytol 190(4):815–817

Maurel C, Reizer J, Schroeder JI, Chrispeels MJ (1993) The vacuolar membrane protein gamma-TIP creates water specific channels in *Xenopus* oocytes. EMBO J 12:2241–2247

Maurel C, Tacnet F, Guclu J, Guern J, Ripoche P (1997) Purified vesicles of tobacco cell vacuolar and plasma membranes exhibit dramatically different water permeability and water channel activity. Proc Natl Acad Sci USA 94:7103–7108

Maurel C, Verdoucq L, Luu DT, Santoni V (2008) Plant aquaporins: membrane channels with multiple integrated functions. Annu Rev Plant Biol 59:595–624

Maurel C (2007) Plant aquaporins: novel functions and regulation properties. FEBS Lett 581: 2227–2236

Mizutani M, Watanabe S, Nakagawa T, Maeshima M (2006) Aquaporin NIP2;1 is mainly localized to the ER membrane and shows root-specific accumulation in *Arabidopsis thaliana*. Plant Cell Physiol 47:1420–1426

Mollapour M, Piper PW (2007) Hog1 mitogen-activated protein kinase phosphorylation targets the yeast Fps1 aquaglyceroporin for endocytosis, thereby rendering cells resistant to acetic acid. Mol Cell Biol 27:6446–6456

Morillon R, Catterou M (2001) Brassinolide may control aquaporin activities in *Arabidopsis thaliana*. Planta 212(2):199–204

Morillon R, Lassalles JP (1999) Osmotic water permeability of isolated vacuoles. Planta 210:80–84

Morillon R, Liénard D, Chrispeels MJ, Lasalles JP (2001) Rapid movements of plants organs require solute-water co-transporters or contractile proteins. Plant Physiol 127:720–723

Mosa KA, Kumar K, Chhikara S, Mcdermott J, Liu Z, Musante C, White JC, Dhankher OP (2012) Members of rice plasma membrane intrinsic proteins subfamily are involved in arsenite permeability and tolerance in plants. Transgenic Res 10:9600–9608

Moshelion M, Becker D, Biela A, Uehlein N, Hedrich R, Otto B, Levi H, Moran N, Kaldenhoff R (2002) Plasma membrane aquaporins in the motor cells of *Samanea saman*: diurnal and circadian regulation. Plant Cell 14:727–739

Munns R, Passioura JB (1984) Effect of prolonged exposure to NaCl on the osmotic pressure of leaf xylem sap from intact transpiring barley plants. Austr J Plant Physiol 26:459–464

Muries B, Faize M, Carvajal M, Martınez-Ballesta MC (2011) Identification and differential induction of the expression of aquaporins by salinity in broccoli plants. Mol Biosyst 7:1322–1335

Navarro JM, Garrido C, Martınez V, Carvajal M (2003) Water relations and xylem transport of nutrients in pepper plants grown under two different salts stress regimes. Plant Growth Regul 41:237–245

Nedjimi B (2009) Salt tolerance strategies of *Lygeum spartum* L.: a new fodder crop for Algerian saline steppes. Flora 204:747–754

Neill S, Barros R, Bright J, Desikan R, Hancock J, Harrison J, Morris P, Ribeiro D, Wilson I (2008) Nitric oxide, stomatal closure, and abiotic stress. J Exp Bot 59:165–176

Niemietz CM, Tyerman SD (1997) Characterization of water channels in wheat root membrane vesicles. Plant Physiol 115:561–567

Niemietz CM, Tyerman SD (2000) Channel-mediated permeation of ammonia gas through the peribacteroid membrane of soybean nodules. FEBS Lett 465:110–114

Opperman CH, Taylor CG, Conkling MA (1994) Root-knot nematode-directed expression of a plant root-specific gene. Science 263:221–223

Paciorek T, Zazımalova E, Ruthardt N, Petrasek J, Stierhof YD (2005) Auxin inhibits endocytosis and promotes its own efflux from cells. Nature 435:1251–1256

Park W, Scheffler BE, Bauer PJ, Campbell BT (2010) Identification of the family of aquaporin genes and their expression in upland cotton (*Gossypium hirsutum* L.). BMC Plant Biol 10:142

Patriarca EJ, Tate R, Ferraioli S, Iaccarino M (2004) Organogenesis of legume root nodules. Int Rev Cytol 234:201–262

Peng YH, Lin WL, Cai WM, Arora R (2007) Overexpression of a Panax ginseng tonoplast aquaporin alters sat tolerance, drought tolerance and cold acclimation ability in transgenic Arabidopsis plants. Planta 226:729–740

Porcel R, Aroca R, Azcon R, Ruiz-Lozano JM (2005) PIP aquaporin gene expression in arbuscular mycorrhizal *Glycine max* and *Lactuca sativa* plants in relation to drought stress tolerance. Plant Mol Biol 60:389–404

Postaire O, Tournaire-Roux C, Grondin A, Boursiac Y, Morillon R, Schaffner AR, Maurel C (2010) A PIP1 aquaporin contributes to hydrostatic pressure-induced water transport in both the root and rosette of Arabidopsis. Plant Physiol 152:1418–1430

Prak S, Hem S, Boudet J, Viennois G, Sommerer N (2008) Multiple phosphorylations in the C-terminal tail of plant plasma membrane aquaporins: role in subcellular trafficking of AtPIP2;1 in response to salt stress. Mol Cell Proteomics 7:1019–1030

Prasad GVR, Coury LA, Finn F, Zeidel ML (1998) Reconstituted aquaporin 1 water channels transport CO_2 across membranes. J Biol Chem 273:33123–33126

Preston GM, Agre P (1991) Isolation of the cDNA for erythrocyte integral membrane protein of 28 kilodaltons: member of an ancient channel family. Proc Natl Acad Sci USA 88:11110–11114

Quigley F, Rosenberg JM, Shachar-Hill Y, Bohnert HJ (2001) From genome to function: the Arabidopsis aquaporins. Genome Biol 3:1–17

Rivers RL, Dean RM, Chandy G, Hall JE, Roberts DM, Zeidel ML (1997) Functional analysis of nodulin 26, an aquaporin in soybean root nodule symbiosomes. J Biol Chem 272:16256–16261

Rodriguez-Gamir J, Ancillo G, Legaz F, Primo-Millo E, Forner-Giner M (2012) Influence of salinity on pip gene expression in citrus roots and its relationship with root hydraulic conductance, transcription and chloride exclusion from leaves. Environ Expt Bot 78:163–166

Ruiz-Lozano JM, Aroca R (2010) Modulation of aquaporin genes by the arbuscular mycorrhizal symbiosis in relation to osmotic stress tolerance. In: Seckbach J, Grube M (eds) Symbioses and stress: joint ventures in biology, cellular origin, life in extreme habitats and astrobiology. Springer, Dordrecht, pp 359–374

Sade N, Gebretsadik M, Seligmann R, Schwartz A, Wallach R, Moshelion M (2010) The role of tobacco aquaporin1 in improving water use efficiency, hydraulic conductivity and yield production under salt stress. Plant Physiol 152:245–254

Sade N, Vinocur BJ, Diber A, Shatil A, Ronen G, Nissan H, Wallach R, Karchi H, Moshelion M (2009) Improving plant stress tolerance and yield production: is the tonoplast aquaporin SlTIP2;2 a key to isohydric to anisohydric conversion? New Phytol 181:651–661

Saito C, Ueda T, Abe H, Wada Y, Kuroiwa T, Hisada A (2002) A complex and mobile structure forms a distinct subregion within the continuous vacuolar membrane in young cotyledons of *Arabidopsis*. Plant J 29:245–255

Sakurai J, Ishikawa F, Yamaguchi T, Uemura M, Maeshima M (2005) Identification of 33 rice S.D. aquaporin genes and analysis of their expression and function. Plant Cell Physiol 46:1568–1577

Sandal NN, Marcker KA (1988) Soybean nodulin 26 is homologous to the major intrinsic protein of the bovine lens fiber membrane. Nucleic Acids Res 16:9347

Sarda X, Tousch D, Ferrare K, Legrand E, Dupuis JM (1997) Two TIP-like genes encoding aquaporins are expressed in sunflower guard cells. Plant J 12:1103–1111

Schaffner AR (1998) Aquaporin function, structure and expression: are there more surprises to surface in water relations. Planta 204:131–139

Schuurmans JA, van Dongen JT, Rutjens BP, Boonman A, Pieterse CM, Borstlap AC (2003) Members of the aquaporin family in the developing pea seed coat include representatives of the PIP, TIP, and NIP subfamilies. Plant Mol Biol 53:633–645

Shatil-Cohen A, Attia Z, Moshelion M (2011) Bundle-sheath cell regulation of xylem-mesophyll water transport via aquaporins under drought stress: a target of xylem-borne ABA? Plant J 67(1):72–80

Shelden MC (2007) A comparison of water stress-induced xylem embolism in two grapevine cultivars, Chardonnay and Grenache, and the role of aquaporins. PhD thesis, University of Adelaide, Adelaide

Siefritz F, Otto B, Bienert GP, van der Krol A, Kaldenhoff R (2004) The plasma membrane aquaporin NtAQP1 is a key component of the leaf unfolding mechanism in tobacco. Plant J 37: 147–155

Siefritz F, Tyree MT, Lovisolo C, Schubert A, Kaldenhoff R (2002) PIP1 plasma membrane aquaporins in tobacco: from cellular effects to function in plants. Plant Cell 14:869–876

Silva C, Martínez V, Carvajal M (2008) Osmotic versus toxic effects of NaCl on pepper plants. Biol Plant 52:72–79

Sorieul M, Santoni V, Maurel C, Luu DT (2011) Mechanisms and effects of retention of over-expressed aquaporin AtPIP2;1 in the endoplasmic reticulum. Traffic 12(4):473–482

Soto G, Alleva K, Mazzella MA, Amodeo G, Muschietti JP (2008) AtTIP1;3 and AtTIP5;1, the only highly expressed *Arabidopsis* pollen-specific aquaporins, transport water and urea. FEBS Lett 582:4077–4082

Srivastava AK, Suprasanna P, Srivastava S, D'Souza SF (2010) Thiourea mediated regulation in the expression profile of aquaporins and its impact on water homeostasis under salinity stress in Brassica juncea roots. Plant Sci 178:517–522

Steudle E, Peterson CA (1998) How does water get through roots? J Exp Bot 49:775–788

Sutka M, Alleva K, Parisi M, Amodeo G (2005) Tonoplast vesicles of *Beta vulgaris* storage root show functional aquaporins regulated by protons. Biol Cell 97:837–846

Sutka M, Li G, Boudet J, Boursiac Y, Doumas P, Maurel C (2011) Natural variation of root hydraulics in Arabidopsis grown in normal and salt-stressed conditions. Plant Physiol 155:1264–1276

Tajkhorshid E, Nollert P, Jensen MO, Miercke LJW, O'Connell J, Stroud RM, Schulten K (2002) Control of the selectivity of the aquaporin water channel family by global orientational tuning. Science 296:525–530

Takano J, Wada M, Ludewig U, Schaaf G, von Wiren N, Fujiwara T (2006) The *Arabidopsis* major intrinsic protein NIP5;1 is essential for efficient boron uptake and plant development under boron limitation. Plant Cell 18:1498–1509

Tanaka Y, Kutsuna N, Kanazawa Y, Kondo N, Hasezawa S, Sano T (2007) Intra-vacuolar reserves of membranes during stomatal closure: the possible role of guard cell vacuoles estimated by 3-D reconstruction. Plant Cell Physiol 48:1159–1169

Terashima I, Ono K (2002) Effects of $HgCl_2$ on CO_2 dependence of leaf photosynthesis: evidence indicating involvement of aquaporins in CO_2 diffusion across the plasma membrane. Plant Cell Physiol 43:70–78

Toole EH, Hendricks SB, Borthwick HA, Toole VK (1956) Physiology of seed germination. Annu Rev Plant Physiol 7:299–324

Tornroth-Horsefield S, Wang Y, Hedfalk K, Johanson U, Karlsson M, Tajkhorshid E, Neutze R, Kjellbom P (2006) Structural mechanism of plant aquaporin gating. Nature 439(7077):688–694

Tournaire-Roux C, Sutka M, Javot H, Gout E, Gerbeau P, Luu DT, Bligny R, Maurel C (2003) Cytosolic pH regulates root water transport during anoxic stress through gating of aquaporins. Nature 425:393–397

Tsuda M, Tyree MT (2000) Plant hydraulic conductance measured by the high pressure flow meter in crop plants. J Exp Bot 51:823–828

Tyerman SD, Bohnert HJ, Maurel C, Steudle E, Smith JA (1999) Plant aquaporins: their molecular biology, biophysics and significance for plant water relations. J Exp Bot 50:1055–1071

Tyerman SD, Niemietz CM, Bramley H (2002) Plant aquaporins: multifunctional water and solute channels with expanding roles. Plant Cell Environ 25:173–194

Tyree MT, Nardini A, Salleo S, Sack L, Omaro BE (2005) The dependence of leaf hydraulic conductance on irradiation during HPFM measurements: any role for stomatal response? J Exp Bot 56:737–744

Tyree MT, Salleo S, Nardini A, Lo Gullo MA, Mosca R (1999) Refilling of embolized vessels in young stems of laurel. Do we need a new paradigm? Plant Physiol 120:11–21

Uehlein N, Lovisolo C, Siefritz F, Kaldenhoff R (2003) The tobacco aquaporin NtAQP1 is a membrane CO_2 transporter with physiological functions. Nature 425:734–737

Uehlein N, Otto B, Hanson DT, Fischer M, McDowell N, Kaldenhoff R (2008) Function of *Nicotiana tabacum* aquaporins as chloroplast gas pores challenges the concept of membrane CO_2 permeability. Plant Cell 20:648–657

Valot B, Dieu M, Recorbet G, Raes M, Gianinazzi S, Dumas-Gaudot E (2005) Identification of membrane-associated proteins regulated by the arbuscular mycorrhizal symbiosis. Plant Mol Biol 59:565–580

Vandeleur R, Niemietz C, Tilbrook J, Tyerman SD (2005) Roles of aquaporins in root responses to irrigation. Plant Soil 274:141–161

Vandeleur RK, Mayo G, Shelden MC, Gilliham M, Kaiser BN, Tyerman SD (2008) The role of PIP aquaporins in water transport through roots: diurnal and drought stress responses reveal different strategies between isohydric and anisohydric cultivars of grapevine. Plant Physiol 149:445–460

Vera-Estrella R, Barkla BJ, Amezcua-Romero JC, Pantoja O (2012) Day/night regulation of aquaporins during the CAM cycle in *Mesembryanthemum crystallinum*. Plant Cell Environ 35(3):485–501

Vera-Estrella R, Barkla BJ, Bohnert HJ, Pantoja O (1999) Salt stress in *Mesembryanthemum crystallinum* L. cell suspensions activates adaptive mechanisms similar to those observed in the whole plant. Planta 207:426–435

Vera-Estrella R, Barkla BJ, Bohnert HJ, Pantoja O (2004) Novel regulation of aquaporins during osmotic stress. Plant Physiol 135:2318–2329

Vera-Estrella R, Bohnert HJ (2011) Physiological roles for the PIP family of plant aquaporins. Plant Cell Monographs 19:193–222

Voicu MC, Cooke JEK, Zwiazek JJ (2009) Aquaporin gene expression and apoplastic water flow in bur oak (*Quercus macrocarpa*) leaves in relation to the light response of leaf hydraulic conductance. J Exp Bot 60(14):4063–4075

Wallace IS, Choi WG, Roberts DM (2006) The structure, function and regulation of thenodulin 26-like intrinsic protein family of plant aqua glyceroporins. Biochim Biophys Acta 1758:1165–1175

Wallace IS, Roberts DM (2004) Homology modeling of representative subfamilies of *Arabidopsis* major intrinsic proteins, classification based on the aromatic/arginine selectivity filter. Plant Physiol 135:1059–1068

Wallace IS, Roberts DM (2005) Distinct transport selectivity of two structural subclasses of the nodulin-like intrinsic protein family of plant aquaglyceroporin channels. Biochemistry 44:16826–16834

Wallace IS, Wills DM, Guenther JF, Roberts DM (2002) Functional selectivity for glycerol of the nodulin 26 subfamily of plant membrane intrinsic proteins. FEBS Lett 523:109–112

Wan XC, Steudle E, Hartung W (2004) Gating of water channels (aquaporins) in cortical cells of young corn roots by mechanical stimuli (pressure pulses): effects of ABA and $HgCl_2$. J Exp Bot 55:411–422

Wan XC (2010) Osmotic effects of NaCl on cell hydraulic conductivity of corn roots. Acta Biochim Biophys Sin 42:351–357

Wang X, Li Y, Ji W, Bai X, Cai H, Zhu D, Sun XL, Chen LJ, Zhu YM (2011) A novel glycine soja tonoplast intrinsic protein gene responds to abiotic stress and depresses salt and dehydration tolerance in transgenic *Arabidopsis thaliana*. J Plant Physiol 168:1241–1248

Weaver CD, Crombie B, Stacey G, Roberts DM (1991) Calcium-dependent phosphorylation of symbiosome membrane-proteins from nitrogen-fixing soybean nodules-evidence for phosphorylation of nodulin-26. Plant Physiol 95:222–227

Weig A, Deswarte C, Chrispeels MJ (1997) The major intrinsic protein family of *Arabidopsis* has 23 members that form three distinct groups with functional aquaporins in each group. Plant Physiol 114:1347–1357

Weig AR, Jakob C (2000) Functional identification of the glycerol permease activity of *Arabidopsis thaliana* NLM1 and NLM2 proteins by heterologous expression in *Saccharomyces cerevisiae*. FEBS Lett 481:293–298

Werner M, Uehlein N, Proksch P, Kaldenhoff R (2001) Characterization of two tomato aquaporins and expression during the incompatible interaction of tomato with the plant parasite *Cuscuta reflexa*. Planta 213:550–555

Witte CP (2011) Urea metabolism in plants. Plant Sci 180:431–438

Wu B, Beitz E (2007) Aquaporins with selectivity for unconventional permeants. Cell Mole Life Sci 64:2413–2421

Yamada S, Komori T, Myers PN, Kuwata S, Kubo T, Imaseki H (1997) Expression of plasma membrane water channel genes under water stress in *Nicotiana* excelsior. Plant Cell Physiol 38:1226–1231

Ye Q, Muhr J, Steudle E (2005) A cohesion/tension model for the gating of aquaporins allows estimation of water channel pore volumes in *Chara*. Plant Cell Environ 28:525–535

Zardoya R (2005) Phylogeny and evolution of the major intrinsic protein family. Biol Cell 97:397–414

Zawonik MS, Ameneiros M, Benavides MP, Vazquez S, Groppa MD (2011) Response to saline stress and aquaporin expression in Azospirillum-inoculated barley seedlings. Appl Microbiol Biotechnol 90:1389–1397

Zelazny E, Borst JW, Muylaert M, Batoko H, Hemminga MA, Chaumont F (2007) FRET imaging in living maize cells reveals that plasma membrane aquaporins interact to regulate their subcellular localization. Proc Natl Acad Sci USA 104:12359–12364

Zhao FJ, Ago Y, Mitani N, Li RY, Su YH, Yamaji N, McGrath SP, Ma JF (2010) The role of the rice aquaporin Lsi1 in arsenite efflux from roots. New Phytol 186(2):392–399

Zhu C, Schraut D, Hartung W, Schäffner AR (2005) Differential responses of maize MIP genes to salt stress and ABA. J Exp Bot 56:2971–2981

Chapter 9
Oilseed Crop Productivity Under Salt Stress

Naghabushana Nayidu, Venkatesh Bollina, and Sateesh Kagale

9.1 Introduction

Food security has become a major and fast growing concern worldwide. It is proposed that there is a need to double the world food production in order to feed the ever increasing population which is set to reach nine billion mark by 2050 (UN 2009). In the current scenario, improving yields in both normal and less productive farm lands including salt affected lands is the only way to address food security concerns, as the amount of unused land available to bring into cultivation is limiting. Among various factors affecting agricultural production, abiotic stress factors are considered to be the main source of yield reduction. Potential yield losses due to individual abiotic stresses are estimated at 17% by drought, 20% by salinity, 40% by high temperature stress, 15% by low temperature stress and 8% by other factors (Ashraf and Harris 2005).

Soil salinity is one of the major abiotic stress factors affecting production and quality of food crops world-wide by limiting growth and development as well as yield potential of crop plants (Bray et al. 2000; Tester and Davenport 2003). More than 20% of the world arable land is now under the threat of salt stress. Agricultural losses due to salinity in Unites States alone are estimated to be about US$12 billion a year, and are expected to rise as soils are expected to be further affected by salinity (Munns 2005; Munns and Tester 2008). In addition to primary salinization of seashore

N. Nayidu • V. Bollina
Saskatoon Research Centre, Agriculture and Agri-Food Canada, 107 Science Place, Saskatoon, SK S7N 0X2, Canada

S. Kagale (✉)
Saskatoon Research Centre, Agriculture and Agri-Food Canada, 107 Science Place, Saskatoon, SK S7N 0X2, Canada

Plant Biotechnology Institute, National Research Council, 110 Gymnasium Place, Saskatoon, SK S7N 0W9, Canada
e-mail: skagale@alumni.uwo.ca

P. Ahmad et al. (eds.), *Ecophysiology and Responses of Plants under Salt Stress*,
DOI 10.1007/978-1-4614-4747-4_9, © Springer Science+Business Media, LLC 2013

salty marshes, a significant portion of cultivated agricultural land is becoming saline due to deforestation, excess irrigation and fertilisation as well as poor drainage (Shannon 1997; Zhu 2007).

9.2 Effect of Salt Stress on Crop Growth and Productivity

Salinity stress adversely affects plant growth and development by inhibiting seed germination (Dash and Panda 2001), seedling growth (Ashraf et al. 2002), enzyme activity (Seckin et al. 2009), DNA, RNA and protein synthesis (Anuradha and Seeta Ram Rao 2001), and cell division (Tabur and Demir 2010). Plants response to salt stress depends on a number of factors, including duration and intensity of stress as well as developmental stage of crop plant (Bray 1997). Yield loss is one of the major yard-sticks applied to measure the impact of salinity stress. Although salt stress affects all growth stages of plant, early plant establishment (germination and seedling development) and the reproductive growth phases are known to be the most sensitive stages in determining crop yield (Barnabás et al. 2008; Cuartero et al. 2006). During germination, increased osmotic pressure of soil solution caused by high salinity restricts the absorption and entry of water into the seed. Additionally, during early developmental stages salt constituents can act as toxins to the embryo and developing young seedlings, thus affecting their growth and establishment. During vegetative and reproductive growth phases, salinity stress is known to adversely affect plant growth and development by inducing water deficit (physiological drought), ion toxicity, nutrient imbalance and oxidative stress (Fig. 9.1; Vinocur and Altman 2005).

Accumulation of ions in the root zone inhibits water absorption leading to the induction of water deficit stress. Salt stress-mediated osmotic effect induces synthesis and accumulation of Abscisic acid (ABA) which when transported to guard cells causes stomata closure (Wilkinson and Davies 2002). Stomata closure reduces transpiration rate and lowers carbon uptake resulting in decreased photosynthetic rate (Chaves et al. 2003; Mullet and Whitsitt 1996). Another factor influencing plant growth response to water deficit is turgor pressure, the force causing plastic enlargement of cells, leaves and stems. Reduced turgor potential due to salinity-mediated water-deficit affects leaf expansion which results in reduced light interception and photosynthesis. During reproductive growth phases salt stress affects growth and yield by delaying onset of flowering as well as reducing the number of flowers (Sinaki et al. 2007).

High deposition of salt in the cytoplasm affects various metabolic processes, including synthesis of proteins, amino acids, nucleic acids, sugars, starch and other organic compounds. Salt stress can cause significant reductions in photosynthetic pigments. Components of photosynthetic electron transport in the chloroplast as well as the enzyme responsible for carbon assimilation are highly sensitive to Na^+ and Cl^- concentrations. Thus salt stress has a major inhibitory effect on photosynthetic apparatus of crop plants.

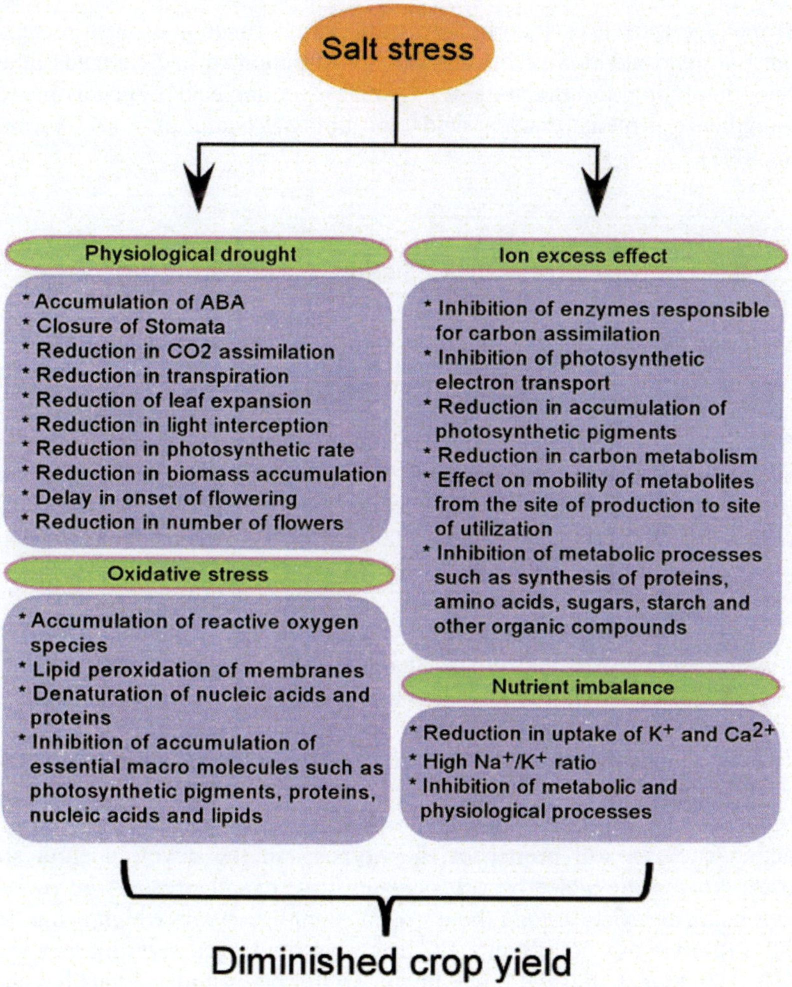

Fig. 9.1 Salt stress responses of crop plants

Since Na^+ and K^+ are similar in chemical nature, excessive accumulation of Na^+ at the root surface affects K^+ uptake as reviewed in Zhu 2007. Additionally, excessive NaCl accumulation in the root zone impedes the uptake of Ca^{2+}, thus causing nutrient deficiency. The increase in Na^+/K^+ ratio resulting due to the elevated salt concentrations at the root zone affects various metabolic and physiological processes including biomass accumulation (Flagella et al. 2002; Yeo 1998).

In addition to the primary effects of salt stress described above, it also can induce secondary effects such as oxidative stress (Dolatabadian and Jouneghani 2009). Salinity stress induces production of reactive oxygen species such as superoxide radicals, hydrogen peroxide and hydroxyl radicals (Verma and Mishra 2005).

Reactive oxygen species can abruptly perturb normal function of plant metabolism through lipid peroxidation of membranes, denaturation of nucleic acids and proteins, and inhibiting accumulation of essential macromolecules such as photosynthetic pigments, proteins, nucleic acids and lipids (Koca et al. 2007; Verma and Mishra 2005).

9.3 Effect of Salt Stress on Oilseed Crop Production

Oilseed crops being major suppliers of edible oil, industrial fuel and animal feed have emerged as one of the most valuable agricultural trade commodities. Global production of major oilseeds during 2009–2010 was reported to be 424.59 million tons from 205.48 million hectares of farmlands (Table 9.1). Currently global oilseed production is dominated by three major crops, including soybean, rapeseed and sunflower, with the world harvested area of 102, 32 and 22 million hectares, respectively, and the seed production about 260, 61 and 31 million tons, respectively (Table 9.1).

Plant salt tolerance is the inherent ability of plants to withstand effect of high salts in the root zone or on the plant surfaces without significant adverse effect on growth, development and yield (Shannon 1997). Using simple convention, salt tolerance can be measured on the basis of two parameters, (1) salinity level causing initial significant reduction in expected yield assessed as the threshold of electrical conductivity of saturated soil paste extract (ECe), and (2) the percentage of yield expected to be reduced for each unit of added salinity, the slope of ECe. The salinity response of oilseed crops varies greatly and generally depends on factors such as soil properties, genotypes and the developmental stages. Although most of the oilseed species are sensitive to salt stress, large variability of salt sensitivity exists among them. As an example, canola which is unaffected by soil salinity upto 10 dS m^{-1} ECe, is classified as a salt tolerant species (Table 9.2); whereas, linseed being unable to tolerate salinity levels beyond 2.0 ds m^{-1} ECe is classified as a salt sensitive species. Soybean, sunflower and safflower are moderately tolerant to salt stress, whereas, peanut is an example for moderately sensitive species (Table 9.2).

Table 9.1 Global oilseed production (2009–2010)

Crop(s)	Area (Million hectares)	Production (Million metric tons)
Major oilseeds	205.48	424.59
Soybean	102.17	260.85
Rapeseed	31.41	60.98
Sunflower seed	22.07	30.39

*Data source: World Agriculture Production, Foreign Agricultural Service, United States Department of Agriculture, December 2011 released

Table 9.2 Salt tolerance of oilseed crops

Crop	Salt tolerance parameters: ECe threshold	Slope % per ds m^{-1}	Rating	Reference
Canola	9.7	14	Tolerant	Francois 1994
Soybean	5.0	20.0	Moderately tolerant	Abel and MacKenzie 1964, Bernstein and Ogata 1966
Sunflower	4.8	5.0	Moderately tolerant	Francois 1996
Safflower	7.5	6.0	Moderately tolerant	Francois and Bernstein 1964
Peanut	3.2	29.0	Moderately sensitive	Shalhevet et al. 1969
Linseed	2.0	10.0	Sensitive	Shannon 1997

Salinity presents potential hazards to oilseed production by affecting both yield and quality of the produce. The most common adverse effect of salinity on oilseed crop is the reduction of seed germination, seedling growth, seedling height, number and size of leaves, reproductive structures, seed number and seed weight as well as deterioration of seed quality (Bybordi 2010; Flagella et al. 2004; Houle et al. 2001; Sinaki et al. 2007; Ulfat et al. 2007; Zheng et al. 1998). A great magnitude of genotypic variation for salt tolerance has been recorded in oilseed crops (Irving et al. 1988; Knowles 1989; Miller 1995), which is greatly affected by climatic and biological factors (Kumar 1995; Minhas et al. 1990). The amphitetraploid Brasssica species such as *B. napus*, *B. carinata* and *B. juncea* are more tolerant to salinity than their progenitors such as *B. rapa*, *B. nigra* and *B. oleracea* (Kumar 1995). Among all Brassica species *B. napus* is the most salt tolerant, whereas, *B. nigra* and *B. rapa* are the most salt sensitive crop species (Kumar 1995).

High salinity affects both content and quality of seed oil (Beke and Volkmar 1995; Francois and Bernstein 1964; Shannon 1997). In oilseed crops, quality of seed oil depends upon the composition of fatty acids such as palmitic, stearic, oleic and linoleic acid. Seed oil with higher percentage of oleic and linoleic acid is considered as the best quality oil (Bergman et al. 2006; Vles and Gottenbos. 1989). Linoleic acid reduces the blood cholesterol level and prevents cardio-vascular diseases. Salinity directly inhibits enzymes such as glyoxysomal catalase, malate synthase, isocitratelyase and oleatedesaturase responsible for fatty acid biosynthesis and modification, which leads to reduction of glycerides, accumulation of more free fatty acids and alteration of fatty acid composition (Bergman et al. 2006; Vles and Gottenbos 1989). Sunflower plants grown on saline irrigation water experienced a progressive increase in oleic acid content which was consistent with decrease in linoleic acid levels, possibly due to inhibition of oleatedesaturase (Di Caterina et al. 2007). Similarly salt stress altered the composition and levels of palmitic, oleic, stearic and linoleic acid in safflower genotypes (Bergman et al. 2006; Vles and Gottenbos. 1989). Overall, salt stress adversely affects both quantity and quality of oilseed crop production.

9.4 Improving Salt Tolerance of Oilseed Crops

Plants have developed numerous defence strategies to overcome salt stress, which can be mainly divided into three main categories: ***Escape***, achieved by inducing early flowering and short life cycle enabling successful completion of reproduction before the onset of severe stress (Mooney et al. 1987); ***Avoidance***, achieved via several mechanisms, such as exclusion of salt ions, maximizing water uptake and minimizing water loss by closing stomata to prevent or decrease the impact of salt stress (Chaves et al. 2003); and ***Tolerance***, when escape or avoidance of salt stress becomes impossible, plants have to adapt to existing salty environment by coordinating both physiological and biochemical alterations at the cellular and/or molecular levels (Morgan 1984). Knowledge gained through extensive research in the past few decades has advanced our understanding of the mechanisms by which salt stress impacts crop growth and productivity, which can be utilized towards engineering genotypes with salt tolerance capability. Salt tolerance being a physiologically complex trait, an integrated approach incorporating genetic and biotechnological modification of existing cultivars is essential to develop salt tolerant genotypes.

9.4.1 Growing Halophytes as an Alternative

Halophytes are plants that naturally grow under high salinity conditions and are therefore tolerant to salt stress (Zhu 2007). Halophytes are widespread among various orders of higher plants which is indicative of a polyphyletic origin of halophytes (Zhu 2007). *Thellungiella halophila*, a Brassicaceae family halophyte, can tolerate NaCl levels upto 500 mM and continue to produce viable seeds (Inan 2004). It has many features similar to Arabidopsis including the floral dip method of transformation and thus can be adopted as a halophytic model for salt-stress tolerance research in Brassicas aimed at understanding complete salt stress pathway and genes associated with it (Inan 2004).

Halophytes have great potential to illuminate specific traits associated with salt tolerance that may be selected in oilseed crops and can be used to domesticate naturally occurring salt tolerant species as crop plants. Some of the potential halophytic oilseed crops are *Kosteletz kyavirginica* (Ruan et al. 2008); *Salvadorapersica* (Reddy et al. 2008); *Salicornia bigelovii* (Glenn et al. 1991) and *Batismaritima* (Marcone 2003). Seeds of various halophytic oilseed crops, such as *Suaeda fruticosa*, *Arathrocnemum macrostachyum*, *S. bigelovii*, *Salicornia brachiata*, *Halogeto nglomeratus*, *Kochia scoparia*, and *Haloxylon stocksii* possess sufficient quantity of high-quality edible oil, with unsaturation of 70–80% (Khan 2006). A desert annual halophyte, *Suaeda aralocaspica*, which has high seed oil content (29% on dry weight basis) and 93% of unsaturated fatty acids (linoleic >68% and oleic >20%) is one of the many promising halophytes for edible oil (Wang et al. 2012). Similarly, *Kochia scoparia*, a mesohalophyte, which can produce

120 kg ha^{-1} seed oil during excessive salt stress condition, is another promising halophyte for edible oil (Salehi et al. 2012). *Halostachys caspica*, a short shrub halophyte can resist upto 700 mM of NaCl (Guan et al. 2010; Liu et al. 2012). Seeds of *Salvadora oleoides* and *Salvadora persica* contain 40–50% seed oil and are a good source of lauric acid (Khan 2006). Purified fat from these species is used for soap- and candle-making and is a potential substitute for coconut oil (Khan 2006). The most productive halophytic species yield upto 20 t/ha of biomass on seawater irrigation, equivalent to conventional crops (Glenn et al. 1999). *Salicornia bigelovii* produces 2 t/ha of seed yield containing 28% oil and 31% protein, which is comparable to soybean seed yield and quality (Glenn et al. 1999). Although several halophytes have been identified as potential oilseed crops, extensive screening of these wild species for quantity and quality of oil is essential before their domestication and cultivation.

In addition to being alternative oilseed crops, halophytes also aid in re-vegetation and remediation of salt affected land (Fogel et al. 2004). Indigenous halophytes can be used to restore overgrazed pasture and reduce reliance on irrigation (Peacock et al. 2003). High yields of biomass and seed can also be obtained by growing halophytes with sea water or brackish water (Ahmed-Hamad and Monsaly 1998). They can be used in various other applications such as recycling saline agricultural wastewater and reclaiming salt-affected soil in arid-zone (Glenn et al. 1999).

9.4.2 Conventional and Molecular Breeding for Enhancing Salt Stress Tolerance

Considerable inter-specific and intra-specific variation exists among most of the oilseed crops including Brassica for salt tolerance, which can be exploited through conventional breeding. Evaluation of several cultivars of Indian mustard, sunflower and safflower for salt tolerance revealed high level of genetic diversity among them (Ahmad et al. 2012; Ghazizade et al. 2012; Hussain et al. 2012), and enabled identification of several physiological traits that can be used as salinity tolerance criteria during oilseed breeding. Salt tolerant cultivars of oilseed crops developed through breeding are summarised in Table 9.3. Most of these cultivars are mainly grown in Indian and Pakistan sub-continent. Like many other traits salinity is also a quantitative character controlled by polygenes, which makes it difficult to breed for salt tolerance (Ashraf and McNeilly 2004; Blumwald et al. 2004; Flowers 2004; Yamaguchi and Blumwald 2005). Quantitative trait loci (QTL) mapping is an important approach to identify the genomic regions that control salinity tolerance traits. Advanced DNA markers techniques, such as AFLP, RFLP, RAPD, SSR, and SNPs are used to identify QTLs. Identification of QTL is an important technique for successful salinity breeding in oilseed crops especially in Brassica species. To date, because of the physiological complexity of the salinity response, no significant QTL with reference to salinity tolerance in Brassica species has been reported, making it

Table 9.3 Cultivars of oilseed species developed through conventional breeding

Oilseed species	Cultivars/lines	Parameter for testing tolerance	References
Brassica napus	Dunkeld (canola), ST9194, Rapora, Mytnitskii, Chisayanatane	Biomass, seed yield, germination	Qasim (2000), Puppala et al. (1999), Pokrovskii (1990)
Brassica juncea	Common Green, Varuna, RH 30, Pusa Bold, Kranti, BM-1, LL-84, P-15, KS-51	Germination, vegetative stage, biomass, seed yield	Kwon et al. (1997), Rai (1977), Kumar (1984), Dhawan et al. (1987), Kumar (1995), Ashraf (1992), Ashraf et al. (1994), Ahmad et al. (2012)
Brassica carinata	C90-1191, P5/80, Yellow Dodella	Germination and seedling growth	Ashraf and Sharif (1997)
Brassica campestris	BSH1	Germination and seed yield	Paliwal (1972), Kumar (1984)
Arachis hypogea	NRCG 2588, 4659, 5513, 6131, 6450, 6820, 6919, 7206, TMV 2 NLM, TG 33, JNDS-2004-15	Seed yield	Singh et al. (2008)
Helianthus annuus	DKS-4040, G-101 and P64A93	Achene yield	Hussain et al. (2012)
Carthamus tinctorius	Golsefid and Isfahan14	germination percentage and seedling growth	Ghazizade et al. (2012)

difficult to characterize the genetic basis of salinity stress tolerance in Brassicas. However, there have been a few attempts looking for QTL for salinity tolerance in other oilseeds and Arabidopsis.

A major QTL related to salt tolerance in soybean was identified in recombinant inbred lines (RILs) derived from a salt tolerant cultivar, S-100 and a susceptible cultivar Tokyo (Lee et al. 2004). The same QTL was identified in a different cross involving F_2 population derived from cultivars, Jackson and a wild soybean accession (Hamwieh and Xu 2008). The identified QTL accounts for 64% of the total variance along with a large dominant effect for salt tolerance (Hamwieh and Xu 2008). Another major QTL was detected on a different linkage group in a cross between soybean cultivar Kefeng No.1 and a salt sensitive cultivar Nannong1138-2 (Chen et al. 2008). Reliable method for validating the effect of a QTL for salt tolerance is by developing near isogenic lines (NILs), since the QTL alleles can be more accurately estimated with a homogeneous genetic background (Salvi and Tuberosa 2005). A significant contribution with regards to identification of a major QTL and development of NILs for salinity tolerance comes from soybean (Hamwieh et al. 2011). Two different RIL populations derived from salt tolerant diverse parents were studied for the salt tolerance. The major QTL identified in RILs of the cross between FT-Abyara and C01 RILs was found to be heterozygous. Further from these residual heterozygous RILs, NILs were developed by selfing. NILs with

FT-Abyara chromosome segment at the QTL regions showed a significantly higher salt tolerance than the lines without FT-Abyara chromosome segment (Hamwieh et al. 2011). The NILs developed for salt tolerance could be used in fine mapping and investigating different mechanisms involved in salt tolerance in soybean.

A wide range of variation was observed for salt tolerance during a germination study involving Arabidopsis accessions (Quesada 2002). Additionally, this study identified six other QTLs, which were contributing for salt tolerance in Arabidopsis accessions. However, accessions which showed salt tolerance during germination stage, failed to tolerate salinity stress during vegetative period, suggesting the mechanisms involved in salt tolerance during early and adult plant stages are different (Quesada 2002).

9.4.3 Genetic Engineering for Enhancing Salt Stress Tolerance

Conventional breeding, although being useful in selection of salt tolerant cultivars/lines, is very time consuming and labour intensive. It can simultaneously allow the transfer of undesirable genes along with desirable ones. Furthermore, reproductive barriers limit transfer of favourable alleles from inter-specific and inter-generic sources. Under such circumstances, genetic engineering has proven to be a powerful means of transferring genes effectively between different crop species.

Metabolic acclimation via the accumulation of compatible solutes is regarded as a basic strategy for the protection and survival of plants in extreme environments (Sakamoto and Murata 2000). Plants tolerate excess of Na^+ by its active uptake and compartmentation in the vacuoles, restriction of Na^+ influx and active efflux (Serrano et al. 1999), osmotic adjustment and accumulation of compatible solutes (Ahmad et al. 2010, Ahmad et al. 2011). Plants accumulate significant amounts of glycinebetaine (betaine), a compatible quaternary amine which is able to restore and maintain the osmotic balance of living cells, in response to high salinity, cold and drought. It is known that glycinebetaine protects plants from salt stress largely through osmotic adjustment (Bhattacharya et al. 2004; Bohnert and Shen 1998; Prasad et al. 2000). Glycinebetaine has also been shown to stabilize membrane integrity and photosynthetic machinery (Deshnium et al. 1995; Hayashi et al. 1997; Huang et al. 2000; Sakamoto and Murata 1998). Exogenous application of osmoregulants, proline or glycinebetaine reduces the harmful effects of salinity on both physiological and growth parameters of canola, suggesting these osmoregulants can be applied where soil sodicity is a problem (Sakr et al. 2012). A choline oxidase gene (*codA*) involved in biosynthesis of glycinebetaine cloned from *Arthrobacter globiformis* transformed into *B. juncea* cv. Pusa Jai kisan (which lacks any means to synthesize glycinebetaine) through *Agrobacterium tumefaciens* mediated transformation significantly enhanced the germination and salt tolerance capability of *B. juncea* (Prasad et al. 2000). Constitutive expression of a bacterial *codA* gene enabled synthesis of substantially higher levels of betaine in transgenic plants (Huang et al. 2000). Cabbage transformants expressing bacterial *bet*A gene conferred enhanced salt tolerance in transgenic plants by protecting important cellular components such as cell membranes and proteins/enzymes (Bhattacharya et al. 2004); these transgenic plants showed better growth response and greater stability in maintaining plant water relations at increased levels of salinity. Furthermore, transgenic *B. campestris* L. spp. *Chinensis* plants expressing *codA* gene accumulated higher betaine which significantly increased net photosynthetic rate under high salinity conditions (100, 200, and 300 mmol/L NaCl, respectively) than the wild-type plants (Wang et al. 2010).

Glyoxalase II, also known as hydroxyacyl-glutathione hydrolase, along with glyoxalase I, constitutes the glyoxalase system, the major function of which is to detoxify by-products of glycolysis. Glyoxalase enzymes have been shown to impart salinity tolerance in crop plants. As an example, overexpression of glyoxalase II gene in transgenic *B. juncea* increased the salt tolerance at germination and delayed leaf senescence as compared to untransformed plants (Saxena et al. 2011).

Pathogenesis-related (PR) proteins apart from being induced in response to pathogen infection are also expressed in response to abiotic stresses. Interestingly, constitutive expression of pea *PR10* gene was found to enhance seed germination and growth of *B. napus* in the presence of 75 mM NaCl (Srivastava et al. 2004). Further research is needed to confirm the actual mechanism by which *PR10* gene helps in the improvement of salt tolerance.

Rise in the cytosolic free calcium concentration is one of the first detectable plants response to sodium stress. Calcium signal serves as a second messenger that turns on the machinery for sodium export and potassium/sodium discrimination. SOS3 (Salt overlay sensitive 3) acts as a sensor for the Ca^{2+} influx which works in a complex with the serine/threonine protein kinase SOS2. Calcium signal generally activates kinase cascade to phosphorylate target proteins such as SOS1. SOS1 is a plasma membrane sodium/proton antiporter that is responsible for removing sodium from the cells (reviewed in Zhu 2002). To-date, 42 SOS mutants have been identified from near-saturated screenings which fall into three main complementation groups. But only three SOS genes (SOS1, SOS2 and SOS3) were found necessary for salt tolerance in Arabidopsis (Zhu et al. 1998). *AtNHX1*, *AtNHX2*, *SsVP2*, *SOD2* (vacuolar Na^+/H^+ antiporter genes) (Yokoi et al. 2002; Gao et al. 2004; Guo et al. 2006) *GmCAX1* (Cation/proton antiporter gene) (Luo et al. 2005); *AVP1* (vacuolar H^+-PPiase gene) (Gaxiola et al. 2001); *AtHKT1* (Sodium and Potassium transporter gene) (Horie et al. 2006); *PgTIP1* (Tonoplast intrinsic protein) (Peng et al. 2007) are some of the genes that confer salt tolerance when over expressed in Arabidopsis. Transgenic *B. napus* plants over-expressing *AtNHX1* were able to grow, flower, and produce seeds in the presence of 200 mM NaCl indicating no effect of high salinity on seed yield and the seed oil quality (Zhang et al. 2001). Transgenic groundnut plants overexpressing the *AtNHX1* gene were more resistant to salt and water deprivation when compared to its wild type counterpart (Asif et al. 2011) implying, overexpression of *AtNHX1* gene not only enhanced the salt tolerance but also drought resistance. Similarly, transgenic *B. juncea* plants over expressing *PgNHX1* (vacuolar Na^+/H^+ antiporter from *Pennisetum glaucum*) survive and produce normal seeds in the presence of upto 300 mM NaCl (Finkelstein et al. 2002; Rajagopal et al. 2006).

LEA (late embryogenesis abundant) proteins in plants are associated with tolerance to osmotic stress resulting from desiccation and salt stress. Transgenic Chinese cabbage (*B. campestris* ssp. *pekinensis*) expressing *B.napus* LEA maintained a higher germination rate than non-transgenic plants under salt-stress condition (Park et al. 2005).

It is now widely known that salt tolerance is a multigenic trait and a multitude of physiological, biochemical and molecular processes influence salt tolerance capability of plants (Zhu 2002). Although surprising, most of the genetically engineered salt tolerant plant species have been developed by altering the levels of single genes. Pyramiding multiple genes involved in conferring salt tolerance is another option that needs to be evaluated in the future.

9.4.4 Agronomic Practices for Management of Salt Stress

Oilseed crops lack the ability to cope up with high soil salinity; additionally, there is little success in developing salt tolerant genotypes through breeding or genetic engineering. In this context, changing some agronomical practices such as exogenous application of proline or glycinebetaine has helped to reduce the harmful effects of salinity on both physiological and reproductive growth parameters in canola (Sakr et al. 2012). Ca^{2+} and Na^{+} ion composition in shoot are more suitable than K^{+} ion composition for selecting the salt tolerant genotypes in *B. napus* (Abbaszadeh et al. 2012). Laser priming of canola seeds helps to reduce the adverse effects of salt stress by increasing the height of plant, number of lateral branches, number of pods and seeds per plant (Mohammadi et al. 2012). In sunflower, plants with greater number and area of leaves can be considered as better salt tolerant hybrids (Hussain et al. 2012). Foliar application of growth regulators, such as ALA (5-aminolevulinic acid), ascorbic acid, pyridoxine, thiamin, α-tocopherol and nicotinamide have potential to enhance tolerance of oilseed crops against salt stress either by decreasing the oxidative damage or by promoting antioxidant activity and increasing the levels of solutes for osmotic adjustment (Hamayun et al. 2010; Sadak et al. 2010; Naeem et al. 2011). Similarly, brassinosteroids, a class of plant polyhydroxy steroids that are similar to animal and insect steroidal hormones, help reduce the inhibitory effect of high salt on seed germination and early seedling development of *B. napus* (Kagale et al. 2007).

9.5 Conclusion and Future Perspective

Salt stress is a significant factor that adversely affects crop growth and productivity as well as quality of agricultural produce. Considering the amount by which food production will have to be increased in the next few decades in order to feed the ever increasing population, there is a need to raise crop varieties that not only with stand high salt conditions but also can maintain optimal yield under adverse conditions. Research conducted in the past two to three decades has unravelled several mechanisms by which the adverse effects of salt stress unfold. During this period several genotypes expressing individual genes having role in ion-homeostasis and osmolyte accumulations have also been genetically engineered. Although salt tolerance is a multigenic complex trait, transgenic approach with candidate genes has proven to be promising. However, the adaptability and suitability of these transgenic genotypes under natural environments and locations where different abiotic stress factors occur simultaneously needs to be tested. Support from plant breeding is also essential for enhancing the adaptability of existing cultivars and developing new genotypes with ability to grow under saline environments. Over the period of time, it has been well understood that salinity tolerance traits are mainly quantitative in nature and are controlled by multiple genes. Complex nature of the traits is likely the reason for little success in breeding and biotechnological approaches towards the improvement of salt tolerance in oilseed crops. Thus, pyramiding several genes or characteristics into a single genotype through conventional breeding or genetic engineering would be a logical way forward for improving the salt tolerance capability of crop plants.

References

Abbaszadeh F, Rameeh V, Cherati A (2012) Salinity stress indices of seed yield and nutrient compositions in rapeseed (*Brassica napus* L.). Int J Biol 4:154–162

Abel GH, MacKenzie AJ (1964) Salt tolerance of soybean varieties (*Glycine max* L. Merrill) during germination and later growth. Crop Sci 4:157–161

Ahmad P (2010) Growth and antioxidant responses in mustard (*Brassica juncea* L.) plants subjected to combined effect of gibberellic acid and salinity. Arch Agro Soil Sci 56:575–588

Ahmad P, Nabi G, Ashraf M (2011) Cadmium-induced oxidative damage in mustard (*Brassica juncea* (L.) Czern. & Coss.) plants can be alleviated by salicylic acid. S African J Bot 77:36–44

Ahmad P, Hakeem RK, Kumar A, Ashraf M, Akram NA (2012) Salt-induced changes in photosynthetic activity and oxidative defense system of three cultivars of mustard (*Brassica juncea* L.). African J Biotechnol 11:2694–2703

Ahmed-Hamad AM, Monsaly HM (1998) Seed soaking presowing in vitamins versus the adverse effects of NaCl salinity on photosynthesis and some related activities of maize and sunflower plants. In: Proceedings of the XIth international congress on photosynthesis, vol IV, Kluwer, Budapest, 17–22 Aug 1998, pp 2617–2620

Anuradha S, Seeta Ram Rao S (2001) Effect of brassinosteroids on salinity stress induced inhibition of seed germination and seedling growth of rice (*Oryza sativa* L.). Plant Growth Regul 33:151–153

Ashraf M, Harris PJC (2005) Abiotic stresses: plant resistance through breeding and molecular approaches. Haworth Press, New York

Ashraf M, McNeilly T (2004) Salinity tolerance in Brassica oilseeds. Cr Rev Plant Sci 23:157–174

Ashraf M, Sharif R (1997) Inter-cultivar variation for salt (NaCl) tolerance in a potential oil-seed crop Ethiopian mustard (*Brassica carinata* A.Br.). Arch Acker Pfl Boden 42:129–136

Ashraf M, Zafar ZU, Tufail M (1994) Intra-specific variation for salt tolerance in a potential oilseed crop, brown mustard (*Brassica juncea* (L.) Czern. and Coss.). Arch Acker Pfl Boden 38:449–458

Ashraf MY, Sarwar G, Ashraf M, Afaf R, Sattar A (2002) Salinity induced changes in α-amylase activity during germination and early cotton seedling growth. Biol Plantarum 45:589–591

Asif MA, Zafar Y, Iqbal J, Iqbal MM, Rashid U, Ali GM, Arif A, Nazir F (2011) Enhanced expression of atnhx1, in transgenic groundnut (*Arachis hypogaea*) improves salt and drought tolerance. Mol Biotechnol 49:250–25

Barnabás B, Jäger K, FehEr A (2008) The effect of drought and heat stress on reproductive processes in cereals. Plant Cell Environ 31:11–38

Beke GJ, Volkmar KM (1995) Mineral composition of flax (*Linum usitatissimum* L.) and safflower (*Carthamus tinctorius* L.) on a saline soil high in sulfate salts. Can J Plant Sci 75:399–404

Bergman J, Flynn N, Carlson C, Wichman G, Dm Kephart K (2006) Registration of 'Montola 2004'safflower. Crop Sci 46:1818–1819

Bernstein L, Ogata G (1966) Effects of salinity on nodulation, nitrogen fixation, and growth of soybean and alfalfa. Agron J 58:201–203

Bhattacharya RC, Maheswari M, Dineshkumar V, Kirti PB, Bhat SR, Chopra VL (2004) Transformation of *Brassica oleracea* var. capitata with bacterial *betA* gene enhances tolerance to salt stress. Sci Hortic 100:215–227

Blumwald E, Grover A, Good AG (2004) Breeding for abiotic stress resistance: challenges and opportunities. In: Proceedings of the 4th international crop science congress, Brisbane

Bohnert HJ, Shen B (1998) Transformation and compatible solutes. Sci Hortic 78:237–260

Bray EA (1997) Plant responses to water deficit. Trends Plant Sci 2:48–54

Bray EA, Bailey-Serres J, Weretilnyk E (2000) Responses to abiotic stress. In: Buchanan B, GWaJ R (eds) Biochemistry and molecular biology of plants. American Society of Plant Physiology, Rockville

Bybordi A (2010) The influence of salt stress on seed germination, growth and yield of canola cultivars. Not Bot Horti Agrobo 38:128–133

Chaves MM, Maroco JP, Pereira JS (2003) Understanding plant responses to drought -from genes to the whole plant. Funct Plant Biol 30:239–264

Chen D, Ahlford A, Schnorrer F, Kalchhauser I, Fellner M, Viràgh E, Kiss I, Syvänen AC, Dickson BJ (2008) High-resolution, high-throughput SNP mapping in *Drosophila melanogaster*. Nat Methods 5:323–329

Cuartero J, Bolarin MC, Asins MJ, Moreno V (2006) Increasing salt tolerance in the tomato. J Exp Bot 57:1045–1058

Dash M, Panda SK (2001) Salt stress induced changes in growth and enzyme activities in germinating *Phaseolus mungo* seeds. Biol Plantarum 44:587–589

Deshnium P, Los DA, Hayashi H, Mustardy L, Murata N (1995) Transformation of Synechococcus with a gene for choline oxidase enhances tolerance to salt stress. Plant Mol Biol 29:897–907

Dhawan RS, Sharma DR, Choudhary JB (1987) Effect of salinity on germination and yield components in three species of *Brassica*. Indian J Agr Sci 57:107–111

Di Caterina R, Giuliani MM, Rotunno T, De Caro A, Flagella Z (2007) Influence of salt stress on seed yield and oil quality of two sunflower hybrids. Ann Appl Biol 151:145–154

Dolatabadian A, Jouneghani RS (2009) Impact of exogenous ascorbic acid on antioxidant activity and some physiological traits of common bean subjected to salinity stress. Not Bot Horti Agrobo 37:165–172

Finkelstein RR, Gampala SSL, Rock CD (2002) Abscisic acid signaling in seeds and seedlings. Plant Cell 14:S15–S45

Flagella Z, Cantore V, Giuliani MM, Tarantio E, De Caro A (2002) Crop salt tolerance: physiological, yield and quality aspects. Rec Res Dev Plant Biol 2:155–186

Flagella Z, Giuliani MM, Rotunno T, Di Caterina R, De Caro A (2004) Effect of saline water on oil yield and quality of a high oleic sunflower (*Helianthus annuus* L.) hybrid. Eur J Agron 21:267–272

Flowers TJ (2004) Improving crop salt tolerance. J Exp Bot 55:1–13

Fogel BN, Crain CM, Bertness MD (2004) Community level engineering effects of *Triglochin maritima* (seaside arrowgrass) in a salt marsh in northern New England, USA. J Ecol 92:589–597

Francois LE (1994) Growth, seed yield, and oil content of canola grown under saline conditions. Agron J 86:233–237

Francois LE, Bernstein L (1964) Salt tolerance of safflower. Agron J 56:38–40

Gao X, Ren Z, Zhao Y, Zhang H (2004) Overexpression of SOD2 increases salt tolerance of *Arabidopsis*. Plant Physiol 133:1873–1881

Gaxiola RA, Soledad JL, Undurraga LM, Dang GJ, Allen SL, Alper GR (2001) Drought and salt tolerant plants result for overexpression of the AVP1 H+ pump. P Natl Acad Sci USA 98:11444–11449

Ghazizade M, Golkar P, Salehinejad F (2012) Effect of salinity stress on germination and seedling characters in safflower (*Carthamus tinctorius* L.) genotypes. Ann Biol Res 3:114–118

Glenn EP, O'Leary JW, Watson MC, Thompson TL, Kuehl RO (1991) *Salicornia bigelovii* Torr: an oilseed halophyte for seawater irrigation. Science 251:1065–1067

Glenn EP, Brown JJ, Blumwald E (1999) Salt tolerance and crop potential of halophytes. Cr Rev Plant Sci 18:227–255

Guan B, Hu Y, Wang Y, Zhang F (2010) Molecular characterization and functional analysis of a vacuolar Na+/H+ antiporter gene (HcNHX1) from *Halostachys caspica*. Mol Biol Rep 38:1889–1899

Guo S, Yin H, Zhang X, Zhao F, Li P, Chen S, Zhao Y, Zhang H (2006) Molecular cloning and characterization of a vacuolar h+ – pyrophosphatase gene, SsVP, from the halophyte *Suaeda salsa* and its overexpression increases salt and drought tolerance of *Arabidopsis*. Plant Mol Biol 60:41–50

Hamayun M, Khan SA, Khan AL, Shin J-H, Ahmad B, Shin D-H, Lee I-J (2010) Exogenous gibberellic acid reprograms soybean to higher growth and salt stress tolerance. J Agric Food Chem 58:7226–7232

Hamwieh A, Xu D (2008) Conserved salt tolerance quantitative trait locus (QTL) in wild and cultivated soybeans. Breeding Sci 58:355–359

Hamwieh A, Tuyen DD, Cong H, Benitez ER, Takahashi R, Xu DH (2011) Identification and validation of a major QTL for salt tolerance in soybean. Euphytica 179:1–9

Hayashi H, Mustardy L, Deshnium P, Ida M, Murata N (1997) Transformation of *Arabidopsis thaliana* with the *codA* gene for choline oxidase; accumulation of glycinebetaine and enhanced tolerance to salt and cold stress. Plant J 12:133–142

Horie T, Horie R, Chan WY, Leung HY, Schroeder JI (2006) Calcium regulation of sodium hypersensitivities of *SOS3* and *Athkt1* mutants. Plant Cell Physiol 47:622–633

Houle G, Morel L, Reynolds CE, Siégel J (2001) The effect of salinity on different developmental stages of an endemic annual plant, *Aster laurentianus* (Asteraceae). Am J Bot 88:62–67

Huang J, Hirji R, Adam L, Rozwadowski KL, Hammerlindl JK, Keller WA, Selvaraj G (2000) Genetic engineering of glycinebetaine production toward enhancing stress tolerance in plants: metabolic limitations. Plant Physiol 122:747–756

Hussain M, Farooq M, Shehzad M, Khan MB, Wahid A, Shabir G (2012) Evaluating the performance of elite sunflower hybrids under saline conditions. Int J Agric Biol 14:131–135

Inan G (2004) Salt cress. A halophyte and cryophyte Arabidopsis relative model system and its applicability to molecular genetic analyses of growth and development of extremophiles. Plant Physiol 135:1718–1737

Irving DW, Shannon MC, Breda VA, Mackey BE (1988) Salinity effects on yield and oil quality of high-linoleate and high-oleate cultivars of safflower. J Agric Food Chem 36:37–42

Kagale S, Divi UK, Krochko JE, Keller W, Krishna P (2007) Brassinosteroid confers tolerance in *Arabidopsis thaliana* and *Brassica napus* to a range of abiotic stresses. Planta 225:353–364

Khan M (2006) Crop diversification through halophyte production on salt-prone land resources. CAB Reviews: Perspectives in Agriculture, Veterinary Science, Nutrition and Natural Resources 1

Knowles PF (1989) Safflower. McGraw-Hill, New York

Koca H, Bor M, Ozdemir F, Turkan I (2007) The effect of salt stress on lipid peroxidation, antioxidative enzymes and proline content of sesame cultivars. Environ Exp Bot 60:344–351

Kumar D (1984) The value of certain plant parameters as an index for salt tolerance in Indian mustard (*Brassica juncea* L.). Plant Soil 79:261–272

Kumar D (1995) Salt tolerance in oilseed Brassicas – present status and future prospects. Plant Breeding Abstracts 65:1439–1447

Kwon TR, Bourne WF, Harris PJC (1997) Growth and ion accumulation of *Brassica juncea* L. cultivars differing in salinity tolerance. Poster paper presented at the 1997 Annual Meeting of the Society of Experimental Biology, University of Kent at Canterbury, 7–10 April 1997; and published as Abstract P8.7 in J Exp Bot 48 (Supplement) 66

Lee GJ, Boerma HR, Villagarcia MR, Zhou X, Carter TE, Li Z, Gibbs MO (2004) A major QTL conditioning salt tolerance in S-100 soybean and descendent cultivars. Theor Appl Genet 109:1610–1619

Liu L, Wang Y, Zeng Y, Haxim Y, Zhang F (2012) Identification and characterization of differentially expressed genes in the halophyte *Halostachys caspica* under salt stress. Plant Cell Tiss Organ Cult. doi:10.1007/s11240-012-0123-z

Luo GZ, Wang HW, Jian H, Tian AG, Wang YJ, Zhang JS, Chen SY (2005) A putative plasma membrane cation/proton antiporter from soybean confers salt tolerance in *Arabidopsis*. Plant Mol Biol 59:809–820

Marcone MF (2003) *Batis maritima* (Saltwort/Beachwort): a nutritious, halophytic, seed bearings, perennial shrub for cultivation and recovery of otherwise unproductive agricultural land affected by salinity. Food Res Int 36:123–130

Miller JF (1995) Inheritance of salt tolerance in sunflower. Helia 18:9–16

Minhas PS, Sharma DR, Khosla BK (1990) Effect of alleviation of salinity stress at different growth stages of Indian mustard (*Brassica juncea*). Indian J Agrl Sci 60:343–346

Mohammadi SK, Shekari F, Fotovat R, Darudi A (2012) Effect of laser priming on canola yield and its components under salt stress. Int Agrophys 26:45–51

Mooney HA, Pearcy RW, Ehleringer J (1987) Plant physiological ecology today. Bioscience 37:18–20

Morgan JM (1984) Osmoregulation and water stress in higher plants. Ann Rev Plant Physiol 35:299–319

Mullet JE, Whitsitt MS (1996) Plant cellular responses to water deficit. Plant Growth Regul 20:119–124

Munns R (2005) Genes and salt tolerance: bringing them together. New Phytol 167:645–663

Munns R, Tester M (2008) Mechanisms of salinity tolerance. Annu Rev Plant Biol 59:651–681

Naeem MS, Rasheed M, Liu D, Jin ZL, Ming DF, Yoneyama K, Takeuchi Y, Zhou WJ (2011) 5-Aminolevulinic acid ameliorates salinity-induced metabolic, water-related and biochemical changes in *Brassica napus* L. Acta Physiol Plant 33:517–528

Paliwal KV (1972) Irrigation with saline water. Indian Agricultural Research Institute Monograph 2, New Delhi

Park BJ, Liu Z, Kanno A, Kameya T (2005) Genetic improvement of Chinese cabbage for salt and drought tolerance by constitutive expression of a *B. napus LEA* gene. Plant Sci 169:553–558

Peacock JM, Ferguson ME, Alhadrami GA, McCann IR, Hajoj AA, Saleh A, Karnik R (2003) Conservation through utilization: a case study of the indigenous forage grasses of the Arabian Peninsula. J Arid Environ 54:15–28

Peng Y, Lin W, Cai W, Arora R (2007) Overexpression of a panax ginseng tonoplast aquaporin alters salt tolerance, drought tolerance and cold acclimation ability in transgenic *Arabidopsis* plants. Planta 226:729–740

Pokrovskii VB (1990) Promising forms of winter swede rape. Selektsiya-i-Semenovodstvo Moskva 4:24–25

Prasad K, Sharmila P, Kumar PA, Saradhi PP (2000) Transformation of *Brassica juncea* (L.) Czern with bacterial codA gene enhances its tolerance to salt stress. Mol Breeding 6:489–499

Puppala N, Flower JL, Poindexter L, Bhardwaj HL (1999) Evaluation of salinity tolerance of canola germination. In: Janick J (ed) Perspectives on new crops and new uses. ASHS Press, Alexandria

Qasim M (2000) Physiological and biochemical studies in a potential oilseed crop canola (*Brassica napus* L.) under salinity (NaCl) stress. PhD thesis, Department of Botany, University of Agriculture, Faisalabad

Quesada V (2002) Genetic architecture of NaCl tolerance in Arabidopsis. Plant Physiol 130:951–963

Rai M (1977) Salinity tolerance in Indian mustard and safflower. Indian J Agr Sci 47:70–73

Rajagopal D, Agarwal P, Tyagi W, Singla-Pareek SL, Reddy MK, Sopory SK (2006) Pennisetum glaucum Na+/H+ antiporter confers high level of salinity tolerance in transgenic *Brassica juncea*. Mol Breeding 19:137–151

Reddy MP, Shah MT, Patolia JS (2008) Salvadorapersica, a potential species for industrial oil production in semiarid saline and alkali soils. Ind Crop Prod 28:273–278

Ruan CJ, Li H, Guo YQ, Qin P, Gallagher JL, Seliskar DM, Lutts S, Mahy G (2008) *Kosteletzkya virginica*, an agroecoengineering halophytic species for alternative agricultural production in China's east coast: Ecological adaptation and benefits, seed yield, oil content, fatty acid and biodiesel properties. Ecol Eng 32:320–328

Sadak MS, Rady MM, Badr NM, Gaballah MS (2010) Increasing sunflower salt tolerance using nicotinamide and α-tocopherol. Int J Acad Res 2:263–270

Sakamoto A, Murata AN (1998) Metabolic engineering of rice leading to biosynthesis of glycinebetaine and tolerance to salt and cold. Plant Mol Biol 38:1011–1019

Sakamoto A, Murata AN (2000) Genetic engineering of glycinebetaine synthesis in plants: current status and implications for enhancement of stress tolerance. J Exp Bot 51:81–88

Sakr MT, El-Sarkassy NM, Fuller MP (2012) Osmoregulators proline and glycine betaine counteract salinity stress in canola. Agron Sustain Dev. doi:10.1007/s13593-011-0076-3

Salehi M, Kafi M, Kiani AR (2012) Salinity and water effects on growth, seed production and oil content of *Kochia scoparia*. J Agron 11:1–8

Salvi S, Tuberosa R (2005) To clone or not to clone plant QTLs: present and future challenges. Trends Plant Sci 10:297–304

Saxena M, Roy SD, Singla-Pareek SL, Sopory SK, Bhalla-Sarin N (2011) overexpression of the glyoxalase ii gene leads to enhanced salinity tolerance in *Brassica juncea*. Open Plant Sci J 5:23–28

Seckin B, Sekmen AH, Türkan I (2009) An enhancing effect of exogenous mannitol on the antioxidant enzyme activities in roots of wheat under salt stress. J Plant Growth Regul 28:12–20

Serrano R, Mulet JM, Rios G, Marquez JA, de Larrinoa IF, Leube MP, Mendizabal I, Pascual-Ahuir A, Proft M, Ros R, Montesinos C (1999) A glimpse of the mechanisms of ion homeostasis during salt stress. J Exp Bot 50:1023–1036

Shalhevet J, Reiniger P, Shimshi D (1969) Peanut response to uniform and non-uniform soil salinity. Agron J 61:384–387

Shannon MC (1997) Adaptation of plants to salinity. Adv Agron 60:75–120

Sinaki JM, Majidi E, Shirani-Rad AL, Moormohammadi G, Zarei G (2007) The effect of water deficit during growth stage of canola (*Brassica napus* L). Am J Agr Environ Sci 2:417–422

Singh AL, Hariprassana K, Solanki RM (2008) Screening and selection of groundnut genotypes for tolerance of soil salinity. Australian J Crop Sci 1:69–77

Srivastava S, Fristensky B, Kav NNV (2004) Constitutive expression of a PR10 protein enhances the germination of *Brassica napus* under saline conditions. Plant Cell Physiol 45:1320–1324

Tabur S, Demir K (2010) Role of some growth regulators on cytogenetic activity of barley under salt stress. Plant Growth Regul 60:99–104

Tester M, Davenport R (2003) Na^+ transport and Na^+ tolerance in higher plants. Ann Bot 91:503–527

U.S. Department of Agriculture, Forest Service Rocky Mountain Research Station, Ogden, UT (2001)

Ulfat M, Athar HR, Ashraf M, Akram NA, Jamil A (2007) Appraisal of physiological and biochemical selection criteria for evaluation of salt tolerance in canola (*Brassica napus* L.). Pakistan J Bot 39:1593–1608

UN (2009) World population prospects: 2008 revision. Population newsletter

Verma S, Mishra SN (2005) Putrescine alleviation of growth in salt stressed *Brassica juncea* by inducing antioxidative defense system. J Plant Physiol 162:669–677

Vinocur B, Altman A (2005) Recent advances in engineering plant tolerance to abiotic stress: achievements and limitations. Curr Opin Biotech 16:123–132

Vles RO, Gottenbos JJ (1989) Nutritional characteristics and food uses of vegetable oils. In: Downey RK, Robbelen G, Ashri A(eds) Oil crops of the world. McGraw-Hill, New York, pp 63–86

Wang Qb, Xu W, Xue QZ, Su WA (2010) Transgenic Brassica chinensis plants expressing a bacterial codA gene exhibit enhanced tolerance to extreme temperature and high salinity. J Zhejiang Univ-Sc B 11:851–861

Wang L, Zhao Z, Zhang K, Tian C (2012) Oil content and fatty acid composition of dimorphic seeds of desert halophyte *Suaeda aralocaspica*. African J Agric Res 7:1910–1914

Wilkinson S, Davies WJ (2002) ABA-based chemical signalling: the co-ordination of responses to stress in plants. Plant Cell Environ 25:195–210

Yamaguchi T, Blumwald E (2005) Developing salt-tolerant crop plants: challenges and opportunities. Trends Plant Sci 10:615–620

Yeo A (1998) Molecular biology of salt tolerance in the context of whole-plant physiology. J Exp Bot 49:915–929

Yokoi S, Cubero FJQB, Ruiz MT, Bressan RA, Hasegawa PM, Pardo JM (2002) Differential expression and function of *Arabidopsis thaliana* NHX Na+/H+ antiporters in the salt stress response. Plant J 30:529–536

Zhang HX, Hodson JN, Williams JP, Blumwald E (2001) Engineering salt-tolerant Brassica plants: characterization of yield and seed oil quality in transgenic plants with increased vacuolar sodium accumulation. P Natl Acad Sci USA 98:12832–12836

Zheng GH, Gao YP, Wilen RW, Gusta LV (1998) Canola seed germination and seedling emergence from pre-hydrated and re-dried seeds subjected to salt and water stresses at low temperatures. Ann Appl Biol 132:339–348

Zhu J (2002) Salt and drought stress signal transduction in plants. Ann Rev Plant Biol 53:247–273

Zhu JK (ed) (2007). Plant salt stress. In: eLS. Wiley, Chichester. http://www.els.net, doi: 10.1002/9780470015902.a0001300.pub2

Zhu BC, Su J, Chan MC, Verma DPS, Fan YL, Wu R (1998) Overexpression of a delta (1)-pyrroline-5-carboxylate synthetase gene and analysis of tolerance to water- and salt-stress in transgenic rice. Plant Sci 139:41–48

Chapter 10
Abscisic Acid and Biomass Partitioning in Tomato Under Salinity

Stella Lovelli, Adriano Sofo, Michele Perniola, and Antonio Scopa

10.1 Introduction

Soil salinity can emerge under different environmental conditions and vary according to chemical, physical and biological properties of the soil itself. Those different conditions can markedly modify the possible indications about the use of the soil, the most suitable techniques to adopt for its agronomic management or the most appropriate actions to apply for its correction and recovery. In particular, an in-depth knowledge of the origin and nature of the processes leading to soil salinization and a clear analysis of the soil changes is very important for a correct soil use planning at land level (Monteleone 2006).

At the same time understanding crop tolerance and adaptation to salinity is very important (Koyro et al. 2012) and forms one of the major research fields on crops in agronomic sector since agricultural productivity is deeply affected by salinization (Yadav et al. 2011). Salinity and drought conditions are the major limiting factors for yields in agriculture (Gregory 2006). Actually they are enormous problems if we consider the global increasing population and climate change. In agriculture it is important and no more postponed to increase productivity also in salinity conditions in order to achieve a sustainable use of environmental resources, to reach food security and finally in order to increase profitability of farms in using production factors.

For classification purposes plants have been organized into two groups: the salt sensitive glycophytes and the salt tolerant halophytes, unluckily all crops are belonging to the first group (Flowers and Flowers 2005). Plants react to stress environmental conditions at different levels, whole plant, root, leaves, reproductive organs, cellular and molecular levels (Jacobsen et al. 2012). Still today any effort to enhance salt tolerance of crops have met low results for the complexity of the crop mechanisms of

S. Lovelli (✉) • A. Sofo • M. Perniola • A. Scopa
Department of Crop Systems, Forestry and Environmental Sciences,
University of Basilicata, Viale dell'Ateneo Lucano, 10, 85100 Potenza, Italy
e-mail: stella.lovelli@unibas.it

P. Ahmad et al. (eds.), *Ecophysiology and Responses of Plants under Salt Stress*,
DOI 10.1007/978-1-4614-4747-4_10,

adaptation to salinity from the physiological and genetic point of view (Flowers and Flowers 2005). In any case crop physiology may explain structure–function relationship of crop characters and modifications in plant function that are caused by stress environmental conditions (Jacobsen et al. 2012; Miflin 2000).

10.2 Tomato Response to Salinity

Salinity is a significant environmental stress for crops. Currently, soil salinization is one of the main causes of crop yield reduction in many areas of the world (Paranychianakis and Chartzoulakis 2005). It was reported that about 20 % of irrigated surface is compromised by increasing salinity according to the United Nations Food and Agriculture Organization (Rozema and Flowers 2008). Soil salinization may arise from intrinsic soil components, use of low quality water for irrigation, or excessive use of fertilizers. The increasing scarcity of good quality water has focused attention on the problem of using brackish waters for irrigation. Solving salt stress problem in agriculture cannot be postponed due to irrigation with saline water and utilization of saline soils to handle the request of the global increasing population (Koushafar et al. 2011; Munns 2002). Agriculture widening to semiarid and arid regions with the practice of modern irrigation will exacerbate secondary salinization because hydrologic balance of the soil between water applied (irrigation and precipitation) and water consumed by crops (transpiration) will change (Chaves et al. 2009). Moreover a number of researchers have suggested that significant impacts of climate change are likely in the Mediterranean area, where in summer season warming greater than the average is expected, with a further increase in heat waves and a significant rainfall reduction (IPCC et al. 2007; Olesen and Bindi 2002; Vitale et al. 2010; Lovelli et al. 2010).

From a physiological point of view the plant response to salinity is complex, since it varies with the species, the salt concentration, the environmental factors and the growth stage. Actually breeding approaches showed that stress tolerance characters are in main part quantitative trait loci (QTLs), this in turn makes genetic selection of these traits very hard (Bartels and Sunkar 2005), even if in some cases stress tolerant genotypes have been likewise obtained, by inserting traits using as genes source wild relatives (Bartels and Sunkar 2005).

Undoubtedly there are a lot of information on plant response to salinity obtained from several researches on different crops made on different approaches, but it is necessary to integrate information regarding aspects of plant salt adaptation derived from physiological studies with those obtained from other approaches since our knowledge for the processes that ensures salinity tolerance is still today unclear (Paranychianakis and Chartzoulakis 2005).

Tomato is a widespread crop in the Mediterranean area where soil salinization is currently a serious problem (Paranychianakis and Chartzoulakis 2005). It is an important annual vegetable crop and it is usually utilised fresh, cooked or after processing (Cuartero and Fernandez-Munoz 1999). Tomato is well adapted to several climates; nevertheless a great part of world tomato production is localized in dry areas such as Mediterranean and California, where cultivation must be necessarily under irrigation (Cuartero and Fernandez-Munoz 1999).

Tomato as crop is classified as "moderately sensitive" to salinity (Foolad 2004) and, undoubtedly, it holds an important position in agriculture section (Koushafar et al. 2011). Water deficit and low water quality are surely the most important factors able to reduce yield and quality of tomato from nutritional value and food safety point of view (Favati et al. 2009; Dorais et al. 2008). Irrigation with saline water may increase sugar and organic acid content of cherry tomatoes (De Pascale et al. 2007) and the flavour of processed tomatoes (Mitchell et al. 1991). All the desirable quality aspects for the processed tomato industry such as dry matter, soluble solids and titratable acidity seem to increase with salinity (Mitchell et al. 1991). From agronomic and physiological point of view as regards salinity response of this crop there are several studies (see review of Cuartero and Fernandez-Munoz 1999, and most recent papers Maggio et al. 2007, Albacete et al. 2008, Perez-Alfocea et al. 2010, Ghanem et al. 2008, Okhovatian-Ardakani et al. 2010, Ghanem et al. 2011b, Lovelli et al. 2012). From the most recent papers it was pointed out that crucial points that are assuming great relevance in the understanding of tomato response to salinity conditions are substantially three:

1. Plant biomass partitioning;
2. ABA signal involved;
3. Tomato root architecture.

Our recent paper (Lovelli et al. 2012) confirmed the critical role for biomass partitioning and for root growth and morphology in tomato process adaptation to salts. Our findings provided important elements for elucidation of crucial mechanisms regarding tomato salt tolerance. Previously it was accepted the idea that in tomato, salinity does not change the usual distribution of dry matter between plant organs (fruits, shoot and root) even when there is a yield decrease (Ehret and Ho 1986). Recently we showed the contrary in agree with other recent papers (Albacete et al. 2008). On tomato we showed the high root-to-shoot ratio under salinity in tomato and the close relationship to high abscisic acid (ABA) root concentration (Lovelli et al. 2012). In tomato under high salinity level, the increase of ABA tissue concentration could regulate plant adaptation processes, such as dry matter partitioning (Albacete et al. 2008) and in particular way the root/shoot ratio (Maggio et al. 2007; Zhang and Blumwald 2001; Lovelli et al. 2012).

10.3 Gas Exchange, Plant Growth and Biomass Partitioning Under Salinity

In general, the negative effect of salt excess in soil water on glycophytic plants is mainly due to three phenomena:

- Osmotic stress, markedly increasing the osmotic potential of soil water stress resulting in a difficulty of plant to uptake water (directly proportional to salt concentration) with consequences similar to those caused by a water deficit (physiological drought);

- Toxic stress, consisting in the toxic effect and denaturing that some excess ions, especially Na^+, cause to cytoplasm enzymatic activities;
- Nutritional stress caused by an unbalanced ion uptake, given the antagonism among certain useful ions against those being in excess in soil water.

These effects change hormonal status and impair plant metabolic processes. As a consequence of those three stresses a reduction of plant growth and yield occurs (Yeo 2007). It was hypothesized that salinity response of tomato, as for other plants, happens in two phases ("biphasic model"; Munns 1993): during the first phase (days to weeks) the osmotic effect is prevalent, while during the second one (weeks to months) growth is controlled by toxic actions of the high salt accumulation in leaf tissues. In other words in plant adaptation to salts it is essential the time scale of the response. During the first phase (osmotic one) plant growth could be hormonal regulated while during the second phase, toxic effect of high salt concentration at tissue level are prevalent on plant growth reduction. Hormonal regulation of growth during the first phase is actually the main field of scientific debate on salinity.

Photosynthesis and the rhythm of cell growth are the first processes to be compromised by salinity (Chaves et al. 2009; Munns et al. 2006). In fact, it is frequently reported that with salt stress as the stomatal resistance rises, due to leaf water potential reduction, photosynthetic assimilation decreases (Prior et al. 1992; Munns 2002; Lovelli et al. 2012; Rivelli et al. 2002). The observed reduction is caused by the effect of salts on each single photosynthetic sub-process (diffusion, photochemical, biochemical processes). The stress determined by the high concentration of solutes in soil water can determine both an increase of stomatal resistance and mesophyll resistance to gas flows, with a subsequent limitation of photosynthetic activity (Flexas et al. 2004, 2007; Lawlor and Cornic 2002). Salt stress effect on photosynthetic non-stomatal components have been studied also, but precise information on the topic are still few (Rivelli et al. 2002; Seemann and Critchely 1985). With "non stomatal limitations" words we usually consider both physical limitations, mesophyll resistances to CO_2 diffusion in the gas and liquid phase, and bio-chemical limitations, mainly carboxylation rate and efficiency, to assimilation rate (Centritto et al. 2003). In addition as regards non-stomatal limitations to photosynthesis, in some cases they may generate confusion of interpretations (Centritto et al. 2003). The difficulty comes from the fact that, being numerous the factors regulating photosynthetic activity, it is particularly difficult to assess whether stomatal or non-stomatal effects prevail in response to salinity. Some studies pointed out that the assimilation activity drop, as a consequence of salt distribution to the crop, should be caused not only by stomatal closure, but mainly by ion actions at biochemical level. Na^+ and Cl^- ions can have a direct effect on photosynthetic apparatus because they reduce the efficiency of ribulose-1 5-bisphosphate carboxylase (Rubisco) in the Calvin's cycle (Bethke and Drew 1992; Martin and Ruiz-Torres 1992). Many studies showed the strict correlations between increased salt concentration, such as Cl^- and photosynthesis decrease (Paranychianakis and Chartzoulakis 2005; Lovelli et al. 2012; Lloyd et al. 1989; Walker et al. 1981; Chartzoulakis et al. 2002). Actually, it was shown that CO_2 concentration in intercellular spaces does not change as an effect of salinity but it stay more or less the same, while stomatal opening decreases; that suggests that both stomata conductance and especially the non stomatal ones are reduced by the salts accumulated in the tissues. Actually understanding the nature of non stomatal limitations of photosynthesis under salinity is an heated field of photosynthesis research (Paranychianakis and Chartzoulakis 2005; Centritto et al. 2003).

Under severe salt stress, photosynthesis of tomato was deeply reduced, so in this way stressed plants had a lower amount of fixed carbon to utilize for plant growth (Lovelli et al. 2012). Lower stomatal conductance and photosynthesis observed in salt stressed tomato plants explain the lower leaf growth and consequently the smaller accumulation of dry matter (Lovelli et al. 2012). During osmotic stress (during the first phase) ABA contributes to salt response through an effective stomatal control (Hassine and Lutts 2010). Indeed, a strong relation between ABA tomato leaf concentration and stomatal conductance occurred (Lovelli et al. 2012).

One consequence of reduced photosynthesis is the overall plant growth reduction, but different parts of the tomato plant grow in different way. In fact we observed an unbalanced growth rhythm of root and shoot under salinity (Lovelli et al. 2012), in particular we showed the high root-to-shoot ratio and the close relationship to high abscisic acid (ABA) root concentration (Lovelli et al. 2012). As said before we refer to the biphasic model of Munns (1993) that considers the physiological and agronomic adaptation of plants to salts as temporal changes in both osmotic and ionic stress (Perez-Alfocea et al. 2010). Actually processes that regulate leaf growth and shoot development under the osmotic phase of salinity are under debate (Albacete et al. 2008), as said before.

During the first osmotic phase it has been hypothesized that inhibition of plant growth could be controlled by hormones or their precursors (Munns and Tester 2008; Lovelli et al. 2012), while later (ionic phase), plant growth is mainly reduced by high leaf salt (Na^+ and Cl^-) build up that in turn involves to whole plant photosynthesis reduction and partly induces premature leaf senescence (Perez-Alfocea et al. 2010).

If from one side stomatal control by ABA increase in leaf tissues was an important research field for long time, actually there are very few data on ABA partitioning among the different plant organs (Assmann 2004; Zhang et al. 2004; Lovelli et al. 2012). Actually the ABA function in growth control is particularly controversy (Albacete et al. 2008), as according to some authors it holds up plant growth (Dodd and Davies 1996, Zhang and Davies 1990), while according to others it favours it (Sharp and LeNoble 2002). Therefore, several contrasting opinions exist on ABA function in the biomass allocation under salt stress (Sachs 2005). Modifications of plant growth under salinity could be controlled by changes in phytohormone tissues concentrations controlling assimilate partitioning from source to sink organs (Perez-Alfocea et al. 2010; van der Werf and Nagel 1996; Hartig and Beck 2006). We known that the ratio between root and shoot dry matter is usually constant, since root system and epigeous plant part grow at the same rate (Lovelli et al. 2012).

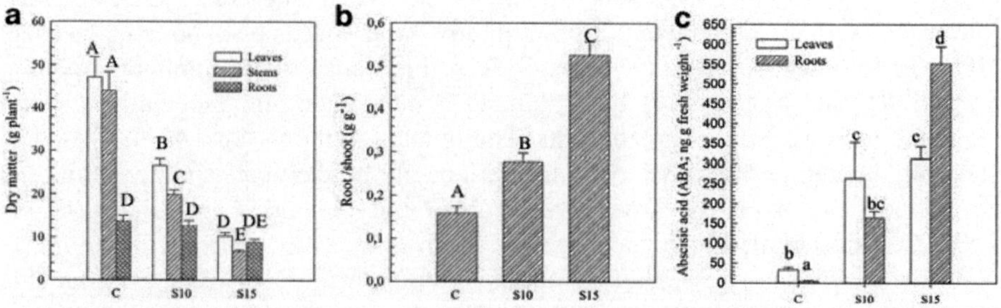

Fig. 10.1 Dry matter (**a**) root shoot ratio (**b**) measured in tomato plants subjected to two levels of salt stress (100 and 150 mM of NaCl, respectively), (**c**) ABA concentration in leaves and roots of tomato plants under two levels of salt stress. Mean value (n=5) within a column followed by different capital and lower case letters are significantly different at $P<0.01$ and $P<0.05$, respectively according to Duncan's multiple range test

The balance of width between each part of the plant is provided by assimilation rate and carbohydrate partitioning, but it can be highly modified by stress conditions (Erice et al. 2010). The ratio between root and shoot dry matter increased with the rise of salt concentration in the nutrient solution. In tomato it was showed that salinity does not affect the usual partitioning of dry matter between fruits, shoot and root even when yield decrease reductions is about close to 25 % of the control (Ehret and Ho 1986). On tomato we showed the contrary (Lovelli et al. 2012). Our results on plant growth are similar to that obtained from other authors always on tomato under salinity (Albacete et al. 2008; Maggio et al. 2007). It is clear now that salinity slows down cellular division and growth (Albacete et al. 2008). It happened that some morphological characters of crucial interest in stress adaptation, such as root growth and stomatal behaviour, have been less studied (Maggio et al. 2007). Sharp and LeNoble (2002) and Spollen et al. (2000) showed that higher root growth in conditions of low water potential is strictly related to ABA increase in the root tissues, and that the keeping of root growth under very negative water potential is controlled by ABA accumulation in roots.

All these results strengthened the idea that photosynthate utilisation rather than its availability is the main factor that limit plant growth under salt stress and assign an important function to hormonal signalling between plant organs (Perez-Alfocea et al. 2010). Moreover channelling assimilates from leaves to the roots (Perez-Alfocea et al. 2010) could be considered a particular choice of the plant, without meaning. At the same time this adaptive plant behaviour allows the plant roots to take out more water and uptake nutrients from the soil and allows to maintain ionic homeostasis, so it have a clear significance from the ecological point of view within salinised environment (Perez-Alfocea et al. 2010). The increase of root/shoot ratio in tomato is likely an effective physiological process that allow to regulate ion increase into tissues under salt stress (Fig. 10.1; Lovelli et al. 2012).

10.4 Role of Abscisic Acid (ABA) and Other Phytohormones Under Salinity

Since plants are sessile organisms, for them having and efficient system of response to the changing environment is crucial for surviving. First discovered group of plant hormones includes auxin, gibberellins (GAs), cytokinins, abscisic acid (ABA) and ethylene. Only recently another group of compounds such as brassinosteroids (BRs), jasmonate (JA), salicylate (SA), strigolactones (SLs), nitric oxide (NO), polyamines, and some oligopeptides were recognized as new families of plant hormones (Javid et al. 2011; Santner and Estelle 2010). Notwithstanding phythormones were studied for many years, interactions that occur between them are still unclear (Ross and O'Neill 2001). Vanguard of the research on this field regards the modalities through which plant hormones are involved in multiple processes and if the so-called "cross-talk" between different hormones results in synergetic or antagonistic interactions in response of plants to abiotic stress (Peleg and Blumwold 2011; Zhu et al. 2012; Gemes et al. 2011).

Under salinity plant response is triggered by osmotic signals (Chaves et al. 2003) or by other compounds (hormones, reactive oxygen species and intracellular second messengers) (Chaves et al. 2009). Surely between them abscisic acid (ABA) has an important function in the whole plant responses to salt stress (Zhang et al. 2006). Generally, ABA operates as a general inhibitor of growth and metabolism, and negatively affects the synthesis of proteins and nucleic acids, even if these actions vary with tissue, developmental stage and the concentrations of this hormone increase substantially under stress conditions (Sofo et al. 2011; Yuan et al. 2011; Kobashi et al. 2001; Srivastava 2002). The changes of the endogenous levels of ABA also stimulate different metabolic and physiological events that increase the level of tolerance to salts (Munns and Tester 2008; Xiong et al. 2002). However, in many stress conditions other hormones (ethylene, cytokinins, auxins) are involved also, and in particular, their biosynthesis could be considered an appropriate indicator of the plant health. Under salinity other hormones such as gibberellins can interact with ABA and other stress metabolites including antioxidants and ROS scavengers (Achard et al. 2006).

Recently it was underlined that all too often ABA is considered "the stress hormones", while other phytohormones such as cytokinins and auxins seems involved in explaining changes in plant biomass partitioning (Albacete et al. 2008; Javid et al. 2011). Notwithstanding it is frequently reported that salinity triggers off ABA synthesis in roots which is relocated to the shoots where it causes stomatal closure (Chaves et al. 2009). ABA can also be produced in leaf cells and then transported in other part of the plant (Wilkinson and Davies 2002). With regards to this aspect, recently it was showed that xylem and apoplastic pH affects ABA movements into plant tissues and in this way it seems to control the levels of ABA reaching the stomata (Jia and Davies 2007). The "alkaline trapping" of ABA may be triggered by salts also (Jia and Davies 2007). ABA concurs in salt response during the osmotic phase through an effective enhancement of stomatal control (Hassine and Lutts 2010). Stomatal control by ABA accumulation in leaf tissues was an important research field for years but data on ABA effects on dry matter partitioning among the different plant organs are lacking (Assmann 2004; Zhang et al. 2004). Modifications of ABA concentrations between leaves and roots may be accountable for the relative changes in growth ratios and biomass partitioning caused by salt stress (Lovelli et al. 2012). Leaf and stem dry matter decrease can be related to a redistribution of photosynthetates to the root system (Maggio et al. 2007) mediated by ABA signaling (Albacete et al. 2008). The few available experimental data on tomato (Albacete et al. 2008; Ghanem et al. 2008) are in agreement with our results (Lovelli et al. 2012), but they disagree with results of other authors (Mulholland et al. 2003; Maggio et al. 2007). In tomato under advanced salinization, the high ABA root levels could regulate organ adaptation, such as dry matter partitioning (Albacete et al. 2008) and usual alteration of the root/shoot ratio (Maggio et al. 2007). It is possible to suppose that similar to its action in the shoot, ABA accumulation may also be useful to keep root growth in salt stressed plants (Albacete et al. 2008). Sharp and LeNoble (2002) and Spollen et al. (2000) clearly showed that higher root growth under low water potential is associated with high ABA levels in the roots, and that the maintaining of root growth under low water potential is controlled by ABA accumulation in roots. Build up of ABA root concentration and root/shoot ratio observed under salt conditions (Lovelli et al. 2012) seems to strength our hypothesis. Some authors (Albacete et al. 2008) also supposed that ABA is related with inhibition of ethylene production, which is sometimes considered a growth inhibitor under stress. It is a clear example of cross-talk between plant hormones during plant response to salt stress, as said before. It could justify how a single hormone produces different effects in different plant organs, in other words in each organ this compound interacts in different modality with the other hormones that are at the sometime present (Ross and O'Neill 2001). Some authors (Ghanem et al. 2011b) observed that the root cytokinin production deeply reduces both ABA and Na^+ build up in the root and other organs without modifying root dry matter under moderate salinity (100 mM NaCl). Another possible explanation could be that given by Zhang et al. (2006). According to this author it is possible to give to ABA a dual function in plant physiological control. That is an inhibitive function when it accumulated at high concentration under stress, and a promoting function when it is at low amount in plant tissues. We observed high ABA root tissue amount in correspondence of high Na^+ and Cl^- root level (Lovelli et al. 2012), so it is possible to hypotize its inhibition function, as other authors reported (Sharp and LeNoble 2002).

In any case, higher root/shoot growth might be interpreted as part of an adaptation behaviour in which plant physiological and metabolic modifications evolve together with plant development, soil salinization and atmospheric parameters during the crop cycle (Maggio et al. 2011). Different tissue root and leaf ABA may have an important function in controlling growth, leaf gas exchange and dry matter partitioning of salinized tomato plants (Lovelli et al. 2012).

10.5 Tomato Root Architecture Modification Under Salinity

Considering its role function in absorbing water and nutrients, the root system is the main part of the plant to meet soil salinity (Ouyang et al. 2007), and likely plays an important role to cope with salts. In particular how salts affects root growth and architecture is of great importance to elucidate mechanisms for plant adaptation process to salinity.

The role of the roots and their function in mediating shoot responses to abiotic stresses such as salinity, was recently emphasised (Ghanem et al. 2011a). Root morphology such as root system architecture should be thoroughly investigated to improve plant development under environmental stress conditions (Ghanem et al. 2011a) because, currently, there are very few information on root architecture/morphology under salt stress (Maggio et al. 2011).

Root growth traits reduction associated to salinity agree with the results of several authors (Schwarz and Grosch 2003; Kafkafi 1996). On tomato we measured a reduction of total root weight and length in salt treatments and a large increase in specific root length (SRL) compared to the control (Lovelli et al. 2012 submitted). Snapp and Shennan (1992) observed no modification of Root Length Density in hydroponically-grown tomato plants under salinity. Recently both a root fresh weight reduction (30 %) was observed on tomato after 3 weeks under saline conditions (Albacete et al. 2008) and a root dry matter reduction under salinity together with a root/shoot increase (Lovelli et al. 2012). According to Cuartero and Fernandez-Munoz (1999) salinity deeply affects root biomass of tomato, but other authors (Abrisqueta et al. 1991) showed that tomato root biomass grown under salinity conditions, have only a delay in reaching a depth of 80 cm and the end root length density is a quarter than in control plants.

Also on other crops there are contrasting results. Considering faba bean growth on salinized soil root length density and root mass density are deeply reduced as effect of salts (Abdelhamid et al. 2010), while on soil-grown alfalfa some authors (Vaughan et al. 2002) showed that root production was stimulated by salinity. These contrasting results may depend also by confusion that comes from heterogeneous growth condition under salinity on soil-grown plants. This is an enormous problem of salinity experiments, each time we want to impose salinity condition in an artificial we may meet difficulties that can generate confusing results. In fact when plants are grown under salinity soil compaction could affect plant growth by causing increased resistance to root penetration and the resulting different mechanisms of salt damage may be very different as the result of the system under which the plants were grown (Tavakkoli et al. 2011). In order to avoid confusing results it is important to separate salt stress from other soil abiotic stress, eliminating soil component and this can be done only growing plant in hydroponics.

In our experiment on hydroponically-growth plants analyzing root length density along the depth we found a significant interaction between salinity and root depth on specific root length (SRL; Lovelli et al. 2012 submitted). A root system with a high SRL in high salinity conditions could be considered an adaptative response

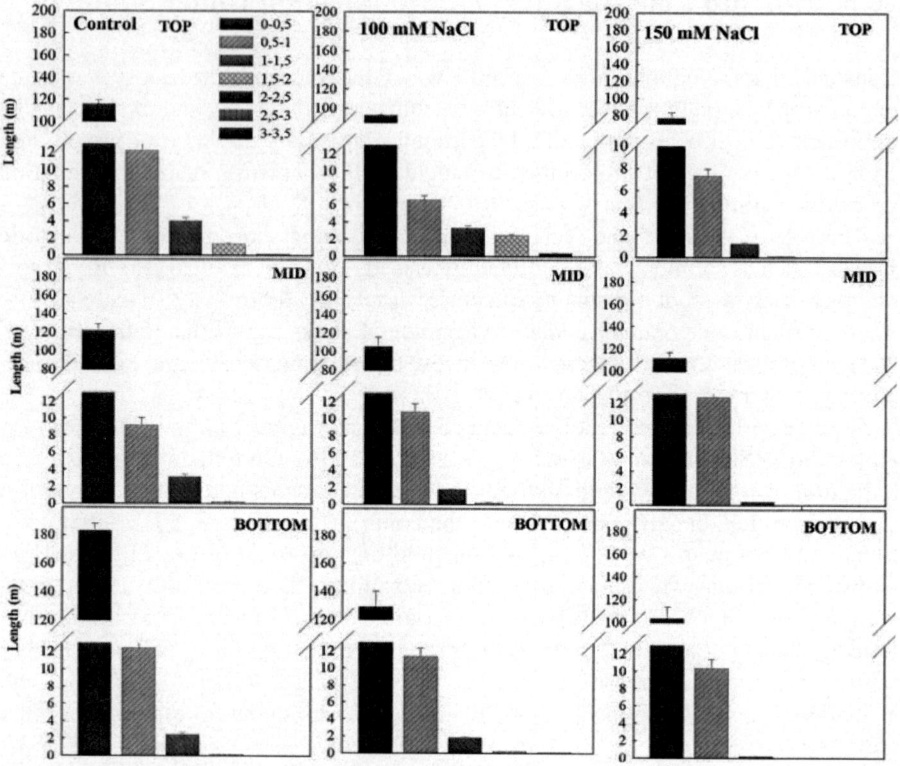

Fig. 10.2 Length of roots belonging to different diameter classes (mm) in the three treatments and at three distant depths (*top*, *mid* and *bottom*). Vertical bars are standard error of the mean

that gives to plants the possibility to growth better in the soil volume (Bazzaz and Morse 1991; Snapp and Shennan 1992), and to delay toxic ions accumulation in plant shoots (Maggio et al. 2007) keeping a right degree of ions homeostasis. Moreover on other crops some authors pointed out that differential rooting was higher in the upper half of the root zone on alfalfa soil grown plants, and that high fibrous rooting in alfalfa is a character to interpret as a salt stress avoidance behaviour (Vaughan et al. 2002). All these recent experiments on root architecture modification under salinity seems to be of great help for elucidation of mechanisms for tomato adaptation to salt stress on whose significance there are still many aspect to clarify.

On tomato it was observed that salt or other abiotic stresses may affect different roots to a different extent (Cuartero and Fernandez-Munoz 1999). Other researches showed that under stress tomato usually grows numerous small lateral feeder roots, which are not present in tomato plants growth in non-stress conditions (Zobel 1975). Moreover in our experiment in salt treatments we observed a particular root diameter distribution. Under severe salt stress we measured a significant amount of tomato roots belonging to the lower diametric class (0–0.5 cm) (Fig. 10.2; Lovelli et al. 2012 submitted). Increased Specific Root Length (SRL) usually associates with low average root diameters (Schwarz et al. 1995; Schwarz and Grosch 2003). These last results are in agreement with other authors (Kurth et al. 1986; Sharp et al. 1990) that observed thinner roots in cotton and maize, respectively, under high level of salinity. In general the increase of Specific Root Length (SRL) under salinity reflects differences in diameter distribution and may be used as an indicator of plant response to management (Basirat et al. 2011) or environmental change (Ostonen et al. 2007).

Moreover modifications in the root class diameter distribution may be considered as a mechanism of adaptation to salinity, thinner roots allow osmotic adjustment without alteration of fixed carbon partitioned to roots (Snapp and Shennan 1992).

10.6 Conclusion and Future Perspective

From recent research activity on tomato, it can be concluded that new elements have emerged that are useful in the elucidation of mechanisms of salt adaptation and tolerance. Source – sink regulation and root-to-shoot signaling are interconnected mechanisms that allow tomato plants to increase salt tolerance since they allow to maintain growth and delay leaf senescence during the first phase of salt response (osmotic one; Perez-Alfocea et al. 2010).

In biomass partitioning plant hormones plays a crucial role. As regards tomato different endogenous ABA at root and leaf level are key aspects in growth control, leaf gas exchange and dry matter partitioning of salt-stressed plants (Lovelli et al. 2012), even if it seems that in the complex plant hormonal network cross-talk between hormones may result in synergetic or antagonic interactions in response to one stress (Peleg and Blumwold 2011).

The role of root architecture in tomato response to salinity is still unclear, but it likely plays an important role. In hydroponically grown tomato plants under high salinity (150 mM NaCl) we observed decrease in root weight, depth and length density but an increase in specific root length, corresponding to an increase in fine roots in the middle part of the root system.

Although this chapter on tomato biomass partitioning under salinity covers only a part of a very complex scientific field of research it is clear that physiological approach is still a powerful tool for analyzing the complex process that is plant adaptation to salts. Several authors (Jacobsen et al. 2012; Boote et al. 2001; Hunt et al. 2003; Martre et al. 2003) think that only integrating knowledge from different approaches (plant physiology, soil science and agrometeorology) into mathematical equations, through models it is possible to forecast plant response and yield in stress conditions. So many efforts would be addressed to create synergies between scientific research groups and to develop a multidisciplinary approach for the salinity stress problematic (Wollenweber et al. 2005) in order to give a further chance to agriculture in areas affected by salinization.

References

Abdelhamid MT, Shokr MMB, Bekheta MA (2010) Growth, root characteristics, and leaf nutrients accumulation of four faba bean (*Vicia faba* L.) cultivars differing in their broomrape tolerance and the soil properties in relation to salinity. Commun Soil Sci Plan 41:2713–2728

Abrisqueta JM, Hernansaenz A, Alarcon JJ, Lozano MA (1991) Root growth dynamics of two tomato genotypes under saline conditions. Suelo y Planta 1:351–361

Achard P, Cheng H, De Grauwe L, Decat J, Schoutteten H, Moritz T, Ven der Straeten D, Peng J, Herberd NP (2006) Integration of plant responses to environmentally activated phytohormonal signals. Science 6:91–94

Albacete A, Ghanem ME, Martinez-Andujar C, Acosta M, Sanchez-Bravo J, Martinez V, Lutts S, Dodd IC, Perez-Alfocea F (2008) Hormonal changes in relation to biomass partitioning and shoot growth impairment in salinized tomato (*Solanum lycopersicum* L.) plants. J Exp Bot 59:4119–4131

Assmann SA (2004) Abscisic acid signal transduction in stomatal responses. In: Davis PJ (ed) Plant hormones: biosynthesis, signal transduction, action. Kluwer, London, pp 291–442

Bartels D, Sunkar R (2005) Drought and salt tolerance in plants. Crit Rev Plant Sci 24:23–58

Basirat M, Malboobi MA, Mousavi A, Asgharzadeh A, Samavat S (2011) Effects of phosphorous supply on growth, phosphate distribution and expression of transporter genes in tomato plants. Aust J Coltural Sci 5(5):537–543

Bazzaz FA, Morse SR (1991) Annual plants: potential for response to multiple stresses. In: Mooney HA, Winner WE, Dell EJ (eds) Response of plants to multiple stresses. Academic, New York, pp 283–305

Bethke PC, Drew MC (1992) Stomatal and non stomatal components to inhibition of photosynthesis in leaves of *Capsicum annum* during progressive exposure to NaCl salinity. Physiol Plant 86:115–123

Boote KJ, Kropff MJ, Bindraban PS (2001) Physiology and modeling of traits in crop plants: implications for genetic improvement. Agric Syst 70:395–420

Centritto M, Loreto F, Chartzoulakis K (2003) The use of low [CO_2] to estimate diffusional and non-diffusional limitations of photosynthetic capacity of salt-stressed olive saplings. Plant Cell Environ 26:585–594

Chartzoulakis K, Loupassaki M, Bertaki M, Androulakis I (2002) Effects of NaCl salinity on growth, ion content and CO_2 assimilation rate of six olive cultivars. Sci Hortic 96:235–247

Chaves MM, Maroco JP, Pereira JS (2003) Understanding plant responses to drought – from genes to the whole plant. Funct Plant Biol 30:239–264

Chaves MM, Flexas J, Pinheiro C (2009) Photosynthesis under drought and salt stress: regulation mechanisms from whole plant to cell. J Exp Bot 103:551–560

Cuartero J, Fernandez-Munoz R (1999) Tomato and salinity. Sci Hortic 78:83–125

De Pascale S, Martino A, Raimondi G, Maggio A (2007) Comparative analysis of water and salt stress-induced modifications of quality parameters in cherry tomato. J Hort Sci Biotechnol 82:283–289

Dodd IC, Davies WJ (1996) The relationship between leaf growth and ABA accumulation in the grass leaf elongation zone. Plant Cell Environ 19:1047–1056

Dorais M, Ehret DL, Papadopoulos AP (2008) Tomato (*Solanum lycopersicum*) health components: from the seed to the consumer. Phytochem Rev 7:231–250

Ehret DL, Ho LC (1986) The effects of salinity on dry matter partitioning and fruit growth in tomatoes grown in nutrient film culture. J Horti Sci 61:361–367

Erice G, Louahlia S, Irigoyen JJ, Sanchez-Diaz M, Avice JC (2010) Biomass partitioning, morphology and water status of four alfalfa genotypes submitted to progressive drought and subsequent recovery. J Plant Physiol 167:114–120

Favati F, Lovelli S, Galgano F, Di Tommaso T, Miccolis V, Candido V (2009) Processing tomato quality as affected by irrigation scheduling. Sci Hortic 122:562–571

Flexas J, Bota J, Loreto F, Cornic G, Sharkey TD (2004) Diffusive and metabolic limitations to photosynthesis under drought and salinity in C3 plants. Plant Biol 6:269–279

Flexas J, Diaz-Espejo A, Galme's J, Kaldenhoff H, Medrano A, Ribas-Carbo M (2007) Rapid variations of mesophyll conductance in response to changes in CO_2 concentration around leaves. Plant Cell Environ 30:1284–1298

Flowers TJ, Flowers SA (2005) Why does salinity pose such a difficult problem for plant breeders. Agric Water Manage 78:15–24

Foolad MR (2004) Recent advances in genetics of salt tolerance and cold tolerance in tomato. Plant Cell Tiss Org 76:101–119

Gemes K, Poor P, Horvath E, Kolbert Z, Szopko D, Szepesi A, Tari I (2011) Cross-talk between salicylic acid and NaCl-generated reactive oxygen species and nitric oxide in tomato during acclimation to high salinity. Physiol Plant 142:179–192

Ghanem ME, Albacete A, Martínez-Andújar C, Acosta M, Romero-Aranda R, Dodd IC, Lutts S, Pérez-Alfocea F (2008) Hormonal changes during salinity-induced leaf senescence in tomato (*Solanum lycopersicum* L.). J Exp Bot 59:3039–3050

Ghanem ME, Albacete A, Smigocki AC, Frebort I, Pospisilova H, Martinez-Andujar C, Acosta M, Sanchez-Bravo J, Lutts S, Dodd IC, Perez-Alfocea F (2011a) Root-synthesized cytokinins improve shoot growth and fruit yield in salinized tomato (*Solanum lycopersicum* L.) plants. J Exp Bot 62:125–140

Ghanem ME, Hichri I, Smigocki AC, Albacete A, Fauconnier ML, Diatloff E, Martinez-Andujar C, Lutts S, Dodd IC, Perez-Alfocea F (2011b) Root-targeted biotechnology to mediate hormonal signalling and improve crop stress tolerance. Plant Cell Rep 30:807–823

Gregory PJ (2006) Food production under poor, adverse climatic conditions. In: Proceedings of the IX ESA Congress, Warsaw, 4–7 Sept 2006, 19pp

Hartig K, Beck E (2006) Crosstalk between auxin, cytokinins and sugars in the plant cell cycle. Plant Biol 8:389–396

Hassine AB, Lutts S (2010) Differential responses of saltbush *Atriplex halimus* L. exposed to salinity and water stress in relation to senescing hormones abscisic acid and ethylene. J Plant Physiol 167:1440–1456

Hunt LA, Reynolds MP, Sayre KD, Rajaram S, White JW, Yan W (2003) Crop modelling and the identification of stable coefficients that may reflect significant groups of genes. Agron J 95:20–31

IPCC, Solomon S, Qin D, Manning M, Chen Z, Marquis M, Averyt KB, Tignor M, Miller HL (2007) IPCC summary for policymakers. In: Climate change 2007: the physical science basis. Contribution of Working Group I to the Fourth Assessment Report of the Intergovernmental Panel on Climate Change. Cambridge University Press, UK/New York

Jacobsen S-E, Jensen CR, Liu F (2012) Improving crop production in the arid Mediterranean climate. Field Crop Res 128:34–47

Javid MG, Sorooshzadeh A, Moradi F, Sanavy SAMM, Allahdadi I (2011) The role of phytohormones in alleviating salt stress in crop plants. Austr J Crop Sci 5:726–734

Jia W, Davies WJ (2007) Modification of leaf apoplastic pH in relation to stomatal sensitivity to root-sourced abscisic acid signals. Plant Physiol 143:68–77

Kafkafi U (1996) Root growth under stress-salinity. In: Waisel Y, Eshel A, Kafkafi U (eds) Plant roots: the hidden half, ed, 2nd edn. Marcel Dekker, New York, pp 375–391

Kobashi K, Sugaya S, Gemma H, Iwahori S (2001) Effect of abscisic acid (ABA) on sugar accumulation in the fresh tissue of peach fruit at the start of the maturation stage. Plant Growth Regul 35:215–223

Koushafar M, Khoshgoftarmanesh AH, Moezzi A, Mobli M (2011) Effect of dynamic unequal distribution of salts in the root environment on performance and crop per drop (CPD) of hydroponic-grown tomato. Sci Hortic 131:1–5

Koyro HW, Ahmad P, Nicole G (2012) Abiotic stress responses in plants: an overview. In: Ahmad P, Prasad MNV (eds) Environmental adaptations and stress tolerance of plants in the era of climate change. Springer Science+Business Media, New York, pp 1–28

Kurth E, Cramer GR, Lauchli A, Epstein E (1986) Effects of NaCl and $CaCl_2$ on cell enlargement and cell production in cotton roots. Plant Physiol 82:1102–1106

Lawlor DW, Cornic G (2002) Photosynthetic carbon assimilation and associated metabolism in relation to water deficits in higher plants. Plant Cell Environ 25:275–294

Lloyd J, Kriedemann PE, Aspinall D (1989) Comparative sensitivity of Prior Lisbon lemon and Valencia orange trees to foliar sodium and chloride concentrations. Plant Cell Environ 12:529–540

Lovelli S, Perniola M, Di Tommaso T, Ventrella D, Moriondo M, Amato M (2010) Effects of rising atmospheric CO_2 on crop evapotranspiration in a Mediterranean area. Agric Water Manage 97:1287–1292

Lovelli S, Scopa A, Perniola M, Di Tommaso T, Sofo A (2012) Abscisic acid root and leaf concentration in relation to biomass partitioning in salinized tomato plants. J Plant Physiol 169:226–233

Maggio A, Raimondi G, Martino A, De Pascale S (2007) Salt stress response in tomato beyond the salinity tolerance threshold. Env Exp Bot 59:276–282

Maggio A, De Pascale S, Fagnano M, Barbieri G (2011) Saline agriculture in Mediterranean environments. Ital J Agron 6:36–43

Martin B, Ruiz-Torres N (1992) Effects of water-deficit stress on photosynthesis, its components and component limitations and on water use efficiency in wheat (*Triticum aestivum* L.). Plant Physiol 100:733–739

Martre P, Porter JR, Jamieson PD, Triboi E (2003) Modeling grain nitrogen accumulation and protein composition to understand the sink/source regulation of nitrogen remobilization for wheat. Plant Physiol 133:1959–1967

Miflin B (2000) Crop improvement in the 21st century. J Exp Bot 51:1–8

Mitchell JP, Shennan C, Grattan SR, May DM (1991) Tomato fruit yields and quality under water deficit and salinity. J Am Soc Hort Sci 116:215–221

Monteleone M (2006) Salinity management in southern Italy irrigation areas. Ital J Agron 1:129–202

Mulholland BJ, Taylor IB, Jackson IC (2003) Can ABA mediate responses of salinity of stressed tomato. Env Exp Bot 50:17–28

Munns R (1993) Physiological processes limiting plant growth in saline soils: some dogmas and hypotheses. Plant Cell Environ 16:15–24

Munns R (2002) Comparative physiology of salt and water stress. Plant Cell Environ 25:239–250

Munns R, Tester M (2008) Mechanisms of salinity tolerance. Annu Rev Plant Biol 59:651–681

Munns R, James RA, Lauchli A (2006) Approaches to increasing the salt tolerance of wheat and other cereals. J Exp Bot 57:1025–1043

Okhovatian-Ardakani AR, Mehrabanian M, Dehghani F, Akbarzadeh A (2010) Salt tolerance evaluation and relative comparison in cuttings of different pomegranate cultivars. Plant Soil Environ 56:176–185

Olesen JE, Bindi M (2002) Consequences of climate change for European agricultural productivity, land use and policy. Eur J Agron 16:239–262

Ostonen I, Püttsepp U, Biel C, Alberton O, Bakker MR, Lõhmus K, Majdi H, Metcalfe D, Olsthoorn AFM, Pronk A, Vanguelova E, Weih M, Brunner I (2007) Specific root length as an indicator of environmental change. Plant Biosyst 141(3):426–442

Ouyang B, Yang T, Li H, Zhang L, Zhang Y, Zhang J, Fei Z, Ye Z (2007) Identification of early salt stress response genes in tomato root by suppression subtractive hybridization and microarray analysis. J Exp Bot 58:507–520

Paranychianakis NV, Chartzoulakis KS (2005) Irrigation of Mediterranean crops with saline water: from physiology to management practices. Agric Ecosyst Environ 106:171–187

Peleg Z, Blumwold E (2011) Hormone balance and abiotic stress tolerance in crop plants. Curr Opin Plant Biol 14:290–295

Perez-Alfocea F, Albacete A, Ghanem ME, Dodd IC (2010) Hormonal regulation of source-sink relations to maintain crop productivity under salinity: a case study of root-to-root signalling in tomato. Funct Plant Biol 37:592–603

Prior LD, Grieve AM, Slavish PG, Gullis PR (1992) Sodium chloride and soil texture interactions in irrigated field grown Sultana grapevines. II. Plant mineral content, growth and physiology. Aust J Agric Res 43:1067–1084

Rivelli AR, Lovelli S, Perniola M (2002) Effects of salinity on gas exchange, water relations and growth of sunflower (*Helianthus annuus* L.). Funct Plant Biol 29:1405–1415

Ross J, O'Neill D (2001) New interactions between classical plant hormones. Trends Plant Sci 6:2–4

Rozema J, Flowers T (2008) Crops for a salinized world. Science 322:1478–1480

Sachs T (2005) Auxin's role as an example of the mechanisms of shoot/root relations. Plant Soil 268:13–19

Santner A, Estelle M (2010) The ubiquitin-proteasome system regulates plant hormone signalling. Plant J 61:1029–1040

Schwarz D, Grosch R (2003) Influence of nutrient solution concentration and root pathogen (*Pythium aphanidermatum*) on tomato root growth and morphology. Sci Hortic 97:109–120

Schwarz D, Heinen M, Van Noordiwijk M (1995) Rooting characteristics of lettuce grown in irrigated sand beds. Plant Soil 176:205–217

Seemann JR, Critchely C (1985) Effects of salt stress on the growth, ion content, stomatal behaviour and photosynthetic capacity of a salt-sensitive species, *Phaseolus vulgaris* L. Planta 164:151–162

Sharp RE, LeNoble ME (2002) ABA ethylene and the control of shoot and root growth under water stress. J Exp Bot 53:33–37

Sharp RE, Hsiao TC, Silk WK (1990) Growth of the maize primary root at low water potentials. II. Role of growth and déposition of hexose and potassium in osmotic adjustment. Plant Physiol 93:1337–1346

Snapp SS, Shennan C (1992) Effects of salinity on root growth and death dynamics of tomato, *Lycopersicon esculentum* Mill. New Phytol 121:71–79

Sofo A, Scopa A, Manfra M, De Nisco M, Tenore G, Troisi J, Fiori RD, Novellino E (2011) *Trichoderma harzianum* strain T-22 induces changes in phytohormone levels in cherry rootstocks (*Prunus cerasus* x *P.canescens*). Plant Growth Regul 65:421–425

Spollen WG, LeNoble ME, Samuels TD, Bernstein N, Sharp RE (2000) Abscisic acid accumulation maintains maize primary root elongation at low water potentials by restricting ethylene production. Plant Physiol 122:967–976

Srivastava LM (2002) Plant, growth and development – hormones and environment. Elsevier Academic Press, San Diego, pp 307–314

Tavakkoli E, Fatehi F, Coventry S, Rengasamy P, McDonald GK (2011) Additive effects of Na^+ and Cl^- ions on barley growth under salinity stress. J Exp Bot 62:2189–2203

Van der Werf A, Nagel OW (1996) Carbon allocation to shoots and roots in relation to nitrogen supply is mediated by cytokinins and sucrose: opinion. Plant Soil 185:21–32

Vaughan LV, MacAdam JW, Smith SE, Dudley LM (2002) Root growth and yield of differing alfalfa rooting populations under increasing salinity and zero leaching. Crop Sci 42:2064–2071

Vitale D, Rana G, Soldo P (2010) Trends and extremes analysis of daily weather data from a site in the Capitanata plain (Southern Italy). Ital J Agron 5:133–143

Walker RP, Torokfalvy E, Scott NS, Kriedemann PE (1981) An analysis of photosynthetic response to salt treatment in Vitis vinifera. Aust J Plant Physiol 8:359–374

Wilkinson S, Davies WJ (2002) ABA-based chemical signalling: the co-ordination of responses to stress in plants. Plant Cell Environ 25:195–210

Wollenweber B, Porter JB, Lubberstedt T (2005) Need for multidisciplinary research towards a second green revolution. Curr Opin Plant Biol 8:337–341

Xiong L, Schumaker KS, Zhu JK (2002) Cell signaling during cold, drought and salt stress. Plant Cell 14(Suppl 1):165–183

Yadav S, Irfan M, Ahmad A, Hayat S (2011) Causes of salinity and plant manifestations to salt stress: a review. J Environ Biol 32:667–685

Yeo AR (2007) Salinity. In: Yeo AR, Flowers TJ (eds) Plant solute transport. Blackwell, Oxford, pp 340–365

Yuan K, Rashotte AM, Wysocka-Diller JW (2011) ABA and GA signaling pathways interact and regulate seed germination and seedling development under salt stress. Acta Physiol Plant 33:261–271

Zhang H-X, Blumwald E (2001) Transgenic salt tolerant tomato plants accumulate salt in the foliage but not in the fruits. Nat Biotech 19:765–768

Zhang J, Davies WJ (1990) Does ABA in the xylem control the rate of leaf growth in soil-dried maize and sunflower plants. J Exp Bot 41:1125–1132

Zhang J, Jia W, Yang J (2004) ABA in plant water stress signalling. Proc Indian Natl Sci Acad B 70:367–377

Zhang J, Jia W, Yang J, Ismail AM (2006) Role of ABA in integrating plant responses to drought and salt stresses. Field Crop Res 97:111–119

Zhu M, Dai S, Chen S (2012) The stomata frontline of plant interaction with the environment-perspectives from hormone regulation. Front Biol 7:96–112

Zobel RW (1975) The genetics of the root development. In: Torrey JG, Clarkson DF (eds) The development and function of roots. Academic, London, pp 261–275

Chapter 11
Phenolic Content Changes in Plants Under Salt Stress

Agnieszka Waśkiewicz, Małgorzata Muzolf-Panek, and Piotr Goliński

11.1 Introduction

11.1.1 Phenolic Compounds: Structures, Biosynthesis and Their Functions in Plants

Phenolics are secondary metabolites of plants, widely spread in nature and studied for a long time (Von Szent-Györgyi 1928). The group contains over 9,000 various compounds differing in structure and molecular weight, and the resulting physico-chemical and biological properties (Crozier et al. 2006). They are synthesized in plants via the shikimic acid metabolic pathway (and manolate in case of flavonoids and stilbenes), which is an endogenously controlled process during developmental differentiation (Crozier et al. 2006), or which can be regulated by exogenous factors such as light, temperature and wounding. In plants the compounds may play a role of signalling molecules, protecting against UV light and pathogens, attracting pollinators, stimulating disease resistance and/or protecting against reactive oxygen species generated when aerobic or photosynthetic metabolism is impaired by various environmental stresses such as salt stress.

Phenolics show a great diversity of their carbon skeleton, from a single hydroxylated aromatic ring to highly complex polymeric substances, e.g. tannins, which results in their division into four major classes: phenolic acids, flavonoids, stilbenes and lignans (Rice-Evans et al. 1996; Jaganath and Crozier 2010).

A. Waśkiewicz • P. Goliński (✉)
Department of Chemistry, Poznań University of Life Sciences,
Wojska Polskiego 75, 60-625 Poznań, Poland
e-mail: piotrg@up.poznan.pl

M. Muzolf-Panek
Department of Food Quality Management, Poznań University of Life Sciences,
Wojska Polskiego 31, 60-624 Poznań, Poland

P. Ahmad et al. (eds.), *Ecophysiology and Responses of Plants under Salt Stress*,
DOI 10.1007/978-1-4614-4747-4_11, © Springer Science+Business Media, LLC 2013

Hydroxybenzoic acid	**R1**	**R2**
p-hydroxybenzoic acid	-H	-H
protocatechuic acid	-H	-OH
gallic acid	-OH	-OH
vanillic acid	$-OCH_3$	-H

Hydroxycinnamic acids	**R1**	**R2**
p-coumaric acid	-H	-H
caffeic acid	-H	-OH
ferulic acid	$-OCH_3$	-H
synapic acid	-OCH3	$-OCH_3$

Fig. 11.1 Phenolic acid structures

Phenolic acids, including hydroxybenzoic acid (C_6–C_1) and hydroxycinnamic acid (C_6–C_3) derivatives (Fig. 11.1), may be present in soluble forms conjugated with sugars or organic acids, as well as bounded to more complex structures such as hydrolysable tannins or lignins (Crozier et al. 2006; Jaganath and Crozier 2010). They are especially widespread in fruit, e.g. berries and apples, vegetables such as onion or raddish, and cereals (Zadernowski et al. 2005; Crozier et al. 2006; Mattila and Kumpulainen 2007).

Flavonoids are a major group of the polyphenolic compounds with a diphenylpropane skeleton (C_6–C_3–C_6), i.e. a three-ring structure: the benzene moiety (A), the γ-pyran ring (C) and the phenyl moiety (B) (Fig. 11.2). According to the oxidation stages of their heterocyclic ring (C), they can be divided into several classes, including flavones, isolavones, flavonols, flavanones, flavan-3-ols, flavan-3,4-diols and anthocyanins (called also anthocyanidins). Structures of various classes of flavonoids are also presented on Fig. 11.2. Other flavonoids are minor compounds in plants, including chalcones, dihydrochalcones, aurones, coumarins and dihydroflavonols (Crozier et al. 2006). The high diversity of flavonoids is a direct effect of the multiple hydroxylation, methoxylation and glycosylation patterns (Bors et al. 1998). Acylation may also occur in the basic skeleton as well as in the glycosyl moiety. The hydroxyl groups play a crucial role in the functional activities of flavonoids. The higher the number of hydroxyl groups in the molecule, the higher the antioxidant activity of the compound is (Rice-Evans et al. 1996; Muzolf et al. 2008). Methoxylation generally reduces the activity of a flavonoid when compared to the native compound (Lemańska et al. 2001; Borkowski et al. 2005). The most commonly present sugar residue in flavonoid glycosides is *D*-glucose attached to carbon at C3, and less frequently at C7, C5, C3′, C4′ or C5′. Flavan-3-ols, also called catechins, do not form glycosides; however, they may be present in the gallate forms, i.e. as esters of gallic acid (C3 position in a C ring), which influences their activity – galloylated catechins exhibit a higher radical scavenging activity than non-galloylated ones (Muzolf et al. 2008). Moreover, catechins may undergo oxidation to chinones in the electron transfer process (Sang et al. 2007; Muzolf-Panek et al. 2008) and further condensation to theaflavins which may polymerize to high-molecular weight compounds – thearubugins. Catechins also form oligomeric and polymeric

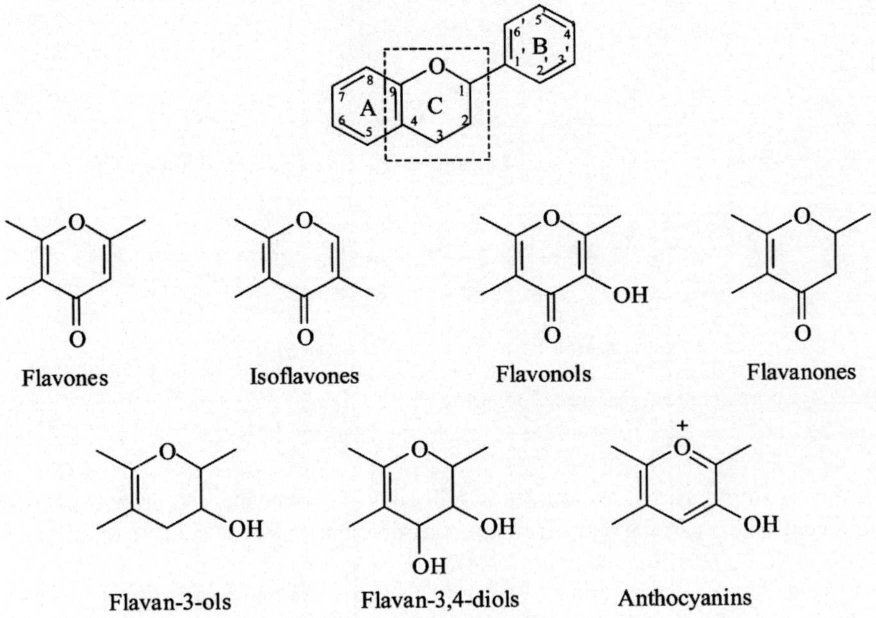

Fig. 11.2 Basic structures of various classess of flavonoids

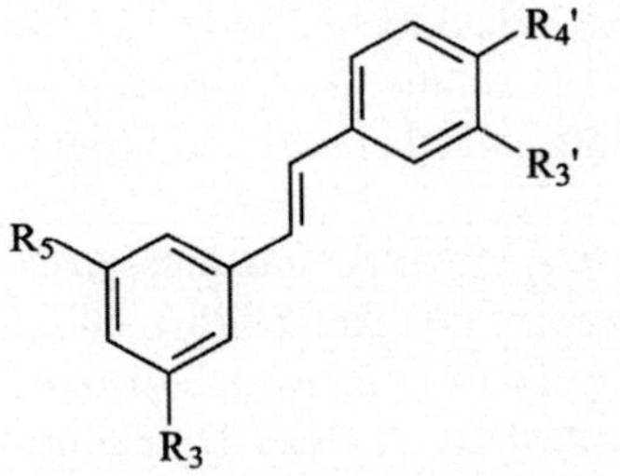

Stilbenes	**R_3**	**R_5**	**$R_{3'}$**	**$R_{4'}$**
trans-resveratrol	-OH	-OH	-H	-OH
trans-piceid	-OGlu	-OH	-H	-OH
pinosylvin	-OH	-OH	-H	-H
piceatannol	-OH	-OH	-OH	-OH
trans-pterostilbene	$-OCH_3$	$-OCH_3$	-H	-OH

Fig. 11.3 Structures of common plant stilbenes

proanthocyanidins, better known as condensed tannins. Flavonoids are widespread in the plant kingdom, as they are found e.g. in large quantities in fruits and vegetables, cocoa, tea, nuts as well as cereal grains, herbs, spices, algae and plants with pharmacological properties, such as ginkgo biloba or Echinacea (Arts et al. 2000a, b; Lee et al. 2010).

Stilbenes – in contrast – are a small group of phenylpropanoids characterized by a 1,2-diphenylethylene skeleton (C_6–C_2–C_6) (Chong et al. 2009). There is a great diversity among stilbenes in terms of structures found in particular plant families; nevertheless, most plant stilbenes are derivatives of the basic unit *trans*-resveratrol (3,5,4-trihydroxy-transstilbene). The basic structure of stilbenes as well as the most common plant stilbenes are shown on Fig. 11.3. Stilbenes are phytoalexins produced by plants in the metabolic pathway as the response to attack of such pathogens as

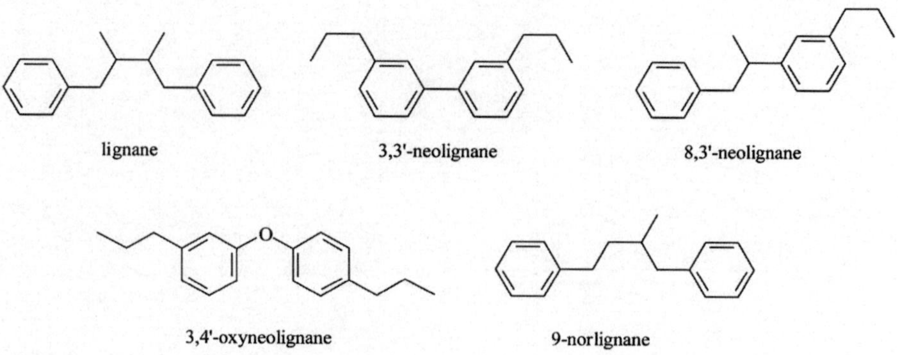

Fig. 11.4 Structures of various groups of lignans

bacteria or fungi. The main sources of stilbenes are soybeans, red grapes, peanuts and recently also various berry fruit are considered as potential sources of stilbenes (Burns et al. 2002; Rimando et al. 2004).

Lignans are a group of plant phenols. Their structure is based on two cinnamic acid units (Fig. 11.4). Oxysubstitution may occur in the aromatic rings, as well as in the side chain. *o*-Methyl and glycosyl residues may also be inserted in lignan structures (Moss 2000). Lignans are one of the groups of phytoestrogens (i.e. estrogen-like phytochemicals). Similarly to other phenolic compounds, they act as antioxidants. The compounds are ubiquitous in flax seed, sesame seed, cereals such as rye, wheat and barley, soybeans, some fruit (apricots, strawberries) and vegetables (broccoli, cabbage) (Smeds et al. 2007).

The biosynthesis of phenolic compounds is a complex network of chemical reactions and runs generally through the shikimate pathway (Winkel-Shirley 2001; Crozier et al. 2006). The shikimic acid pathway is derived from the metabolism of carbohydrates and glycolysis and starts with the biosynthesis of 3-deoxy-D-arabino-heptulosonic acid 7-phosphate (DAHP) by the corresponding synthase (DAHPS, EC 4.1.2.15). The following reactions, including shikimic acid formation, lead to the biosynthesis of the aromatic amino acid L-phenylalanine. This was the first step in the biosynthesis of phenolic compounds. The second stage is initiated with the deamination of L-phenylalanine by phenylalanine ammonia lyase (PAL, EC 4.3.1.5) leading to the formation of cinnamic acid and the following phenylpropanoid pathway (Davis and Shwinn 2006; MacDonald and D'Cunha 2007). The pathway continues with the hydroxylation of the latter by a P450 enzyme (Ehlting et al. 2006), cinnamate-4-hydroxylase (C4H, 1.14.13.11), to form *p*-coumaric acid, which in turn is metabolized to *p*-coumaroyl-CoA by *p*-coumarate: CoA ligase (4CL, EC 6.2.1.12). *p*-Coumaric acid is also a precursor of salicylic acid and many phenolic acids, such as caffeic, ferulic and synapic acids, and further reactions lead to the formation of lignins. *p*-Coumaroyl-CoA is a pivotal compound which gives rise to many flavonoids and stilbenes. The three-ring flavonoid structure formation is followed by the condensation of *p*-coumaroyl-CoA with three malonyl-CoA units catalysed by chalcone synthase (CHS, EC 2.3.1.74). *p*-Coumaroyl-CoA is a basic compound for the phenyl B-ring and the bridge synthesis. The benzene ring A (Fig. 11.1) is derived from carbons of three malonyl-CoA residues. Altogether, the reactions lead to the biosynthesis of naringenin-chalcone followed by the stereospecific conversion to naringenin by chalcone isomerase (CHI, EC 5.5.1.6). Further steps, starting from naringenin formation, are specific for various flavonoid classes (Punyasiri et al. 2004; Crozier et al. 2006; Togami et al. 2006). Stilbenes such as trans-resveratrol also originate from the condensation of *p*-coumaroyl-CoA with three malonyl-CoA units (Chong et al. 2009) and a key enzyme in this reaction is stilbene synthase (SS, EC 2.3.1.95). The biosynthesis of lignans proceeds through the formation of caffeic acid derivatives, from *p*-coumaroyl-CoA, starting from caffeoyl quinic acid to coniferyl and synapic alcohols (lignans subunits) (Suzuki et al. 2002; Umezawa 2003; Schmidt et al. 2006).

Phenolics serve diverse functions crucial for plant growth and development (Gould and Lister 2006). Phenolics, mainly in the form of lignins, are components of the cell-wall structure, constituting a mechanical and microbiological barrier (Cvikrova et al. 2006). A good example here is the synthesis of cell-wall bound phenolics formed in response to pathogen attack (Soylu 2006).

One of the most obvious functions of phenolic compounds is their contribution to flower and fruit colours (Crozier et al. 2006). Anthocyanins, the most colourful among phenolic compounds, are red and blue pigments attracting insect pollinators and other fruit-eating animals, facilitating seed dispersal. The function of colorants, rather as co-pigments than the main pigments, is also ascribed to flavonols and flavones, responsible – depending on their structures – for white, yellow and ivory colours in plant tissues. It is also worth mentioning that yellow pigmentation is the result of the presence of chalcones.

Flavonoids are important compounds for plant reproduction, contributing not only to colour formation to attract pollinators, but they also affect pollen germination and tube growth (Gould and Lister 2006; reviewed by Amalesh et al. 2011). Polyphenolic compounds are present in most plant seeds, protecting them against microbial pathogens and insect pests and participating in seed maturation and dormancy (Routaboul et al. 2006; reviewed by Amalesh et al. 2011).

Flavonoid compounds facilitate the establishment of a symbiotic relationship between soil bacteria, *Rhizobia*, and plants (reviewed by Lattanzio et al. 2006; Shaw et al. 2006). They play a role of chemical signals which are recognized by *Rhizobia*, which initiates nodule meristem formation and nitrogen fixation (reduction of atmospheric nitrogen to ammonia) (Begum et al. 2001; Fox et al. 2001; Sundaravarathan and Kannaiyan 2002; Cooper 2004; Eckardt 2006).

Phenolic compounds, such as hydroxybenzoic aids, some flavonols and (iso-) flavones, may influence positively spore germination and hyphal growth of arbuscular mycorrhizal fungi (Scervino et al. 2005; reviewed by Lattanzio et al. 2006; Kensuke et al. 2007; Scervino et al. 2007; Carlsen et al. 2008; Ponce et al. 2009).

Unlike the basic metabolism, including anabolic and catabolic processes, which is necessary for plant cell maintenance and proliferation, secondary metabolism, gives rise to a broad range of compounds, which play a crucial role in plant survival under various environmental conditions (Lattanzio et al. 2006; Drzewiecka et al. 2011).

When plants are exposed to pathogenic microorganisms they produce various compounds known as phytoalexins, i.e. low molecular weight substances synthesized by a plant in response to pathogen attack. This requires de novo expression of enzymes involved in the biosynthetic pathway. Studies of Fofana et al. (2005) on the role of flavonoid phytoalexin production in the induced resistance of cucumber against powdery mildew (caused by *Podosphaera xanthii*) showed that the resistance could be suppressed through down-regulation of CHS and disruption of the flavonoid pathway, which strongly suggests that flavonoid phytoalexins play a key role in determining the outcome of plant–powdery mildew interactions. The compound functions are mainly ascribed to the group of isoflavonoids (reviewed by Amalesh et al. (2011); Zabala et al. (2006)).

Some of the phenolic compounds may act as allelochemicals, in view of the fact that allelopathy is a biological phenomenon, in which the growth, survival and reproduction of one plant species is biochemically influenced by another and the compounds released to the environment by the plant are called allelochemicals. Up to now studies revealed that catechin, gallic acid, protocatechuic acid, chlorogenic acid, caffeic acid, hydroquinone, 5,7,4′-trihydroxy-3′,5′-dimethoxyflavone, stilbenes (such as prenyl-trans-stilbene) and other phenolic compounds may inhibit the growth of neighboring plants (Weir et al. 2003; Kong et al. 2004; Weidenhamer and Romeo 2004; reviewed by Bais et al. 2006; Li et al. 2010; Lobo et al. 2010; Scognamiglio et al. 2012).

The toxic effect of reactive oxygen species (ROS) on plants and other living organisms is quite well-known (Waris and Ahsan 2006). In plants one of the main sources of ROS formation is nothing other than the photosynthetic electron transport chain. Also various environmental stresses, such as exposure to toxic metals, temperature, UV radiation, water deficit and/or salinity, may induce the cascade of ROS production (Dat et al. 2000; Desikan et al. 2005; Koyro et al. 2012). Thus, plants have evolved various defence mechanisms, including phenolic compound biosynthesis, against these toxic species. Flavonoids and other phenolic compounds may scavenge free radicals and other active species through the hydrogen atom and/or electron donation. However, the efficiency of the antioxidant activity is highly structure-dependent (Rice-Evans et al. 1996) as well as pH-dependent (Lemańska et al. 2001; Borkowski et al. 2005; Muzolf et al. 2008), as it was mentioned earlier in this chapter.

Plants are constantly exposed to UV-B radiation, which may negatively affect their photosynthesis, transpiration, pollination and causes cellular damage, including DNA alterations. To protect themselves from these harmful effects, plants synthesize flavonoids (such as anthocyanins, flavonols – mainly in the form of glycosides and methylated glycosides) and phenolic acids (such as caffeic acid, *p*-coumaric acid and ferulic acid). These phenolic compounds absorb UV-B light and are mainly accumulated in the epidermal cells to protect photosynthetic tissues (Bourchard et al. 2000; Gould and Lister 2006; Luthria et al. 2006; Pinter et al. 2007; Tsormpatsidis et al. 2008). An increased flavonoid content was also observed in hairs and epicuticular wax (Gould and Lister 2006). A similar situation is observed in toxigenic fungi – present in cereals and cereal-derived food products and/or feed components – able to form a group of phenolic compounds which protect them against this denaturating light (Goliński et al. 1995), hindering prevention of mycotoxin formation by toxigenic microorganisms.

In the aftermath of industrial development the soil is getting increasingly polluted with heavy metals (Michalak 2006). These toxic elements cause the inhibition of plant growth, reduce biomass and finally lead to plant death. The negative effect of heavy metals on plants is also related with reactive oxygen species generation through the oxidative stress induction. Phenolics participate in the plant resistance to the harmful effects of heavy metals in soil by various mechanisms, including metal chelation (Gould and Lister 2006; Michalak 2006; Keilig and Ludwig-Müller 2009). Pawlak-Sprada et al. (2011a) revealed that the treatment of soybean and lupine with high levels of cadmium and lead causes an increase in PAL activity – the principal enzyme of the phenylpropanoid pathway. Further studies of the same authors showed that plant exposure to cadmium and lead increases the level of isoflavonoids, especially 2′-hydroxygenistein and its malonylated glucoside (Pawlak-Sprada et al. 2011b). Chelated heavy metals, similarly to other toxic stressors (such as mycotoxins), are bound e.g. in the conducting tissue (xylem), so they are no longer mobile and are not transported with the nutrient solution to ribosomes, protecting in this way against macromolecule denaturation, genetic information degradation, cell apoptosis and in the consequence – plant death (Waśkiewicz et al. 2010; Drzewiecka et al. 2011; Karolewski et al. 2011).

The most common plant stressors are temperature (low or high) and a lack of water. Chilling as well as heat stresses promote the formation of phenolics in plants (Kirakosyan et al. 2004; Crifò et al. 2011), through the stimulation of their synthesis (increased PAL activity) and inhibition of their oxidation (decreased peroxidase and polyphenol oxidase activities) (Rivero et al. 2001). Water deficit is also a factor inducing phenolic production (Kirakosyan et al. 2004). It induces the shikimate pathway and decreases the production of degradation-related enzymes, including polyphenol oxidase in water stress-tolerant plants (Sánchez-Rodríguez et al. 2011).

Salt stress is one of the most important factors limiting plant growth and yield. Phenolic compounds may influence plant resistance to salt stress, which will be discussed in detail later in this chapter.

11.2 Oxidative Stress: Definition and Factors Involved in Its Induction

Most environmental stresses affect the production of active oxygen species in plants, causing oxidative stress, which is defined as a shift of the balance between prooxidative and antioxidative reactions in favour of the former, which seems to be a common denominator of the action of various agents on living organisms, creating the potential for organic damage. Pro-oxidants are by definition free radicals, atoms or clusters of atoms with a single unpaired electron. Physiological concentrations of pro-oxidants are determined both by internal and external factors. Antioxidants are chemical compounds (enzymatic and non-enzymatic subtypes) that can bind to free radicals and thus prevent from damaging healthy cells. The reduction of oxygen to form superoxide, hydrogen peroxide and hydroxyl radicals is the principle mechanism of oxygen activation in most biological systems. However, in photosynthetic plants, the formation of singlet oxygen by the photosystems is of prime importance. Activated oxygen is often formed as a component of metabolism to enable "complex" chemical reactions, such as oxidation of xenobiotics or polymerisation of lignin, but in other instances activated oxygen is formed by the dysfunctioning of enzymes or electron transport systems, as a result of perturbations in metabolism caused by chemical or environmental stress.

11.3 Salt Stress in Plant

Abiotic stresses are major constraints to agricultural production worldwide. Plants have an inbuilt mechanism to respond to fluctuations in circadian and seasonal environmental conditions (Khan et al. 2011). Among abiotic environmental stresses, high salinity stress is the most severe, which impairs crop production over at least 20 % of irrigated land worldwide (Zhao et al. 2007; Kumar et al. 2010; Tavakkoli et al. 2011). Salt stress in soil or water is one of the major stresses especially in arid and semi-arid regions, where salinity strongly limits crop development (Allakhverdiev et al. 2000; Koca et al. 2007; Munns and Tester 2008).

Among various sources of soil salinity, irrigation combined with poor drainage is the most serious one (Zhu 2007). Soil type and environmental factors, such as vapour, radiation and temperature, may further alter salt tolerance (Chinnusamy et al. 2005). Actually in fields salt levels fluctuate seasonally and spatially, and this variation will occur due to the circumstances influencing each particular plant (Estes 2002). In addition, the continuous use of the same soil for growing vegetables results in an increase of salinization.

Damage caused by high salinity is often associated with three different mechanisms (Levinsh 2006), the first being ion toxicity caused by an excessive accumulation of Na^+ and Cl^- in the cytoplasm, leading to an ionic imbalance which can be counteracted by an increased transport intensity of the ions to the vacuole. In the second mechanism, even if massive ion compartmentation occurred in the vacuole, the cytosol water potential must be lowered to balance a low external water potential, thus allowing water intake in the plant cell and preventing macromolecule damage. In the third mechanism a high cellular NaCl concentration causes an increased formation of ROS (Hernandez and Almansa 2002), which is considered the primary event under a variety of stress conditions (Noctor and Foyer 1998).

11.4 Tolerance to Salinity Stress

Tolerance to salinity stress has often been associated with oxidative stress, since one of the consequences of exposure to salinity is the production of ROS, such as superoxide radicals ($^{\bullet}O_2^-$), hydrogen peroxide (H_2O_2) and hydroxyl radicals ($^{\bullet}OH$) (Asada 2006; Ashraf and Foolad 2007; Ashraf 2009). The generation of ROS in plants under oxidative stress at different locations in the plant cell (mitochondria, chloroplast, peroxisome and nucleus) causes injury and cell death (Mano 2002). On the other hand, ROS play a vital role in intracellular redox signaling, activating antioxidant resistance mechanisms. Thus, it is a surviving response for plants to control the concentration of ROS (Khan et al. 2011). These species cause oxidative damage to different cellular components including membrane lipids, proteins and nucleic acids (Valko et al. 2006). Plants possess efficient systems for scavenging ROS that protect them from destructive oxidative reactions (Eryilmaz 2006; Sharma et al. 2010).

The major ROS-scavenging mechanisms in plants are connected with superoxide dismutase (SOD), catalase (CAT), peroxidase (POD), glutathione reductase (GR), and the non-enzymic constituents such as ascorbic acid, glutathione, tocopherols, carotenoids, and phenolic compounds, which help in the deactivation of active oxygen species in multiple redox reactions, thereby contributing to the protective system against oxidative stress (Lee et al. 2001). The ROS scavengers can increase the plant resistance to salinity stress (Foyer et al. 1997; Navari-Lazzo and Quartacci 2001; Vinocur and Altman 2005).

Deleterious effects of salinity on plant growth are associated with: (1) lowering of the water potential, (2) direct toxicity of any Na and Cl absorbed, and (3) interference with the uptake of essential nutrients, or (4) a combination of these factors (Ahmad and Sharma 2008). Plants are either dormant during the salt episode or they must adjust to the new environmental conditions. Their responses depend on the species, genotype, as well as length and severity of the salinity, the age and stage of development, the organ and the cell type as well as the sub-cellular compartment (Yokoi et al. 2002).

Salt tolerance depends also on such factors as morphology, regulation of transpiration, control of ion movement, membrane characteristics, tolerating high Na/K ratios in the cytoplasm and salt glands (Flowers and Flowers 2005).

11.5 Halophytes and Glycophytes

The adaptations required to survive in salt-affected soils are the same in all plants and depend on the plant genotype, determining morphological, biochemical and physiological mechanisms providing for plant growth under previously unfavorable conditions (Dajic 2006). Plants are generally divided into two groups with respect to salt tolerance, i.e. halophytes and glycophytes, of which the former can tolerate high levels of salinity and complete their life cycle in soils with salt concentrations above 200 mM NaCl (Flowers and Colmer 2008) that may be potentially useful for economic (oilseed, forage, production of metabolites) applications (Single et al. 1996). Some species of halophytes even require salt concentrations of this order for

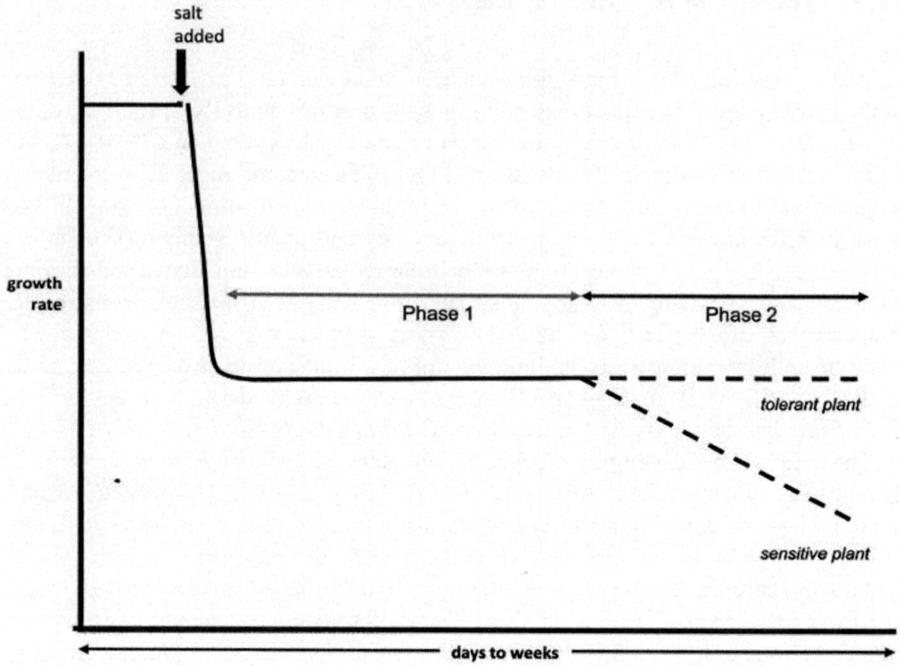

Fig. 11.5 Scheme of the two-phase growth response to salinity (Adapted from Munns et al. 1995)

optimal growth, and grow poorly without it. In contrast, growth of glycophytes is markedly reduced even at low-salt concentrations and they cannot complete their life cycle in saline soils (Ellouzi et al. 2011).

Halophytes must balance their requirement for salts needed for osmotic adjustment with their growth rate. Regulating transpiration plays an important part in this process, as it is the transpiration stream that carries ions between roots and shoots. Consequently, factors that influence the rate of water loss by plants are important in salt tolerance. Many halophytes show morphological adaptations associated with limiting transpiration, such as reduced leaves (which may be 'fleshy' or 'succulent') (Huchzermeyer et al. 2004; Flowers and Flowers 2005).

Halophytes, although widespread taxonomically, are relatively rare amongst the 250,000 species of flowering plants and virtually all of our crops are glycophytes (Sairam and Tyagi 2004). There is, however, considerable variability in the tolerance of these glycophytes to salt, as variation occurs both between and within species. Glycophytes, if they are to survive salinity, must adapt to external salt concentrations; they face the same problems as those faced by halophytes. Measurements of ion contents in plants under salt stress revealed that halophytes accumulate salts, whereas glycophytes tend to exclude the salts (Zhu 2007).

There are two phases in plant response to salt (Fig. 11.5). Plants sensitive or tolerant to salinity differ in the rate at which salt reaches toxic levels in leaves.

The timescale is days, weeks or months, depending on the species and the salinity level. The first phase is a response to the changes in water potential by lowering the external water potential by salt. These initial effects of salinity (phase 1) are likely to be the same for cultivars of different salt tolerance. Only when ions are accumulated over time (phase 2), do actual differences in salt tolerance appear. Sensitive cultivars (glycophytes) accumulate ions more quickly than tolerant cultivars (halophytes) and this ion accumulation leads to leaf death and, progressively, death of the plant (Munns 2002).

11.6 Changes of Phenolic Compound Content in Various Plants Under Salt Stress

Salt stress, as it was mentioned above, causes a limitation of photosynthesis, which results in excessive ROS production. To adapt to these harmful environmental conditions plants induce the synthesis of various secondary metabolites, such as e.g. phenolic compounds.

11.6.1 Fruit and Vegetables

Although extensive research has been carried out on the effect of salinity on physiological functions and antioxidant enzyme responses, there is relatively little data available, to the best of the authors' knowledge, on the influence of salt stress on phenolic content in vegetables and fruit. Tables 11.1 and 11.2 present recent studies on phenolic content changes under salt stress in edible plants. Based on the data collected in the tables it may be stated that the reviewed studies on the direction of phenolic changes upon plant treatment with high doses of salt give inconclusive results. It is also difficult to directly compare the results of various studies, since various experimental conditions are applied by researchers (source of salt stress, salt doses and time of treatment) and the phenolic profile was investigated in different parts of plants (leaves, roots, fruit) (Keutgen and Paweltzik 2008; Mohamed and Aly 2008; Telesiński et al. 2008; Lopez-Berenguer et al. 2009; Noreen and Ashraf 2009; Chisari et al. 2010; Dkhil and Denden 2010; Frary et al. 2010; Mahmoudi et al. 2010; Neves et al. 2010; Tiwari et al. 2010; Demiral et al. 2011; Yiu et al. 2012; Petridis et al. 2012; Rezazadeh et al. 2012). Moreover, the initial level of total phenolic acids and/or flavonoids varies widely among species and cultivars (Keutgen and Paweltzik 2008; Frary et al. 2010; Mahmoudi et al. 2010). It is also worth noting that very few authors studied the effect of salinity on bioactive compounds at different ripening phases of plants (Navarro et al. 2006; Lopez-Berenguer et al. 2009).

Table 11.1 Total phenolic content in fruit and vegetables exposed to salt stress

Species	Total phenolics	Source and time of salt exposure	Response to increasing salt stress	Changes in phenolic content	References
Fruit					
Mulberry	[g/kg DW]	Saline water ($NaCl$, Na_2SO_4 and $CaCl_2$); 60 days			Agastian et al. 2000
Morus alba L.					
Genotypes:					
M-5			Increase at low salinity (up to 2 mS/cm) and decrease at higher salinity (4 and 12 mS/cm)	8.86–5.70	
BC2-59 (salt-tolerant)				8.30–11.12	
S30 (salt-tolerant)			Increase at low salinity (up to 4 mS/cm) and decrease at higher salinity (8 and 12 mS/cm)	8.73–13.20	
Strawberries	[mg GAE/100 g FW]	NaCl; 4 months			Keutgen and Paweltzik 2008
Fragaria ananassa					
Varieties:					
Elsanta (salt-sensitive)			Increase	236.65–280.77	
Korona (less sensitive)			Slight increase	276.95–281.55	
Vegetables					
Tomato	[mg GAE/kg FW]	NaCl; 21 days	Decrease in 60 % of the lines	min. 3.0-fold	Frary et al. 2010
Solanum pennellii			Increase in 38 % of the lines	max. 3.3-fold	
Introgression lines					
Parental lines:					
S. lycopersicum cv. M82 (salt-sensitive)			Decrease	No numbers available	
S. pennellii LA716 (salt-tolerant)			Increase	max. 2.4-fold	
Soybean	[mg FA/g DW]	NaCl; 2 days	Increase	50 %	Neves et al. 2010
Glycine max L. Merrill					
Romaine lettuce	[μg C/g FW]	Lack of data	Decrease et lower salinity and unaffected at higher salinity	No numbers available	Chisari et al. 2010
L. sativa cv. Duende					

Romaine lettuce *L. sativa* var. longifolia (salt-sensitive)	[mg/g FW]	NaCl or Na_2SO_4; 12 days	Increase after 100 mM NaCl treatment Decrease after 77 mM Na_2SO_4 treatment	0.74–0.95 0.74–0.35	Mahmoudi et al. 2010
Verte lettuce *L. Sativa* var. Verte de Cobham (salt-tolerant)			Increase after 100 mM NaCl treatment Decrease after 77 mM Na_2SO_4 treatment	0.61–1.04 0.61–0.58	
Romaine lettuce *Lactuca sativa* L. (salt-sensitive)	[mg GAE/g DW]	NaCl; 2 days	Decrease	12.0–9.5	Kim et al. 2008
Radish sprouts	[mg GAE/100 g FW]	NaCl; 3, 5 and 7 days	Decrease with salt stress enhancement up to 50 mM NaCl; increase at 100 mM of NaCl compared to control	No numbers available	Yuan et al. 2010
Pepper fruit *Capsicum annuum L.*	[g p-coumaric acid/kg DW]	NaCl	Increase in red pepper fruits and no effect at the green stage and turning stage	5.32–5.45	Navarro et al. 2006
Pepper (leaves) *Capsicum annuum* L.	[mg GAE/g DW]	NaCl; 3, 6 and 9 days	Decrease	No numbers available	Yiu et al. 2012
Pea *Pisum sativum* L.	[mg GAE/g FW]	NaCl; 20 days	Increase	No numbers available	Noreen and Ashraf 2009
Onion *Allium cepa* L.	[mg PG/g DW]	Seawater	Decrease	39.68–34.51	Mohamed and Aly 2008
Olive tree *Olea europea* cv Gemlik	[mg GAE/g FW]	NaCl; a month	Decrease at salinity op t 8 dS/m and increase at 12 dS/m	No numbers available	Demiral et al. 2011
Olive tree (leaves) *Olea europea*	[mg GAE/g DW]	NaCl; 5 months	Increase	Up to 129 %	Petridis et al. 2012
Olive tree (roots) *Olea europea*	[mg GAE/g DW]	NaCl; 5 months	Increase at middle level of salinity (75 mM), decrease at higher salinity (125 mM)	No numbers available	Petridis et al. 2012
Okra *Abelmoschus esculentus* L.	[mg GAE/g FW]	NaCl; 10 days	Increase	18.5–21.0	Dkhil and Denden 2010

(continued)

Table 11.1 (continued)

Species	Total phenolics	Source and time of salt exposure	Response to increasing salt stress	Changes in phenolic content	References
Cucumber	[mg catechol/100 g]	NaCl, Na_2CO_3 and K_2SO_4; 60 days	Increase	1.80–2.44	Tiwari et al. 2010
Bean *Phaseolus vulgaris* L.	[mg GAE/g FW]	NaCl; 14, 21 and 28 days	Decrease	0.148–0.058	Telesiński et al. 2008
Artichoke (leaves) *Cynara scolymus* L.	[mg GAE/g DW]	NaCl; 120 days	Increase at moderate salinity (6.5 ds/m), decrease at higher salinity (29 dS/m)	50.10–70.22–10.18	Rezazadeh et al. 2012
Almond (root and leaves)	[mg GAE/kg FW]	NaCl; 30 days		No numbers available	Zrig et al. 2011
Genotypes:					
Bitter almonds			Increase at low salinity, decrease at moderate salinity – roots		
Prunus amygdalus			Unchanged – leaves		
GF677			Decrease – roots		
			Increase – leaves		
Garnem GN15			Increase at low salinity, decrease at moderate salinity – roots		
			Decrease – leaves		

FA ferulic acid, *GAE* gallic acid equivalent, *QE* quercetin equivalent, *PG* pyrogallol, *C* catechin

Table 11.2 Flavonoid and phenolic acid contents in vegetables exposed to salt stress

Species	Phenolic compounds	Source and time of salt exposure	Response to increasing salt stress	Changes in phenolic content	References
Tomato *Solanum pennellii*	Flavonoids [mg EC/kg FW]	NaCl; 21 days	Increase in 74 % of the lines	max. 4.0-fold	Frary et al. 2010
Introgression lines			Decrease in 22 % of the lines	min 3.2-fold	
Parental lines:					
S. lycopersicum cv. M82 (salt-sensitive)			Slight increase	1.3-fold	
S. pennellii LA716 (salt-tolerant)			Increase	2.4-fold	
Romaine lettuce	Flavonoids	NaCl or Na_2SO_4; 12 days	Increase after 100 mM NaCl treatment	0.04–0.58	Mahmoudi et al. 2010
L. sativa var. longifolia (salt-sensitive)			Increase after 77 mM Na_2SO_4 treatment	0.04–0.19	
Verte lettuce			Increase after 100 mM NaCl treatment	0.06–0.25	
L. Sativa var. Verte de			Increase after 77 mM Na_2SO_4 treatment	0.06–0.16	
Romaine lettuce	Phenolic acids [mg/g FW]	NaCl or Na_2SO_4; 12 days	Decrease after 100 mM NaCl treatment	0.70–0.37	Mahmoudi et al. 2010
L. sativa var. longifolia (salt-sensitive)			Decrease after 77 mM Na_2SO_4 treatment	0.70–0.15	
Verte lettuce			Increase after 100 mM NaCl treatment	0.55–0.79	
L. Sativa var. Verte de			Decrease after 77 mM Na_2SO_4 treatment	0.55–0.42	
Onion *Allium cepa* L.	Flavonoids [mg QE/g DW]	Seawater	Increase at low salinity and no effect at higher salinity	30.08–34.08–29.96	Mohamed and Aly 2008
Broccoli *Brassica oleracea* L. var. *italica* acv. Marathon	Flavonoids [mg QE/100 g FW]	NaCl; 11 weeks	Decrease		Lopez-Berenguer et al. 2009
Bean *Phaseolus vulgaris* L.	Flavonoids [mg QE/g FW]	NaCl; 14, 21 and 28 days	Decrease	1.098–0.235	Telesiński et al. 2008
Artichoke (leaves) *Cynara scolymus* L.	Flavonoids [mg QE/g DW]	NaCl; 120 days	Increase at moderate salinity (6.5 dS/m), decrease at higher salinity (29 dS/m)	35.10–49.15–28.10	Rezazadeh et al. 2012

The antioxidant response of plants is strongly dependent on the dose and time of salt treatment, as well as the source of salt stress. Most studies focused on NaCl-induced salt stress in plants (Keutgen and Paweltzik 2008; Kim et al. 2008; Telesiński et al. 2008; Lopez-Berenguer et al. 2009; Noreen and Ashraf 2009; Frary et al. 2010; Petridis et al. 2012; Rezazadeh et al. 2012). Telesiński et al. (2008) revealed that the content of chlorides which increased in bean plant tissues with an increasing NaCl concentration in the soil (0, 10, 30 and 50 mM/kg) correlates negatively with the flavonoid as well as total phenol contents after 28 days of the experiment. However, despite a decreasing total phenolic content at day 14 of the experiment the level of flavonoids increased with enhanced salinity. Total phenolic content and the concentration of flavonoids were stable under salt stress enhancement at day 21 of the study (Telesiński et al. 2008). Simultaneously, the authors showed that the activities of antioxidant enzymes, such as catalase and peroxidase, rose with an increasing salt stress in bean plants. This increase was observed at day 14 for both enzymes, as well as days 21 and 28 for peroxidase only. The activity of catalase at days 21 and 28 fluctuated, as it initially declined and then rose at the dose of 50 mM of NaCl/kg soil. The results of a study by Kim et al. (2008) on the influence of increasing salinity on phenolic compound levels in Romaine lettuce during short- (2-day) and long-term (15-day) treatments showed that phenolic content decreased with short-term salt exposure to a high NaCl concentration (up to 1,000 mM of NaCl), whereas there were no significant differences among samples exposed to long-term irrigation at relatively low and moderate NaCl levels (up to 200 mM). In contrast to the results reported by Kim et al. (2008), Mahmoudi et al. (2010) indicated that after 12 days of 100 mM NaCl treatment of Romaine lettuce the total phenolic content increased about 1.3-times, which coincided with a 14.5-fold increase of flavonoids. However, when treating the plant with 77 mM Na_2SO_4 a 0.5-fold decrease in total phenolic content was observed together with a 4.5-fold increase in the flavonoid level (Mahmoudi et al. 2010). The differences in phenolic compound fractions of Romaine lettuce under various sources of salt stress may suggest that plants synthesized flavonoids and other phenolic compounds differentially to adapt to various salinities.

The effect of various levels of salt stress on the phenolic antioxidant system in plants was also investigated by Agastian et al. (2000), Yuan et al. (2010), Biteur et al. (2011), and Rezazadeh et al. (2012). While at a relatively low salt treatment total phenolic content decreased in all among the analysed mulberry genotypes (Agastian et al. 2000) and increased at higher salinity, the direction of these changes was opposite in radish sprouts (Yuan et al. 2010). It is worth noting that mulberry was exposed to a long-term salt stress, whereas radish to a short-term one. The study of Rezazadeh et al. (2012) on the effect of salinity on the phenolic content in artichoke gave similar results as those recorded by Yuan et al. (2010). At moderate salinity total phenolic as well as flavonoid contents increased significantly, whereas a further increase in salinity resulted in a reduction of phenolic compound levels in artichoke leaves. The observed changes coincided with the changes of chlorogenic acid content.

Many authors showed different responses of various salt-sensitive plants to stress (Keutgen and Paweltzik 2008; Frary et al. 2010; Mahmoudi et al. 2010).

Total phenolic content in a strawberries cv. Korona, which is less sensitive to higher NaCl concentrations in soil, increased up to 10 % and 16 % for 40 and 80 mM NaCl, respectively (Keutgen and Paweltzik 2008). For cv. Elsanta, which is salt-sensitive, this increase was even more significant (14 % and 23 %). It should be stressed here that the mean initial level of phenolics in cv. Korona was higher when compared to cv. Elsanta (Keutgen and Paweltzik 2008). The content of anthocyanins in strawberries under various salt stress conditions was also monitored by Keutgen and Paweltzik (2008). The anthocyanin concentration changes in the plants coincided with the total phenolic content changes upon NaCl treatment. However, the rate of anthocyanin increase was higher for both Korona and Elsanta than the total phenolic increase rate. Since the level of sodium dismutase (SOD) was also elevated in the samples treated with higher amounts of NaCl (Keutgen and Paweltzik 2008), it may be supposed that oxidative stress induced as a result of salinity enhancement stimulated the antioxidant defense mechanisms, such as the synthesis of phenolics in plants and a higher antioxidant enzyme activity. This effect was more pronounced in plants exhibiting high sensitivity to salt stress. Frary et al. (2010) also reported that various salt-sensitive lines of tomatoes have different antioxidant profiles in leaves under salt stress (max. 150 mM of NaCl treatment for 21 days). However, these differences in total antioxidant contents (total phenolics and flavonoids) between the two parental lines were opposite to those reported by Keutgen and Paweltzik (2008). According to Table 11.1 the total phenolic level decreased in salt-sensitive plants and significantly increased in salt-tolerant plants under increased salinity. Also, in the contrast to the results presented by Keutgen and Paweltzik (2008) the mean initial level of phenolics and flavonoids in cultivated salt-sensitive (*S. Lycopersicum*) tomatoes was higher when compared to the wild salt-tolerant line (*S. pennellii*) (Frary et al. 2010). Differences in the phenolic content changes between plants of various sensitivity to salt stress were also examined by Mahmoudi et al. (2010). The authors investigated contents of total phenolics, flavonoids and phenolic acids in two varieties of lettuce, i.e. Verte (salt-tolerant) and Romaine (salt-sensitive) after 12 days of 100 mM NaCl treatment. Both lettuce varieties contained mainly phenolic acids and a minor flavonoid fraction under normal growth conditions. Moreover, the salt-tolerant lettuce cv. Verte contained lower levels of both phenolic acids and total phenolics than cv. Romaine in the absence of salinity treatment. However, both lettuce varieties exhibited different changes in bioactive compound levels upon salt stress. While a significant increase was found in the flavonoid fraction for both plants (14.5- and 4.2-fold for Romaine and Verte, respectively), an increase in phenolic acid content was observed only for cv. Verte (1.4-fold). The phenolic acid content in Romaine lettuce decreased 0.5-times after salt treatment. Nevertheless, the total phenolic content was significantly higher in both lettuce varieties after NaCl exposure, with a more pronounced increase in the salt-tolerant plant than in the salt-sensitive Romaine lettuce.

The discrepancies between results of various studies resulted also from the type of the samples, i.e. the analysed part of the plant. As it was shown by many authors, phenolic profile changes upon salt stress varied widely among tissue type (fruit, leaves, roots) (Navarro et al. 2006; Lopez-Berenguer et al. 2009; Neves et al. 2010;

Yiu et al. 2012; Zrig et al. 2011; Petridis et al. 2012). While total phenolic levels increased after exposure to NaCl in red pepper fruit (Navarro et al. 2006), their content decreased in pepper leaves (Yiu et al. 2012). Lopez-Berenguer et al. (2009) reported that the sinapic acid derivative level increased in broccoli inflorescences upon a 40 mM salt treatment, whereas an approx. 30 % reduction was observed in the leaves. According to the results of the study by Zrig et al. (2011), total phenolic response to salinity differed significantly between roots and leaves of bitter almonds. At a relatively low NaCl dose the total phenolic content was higher in roots when compared to the control, while at higher salt concentrations the antioxidant level diminished. Similar tendency was observed in the studies of Petridis et al. (2012) on the total phenolic content changes in olive tree roots after salinity. In almond leaves the content of phenolics was stable regardless of salinity (Zrig et al. 2011). Other almond genotypes showed various phenolic responses to NaCl exposure (Zrig et al. 2011). Based on the results of Petridis et al. (2012) it was concluded that phenolic content increases with increasing salinity up to 129 % in leaves of olive tree compared to control sample.

It was observed that the antioxidant response of plants to salinity depends strongly on their maturity stage (Navarro et al. 2006; Lopez-Berenguer et al. 2009). Based on the results of the study on the total phenolic content in green, turning (one half green skin and the other half red) and red pepper fruit under a 3-day salt treatment it was concluded that the phenolic level increased with salinity in red fruit, but was unchanged or slightly decreased in green and turning fruit, respectively (Navarro et al. 2006). Lopez-Berenguer et al. (2009) investigated the content of flavonoids, caffeoylquinic acid derivatives as well as sinapic and ferulic acid derivatives in young and old leaves of broccoli exposed to an 11-week NaCl-induced salt stress. The authors showed that both maturity states of leaves were characterized by similar phenolic profiles under control conditions, with flavonoids as a major fraction. After salt treatment the flavonoid content decreased by approx. 60 % in both types of leaves. However, while the content of sinapic acid derivatives decreased in old leaves with an increase in salinity, they were unchanged in the young leaves.

Finally, in order to induce higher salt tolerance to plants various external antioxidants were applied (Mohamed and Aly 2008; Yiu et al. 2012). Yiu et al. (2012) reported a lower polyphenol content in sweet pepper leaves after a 9-day 150 mM NaCl treatment. However, while applying catechin as a compound in the irrigation solution, total polyphenol content increased by approx. 23 % in comparison to plants treated with only NaCl. Nevertheless, pepper plants growing under normal conditions still showed higher phenolic content. Similar results were reported by Mohamed and Aly (2008) with α-tocopherol spraying during onion growth under salt stress. Altogether, the results indicate that an antioxidant application could be used to alleviate the harmful effect of salt stress on phenolic content in some plants.

Most studies focused on experiments carried out in greenhouses, where all experimental conditions could be monitored during the entire experimental period. However, outdoor climatic conditions are not stable. Thus, many factors may influence the overall plant response to salt stress observed at outdoor cultivation.

Rezazadeh et al. (2012) investigated the effect of salinity on the artichoke phenolic level during field and greenhouse trials. The authors observed that plant response to salt stress was the same in the field and pot experiments. The contents of both flavonoids and total phenols increased at moderate salinity (6.5 and 6.9 dS/m in the field and greenhouse trials, respectively).

Altogether, these results indicate the importance of plant tissue (fruit vs. leaves), species as well as salt doses, time of salinity exposure and initial levels of phenolics in determining phenolic response to salinity.

It is important to note that the numerical values from Tables 11.1 and 11.2 could not be compared directly, since various phenolic content determination methods as well as standard compounds were used. Moreover, not all authors expressed their values in terms of dry weight, on which reliable statements on the cases of a higher metabolite concentration under salt stress could be drawn. Thus, in some studies it is not clear whether the higher phenolic level for the stressed plants indeed results from a higher overall amount of the metabolites, or whether it is due to a putative increase in the concentration that is an effect of biomass reduction (Khan et al. 2011).

11.6.2 Cereals

Soil salinity affects large areas of agriculturally utilized land worldwide, causing significant reductions in crop yields (Schleiff 2008; Tavakkoli et al. 2011). Genetic variations among various crop plants are useful in providing a valuable tool in the selection of cultivars with desirable traits (Misra and Dwivedi 2004).

Many researchers showed different responses of various cereals to salinity (Ali and Abbas 2003; Lilia et al. 2005; Kattab 2007; Chutipaijit et al. 2009; Hichem et al. 2009; Ashraf et al. 2010; Daiponmak et al. 2010; Danai-Tambhale et al. 2011). Daiponmak et al. (2010) tested six Thai rice cultivars for their anthocyanin cyanidin-3-glucoside content and the antioxidant activity under salinity stress (60 mM NaCl, for 11 days). At a low salt treatment the total phenolic content increased in leaves of almost all genotypes, as in two of them a significant increase (from 0.83 to 1.49 and from 0.41 to 0.73 mg GAE/g FW) was observed, in another two it was a moderate increase (from 0.50 to 0.73 and from 0.44 to 0.72 mg GAE/g FW) and in another two only a slight increase (from 0.50 to 0.56 and from 0.55 to 0.57 mg GAE/g FW) when compared to the control. The reported results suggest that an increase of the total phenolic content in rice genotypes – as a result of salt stress – protects plants from oxidative damage. In other studies four varieties of *indica* rice, differing in salt sensitivity, were used for a comparative study of defense systems in response to salinity (100 mM NaCl, for 4 days) (Chutipaijit et al. 2009). It was shown that the total flavonoid content increased in salt-stressed seedlings of salt-tolerant rice varieties, by 6.34–7.31 % and 1.72–3.48 % in salt-sensitive plants, which indicates that probably flavonoids – similarly to proline compounds – serve a protective role under stress conditions. The above mentioned compounds, accumulated in plant tissue, protect plants against damages caused by free radicals.

The effect of stress at various doses of salt (50, 100, 150, 200 and 300 mM NaCl) on the phenolic content in two popular scented non-basmati type indica rice cultivars was investigated by Danai-Tambhale et al. (2011). The results, recorded at day 21 after germination, indicate an increase in the total polyphenol content with salt stress progress in both cultivars (from 11.58 to 26.32 mg/g FW for Indrayani and from 19.74 to 23.01 mg/g FW for Ambemohar). Similar results were presented by Hichem et al. (2009) when testing two forage maize cultivars under 0, 34, 68 and 102 mM NaCl supplementation for 6 weeks under glasshouse conditions. In both cultivars the amount of total polyphenols increased with salinity, but decreased with leaf tissue senescence (young leaves>mature leaves>senescent leaves). In the tested cultivars the concentration of total flavonoids, proanthocyanins and anthocyanins, as well as its variation in leaves of different age was similar and well-correlated to total polyphenol content. In contrast to the above results, a significantly lower level (mg/g FW) of total polyphenols was found in seedlings of two different canola cultivars (Serw and Pactol) grown under saline conditions (200 mM NaCl) for 3 weeks, amounting to 7.17 and 8.31 for Serw and Pactol, respectively, when compared to that of the control nonstressed plants at 20.80 and 12.68 (Kattab 2007).

In other studies on the effects of salt treatment (70 and 140 mM NaCl) on the expression of phenolics in two barley cultivars (Arig 8 and Acsad 1230), both in roots and leaves during 4 and 20 weeks (Lilia et al. 2005), it was reported that Ascad 1230 was less sensitive to salt stress than Arig 8. At a lower salt supplementation (70 mM NaCl) both cultivars exhibited comparable levels of phenolic compounds, while at 140 mM NaCl the metabolite levels were different. A significant accumulation of benzoic acid, as well as several ferulic acid, apigenic and luteolin derivatives was detected in leaf extracts of both cultivars, while in case of roots the phenolic profile indicated a higher accumulation of benzoic acid and *p*-coumaric acid derivatives. Moreover, Ashraf et al. (2010) reported the effect of root zone salinity (0 and 150 mM NaCl) on two bread wheat cultivars (salt-tolerant and salt-sensitive) at different growth stages (vegetative, booting and reproductive). Phenolic contents in leaves in both cultivars significantly decreased only at the booting stage, while at the other growth stages such an inhibitory effect of salt stress on leaf phenolic content was not observed.

Ali and Abbas (2003) examined the effect of different levels of phenylurea (0.1, 1, 3, 5 ppm) on germination, seedling growth as well as the contents of total phenolic compounds, flavonoids and antioxidant enzymes in barley grown under salt stress (50 and 100 mM NaCl) for 10 days. The results indicated that phenylurea application resulted in a significant increase in the growth rate of shoots and roots and a reduction of the total phenolic compound concentration (from 0.49 to 0.11 and 0.50–0.16 mg/g FW, respectively for 50 and 100 mM NaCl) and flavonoids (from 0.40 to 0.28 and 0.44–0.16) only in shoots. In contrast, a slight increase of the metabolite levels was observed in roots.

11.6.3 Spices, Herbs and Medicinal Plants

The effects of various salt treatments on the phenolic content were also investigated in such plants as rosemary, black cumin, mint and basil (Bourgou et al. 2010; Oueslati et al. 2010; Kiarostami et al. 2010; Zahedi et al. 2011; Mehrizi et al. 2012).

Rosemary is a good source of phenolic antioxidant compounds (2–3 % rosemarinic, chlorogenic and caffeic acids), able to scavenge ROS (Zheng and Wang 2001; Almela et al. 2006), while stress caused by NaCl induced an accumulation of proline, total phenolics and other antioxidants in rosemary (*Rosmarinus officinalis*). The antioxidant activity and total phenolic content increased in salinized plants (50, 100, and 150 mM NaCl, for 3 weeks) (Kiarostami et al. 2010). Phenolic content (in mg/g) in leaves significantly increased at 50 mM NaCl (from 0.91 to 1.59), increased to 4.22 at 100 mM NaCl and decreased at 150 mM NaCl (to 3.47). The above observation of an increase in total phenolic content and the antioxidant activity of rosemary under salinity can suggest a reduction of oxidative stress and in the consequence may be an interesting and valuable tip for the production of antioxidant compounds in rosemary plants under salt stress. In other studies on rosemary the effects of salinity (0, 50 and 100 mM NaCl, for 3 months) and various levels of copper (0, 0.5 and 1.0 μM Cu^{2+} as $CuSO_4$) on total phenolic content were investigated (Mehrizi et al. 2012). Copper plays an important role in the synthesis of phenolic compounds and its deficiency can decrease phenolic level in plants (Dicko et al. 2006), which is in agreement with data on the effect of some metals on the oxidative stress (Drzewiecka et al. 2011). Regardless of copper concentration, the highest total phenolic content (TPC) was found in the leaves of plants grown under the 100 mM NaCl treatment. The copper nutrition effect on total phenolic content and its interaction with salinity were significant. Plants exposed to 0.5 and 1.0 μM Cu^{2+} accumulated, respectively, 12 and 37 % total phenolics more than those grown in the Cu-free treatement (at 50 mM NaCl) and the above values dropped to 12 % and 15 %, respectively at higher salinity (at 100 mM NaCl).

The content of phenolics in different parts of *Mentha pulegium* L. plants under various salt stress levels (25, 50, 75 and 100 mM NaCl) after 2 weeks was also monitored (Oueslati et al. 2010). Phenolic content increased in leaf methanolic extracts up to 8.17 mg GAE/g DW at 100 mM NaCl, which was about 3.5 times higher than the total polyphenol content in leaves of the control (2.41 mg GAE/g DW). Interesting results were reported at 0 and 100 mM NaCl in root extracts concerning polyphenol content – 8.89 and 6.17 mg GAE/g DW, respectively, as compared to the amount observed at 25 and 50 mM NaCl (3.80 and 3.17 mg GAE/g DW).

Moreover, the effect of the type of salt (KCl, K_2SO_4, $NaHCO_3$, Na_2SO_4, $CaCl_2$, NaCl, Na_2CO_3) and its different levels (2, 4, 6, 8 and 10 dS/m) on seed germination and growth of basil (*Ocimum basilicum* L.) was investigated (Zahedi et al. 2011). The results show that vegetative growth was reduced with an increasing salinity level and the most negative effect was related to Na_2SO_4 and $CaCl_2$, while germination reduction was related to salinity that induced disturbance of metabolic processes leading to an increase in phenolic compounds (Ayaz et al. 2000).

Bourgou et al. (2010) evaluated in their studies the effect of saline conditions (0, 20, 40, and 60 mM NaCl, after 12 weeks) on fruit yield, fatty acid and essential oil compositions and phenolic content in seeds of black cumin (*Nigella sativa*), an aromatic and medicinal plant belonging to the Ranunculacea family, with seeds traditionally used in several countries for culinary and medicinal purposes. In contrast to the results of Kiarostami et al. (2010) and Mehrizi et al. (2012) the total polyphenol content in black cumin seeds decreased by 30 %, 54 %, and 61 % as a response to 20, 40, and 60 mM NaCl, respectively, compared to the control (11.69 mg GAE/g DW). In this experiment flavonoid concentrations were also reduced, with the lowest values observed at 40 mM NaCl.

11.6.4 Other Plants

Salinity-induced oxidative damage and differential response of enzymatic and non-enzymatic antioxidants in *Colubrina asiatica*, treated with different concentrations (100, 200 and 300 mM, for 8 weeks) of sodium chloride solutions were examined by Sonar et al. (2011). Since their leaves contain saponin this plant is of some economic value and it is used as a soap substitute, to prepare bowels, used to wash and for food production, medicine and as tooth cleaners (Richardson et al. 2000; Uphoff et al. 2001). Activities of enzymatic antioxidants (catalase, peroxidase) both in leaves and roots were significantly increased by salt stress, with a maximum induction at high salinity (200 and 300 mM NaCl treatment). In the case of non-enzymatic antioxidants (polyphenols) their contents also increased with an increasing NaCl treatment (from 24 to 38 mg GAE/g DW), which together were found to be positively correlated also with the parameters of oxidative damage.

Salah et al. (2011) tested two *Medicago ciliaris* lines (a salt-tolerant and salt-sensitive) with contrasting responses to 100 mM NaCl, for 3 weeks. *Medicago ciliaris* belongs to legumes, critical components of natural ecosystems and agriculture, which have recently been the subject of several studies highlighting their benefits and responses to cultivation under saline conditions. Salt stress did not affect contents of total phenolics, flavonoids and tannins in leaves of the tolerant line. Levels of these biochemical components were initially higher in leaves of the salt-sensitive line than in the salt–tolerant line in the absence of salt, but then decreased by 18 %, 50 % and 50 % for total phenolics, flavonoids and tannins, respectively. Levels of phenolics and tannins in roots were higher in the salt-tolerant line than in the salt-sensitive line without salt treatment. After salt application, the contents of root phenolics and root tannins declined proportionately in both lines, while root flavonoid level dropped down to 50 % of the nonstress level in the salt-tolerant, and finally to 25 % in the salt-sensitive line.

Morinda citrifilia, a plant used for traditional food and folk medicine in Polynesia for over 2,000 years, contains several medicinally active compounds, e.g. anthraquinones, terpenoids, alkaloids, sitosterol, carotene, flavones and glycosides (Ahmed et al. 2008), that exhibit various therapeutic effects such as antibacterial, antiviral and anticancer activities (Wang et al. 2002). In this experiment *Morinda citrifolia* adventitious roots were cultured in different salt strengths (0.25, 0.50, 0.75, 1.0, 1.5 and 2.0) of Murashige and Skoog (MS) medium supplemented with 5 mg/L indole butyric acid and 30 g/L sucrose suitable for the production of both biomass and secondary metabolites (Abdullahil et al. 2010). After 4 weeks the recorded results indicated that the best media for the production of total phenolics (24.45 mg/g DW) was 2.0 MS, while in the case of flavonoid content it was 0.25 MS (14.28 mg/g DW).

11.6.5 Halophytes

In the last few years a correlation between the antioxidant capacity and salt tolerance has been found in different halophytic plant species, including *Centaurea tuzgoluensis* (Yıldıztugay et al. 2011), *Plantago maritima* (Sekmen et al. 2004), *Cakile maritima* (Amor et al. 2006) *Aegiceras corniculatum* (Parida et al. 2004), *Mesembryanthemum edule* (Falleh et al. 2012) and *Gypsophila oblanceolata* BARK (Sekmen et al. 2012).

Two *Mesembryanthemum edule* L. provenances, which differed in their climatic conditions, i.e. Jerba and Bizerte, were investigated by Falleh et al. (2012). This plant is an edible and medicinal halophyte and contains high levels of polyphenolic compounds, notably procyanidins and propelargonidins. Under salt treatment (0, 300 and 600 mM NaCl) after 67 days polyphenol content significantly decreased (from 3.00 to 2.11 mg/g DW for Jerba and 4.03–2.50 mg/g DW for Bizerte) and salt treatment caused differential responses of major phenolics, as reyotrin tended to disappear upon salinity, while isorhamnoside rutinol accumulated in salt-treated plants.

In contrast to the above results Parida et al. (2004) reported that salt-induced biochemical changes were observed in hydroponically grown plants of a salt secrector mangrove, *Aegiceras corniculatum* (Myrsinaceae). *Aegiceras corniculatum* is one of the minor components of the tropical mangrove ecosystem, with characteristic salt glands in the leaves for the secretion of excess salt, showing tolerance to salinity gradients. All the experiments were conducted up to 30 days at 0 and 250 mM NaCl treated plants. Polyphenol content increased significantly with the duration of NaCl treatment, and varied from 2.4 in the control to 4.6 mg/g DW in 250 mM NaCl treated plants.

Ksouri et al. 2007 analyzed two Tunisian accessions (Jerba and Tabarka) of *Cakile maritima*, a local oilseed halophyte exhibiting potential for secondary metabolite biosynthesis. Three-week-old plants were subjected to 0, 100 and 400 mM NaCl for 28 days under glasshouse conditions. Total polyphenol analyses showed that in the absence of NaCl both accessions accumulated polyphenols at the same concentrations in their leaves. However, when salt-challenged Jerba presented a significant increase in polyphenol accumulation (56 % at 100 mM NaCl, and 30 % at 400 mM), while this parameter decreased in Tabarka. In further studies

Ksouri et al. (2008) assessed the phenolic content and the antioxidant activity of some Tunisian halophytes (*Cakile maritima*, *Limoniastrum monopetalum*, *Mesembryanthemum crystallinum*, *M. edule*, *Salsola kali* and *Tamarix gallica*) depending on biological (species, organ and developmental stage), environmental, and technical (extraction solvent) factors. For example, in *T. gallica* a comparison between leaves and flowers showed that both phenolic content and the antioxidant activity were organ-dependent. Flower extracts were characterized by higher polyphenol contents (70.56 mg GAE/g DW), when compared to leaf extracts (20.69 mg GAE/g DW). Presented data appeared strictly dependent on a number of biotic (species, organ and physiological stage) and abiotic (environmental, handling, solvent extraction) factors. Taken together; this information may confirm the interesting potential of halophytes as a valuable source for natural antioxidant molecules.

11.7 Conclusion and Future Perspective

In conclusion phenolics show a great diversity of their carbon skeleton with consequence of different classes of the compounds followed by the metabolites diverse functions crucial for plant growth and development. Their biosynthesis is not only the response of the plant to unfavorable conditions of the oxidative stress, but also the compounds may positively influence for example spore germination and hyphal growth of arbuscular mycorrhizal fungi. Experiments in the topic are spread all over the world and generated this way results and in the consequence knowledge is the potent tool protecting plants developing under stress conditions. Considering all the above, overestimation of the phenolics importance is out of the discussion and further studies in the field of the group of the metabolites is of prime concern.

References

Abdullahil MB, Lee EJ, Paek KI (2010) Medium salt strength induced changes in growth, physiology and secondary metabolite content in adventitious roots of *Morinda citrifolia*: the role of antioxidant enzymes and phenylalanine ammonia lyase. Plant Cell Rep 29:685–694

Agastian P, Kingsley SJ, Vivekanandan M (2000) Effect of salinity on photosynthesis and biochemical characteristics in mulberry genotypes. Photosynthetica 38:287–290

Ahmad P, Sharma S (2008) Salt stress and phyto-biochemical responses of plants – a review. Plant Soil Environ 54:89–99

Ahmed S, Hahn EJ, Paek KY (2008) Aeration volume and photosynthetic photon flux affect cell growth and secondary metabolite contents in bioreactor cultures of *M. citrifolia*. J Plant Biol 51:209–212

Ali RM, Abbas HM (2003) Response of salt stressed barley seedlings to phenylurea. Plant Soil Environ 49:158–162

Allakhverdiev SI, Sakamoto A, Nishiyama Y, Inaba M, Murata N (2000) Ionic and osmotic effects of NaCl-induced inactivation of photosystems I and II in *Synechococcus* sp. Plant Physiol 123:1047–1056

Almela L, Sanchez-Munoz B, Fernandez-Lopez JA, Rocaa MJ, Virginia R (2006) Liquid chromatograpic-mass spectrometric analysis of phenolics and free radical scavenging activity of rosemary extract from different raw material. J Chromatogr 7:221–229

Amalesh S, Das G, Das SK (2011) Role of flavonoids in plants. Int J Pharm Sci Tech 6:12–35

Amor NB, Jimenez A, Megdiche W, Lundqvist M, Sevilla F, Abdelly C (2006) Response of antioxidant systems to NaCl stress in the halophyte *Cakile maritima*. Physiol Plant 126:446–457

Arts IC, van de Putte B, Hollman PC (2000a) Catechin contents of foods commonly consumed in The Netherlands. 1. Fruits, vegetables, staple foods, and processed foods. J Agric Food Chem 48:1746–1751

Arts IC, van de Putte B, Hollman PC (2000b) Catechin contents of foods commonly consumed in The Netherlands. 2. Tea, wine, fruit juices, and chocolate milk. J Agric Food Chem 48:1752–1757

Asada K (2006) Production and scavenging of reactive oxygen species in chloroplasts and their functions. Plant Physiol 141:391–396

Ashraf M (2009) Biotechnological approach of improving plant salt tolerance using antioxidants as markers. Biotechnol Adv 27:84–93

Ashraf M, Foolad MR (2007) Improving plant abiotic-stress resistance by exogenous application of osmoprotectants glycinebetaine and proline. Environ Exp Bot 59:206–216

Ashraf MA, Ashraf M, Ali AQ (2010) Response of two genetically diverse wheat cultivars to salt stress at different growth stages: leaf lipid peroxidation and phenolic contents. Pak J Bot 42:559–565

Ayaz FA, Kadioglu A, Turgut R (2000) Water stress effects on the content of low molecular weight carbohydrates and phenolic acids in *Ctenanthe setosa* (Rose.) Eichler. Can J Plant Sci 80:373–378

Bais HP, Weir TL, Perry LG, Gilroy S, Vivanco JM (2006) The role of root exudates in Rhizosphere interactions with plants and other organisms. Ann Rev Plant Biol 57:233–266

Begum AA, Leibovitch S, Migner P, Zhang F (2001) Specific flavonoids induced nodes of *Rhizobium leguminosarum* increased pea (*Pisum sativum* L.) and lentil (*Lens culinaris* L.) nodulation in controlled growth chamber environment. J Exp Bot 52:1537–1543

Borkowski T, Szymusia H, Gliszczyńska-Świgło A, Rietjens IMCM, Tyrakowska B (2005) Radical scavenging capacity of wine anthocyanins is strongly pH-dependent. J Agric Food Chem 53:5526–5534

Bors W, Heller W, Michel C (1998) The chemistry of flavonoids. In: Rice-Evans CA, Pecker L (eds) Flavonoids in health and disease. Marcel Dekker, New York, pp 11–136

Bourchard P, Bilger W, Weissenbock G (2000) Contribution of hydroxycinnamates and flavonoids to epidermal shielding of UV-A and UV-B radiation in developing rye primary leaves as assessed by ultravioletinduced chlorophyll fluorescence measurements. Plant Cell Environ 23:1373–1380

Bourgou S, Bettaieb I, Saidani M, Marzouk B (2010) Fatty acids, essential oil, and phenolics modifications of black cumin fruit under NaCl stress conditions. J Agric Food Chem 58:12399–12406

Burns J, Yokota T, Ashihira H, Lean MEJ, Crozier A (2002) Plant foods and herbal sources of resveratrol. J Agric Food Chem 50:3337–3340

Carlsen SCK, Understrup A, Fomsgaard IS, Mortensen AG, Ravnskov S (2008) Flavonoids in roots of white clover: interaction of arbuscular mycorrhizal fungi and a pathogenic fungus. Plant Soil 302:33–43

Chinnusamy V, Jagendorf A, Zhu J-K (2005) Understanding and improving salt tolerance in plants. Crop Sci 45:437–448

Chisari M, Todaro A, Barbagallo RN, Spagna G (2010) Salinity effects on enzymatic browning and antioxidant capacity of fresh-cut baby Romaine lettuce (*Lactuca sativa* L. Cv. Duende). Food Chem 119:1502–1506

Chong J, Poutaraud A, Hugueney P (2009) Metabolism and roles of stilbenes in plants. Plant Sci 177:143–155

Chutipaijit S, Cha-Um S, Sompornpailin K (2009) Differential accumulations of proline and flavonoids in indica rice varieties against salinity. Pak J Bot 41:2497–2506

Cooper JE (2004) Multiple responses of rhizobia to flavonoids during legume root infection. Adv Bot Res 41:1–62

Crifò T, Puglisi I, Petrone G, Recupero GR, Piero ARL (2011) Expression analysis in response to low temperature stress in blood oranges: implication of the flavonoid biosynthetic pathway. Gene 476:1–9

Crozier A, Jaganath IB, Clifford MN (2006) Phenols, polyphenols and tannins: an overview. In: Crozier A, Ashihara H (eds) Plant secondary metabolites: occurrence, structure and role in the human diet. Blackwell, Oxford, pp 1–24

Cvikrova M, Mala J, Hrubcova M, Eder J (2006) Soluble and cell wall-bound phenolics and lignin in *Ascocalyx abietina* infected Norway spruces. Plant Sci 170:563–570

Daiponmak W, Theerakulpisut P, Thanonkao P, Vanavichit A, Prathepha P (2010) Changes of anthocyanin cyanidin-3-glucoside content and antioxidant activity in Thai rice varieties under salinity stress. Sci Asia 36:286–291

Dajic Z (2006) Salt stress. In: Madhava RKV, Raghavendra AS, Janardhan RK (eds) Physiology and molecular biology of stress tolerance in plants. Kluwer, Dordrecht, pp 41–101

Danai-Tambhale S, Kumar V, Shriram V (2011) Differential response of two scented indica rice (*Oryza sativa*) cultivars under salt stress. J Stress Physiol Biochem 7:387–397

Dat J, Vandenabeele S, Vranova E, Van Montagu M, Inze D, Van Breusegem F (2000) Dual action of the active oxygen species during plant stress responses. Cell Mol Life Sci 57:779–795

Davis K, Shwinn K (2006) Molecular biology and biotechnology of flavonoid biosynthesis. In: Andersen OM, Markham KR (eds) Flavonoids. Chemistry, biochemistry and applications. CRC Press/Taylor and Francis, Boca Raton, pp 143–218

Demiral MA, Aktas Uygun D, Uygun M, Kasirga E, Karagozler AA (2011) Biochemical response of *Olea europaea* cv. Gemlik to short-term salt stress. Turk J Biol 35:433–442

Desikan R, Hancock J, Neil S (2005) Reactive oxygen species as signalling molecules. In: Smirnoff N (ed) Antioxidants and reactive oxygen species in plants. Blackwell, Oxford, pp 169–196

Dicko MH, Gruppen H, Traoré AS, Voragen AGJ, van Berkel WJH (2006) Phenolic compounds and related enzymes as determinants of sorghum for food use. Biotechnol Mol Biol Rev 1:21–38

Dkhil BB, Denden M (2010) Salt stress induced changes in germination, sugars, starch and enzyme of carbohydrate metabolism in *Abelmoschus esculentus* L. (Moench.) seeds. Afr J Agric Res 5:1412–1418

Drzewiecka K, Mleczek M, Waśkiewicz A, Goliński P (2011) Oxidative stress and phytoremediation. In: Ahmad P, Prasad MNV (eds) Abiotic stress responses in plants. Springer, New York, pp 425–449

Eckardt NA (2006) The role of flavonoids in root nodule development and auxin transport in *Medicago truncatula*. Plant Cell 18:1539–1540

Ehlting J, Hamberger B, Million-Rousseau R, Werck-Reichhart D (2006) Cytochromes P450 in phenolic metabolism. Phytochem Rev 5:239–270

Ellouzi H, Hamed KB, Cela J, Munne-Bosch S, Abdelly C (2011) Early effects of salt stress on the physiological and oxidative status of *Cakile maritima* (halophyte) and *Arabidopsis thaliana* (glycophyte). Physiol Plant 142:128–143

Eryilmaz F (2006) The relationships between salt stress and anthocyanin content in higher plants. Biotechnol Biotechnol Equipment 20:47–52

Estes M (2002) A review of salinity stress on plants. Available at: http://www.bio.davidson.edu/people/kabernd/seminar/2002/stress/Salinity%20Paper.htm. Accessed Dec 2011

Falleh H, Jalleli I, Ksouri R, Boulaaba M, Guyot S, Magné M, Abdely C (2012) Effect of salt treatment on phenolic compounds and antioxidant activity of two *Mesembryanthemum edule* provenances. Plant Physiol Biochem 52:1–8

Flowers TJ, Colmer TD (2008) Salinity tolerance in halophytes. New Phytol 179:945–963

Flowers TJ, Flowers SA (2005) Why does salinity pose such a difficult problem for plant breeders? Agric Water Manag 78:15–24

Fofana B, Benhamou N, McNally DJ, Labbé C, Séguin A, Bélanger RR (2005) Suppression of induced resistance in cucumber through disruption of the flavonoid pathway. Phytopathology 95:114–123

Fox JE, Starcevic M, Kow KY, Burow ME (2001) Nitrogen fixation: endocrine disrupters and flavonoid signalling. Nature 413:128–129

Foyer CD, Changey RL, White MC (1997) The physiology of metal toxicity in plants. Ann Rev Plant Physiol 29:11–566

Frary A, Gol D, Keles D, Okmen B, Pinar H, Sigva HO, Yemenicioglu A (2010) Salt tolerance in *Solanum pennellii*: antioxidant response and related QTL. Plant Biol 10:58–73

Goliński P, Krick TP, Blanchette RA, Mirocha CJ (1995) Chemical characterization of red pigment (5,8-dihydroxy-2,7-dimethoxy-1.4-naphthalenedione) produced by *Arthrographis cuboidea* in pink stained wood. Holzforschung 49:407–410

Gould KS, Lister C (2006) Flavonoid functions in plants. In: Andersen OM, Markham KR (eds) Flavonoids: chemistry, biochemistry and applications. CRC Press/Taylor and Francis, Boca Raton, pp 397–442

Hernandez JA, Almansa MS (2002) Short-term effects of salt stress on antioxidant systems and leaf water relations of pea leaves. Physiol Plant 115:251–257

Hichem H, Mounir D, Naceur EA (2009) Differential responses of two maize (*Zea mays* L.) varieties to salt stress: changes on polyphenols composition of foliage and oxidative damages. Ind Crops Prod 30:144–151

Huchzermeyer B, Hausmann N, Paquet-Durant F, Koyro H-W (2004) Biochemical and physiological mechanisms leading to salt tolerance. Tropic Ecol 45:141–150

Joganath IB, Crozier A (2010) Dietary flavonoids and phenolic compounds. In: Fraga G (ed) Plant phenolics and human health. Wiley, Hoboken, pp 1–50

Karolewski Z, Waśkiewicz A, Irzykowska L, Bocianowski J, Kostecki M, Golinski P, Knaflewski M, Weber Z (2011) Fungi presence and their mycotoxins distribution in asparagus spears. Pol J Environ Stud 20:911–919

Kattab H (2007) Role of glutathione and polyadenylic acid on the oxidative defense systems of two different cultivars of canola seedlings grown under saline condition. Aust J Basic Appl Sci 1:323–334

Keilig K, Ludwig-Müller J (2009) Effect of flavonoids on heavy metal tolerance in *Arabidopsis thaliana* seedlings. Bot Stud 50:311–318

Kensuke K, Norihisa M, Kazuo S, Taizo H (2007) Flavonoids induce germination of basidiospores of the ectomycorrhizal fungus *Suillus bovinus*. Mycorrhiza 17:563–570

Keutgen AJ, Paweltzik E (2008) Quality and nutritional value of strawberry fruit under long term salt stress. Food Chem 107:1413–1420

Khan TA, Mazid M, Mohammad F (2011) Status of secondary plant products under abiotic stress: an overview. J Stress Physiol Biochem 7:75–98

Kiarostami Kh, Mohseni R, Saboora A (2010) Biochemical changes of R*osmarinus officinalis* under salt stress. J Stress Physiol Biochem 6:114–122

Kim H-J, Fonseca JM, Choi J-H, Kubota C, Kwon DY (2008) Salt in irrigation water affects the nutritional and visual properties of Romaine lettuce (*Lactuca sativa* L.). J Agric Food Chem 56:3772–3776

Kirakosyan A, Kaufman P, Warber S, Zick S, Aaronson K, Bolling S, Chang SC (2004) Applied environmental stresses to enhance the levels of polyphenolics in leaves of hawthorn plants. Physiol Plant 121:182–186

Koca M, Bor M, Ozdemir F, Turkan I (2007) The effect of salt stress on lipid peroxidation, antioxidative enzymes and proline content of sesame cultivars. Environ Exp Bot 60:344–351

Kong C, Liang W, Xu X, Hu F, Wang P, Jiang Y (2004) Release and activity of allelochemicals from allelopathic rice seedlings. J Agric Food Chem 52:2861–2865

Koyro H-W, Ahmad P, Geissler N (2012) Abiotic stress responses in plants: an overview. In: Ahmad P, Prasad MNV (eds) Environmental adaptations and stress tolerance 1 of plants in the era of climate change. Springer, New York, pp 1–28

Ksouri R, Megdiche W, Debez A, Falleh H, Grignon C, Abdelly C (2007) Salinity effects on polyphenol content and antioxidant activities in leaves of the halophyte *Cakile maritime*. Plant Physiol Biochem 45:244–249

Ksouri R, Megdiche W, Falleh H, Trabelsi N, Boulaaba M, Smaoui M, Abdelly C (2008) Influence of biological, environmental and technical factors on phenolic content and antioxidant activities of Tunisian halophytes. C R Biologies 331:865–873

Kumar V, Shriram V, Kavi Kishor PB, Jawali N, Shitole MG (2010) Enhanced proline accumulation and salt stress tolerance of transgenic *indica* rice by over expressing P5CSF129A gene. Plant Biotechnol Rep 4:37–48

Lattanzio V, Lattanzio VMT, Cardinali A (2006) Role of phenolics in the resistance mechanisms of plants against fungal pathogens and insects. In: Imperato F (ed) Phytochemistry: Advances in Research. Research Signpost, Trivandrum, Kerala pp 23–67

Lee DH, Kim YS, Lee CB (2001) The inductive responses of the antioxidant enzymes by salt stress in the rice (*Oryza sativa* L.). Plant Physiol 158:735–747

Lee T-T, Huang C-C, Shieh X-H, Chen C-L, Chen L-J, Yu B (2010) Flavonoid, phenol and polysaccharide contents of *Echinacea Purpurea* L. and its immunostimulant capacity *in vitro*. Int J Environ Sci Dev 1:5–9

Lemańska K, Szymusiak H, Tyrakowska B, Zieliński R, Soffers A, Ritjens IMCM (2001) The influence of pH on antioxidant properties and the mechanism of antioxidant action of hydroxyflavones. Free Radic Biol Med 31:869–881

Levinsh G (2006) Biological basis of biological diversity: physiological adaptations of plants to heterogeneous habitats along a sea coast. Acta Univ Latv 710:53–79

Li Z-H, Wang Q, Ruan X, Pan C-D, Jiang D-A (2010) Phenolics and plant allelopathy. Molecules 15:8933–8952

Lilia E, Salah R, Hajer SA, Abderrazak D, Majida EH, Ismail EH (2005) Effect of salt treatment on the expression of phenolics and peroxidase activity assessed in two barley cultivars Acsad 1230 and Arig 8. J Agron 4:196–202

Lobo LT, da Silva GA, de Freitas MCC, Souza Filho APS, da Silva MN, Arruda AC, Guilhon GMSP, Santos LS, Santos AS, Arruda MSP (2010) Stilbenes from *Deguelia rufescens* var. *urucu* (Ducke) A. M. G. Azevedo leaves: effects on seed germination and plant growth. J Braz Chem Soc 21:1838–1844

Lopez-Berenguer C, del Carmen M, Martinez-Ballesta C, Moreno DA, Carvajal M, Garcia-Viguera C (2009) Growing hardier crops for better health: salinity tolerance and the nutritional value of broccoli. J Agric Food Chem 57:572–578

Luthria DL, Mukhopadhyay S, Krizek DT (2006) Content of total phenolics and phenolic acids in tomato (*Lycopersicon esculentum* Mill.) fruits as influenced by cultivar and solar UV radiation. J Food Comp Anal 19:771–777

MacDonald MJ, D'Cunha GB (2007) A modern view of phenylalanine ammonia lyase. Biochem Cell Biol 85:273–282

Mahmoudi H, Huang J, Gruber MY, Kaddour R, Lachaal M, Ouerghi Z, Hannoufa A (2010) The impact of genotype and salinity on physiological function, secondary metabolite accumulation, and antioxidative responses in lettuce. J Agric Food Chem 58:5122–5130

Mano J (2002) Early events in environmental stresses in plants. Induction mechanisms ofoxidative stress. In: Inze D, Van Montagu M (eds) Oxidative stress in plants. Taylor and Francis, London, pp 217–246

Mattila P, Kumpulainen J (2007) Determination of free and total phenolic acids in plant-derived foods by HPLC with diode-array detection. J Agric Food Chem 50:3660–3667

Mehrizi MH, Shariatmadari H, Khoshgoftarmanesh AH, Dehghani F (2012) Copper effects on growth, lipid peroxidation, and total phenolic content of rosemary leaves under salinity stress. J Agric Sci Tech 14:205–212

Michalak A (2006) Phenolic compounds and their antioxidant activity in plants growing under heavy metal stress. Pol J Environ Stud 15:523–530

Misra N, Dwivedi UN (2004) Genotypic differences in salinity tolerance of green gram cultivars. Plant Sci 166:1135–1142

Mohamed AA, Aly AA (2008) Alterations of some secondary metabolites and enzymes activity by using exogenous antioxidant compound in onion plants grown under seawater salt stress. Am Eurasian J Sci Res 3:139–146

Moss GP (2000) Nomenclature of lignans and neolignans (IUPAC recommendations 2000). Pure Appl Chem 72:1493–1523

Munns R (2002) Comparative physiology of salt and water stress. Plant Cell Environ 25:239–250

Munns R, Tester M (2008) Mechanisms of salinity tolerance. Ann Rev Plant Biol 59:651–681

Munns R, Schachtman D, Condon A (1995) The significance of a two-phase growth response to salinity in wheat and barley. Aust J Plant physiol 22:561–569

Muzolf M, Szymusiak H, Gliszczyńska-Świgło A, Rietjens IM, Tyrakowska B (2008) pH-dependent radical scavenging capacity of green tea catechins. J Agric Food Chem 56:816–823

Muzolf-Panek M, Gliszczyńska-Świgło A, de Haan L, Aarts JMMJG, Szymusiak H, Vervoort JM, Tyrakowska B, Rietjens IMCM (2008) Role of catechin quinones in the induction of EpRE-mediated gene expression. Chem Res Toxicol 21:2352–2360

Navari-Lazzo F, Quartacci MF (2001) Phytoremedition of metals: tolerance mechanisms against oxidative stress. Minerva Biotechnol 13:73–83

Navarro JM, Folres P, Garrido C, Martinez V (2006) Changes in the contents of antioxidant compounds in pepper fruits at different ripenning stages, as affected by salinity. Food Chem 96:66–73

Neves GYS, Marchiosi R, Ferrarese MLL, Siqueira-Soares C, Ferrarese-Filho O (2010) Root growth inhibition and lignification induced by salt stress in soybean. J Agron Crop Sci 196:467–473

Noctor G, Foyer CH (1998) Ascorbate and glutathione: keeping active oxygen under control. Ann Rev Plant Physiol Plant Mol Biol 49:623–647

Noreen Z, Ashraf M (2009) Assessment of variation in antioxidative defense system in salt-treated pea (*Pisum sativum*) cultivars and its putative use as salinity tolerance markers. J Plant Physiol 166:1764–1774

Oueslati S, Karray-Bouraoui N, da Attia H, Rabhi M, Ksouri R, Lachaal M (2010) Physiological and antioxidant responses of *Mentha pulegium* (Pennyroyal) to salt stress. Acta Physiol Plant 32:289–296

Parida AK, Das AB, Sanadac Y, Mohantyb P (2004) Effects of salinity on biochemical components of the mangrove, *Aegiceras corniculatum*. Aquat Bot 80:77–87

Pawlak-Sprada S, Arasimowicz-Jelonek M, Podgórska M, Deckert J (2011a) Activation of phenylpropanoid pathway in legume plants exposed to heavy metals. Part I. Effects of cadmium and lead on phenylalanine ammonia-lyase gene expression, enzyme activity and lignin content. Acta Biochim Pol 58:211–216

Pawlak-Sprada S, Stobiecki M, Deckert J (2011b) Activation of phenylpropanoid pathway in legume plants exposed to heavy metals. Part II. Profiling of isoflavonoids and their glycoconjugates induced in roots of lupine (*Lutinus lupeus*) seedling treated with cadmium and lead. Acta Biochim Pol 58:217–223

Petridis A, Therios I, Samouris G, Tananaki C (2012) Salinity-induced changes in phenolic compounds in leaves and roots of four olive cultivars (*Olea europaea* L.) and their relationship to antioxidant activity. Environ Exp Bot 79:37–43

Pinter J, Kosa E, Hadi G, Hegyi Z, Spitkó T, Tóth Z, Szigeti Z, Páldi E, Marton LC (2007) Effect of increased UV-B radiation on the anthocyanin content of maize (*Zea mays* L.) leaves. Acta Agronomica Hugarica 55:7–17

Ponce MA, Bompadre MJ, Scervino JM, Ocampo JA, Chaneton EJ, Godeas AM (2009) Flavonoids, benzoic acids and cinnamic acids isolated from shoots and roots of Italian rye grass (*Lolium multiflorum* Lam.) with and without endophyte association and arbuscular mycorrhizal fungus. Biochem Syst Ecol 37:245–253

Punyasiri PAN, Abeysinghe ISB, Kumar V, Treutter D, Duy D, Gosch C, Martens S, Forkmann G, Fischer TC (2004) Flavonoid biosynthesis in the tea plant *Camellia sinensis*: properties of enzymes of the prominent epicatechin and catechin pathways. Arch Biochem Biophys 431:22–30

Rezazadeh A, Ghasemneshaz A, Barani M, Telmadarrehei T (2012) Effect of salinity on phenolic composition and antioxidant activity of artichoke (*Cynara scolymus* L.) leaves. Res J Med Plant 6:245–252

Rice-Evans C, Miller NJ, Paganga G (1996) Structure-antioxidant activity relationships of flavonoids and phenolic acids. Free Radic Biol Med 20:933–956

Richardson JE, Fay MF, Cronk QCB, Bowman D, Chase MW (2000) A phylogenetic analysis of Rhamnaceae using rbcL and trnL-F plastid DNA sequences. Am J Bot 87:1309–1324

Rimando AM, Kalt W, Magee JB, Dewey J, Ballington JR (2004) Resveratrol, pterostilbene, and piceatannol in *Vaccinium* berries. J Agric Food Chem 52:4713–4719

Rivero R, Ruiz JM, Garcia PC, Lopez-Lefebre LR, Sanchez E, Romero L (2001) Resistance to cold and heat stress: accumulation of phenolic compounds in tomato and watermelon plants. Plant Sci 160:315–321

Routaboul JM, Kerhoas L, Debeaujon I, Pourcel L, Caboche M, Einhorn J, Lepiniec L (2006) Flavonoid diversity and biosynthesis in seed of *Arabidopsis thaliana*. Planta 224:96–107

Sairam RK, Tyagi A (2004) Physiology and molecular biology of salinity stress tolerance in plants. Curr Sci 86:407–421

Salah IB, Mahmoudi H, Gruber M, Slatni T, Boulaaba T, Gandour M, Messedi D, Hamed KB, Ksouri B, Hannoufa A, Abdelly C (2011) Phenolic content and antioxidant activity in two contrasting Medicago ciliaris lines cultivated under salt stress. Biologia 66:813–820

Sánchez-Rodríguez E, Moreno DA, Ferreres F, Rubio-Wilhelmi Mdel M, Ruiz JM (2011) Differential responses of five cherry tomato varieties to water stress: changes on phenolic metabolites and related enzymes. Phytochemistry 72:723–729

Sang S, Yang I, Buckley B, Ho C-T, Yang CS (2007) Autoxidative quinone formation in vitro and metabolite formation in vivo from tea polyphenol (−)-epigallocatechin-3-gallate: studies by real-time mass spectrometry combined with tandem mass ion mapping. Free Radic Biol Med 43:362–371

Scervino JM, Ponce MA, Erra-Bassells R, Vierheilih H, Ocampo JA, Godeas A (2005) Flavonoids exhibit fungal species and genus specific effects on the presymbiotic growth of Gigaspora and Glomus. Mycol Res 109:789–794

Scervino JM, Ponce MA, Erra-Bassells R, Bompadre J, Vierheilig H, Ocampo J, Godeas A (2007) The effect of flavones and flavonols on colonization of tomato plants by arbuscular mycorrhizal fungi of the genera *Gigaspora* and *Glomus*. Can J Microbiol 53:702–709

Schleiff U (2008) Analysis of water supply of plants under saline soil conditions and conclusions for research on crop salt tolerance. J Agric Crop Sci 194:1–8

Schmidt TJ, Hemmati S, Fuss E, Alfermann AW (2006) A combined HPLC-UV and HPLC-MS method for the identification of lignans and its application to the lignans of *Linum usitatissimum* L and L. bienne Mill. Phytochem Anal 17:299–311

Scognamiglio M, Esposito A, D'Abrosca B, Pacifico S, Fiumano V, Tsafantakis N, Monaco P, Fiorentino A (2012) Isolation, distribution and allelopathic effect of caffeic acid derivatives from *Bellis perennis* L. Biochem Syst Ecol 43:108–113

Sekmen AH, Ozdemir F, Turkan I (2004) Effects of salinity, light, and temperature on seed germination in a Turkish endangered halophyte, *Kalidiopsis wagenitzii* (Chenopodiaceae). Isr J Plant Sci 52:21–30

Sekmen AH, Turkan I, Tanyolac ZO, Ozfidan C, Dinc A (2012) Different antioxidant defense responses to salt stress during germination and vegetative stages of endemic halophyte *Gypsophila oblanceolata* BARK. Environ Exp Bot 77:63–76

Sharma I, Pati PK, Bhardwaj R (2010) Regulation of growth and antioxidant enzyme activities by 28-homobrassinolide in seedlings of *Raphanus sativus L.* under cadmium stress. Indian J Biochem Biophys 47:172–177

Shaw LJ, Morris P, Hooker JP (2006) Perception and modification of plant flavonoid signals by rhizosphere microorganisms. Environ Microbiol 8:1867–1880

Single RS, Glenn EP, Squires V (1996) Growth performance of lambs fed mixed diets containing halophyte ingredients. Anim Feed Sci Technol 63:137–148

Smeds AI, Eklund PC, Sjöholm RE, Willför SM, Nishibe S, Deyama T, Holmbom BR (2007) Quantification of a broad spectrum of lignans in cereals, oilseeds, and nuts. J Agric Food Chem 55:1337–1346

Sonar BA, Desai Nivas M, Gaikwad DK, Chavan PD (2011) Assessment of salinity-induced antioxidative defense system in *Colubrina asiatica* brong. J Stress Physiol Biochem 7:193–200

Soylu S (2006) Accumulation of cell-wall bound phenolic compounds and phytoalexin in *Arabidopsis thaliana* leaves following inoculation with pathovars of *Pseudomonas syringae*. Plant Sci 170:942–952

Sundaravarathan S, Kannaiyan S (2002) Role of plant flavonoids as signal molecules to Rhizobium. In: Kannaiyan S (ed) Biotechnology of biofertilizers. Kluwer, The Netherlands, pp 144–164

Suzuki S, Umezawa T, Shimada M (2002) Stereochemical diversity in lignan biosynthesis of *Arctium lappa* L. Biosci Biotechnol Biochem 66:1262–1269

Tavakkoli E, Fatehi F, Coventry S, Rengasamy P, McDonald GK (2011) Additive effects of Na^+ and Cl^- ions on barley growth under salinity stress. J Exp Bot 62:2189–2203

Telesiński A, Nowak J, Smolik B, Dubowska A, Skrzypiec N (2008) Effect of soil salinity on activity of antioxidant enzymes and content of ascorbic acid and phenols in bean (*Pheseolus vulgaris* L.) plants. J Elementol 13:401–409

Tiwari JK, Munshi AD, Kumar R, Pandey RN, Arora A, Bhat JS, Sureja AK (2010) Effect of salt stress on cucamber: Na+– K+ratio, osmolyte concentration, phenols and chlorophyll content. Acta Physiol Plant 32:103–114

Togami J, Tamura M, Ishiguro K, Hirose C, Okuhara H, Ueyama Y, Nakamura N, Yonekura-Sakakibara K, Fukuchi-Mizutani M, Suzuki K, Fukus Y, Kusumi T, Tanaka Y (2006) Molecular characterization of the flavonoid biosynthesis of *Verbena hybrida* and the functional analysis of verbena and *Clitoria ternatea F3_5_H* genes in transgenic verbena. Plant Biotechnol 23:5–11

Tsormpatsidis E, Henbest RGC, Davis FJ, Battey NH, Hadley P, Wagstaffe A (2008) UV irradiance as a major influence on growth, development and secondary products of commercial importance in Lollo Rosso lettuce 'Revolution' grown under polyethylene films. Environ Exp Bot 63:232–239

Umezawa T (2003) Diversity in lignan biosynthesis. Phytochem Rev 2:371–390

Uphoff RA, Van den Maagdenberg AM, Roon KI, Ferrari MD, Frantus RR (2001) The impact of pharmacogenetics for migraine. Eur J Pharmacol 9:1–10

Valko M, Rhodes CJ, Moncol J, Izakovic M, Mazur M (2006) Free radicals, metals and antioxidants in oxidative stress-induced cancer. Chem Biol Interact 160:1–40

Vinocur B, Altman A (2005) Recent advances in engineering plant tolerance to abiotic stress: achievements and limitations. Curr Opin Biotechnol 16:123–132

Von Szent-Györgyi A (1928) Observations on the function of peroxidase systems and the chemistry of the adrenal cortex. Description of a new carbohydrate derivative. Biochem J 22:1387–1409

Wang MY, Brett JW, Jensen CJ, Nowicki D, Su C, Paul AK, Anderson G (2002) *M. citrifolia* (Noni): a literature review and recent advances in noni research. Acta Pharmacol Sin 23:1127–1141

Waris G, Ahsan H (2006) Reactive oxygen species: role in the development of cancer and various chronic conditions. J Carcinog 5:14–22

Waśkiewicz A, Irzykowska L, Bocianowski J, Karolewski Z, Kostecki M, Weber Z, Goliński P (2010) Occurrence of *Fusarium* fungi and mycotoxins in marketable asparagus spears. Pol J Environ Stud 49:367–372

Weidenhamer JD, Romeo JT (2004) Allelochemicals of *Polygonella myriophylla*: chemistry and soil degradation. J Chem Ecol 20:1067–1082

Weir WL, Bais HP, Vivanco JM (2003) Intraspecific and interspecific interactions mediated by a phytotoxin, (–)-catechin, secreted by the roots of *Centaurea maculosa* (spotted knapweed). J Chem Ecol 29:2397–2412

Winkel-Shirley B (2001) Flavonoid biosynthesis. A colorful model for genetics, biochemistry, cell biology, and biotechnology. Plant Physiol 126:485–493

Yıldıztugay E, Sekmen AH, Turkan I, Kucukoduk M (2011) Elucidation of physiological and biochemical mechanisms of an endemic halophyte *Centaurea tuzgoluensis* under salt stress. Plant Physiol Biochem 49:816–824

Yiu J-C, Tseng M-J, Liu C-J, Kuo C-T (2012) Modulation of NaCl stress in *Capsicum annuum* L. seedlings by catechin. Sci Hortic 134:200–209

Yokoi S, Quintero FJ, Cubero B, Ruiz MT, Bressan RA, Hasegawa PM, Pardo JM (2002) Differential expression and function of *Arabidopsis thaliana* NHX Na+/H+antiporters in the salt stress response. Plant J 30:529–539

Yuan G, Wang X, Guo R, Wang Q (2010) Effect of salt stress on phenolic compounds, glucosinolates, myrosinase and antioxidant activity in radish sprouts. Food Chem 121:1014–1019

Zabala G, Zou J, Tuteja J, Gonzalez DO, Clough SJ, Vodkin LO (2006) Transcriptome changes in the phenylpropanoid pathway of *Glycine max* in response to *Pseudomonas syringae* infection. Plant Biol 6:26–44

Zadernowski R, Naczk M, Nesterowicz J (2005) Phenolic acid profiles in some small berries. J Agric Food Chem 53:2118–2124

Zahedi SM, Nabipour M, Azizi M, Gheisary H, Jalali M, Amini Z (2011) Effect of kinds of salt and its different levels on seed germination and growth of basil plant. World Appl Sci J 15:1039–1045

Zhao J, Ren W, Zhi D, Wang L, Xia G (2007) Arabidopsis DREB1A/CBF3 bestowed transgenic tall rescue increased tolerance to drought stress. Plant Cell Rep 26:1521–1528

Zheng W, Wang SY (2001) Antioxidant activity and phenolic compounds in selected herbs. J Agric Food Chem 49:5165–5170

Zhu J-K (2007) Plant salt stress. In: Encyclopedia of Life Sciences. John Wiley & Sons, Ltd West Sussex, pp 1–3

Zrig A, Tounekti T, Vadel AM, Mohamed HB, Valero D, Serrano M, Chtara C, Khemira H (2011) Possible involvement of polyphenols and polyamines in salt tolerance of almond rootstocks. Plant Physiol Biochem 49:1313–1322

Chapter 12
Polyamines and Their Roles in the Alleviation of Ion Toxicities in Plants

Stanley Lutts, Jean-François Hausman, Muriel Quinet, and Isabelle Lefèvre

12.1 Introduction

Polyamines (PAs) are small aliphatic amines that are ubiquitous to all living organisms. In plants, the most abundant polyamines are putrescine (Put, 1,4-diaminobutane), spermidine (Spd, *N*-3-aminopropyl-1,4-diaminobutane) and spermine (Spm, bis(*N*-3-aminopropyl)-1,4-diaminobutane) in amounts varying from micromolar to more than millimolar. Cadaverine (Cad, 1,5-diaminopentane) was also reported in some higher plants, in particular Gramineae, Leguminoseae and Solanaceae. Some unusual polyamines with longer amine chain have also been detected (Bagni and Tassoni 2001). Plant polyamines have been suggested to play important roles in morphogenesis, growth, embryogenesis, organ development, leaf senescence, biotic and abiotic stress responses (Martin-Tanguy 1997; Walden et al. 1997; Kakkar and Sawhney 2002; Kuznetsov et al. 2006; Kusano et al. 2007; Groppa and Benavides 2008; Kusano et al. 2008). Despite the fact that their possible implications in the latest processes have long been reviewed, their exact functions and their way of action remain elusive.

S. Lutts (✉) • M. Quinet
Groupe de Recherche en Physiologie végétale (GRPV),
Earth and Life Institute – Agronomy, Université catholique de Louvain,
Croix du Sud 4-5 bte L7.07.13, Louvain-la-Neuve 1348, Belgium
e-mail: stanley.lutts@uclouvain.be

J.-F. Hausman
Department EVA, Centre de Recherche Public – Gabriel Lippmann,
41, rue du Brill, Belvaux L-4422, Luxembourg

I. Lefèvre
Groupe de Recherche en Physiologie végétale (GRPV),
Earth and Life Institute – Agronomy, Université catholique de Louvain,
Croix du Sud 4-5 bte L7.07.13, Louvain-la-Neuve 1348, Belgium

Department EVA, Centre de Recherche Public – Gabriel Lippmann,
41, rue du Brill, Belvaux L-4422, Luxembourg

P. Ahmad et al. (eds.), *Ecophysiology and Responses of Plants under Salt Stress*,
DOI 10.1007/978-1-4614-4747-4_12,

Ion toxicity is a specific stress condition occurring when an essential element is absorbed and accumulates within plant tissues in excess, leading to metabolic impairment, or when non-essential elements become fully available in the soil and are absorbed by the transpiration stream. Metabolic disorders resulting from ion toxicity are a consequence of a direct toxicity of accumulated ions, induced deficiencies in other essential elements or alterations in the plant water status. Salinity, heavy metals and aluminium are the most important ion toxicities affecting plant growth and development in both field and natural environment. Under flood conditions, iron excess resulting from reduction of ferric to ferrous form may also affect plant development. Under field conditions, excessive application of fertilizers may induce transient ion toxicities, especially at the seedling stage. Behaving at physiological pH as polycations, polyamines were reported to be involved in the plant response to most ion toxicities, even if their underlying physiological functions remain unclear in relation to the complexity of the environmental constraint which may vary in field conditions over a small scale and involve several toxic elements, but also because of the multifarious roles of polyamines in plant growth and developmental processes.

12.2 Metabolic Pathways: Specificities of Polyamine Metabolism in Plants

In plants, PAs have not only been localized in the cytoplasm, but also in organelles such as vacuoles, mitochondria, chloroplasts and nucleus (Kumar et al. 1997). In plants, PAs are synthesized from two major alternative pathways, which are represented in Fig. 12.1. Starting from arginine, the diamine putrescine may be formed via ornithine by arginase (EC 3.5.3.1) and ornithine decarboxylase (ODC, EC 4.1.1.17). Putrescine can also be synthesized via agmatine by three sequential reactions catalysed by arginine decarboxylase (ADC, EC 4.1.1.19), agmatine iminohydrolase (AIH, EC 3.5.3.12), and *N*-carbamoylputrescine amidohydrolase (CPA, EC 3.5.1.53), respectively. As far as we know, this previous pathway is specific to plants. The existence of these alternative biosynthetic pathways may be explained by the differential localization of the two enzymes ADC and ODC, ADC being localized in thylakoid membranes of chloroplasts and ODC being localized in the nucleus and tightly associated with chromatin, resulting in the specific regulation of different plant processes. Putrescine can also be directly converted from agmatine by the action of agmatinase (EC 3.5.3.11). Ornithine can be converted back to citrulline by transcarbamoylase activity (EC 2.1.3.3) and then arginine by the action of ARS synthase (EC 6.3.4.5) and ARS lyase (EC 4.3.2.1). In *Sesamum* leaves and *Helianthus tuberosus* tuber explants, a citrulline decarboxylase activity which forms *N*-carbamoylputrescine from citrulline was also detected (Bagni and Tassoni 2001). Spd and Spm are synthesized from Put by the addition of aminopropyl groups, transferred from decarboxylated S-adenosyl methionine (SAM) in reactions catalysed by Spd synthase (EC 2.5.1.16) and Spm synthase (EC 2.5.1.22). Decarboxylated SAM is produced from SAM by SAM decarboxylase (SAMDC, EC 4.1.1.50).

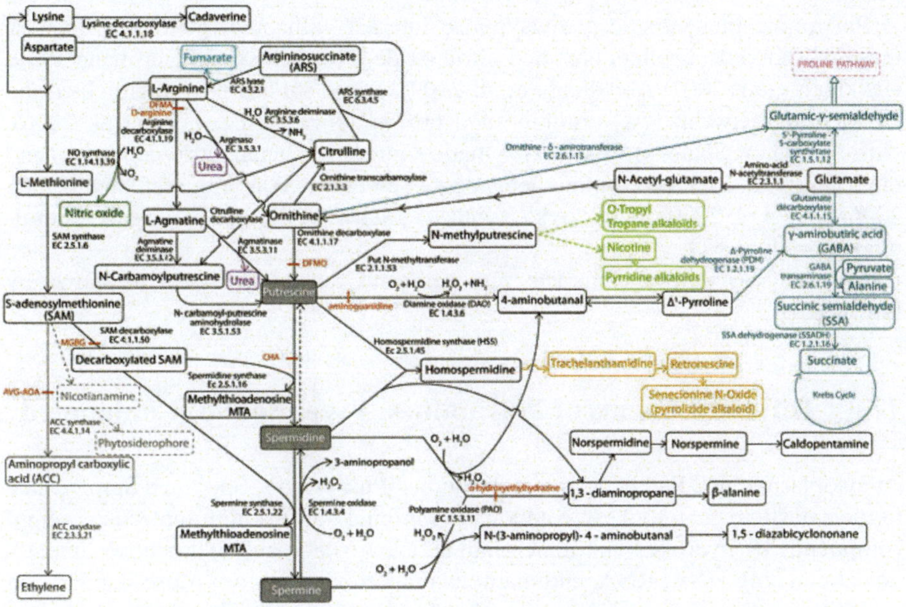

Fig. 12.1 Polyamine biosynthetic and catabolic pathway, and linkage to other related pathways

The diamine Cad is derived from lysine via a reaction of lysine decarboxylase (LDC, EC 4.1.1.18), an enzyme mainly localised in chloroplasts. The biosynthesis of the aromatic amines phenylethylamine and tyramine has not yet been described.

Polyamine catabolism in plants is catalysed by two oxidative enzymes, copper-containing diamine oxidase (DAO, EC 1.4.3.6) and flavoprotein-dependant polyamine oxidase (PAO, EC 1.5.3.11). Both enzymes are localized in the cell walls. DAO catalyses the oxidation of Put to 4-aminobutanal with concomitant production of NH_3 and H_2O_2, and the resulting aldehyde is further metabolised to γ-aminobutyric acid via Δ^1-pyrroline. PAO catalyses the conversion of Spd and Spm to 4-aminobutanal and N-(3-aminopropyl)-4-aminobutanal, respectively, along with the production of 1,3-diaminopropane and H_2O_2. A backcross conversion of Spm to Spd, and from Spd to Spm similar to that found in animals, has recently been demonstrated (Tavladoraki et al. 2006; Moschou et al. 2008c).

As shown in Fig. 12.1, ethylene and PAs share a common precursor, SAM, which can be metabolised successively to 1-amino cyclopropane-1-carboxylic acid (ACC) and ethylene. The presence of this common precursor can be a cause for reciprocal relationships between PAs and ethylene. Spm has been shown to inhibit the enzymes of ethylene biosynthesis, ACC oxidase in particular, (Locke et al. 2000). Conversely, ethylene has been shown to inhibit the activity of ADC and ODC, but increase DAO's one, an enzyme involved in PAs catabolism (Aziz et al. 1997; Torrigiani et al. 2003). Polyamines also share this common precursor SAM with nicotianamine and phytosiderophores in graminaceous species (Fig. 12.1). Competition between PAs and phytosiderophores has recently been reported (Tari et al. 2006).

Polyamine biosynthetic pathways also interact with other pathways; such as those of alkaloids, proline, urea and nitric oxide (NO) (Ober and Hartmann 1999; Häkkinen et al. 2007; Abdelhady et al. 2009). Some studies have highlighted the interactions between PAs, ornithine and proline (Jiménez-Bremont et al. 2006). Nitric oxide originates from two enzymatic sources in plants, nitric-dependent and the recently discovered arginine-dependant (Tun et al. 2006; Yamasaki and Cohen 2006; Gao et al. 2009). Nitric oxide synthase (NOS, EC 1.14.13.39) catalyses arginine to produce NO.

PA catabolic pathway interacts with pathways related to GABA as shown by Fig. 12.1.

12.3 Different Forms of Polyamines: Free, Bound, Conjugated

At cytoplasmic pH, PAs behave as polycations (Put^{2+}, Cad^{2+}, Spd^{3+} and Spm^{4+}). Due to their positive charge, these compounds can bind various high molecular weight compounds by hydrogen and ionic binding, electrostatic and hydrophobic interactions, including DNA, RNA, chromatin and proteins, which can cause stabilization or destabilization of these large molecules. Furthermore, they can covalently conjugate to endoglutamines of proteins by action of transglutaminase (TGase, EC 2.3.2.13) (Serafini-Fracassini et al. 1995). These molecules can also be conjugated by transferase to low molecular weight compounds like hydroxycinnamic acids, especially in Solanaceae, forming phenol amides. Put mainly forms monomers (PCA-soluble fraction) with coumaric acid, caffeoyl acid or ferulic acid, but can also conjugate dimmers of these hydroxycinammic acids (PCA-insoluble fraction) (Martin-Tanguy 1997; Bagni and Tassoni 2001). Thus, these ligands are of particular importance for the regulation of PAs concentration inside the cell, but are also implicated in a myriad of fundamental cellular processes, including regulation of gene expression, translation, cell proliferation, modulation of cell signalling, and membrane stabilization (Martin-Tanguy 1997). Also, Put can be methylated to form methylputrescine, the first specific precursor of nicotine and tropane alkaloids (Hashimoto et al. 1989). The balance of free and conjugated PAs is critical for different developmental processes and the relative proportions of each form may vary among different species as reviewed by Bagni and Tassoni (2001) and Martin-Tanguy (1997).

12.4 Polyamine Transport

Application of exogenous PAs, labelled or not ([^{14}O]Put, [^{14}C]Put and [^{14}C]Spd), at the whole plant or tissue level, gave evidences of their absorption and translocation in plants, their translocation across tonoplast and plasma membrane (Antognoni et al. 1998).

At the whole plant level, PAs have been identified in xylem as well as in phloem, but only under their free form (Friedman et al. 1986; Antognoni et al. 1998) and a long-distance transport has been displayed by Rabiti et al. (1989) and by Caffaro et al. (1993). Transport rate may be plant age-dependent and contents in xylem appear much lower than those in phloem (Antognoni et al. 1998). Differential impact of some abiotic stress on Put, Spd and Spm translocation from root to shoot suggests that such translocation does not occur passively through the transpiration stream but that it may involve specific transporters contributing to xylem loading (Ndayiragije and Lutts 2006a).

Nevertheless, despite the fact that long-distance translocation of PAs and transport between organelles and cells has been displayed for numerous years in plants, few studies have focused on the mechanisms involved in these transports. Several PA transport systems have been proposed in plants, but none have been identified at the molecular level. Some results from uptake-experiments of Put and Spd into carrot cells suggested that the entry of PAs into the cells is driven by a transmembrane electrical gradient, with a possible antiport mechanism between external and internal PAs (Pistocchi et al. 1987). Other results suggested that in maize roots, the bulk of exogenously applied Put is transported across the plasma membrane by a carrier-mediated process (DiTomaso et al. 1992a).

Whether single or different carriers for PAs exist, as well as the regulation of PAs allocation, have yet to be elucidated.

12.5 Modelling Plant Polyamine Status Through the Use of Specific Inhibitors and Transgenic Approaches

12.5.1 Specific Inhibitors

Irreversible inhibitors of the enzymes of PA metabolism have been developed initially to investigate PA biosynthesis and catabolism during plant development and their possible functions in growth and differentiation processes as reviewed by Bais and Ravishankar (2002). Their use to induce disturbances of cellular PA homeostasis by partial inhibition of their metabolic pathway have enhanced our understanding of PA regulation in response to abiotic stresses. The site of inhibition on PA metabolism is presented in Fig. 12.1.

DL-α-difluoromethylarginine (DFMA) and D-arginine are irreversible inhibitors of ADC activity, while DL-α-difluoromethylornithine (DFMO) is an irreversible inhibitor of ODC. The ability to carry out the selective inhibition of ADC or ODC with these compounds has provided valuable insights regarding the metabolic control of Put synthesis in response to stress or hormonal induction of decarboxylase activity. The increase in Put content occurring in stressed organs can be blocked by DFMA, but not by DFMO which suggests that Put synthesis is primarily the result of increased ADC activity rather than ODC activity in numerous stressed plant species (Benavides et al. 1997; Hao et al. 2005; Liu et al. 2005, 2006a). This, however, is not a general rule, as demonstrated in species treated with heavy metals in which an increase of ODC activity is largely involved in the increase of Put content (Wang and Kao 2006; Groppa et al. 2008).

Put incorporation into Spd and Spm can be blocked by methylglyoxal(bis-guanilhydrazone) (MGBG), an inhibitor of SAM decarboxylase, while Spd is inhibited by cyclohexylammonium sulphate (CHA). Diamine oxidase and PAO activity can be inhibited by aminoguanidine and α-hydroxyethylhydrazine, respectively (Fig. 12.1) (Aziz et al. 1998; Maiale et al. 2004; Ndayiragije and Lutts 2006b).

A modification in PAs titers can be related to a modification in PAs biosynthesis or in PAs catabolism. By compiling the experimental use of various inhibitors, the importance of each metabolic step of the PA pathway in the stress response can be pointed out. For example, D-arginine and CHA application showed that an increase of Put in response to NaCl can be related to an increase of PAs biosynthesis, although a parallel decrease of their catabolism is not excluded (Ndayiragije and Lutts 2006b).

Incorporation of inhibitors can give some clues about a putative beneficial or deleterious effect of PAs on plant growth and can help to understand the underlying mechanisms of PAs in plant response to ionic stress by comparing responses of plants organs or tissue differing in their PAs titer and/or ratios. For example, callus treatment with D-arginine led to more serious growth reduction than no treatment under salt stress (Liu et al. 2006b). Incorporation of D-arginine to rice culture alleviates the toxic effect of $AlCl_3$ on root growth (Wang and Kao 2006). Application of CHA, which induced an increase of Put content, increased salt deleterious effect at both root and shoot level (Ndayiragije and Lutts 2006b). Spermidine and Spm reduction induced by MGBG were associated with growth reduction in barley treated with salt (Liu et al. 2006b).

These inhibitors also help to understand interconnection between PAs pathway and proline one as they share common precursor and product (Fig. 12.1). By modulating the activity of ADC or DAO with DFMA or aminoguanidine, respectively, some authors have demonstrated the contribution of increasing Put synthesis or degradation under salt stress to GABA or proline formation (Aziz et al. 1998; Xing et al. 2007; Su and Bai 2008). Inhibitors of both PA and ethylene synthesis make it possible to probe the interaction of PAs with ethylene in a variety of experimental procedures, considering the fact that they share a common precursor (Fig. 12.1). Investigation with the use of aminoethoxyvinylglycine (AVG) or aminooxyacetic acid (AOA) (Aziz et al. 1997; Bar et al. 1998), both inhibitors of ethylene synthesis, demonstrated the competition between Spd, Spm and ethylene pathways, and the effect of the modulation of their regulation on plant growth in response to ionic stress (Bar et al. 1998). The use of AVG and MGBG also confirms a competition between PAs and phytosiderophore synthesis, which share a common SAM precursor (Fig. 12.1), that modulates Cu^{2+} accumulation in wheat (Tari et al. 2006).

Nevertheless, use of inhibitors as a tool to modify plant metabolism suffers from a number of limitations. Some doubts are raised about the specificity of some inhibitors. (Suzuki 1996; Scaramagli et al. 1999; Hao et al. 2005; Maiale et al. 2004). However, the effect of some other molecules as putative inhibitor of PAO has been investigated and some synthetic amines, PA analogues or oligamine appears to be good candidates in further assays (Cona et al. 2004; Maiale et al. 2008). Moreover, some authors had evidenced the possible interaction between the inhibitor and the stressing agent. Wang and Kao (2006) have suggested a possible binding of Al to the carboxyl group of D-arginine or methylornithine, which in turn reduce the uptake of Al.

12.5.2 Transgenic Approaches

Transgenics offer a convenient tool for studying the involvement of PAs not only in plant growth and development but also in response to ion toxicities. Similarly, identification of mutants affected in PA biosynthesis, especially in the model plant species *Arabidopsis thaliana*, allowed researchers to gain more information about their implication in growing processes. However, the resulting abnormal phenotypes of those mutants or transgenic lines are not always suitable for further test aiming to quantify their level of resistance to ion toxicities. considering the numerous functions of PAs and interaction with plant growth regulators, the only reliable strategy to analyse the impact of gene transfer affecting PA metabolism on stress response consists in the association of transferred coding sequences with inducible promoter (Masgrau et al. 1997). The overexpression of introduced gene might, in some cases, repress the expression of the corresponding endogenous gene.

The use of the constitutive promoter to drive gene expression does not constitute the optimal solution since endogenous polyamines usually accumulated to higher amounts in specific tissues and cell types (Watson et al. 1998; Clay and Nelson 2005). The consequence of overexpression of a given gene may widely vary depending on the considered organs: as an example, Thu-Hang et al. (2002) transferred SAMDC from *Datura stramonium* to rice and concluded that consequences were completely different in leaves and seeds from both a quantitative and qualitative point of view.

Finally, because of numerous post-transcriptional and post-translational regulatory processes, the overexpression of a given gene involved in the synthesis of a given polyamine does not always lead to accumulation in this polyamine in the free soluble form (Mayer and Michael 2003).

An ADC-coding gene under the control of an ABA-inducible promoter has been integrated in rice and transgenic plants exhibited a higher ADC activity in salt stress conditions (Roy and Wu 2001). However, transgenic plants remained unaffected for ODC, Spd synthase or SAMDC activities, therefore suggesting that components of PA pathways could be individually manipulated. Kasinathan and Wingler (2004) identified two mutants in *A. thaliana* exhibiting constitutively low levels of ADC: plants were unable to accumulate PAs in salt stress conditions and are hypersensitive to low NaCl dose, thus demonstrating the positive role of ADC in salt stress resistance. However, overexpression of ADC in transgenic lines do not allow to increase salt-resistance in *A. thaliana* and only leads to Put accumulation while Spd and Spm titers remain unaffected: transgenic plants usually present a dwarf phenotype with numerous morphological abnormalities (Alcazar et al. 2005).

Overexpression of Spm synthase encoding genes afford protection against numerous abiotic stresses, as recorded by Kasukabe et al. (2004) in *A. thaliana* and by Wen et al. (2008) in European pear (*Pyrus communis*).This gene therefore constitute a good candidate for future improvement of cross resistance against different types of stress in plants, although additional works are required to more precisely identify the impact of Spd in stressed tissues. Some experimental evidences based on microarray approaches suggest that Spd accumulation could trigger the expression of numerous stress-related genes, including transcription factors encoding genes. Surprisingly, overexpression of SAMDC is also leading to Spd accumulation but was not reported to afford similar advantages in terms of resistance to abiotic stresses. Manipulation of PA biosynthetic pathway in transgenic plants was also shown to provide protection not only to abiotic but also to biotic stresses (Hussain et al. 2011).

12.6 Responses to Ion Toxicities

12.6.1 Salt Stress

More than 800 million hectares of land worldwide are affected by salinity, which accounts for nearly 7 % of the world's total land area (FAO 2008). Some of these salt affected lands have arisen from natural causes (weathering of parental rocks, oceanic salts carried by wind etc.), but a significant proportion of recently cultivated agricultural lands has become saline owing to land clearing or irrigation, both of which causes water tables to rise and concentrate the salts in the root zone (Munns and Tester 2008).

It has been demonstrated that exogenous PAs can mitigate the deleterious effects of salt stress on seed germination (Zhu et al. 2006; Ali 2000; Ali et al. 2009; Chai et al. 2010) and may contribute to drastic increase in root and shoot elongations and fresh weights of the plants growing in saline conditions (Cavusoglu et al. 2007). The influence of exogenous PAs and their endogenous concentrations on the one hand, and the underlying cause of the registered improvement on the other hand, still remain a matter of debate. Some authors, for example, reported that exogenously-applied Put on salt-treated plants is hardly converted to Spd (Sarjala et al. 1997). Similarly, exogenous application of PAs may induce synthesis of uncommon PAs such as homospermidine (Larher et al. 1998). According to Roychoudhury et al. (2011), exogenously applied Spd and Spm may prevent salt-induced cellular damages in rice and help to maintain a proper K^+/Na^+ balance. Some authors, however, concluded that in some experimental systems, exogenously applied Spd and Spm could be immediately converted in diaminopropane as a consequence of PA oxidase activity and the response of salt-treated plants to a given exogenous PA should therefore not be regarded as a result of this unique compound (Bouchereau et al. 1999; Takahashi and Kakehi 2010).

Total PAs have been shown to increase in response to salinity (Santa-Cruz et al. 1998; Jiménez-Bremont et al. 2007; Upreti and Murti 2010) Nevertheless, there are numerous conflicting data concerning (1) the nature of accumulated PAs (2) the enzymes involved in PA synthesis in response to stress and (3) the nature of the initial signal triggering PA accumulation.

In numerous cases, ADC was directly involved in salt-stress-induced PAs over-synthesis (Hummel et al. 2004; Legocka and Kluk 2005). Transcriptional regulation of *adc* genes has been reported to occur under salt stress (Hao et al. 2005). Similarly, Chattopadhayay et al. (1997) demonstrated that ADC activity was higher in salt-resistant than in salt sensitive rice cultivars. Most plant species, however, contain several genes coding for arginine decarboxylase and these genes could be differently regulated in salt stress conditions (Mo and Pua 2002; Urano et al. 2003; Hummel et al. 2004; Quinet et al. 2010). According to Kumria and Rajam (2002), overexpression of ODC in tobacco resulted in increased salt resistance in this species. Activities of both the enzymes ODC and ADC were higher in salt-resistant Pokkali than in salt-sensitive IKP cultivar of rice, especially in the roots of salt-treated plants (Quinet et al. 2010). However, ADC and ODC activation leading to Put accumulation does not appear a sufficient step for salt tolerance improvement since several studies reported that such an improvement is a direct consequence of a stress-induced increase in (Spm + Spd/Put) ratio (Bar et al. 1996; El-Shintinawy 2000; Zapata et al. 2004). According to Santa-Cruz et al. (1998), Spd and Spm synthesis are therefore of a crucial importance and could be directly affected by light environment which directly influence both Spd- and Spm synthase activities. In maize, salinity is also triggering *Zmspds* genes (*Zmspds1* being hyperosmotic responsive while *Zmspds2* is NaCl responsive) (Jiménez-Bremont et al. 2007). Quinet et al. (2010) found that salt stress also increased DAO and PAO activities in the roots of salt-resistant Pokkali and in the shoots of salt-sensitive IKP, while Roy et al. (2005) found that the PAO activity was higher in salt-sensitive rice cultivars than in salt-tolerant ones. They also showed that majority of PAs bound to plasma membrane in the root of tolerant cultivars were found to be Spd and Spm, whereas the plasma membrane of salt sensitive rice cultivars contain only Put.

A protective role of PAs on photosystems has also been reported in salt-stressed rice (Chattopadhayay et al. 2002). Polyamines were recently demonstrated to alleviate the detrimental effects of increasing salinity on chromosomal aberrations (Tabur and Demir 2010). Polyamines have been demonstrated to act on ATPase activities and to be involved in regulation of several cation channels which could be involved in toxic Na^+ compartmentation at the cell level (Zhao et al. 2007; Shabala et al. 2007). Putrescine was recently proposed as a key component regulating salt excretion to epidermal trichomes in the halophyte species *Atriplex halimus* (Ben Hassine et al. 2009). (Ndayirajige and Lutts 2006a, b) demonstrated that exogenous application of PAs improved discrimination among monovalent cations at the root level of salt-stressed rice, especially at the xylem loading sites. Exogenous Put also improved CO_2 assimilation, at least partly as a consequence of an increased stomatal conductance, and also increased pollen viability allowing higher yields under moderate salinity (Ndayirajige and Lutts 2007).

Beside accumulation of toxic ions (namely Na^+ and Cl^-), salt stress also induces deficiencies in essential elements. A decrease in K^+ concentration was reported as an efficient signal triggering Put synthesis (Shabala et al. 2007). Similar observations have been reported for salt-induced deficiencies in Mg^{2+} (Geny et al. 1997), phosphate (Shih and Kao 1996) and Ca^{2+} (Anjum 2010). According to Watson and Malmberg (1996), a decrease in K^+ induced an increase in ADC activity occurring as a consequence of a post-translational activation process: although Put increases in response to K^+ deficiency, the endogenous concentration of Spd and Spm was reported to decrease under these conditions. According to Hummel et al. (2004), root elongation increased in response to agmatine accumulation occurring as a result of ADC activation: this accumulation could increase root elongation allowing the root system to explore soil profile at deeper depth in case of ion deficiency.

According to Xing et al. (2007), there is a positive correlation between the ability of plants to recover after exposure to salt stress and a decrease in GABA concentration issued from PA oxidation, thus suggesting that this compound could be involved in triggering of the plant defence mechanisms.

Salt stress is a complex environmental constraint which includes both an ionic and an osmotic component. The latter component is mainly due to decrease in the external water potential which, in turns, compromises water uptake by stressed plants. Accordingly, the putative involvement of PAs in plant response to water shortage is a matter of great interest. Osmo-protecting compounds such as proline may contribute to osmotic adjustment (Kishor et al. 2005). Polyamines and proline share common precursor and product (see Sect. 2). By blocking Put synthesis with DFMA or aminoguanidine in tomato leaf discs treated with salt, proline accumulation was also inhibited (Aziz et al. 1998). An increase of Put degradation via DAO activity under salt stress contributes to GABA and proline synthesis (Lin and Kao 2002; Xing et al. 2007; Su and Bai 2008). Polyamines may also interact in a very complex way in stomatal regulation. Liu et al. (2000) and Zhao et al. (2007) consider that all PAs may interact with Ca^{2+} channels involved in turgor regulation of gard cell and such an interaction may involve inositol 1,4,5-triphosphate as key components (Wilson et al. 2009). Peremarti et al. (2009) also demonstrated that a transgenic rice overexpressing SAMDC from *Datura stramonium* is more resistant to drought, especially during the recovery period occurring subsequently to stress relief.

12.6.2 Heavy Metals

Heavy metals are elements which density is higher than 5.0 or 6.0 $g\,cm^{-3}$, such as Cd, Cr, Hg, Pb, Al, Ag, Sn etc. or elements with a high atomic mass given dense sulphurs of low solubility (di Toppi and Gabbrielli 1999; Girard et al. 2005). Some of them are essential elements for all living organisms, including plants (Cu, Zn, Mn etc.) while others do not exert any known function in plant metabolism (Cd, Hg, Ni, Pb, Cr etc.). Some of these elements may create long term problems not only because they accumulate in organisms, and thus circulate in food chains, but also because they remain in the ecosystems in potentially dangerous concentrations, especially in sediments. The soil covering ore-bearing rock contains heavy metals (mainly Zn, Pb, Ni, Cr, Cu, Co) and metalloids (Mn, Cd, As, Se). Heavy metal contamination also occurs in industrial areas where the source includes road transport, refuse dumps and sewage sludge. Emission of dust from the metal processing industries is another source of contamination (especially Cd, Zn, Fe, Pb, Cu, Cr, Hg, Ni).

Cadmium is one of the major heavy metals that show phytotoxicity even at a low concentration. This metal can result in the damage of cell membrane by lipid peroxidation (Shah et al. 2001), inhibits growth and even causes plant death (di Toppi and Gabbrielli 1999; Aravind and Prasad 2003). Cadmium stress was reported to induce changes in PAs levels (Groppa et al. 2001, 2003), antioxidant enzyme activities and membrane permeability (Shah et al. 2001). Spermidine and Spm, but not Put, were effective in reducing $CdCl_2$-induced toxicity in rice and this protection is most likely related to the avoidance of H_2O_2 generation but also to the reduction of Cd uptake (Hsu and Kao 2007). In other experimental systems, Put was found to be the main PA overproduced in response to Cd toxicity (Groppa et al. 2003, 2008; Geuns et al. 1997; Balestrasse et al. 2005). Exogenous Spm reduced Cd-induced oxidative stress in wheat and induction of NO by this PA could play an important adaptive role in this respect (Groppa et al. 2008). In the specific case of Cd, both ADC and ODC are involved in Put overproduction and accumulation of this diamine could also partly result from DAO inhibition (Balestrasse et al. 2005). The consequences of Cd accumulation on PAs metabolism may also be influenced by the simultaneous presence of other ionic compounds, as it frequently occurs in dry areas of developing countries where the lack of wastewater treatment facilities may be responsible for considerable loads of both salt and heavy metals on agriculture soils. Lefèvre et al. (2009b) demonstrated that free Put, Spd and Spm increased in response to Cd + NaCl or Cd + KCl while only Put increased in response to Cd + $NaNO_3$.

Polyamines also accumulated in response to Zn and the conjugated fractions appear to assume key (although still unknown) functions in plant response to this element (Todeschini et al. 2007; Lingua et al. 2008). Zinc lead to overexpression of *PaADC* and *PaODC* in poplar shoots and both free and conjugated PAs accumulated in response to this metal (Franchin et al. 2007). Castiglione et al. (2009) recently established a strong correlation between the ability of poplar clones to cope with high amounts of Zn and Cu within roots on the one hand, and leaf accumulation of conjugated PAs on the other hand, suggesting that these compounds could play a role in root-to-shoot signalling. In Cu-treated carrot cell suspension, Put also accumulated in bound fractions (Gorecka et al. 2007). Light conditions and sugar synthesis were reported to be necessary for Cu-induced upregulation of PAs biosynthetic pathway (Lin and Kao 1999; Franchin et al. 2007; Lingua et al. 2008). Both Put and Spd were shown to decrease in radish seedlings exposed to Cu stress, although exogenous epibrassinolide mitigated the Cu-induced changes, therefore suggesting a link between PAs and brassinolides pathway in stress conditions (Choudhary et al. 2010).

Polyamines have also been shown to accumulate in response to Pb (Zacchini et al. 2003), Hg (Ding et al. 2010) and As (Mascher et al. 2002). Some heavy metals (especially Cu, Cd and Zn) could directly bind to PA groups to form Schiff base complexes in artificial experimental systems and this property was recently used for the synthesis of resins used for purification of heavy-metal containing industrial water (Kolodynska et al. 2008; Keypour et al. 2008). Since PAs exhibit a polycationic charge at physiological pH, such protective functions by direct chelating properties appear unlikely in plant cell, although the presence of additional anionic groups, such as phosphate, may directly influence the binding properties of heavy metals on amine groups of purine basis (Bregier-Jarzebowska et al. 2009). According to Ding et al. (2010) Hg stress disturbed the activities of ADC, ODC and PAO in water hyacinth leaves, and exogenous Spd can alleviate the metabolic disturbance of PAs caused by Hg.

In numerous polluted areas, several heavy metals are usually simultaneously present. Polyamine metabolism constitutes a promising target to confer cross-resistance to several pollutants since Wen et al. (2010) demonstrated that a transgenic pear (*Pyrus communis* L.) overexpressing apple *SPDS* (*MdSPDS1*) exhibited improved tolerance to $CdCl_2$, $PbCl_2$ and $ZnCl_2$ comparatively to wild type. Since PAs are involved in both plant growth regulation (see Sect. 8) and tolerance to heavy metals, they received attention in the framework of phytoremediation aiming at decontaminating polluted soil with high biomass producing plants. Shevyakova et al. (2011) recently demonstrated that spraying leaves of adult rape plants (*Brassica napus*) with Put markedly reduced the toxic effect of Ni on root growth, enhanced leaf supply with Fe and increased Ni content in young leaves by 2.5 times. Cicatelli et al. (2010) also demonstrated that arbuscular mycorrhizal fungi restore normal growth in white poplar clone exposed to heavy metal-contaminated soil and that this is associated with upregulation of PA biosynthetic gene expression.

12.6.3 Iron Toxicity

Iron is present in aerobic soils as Fe^{3+} but a large part of this element could be reduced to Fe^{2+} in waterlogged soils characterized by anaerobic conditions and low pH. Iron toxicity consequently appears in plants due to excessive Fe^{2+} absorption by roots. Since it concerns mainly (if not only) rice production, iron toxicity is very often a neglected ionic toxicity, although it constitutes a widespread nutrient disorder affecting the growth of wetland rice in Asia, Africa and South America. Majerus et al. (2007) demonstrated that PAs concentrations are higher in iron-resistant cultivar TOG7105 than in the iron-sensitive IRGC104047, one of the most obvious differences being the high concentration of Put in shoot of the former comparatively to the latter. Such an increase in Put occurred concomitantly with an iron-induced oversynthesis of ethylene, suggesting that PA synthesis from methionine was more efficient in the resistant than in the sensitive cultivar. In an in vitro system, it has also been reported that PAs may protect phospholipids vesicles or isolated DNA (Pedreno et al. 2005) from the deleterious effect of iron excess leading to PAs synthesis.

12.6.4 *Aluminium*

Aluminium (Al) belongs to the non-essential metals. Under neutral soil conditions, it exists in the non-phytotoxic insoluble form, whereas acidification of soil and soil water below pH dramatically enhances the release of the phytotoxic Al ion. Acid soils occupy up to 40 % of world arable lands and Al phytotoxicity may be considered as one of the major limiting factors of crop productivity in the world.

According to Wang and Kao (2006), Put accumulation in rice is the main factor causing root growth inhibition under Al stress. According to these authors, lignification could be responsible for both Put and $AlCl_3$-inibited growth of rice roots, although a direct impact of PAs on mitotic index independently of lignification process has also been demonstrated (Mahajan et al. 2009).

Putrescine accumulation in cell suspensions exposed to Al occurs concomitantly with exudation of organic acids (such as succinate, malate or oxalate) which play a key role in Al resistance, but no experimental evidence demonstrates a causal link between these events and exogenous succinate has no impact on PAs titers (Minocha and Long 2004). Aluminium strongly stimulates ADC activities in *Picea rubens* but has only a small impact on Spd and Spm concentrations (Minocha et al. 1996). Some authors postulated that Put accumulation is a direct consequence of Al-induced deficiencies in Ca concentrations and that this diamine exhibiting two positive charges at cellular pH could substitute with Ca^{2+} in various interactions with cellular components (Minocha et al. 1996). Mohapatra et al. (2010) generated poplar transgenic cell lines overexpressing mouse ODC and reported that oxidation of accumulated Put induced oxidative stress in response to Ca^{2+} deficiencies, although these transgenic lines also accumulated lower concentration of Al^{3+} on Al-enriched media.

Beside Put, Spd could assume protecting functions in response to Al toxicity: overexpression of Spd synthase (*MdSPDS1*) in *Pyrus communis* clearly improves plant tolerance to 30 μM $AlCl_3$ via ameliorating oxidative status through stimulation of superoxide dismutase and glutathione reductase (Wen et al. 2009).

12.7 Involvement of Polyamines in Sensing and Signalling

Ionic stresses can lead to a range of different constraints in plants, which have to cope with all these environmental signals. Plant adaptation to environmental stresses is dependent upon the activation of cascades of molecular networks involved in stress perception, signal transduction, and the expression of specific stress-related genes and metabolites which are described thereafter and in Fig. 12.2.

High salinity causes both (1) an ionic stress linked to the toxic effects of excess ions, mainly Na^+ and Cl^-; (2) an osmotic stress associated with low water potential of the root medium which creates a water deficit within the plant; (3) a nutritional

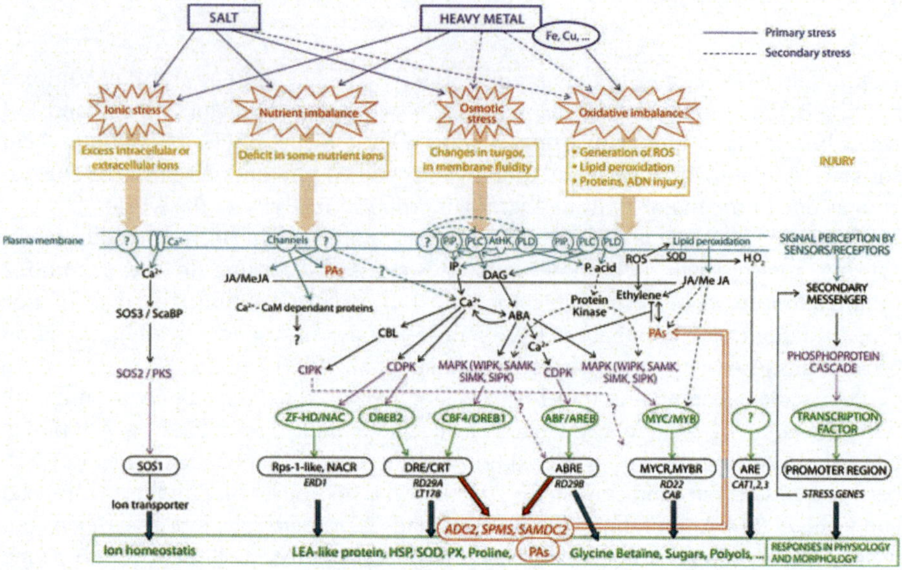

Fig. 12.2 Ionic stress sensing and signalling. The model begins with the perception of signals from environments by receptors, which then results in the activation of secondary messengers. An input signal can be osmotic stress (e.g., turgor changes) or derived from osmotic stress injury. These latest can modify membrane fluidity, and these alterations can be perceived by cells via sensory proteins embedded in membranes, while the way they perceive the signal remains unknown (Munnik and Meijer 2001; Los and Murata 2004). The histidine kinase AtHK1, identified by Urao et al. (1999), has been shown to be inactivated in reposne to high salinity.Membrane phospholoipids constitue a system that generate important signalling molecules, such as IP_3, phosphatidic acid (P. acid) and diacylglycerol (DAG) (Kacperska 2004). Transient IP_3 signals, involved in the opening of Ca channels, are generated by the phospholipase C (PLC)-mediated hydrolysis of phosphatidylinositol-4,5-biphosphate (PIP_2). The enzyme is activated by changes in cytosolic Ca in response to osmotic stress. Input signal can also be an excess or deprivation of ions which can be perceived by ion transporters or proteins. These proteins are supposed to be histidine kinase, while they have not been identified yet. Proteins then transmit the signals into the cell through the signal transduction pathways leading to gene expression and biochemical changes. The redox status has also been proposed as a stress-sensing system. Accumulation of H_2O_2, together with changes in the thiol disulfide status of the cell, provides the redox signal leading to changes in gene expression (Kacperska 2004). Secondary messengers can modulate intracellular Ca^{2+}-levels, often initiating a protein phosphorylation cascade that finally targets proteins directly involved in cellular protection or transcription factors controlling specific sets of stress-regulated genes. One of the major salt detoxyfing mechanisms in the cell is calcium activated SOS3-SOS2 protein complex, which activates SOS1, a Na^+/H^+-antiporter on the plasma membrane responsible for extrusion of Na^+ out of the cell. At the same time SOS3-SOS2 complex is also involved in inhibiting Na^+ transporter (Mahajan et al. 2008). SOS pathway appeared to be specific to high cytosolic Na^+ level. It is probably one of the most elucidated pathways. Osmotic stress may trigger both ABA-dependant and ABA-independent pathways (Xiong and Zhu 2002; Xiong et al. 2002). Second messengers modulate intracellular Ca^{2+} and/or ABA levels, often initiating a protein phosphorylation cascade (MAPK, CDPK, SOS2/PKS) that finally targets proteins directly involved in cellular protection or transcription factors controlling specific sets of stress regulated genes (Chinnusamy et al. 2004; Shao et al. 2007). Induction of the jasmonate pathway is related to heavy metal stress and nutrient deficency. They act as signal molecules and can increase ethylene concentration by stimulating the activity of ACC synthase and oxidase (Maksymiec 2007). Transcription factors are proteins with a DNA domain

imbalance caused by reduced nutrient uptake and/or transport to the shoot (Bartels and Sunkar 2005). Heavy metals also induce an ionic stress and a nutritional imbalance. They are present in the environment at such low concentrations that their effect on the substrate's osmotic potential is negligible. However, several studies have shown that toxic levels of these ions induce a secondary water stress (Poschenrieder and Barceló 1999 and references therein). Some metals such as Fe, Cu, Cr, Va and Co are able to induce oxidative stress via Fenton- Haber-Weiss-type reactions that produce reactive oxygen species (ROS). Salt stress also can however induce secondary oxidative stress by changes in the redox balance which occurs as a result of unspecific cell degradation processes (Møller 2001). All these constraints generated by excessive concentrations of ions can induce plant injuries by disruption of intracellular ion homeostasis, membrane dysfunction, photosynthesis impairment and inhibition of metabolic activity resulting in growth and yield reduction. The relative contribution of each different constraint to growth inhibition is difficult to assess, as many factors are involved such as ion concentrations, duration of exposure, plant species, cultivars, plant development stage and organ, environmental conditions etc. Progress has been made in identifying components of signalling pathways involved in salt and osmotic stresses, while those involved in heavy metals response are still scarce. Nevertheless, cell signalling still remains a puzzle, as primary target of toxicity, degree of response specificity, sensing and signalling events that lead to transcriptional activation are left mysterious.

The perception of the signal generates second messengers (such as inositol phosphates (IP), ROS, ABA, JA, Ca^{2+}). Calcium is an ubiquitous messenger elicited by numerous abiotic as well as developmental, hormonal and biotic stress cues (For reviews see Lecourieux et al. 2006; Hong-Bo et al. 2008a, b). Abscissic acid plays a major role in plant response to stress as it accumulates rapidly, mediate long-distance signal and also the expression of some genes responsive to abiotic constraints (Zhang et al. 2006; Wasilewska et al. 2008).

Fig. 12.2 (continued) that binds to the *cis*-acting elements present in the promoter of a target gene. They induce or repress the activity of the RNA polymerase, thus regulating gene expression. Transcription factors can be grouped into families according to their DNA-binding domain. A group of genes controlled by a certain type of transcription factor is known as a regulon. In the plant response to abiotic stresses, at least four different regulons can be identified: (1) CGF/DREB regulon; (2) NAC (NAM, ATAF and CUC), and ZF-HD (Zinc-finger homeodomain) regulon; (3) AREB/ABF (ABA-responsive element-binding protein/ABA-biding factor) regulon; (4) MYC (myelocytomatosis oncogene)/MYB (myeloblastosis oncogene) regulon. The first two regulons are ABA independent, and the last two are ABA dependent (Zhu 2001; Cheong et al. 2003; Agarwal et al. 2006; Kim 2006; Nakashima and Yamaguchi-Shinozaki 2006; Saibo et al. 2009). MYC/MYB regulon are also regulated by jasmonic acid (Abe et al. 2003). A recently discovered sensor, CBL (calcineurin B-like) protein, interact with a class of kinases known as CBL-interacting protein kinase (CIPK) to transduce the signal (Li et al. 2009). Mutations in these calcium sensors like *CBL1* and their interacting protein kinases have been shown to cause aberrations in the expression of some of the major stress responsive genes like *RD29* and *RD22* (Cheong et al. 2003; Mahajan and Tuteja 2005). The product of the stress-regulated genes may participate in the production of transcription factors and regulatory molecules like the plant hormones ABA, ethylene and salicylic acid (SA). Some of these regulatory molecules can, in turn, initiate a second round of circulation (Xiong and Zhu 2002; Xiong et al. 2002; Shao et al. 2007)

Despite contributing to stress damage, ROS are key components to many biological processes, notably to cell signalling pathway (Foyer and Noctor 2005; Fujita et al. 2006; Mittler et al. 2004). ROS may be sensed directly by key signalling proteins such as tyrosine phosphatase through oxidation of conserved cysteine residues (reviewed by Xiong et al. 2002), but ROS can also activate signal cascades via Ca^{2+} or ABA (Zhang et al. 2006). The chemical identity and the subcellular source of the ROS that accumulate during ionic stress could dictate the expression pattern of specific sets of genes and the induction of certain defence mechanisms.

Growth regulatory hormones have been shown to elicit tissue specific changes in PAs metabolism, despite the fact that precise mechanisms remain unknown. Polyamines have been shown to be the downstream components in ABA signalling (Liu et al. 2005). Moreover, some of the genes involved in polyamine biosynthesis (*ADC2*, *SPMS*, *SAMDC2*) are inducible by the ABA-dependent response of ionic stresses (Urano et al. 2003). These results are emphasized by the fact that stress – (DRE/CRT) and ABA-responsive elements (ABRE and/or ABRE-related motifs) are present in the promoters of these genes (Alcázar et al. 2006) (Fig. 12.2).

The initial signal triggering PAs oversynthesis in plants submitted to ionic stresses remains unknown. Despite the fact that ABA response is linked to osmotic stress, and that Erdei et al. (1996) postulated that an osmotic signal linked to the decrease in external osmotic potential is required for polyamine accumulation, this abiotic stress does not seem to be the only signal of PAs biosynthesis in short term response. As a matter of fact, Lefèvre et al. (2001) have shown that increase of ion and PA contents in rice subjected to different salt stresses occurred rapidly after 3 h independently of any change in the plant water status, while exposure to PEG during 12 h which induce a strong decrease in shoot osmotic potential and water content had only a limited impact on PAs content.

Ion imbalance and some nutrient elements deficiency generated by excess of some ions (Fig. 12.1) were reported to also generate increase of PAs (Sudha and Ravishankar 2002). Putrescine accumulation was notably reported in response to K or Mg deficiency. Calcium influence the PA transport through a cascade pathway involving protein kinase and phosphate activities.

Induction of the jasmonate pathway is related to heavy metal stress (Fig. 12.1) (Maksymiec 2007). Some studies have also highlighted a role of jasmonate in nutrient signalling (Armengaud et al. 2004). A strong up-regulation of *ADC2* occurs in response to a K^+-deficiency, often reported beside accumulation of toxic ions (see Sect. 6) (Armengaud et al. 2004). *ADC2* is reported to be a JA-responsive gene for various stresses (Goossens et al. 2003; Perez-Amador et al. 2002; Urano et al. 2003). Jasmonic acid and its methyl ester methyl jasmonate (MeJA) induce increase in free Put, soluble conjugated Put, Spd, Spm and N-methylPut in hairy roots of *Hyoscyamus muticus* (Biondi et al. 2000). Methyl jasmonate elicits massive accumulation of caffeoylputrescine in tomato leaves (Chen et al. 2006). This molecule induces a strong and rapid accumulation of N-methylPut in tobacco cells, both as free and soluble–conjugated forms (Goossens et al. 2003). It induces ADC, ODC and SAMDC genes (Goossens et al. 2003). Xu et al. (2004) demonstrated that *NtODC1* in tobacco is the gene encoding for ODC that is induced by MeJA. Sequence analysis revealed that *ODC1* promoter fragment contains both G-box and TGACG-motif elements similar to those found in other MeJA-responsive genes.

Polyamine catabolism produces H_2O_2 (see Sect. 2). Some authors have shown that a higher ROS production can be correlated with low PA biosynthetic activities or high PA catabolic activities Moschou et al. 2008b), indicating a possible feedback on gene regulation. Up-regulation of Spd synthase activity in transgenic *Arabidopsis thaliana* displayed a remarkable increase of ADC activity, Spd and Spm content in leaves, indicating that Spd synthase plays an important role in regulating the cellular levels of PAs (Kasukabe et al. 2004).

12.8 Polyamine Functions

Polyamines assume a myriad of biological functions during plant growth and development. Indeed, PA metabolism directly impacts on seed germination (Benavides et al. 1997), seedling growth (Urano et al. 2005), flowering processes (Gomez-Jimenez et al. 2010), fruit maturation (Mattoo et al. 2007) and senescence (Serafini-Fracassini et al. 2010). Polyamines are directly involved in both sexual (Falasca et al. 2010; Del Duca et al. 2010) and vegetative reproduction (Vusoku et al. 2006). Beside their involvement as key regulatory molecules in unstressed conditions, PAs may also act as protective molecules to repair or prevent stress-induced injuries through direct interaction with numerous cell structures and biochemical pathways.

12.8.1 Protein Protection

Since PAs are present in all cell compartments, they are thus able to interact with negatively-charged radicals of many enzymes and may help to stabilize their structure under ionic stress conditions. Spermine has been shown to afford protection to protein kinases in various plant species (Kuznetsov and Shevyakova 1997; Chang and Kang 1999; Smith and Maravolo 2004). According to Kuehn et al. (1983), both Spm and Spd directly act on phosphorylation of nuclear proteins in *Physarum polycephalum*. Smith and Maravolo (2004) also confirmed that Spm activates protein kinase in *Marchantia polymorpha*. However, Ye et al. (1994) showed that Put may also efficiently contribute to protein phosphorylation. Beside their positive impact on phosphorylating processes, PAs could also inhibit phosphatase activities (Guo and Roux 1995).

Specific interactions between PAs and target protein depend on acyltransferase (glutamyl-peptide γ-glutamyltransferase; EC 2.3.2.13) (Serafini-Fracassini et al. 1995). Polyamines may specifically interact with enzymes involved in cell cycle regulation (Del Duca et al. 2000a). Because of their positive charges, PAs may substitute to Mg^{2+} as cofactors of several key proteins, as reported by Athwal and Huber

(2002) for protein 14-3-3 where PAs bind at specific sites on loop n°8 and modify both the structure and the hydrophobicity of this regulatory protein.

Polyamines assume key protecting roles of several chloroplastic proteins, especially Rubisco, and thus help to maintain an efficient photosynthesis under stress conditions (Del Duca et al. 2000b). Besford et al. (1993) established that exogenous Spd and Spm effectively delayed the loss of D1, D2, cytochrome b6f protein components of thylakoid membrane and the stromal protein Rubisco (large subunit) and chlorophyll from osmotically treated oat leaf. Polyamines could interact with numerous proteins of PSII (H-bonding) through polypeptides C=O, C – N and N – H groups (Szalai et al. 1997; Kotzabasis et al. 1999; Legocka and Zajchert. 1999). It has been demonstrated however that some amines may thus cause the release of extrinsic polypeptides of 17, 23 and 33 kDa associated with the oxygen evolving complex and may have deleterious impact on PSII (Beauchemin et al. 2007) although this mainly occur for monoamines but not for naturally-occurring biogenic PAs (Hamdani et al. 2009).

Polyamines were also reported to have similar protective effects on proteins integrated in the respiratory chain (Torrigiani et al. 1986). Spermidine also protects superoxide dismutase (SOD) and ascorbate peroxidase (APX), allowing an appropriate response of stressed tissues to secondary oxidative stress resulting from toxic ion overload (Ahn and Jin 2004; see Sect. 8.2).

Beside specific interactions with given proteins, polyamines were also reported to amplify protein synthesis and to hamper protein degradation through protease inhibition (Scaramagli et al. 1999) and these properties could partly account for the antisenescing impact of PAs on stressed plant tissues.

12.8.2 Free Radical Scavenger and Protection Against Oxidative Stress

Oxidative stress is an important component of all types of ion toxicities. Since the early work of Drolet et al. (1986), PAs were demonstrated to act as free radical scavenger and could therefore efficiently prevent ROS – induced damages on plant cells (Hussain et al. 2011). Such property explains the positive impact of PAs on other ROS-inducing environmental constraints such as ozone (Langebartels et al. 1991) and paraquat application (Kurepa et al. 1998; Zheleva et al. 1994). Wang et al. (2007) consider that the protective impact of PAs against oxidative stress is the main component of resistance to Cu excess in *Nymphoides pelatum*. Similarly, Wen et al. (2010) demonstrated that Spd-levels are involved in enhanced heavy metal tolerance in *Pyrus communis* by exerting an antioxidant activity.

It still remains to be determined which fraction of PA displays such an antioxidative property: according to Kubis (2005), free Spd and Spm may directly detoxify superoxide anions while other authors consider that conjugated PAs are mainly responsible for this property through their hydrocynamic groups (Langebartels et al. 1991). Papadakis and Roubelakis-Angelakis (2005) demonstrated that PAs may inhibit ROS synthesis through inhibition of NADPH oxidase in mitochondria. Alternatively, PAs also stimulated the synthesis of antioxydants such as ascorbate and reduced glutathione in response to low temperatures and water stress in *Cicer arietinum* (Nayyar and Chander 2004) and to Cd-induced phytotoxicity in wheat roots (Groppa et al. 2008).

It has also been demonstrated that PAs may stimulate antioxidative enzymes such as peroxidase and glutathione reductase (Tang et al. 2004; Verma and Mishra 2005). According to Poduslo and Curran (1996), Put may form a complex with SOD, which could be targeted to the main sites of ROS production more efficiently than the isolated enzyme. Similarly, chilling-induced increase in PAs content in cold-resistant cucumber is directly related to maintenance of SOD, peroxidase and catalase activities, while all of these enzyme activities are inhibited in cold-sensitive genotypes unable to display efficient PAs accumulation (Zhang et al. 2009). According to Ye et al. (1997), resistance to oxidative stress in wheat is directly related to the stress-induced stimulation of both ADC and ODC activities leading to increased concentration of endogenous Put. Nevertheless, according to these authors, the response to a similar increase in Put in terms of resistance to secondary oxidative stress widely differs among cultivars.

As previously stated (see Sect. 2), PA catabolism itself produces H_2O_2 and poplar transgenic cell lines overexpressing mouse *ODC* gene suffers from numerous abnormalities as a direct consequence of oxidative stress resulting from Put oxidation (Mohapatra et al. 2009): these data therefore demonstrate that PAs synthesis and degradation need to be tightly regulated in stresses plant tissues.

12.8.3 Stabilization of Biological Membranes

Polyamines stabilize membrane biological structures and therefore contribute to the maintenance of cell functions and exchanges (Bouchereau et al. 1999; Groppa et al. 2008). Such a protective effect could be directly linked to their involvement as free radical scavengers or as regulatory molecules triggering antioxidative defense mecanisms (see Sect. 8.2.) since biological membranes are especially sensitive to the deleterious impact of ROS.

A close association between biological membranes and PAs is however frequently reported in the literature and could be part of a protective mechanism as first reported by Besford et al. (1993) in osmotically-stressed *Avena sativa*. This property may be partly explained by the positive charges of PAs, allowing them to substitute with Ca^{2+} in their stabilizing interaction with negatively charged phospholipids. Protection thus appears to be more efficient for Spd exhibiting three positive charges at cellular pH than for Put which only presents two positive charges, the latter having only minor effect on plasma membrane stability under stress conditions (Majewska-Sawka et al. 1998).

Similar protective functions of PAs have also been recorded on thylakoid membranes (Velikova et al. 1998; Srivastava et al. 1995). Interactions of PAs with light harvesting complexes (LHC) of photosynthetic tissues could have a direct impact on stress sensitivity: according to Sfichi et al. (2004), Put allows to reduce the size of LHCII while Spm increases it.

Beside phospholipids, PAs were also found to physically interact with membrane proteins through hydrophobic interactions involving both amino – and imino groups (Serafini-Fracassini et al. 1995). Protein PotD from *Escherichia coli* was the first one to be isolated and purified from microsomal fractions and it appears to be involved in complex regulation of transport systems across the plasmamembrane (Pistocchi et al. 1993). Soon after, a similar protein was isolated from *Cucurbita pepo* (Tassoni et al. 1996): it belongs to two different higher protein complexes of 66 and 44 kD (Tassoni et al. 1998) and fixes only Spd but not Put. Similar proteins were also isolated from maize but could be involved in Spd translocation across biological membranes rather than in protective mechanisms (Tassoni et al. 2002). Other interactions between PAs and membrane proteins may involve H^+-ATPase, which may contribute to Na^+ compartmentation in salt stressed-rice (Roy et al. 2005) and barley (Zhao and Qin 2005), although covalently conjugated PAs may also have a positive impact on tonoplastic H^+-ATPase from osmotically-stressed plants (Liu et al. 2004).

12.8.4 Regulation of Mineral Nutrition and Ion Homeostasis

Although ion toxicities have an impact on PA biosynthesis (see Sect. 6), some experimental evidences also demonstrate that PAs may conversely have a strong impact on plant nutrition and ion distribution, not only among plant organs within the plant but also within cell compartments. Polyamine synthesis offers an efficient strategy for charge homeostasis within cells since they constitute the results of a metabolic pathway, which could be regulated according to cellular nutritional status (Watson and Malmberg 1996). Accordingly, PAs (and especially Put) accumulation was reported to occur as a result of K deficiency (Jokela et al. 1997; Watson and Malmberg 1996) and it has been hypothesized that the diamine may substitute to K in some of its structural functions.

By reinforcing barrier effects of Casparian bands, exogenous Spd inhibits Na^+ transport from roots to shoots under conditions of high salinity and thus attenuates salt injuries in barley seedlings (Zhu et al. 2006). Exogenous Put and Spd were shown to improve growth of salt-treated rice in relation to an increase in K^+/Na^+ ratio in both roots and shoots (Krishnamurthy 1991; Ndayiragije and Lutts 2006a, 2007) and a similar effect was detected in undifferentiated calli (Ndayiragije and Lutts 2006b). Exogenous Spd and Spm also reduce salt-induced Mg deficiencies in salt-treated rice (Chattopadhayay et al. 2002) and Cu accumulation in *Nymphoides pelatum* exposed to this heavy metal (Wang et al. 2007).

Tamai et al. (1999) demonstrated that Put influence K^+ partitioning between roots and shoots in barley. Ben Hassine et al. (2009) also confirmed that endogenous PAs (Spd and Spm) are involved in the salt excretion process in epidermal trichomes of the xerohalophyte species *Atriplex halimus*. Some of these data suggest that PA could interfere with ion channels and transporters at the plant cell level. Indeed, it has been recently demonstrated that in plant cells, elevated PA levels in the cell cytosol modulates the activity of plasma membrane ion channels and improve ionic relation. Polyamines may activate H^+-ATPase through stabilisation of the regulatory 14-3-3 protein fixation on the enzyme (Garufi et al. 2007).

Shabala et al. (2007) used patch-clamp measurements on protoplasts and reported that micromolar concentrations of exogenous PAs are efficient in preventing NaCl-induced K^+ efflux. However, those externally applied PAs should be transported across the plasma membrane and exert their inhibitory effects on K^+-efflux and non-selective cation channel (NSCC)-mediated currents from the cytosolic side only. Since those NSCC are often cited as a major route for Na^+ uptake (Demidchik and Tester 2002), the fact that Put and Spm block NSCC may help to reduce the magnitude of NaCl-induced plasma membrane depolarization. Zhao et al. (2007) similarly confirm that PA significantly blocks the inward K^+ and Na^+ currents in root epidermal and cortical cells. These blocking effects of PAs were increased with increasing polyaction charge. In root xylem parenchyma, the inward K^+ currents were blocked by extracellular Spd while the outward K^+ currents were enhanced. According to these authors, PA improve K^+/Na^+ homeostasis in barley and should consequently be regarded as a self-protecting response. Vacuolar sequestration of Na^+ is also an important component of salt stress tolerance and Brüggemann et al. (1998) demonstrated that Spd and Spm inhibit fast activating vacuolar currents in tonoplast.

Polyamines may also regulate Ca^{2+}-permeable channels, and thus activate the K^+ inward at the plasma membrane in guard cells by raising cytoplasmic Ca^{2+} concentration, which would stimulate stomatal closure (Liu et al. 2000; Oliver et al. 2000; Yamaguchi et al. 2007). Spermine having a higher net charge has been shown to be more efficient than Put and Spd in this regulation. Putrescine is also suggested to indirectly regulate Ca^{2+}-channel *via* its catabolism and resulting H_2O_2 production (see Sect. 2).

12.8.5 Interaction with Cell Wall Components and Involvement in Lignification Processes

Polyamines are associated with primary cell wall and may directly interact with their pectic components. According to Messiaen et al. (1997), PA fixation on cell wall is a direct consequence of their charges at physiological pH and selectivity is an outcome of electrostatic interaction. Accordingly, Put is fixed only on type I rhamnogalacturonane. Polyamine may bind to polygalactutonic acid in a competitive way with both monovalent (Na^+) and divalent (Ca^{2+}) cations. According to Messiaen and Van Cutsem (1999), PAs may have an impact on polygalacturonic acid conformation and could also influence transduction signal associated with the mobilisation of pectic fragments.

Polyamines also influence pectine methylesterase activities (Charnay et al. 1992) and lignin synthesis (Angelini et al. 1993). It is indeed well established that diamine and PA oxidases present an apoplastic distribution and that H_2O_2 issued from PA oxidation is used for further lignin synthesis by associated peroxydases or for the establishment of cross-linking between extensin and phenolics associated with cell wall polysaccharides (Su et al. 2005; Fincato et al. 2011). Rea et al. (1998) demonstrated that apoplastic CuAO are directly involved in periderm formation in response to pathogenesis. Cross-linking between cell wall polymers appears to be affected by inhibitors of CuAO (Wisniewski et al. 2000). According to Laurenzi et al. (2001), those enzymes are associated with extension in intercellular spaces, suggesting that PA metabolism is directly influencing cell wall rigidification. Conversely, Spm was shown to inhibit polygalacturonase activity involved in pectin denaturation (Sitrit and Bennett 1998). All ionic toxicities involve a water stress component and PA-induced modifications of cell wall rheological properties may thus display a strong impact on relationship between cell turgor upholding and cell elongation processes. Since cell wall could constitute a site of sequestration for toxic divalent cations such as Pb or Zn, it could be hypothesized that PA accumulation could somewhat interfere with cell wall binding capacities.

12.8.6 Interaction with DNA and Regulation of Cell Cycle

One of the major symptoms of ion toxicity consists in plant growth inhibition, which may itself be triggered by a decrease in the cell division rate and/or inhibition of cell elongation. It also frequently involves specific gene expression and interactions between PAs and nucleic acid are of crucial importance in this respect. Polyamines could indeed be directly associated with DNA and chromatin and there is a direct correlation between ploidy level and nuclear amounts of PAs (Figueras et al. 1990). Polyamines directly influence DNA conformation as well as interactions between DNA and associated proteins (Vandenroeck et al. 1994; Hobbs et al. 2002). Polyamines were reported to stabilize DNA and to protect it from stress-induced damages, including thermal denaturation (Ha et al. 1998; Pagoria and Maravolo 2005; Terui et al. 2005). According to Liu et al. (2006c), the conjugated fraction of PAs assumes key functions in this respect. Because of its small size, Put may fix on both major and minor grooves of DNA while Spd and Spm are preferentially fixed on major grooves (Bryson and Greenall 2000). Polyamines may also interact with mRNA, tRNA and rRNA and could therefore have a positive impact on protein synthesis, independently from their impact on gene expression and chromatin stabilisation. Spermidine stabilizes ribosome structure and reduces the occurrence of dissociations of the two subunits (Mulo et al. 1998; Igarashi et al. 2006).

Self-assembly of PAs with phosphate ions has been demonstrated (D'Agostino and Di Luccia 2002): the intercalation of phosphate anion between the N-terminal ends of two PAs determines, through electrostatic interactions, the formation of basic cyclical structures that further aggregate into supramolecular complexes, thus producing aggregates that interact with genomic DNA and protect it from nuclease activities, from thermal denaturation or even from direct interactions with toxic ions such as Na^+ (D'Agostino et al. 2005, 2006; Di Lucia et al. 2009). As a consequence, PA could directly influence binding of transcription factors on DNA and may therefore indirectly control specific gene expression (Hiraga et al. 2000; Kasukabe et al. 2004; Sudha and Ravishankar 2002).

Some isoforms of ODC are directly associated with chromatin while ADC mainly presents a cytosolic localisation (Acosta et al 2005). Activity of ODC thus directly influences the rate of cell division in meristematic tissues. Not only Put but also Spd appears to assume key functions in this respect and activation of Spd synthase in actively growing tissues has also been reported (Cvikrova et al. 1999; Paschalidis and Roubelakis-Angelakis 2005). Some uncommon aromatic monoamine, such as tyramine, has also been shown to influence the rate of cell division in meristematic tissues (Martin-Tanguy and Carré 1993). An increase in PAs synthesis is usually recorded during the G1 phase of the cell cycle (Bueno et al. 1993). Stress-induced inhibition of cell division has been reported as a consequence of a decrease in Spm concentration (Hussain et al. 2011). Similarly, Fowler et al. (1996) demonstrated that cyclin required for mitosis coordination is activated by PAs.

12.8.7 Involvement in Abiotic Stress Signalling and in Gene Expression

As discussed in Sect. 7, the concentration-dependent imbalance in the antioxidant status can mediate a signal in plants (Foyer and Noctor 2005; Scandalios 2005). Among the different ROS generated in response to ionic stresses, H_2O_2 has been recognized as the most potent signalling ROS in plants owing to its relative stability and ability to diffuse through membranes (Van Breusegem et al. 2008). Evidences are provided that H_2O_2 issued from the degradation of Put by DAO is involved in ABA-induced stomatal closure in *Vicia faba* leaves (An et al. 2008). These authors have proposed a model of H_2O_2 generation by Put; DAO is activated in response to ABA which produces H_2O_2 by catalysing Put oxidation. The increase H_2O_2 production may activate the Ca^{2+} channel, resulting in an increased Ca^{2+} level in guard cells. Therefore, a Ca^{2+} messenger mediates ABA signalling processes and induces stomatal closure.

By overexpressing or downregulating apoplastic PAO, or downregulating SAMDC in tobacco, some authors suggest that apoplastic H_2O_2 generation induces either the expression of effector stress-responsive genes or the programmed cell death syndrome, depending on a specific H_2O_2 threshold linked to stress magnitude (Rea et al. 2004; Moschou et al. 2008a, b). The apoplastic catabolism of PAs could transduce more efficiently H_2O_2-derived signals, as apoplast hold weak antioxidant potential. Moschou et al. (2008b) suggest that when PA anabolism predominates over PA catabolism, programmed cell death fails to occur, whereas when the opposite occurs, programmed cell death is induced.

Nitric oxide, a highly reactive gaseous molecule that shares a common precursor with PAs (see Sect. 2), was shown to play a key role as a signalling molecule in plant responses to biotic and abiotic stresses through activation of various defence genes (Delledonne et al. 2001) or induction of a dose-dependent stomatal closure (Neill et al. 2002). Tun et al. (2006) presented evidences that Spd and Spm induce NO production in various tissues of *Arabidopsis thaliana* seedlings without apparent lag phase, whereas Put and arginine have little or no effect. Nitric oxide was found to be significantly increased in the roots of both Cd and PAs treated wheat, especially when exposed to Spm (Groppa et al. 2008). Cross-talk was demonstrated between PAs and NO in cucumber leaves under drought stress, suggesting that NO may act downstream of PAs (Arasimowicz-Jelonek et al. 2009). Although exogenous PAs did not affect NO production in well-watered cucumber seedlings, treatment of seedlings with PAs prior to imposition of water deficiency triggers NO production, Spd and Spm being more efficient than Put (Arasimowicz-Jelonek et al. 2009). PAs modulate NO production at least by the arginine-dependant pathway, which has been reported to be the main pathway of NO synthesis in water stress signalling (Hao et al. 2008).

Kasukabe et al. (2004) evidenced the role of Spd as a signalling molecule in plant response to stress in transgenic *Arabidopsis thaliana* over-expressing Spd synthase gene. Results from microarray analysis revealed the up-regulation of several stress-related genes in the transgenic Arabidopsis compared to the wild type linked to an increase in Spd and Spm content, more specifically genes encoding DREB transcription factors, DREB1A, DREB1B, and DREB2B. These DREB transcription factors regulate the expression of several stress-responsive genes by binding to the *cis*-acting DRE motif in the promoter region of the target genes. Kusano et al. (2008) demonstrated that Put activates transcription of genes coding for PAs absorption from the external medium. Terui et al. (2004) showed that more than 300 genes are specifically induced by PAs in *Escherichia coli* and that among them, 30 % are coding for transcription factors. In tomato fruits, more than 1,000 genes are directly regulated by PAs: they include gene coding for transcription factors, modulators of signal transduction, genes involved in amino acids and ethylene synthesis, isoprenoid pathway and flavonoid biosynthesis (Handa and Mattoo 2010). According to these authors, genes over-expressed in the presence of Spm and Spd are often down-regulated by Put. Transgenic tomato plants overexpressing a yeast SAMDC displayed a strong modification in the transcriptomic profile (Mehta et al. 2002).

12.8.8 Anti-senescing Properties

Like other abiotic stresses, ion toxicities hasten senescence in relation to a decrease in cell membrane stability, increase in protein degradation, photosynthesis inhibition and decrease in chlorophyll concentration (Lutts et al. 1996; Ghanem et al. 2009; Lefèvre et al. 2009a). Polyamines clearly exhibit antisenescing properties, which could be due to the fact that PAs oversynthesis reduces the production of the senescing hormone ethylene as a consequence of the competitivity of their metabolic pathways. Antisenescing properties of PAs could also be due to the protection afforded by these compounds to various cellular components (see Sects. 8.1, 8.3, 8.5, 8.6) as well as to their antioxidative properties (see Sect. 8.2) (Larher et al. 1998; Tassoni et al. 2006; Malik and Singh 2005). Antisenescing properties of PAs have been related to their impact on proteolytic, amylasic or nucleasic activities (Sung et al. 1994). Several PAs, including uncommon cadaverine, may drastically reduce endopeptidase (Desjouis et al. 1996), RNase (Galston and Kaur-Sawhney 1990) and lipoxygenase activities (Borrell et al. 1997). According to Antognoni et al. (1998), oxidation of PAs and mobilisation of Put through phloem could be observed during the first steps of senescence processes while Galston and Kaur-Sawhney (1990) reported that senescence could be directly linked to ADC degradation.

According to Serafini-Fracassini et al. (2010), antisenescing properties of Spd is directly associated with an inhibition of DNA fragmentation and an increase in chloroplast viability. Both Spd and Spm may closely associate with light harvesting complex, protect PSII from degradation and stabilize chlorophyll through the establishment of coordination links with central Mg (Della Mea et al. 2004).

12.9 Conclusions and Future Perspectives

As presented in this review, PAs assume numerous functions in sensing and signalling in response to ion toxicities, as well as in repairing or preventing stress-induced injuries. Different tools involving specific inhibitors of PA synthesis, exogenous application of PAs as well as transgenic approaches considerably improved our knowledge about PA functions. However, albeit numerous studies have been performed these last years to elucidate their way of action, numerous data are still lacking, such as the nature of the initial signal triggering PA accumulation, the nature of the accumulated PAs, their form (free, bound or conjugated) and the precise mechanisms of PA interaction with other molecules. The diversity of PA natural content between different species and cultivars also complicates the different approaches used to understand their metabolism.

An integrative approach of PA synthesis involving an insight into the identification of genes underlying the differential regulation of polyamine levels by traditional quantitative trait locus (QTL) mapping and cloning, as well as by transcriptomic analysis, coupled to a proteomic and metabolomic analysis will lead to a better overview of PA metabolism. Such tools could improve the understanding of some remaining missing data. Polyamine localization inside the cells and transport mechanisms also need to be elucidated. In this way, PA research still remains a challenge.

Yet, PA synthesis appears as a promising criterion to evaluate plant tolerance to mineral toxicities, and may probably be used in the future for selection of cultivated plants exhibiting a higher tolerance, assuming that the remaining questions previously mentioned would be resolved.

Acknowledgements This work was partly supported by the National Research Fund, Luxembourg, and the Ministry of Finances, Luxembourg.

References

Abdelhady MIS, Beuerle T, Ober D (2009) Homospermidine in transgenic tobacco results in considerably reduced spermidine levels but is not converted to pyrrolizidine alkaloid precursors. Plant Mol Biol 71:145–155

Abe H, Urao T, Ito T, Seki M, Shinozaki K, Yamagushi-Shinozaki K (2003) *Arabidopsis* AtMYC2 (bHLH) and AtMYB2 (MYB) function as transcription activators in abscisic acid signaling. Plant Cell 15:69–78

Acosta C, Perez-Amador MA, Carbonell J, Granell A (2005) The two ways to produce putrescine in tomato are cell-specific during normal development. Plant Sci 168:1053–1057

Agarwal PK, Agarwal P, Reddy MK, Sopory SK (2006) Role of DREB transcription factors in abiotic and biotic stress tolerance in plants. Plant Cell Rep 25:1263–1274

Ahn SM, Jin CD (2004) Protective role of exogenous spermidine against paraquat toxicity in radish chloroplasts. J Plant Biol 47:338–347

Anjum MA (2010) Response of Cleopatra mandarin seedlings to a polyamine-biosynthesis inhibitor under salt stress. Acta Physiol Plant 32:951–959

Alcazar R, Garcia-Martinez JL, Cuevas JC, Tiburcio AF, Altabella T (2005) Overexpression of *ADC2* in *Arabidopsis* induces dwarfism and late-flowering through GA deficiency. Plant J 43: 425–436

Alcázar R, Marco F, Cuevas JC, Patron M, Ferrando A, Carrasco P, Tiburcio AF, Altabella T (2006) Involvement of polyamines in plant response to abiotic stress. Biotechnology Letters 28:1867–1876

Ali RM (2000) Role of putrescine in salt tolerance of *Atropa belladonna* plant. Plant Science 152:173–179

Ali RM, Abbas HM, Kamal RK (2009) The effects of treatment with polyamines on dry matter and some metabolites in salinity – stressed chamomile and sweet majoram seedlings. Plant Soil Environ 55:477–483

An Z, Jing W, Liu Y, Zhang W (2008) Hydrogen peroxide generated by copper amine oxidase is involved in abscisic acid-induced stomatal closure in *Vicia faba*. J Exp Bot 59:815–825

Angelini R, Bragaloni M, Federico R, Infantino A, Portapuglia A (1993) Involvement of polyamines, diamine oxidase and peroxidase in resistance of chickpea to *Ascochyta rabiei*. J Plant Physiol 142:704–709

Antognoni F, Fornalè S, Grimmer C, Komor E, Bagni N (1998) Long-distance translocation of polyamines in phloem and xylem of *Ricinus communis* L. plants. Planta 204:520–527

Arasimowicz-Jelonek M, Floryszak-Wieczorek J, Kubis J (2009) Interaction between polyamine and nitric oxide signaling in adaptive responses to drought in cucumber. J Plant Growth Regul 28:177–186

Aravind P, Prasad MNV (2003) Zinc alleviates cadmium-induced oxidative stress in *Ceratophyllum demersum* L.: a free floating freshwater macrophyte. Plant Physiol Biochem 41:391–397

Armengaud P, Breitling R, Amtmann A (2004) The potassium-dependent transcriptome of *Arabidopsis* reveals a prominent role of jasmonic acid in nutrient signalling. Plant Physiol 136:2556–2576

Athwal GS, Huber SC (2002) Divalent cations of polyamines bind to loop 8 of 14-3-3 proteins, modulating their interaction with posphorylated nitrate reductase. Plant J 29:119–129

Aziz A, Martin-Tanguy J, Larher F (1997) Plasticity of polyamine metabolism associated with high osmotic stress in rape leaf discs and with ethylene treatment. Plant Growth Regul 21:153–163

Aziz A, Martin-Tanguy J, Larher F (1998) Stress-induced changes in polyamine and tyramine levels can regulate proline accumulation in tomato leaf discs treated with sodium chloride. Physiol Plant 104:195–102

Bagni N, Tassoni A (2001) Biosynthesis, oxidation and conjugation of aliphatic polyamines in higher plants. Amino Acids 20:301–317

Bais HP, Ravishankar GA (2002) Role of polyamines in the ontogeny of plants and their biotechnological applications. Plant Cell Tissue Organ Cult 69:1–34

Balestrasse KB, Gallego SM, Benavides MP, Tomaro ML (2005) Polyamines and proline are affected by cadmium stress in nodules and roots of soybean plants. Plant Soil 270:343–353

Bar Y, Apelbaum A, Kafkafi U, Goren R (1996) Polyamines in chloride-stressed citrus plants: alleviation of stress by nitrate supplementation via irrigation water. J Am Soc Horticult Sci 121:507–513

Bar Y, Apelbaum A, Kafkafi U, Goren R (1998) Ethylene association with chloride stress in citrus plants. Sci Hortic 73:99–109, Amsterdam

Bartels D, Sunkar R (2005) Drought and salt tolerance in plants. Crit Rev Plant Sci 24:23–58

Beauchemin R, Harnois J, Rouillon R, Tajmir-Riahi HA, Carpentier R (2007) Interaction of polyamines with proteins of photosystem II: cation binding and photosynthetic oxygen evolution. J Mol Struct 833:1–3

Ben Hassine A, Ghanem ME, Bouzid S, Lutts S (2009) Abscisic acid has contrasting effects on salt excretion and polyamine concentrations of an inland and a coastal population of the Mediterranean xero-halophyte species *Atriplex halimus*. Ann Bot 104:925–936

Benavides MP, Aizencang G, Tomaro ML (1997) Polyamines in *Helianthus annuus* L. during germination under salt stress. J Plant Growth Regul 16:205–211

Besford TR, Richardson CM, Campos JL, Tiburcio AF (1993) Effect of polyamines on stabilization of molecular-complexes in thylakoid membranes of osmotically stressed oat leaves. Planta 189:201–206

Bregier-Jarzebowska R, Gasowska A, Jastrzab R, Lomozik L (2009) Noncovalent interactions and coordination reactions in the systems consisting of copper(II) ions, aspartic acid and diamines. J Inorg Biochem 103:1228–1235

Biondi S, Fornale S, Oksman-Caldentey KM, Eeva M, Agostani S, Bagni N (2000) Jasmonates induce over-accumulation of methylputrescine and conjugated polyamines in *Hyoscyamus muticus* L. root cultures. Plant Cell Rep 19:691–697

Borrell A, Carbonell L, Farras R, Puig-Parellada P, Tiburcio AF (1997) Polyamines inhibit lipid peroxidation in senescing oat leaves. Physiol Plant 99:385–390

Bouchereau A, Aziz A, Larher F, Martin-Tanguy J (1999) Polyamines and environmental challenges: recent development. Plant Sci 140:103–125

Brüggemann LI, Pottosin II, Schönknecht G (1998) Cytoplasmic polyamines block the fast-activating vacuolar cation channel. Plant J 16:101–105

Bryson K, Greenall RJ (2000) Binding sites of the polyamines putrescine, cadaverine, spermidine and spermine on A- and B-DNA located by simulated annealing. J Biomol Struct Dyn 18:393–412

Bueno M, Garrido D, Matilla A (1993) Gene-expression induced by spermine in isolated embryonic axes of chickpea seeds. Physiol Plant 87:381–388

Caffaro S, Scaramagli S, Antognoni F, Bagni N (1993) Polyamine content and translocation in soybean plants. J Plant Physiol 141:563–568

Castiglione S, Todeschini V, Franchin C, Torrigiani P, Gastaldi D, Cicatelli A, Rinaudo C, Berta G, Biondi S, Lingua G (2009) Clonal differences in survival capacity, copper and zinc accumulation, and correlation with leaf polyamine levels in poplar: a large-scale field trial on heavily polluted soil. Environ Pollut 157:2108–2117

Cavusoglu K, Kilic S, Kabar K (2007) Some morphological and anatomical observations during alleviation of salinity (NaCl) stress on seed germination and seedling growth of barley by polyamines. Acta Physiol Plant 29:551–557

Chai YY, Jiang CD, Shi L, Shi TS, Gu WB (2010) Effects of exogenous spermine on sweet sorghum during germination under salinity. Biol Plant 54:145–148

Chang SC, Kang BG (1999) Effects of spermine and plant hormones on nuclear protein phosphorylation in Ranunculus petioles. J Plant Physiol 154:463–470

Charnay D, Nari J, Noat G (1992) Regulation of plant cell-wall pectin methyl esterase by polyamines – interactions with the effect of metal ions. Eur J Biochem 205:711–714

Chattopadhayay MK, Gupta S, Sengupta DN, Ghosh B (1997) Expression of arginine decarboxylase in seedlings of indica rice (*Oryza sativa* L.) cultivars as affected by salinity stress. Plant Mol Biol 34:477–483

Chattopadhayay MK, Tiwari BS, Chattopadhayay G, Bose A, Sengupta DN, Ghosh B (2002) Protective role of exogenous polyamines on salinity-stressed rice (*Oryza sativa*) plants. Physiol Plant 116:192–199

Chen H, Jones AD, Howe GA (2006) Constitutive activation of the jasmonate signalling pathway enhances the production of secondary metabolites in tomato. FEBS Lett 580:2540–2546

Cheong YH, Kim KN, Pandey GK, Gupta R, Grant JJ, Luan S (2003) CBL1, a calcium sensor that differentially regulates salt, drought, and cold responses in *Arabidopsis*. Plant Cell 15: 1833–1845

Chinnusamy V, Schumaker K, Zhu JK (2004) Molecular genetic perspectives on cross-talk and specificity in abiotic stress signalling in plants. J Exp Bot 55:225–236

Choudhary SP, Bhardwaj R, Gupta DT, Dutt P, Gupta RK, Biondi S, Kanwar M (2010) Epibrassinolide induces changes in indole-3-acetic acid, abscisic acid and polyamine concentrations and enhances antioxidant potential of radish seedlings under copper stress. Physiol Plant 140:280–296

Cicatelli A, Linhua G, Todeschini V, Biondi S, Torrigiani P, Castiglione S (2010) Arbuscular mycorrhizal fungi restore normal growth in a white poplar clone grown on heavy metal-contaminated soil, and this is associated with upregulation of foliar metalothionein and polyamine biosynthetic gene expression. Ann Bot-London 106:791–802

Clay NK, Nelson T (2005) *Arabidopsis* thickvein mutation affects vein thickness and organ vascularization, and resides in a provascular cell-specific spermine synthase involved in vein definition and in polar auxin transport. Plant Physiol 138:767–777

Cona A, Manetti F, Leone R, Corelli F, Tavladoraki P, Polticelli F, Botta M (2004) Molecular basis for the binding of competitive inhibitors of maize polyamine oxidase. Biochem-US 43:3426–3435

Cvikrova M, Binarova P, Cenklova V, Eder J, Machackova I (1999) Reinitiation of cell division and polyamine and aromatic monoamine levels in alfalfa explants during the induction of somatic embryogenesis. Physiol Plant 105:330–337

D'Agostino L, Di Luccia A (2002) Polyamines interact with DNA as molecular agregates. Eur J Biochem 269:4317–4325

D'Agostino L, di Pietro M, Di Luccia A (2005) Nuclear aggregates of polyamines are supramolecular structures that play a crucial role in genomic DNA protection and conformation. FEBS J 272:3777–3787

D'Agostino L, di Pietro M, Di Luccia A (2006) Nuclear agragates of polymaines. IUBMB Life 58:75–82

Del Duca S, Cai G, Di Sandro A, Serafini-Fracassini D (2010) Compatible and self-incompatible pollination in *Pyrus communis* displays different polyamine levels and transglutaminase activity. Amino Acids 38:659–667

Del Duca S, Creus JA, D'Orazi D, Dondini L, Bregoli AM, Serafini-Fracassini D (2000a) Tuber vegetative stages and cell cycle in *Helianthus tuberosus*: protein pattern and their modification by spermidine. J Plant Physiol 156:17–25

Del Duca S, Dondini L, Della Mea M, de Munoz P, Serafini-Fracassini D (2000b) Factors affecting transglutaminase activity catalysing polyamine conjugation to endogenous substrates in the entire chloroplast. Plant Physiol Biochem 38:429–439

Della Mea M, Di Sandro A, Dondini L, Del Duca S, Vantini F, Bergamini C, Bassi R, Serafini-Fracassini D (2004) A Zea mays 39-kDa thylakoid transglutaminase catalyses the modification by polyamines of light-harvesting complex II in a light-dependent way. Planta 219:754–764

Delledonne M, Zeier J, Marocco A, Lamb C (2001) Signal interactions between nitric oxide and reactive oxygen intermediates in the plant hypersensitive disease resistance response. Proc Natl Acad Sci U S A 98:13454–13459

Demidchik V, Tester M (2002) Sodium fluxes through nonselective cation channels in the plasma membrane of protoplast from *Arabidopsis* roots. Plant Physiol 128:379–387

Desjouis M, LeDily F, Boucaud J (1996) Evidence for a polyamine-mediated control of soluble nitrogen mobilization during post-clipping regrowth of white clover (*Trifolium repens* L.). J Plant Growth Regul 19:257–264

Di Lucia A, Picariello G, Iacomino G, Formisano A, Paduano L, D'Agostino L (2009) The in vitro aggregates of polyamines. FEBS J 276:2324–2335

di Toppi LS, Gabbrielli R (1999) Response to cadmium in higher plants. Environ Exp Bot 41:105–130

DiTomaso JM, Hart JJ, Kochian LV (1992a) Transport kinetics and metabolism of exogenously applied putrescine in roots of intact maize seedlings. Plant Physiol 98:611–620

Ding C, Shi G, Xu X, Yang H, Xu Y (2010) Effect of exogenous spermidine on polyamine metabolism in water hyacinth leaves under mercury stress. J Plant Growth Regul 60:61–67

Drolet G, Dumbroff EB, Legge RL, Thompson JE (1986) Radical scavenging properties of polyamines. Phytochemistry 25:367–371

El-Shintinawy F (2000) Photosynthesis in two wheat cultivars differing in salt susceptibility. Photosynthetica 38:615–620

Erdei L, Szegletes Z, Barabas K, Pestenacz A (1996) Responses in polyamine titer under osmotic and salt stress in sorghum and maize seedlings. J Plant Physiol 147:599–603

Falasca G, Franceschetti M, Bagni N, Altamura MM, Biasi R (2010) Polyamine biosynthesis and control of the development of functional pollen in kiwifruit. Plant Physiol Biochem 48: 565–573

Friedman R, Levin N, Altman A (1986) Presence and identification of polyamines in xylem and phloem exudates of plants. Plant Physiol 82:1154–1157

Figueras X, Gendy CA, Pinol MT, Van KTT, Tiburcio AF (1990) Polyamine content in relation to ploidy level and to different organs of *Nicotiana plumbaginifolia*. Plant Cell Physiol 31:823–828

Fincato P, Moschou PN, Spedaletti V, Tavazza R, Angelini R, Federico R, Roubelakis-Angelakis KA, Tavladoraki P (2011) Functional diversity inside the *Arabidopsis* polyamine oxidase gene family. J Exp Bot 62:1155–1168

Fowler MR, Kirby MJ, Scott NW, Slater A, Eliott MC (1996) Polyamine metabolism and gene regulation during the transition of autonomous sugar beet cells in suspension culture from quiescence to division. Physiol Plant 98:439–446

Foyer CH, Noctor G (2005) Redox homeostasis and antioxidant signaling: a metabolic interface between stress perception and physiological responses. Plant Cell 17:1866–1875

Franchin C, Fossati T, Pasquini E, Lingua G, Castiglione S, Torrigiani P, Biondi S (2007) High concentrations of zinc and copper induce differential polyamine responses in micropropagated white poplar (*Populus alba*). Physiol Plant 130:77–90

Fujita M, Fujita Y, Noutoshi Y, Takahashi F, Narusaka Y, Yamaguchi-Shinozaki K, Shinozaki K (2006) Crosstalk between abiotic and biotic stress responses: a current view from the points of convergence in the stress signaling networks. Curr Opin Plant Biol 9:436–442

Galston AW, Kaur-Sawhney RK (1990) Polyamines in plant physiology. Plant Physiol 94:406–410

Gao HJ, Yang HQ, Wang JX (2009) Arginine metabolism in roots and leaves of apple (*Malus domestica* Borkh.): the tissue-specific formation of both nitric oxide and polyamines. Sci Hortic 119:147–152, Amsterdam

Garufi A, Visconti S, Camoni L, Aducci P (2007) Polyamines as physiological regulators of 14-3-3 içnteraction with the plasma memebrane H^+-ATPase. Plant Cell Physiol 48:434–440

Geny L, Broquedis M, MartinTanguy J, Soyer JP, Bouard J (1997) Effects of potassium nutrition on polyamine content of various organs of fruiting cuttings of *Vitis vinifera* L cv Cabernet Sauvignon. Am J Enol Viticult 48:85–92

Geuns JMC, Cuypers AJF, Michiels T, Colpaert JV, VanLaere A, VandenBroeck KAO, Vandecasteele CHA (1997) Mung bean seedlings as bio-indicators for soil and water contamination by cadmium. Sci Total Environ 203:183–197

Ghanem ME, van Elteren J, Albacete A, Quinet M, Martinez-Andujar C, Kinet JM, Perez-Alfocea F, Lutts S (2009) Impact of salinity on early reproductive physiology of tomato (*Solanum lycopersicum*) in relation to a heterogeneous distribution of toxic ions in flower organs. Funct Plant Biol 36:125–136

Girard MC, Walter C, Rémy JC, Berthelin J, Morel JL (2005). Présence et impact des elements en trace dans les sols. In Sols et environnement: cours, exercices et études de cas. Dunod, Paris

Gomez-Jimenez MC, Paredes MA, Gallardo M, Fernandez-Garcia N, Olmos E, Sanchez-Calle IM (2010) Tissue-specific expression of olive S-adenosyl methionine decarboxylase and spermidine systhase genes and polyamine metabolism during flower opening and early fruit development. Planta 232:629–647

Goossens A, Hakkinen ST, Laakso I, Seppanen-Laakso T, Biondi S, De Sutter V, Lammertyn F, Nuutila AM, Soderlund H, Zabeau M, Inzé D, Oksman-Caldentey KM (2003) A functional genomics approach toward the understanding of secondary metabolism in plant cells. Proc Natl Acad Sci U S A 100:8595–8600

Gorecka K, Cvikrova M, Kowalska U, Eder J, Szafranska K, Gorecki R, Janas KM (2007) The impact of Cu treatment on phenolic and polyamine levels in plant material regenerated from embryos obtained in anther culture of carrot. Plant Physiol Biochem 45:54–61

Groppa MD, Benavides MP (2008) Polyamines and abiotic stress: recent advances. Amino Acids 34:35–45

Groppa MD, Benavides MP, Tomaro ML (2003) Polyamine metabolism in sunflower and wheat leaf discs under cadmium or copper stress. Plant Sci 164:293–299

Groppa MD, Rosales EP, Iannone MF, Benavides MP (2008) Nitric oxide, polyamines and Cd-induced phytotoxicity in wheat roots. Phytochemistry 69:2609–2615

Groppa MD, Tomaro ML, Benavides MP (2001) Polyamines as protectors against cadmium or copper-induced oxidative damage in sunflower leaf discs. Plant Sci 161:481–488

Guo YL, Roux SJ (1995) Partial-purification and characterization of an enzyme from pea nuclei with protein-tyrosine-phosphatase activity. Plant Physiol 107:167–175

Ha HC, Sirisoma NS, Kuppusamy P, Zweier JL, Woster PM, Casero RA (1998) The natural polyamine spermine functions directly as a free radical scavenger. Proc Natl Acad Sci USA 95:11140–11145

Häkkinen ST, Tilleman S, Świątek A, De Sutter V, Rischer H, Vanhoutte I, Van Onckelen H, Hilson P, Inzé D, Oksman-Caldentey KM, Goossens A (2007) Functional characterisation of genes involved in pyridine alkaloid biosynthesis in tobacco. Phytochemistry 68:2778–2785

Hamdani S, Tajmir-Riahi HA, Carpentier R (2009) Methylamine interaction with proteins of photosystem II: a comparison with biogenic polyamines. J Photochem Photobiol B 96:201–206

Handa AK, Mattoo AK (2010) Differential and functional interactions emphasize the multiple roles of polyamines in plants. Plant Physiol Biochem 48:540–546

Hao YJ, Kitashiba H, Honda C, Nada K, Moriguchi T (2005) Expression of arginine decarboxylase genes in apple cells and stressed shoots. J Exp Bot 56:1105–1115

Hao GP, Xing Y, Zhang JH (2008) Role of nitric oxide dependence on nitric oxide synthase-like activity in the water stress signaling of maize seedling. J Integr Plant Biol 50:435–442

Hashimoto T, Yukimune Y, Yamada Y (1989) Putrescine and putrescine N-methyltransferase in the biosynthesis of tropane alkaloids in cultured roots of *Hyoscyamus albus*.2. incorporation of labeled precursors. Planta 178:131–137

Hiraga S, Ito Y, Yamakawa H, Ohtsubo N, Seo S, Mitsuhara I, Matsui H, Honma M, Ohashi Y (2000) An HR-induced tobacco peroxidase gene is responsive to spermine, but not to salicylate, methyl jasmonate, and etephon. Mol Plant Microbe Interact 13:210–216

Hobbs CA, Paul BA, Glimour SK (2002) Deregulation of polyamine biosynthesis alters intrinsic histone acetyltransferase and deacetylase activities in murine skin and tumors. Cancer Res 62: 67–74

Hong-Bo S, Li-Ye C, Ming-An S (2008a) Calcium as a versatile plant signal transducer under soil water stress. Bioessays 30:634–641

Hong-Bo S, Li-Ye C, Ming-An S, Shi-Qing L, Ji-Cheng Y (2008b) Bioengineering plant resistance to abiotic stresses by the global calcium signal system. Biotechnol Adv 26:503–510

Hsu YT, Kao CH (2007) Cadmium-induced oxidative damage in rice leaves is reduced by polyamines. Plant Soil 291:27–37

Hummel I, Gouesbet G, El Amrani A, Ainouche A, Couée I (2004) Characterization of the two arginine decarboxylase (polyamine biosynthesis) paralogues of the endemic subantarctic cruciferous species *Pringlea antiscorbutica* and analysis of their differential expression during development and response to environmental stress. Gene 342:199–209

Hussain SS, Ali M, Ahmad M, Siddique KHM (2011) Polyamines: natural and engineered abiotic and biotic stress tolerance in plants. Biotechnol Adv 29:300–311

Igarashi K, Ueda S, Yoshida K, Kashiwagi K (2006) Polyamines in renal failure. Amino Acids 31: 477–483

Jiménez-Bremont JF, Becerra-Flora A, Hernández-Lucero E, Rodríguez-Kessler M, Acosta-Gallegos JA, Ramírez-Pimentel JG (2006) Proline accumulation in two bean cultivars under salt stress and the effect of polyamines and ornithine. Biologia Plant 50:763–766

Jiménez-Bremont JF, Ruiz OA, Rodríguez-Kessler M (2007) Modulation of spermidine and spermine levels in maize seedlings subjected to long-term salt stress. Plant Physiol Biochem 45:812–821

Jokela A, Sarjala T, Kaunisto S, Huttunen S (1997) Effects of foliar potassium concentration on morphology, ultrastructure and polyamine concentrations of Scot pine needles. Tree Physiol 17:677–685

Kacperska A (2004) Sensor types in signal transduction pathways in plant cells responding to abiotic stressors: do they depend on stress intensity? Physiol Plant 122:159–168

Kakkar RK, Sawhney VK (2002) Polyamine research in plants – a changing perspective. Physiol Plant 116:281–292

Kasinathan V, Wingler A (2004) Effect of reduced arginine decarboxylase activity on salt tolerance and on polyamine formation during salt stress in *Arabidopsis thaliana*. Physiol Plant 121: 101–107

Kasukabe Y, He L, Nada K, Misawa S, Ihara I, Tachibana S (2004) Overexpression of spermidine synthase enhances tolerance to multiple environmental stresses and up-regulates the expression of various stress-regulated genes in transgenic *Arabidopsis thaliana*. Plant Cell Physiol 45: 712–722

Keypour H, Azadbakht R, Salehzadeh S, Khanmohammadi H, Khavasi H, Adams H (2008) Synthesis, crystal structure and spectroscopic properties of some cadmium(II) complexes with three polyamine and corresponding macroacyclic Schiff base ligands. J Organomet Chem 693:2237–2243

Kim SY (2006) The role of ABF family bZIP class transcription factors in stress response. Physiol Plant 126:519–527

Kishor PBK, Sangam S, Amrutha RN, Laxmi PS, Naidu KR, Rao KRSS, Rao S, Reddy KJ, Theriappan P, Sreenivasulu N (2005) Regulation of proline biosynthesis, degradation, uptake and transport in higher plants: its implications in plant growth and abiotic stress tolerance. Curr Sci 88:424–438, India

Kolodynska D, Hubicki Z, Geca M (2008) Application of a new-generation complexing agent in removal of heavy metal ions from aqueous solutions. Ind Eng Chem Res 47:3192–3199

Kotzabasis K, Navakoudis E, Tsolakis G, Senger H, Dornemann D (1999) Characterization of thye photoreceptors responsible for the regulation of the intracellular polyamine level and the putative participation of heterotrimeric G-proteins in the signal transduction chain. J Photochem Photobiol B 50:38–44

Krishnamurthy K (1991) Amelioration of salinity effect in salt tolerant rice (*Oryza sativa* L) by foliar application of putrescine. Plant Cell Physiol 32:699–703

Kuehn GD, Atmar VJ, Daniels GR (1983) Polyamine-dependent protein-kinase from the slime-mold *Physarum polycephalum*. Methods Enzymol 94:147–154

Kubis J (2005) The effect of exogenous spermidine on superoxide dismutase activity, H_2O_2 and superoxide radical level in barley leaves under water deficit conditions. Acta Physiol Plant 27: 289–295

Kumar A, Altabella T, Taylor MA, Tiburcio AF (1997) Recent advances in polyamine research. Trends Plant Sci 2:124–130

Kumria R, Rajam MV (2002) Ornithine decarboxylase transgene in tobacco affects polyamines, in vitro-morphogenesis and response to salt stress. J Plant Physiol 159:983–990

Kurepa J, Smalle J, Van Montagu M, Inzé D (1998) Polyamines and paraquat toxicity in *Arabidopsis thaliana*. Plant Cell Physiol 39:987–992
Kusano T, Berberich T, Tateda C, Takahashi Y (2008) Polyamines: essential factors for growth and survival. Planta 228:367–381
Kusano T, Yamaguchi K, Berberich T, Takahashi Y (2007) Advances in polyamine research in 2007. J Plant Res 120:345–350
Kuznetsov VV, Radyukina NL, Shevyakova NI (2006) Polyamines and stress: biological role, metabolism, and regulation. Russ J Plant Physiol 53:583–604
Kuznetsov VV, Shevyakova NI (1997) Stress responses of tobacco cells to high temperature and salinity. Proline accumulation and phosphorylation of polypeptides. Physiol Plant 100:320–326
Langebartels C, Kerner K, Leonardi S, Schraudner M, Trost M, Heller W, Sandermann H (1991) Biochemical-plant responses to ozone.1. differential induction of polyamine and ethylene biosynthesis in tobacco. Plant Physiol 95:882–889
Larher F, Aziz A, Deleu C, Lemesle P, Ghaffar A, Bouchard F, Plasman M (1998) Suppression of the osmoinduced proline response of rapeseed leaf discs by polyamines. Physiol Plant 102:139–147
Laurenzi M, Tipping AJ, Marcus SE, Knox JP, Federico R, Angelini R, McPherson MJ (2001) Analysis of the distribution of copper amine oxidase in cell walls of legume seedlings. Planta 214:37–45
Lecourieux D, Ranjeva R, Pugin A (2006) Calcium in plant defence-signalling pathways. New Phytol 171:249–269
Lefèvre I, Correal E, Faz-Cano A, Zanuzzi A, Lutts S (2009a) Structural development, water status, pigment concentrations, and oxidative stress of *Zygophyllum fabago* seedlings in relation to cadmium distribution in the shoot organs. Int J Plant Sci 170:226–236
Lefèvre I, Gratia E, Lutts S (2001) Discrimination between the ionic and osmotic components of salt stress in relation to free polyamine level in rice (*Oryza sativa*). Plant Sci 161:943–952
Lefèvre I, Marchal G, Meerts P, Corréal E, Lutts S (2009b) Chloride salinity reduces cadmium accumulation by the Mediterranean halophyte species *Atriplex halimus* L. Environ Exp Bot 65:142–152
Legocka J, Kluk A (2005) Effect of salt and osmotic stress on changes in polyamine content and arginine decarboxylase activity in *Lupinus luteus* seedlings. J Plant Physiol 162:662–668
Legocka J, Zajchert I (1999) Role of spermidine in the stabilization of the apoprotein of the light-harvesting chlorophyll a/b-protein complex of photosystem II during leaf senescence process. Acta Physiol Plant 21:127–132
Li M, Hong Y, Wang X (2009) Phospholopase D- and phosphatidic acid-mediated signaling in plants. Biochim Biophys Acta 1791:927–935
Lin CC, Kao CH (1999) Excess copper induces an accumulation of putrescine in rice leaves. Bot Bull Acad Sinica 40:213–218
Lin CC, Kao CH (2002) NaCl-induced changes in putrescine content and diamine oxidase activity in roots of rice seedlings. Biologia Plant 45:633–636
Lingua G, Franchin C, Todeschini V, Castiglione S, Biondi S, Burlando B, Parravicini V, Torrigiani P, Berta G (2008) Arbuscular mycorrhizal fungi differentially affect the response to high zinc concentrations of two registered poplar clones. Environ Pollut 153:137–147
Liu HP, Dong BH, Zhang YY, Liu ZP, Liu YL (2004) Relationship between osmotic stress and the levels of free, conjugated and bound polyamines in leaves of wheat seedlings. Plant Sci 166: 1261–1267
Liu K, Fu HH, Bei QX, Luan S (2000) Inward potassium channel in guard cells as a target for polyamine regulation of stomatal movements. Plant Physiol 124:1315–1325
Liu J, Jiang MY, Zhou YF, Liu YL (2005) Production of polyamines is enhanced by endogenous abscisic acid in maize seedlings subjected to salt stress. J Integr Plant Biol 47:1326–1334
Liu JH, Nada K, Honda C, Kitashiba H, Wen XP, Pang XM, Moriguchi T (2006a) Polyamine biosynthesis of apple callus under salt stress: importance of the arginine decarboxylase pathway in stress response. J Exp Bot 57:2589–2599
Liu J, Yu BJ, Liu YL (2006b) Effects of spermidine and spermine levels on salt tolerance associated with tonoplast H^+-ATPase and H+-PPase activities in barley roots. Plant Growth Regul 49: 119–126

Liu H, Ji X, Yu B, Liu Y (2006c) Relationship between osmotic stress and polyamines conjugated to the deoxyribonucleic acid-protein in wheat seedling roots. Science in China: Series C Life Sciences 49:12–17

Locke JM, Bryce JH, Morris PC (2000) Contrasting effects of ethylene perception and biosynthesis inhibitors on germination and seedling growth of barley (*Hordeum vulgare* L.). J Exp Bot 51:1843–1849

Los DA, Murata N (2004) Membrane fluidity and its roles in the perception of environmental signals. Biochimica et Biophysica Acta - Biomembranes 1666:142–157

Lutts S, Kinet JM, Bouharmont J (1996) NaCl-induced senescence in leaves of rice (*Oryza sativa* L.) cultivars differing in salinity resistance. Ann Bot-London 78:389–398

Mahajan S, Tuteja N (2005) Cold, salinity and drought stresses: An overview. Archives of Biochemistry and Biophysics 444:139–158

Mahajan S, Pandey G, Tuteja N (2008) Calcium- and salt stress signaling in plants: shedding light on SOS pathway. Arch Biochem Biophys 471:146–158

Mahajan A, Sharma S (2009) Antagonistic effect of polyamines on ABA-induced suppression of mitosis in *Allium cepa* L. Indian J Exp Biol 47:136–139

Maiale SJ, Marina M, Sánchez DH, Pieckenstain FL, Ruiz OA (2008) In vitro and in vivo inhibition of plant polyamine oxidase activity by polyamine analogues. Phytochemistry 69:2552–2558

Maiale S, Sanchez DH, Guirado A, Vidal A, Ruiz OA (2004) Spermine accumulation under salt stress. J Plant Physiol 161:35–42

Majerus V, Bertin P, Lutts S (2007) Effects of iron toxicity on osmotic potential, osmolytes and polyamines concentrations in the African rice (*Oryza glaberrima* Steud.). Plant Sci 173: 96–105

Majewska-Sawka A, Butowt R, Niklas A (1998) Do polyamines release membrane-bound calcium in sugar beet protoplasts? J Plant Physiol 153:247–250

Maksymiec W (2007) Signaling responses in plants to heavy metal stress. Acta Physiol Plant 29:177–187

Malik AU, Singh Z (2005) Pre-storage application of polyamines improves shelf-life and fruit quality of mango. J Hortic Sci Biotech 80:363–369

Martin-Tanguy J (1997) Conjugated polyamines and reproductive development: biochemical, molecular and physiological approaches. Physiol Plant 100:675–688

Martin-Tanguy J, Carré M (1993) Polyamines in grapevine microcuttings cultivated in-vitro - effects of amines and inhibitors of polyamine biosynthesis on polyamine levels and microcutting growth and development. Plant Growth Regul 13:269–280

Mascher R, Lippmann B, Holzinger S, Bergmann H (2002) Arsenate toxicity: effects on oxidative stress response molecules and enzymes in red clover plants. Plant Sci 163:961–969

Masgrau C, Altabella T, Farras R, Flores D, Thompson AJ, Besford RT, Tiburcio AF (1997) Inducible overexpression of oat arginine decarboxylase in transgenic tobacco plants. Plant J 11:465–473

Mattoo K, Chung SH, Goyal RK, Fatima T, Solomos T, Srivastava A, Handa AK (2007) Overaccumulation of higher polyamines in ripening transgenic tomato fruits revives metabolic memory, upregulates anabolism-related genes, and positively impacts nutritional quality. J AOAC Int 90:1456–1464

Mayer MJ, Michael AJ (2003) Polyamine homeostasis in transgenic plants overexpressing ornithine decarboxylase includes ornithine limitation. J Biochem 134:765–772

Mehta RA, Cassol T, Li N, Ali N, Handa AK, Mattoo AK (2002) Engineered polyamine accumulation in tomato enhances phytonutrient content, juice quality, and vine life. Nat Biotechnol 20: 613–618

Messiaen J, Cambier P, Van Cutsem P (1997) Polyamines and pectins: I. Ion exchange and selectivity. Plant Physiol 113:387–395

Messiaen J, Van Cutsem P (1999) Polyamines and pectins. II. Modulation of pectic-signal transduction. Planta 208:247–256

Minocha R, Long S (2004) Effects of aluminum on organic acid metabolism and secretion by red spruce cell suspension cultures and the reversal of A1 effects on growth and polyamine metabolism by exogenous organic acids. Tree 24:55–64

Minocha R, Shortle WC, Coughlin DJ, Minocha SC (1996) Effects of aluminum on growth, polyamine metabolism, and inorganic ions in suspension cultures of red spruce (*Picea rubens*). Can J Forest Res 26:550–559

Mittler R, Vanderauwera S, Gollery M, Van Breusegem F (2004) Reactive oxygen gene network of plants. Trends Plant Sci 9:490–498

Mo H, Pua EC (2002) Up-regulation of arginine decarboxylase gene expression and accumulation of polyamines in mustard (*Brassica juncea*) in response to stress. Physiol Plant 114:439–449

Mohapatra S, Cherry S, Minocha R, Majumdar R, Thangavel P, Long S, Minocha SC (2010) The response of high and low polyamine-producing cell lines to aluminum and calcium stress. Plant Physiol Biochem 48:612–620

Mohapatra S, Minocha R, Long S, Minocha SC (2009) Putrescine overproduction negatively impacts on the oxidative state of poplar cells in culture. Plant Physiol Biochem 47:262–271

Møller IM (2001) Plant mitochondria and oxidative stress: electron transport, NADPH turnover, and metabolism of reactive oxygen species. Ann Rev Plant Physiol 52:561–591

Moschou PN, Delis ID, Paschalidis KA, Roubelakis-Angelakis KA (2008a) Transgenic tobacco plants overexpressing polyamine oxidase are not able to cope with oxidative burst generated by abiotic factors. Physiol Plant 133:140–156

Moschou PN, Paschalidis KA, Delis ID, Andriopoulou AH, Lagiotis GD, Yakoumatis DI, Roubelakis-Angelakis KA (2008b) Spermidine exodus and oxidation in the apoplast induced by abiotic stress is responsible for H_2O_2 signatures that direct tolerance responses in tobacco. Plant Cell 20:1708–1724

Moschou PN, Sanmartin M, Andriopoulou AH, Rojo E, Sanchez-Serrano JJ, Roubelakis-Angelakis KA (2008c) Bridging the gap between plant and mammalian polyamine catabolism: a novel peroxisomal polyamine oxidase responsible for a full back-conversion pathway in *Arabidopsis*. Plant Physiol 147:1845–1857

Mulo P, Eloranta T, Aro EM, Maenpaa P (1998) Disruption of a spe-like open reading frame alters polyamine content and psbA-2 mRNA stability in the cyanobacterium *Synechocystis* sp. PCC 6803. Botanica Acta 111:71–76

Munnik T, Meijer HJC (2001) Osmotic stress activates distinct lipid and MAPK signalling pathways in plants. FEBS Lett 498:172–178

Munns R, Tester M (2008) Mechanisms of salinity tolerance. Ann Rev Plant Physiol 59:651–681

Nakashima K, Yamaguchi-Shinozaki K (2006) Regulons involved in osmotic stress-responsive and cold stress-responsive gene expression in plants. Physiol Plant 126:62–71

Nayyar H, Chander S (2004) Protective effects of polyamines against oxidative stress induced by water and cold stress in chickpea. J Agron Crop Sci 190:355–365

Ndayiragije A, Lutts S (2006a) Exogenous putrescine reduces sodium and chloride accumulation in NaCl-treated calli of the salt-sensitive rice cultivar I Kong Pao. Plant Growth Regul 48: 51–63

Ndayiragije A, Lutts S (2006b) Do exogenous polyamines have an impact on the response of a salt-sensitive rice cultivar to NaCl? J Plant Physiol 163:506–516

Ndayirajige A, Lutts S (2007) Long term exogenous putrescine applicatrion improves grain yield of a salt-sensitive rice cultivar exposed to NaCl. Plant Soil 291:225–238

Neill SJ, Desikan R, Clarke A, Hurst RD, Hancock JT (2002) Hydrogen peroxide and nitric oxide as signalling molecules in plants. J Exp Bot 53:1237–1247

Ober D, Hartmann T (1999) Homospermidine synthase, the first pathway-specific enzyme of pyrrolizidine alkaloid biosynthesis, evolved from deoxyhypusine synthase. Proc Natl Acad Sci USA 96:14777–14782

Oliver D, Baukrowitz T, Fakler B (2000) Polyamines as gating molecules of inward-rectifier K^+ channels. Eur J Biochem 267:5824–5829

Pagoria DA, Maravolo NC (2005) DNA fragmentation in *Marchantia polymorpha* Thalli in response to spermine treatment. Int J Plant Sci 166:589–594

Papadakis AK, Roubelakis-Angelakis KA (2005) Polyamines inhibit NADPH oxidase-mediated superoxide generation and putrescine prevents programmed cell death induced by polyamine oxidase-generated hydrogen peroxide. Planta 220:826–837

Paschalidis KA, Roubelakis-Angelakis KA (2005) Spatial and temporal distribution of polyamine levels and polyamine anabolism in different organs/tissues of the tobacco plant. Correlations with age, cell division/expansion, and differentiation. Plant Physiol 138:142–152

Pedreno E, Lopez-Contreras AJ, Cremades A, Penafiel R (2005) Protecting or promoting effects of spermine on DNA strand breakage induced by iron or copper ions as a function of metal concentration. J Inorg Biochem 99:2074–2080

Peremarti A, Bassie L, Christou P, Capell T (2009) Spermine facilitates recovery from drought but does not confer drought tolerance in transgenic rice plants expressing *Datura stramonium* S-adenosylmethionine decarboxylase. Plant Mol Biol 70:253–264

Perez-Amador MA, Leon J, Green PJ, Carbonell J. (2002) Induction of the arginine decarboxylase ADC2 gene provides evidence for the involvement of polyamines in the wound response in Arabidopsis. Plant Physiology 130:1454–1463

Pistocchi R, Bagni N, Creus JA (1987) Polyamine uptake in carrot cell-cultures. Plant Physiol 84:374–380

Pistocchi R, Kashiwagi K, Miyamoto S, Nukui E, Sadakata Y, Kobayashi H, Igarashi K (1993) Characteristics of the operon for a putrescine transport-system that maps at 19 minutes on the *Escherichia-coli* chromosome. J Biol Chem 268:146–152

Poduslo JF, Curran GL (1996) Increased permeability of superoxide dismutase at the blood-nerve and blood–brain barriers with retained enzymatic activity after covalent modification with the naturally occurring polyamine, putrescine. J Neurochem 67:734–741

Poschenrieder C, Barceló J (1999) Water relations in heavy metal stressed plants. In: Prasad MNV, Hagemeyer J (eds) Heavy metal stress in plants, from molecules to ecosystems. Springer Verlag, Berlin, pp 207–229

Quinet M, Ndayiragije A, Lefèvre I, Lambillotte B, Dupont-Gillain CC, Lutts S (2010) Putrescine differently influences the effect of salt stress on polyamine metabolism and ethylene synthesis in rice cultivars differing in salt resistance. J Exp Bot 61:2719–2733

Rabiti A, Pistocchi R, Bagni N (1989) Putrescine uptake and translocation in higher plants. Physiol Plant 77:225–230

Rea G, de Pinto MC, Tavazza R, Biondi S, Gobbi V, Ferrante P, De Gara L, Federico R, Angelini R, Tavladoraki P (2004) Ectopic expression of maize polyamine oxidase and pea copper amine oxidase in the cell wall of tobacco plants. Plant Physiol 134:1414–1426

Rea G, Laurenzi M, Tranquilli E, D'Ovidio R, Federico R, Angelini R (1998) Developmentally and wound-regulated expression of the gene encoding a cell wall copper amine oxidase in chickpea seedlings. FEBS Lett 437:177–182

Roy P, Niyogi K, SenGupta DN, Ghosh B (2005) Spermidine treatment to rice seedlings recovers salinity stress-induced damage of plasma membrane and PM-bound H^+-ATPase in salt-tolerant and salt-sensitive rice cultivars. Plant Sci 168:583–591

Roy M, Wu R (2001) Arginine decarboxylase transgene expression and analysis of environmental stress tolerance in transgenic rice. Plant Sci 160:869–875

Roychoudhury A, Basu S, Sengupta DN (2011) Amelioration of salinity stress by exogenously applied spermidine or spermine in three varieties of indica rice differing in their level of salt toleranbce. J Plant Physiol 68:317–328

Saibo NJM, Lourenço T, Oliveira MM (2009) Transcription factors and regulation of photosynthetic and related metabolism under environmental stresses. Annals of Botany 103:609–623

Santa-Cruz A, Perez-Alfocea F, Caro M, Acosta M (1998) Polyamines as short-term salt tolerance traits in tomato. Plant Sci 138:9–16

Sarjala T, Haggman H, Aronen T (1997) Effect of exogenous polyamines and inhibitors of polyamine biosynthesis on growth and free polyamine contents of embryogenic Scots pine callus. J Plant Physiol 150:597–602

Scandalios JG (2005) Oxidative stress: molecular perception and transduction of signals triggering antioxidant gene defenses. Braz J Med Biol Res 38:995–1014

Scaramagli S, Biondi S, Torrigiani P (1999) Methyl(bis-guanylhydrazone) inhibition of organogenesis is not due to S-adenosylmethionine decarboxylase inhibition/polyamine depletion in tobacco thin layers. Physiol Plant 107:353–360

Serafini-Fracassini D, Delduca S, Beninati S (1995) Plant transglutaminases. Phytochemistry 40: 355–365

Serafini-Fracassini D, Di Sandro A, Del Duca S (2010) Spermine delays leaf senescence in *Lactuca sativa* and prevents the decay of chloroplast photosystems. Plant Physiol Biochem 48:602–611

Sfichi L, Ioannidis N, Kotzabasis K (2004) Thylakoid-associated polyamines adjust the UV-B sensitivity of the photosynthetic apparatus by meansq of light-harvesting complex II changes. Photochem Photobiol 80:499–506

Shabala S, Cuin TA, Pottosin I (2007) Polyamines prevent NaCl-induced K^+ efflux from pea mesophyll by blocking non-selective cation channels. FEBS Lett 581:1993–1999

Shah K, Kumar RG, Verma S, Dubey RS (2001) Effect of cadmium on lipid peroxidation, superoxide anion generation and activities of antioxidant enzymes in growing rice seedlings. Plant Sci 161:1135–1144

Shao HB, Guo QJ, Chu LY, Zhao XN, Su ZL, Hu YC, Cheng JF (2007) Understanding molecular mechanism of higher plant plasticity under abiotic stress. Colloids Surf B 54:37–45

Shevyakova NIO, Il'Ina EN, Stetsenko LA, Kuznetsov VV (2011) Nickel accumulation in rape shoots (*Brassica napus* L.) increased by putrescine. Int J Phytoremediat 13:345–356

Shih CY, Kao CH (1996) Growth inhibition in suspension-cultured rice cells under phosphate deprivation is mediated through putrescine accumulation. Plant Physiol 111:721–724

Sitrit Y, Bennett AB (1998) Regulation of tomato fruit polygalacturonase mRNA accumulation by ethylene: a re-examination. Plant Physiol 116:1145–1150

Smith MA, Maravolo NC (2004) The influence of spermine on the in situ expression of a protein kinase associated with senescence in *Marchantia polymorpha* thalli. Int J Plant Sci 165: 745–751

Srivastava HS, Ormrod DP, Hale BA (1995) Polyamine mediated modifications of bean leaf response to nitrogen-dioxide. J Plant Physiol 146:313–317

Su GX, An ZF, Zhang WH, Liu YL (2005) Light promotes the synthesis of lignin through the production of H_2O_2 mediated by diamine oxidases in soybean hypocotyls. J Plant Physiol 162:1297–1303

Su GX, Bai X (2008) Contribution of putrescine degradation to proline accumulation in soybean leaves under salinity. Biol Plantarum 52:796–799

Sudha G, Ravishankar GA (2002) Involvement and interaction of various signaling compounds on the plant metabolic events during the defense response, resistance to stress factors, formation of secondary metabolites and their molecular aspects. Plant Cell Tiss Org 71:181–212

Sung HI, Liu LF, Kao CH (1994) Putrescine accumulation is associated with growth-inhibition in suspension-cultured rice cells under potassium-deficiency. Plant Cell Physiol 35: 313–316

Suzuki Y (1996) Purification and characterization of diamine oxidase from *Triticum aestivum* shoots. Phytochemistry 42:291–293

Szalai G, Janda T, Bartok T, Paldi E (1997) Role of light in changes in free amino acid and polyamine contents at chilling temperature in maize (*Zea mays*). Physiol Plantarum 101:434–438

Tabur S, Demir K (2010) Protective roles of exogenous polyamines on chromosomal aberrations in *Hordeum vulgare* exposed to salinity. Biologia 65:947–953

Takahashi T, Kakehi JI (2010) Polyamines: ubiquitous polycations with unique roles in growth and stress responses. Ann Bot-London 105:1–6

Tamai T, Inoué M, Sugimoto T, Sueyoshi K, Shiraishi N, Oji Y (1999) Ethylene-induced putrescine accumulation modulates K^+ partitioning between roots and shoots in barley seedlings. Physiol Plantarum 106:296–301

Tang W, Newton RJ, Outhavong V (2004) Exogenously added polyamines recover browning tissues into normal callus cultures and improve plant regeneration in pine. Physiol Plant 122: 386–395

Tari I, Csiszar J, Gemes K, Szepesi A (2006) Modulation of Cu^{2+} accumulation by (aminoethoxyvinyl)glycine and methyloxal bis(guanylhydrazone), the inhibitors of stress ethylene and polyamine synthesis in wheat genotypes. Cereal Res Commun 34:989–996

Tassoni A, Accettulli P, Bagni N (2006) Exogenous spermidine delays senescence of *Dianthus caryophyllus* flowers. Plant Biosyst 140:107–114

Tassoni A, Antognoni F, Bagni N (1996) Polyamine binding to plasma membrane vesicles isolated from zucchini hypocotyms. Plant Physiol 110:817–824

Tassoni A, Antognoni F, Battostini ML, Sanvido O, Bagni N (1998) Characterization of spermidine binding to solubilized plasma membrane proteins from succhini hypocotyls. Plant Physiol 117:971–977

Tassoni A, Napier RM, Franschetti M, Venis MA, Bagni N (2002) Spermidine-bionding proteins. Purification and expression analysis in maize. Plant Physiol 128:1303–1312

Tavladoraki P, Rossi MN, Saccuti G, Perez-Amador MA, Polticelli F, Angelini R, Federico R (2006) Heterologous expression and biochemical characterization of a polyamine oxidase from *Arabidopsis* involved in polyamine back conversion. Plant Physiol 141:1519–1532

Terui Y, Ohnuma M, Hiraga K, Kawashima E, Oshima T (2005) Stabilization of nucleic acids by unusual polyamines produced by an extreme thermophile, *Thermus thermophilus*. Biochem J 388:427–433

Thu-Hang P, Bassie L, Safwat G, Trung-Nghia P, Christou P, Capell T (2002) Expression of heterologous S-adenosylmethionine decarboxylase cDNA in plants demonstrates that changes in S-adenosyl-L-methionine decarboxylase activity determine levels of the higher polyamines spermidine and spermine. Plant Physiol 129:1744–1754

Todeschini V, Franchin C, Castiglione S, Burlando B, Biondi S, Torrigiani P, Berta G, Lingua G (2007) Responses to copper of two registered poplar clones inoculated or not with arbuscular mycorrhizal fungi. Caryologia 60:146–155

Torrigiani P, Scaramagli S, Castiglione S, Maddalena A, Biondi S (2003) Downregulation of ethylene production and biosynthetic gene expression is associated to changes in putrescine metabolism in shoot-forming tobacco thin layers. Plant Sci 164:1087–1094

Torrigiani P, Serafini-Fracassini D, Biondi S, Bagni N (1986) Evidence for the subcellular-localization of polyamines and their biosynthetic-enzymes in plant cells. J Plant Physiol 124:23–29

Tun NN, Santa-Catarina C, Begum T, Silveira V, Handro W, Floh EIS, Scherer GFE (2006) Polyamines induce rapid biosynthesis of nitric oxide (NO) in *Arabidopsis thaliana* seedlings. Plant Cell Physiol 47:346–354

Upreti KK, Murti GSR (2010) Response of grape rootstocks to salinity: changes in root growth, polyamines and abscisic acid. Biologia Plant 54:730–734

Urano K, Hobo T, Shinozaki K (2005) *Arabidopsis* ADC genes involved in polyamine biosynthesis are essential for seed development. FEBS Lett 579:1557–1564

Urano K, Yoshiba Y, Nanjo T, Igarashi Y, Seki M, Sekigushi F, Yamagushi-Shinozaki K, Shinozaki K (2003) Characterization of *Arabidopsis* genes involved in biosynthesis of polyamines in abiotic stress responses and developmental stages. Plant Cell Environ 26:1917–1926

Urao T, Yakubov B, Satoh R, Yamaguchi-Shinozaki K, Seki M, Hirayama T, Shinozaki K (1999) A transmembrane hybrid-type histidine kinase in *Arabidopsis* functions as an osmosensor. Plant Cell 11:1743–1754

Van Breusegem F, Bailey-Serres J, Mittler R (2008) Unraveling the tapestry of networks involving reactive oxygen species in plants. Plant Physiol 147:978–984

Vandenroeck D, Vanderstraeten D, Vanmontagu M, Caplan A (1994) A group of chromosomal-proteins is specifically released by spermine and loses DNA-binding activity upon phosphorylation. Plant Physiol 106:559–566

Velikova VB, Yordanov IT, Georgieva KM, Tsonev TD, Goltsev V (1998) Effects of exogenous polyamines applied separately and in combination with simulated acid rain on functional activity of photosynthetic apparatus. J Plant Physiol 153:299–307

Verma S, Mishra SN (2005) Putrecine alleviation of growth in salt stressed *Brassica juncea* by inducing antioxidative defense system. J Plant Physiol 162:669–677

Vusoku J, Jokela A, Laara E, Saaskilathi M, Muilu R, Sutela S, Altabella T, Sarjala T, Hagglman H (2006) Consistency of polyamines profiles and expression of arginine decarboxylase in mitosis during zygotic embryogenesis of Scot pine. Plant Physiol 142:1027–1038

Walden R, Cordeiro A, Tiburcio AF (1997) Polyamines: small molecules triggering pathways in plant growth and development. Plant Physiol 113:1009–1013

Wang JW, Kao CH (2006) Aluminium-inhibited root growth of rice seedlings is mediated through putrescine accumulation. Plant Soil 288:373–381

Wang X, Shi G, Xu Q, Hu J (2007) Exogenous polyamines enhance copper tolerance of *Nymphoides peltatum*. J Plant Physiol 164:1062–1070

Wasilewska A, Vald F, Sirichandra C, Redko Y, Jammes F, Valon C, Frey N, Leung J (2008) An update on abscissic acid signaling in plants and more…. Mol Plant 1:198–217

Watson MB, Emory KK, Piatak RM, Malmberg RL (1998) Arginine decarboxylase (polyamine synthesis) mutants of *Arabidopsis thaliana* exhibit altered root growth. Plant J 13:231–239

Watson MB, Malmberg RL (1996) Regulation of *Arabidopsis thaliana* (L) Heynh arginine decarboxylase by potassium deficiency stress. Plant Physiol 111:1077–1084

Wen XP, Ban Y, Inoué H, Matsuda N, Moriguchi T (2009) Aluminium tolerance in a spermidine synthase-overexpressing transgenic European pear is correlated with the enhanced level of spermidine via alleviating oxidative status. Environ Exp Bot 66:471–478

Wen XP, Ban Y, Inoué H, Matsuda N, Moriguchi T (2010) Spermidine levels are implicated in heavy metal tolerance in a *spermidine synthase* overexpressing transgenic European pear by exerting antioxidant activities. Transgenic Res 19:91–103

Wilson PB, Estavillo GM, Field KJ, Pornsiriwong W, Carroll AJ, Hiowell KA, Woo NS, Lake JA, Smith SM, Millar AH, von Caemmerer S, Pogson BJ (2009) The nucleotidase/phosphatase SAL1 is a negative regulator of drought tolerance in *Arabidopsis*. Plant J 58:299–317

Wen XP, Pang XM, Matsuda N, Kita M, Inoué H, Hao YJ, Honda C, Moriguchi C (2008) Overexpression of the apple spermidine synthase gene in pear confers multiple abiotic stress tolerance by altering polyamines titers. Transgenic Res 17:251–263

Wisniewski JP, Rathbun EA, Knox JP, Brewin NJ (2000) Involvement of diamine oxidase and peroxidase in insolubilization of the extracellular matrix: implication for pea nodule initiation by *Rhizobium leguminosarum*. Mol Plant Microbe Interact 13:413–420

Xing SG, Jun BY, Hau ZW, Liang LY (2007) Higher accumulation of γ-aminobutyric acid induced by salt stress through stimulating the activity of diamine oxidases in *Glycine max* (L.) Merr. roots. Plant Physiol Biochem 45:560–566

Xiong L, Schumacher KS, Zhu JK (2002) Cell signaling during cold, drought and salt stress. Plant Cell 14:S165–S285

Xiong L, Zhu JK (2002) Molecular and genetic aspects of plant responses to osmotic stress. Plant Cell Environ 25:131–139

Xu BF, Sheehan MJ, Timko MP (2004) Differential induction of ornithine decarboxylase (ODC) gene family members in transgenic tobacco (*Nicotiana tabacum* L. cv. Bright Yellow 2) cell suspensions by methyl-jasmonate treatment. Plant Growth Regul 44:101–116

Yamaguchi K, Takahashi Y, Berberich T, Imai A, Takahashi T, Michael AJ, Kusano T (2007) A protective role for the polyamine spermine against drought stress in *Arabidopsis*. Biochem Biophys Res Comm 352:486–490

Yamasaki H, Cohen MF (2006) NO signal at the crossroads: polyamine-induced nitric oxide synthesis in plants? Trends Plant Sci 11:522–524

Ye XS, Avidiushko SA, Kuc J (1994) Effects of polyamines on in-vitro phoshorylation of soluble and plasma-membrane proteins in tobacco, cucumber and *Arabidopsis thaliana*. Plant Sci 97:109–118

Ye B, Muller HH, Zhang J, Gressel J (1997) Constitutively elevated levels of putrescine and putrescine-generating enzymes correlated with oxidant stress resistance in *Conyza bobariensis* and wheat. Plant Physiol 115:1443–1451

Zacchini M, Rea E, Tullio M, de Agazio M (2003) Increased antioxidative capacity in maize calli during and after oxidative stress induced by a long lead treatment. Plant Physiol Biochem 41:49–54

Zapata PJ, Serrano M, Pretel MT, Amorós A, Botella MA (2004) Polyamines and ethylene changes during germination of different plant species under salinity. Plant Sci 167:781–788

Zhao FG, Qin P (2005) Protective effects of exogenously fatty acids on root tonoplast function against salt stress in barley seedlings. Environ Exp Bot 53:215–223

Zhao FG, Song CP, He JQ, Zhu H (2007) Polyamines improve K^+/Na^+ homeostasis in barley seedlings by regulating root ion channel activities. Plant Physiol 145:1061–1072

Zhang J, Jia W, Yang J, Ismail AM (2006) Role of ABA in integrating plant responses to drought and salt stresses. Field Crop Res 97:111–119

Zhang WP, Jiang BA, Li WG, Song H, Yu YS, Chen JF (2009) Polyamines enhance chilling tolerance of cucumber (*Cucumis sativus* L.) through modulatinfg antioxidative system. Sci Hortic 122:200–208, Amsterdam

Zheleva D, Tsonev V, Sergiev I, Karanov E (1994) Protective effects of exogenous polyamines against atrazine in pea plants. Plant Growth Regul 13:203–211

Zhu JK (2001) Cell signalling under salt, water and cold stresses. Curr Opin Plant Biol 4:401–406

Zhu H, Ding GH, Fang K, Zhao FG, Qin P (2006) New perspective on the mechanism of alleviating salt stress by spermidine in barley seedlings. Plant Growth Regul 49:147–156

Chapter 13
Role of Polyamines in Alleviating Salt Stress

Dessislava Todorova, Zornitsa Katerova, Iskren Sergiev, and Vera Alexieva

13.1 Introduction

Abiotic and biotic stresses cause alterations in the normal physiological processes of all plants, including the economically important crops. Plant damage and decrease in their productivity take place most often due to naturally occurring unfavorable factors of the environment (natural stress factors). These include extreme temperatures; water deficit or abundance; increased soil salinity; high solar irradiance; early autumn or late spring ground frosts; pathogens, etc. Along with these factors, plants are imposed to a large scale of new stressors related to human activity (anthropogenic stress factors) including, toxic pollutants such as pesticides, noxious gasses (SO_2, NO, NO_2, NO_x, O_3 and photochemical smog); photooxidants; soil acidification and mineral deficit due to acid rains; overdoses of fertilizers; heavy metals; intensified UV-B irradiation, etc. (Fig. 13.1). All these stresses cause an increased production of reactive oxygen species (ROS) in plants that alter their normal physiological functions, decrease the biosynthetic capacity of plant organisms, and cause damages which may lead to plant death (Mittler 2002; Ahmad et al. 2008; Gill and Tuteja 2010b; Potters et al. 2010).

13.1.1 Reactive Oxygen Species

Independently of the type of stress (natural or anthropogenic), the accumulation of ROS is an undeniably established fact. Currently, overproduction of more than ten

D. Todorova • Z. Katerova • I. Sergiev • V. Alexieva (✉)
Institute of Plant Physiology and Genetics, Bulgarian Academy of Sciences,
Acad. G. Bonchev Str, Bl. 21, 1113 Sofia, Bulgaria
e-mail: verea@bio21.bas.bg

P. Ahmad et al. (eds.), *Ecophysiology and Responses of Plants under Salt Stress*,
DOI 10.1007/978-1-4614-4747-4_13,

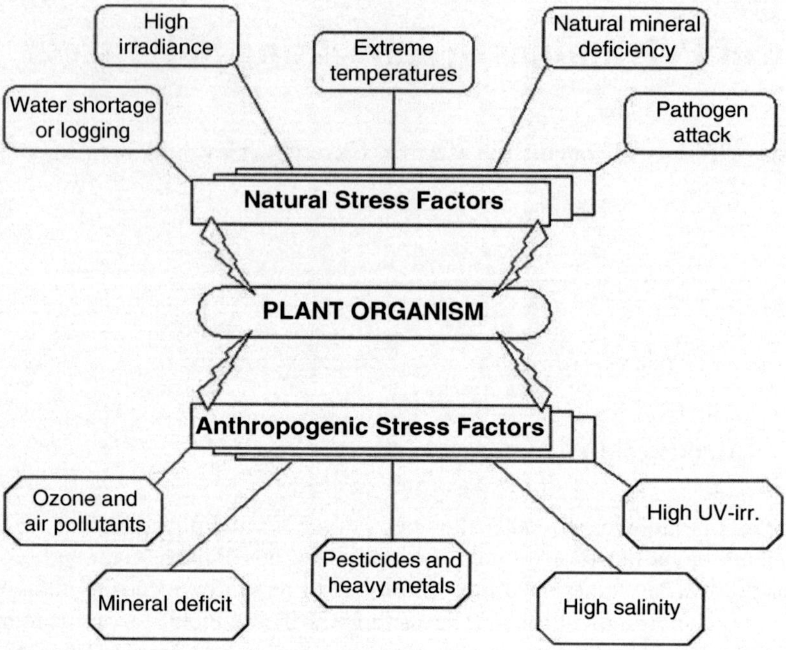

Fig. 13.1 Natural and anthropogenic stress factors

Table 13.1 Reactive oxygen species

Free Radicals		Nonradicals	
Superoxide radical	$O_2^{\bullet -}$	Hydrogen peroxide	H_2O_2
Hydroxyl radical	$OH^{\bullet}$	Hypobromous acid	HOBr
Hydroperoxyl radical	$HO_2^{\bullet}$	Hypochlorous acid	HOCl
Carbonate radical	$CO_3^{\bullet -}$	Ozone	O_3
Peroxyl radical	$RO_2^{\bullet}$	Singlet oxygen	$O_2{}^1\Delta g$
Alkoxyl radical	$RO^{\bullet}$	Organic peroxides	ROOH
Carbon dioxide radical	$CO_2^{\bullet -}$	Peroxynitrite	$ONOO^-$
Singlet radical	$O_2{}^1\Sigma g^+$	Peroxynitrate	O_2NOO^-
		Peroxynitrous acid	ONOOH
		Peroxomonocarbonate	$HOOCO_2^-$

oxygen-containing molecules and radicals (Table 13.1) are known to induce oxidative stress. However, most detrimental to all biological systems are $O_2^{\bullet -}$, H_2O_2, $OH^{\bullet}$, $^1O_2$ (Halliwell 2006). In plants, ROS are generated mainly as by-products of various processes requiring high metabolic activity or high rate of electron flow via electron-transport chains. The major targets of deleterious ROS action are cellular macromolecules as phospholipids, proteins, and nucleic acids.

13.1.2 Plant Defense Systems

During the phylogenesis, plants have developed a complex of antioxidant protective systems in order to cope with all destructive effects of the unfavourable environmental conditions (Fig. 13.2). In general, the plant antioxidative systems can be divided in to, (a) Enzymatic antioxidants including superoxide dismutase (EC 1.15.1.1), catalase (EC 1.11.1.6), guaiacol peroxidase (EC 1.11.1.7) and enzymes belonging to the ascorbate-glutathione cycle – ascorbate peroxidase (EC 1.11.1.11), glutathione peroxidase (EC 1.11.1.9), monodehydroascorbate reductase (EC 1.6.5.4), dehydroascorbate reductase (EC 1.8.5.1), glutahtione reductase (EC 1.6.4.2.) and glutathione-S-transferase (EC 2.5.1.18). (b) Non-enzymatic antioxidants including lipid-soluble, membrane associated antioxidants – α-tocopherol, β-carotene, which directly quench free radicals of lipid peroxidation (triplet chlorophyll and $^1O_2$). Water-soluble antioxidants – glutathione and ascorbate, taking part in the detoxification of $O^{\bullet}_2{}^-$ and H_2O_2; polyphenols (flavonoids, tannins and anthocyanins), proteinaceus thiols, proline, glycinebetaine, and *polyamines.*

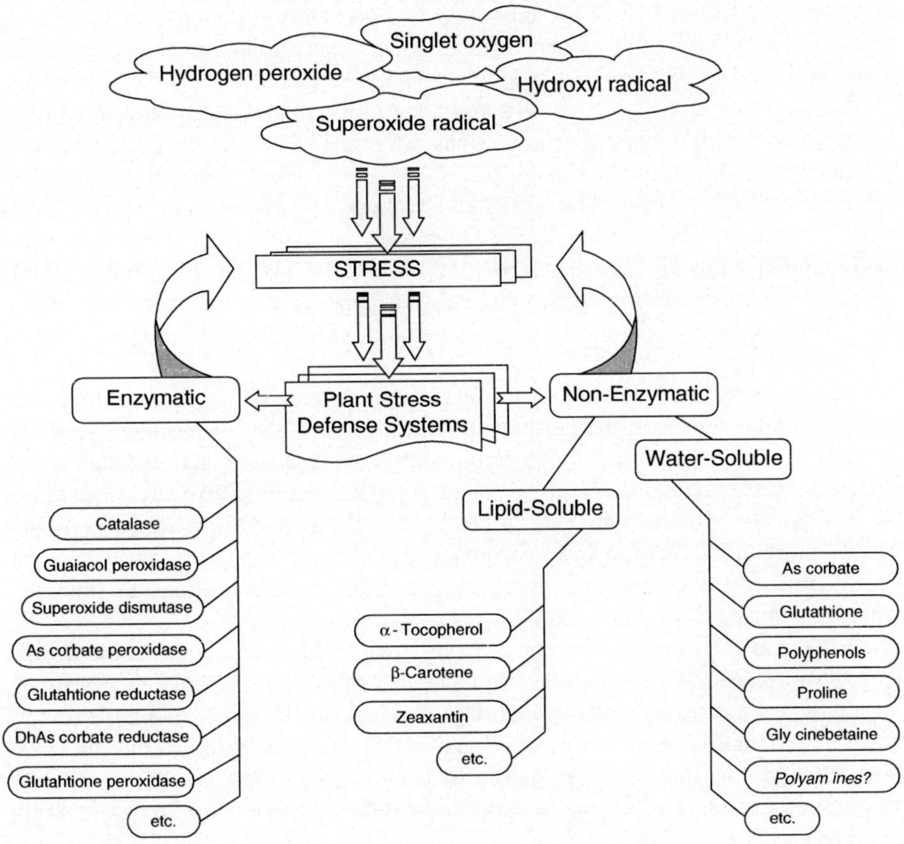

Fig. 13.2 Reactive oxygen species and plant defense systems

By activation of some or all of these systems, the plants are capable of overcoming the oxidative stress. However, in the case of prolonged or acute short stress, the capacity of the defense systems becomes exhausted or overloaded and this leads to considerable damages and even to plant death.

13.1.3 Salinity Stress

Salinity of soil as a result from increased quantity of cations (Na^+, K^+, Mg^{2+}, Ca^{2+}) and anions (Cl^-, SO_4^{2-}, HCO_3^-) which are originated from more water-soluble salts such as $NaSO_4$, $NaHCO_3$, NaCl, and $MgCl_2$ as well as less water-soluble salts such as $CaSO_4$, $MgSO_4$, and $CaCO_3$, is one of the major abiotic stresses that reduce plant growth and productivity of many crops worldwide. Salinity may originate from natural factors (for example mineral erosion in soil), but also may result from human activities (as irrigation with high mineralized water and/or ineffective draining of the irrigated area, application of fertilizers, etc.). Most often soil salinity is enhanced by the presence of NaCl in water and soil, although other salts also can be involved. Detrimental effects of NaCl are complex, and include ion toxicity, hyperosmotic stress, and may induce subsequent stresses such as nutritional imbalances and oxidative stress (Zhu 2001).

The metabolic toxicity of Na^+ is mainly due to its capability to compete with K^+ for target sites with important cell functions. More than 50 enzymes are activated by K^+ and elevated levels of Na^+ or high Na^+/K^+ ratio can disrupt a number of enzymatic processes in the cytoplasm. The protein biosynthesis also requires appropriate levels of K^+ for proper tRNA binding to the ribosomes, and high levels of Na^+ can negatively influence this process (Tester and Davenport 2003; Bartels and Sunkar 2005).

Increased concentration of Na^+ can cause a hyperosmotic stress by preventing plant water uptake which results in the so-called "physiological drought" (Turkan and Demiral 2008).

Elevated concentration of Na^+ can cause a deficit of nutrition elements by inhibiting their intake through the plasmalemma transporters (for instance the selective K^+ channels) in the root cells. The salinity can lead to a hormonal disbalance and increased ROS production which can result to additional oxidative stress (Azevedo Neto et al. 2008; Turkan and Demiral 2008). Like drought stress salinity can induce accumulation of ABA followed by stomatal closure and respective decrease in the CO_2/O_2 ratio in leaves and inhibits the CO_2 fixation. These conditions are prerequisite for enhanced ROS production.

To unscramble the salinity stress from other plant stresses (as drought and osmotic stress) is difficult because the increased salts alter the ionic chemical balance in plants and affect water availability to plants, and similarly to other stressors may cause oxidative stress via production of ROS. The plant responses to salt stress is documented as multigenic in nature as adaptation to high salt levels involves osmotic adjustment, toxic ions compartmentation and oxidative stress tolerance (Turkan and Demiral 2008).

The investigation on salt stress and modulation of plant salinity tolerance in various cultivars has been extensively carried out using different approaches (Georgiev and Atkins 1993; Tanaka et al. 1999; Apse and Blumwald 2002; Garratt et al. 2002; Rios-Gonzalez et al. 2002; Borsani et al. 2003; Badawi et al. 2004; Flowers 2004; Brankova et al. 2007; Ivanova et al. 2008; Nenova 2008; Ogawa and Mitsuya 2012).

13.2 Polyamines: Chemistry and Metabolism

The triamine spermidine (Spd) and tetraamine spermine (Spm), as well as their precursor the diamine putrescine (Put) are the major polyamines (PAs) which are constitutive for all plant species. Polyamines are organic low-weight molecules with straight-chained C_3-C_{15} aliphatic structure with at least two primary amino groups and one or more internal imino groups (Edreva 1996; Groppa and Benavides 2008; Gill and Tuteja 2010a). Besides putrescine, spermine and spermidine, which are common for all plant species, there are also unusual polyamines which occur only in distinct plant species (i.e. diamines cadaverine and 1,3-diaminopropane) or synthesized under certain conditions (i.e. norspermine, norspermidine, thermospermine and caldopentamine) (Table 13.2).

Generally, polyamine biosynthesis in plants can be described as a two-phase process – the first stage is the biosynthesis of diamines, and the second stage is spermidine and spermine biosynthesis (Fig. 13.3). The putrescine is synthesized through decarboxylation of L-arginine to agmatine by arginine decarboxylase (ADC – E.C.4.1.1.19), followed by hydrolysis and deamination of agmatine by agmatine iminohydrolase (AIH – E.C. 3.5.3.12) and formation of N-carbamoylputrescine. N-carbamoylputrescine is then subjected to hydrolysis, deamination and decarboxylation by N-carbamoylputrescine amidohydrolase (CPA – E.C. 3.5.1.53) and the final product is putrescine. A parallel pathway for putrescine synthesis is the

Table 13.2 Naturally occurring polyamines in plants

Name	Structure
1,3-Diaminopropane	$H_2N(CH_2)_3NH_2$
Putrescine	$H_2N(CH_2)_4NH_2$
Cadaverine	$H_2N(CH_2)_5NH_2$
Norspermidine	$H_2N(CH_2)_3NH(CH_2)_3NH_2$
Spermidine	$H_2N(CH_2)_3NH(CH_2)_4NH_2$
Homospermidine	$H_2N(CH_2)_4NH(CH_2)_4NH_2$
Aminopropylcadaverine	$H_2N(CH_2)_3NH(CH_2)_5NH_2$
Norspermine	$H_2N(CH_2)_3NH(CH_2)_3NH(CH_2)_3NH_2$
Spermine	$H_2N(CH_2)_3NH(CH_2)_4NH(CH_2)_3NH_2$
Thermospermine	$H_2N(CH_2)_3NH(CH_2)_3NH(CH_2)_4NH_2$
Homospermine	$H_2N(CH_2)_3NH(CH_2)_4NH(CH_2)_4NH_2$
Canavalmine	$H_2N(CH_2)_4NH(CH_2)_3NH(CH_2)_4NH_2$

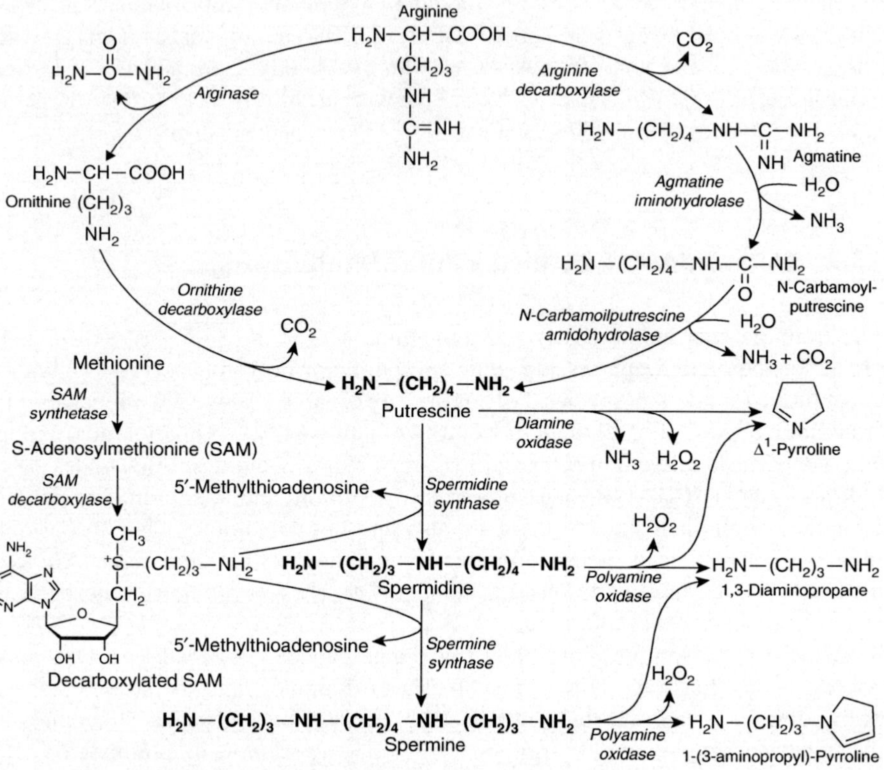

Fig. 13.3 Polyamine metabolic pathways

direct decarboxylation of L-ornithine, catalyzed by ornithine decarboxylase (ODC – E.C.4.1.1.17). The polyamines spermidine and spermine are synthesised by incorporation of an aminopropyl residue from decarboxylated S-adenosylmethionine to putrescine or spermidine – this step is catalyzed by the enzymes spermidine synthase (SPDS – E.C.2.5.1.16) or spermine synthase (SPMS – E.C.2.1.5.22) respectively. The necessary for polyamine biosynthesis decarboxylated S-adenosylmethionine is formed by decarboxylation (S-adenosylmethionine decarboxylase SAMDC – E.C.4.1.1.50) of S-adenosylmethionine (SAM) which is a common precursor of polyamines and ethylene (Slocum 1991).

The polyamine degradation is realized through oxidative deamination catalyzed by aminooxidases, which are copper-containing diamine oxidases (DAO – E.C.1.4.3.6) and flavoprotein-containing polyamine oxidases (PAO – E.C.1.5.3.14). DAO oxidize the primary amino groups of polyamines. The oxidative deamination of putrescine produces Δ^1-pyrroline, H_2O_2 and NH_3. PAO oxidize the secondary amino groups of polyamines and the final products of the process are Δ^1-pyrroline (from Spd oxidation) or 1-(3-aminopropyl)-pyrroline (from Spm oxidation), along with 1,3-diaminopropane and H_2O_2 (Federico and Angelini 1991).

Under physiological pH conditions polyamines bear positive charge and may conjugate with other negatively charged organic molecules like phenolic acids, proteins, phospholipids or nucleic acids. Thus polyamines in higher plants could be present in free, soluble conjugated and insoluble bound forms. The interaction with macromolecules and cell substructures allow polyamines to participate in a number of important growth and developmental processes in plants (regulation of gene expression; cell division, growth, and differentiation; cell, organ, and tissue senescence; dormancy breaking of tubers and germination of seeds; stimulation, support and development of flower buds; embryo- and organogenesis; fruit set, growth and ripening; plant morphogenesis; etc.). Additionally, PAs act in concert with light and phytohormones and are considered as plant endogenous growth regulators with hormone-like properties. Because of their polycationic nature, polyamines possess free radical scavenging features and antioxidant activity and may confer plant tolerance to different biotic and abiotic stresses (Groppa and Benavides 2008; Gill and Tuteja 2010a).

13.3 Polyamines and Salinity Stress Tolerance in Plants

The participation of polyamines in the scavenging of free radicals, antioxidant activity and modulation of plant stress tolerance to various abiotic stresses has been extensively studied. A number of authors have reported the relationship of PA metabolism with plant responses to nutrient deficiency in the environment (Sarjala and Kaunisto 2002; Camacho-Cristobal et al. 2005), low or high temperature stress (Pillai and Akiyama 2004; Todorova et al. 2007), drought (Yamaguchi et al. 2007; Todorova et al. 2008), UV treatment (Smith et al. 2001; An et al. 2004; Zacchini and de Agazio 2004; Katerova and Todorova 2009), excess of heavy metals (Groppa et al. 2001, 2003; Zhao et al. 2008), and herbicide treatments (Benavides et al. 2000; Durán-Serantes et al. 2002; Szigeti and Lehoczki 2003; Cuevas et al. 2004; Deng 2005). The cited extensive investigations clearly demonstrate that polyamines play a pivotal role in conferring plant stress tolerance under unfavorable environmental conditions.

13.3.1 Endogenous Polyamine Concentrations

Similarly to other stresses, investigations with different plant species have shown that alteration of polyamine concentrations occur in response to salinity. The effect of saline conditions on endogenous polyamines showed that polyamine concentrations altered in different manner depending on plant species and cultivars, plant organ and developmental stage of tissues, duration and intensity of stress treatment (Liu et al. 2007). Although some authors have found a decrease in endogenous polyamine concentrations under salinity stress (Benavides et al. 1997; Liu et al. 2008; Legocka and Sobieszczuk-Nowicka 2012), mainly an accumulation of PAs due to salt stress has been reported in plant tissues (Kakkar et al. 2000; Upreti and Murti 2010; Iqbal and Ashraf 2012). Lefèvre et al. (2001) reported that short-term salt stress (50/100 mM NaCl or KCl) markedly increased Put (putrescine) concentrations of roots for the salt-resistant rice cultivar Pokkali compared with the sensitive one (IKP). The authors did not observe a clear relationship between the mean level of salinity resistance and the endogenous amounts of spermidine or spermine. Shoot Put concentration of the salt-resistant cultivar was only slightly increased after exposure to both salt stresses. Therefore, the authors suggested that physiological significance of Put accumulation may depend on the organ considered. It was also found that in salt-stressed heterotrophic maize callus culture total PA content and especially Put amount was higher in the resistant cultivar than in the salt-sensitive one (Willadino et al. 1996).

Most common PAs abundance (and especially of higher polyamine spermidine and spermine) was associated with increased stress tolerance to salinity (Ahmad et al. 2009; Ben et al. 2009; Yamamoto et al., 2011; Alet et al. 2011, 2012). Zapata et al. (2003, 2004, 2008) studied the effect of salinity on polyamine levels in different plant species subjected to 100 mM or 150 mM NaCl. They found that in general Put decreased, while Spd and Spm increased in saline-stressed plants (Zapata et al. 2003, 2004). Additionally the ratio (Spd + Spm)/Put increased with salinity (Zapata et al. 2004), which correlated with the idea of a protective role of higher PAs against salt stress. Similarly, El-Shintinawy (2000) also pointed out that salinity significantly enhanced the augmentation of Spm and Spd accompanied with a reduction in Put amount in wheat cultivars. In their study Tassoni et al. (2008) analyzed polyamine metabolism in *Arabidopsis thaliana* (ecotype Columbia) inflorescences and stalks collected from plants germinated and grown under increasing salt-stress conditions (0–75 mM NaCl). The authors found that free spermidine was the most abundant polyamine and its levels, as well as those of free spermine, increased with salt concentration, supporting the hypothesis for a specific role of higher polyamines in the tolerance to salt stress. In experiment with barley seedlings Zhao et al. (2003) have examined the effect of wide range (0–300 mmol/L) of NaCl on free and bound polyamines and have found that polyamine content raised up in dose-dependent manner in relation to salt concentrations, but amount of bound fractions were reduced by highest concentration. Additionally, they found a positive correlation between the plant growth rate and the ratio of bound PA/free PA levels. The authors suggested that under salt stress, the balance between free PA and bound PA content in roots is important for salt tolerance of barley seedlings. In order to examine the plant responses to salt stress (0, 50, and 100 mM NaCl), Kim et al. (2010) investigated Chinese cabbage seedlings grown as a hydroponic culture. They found that spermidine content decreased as salinity increased, but spermine content increased, which also correlated with proposed idea of polyamine protection under saline environment. Jouve et al. (2004) have shown that *Populus tremula* was able to cope with up to 150 mM NaCl mainly due to accumulation of spermine as well as compatible solutes (sucrose, proline, mannitol and raffinose). Ghosh et al. (2011) found that salt induction had changed the nature of polyamine content in time-dependent manner. Accumulation of polyamines in both rice varieties (Pokkali and Nonabokra) amplified with increased duration of NaCl treatment. In other experiment with three rice cultivars differing in their salt-sensitivity Roychoudhury et al. (2008) found that salt-tolerant plants (Nonabokra) accumulate less Put than salt-sensitive cultivars (M-1-48) after salinity treatment for 48 h, which was comparable with less significant reduction of growth rate. Correspondingly, Spd and Spm synthesis in the salt-tolerant variety is stimulated to stabilize membrane systems. The effect of salt stress (50, 100 and 150 mM NaCl) on the endogenous levels of free, bound and total polyamines was studied in root tissues of salt tolerant (Coban) and salt sensitive (Sanbro) cultivars of sunflower (*Helianthus annuus* L.) plants for the 5th, 15th and 25th days of the growth periods (Mutlu and Bozcuk 2007). The amounts of free, bound and total Spm increased in root tissues of sunflower plants, while the levels of other polyamine titers were either decreased in general or remained unchanged significantly. The authors suggested that increase in some polyamine concentrations in sunflower root tissues under salinization implies their possible role in diminishing the injurious effect of salinity stress. The salt tolerance in sunflower plants was related to the excessive augmentation of total polyamines in root tissues of salt tolerant cultivar Coban under saline condition (Mutlu and Bozcuk 2007). A marked increase was noted in free Spd and Spm, soluble conjugated and insoluble bound Put, Spd and Spm contents in the roots of cucumber (*Cucumis sativus* L) cultivar Changchun mici (comparatively tolerant to high salinity) than Jinchun No. 2 under short-term salt stress (Duan et al. 2008). Hummel et al. (2004) observed the reduction in polyamines measured in shoots and partially in roots during long term exposure of *Pringlea antiscorbutica* seedlings to severe salinity (300 mM NaCl). Only spermine content was increased in roots after salt stress. The authors concluded that major effect of saline stress was the modification of polyamine distribution between roots and shoots. Higher Spm content in roots was a developmental response to stress and its accumulation in roots facilitated reinitiation of root growth (Hummel et al. 2004).

In general one possible mechanism is assumed which explain the participation of endogenous polyamines (primarily spermine) in plant salt tolerance. In a number of articles where the protective role of endogenous higher polyamines has been shown, it was proposed that under salinity conditions PAs regulated the activity of plasma membrane H^+-ATPase. Janicka-Russak et al. (2010) suggested that the decrease in PAs could result in the removal of excess cations from the cell. This action of endogenous PAs contributes to ionic homeostasis by modification of the plasma membrane H^+-ATPase and the vacuolar H^+-ATPase activities in cucumber roots treated with NaCl. Similarly, Roy et al. (2005), showed that root plasma membranes of rice salt-tolerant cultivars Nonabokra and Pokkali were rich in Spm and Spd, whereas the root plasma membranes of sensitive cultivars (M-1-48 and IR8) were rich in Put only. After treatment of barley seedlings with different concentrations of NaCl (0–300 mM) for 3 days Liu et al. (2006a) have shown that Put content declined, while spermine and spermidine were amplified in roots. The polyamines, especially Spd, enhanced the activities of tonoplast H^+-ATPase and H^+-PPase, which are capable to create the proton gradient across the vacuole membrane and to pump in the excessive cytosol Na^+ into the vacuole and to rebuild the relative ion and pH balance. Thus tonoplast H^+-ATPase and H^+-PPase play crucial roles in plant salt adaptation and maintenance of their activities higher is essential for plant growth and survival under salt stress. The authors suggested that the conversion of Put to Spd and Spm and maintenance of higher levels of Spd and Spm were necessary for plant salt tolerance through enhanced tonoplast H^+-ATPase and H^+-PPase activities in roots (Liu et al. 2006a).

13.3.2 Biosynthetic Polyamine Enzymes

It is well known that plant ability to control stress, including salinity, is linked to their ability to synthesize PAs (Kasinathan and Wingler 2004; Hamdani et al. 2011). When Kasinathan and Wingler (2004) analyzed polyamine concentration and salt stress tolerance in two *Arabidopsis thaliana* mutants, *spe1-1* and *spe2-1* with reduced activity of ADC, they showed that polyamine accumulation depends on acclimation to salinity and that decreased polyamine formation leads to reduced salt tolerance. Similarly, a marked increase was reported in ADC, ODC, SAMDC and DAO activities in the roots of comparatively tolerant to high salinity cucumber cultivar Changchun mici than in cv. Jinchun No. 2 under short-term salt stress (Duan et al. 2008). It was reported that ADC activity in NaCl-stressed heterotrophic maize callus culture is also significantly incremented, especially for the salt-resistant calluses, which was believed to be related with the rise in putrescine (Willadino et al. 1996). However, long-term (salt treatment 21 days) led to reduced activities of ADC and SAMDC in salt tolerant (var Giza) and salt sensitive (var El Paso) *Oryza sativa* cultivars (Maiale et al. 2004). The activity of SPDS was reduced in the salt tolerant rice cultivar but not in the salt sensitive variety and the authors suggested that SPDS activity has important role for plants subjected to salt stress.

The expression of several genes involved in PA biosynthesis (*ZmODC*, *ZmSPDS2A*, *ZmSPDS2B*, *AtADC2*, *AtSPMS*, *SPMS*, *SAMDC1*, *SAMDC2*) is up-regulated in the presence of salt stress (Li and Chen 2000a; Urano et al. 2003, 2004; Rodríguez-Kessler et al. 2006) in different plants (*Zea mays*, *Arabidopsis thaliana*, *Oryza sativa*). Rodríguez-Kessler et al. (2006) described a spermidine synthase cDNA from *Zea mays* leaves (*ZmSPDS2A*) and discussed the transcriptional regulation of the corresponding gene and other genes related to PA biosynthesis (*ZmODC*, *ADC* and *SAMDC*) under salinity. *ZmSPDS2A* was equally expressed in leaves, stem and roots, however the transcripts of other genes involved in PA biosynthesis (*ZmODC*, *ADC* and *SAMDC*) showed tissue-specific regulation. It was reported that in maize only *ZmODC*, *ZmSPDS2A* and the identified second transcript encoding a spermidine synthase (*ZmSPDS2B*) were up-regulated by salt stress. In *Oryza sativa* L. seedlings, the expression of the *SAMDC1* gene was considerably induced by salinity (Li and Chen 2000a). The transcript levels of *SAMDC1* in two rice varieties differing in salt tolerance was higher in the salt-tolerant japonica rice variety Lansheng, than in the salt-sensitive one (variety 77–170), and occurred more quickly when both varieties were exposed to low salt conditions. The results suggested that the expression of the *SAMDC1* gene in seedlings is positively correlated with the salt tolerance of rice. Hao et al. (2005) studied two SAMDC c DNAs isolated from apple (*MdSAMDC1* and *MdSAMDC2*) and found that *MdSAMDC2* was positively induced by salt stresses, but *MdSAMDC1* was not activated. In *Arabidopsis* the polyamine biosynthetic gene *ADC1* has been found to be expressed in all tissue tested, while *ADC2* is mainly expressed in siliques and cauline leaves, and is induced upon various stress types, including salinity (Urano et al. 2003, 2004; Bagni et al. 2006). When analyzing the expression profiles of genes responsible for PA biosynthesis in *Arabidopsis thaliana* (two genes for *ADC*, four genes for *SAMDC*, four genes for *SPDS*, and two genes for *SPMS*) under various abiotic stress conditions, Urano et al. (2003), found that *AtADC2* and *AtSPMS* mRNAs (encoding ADC and spermine synthase) increased markedly in response to NaCl. Stress-inducible accumulation of *AtADC2* mRNA correlated with putrescine accumulation under NaCl treatment. Normally ODC activity is involved in Put biosynthesis but in *Arabidopsis* there is no detectable ODC activity (Hanfrey et al. 2001) and ADC is considered as a key enzyme in PA biosynthesis in *Arabidopsis*. The role of stress-inducible *AtADC2* gene in *Arabidopsis* was analyzed through a *Ds* insertion mutant of *AtADC2* gene (*adc2-1*) under salinity conditions (Urano et al. 2004). In the *adc2-1* mutant, which was more sensitive to salinity than the control plants, free Put content was significantly reduced (compared with the control plants) and did not increase under salt stress. The salinity sensitive phenotype of *adc2-1* was recovered by the addition of exogenous Put. The authors concluded that endogenous Put plays an important role in salt tolerance in *Arabidopsis* and *AtADC2* is a key gene for its production under salt stress and normal conditions.

It was reported that *MdADC* expression and ADC activity decreased in apple (*Malus sylvestris* (L.) Mill. var. domestica) in vitro callus experiencing salt recovery, and increased when the callus was subjected to successive salt stress (Liu et al. 2006b). Under the same conditions ODC activity showed a pattern opposite to that of ADC. Moreover, treatment with the ADC inhibitor (D-Arginine) caused serious growth impairment under salt stress. The authors' results clearly showed that the ADC pathway is involved in the salt stress response.

When the localization of *SPMS* gene promoter activity was examined using the *SPMS* promoter-β-glucuronidase (*GUS*) gene expressed in transgenic *Arabidopsis* plants the gene was detected in almost all organs during all developmental growth stages (Sagor et al. 2011). However, *ACL5* (encoding thermospermine synthase) promoter activity was predominantly observed in the vascular systems. The authors reported that upon high salt stress *SPMS* promoter activity was strongly detected in all organs except cotyledons, whereas the *ACL5* promoter activity was reduced which is consistent with the decreased levels of *ACL5* transcripts. The results showed that *SPMS* expression differs from that of *ACL5* in respect to tissue specificity and salt stress response, suggesting relevant differences in Spm and thermospermine functions (Sagor et al. 2011).

During the last decade number of authors reported enhanced salinity tolerance in different transgenic plants engineered to overproduce polyamines by overexpression of the genes involved in PAs biosynthesis (Table 13.3).

In *Oryza sativa*, the overproduction of polyamines was achieved by overexpression of *ADC* from *Avena sativa* and *SAMDC* under the control of ABA inducible

Table 13.3 Salt stress tolerance in transgenic plants engineered to overproduce polyamines

Gene	Gene source	Transgenic plant	PA overproduction	Reference
ADC	Oat	Rice	Put	Roy and Wu (2001)
ADC	Oat	Eggplant	Put, Spd (particularly conjugated forms) and free Spm fraction	Prabhavathi and Rajam (2007)
ODC	Mouse	Tobacco	Put	Kumria and Rajam (2002)
SAMDC	Tritordeum	Rice	Spm and Spd	Roy and Wu (2002)
SAMDC	Human	Tobacco	Spd, Put, especially conjugated fraction	Waie and Rajam (2003)
SAMDC	Carnation	Tobacco	Put, Spd and Spm	Wi et al. (2006)
SAMS (SsSAMS2)	Suadea salsa	Tobacco	Free Spm, Spd and Put	Qi et al. (2010)
SPDS (MdSPDS1)	Apple	European pear	Spd	He et al. (2008)
MdSPDS1	Apple	European pear	Spd	Wen et al. (2008)
MdSPDS1	Apple	Tomato	Spd and Spm	Neily et al. (2011)
SPDS	Fig leaf gourd	Arabidopsis	Spd, especially conjugated fraction	Kasukabe et al. (2004)
SPDS (FSPD1)	Fig leaf gourd	Sweet potato	Spd	Kasukabe et al. (2006)

promoter and it was shown that transgenic plants had higher salinity tolerance as compared to non-transformed plants (Roy and Wu 2001, 2002). The transgenic *Nicotiana tabacum* plants, overexpressing the putrescine synthesis gene *ODC* from mouse possessed enhanced salt tolerance (Kumria and Rajam 2002). The introduction of *SPDS* gene into *Arabidopsis* and *Ipomoea batatas* cv. Kokei 14 led to the enhanced tolerance against multiple abiotic stresses including salinity (Kasukabe et al. 2004, 2006). Transgenic eggplants overexpressing *ADC* gene from *Avena sativa* accumulated polyamines and exhibited an increased tolerance to multiple abiotic stresses including salinity (Prabhavathi and Rajam 2007). The transgenic tobacco overexpressing human *SAMDC* gene has showed increased PA levels (especially conjugated Put and Spd fractions) and tolerance to salinity (Waie and Rajam 2003). Salt-induced damage was attenuated in the transgenic *Nicotiana tabacum* L. plants overexpressing *SAMDC* from *Dianthus caryophyllus* L. flower (Wi et al. 2006). Recently, it was reported that salt stress induced damage in tobacco transgenic plants overexpressing the suadea salsa full-length S-adenosylmethionine synthetase (*SsSAMS2*) gene under the control of cauliflower mosaic virus 35S promoter, was attenuated (Qi et al. 2010). The authors showed that *SsSAMS2* overexpression in transgenic tobacco plants leads to increase in free Spm, Spd and Put content, and as a result promotes salt tolerance. Transgenic European pear (*Pyrus communis* L. 'Ballad') and tomato overexpressing apple *SPDS* (*MdSPDS1*) performed attenuated susceptibility to NaCl stress in relation to the wild plant (Wen et al. 2008; Neily et al. 2011). Additionally, He et al. (2008) documented that the same transformation led to enhanced enzymatic (SOD, APX, MDHAR, GR) and non-enzymatic antioxidant capacity in response to salinity. The overexpression of fig leaf gourd (*Cucurbita ficifolia*) *SPDS* in *A. thaliana* enhanced the tolerance of transgenic plants to various types of environmental stresses, including salinity, suggesting the key role for spermidine as regulator in stress signaling pathways (Kasukabe et al. 2004). The authors suggested that Spd could be involved in stress tolerance phenomenon as a direct stress-protecting compound and as a stress-signaling regulator.

13.3.3 Aminooxidase(s)

Under normal and stress conditions PA content in plant cells depends not only on their biosynthesis and transport but also on their catabolism where aminooxidases (copper-containing DAO and flavin-containing PAO) are involved. PAOs are grouped into families, which are participating either in terminal catabolism or back-conversion of polyamines. DAOs exhibit high affinity for diamines, while PAO oxidize secondary amine groups from Spd and Spm (Alcazar et al. 2006). However, it was noted that in soybean plants CuAO preferred Cad over Put and the catabolic enzyme was active even under salinity (Campestre et al. 2011). DAO and PAO are localized in the cytoplasm and cell walls where they provide H_2O_2 required for suberinization, lignification and formation of cross-bridges between the components of the cell wall that confer cell wall-stiffening (Cona et al. 2003; Kuznetsov and Shevyakova 2007). In contrast to DAO, which are known to occur in high amounts in dicotyledonous, PAO occur in high levels in monocotyledonous plants (Šebela et al. 2001; Cona et al. 2006). However, Maiale et al. (2004) did not detect PAO activity in crude extracts of salt tolerant (var Giza) and salt sensitive (var El Paso) *Oryza sativa* cultivars subjected to long-term salt stress.

The activities of CuAO and PAO lead to a reduction in the PA levels. Additionally, along with the other products, CuAO and/or DAO generate H_2O_2, which is known to act as a signal molecule in low concentrations, and might coordinate adaptation processes in plants (Kuznetsov and Shevyakova 2007; Alcazar et al. 2010; Campestre et al. 2011). Therefore the PA catabolism and the associated H_2O_2 production are important in the induction of abiotic stress (including salinity) tolerance in plants. Moschou et al. (2008) revealed the importance of the H_2O_2 derived from PA catabolism in the induction of salinity-induced tolerance in tobacco (*Nicotiana tabacum* cv Xanthi) using transgenic plants overexpressing or downregulating apoplastic PAO. In addition, Rodríguez et al. (2009) reported that PA oxidation by the activity of PAO could be the major source contributing to ROS production involved in maize leaf elongation under salinity. The authors showed that maximal activity of PAO was not affected by salinity, and suggested that the enzyme in *Zea mays* is tolerant to salt stress. As expected, the total activity of PAO was found to be up to 20-fold higher than that of the CuAO.

Campestre et al. (2011) examined the possible relationship between polyamine catabolism mediated by CuAO and the elongation of *Glycine max* L. hypocotyls from plants exposed to NaCl. Salinity caused a rise in CuAO activity in segments of the hypocotyl elongation zone in in vitro and in vivo experiments. The expression of *GmCuAO1* gene (found by Delis et al. 2006 predominantly in tissues characterized by rapid extension growth, as the apical segments of etiolated hypocotyls) was detected in the elongation zone but, surprisingly, there was no change in its expression level under salinity (Campestre et al. 2011). The authors conclude that the higher CuAO activity found under salt stress could not be attributed to a higher expression of *GmCuAO1* gene.

Classical approaches, using inhibitors of enzymes involved in polyamine metabolism, pointed to their possible role in plant adaptation to NaCl stress. Using a DAO specific inhibitor aminoguanidine and exogenously applied Put, Su and Bai (2008) reported that a 15–20 % of proline accumulated in *Glycine max* leaves under salt stress (50–150 mM NaCl) could result from PA (Put) degradation products. Additionally, the activity of DAO was also enhanced. Shevyakova et al. (2006) suggested that salinity (400 mM NaCl) induced peroxide in the leaves of the halophyte *Mesembryanthemum crystallinum* L. was produced via DAO activity. This assumption was supported in experiments with addition of the inhibitor aminoguanidine to the root medium in the presence of NaCl. In addition, their study indicated that activation of DAO combined with H_2O_2–peroxides reaction in leaves and roots is implicated in the regulation of free and conjugated PA concentrations under salinity. In a study conducted with similar model system Kuznetsov et al. (2007) also noted enhanced DAO activity, which correlated with the increased H_2O_2 production in the presence of NaCl. Salinity (50, 100, 150 mM NaCl) strongly increased the activity of DAO in *Glycine max* (cv. Suxie-1) roots (Xing et al. 2007). Accumulation of γ-aminobutyric acid (GABA) levels also increased with the increasing NaCl concentrations. Expectedly, PA degradation under NaCl treatment led to reduced amounts of free Put, Cad and Spd. The authors observed a close correlation between the alterations in DAO activity and GABA accumulation also after treatment with the aminoguanidine inhibitor. It was suggested that higher GABA accumulation induced by treatment with NaCl might result from PA degradation, concluding that PAs could execute their functions through formation of GABA under salinity. Soybean plants treated with N,N^1-diaminoguanidine (an inhibitor of CuAO) showed a decrease in CuAO and significant reduction of reactive oxygen species in the elongation zone, even under salinity (Campestre et al. 2011). N,N^1-diaminoguanidine ceased the increased hypocotyl length in control and salinity exposed plants. The authors suggested that the activity of CuAO may be is partly contributing to the hypocotyl growth under NaCl stress, via generation of H_2O_2 by polyamine catabolism – thus reinforcing the importance of polyamine catabolism and H_2O_2 production in the induction of salt tolerance in plants. Supporting the idea of Moschou et al. (2008), Rodríguez et al. (2009) and Campestre et al. (2011) for the important role of H_2O_2 produced by PAO and/or DAO, Shoresh et al. (2011) noted that under salinity PAO activity was highly increased in the leaf-growing zone in *Zea mays*, possibly providing the ROS required for elongation. The authors suggested that the enhanced PAO activity was an adaptive mechanism directed to restore the normal levels of ROS at the expansion zone where NADPH oxidase could no longer provide the required ROS for growth.

It was reported that under 0–200 mM NaCl PAO activity was increased in the roots of barley seedlings (Zhao et al. 2003). Interestingly DAO activity was also increased in transgenic eggplants (overexpressing oat *ADC* gene) that exhibited an increased tolerance levels to multiple abiotic stresses including salinity (Prabhavathi and Rajam 2007). Duan et al. (2008) also reported that under short-term salt stress DAO activity increased along with the biosynthetic PA enzymes (ADC, ODC, SAMDC) in the roots of comparatively tolerant to high salinity cucumber cultivar Changchun mici than in cv. Jinchun No. 2.

The polyamine metabolic pathway is interconnected with other metabolic routes engaged in the formation of different signaling molecules (as H_2O_2 and NO) and metabolites (ethylene, SAM, proline, GABA) which are important in plant stress (including salinity) responses (Kuznetsov et al. 2007; Xing et al. 2007; Su and Bai 2008; Alcazar et al. 2010).

13.4 Application of Polyamines for Enhanced Salinity Stress Tolerance

Large number of evidence suggested that exogenous application of PAs could be used as promising tool to enhance plant tolerance under salt stress conditions. Ali (2000) showed that exogenous Put reduced the net accumulation of Na^+ and Cl^- ions in different organs of *Atropa belladonna* subjected to salinity stress. Put alleviated the negative effect of NaCl during germination and early seedling growth and increased endogenous Put of *A. belladonna*. Verma and Mishra (2005) also reported that Put counteracted the salinity induced decrease in seedling growth and biomass accumulation, and increased the activity of antioxidant enzymes and carotenoids in leaf tissues of salt stressed *Brassica juncea* seedlings. Similar results were reported by Tang and Newton (2005), who showed that polyamines (mainly Put) lessened salt-induced oxidative damage by increasing the activities of antioxidant enzymes and decreasing lipid peroxidation in callus and seedlings of Virginia pine. Putrescine (0.5 mM) was completely able to amend the toxic effects of salt stress (100 mM NaCl) on electrolyte leakage and lipid peroxidation and partially on relative water content in chickpea plants (Sheokand et al. 2008). Additionally, Put had a positive effect on antioxidant enzyme activities under salt stress. In other study (Shi et al. 2008), exogenous putrescine (100 μM) was added to nutrient solution 3 days before cucumber (*Cucumis sativus* L. cv. "Jinyan No.4") seedlings were exposed to 100 mM NaCl treatment. Putrescine considerably diminished the negative effects of NaCl on root growth through decreasing of Na^+ uptake and increasing of potassium accumulation in roots. Demetriou et al. (2007) reported that enhanced salinity provoked negative changes in the photosynthetic apparatus of green alga *Scenedesmus obliquus* by affecting both its structure and function: 10‰ NaCl increased the effective antenna size and decreased the active reaction center population, accompanied by a significant inhibition of the photosynthetic rate. Exogenously added Put altered these negative changes, by inducing the reciprocal reorganization of the photosynthetic apparatus (Demetriou et al. 2007). Increased salinity (150 mM NaCl) provoked a negative modification in plasma membrane phospholipids and polyamine treatment (particularly putrescine and spermidine) caused a favorable effect in maintaining of plasma membrane stability and functions of wheat roots under surplus salinity (Mansour et al. 2002).

The effects of higher polyamines (Spd and Spm) on physiological and biochemical changes in 12-day-old rice seedlings (salt-tolerant Pokkali and salt-sensitive M-1–48) were investigated during salinity stress (Chattopadhayay et al. 2002). Both polyamines considerably prevented the electrolyte leakage, amino acid and chlorophyll loss, and inhibition of photochemical reactions of photosynthesis, as well as the downregulation of chloroplast-encoded genes like *psbA*, *psbB*, *psbE* and *rbcL* induced by the salinity stress. In similar experiments with roots of salt-tolerant (Nonabokra) and salt-sensitive (M-1-48) rice cultivars treated with 150 mM NaCl and 1 mM Spd, Gupta et al. (2012b) for a first time showed that polyamines participate in the salt stress signaling due to spermidine-mediated phosphorylation and activation of 42 kDa Ca^{2+}-independent non-MAPK protein kinase. This fact allowed the authors to presume that this protein kinase plays a key role in the activation of different stress regulatory biomolecules, indicating its importance in salinity mediated signal transduction (Gupta et al. 2012a, b). In other study with rice seedlings (salt-tolerant Pokkali and salt-sensitive KDML 105), the application of 1 mM spermidine helps the plant to withstand the negative effect of NaCl (Saleethong et al. 2011). The authors concluded that exogenous spermidine enhanced the salinity tolerance of rice by stabilizing membrane, scavenging free radicals and maintaining K^+/Na^+ balance.

Salt stress reduced all evaluated growth parameters and yield components, content of leaf pigments, carbohydrates, protein, spermidine and spermine as well as amylase activity of *Vigna sinensis* plants (Alsocari 2011). Exogenous application of spermine mitigated the deleterious effects of salinity stress on growth and yield of the stressed plants. The protective effect of spermine on *V. sinensis* plants appeared to be mainly due to the increased chlorophyll and protein content and endogenous polyamines (Alsocari 2011). Similar results were reported by Sakr and El-Metwally (2009) who showed that exogenous spermine partially alleviated the harmful effect of soil salinization on wheat growth and yield. Spermidine application in salinized nutrient solution during short-time stress resulted in alleviation of the salinity-induced membrane damage in the roots and plant growth and photosynthesis inhibition, together with an increase in polyamine and proline contents and antioxidant enzyme activities in the roots of salt-sensitive cucumber cultivar Jinchun No. 2 (Duan et al. 2008). Recently, Roychoudhury et al. (2011) provided an additional information concerning protective effect of exogenous application of spermidine and spermine on NaCl-treated rice seedlings. The authors showed that mitigation of inhibitory effect of salinity stress was conferred by preventing growth inhibition, averting different forms of cellular injuries, maintaining K^+/Na^+ balance or increasing the level of compatible osmolytes and activity of antioxidant enzymes.

Similarly to endogenous polyamines, the beneficial effect of the exogenous polyamines is related to the improvement of the ion balance in salt-treated cells due to their polycationic nature. As was illustrated in series of experiments, the application of 1 mM Put decreased the Na^+ accumulation of rice calli and rice plants exposed to salt stress (Ndayiragije and Lutts 2006, 2007). The protective effect of exogenous polyamines under high salinity is associated mainly with improving of the K^+/Na^+ homeostasis through restricting the Na^+ influx into roots and thus preventing the loss of K^+ from cells (Zhao et al. 2007). Zhao and Qin (2004) also proposed that one of the possible mechanisms involved in the alleviating of salt injury in barley seedlings by PAs application was to maintain tonoplast integrity and function under saline conditions. Furthermore, Zhu et al. (2006) suggested that exogenous Spd inhibits Na^+ transport from roots to shoots under conditions of high salinity by reinforcing the barrier effects of Casparian bands, which are beneficial for attenuating the salt injuries in barley seedlings. Salinity stress severely inhibited the H^+-ATPase activity in rice plants, but spermidine treatment significantly recovered its activity (Roy et al. 2005). Shabala et al. (2007) showed that application of polyamines in micromolar concentrations was efficient in preventing NaCl-induced K^+ efflux from the pea mesophyll. The authors suggested that polyamines may directly block the K^+ efflux through non-selective cation channels and activate the plasma membrane H^+-ATPase, so restoring the membrane potential (Shabala et al. 2007). Moreover, maintaining of cytosolic K^+/Na^+ balance seems to be the major beneficial effect of polyamine regulation of membrane transport activity (Shabala and Cuin 2008). However, preliminary application of 1 mM Spm, Spd or Put prevented NaCl-induced K^+ leak only in the mature root zone of hydroponically grown maize and Arabidopsis. In contrast, in the distal elongation root zone, PA pre-treatment resulted in an even larger NaCl-induced K^+ efflux, so PAs affect the cell membrane transporters in a highly specific way (Pandolfi et al. 2010). The authors concluded that the ameliorative affect of PAs is a result of combination of several issues which probably incorporate PA transport, accumulation and metabolism in cell, and the functional expression of specific target proteins or signaling components (Pandolfi et al. 2010).

13.5 Conclusion and Future Perspective

The importance, actuality and inevitability of the salinity stress, as well as its negative impact on the physiological processes and plant productivity have drawn the attention of a number of investigators to study the problems of salt stress. The research in this area is routed mainly into five directions:

- Studies on the physiological response of the plant organism subjected to salinity stress
- Comparison of the effects of salt stress on several plant organisms, differing in their species, variety or genetically determined resistance to the stress factor

- Searching for possibilities to decrease the unfavorable consequences caused by a salinity by induction of adaptation or by means of application of xenobiotics, including polyamines
- Selection using cell cultures and conventional plant breeding
- Genetic engineering for improved stress tolerance

One popular procedure for increasing plant salt tolerance is the using of exogenous application of different chemicals. For now the polyamine large-scale agricultural benefit application is rather limited.

Alterations in polyamine content, expression of genes and/or activity of polyamine biosynthetic enzymes due to salt stress have been explored in a large number of plant species, for example rice (Krishnamurthy and Bhagwat 1989; Chattopadhyay et al. 1997; Quinet et al. 2010), tomato (Santa-Cruz et al. 1997a, b, 1998; Botella et al. 2000), wheat (Li and Chen 2000b), sunflower (Alvarez et al. 2003), *Fraxinus angustifolia* (Tonon et al. 2004), Arabidopsis (Urano et al. 2004), *Lotus glaber* (Sanchez et al. 2005) and *Lupinus luteus* (Legocka and Kluk 2005). Most of the studies demonstrate that polyamine abundance is associated with enhanced plant salinity tolerance. Hence, perspective and modern strategy for improvement of plant tolerance to cope with salinity is the using of transgenic and molecular genetic approaches to increase cellular PAs levels. Current studies with gain- or loss-of-function mutants have also suggested a close relationship between plant stress tolerance and the level of endogenous PAs in plants (Alcazar et al. 2006). Therefore, screening of varieties that maintain high level of endogenous polyamines is promising approach to enhance salinity tolerance.

Acknowledgments The authors express their gratitude to the National Science Fund of Republic Bulgaria – Grant DMU03/60.

References

Ahmad P, Sarwat M, Sharma S (2008) Reactive oxygen species, antioxidants and signaling in pants. J Plant Biol 51:167–173

Ahmad MSA, Ali Q, Ashraf M, Haider MZ, Abbas Q (2009) Involvement of polyamines, abscisic acid and anti-oxidative enzymes in adaptation of Blue Panicgrass (*Panicum antidotale* Retz.) to saline environments. Environ Exp Bot 66:409–417

Alcazar R, Altabella T, Marco F, Bortolotti C, Reymond M, Knocz C, Carrasco P, Tiburcio AF (2010) Polyamines: molecules with regulatory functions in plant abiotic stress tolerance. Planta 231:1237–1249

Alcazar R, Marco F, Cuevas JC, Patron M, Ferrando A, Carrasco P, Tiburcio AF, Altabella T (2006) Involvement of polyamines in plant response to abiotic stress. Biotech Lett 28:1867–1876

Alet A, Sánchez D, Ferrando A, Fernandez-Tiburcio A, Alcazar R, Cuevas JC, Altabella T, Pico FM, Carrasco P, Menéndez AB, Ruiz OA (2011) Homeostatic control of polyamine levels under long-term salt stress in Arabidopsis. Changes in putrescine content do not alleviate ionic toxicity. Plant Signal Behav 6:1–6

Alet A, Sánchez DH, Cuevas JC, Marina M, Carrasco P, Altabella T, Tiburcio AF, Ruiz OA (2012) New insights into the role of spermine in *Arabidopsis thaliana* under long-term salt stress. Plant Sci 182:94–100

Ali RM (2000) Role of putrescine in salt tolerance of *Atropa belladonna* plant. Plant Sci 152:173–179

Alsocari SS (2011) Synergistic effect of kinetin and spermine on some physiological aspects of seawater stressed *Vigna sinensis* plants. Saudi J Biol Sci 18:37–44

Alvarez I, Tomaro ML, Benavides MP (2003) Changes in polyamines, proline and ethylene in sunflower calluses treated with NaCl. Plant Cell Tiss Organ Cult 74:51–59

An LZ, Liu GX, Zhang MX, Chen T, Liu YH, Feng HY, Xu SJ, Qiang WY, Wang XL (2004) Effect of enhanced UV-B radiation on polyamine content and membrane permeability in cucumber leaves. Russ J Plant Physiol 51:658–662

Apse MP, Blumwald E (2002) Engineering salt tolerance in plants. Curr Opin Biotech 13:146–150

Azevedo Neto AD, Gomes-Filho E, Prisco JT (2008) Salinity and oxidative stress. In: Khan NA, Singh S (eds) Abiotic stress and plant responses. I. K. International Publishing House, New Delhi/Bangalore/Mumbai, pp 57–82

Badawi GH, Yamauchi Y, Shimada E, Sasaki R, Naoyoshi K, Tanaka K, Tanaka K (2004) Enhanced tolerance to salt stress and water deficit tolerance by overexpressing superoxide dismutase in tobacco (*Nicotiana tabacum*). Plant Sci 166:919–928

Bagni N, Ruiz-Carrascoa K, Franceschetti M, Fornalè S, Fornasiero RB, Tassoni A (2006) Polyamine metabolism and biosynthetic gene expression in *Arabidopsis thaliana* under salt stress. Plant Physiol Bioch 44:776–786

Bartels D, Sunkar R (2005) Drought and salt tolerance in plants. Critical Rev Plant Sci 24:23–58

Ben HA, Ghanem ME, Bouzid S, Lutts S (2009) Abscisic acid has contrasting effects on salt excretion and polyamine concentrations of an inland and a coastal population of the Mediterranean xero-halophyte species *Atriplex halimus*. Ann Bot 104:925–936

Benavides MP, Aizencang G, Tomaro ML (1997) Polyamines in *Helianthus annuus* L. during germination under salt stress. J Plant Growth Regul 16:205–211

Benavides MP, Gallego SM, Comba ME, Tomaro ML (2000) Relationship between polyamines and paraquat toxicity in sunflower leaf discs. Plant Growth Regul 31:215–224

Borsani O, Valpuesta V, Botella MA (2003) Developing slat tolerant plants in a new century: a molecular biology approach. Plant Cell Tissue Organ Cult 73:101–115

Botella MA, del Amor F, Amoros A, Serrano M, Martinez V, Cerda A (2000) Polyamine, ethylene and other physico-chemical parameters in tomato (*Lycopersicon esculentum*) fruits as affected by salinity. Physiol Plant 109:428–434

Brankova L, Ivanov S, Alexieva V (2007) The induction of microsomal NADPH:cytochrome P450 and NADH:cytochrome b5 redictases by long-term salt treatment of cotton (*Gossypium hirsutum* L.) and bean (*Phaseolus vulgaris* L.) plants. Plant Physiol Biochem 45:691–695

Camacho-Cristobal JJ, Maldonado JM, Gonzales-Fontes A (2005) Boron deficiency increases putrescine levels in tobacco plants. J Plant Physiol 162:921–928

Campestre MP, Bordenave CD, Origone AC, Menéndez AB, Ruiz OA, Rodríguez AA, Maiale SJ (2011) Polyamine catabolism is involved in response to salt stress in soybean hypocotyls. J Plant Physiol 168:1234–1240

Chattopadhayay MK, Tiwari BS, Chattopadhyay G, Bose A, Sengupta DN, Ghosh B (2002) Protective role of exogenous polyamines on salinity-stressed rice (*Oryza sativa*) plants. Physiol Plantarum 116:192–199

Chattopadhyay MK, Gupta S, Sengupta DN, Ghosh B (1997) Expression of arginine decarboxylase in seedlings of indica rice (*Oryza sativa* L.) cultivars as affected by salinity stress. Plant Mol Biol 34:477–483

Cona A, Cenci F, Cervelli M, Federico R, Mariottini P, Moreno S, Angelini R (2003) Polyamine oxidase, a hydrogen peroxide-producing enzyme, is up-regulated by light and down-regulated by auxin in the outer tissues of the maize mesocotyl. Plant Physiol 131:803–813

Cona A, Rea G, Angelini R, Federico R, Tavladoraki P (2006) Functions of amine oxidases in plant development and defence. Trends Plant Sci 11:80–88

Cuevas JC, Sánchez DH, Marina M, Ruiz OA (2004) Do polyamines modulate the *Lotus glaber* NADPH oxidation activity induced by the herbicide methyl viologen? Func Plant Biol 31: 921–928

Delis C, Dimou M, Flemetakis E, Aivalakis G, Katinakis P (2006) A root- and hypocotyls-specific gene coding for copper-containing amine oxidase is related to cell expansion in soybean seedlings. J Exp Bot 57:101–111

Demetriou G, Neonaki C, Navakoudis E, Kotzabasis K (2007) Salt stress impact on the molecular structure and function of the photosynthetic apparatus – the protective role of polyamines. Biochim Biophys Acta 1767:272–280

Deng F (2005) Effects of glyphosate, chlorsulfuron, and methyl jasmonate on growth and alkaloid biosynthesis of jimsonweed (*Datura stramonium* L.). Pest Biochem Physiol 82:16–26

Duan J, Li J, Guo S, Kang Y (2008) Exogenous spermidine affects polyamine metabolism in salinity-stressed *Cucumis sativus* roots and enhances short-term salinity tolerance. J Plant Physiol 165:1620–1635

Durán-Serantes B, González L, Reigosa MJ (2002) Comparative physiological effects of three allelochemicals and two herbicides on *Dactylis glomerata*. Acta Physiol Plant 24:385–392

Edreva A (1996) Polyamines in plants. Bulg J Plant Physiol 22:73–101

El-Shintinawy F (2000) Photosynthesis in two wheat cultivars differing in salt susceptibility. Photosynthetica 38:615–620

Federico R, Angelini R (1991) Polyamine catabolism in plants. In: Slocum RD, Flores HE (eds) Biochemistry and physiology in plants. CRC Press, Boca Raton/Ann Arbor/London, pp 41–56

Flowers TJ (2004) Improving crop salt tolerance. J Exp Bot 55:307–319

Garratt LC, Janagoudar BS, Lowe KC, Anthony P, Power JB, Davey MR (2002) Salinity tolerance and antioxidant status in cotton cultures. Free Radic Biol Med 33:502–511

Georgiev GI, Atkins CA (1993) Effect of salinity on N_2 fixation, nitrogen metabolism and export and diffusive conductance of cowpea root nodules. Symbiosis 15:239–255

Ghosh N, Adak MK, Ghosh PD, Gupta S, Sen Gupta DN, Mandal C (2011) Differential responses of two rice varieties to salt stress. Plant Biotechnol Rep 5:89–103

Gill SS, Tuteja N (2010a) Polyamines and abiotic stress tolerance. Plant Signal Behav 5:26–33

Gill SS, Tuteja N (2010b) Reactive oxygen species and antioxidant machinery in abiotic stress tolerance in crop plants. Plant Physiol Biochem 48:909–930

Groppa MD, Benavides MP (2008) Polyamines and abiotic stress: recent advances. Amino Acids 1:35–45

Groppa MD, Tomaro ML, Benavides MP (2001) Polyamines as protectors against cadmium or copper-induced oxidative damage in sunflower leaf discs. Plant Sci 161:81–88

Groppa MD, Benavides MP, Tomaro ML (2003) Polyamine metabolism in sunflower and wheat leaf discs under cadmium and copper stress. Plant Sci 164:293–299

Gupta B, Gupta K, Sengupta DN (2012a) Spermidine-mediated in vitro phosphorylation of transcriptional regulator OSBZ8 by SNF1-type serine/threonine protein kinase SAPK4 homolog in indica rice. Acta Physiol Plant. doi:10.1007/s11738-012-0929-7

Gupta K, Gupta B, Ghosh B, Sengupta DN (2012b) Spermidine and abscisic acid-mediated phosphorylation of a cytoplasmic protein from rice root in response to salinity stress. Acta Physiol Plant 34:29–40

Halliwell B (2006) Reactive species and antioxidants. Redox biology is a fundamental theme of aerobic life. Plant Physiol 141:312–322

Hamdani S, Yaakoubi H, Carpentier R (2011) Polyamines interaction with thylakoid proteins during stress. J Photochem Photobiol B Biol 104:314–319

Hanfrey C, Sommer S, Mayer MJ, Burtin D, Michael AJ (2001) Arabidopsis polyamine biosynthesis: absence of ornithine decarboxylase and the mechanism of arginine decarboxylase activity. Plant J 27:551–560

Hao Y-J, Zhang Z, Kitashiba H, Honda C, Ubi B, Kita M, Moriguchi T (2005) Molecular cloning and functional characterization of two apple S-adenosylmethionine decarboxylase genes and their different expression in fruit development, cell growth and stress responses. Gene 350:41–50

He L, Ban Y, Inoue H, Matsuda N, Liu J, Moriguch T (2008) Enhancement of spermidine content and antioxidant capacity in transgenic pear shoots overexpressing apple spermidine synthase in response to salinity and hyperosmosis. Phytochem 69:2133–2141

Hummel I, Quemmerais F, Gouesbet G, El Amrani A, Frenot Y, Hennion F, Couée I (2004) Characterization of environmental stress responses during early development of *Pringlea antiscorbutica* in the field at Kerguelen. New Phytol 162:705–715

Iqbal M, Ashraf M (2012) Gibberellic acid mediated induction of salt tolerance in wheat plants: growth, ionic partitioning, photosynthesis, yield and hormonal homeostasis. Environ Exp Bot. doi:10.1016/j.envexpbot.2010.06.002

Ivanova A, Seizova K, Kostova I, Evstatieva L, Stefanov K (2008) Effect of soil salinity on the chemical composition of two populations of *Tribulus terrestris* from Bulgaria. Compt Rend Acad Bulg Sci 61:55–62

Janicka-Russak M, Kabala K, Mlodzinska E, Klobus G (2010) The role of polyamines in the regulation of the plasma membrane and the tonoplast proton pumps under salt stress. J Plant Physiol 167:261–269

Jouve L, Hoffmann L, Hausman J-F (2004) Polyamine, carbohydrate, and proline content changes during salt stress exposure of aspen (*Populus tremula* L.): involvement of oxidation and osmoregulation metabolism. Plant Biol 6:74–80

Kakkar RK, Bhaduri S, Rai VK, Kumar S (2000) Amelioration of NaCl stress by arginine in rise seedlings: changes in endogenous polyamines. Biol Plant 43:419–422

Kasinathan V, Wingler A (2004) Effect of reduced arginine decarboxylase activity on salt tolerance and on polyamine formation during salt stress in *Arabidopsis thaliana*. Physiol Plant 121: 101–107

Kasukabe Y, He L, Nada K, Misawa S, Ihara I, Tachibana S (2004) Overexpression of spermidine synthase enhances tolerance to multiple environmental stress and upregulates the expression of various stress-regulated genes in transgenic *Arabidopsis thaliana*. Plant Cell Physiol 45:712–722

Kasukabe Y, He L, Watakabe Y, Otani M, Shimada T, Tachibana S (2006) Improvement of environmental stress tolerance of sweet potato by introduction of genes for spermidine synthase. Plant Biotechnol 23:75–83

Katerova ZI, Todorova D (2009) Endogenous polyamines lessen membrane damages in pea plants provoked by enhanced ultraviolet-C radiation. Plant Growth Regul 57:145–152

Kim J-S, Shim I-S, Kim M-J (2010) Physiological response of Chinese cabbage to salt stress. Kor J Hort Sci Technol 28:343–352

Krishnamurthy R, Bhagwat KA (1989) Polyamines as modulators of salt tolerance in rice cultivars. Plant Physiol 91:500–504

Kumria R, Rajam MV (2002) Ornithine decarboxylase transgene in tobacco affects polyamines, in vitro-morphogenesis and response to salt stress. J Plant Physiol 159:983–990

Kuznetsov VV, Shevyakova NI (2007) Polyamines and stress tolerance of plants. Plant Stress 1:50–71

Kuznetsov V, Shorina M, Aronova E, Stetsenko L, Rakitin V, Shevyakova N (2007) NaCl- and ethylene-dependent cadaverine accumulation and its possible protective role in the adaptation of the common ice plant to salt stress. Plant Sci 172:363–370

Lefèvre I, Gratia E, Lutts S (2001) Discrimination between the ionic and osmotic components of salt stress in relation to free polyamine level in rice (*Oryza sativa*). Plant Sci 16:943–952

Legocka J, Kluk A (2005) Effect of salt and osmotic stress on changes in polyamine content and arginine decarboxylase activity in *Lupinus luteus* seedlings. J Plant Physiol 162:662–668

Legocka J, Sobieszczuk-Nowicka E (2012) Sorbitol and NaCl stresses affect free, microsome-associated and thylakoid-associated polyamine content in *Zea mays* and *Phaseolus vulgaris*. Acta Physiol Plant 34:1145–1151

Li Z-Y, Chen S-Y (2000a) Differential accumulation of the S-adenosylmethionine decarboxylase transcript in rice seedlings in response to salt and drought stresses. Theor Appl Genet 100:782–788

Li ZY, Chen SY (2000b) Isolation and characterization of a salt- and drought-inducible gene for S-adenosylmethionine decarboxylase from wheat (*Triticum aestivum* L.). J Plant Physiol 156:386–393

Liu J, Yu B-J, Liu Y-L (2006a) Effects of spermidine and spermine levels on salt tolerance associated with tonoplast H^+-ATPase and H^+-PPase activities in barley roots. Plant Growth Regul 49:119–126

Liu J-H, Nada KS, Da CH, Shiba HK, Wen X-P, Pang X-M, Moriguch T (2006b) Polyamine biosynthesis of apple callus under salt stress: importance of the arginine decarboxylase pathway in stress response. J Exp Bot 57:2589–2599

Liu J-H, Kitashiba H, Wang J, Ban Y, Moriguchi T (2007) Polyamines and their ability to provide environmental stress tolerance to plants. Plant Biotechnol 24:117–126

Liu J-H, Inoue H, Moriguchi T (2008) Salt stress-mediated changes in free polyamine titers and expression of genes responsible for polyamine biosynthesis of apple in vitro shoots. Environ Exp Bot 62:28–35

Maiale S, Sánchez DH, Guirado A, Vidal A, Ruiz OA (2004) Spermine accumulation under salt stress. J Plant Physiol 161:35–42

Mansour MMF, Salama KHA, Al-Mutawa MM, Abou Hadid F (2002) Effect of NaCl and polyamines on plasma membrane lipids of wheat roots. Biol Plant 45:235–239

Mittler R (2002) Oxidative stress, antioxidants and stress tolerance. Trends Plant Sci 7:405–410

Moschou PN, Paschalidis KA, Delis ID, Andriopoulou AH, Lagiotis GD, Yakoumakis DI, Roubelakis-Angelakis KA (2008) Spermidine exodus and oxidation in the apoplast induced by abiotic stress is responsible for H_2O_2 signatures that direct tolerance responses in tobacco. Plant Cell 20:1708–1724

Mutlu F, Bozcuk S (2007) Salinity-induced changes of free and bound polyamine levels in sunflower (*Helianthus annuus* L.) roots differing in salt tolerance. Pak J Bot 39:1097–1102

Ndayiragije A, Lutts S (2006) Exogenous putrescine reduces sodium and chloride accumulation in NaCl-treated calli of the salt-sensitive rice cultivar I Kong Pao. Plant Growth Regul 48:51–63

Ndayiragije A, Lutts S (2007) Long term exogenous putrescine application improves grain yield of a salt-sensitive rice cultivar exposed to NaCl. Plant Soil 291:225–238

Neily MH, Baldet P, Arfaoui I, Saito T, Qiu-li L, Asamizu E, Matsukura C, Moriguchi T, Ezura H (2011) Overexpression of apple spermidine synthase 1 (MdSPDS1) leads to significant salt tolerance in tomato plants. Plant Biotechnology 28:33–42

Nenova V (2008) Growth and mineral concentration of pea plants under different salinity levels and iron supply. Gen Appl Plant Physiol 34:189–202

Ogawa S, Mitsuya S (2012) S -methylmethionine is involved in the salinity tolerance of Arabidopsis thaliana plants at germination and early growth stages. Physiol Plant 144:13–19

Pandolfi C, Pottosin I, Cuin T, Mancuso S, Shabala S (2010) Specificity of polyamine effects on NaCl-induced ion flux kinetics and salt stress amelioration in plants. Plant Cell Physiol 51: 422–434

Pillai MA, Akiyama T (2004) Differential expression of an S-adenosyl-L-methionine decarboxylase gene involved in polyamine biosynthesis under low temperature stress in japonica and indica rice genotypes. Mol Gen Genom 271:141–149

Potters G, Horemans N, Jansen MAK (2010) The cellular redox state in plant stress biology – a charging concept. Plant Physiol Biochem 48:292–300

Prabhavathi VR, Rajam MV (2007) Polyamine accumulation in transgenic eggplant enhances tolerance to multiple abiotic stresses and fungal resistance. Plant Biotechnol 24:273–282

Qi Y-C, Wang F-F, Zhang H, Liu W-Q (2010) Overexpression of suadea salsa *S*-adenosylmethionine synthetase gene promotes salt tolerance in transgenic tobacco. Acta Physiol Plant 32:263–269

Quinet M, Ndayiragije A, Lefèvre I, Lambillotte B, Dupont-Gillain CC, Lutts S (2010) Putrescine differently influences the effect of salt stress on polyamine metabolism and ethylene synthesis in rice cultivars differing in salt resistance. J Exp Bot 61:2719–2733

Rios-Gonzalez K, Erdei L, Lips SH (2002) The activity of antioxidant enzymes in maize and sunflower seedlings as affected by salinity and different nitrogen sources. Plant Sci 162: 923–930

Rodríguez AA, Maiale SJ, Menéndez AB, Ruiz OA (2009) Polyamine oxidase activity contributes to sustain maize leaf elongation under saline stress. J Exp Bot 60:4249–4262

Rodríguez-Kessler M, Alpuche-Solís AG, Ruiz OA, Jiménez-Bremont JF (2006) Effect of salt stress on the regulation of maize (*Zea mays* L.) genes involved in polyamine biosynthesis. Plant Growth Regul 48:175–185

Roy M, Wu R (2001) Arginine decarboxylase transgene expression and analysis of environmental stress tolerance in transgenic rice. Plant Sci 160:869–875

Roy M, Wu R (2002) Overexpression of S-adenosyl methionine decarboxylase gene in rice increases polyamine level and enhances sodium chloride-stress tolerance. Plant Sci 163:987–992

Roy P, Niyogi K, Sen Gupta DN, Ghosh B (2005) Spermidine treatment to rice seedlings recovers salinity stress-induced damage of plasma membrane and PM-bound H^+-ATPase in salt-tolerant and salt-sensitive rice cultivars. Plant Sci 168:583–591

Roychoudhury A, Basu S, Sarkar SN, Sengupta DN (2008) Comparative physiological and molecular responses of a common aromatic indica rice cultivar to high salinity with non-aromatic indica rice cultivars. Plant Cell Rep 27:1395–1410

Roychoudhury A, Basu S, Sengupta DN (2011) Amelioration of salinity stress by exogenously applied spermidine or spermine in three varieties of indica rice differing in their level of salt tolerance. J Plant Physiol 168:317–328

Sagor GHM, Yamaguchi K, Watanabe K, Berberich T, Kusano T, Takahashi Y (2011) Spatio-temporal expression analysis of Arabidopsis thaliana spermine synthase gene promoter. Plant Biotechnol 28:407–411

Sakr MT, El-Metwally MA (2009) Alleviation of the harmful effects of soil salt stress on growth, yield and endogenous antioxidant content of wheat plant by application of antioxidants. Pak J Biol Sci 12:624–630

Saleethong P, Sanitchon J, Knog-ngern K, Theerakulpisut P (2011) Pretreatment with spermidine reverse inhibitory effects of salt stress in two rice (*Oriza sativa* L.) cultivars differing in their tolerance. Asian J Plant Sci 10:245–254

Sanchez DH, Cuevas JC, Chiesa MA, Ruiz OA (2005) Free spermidine and spermine content in *Lotus glaber* under long-term salt stress. Plant Sci 168:541–546

Santa-Cruz A, Acosta M, Perez-Alfocea F, Bolarin MC (1997a) Changes in free polyamine levels induced by salt stress in leaves of cultivated and wild tomato species. Physiol Plant 101:341–346

Santa-Cruz A, Estan MT, Rus A, Bolarin MC, Acosta M (1997b) Effects of NaCl and mannitol iso-osmotic stresses on the free polyamine levels in leaf discs of tomato species differing in salt tolerance. J Plant Physiol 151:754–758

Santa-Cruz A, Perez-Alfocea F, Caro M, Acosta M (1998) Polyamines as short-term salt tolerance traits in tomato. Plant Sci 138:9–16

Sarjala T, Kaunisto S (2002) Potassium nutrition and free polyamines of *Betula pendula* Roth and *Betula pubescens* Ehrh. Plant Soil 238:141–149

Šebela M, Radová A, Angelini R, Tavladoraki P, Frébort I, Pec P (2001) FAD-containing polyamine oxidases: a timely challenge for researchers in biochemistry and physiology of plants. Plant Sci 160:197–207

Shabala S, Cuin TA (2008) Potassium transport and plant salt tolerance. Physiol Plant 133: 651–669

Shabala S, Cuin TA, Pottosin I (2007) Polyamines prevent NaCl-induced K^+ efflux from pea mesophyll by blocking non-selective cation channels. FEBS Lett 581:1993–1999

Sheokand S, Kumari A, Sawhney V (2008) Effect of nitric oxide and putrescine on antioxidative responses under NaCl stress in chickpea plants. Physiol Mol Biol Plant 14:355–362

Shevyakova NI, Rakitin VY, Stetsenko LA, Aronova EE, Kuznetsov VV (2006) Oxidative stress and fluctuations of free and conjugated polyamines in the halophyte *Mesembryanthemum crystallinum* L. under NaCl salinity. Plant Growth Regul 50:69–78

Shi K, Huang YY, Xia XJ, Zhang YL, Zhou YH, Yu JQ (2008) Protective role of putrescine against salt stress is partially related to the improvement of water relation and nutritional imbalance in cucumber. J Plant Nutr 31:1820–1831

Shoresh M, Spivak M, Bernstein N (2011) Involvement of calcium-mediated effects on ROS metabolism in the regulation of growth improvement under salinity. Free Radical Biol Med 51:1221–1234

Slocum R (1991) Polyamine Biosynthesis in plants. In: Slocum RD, Flores HE (eds) Biochemistry and physiology of polyamines in plants. CRC Press, Boca Raton/Ann Arbor/London, pp 23–40

Smith J, Burrit D, Bannister P (2001) Ultraviolet-B radiation leads to a reduction in free polyamines in *Phaseolus vulgaris* L. Plant Growth Regul 35:289–294

Su GX, Bai X (2008) Contribution of putrescine degradation to proline accumulation in soybean leaves under salinity. Biol Plant 52:796–799

Szigeti Z, Lehoczki E (2003) A review of physiological and biochemical aspects of resistance to atrazine and paraquat in Hungarian weeds. Pest Manag Sci 59:451–458

Tanaka Y, Hibino T, Hayashi Y, Tanaka A, Kishitani S, Takabe T, Yokota S, Takabe T (1999) Salt tolerance of transgenic rice overexpressing yeast mitochondria Mn-SOD in chloroplast. Plant Sci 148:131–138

Tang W, Newton RJ (2005) Polyamines reduce salt-induced oxidative damage by increasing the activities of antioxidant enzymes and decreasing lipid peroxidation in Virginia pine. Plant Growth Regul 46:31–43

Tassoni A, Franceschetti M, Bagni N (2008) Polyamines and salt stress response and tolerance in *Arabidopsis thaliana* flowers. Plant Physiol Biochem 46:607–613

Tester M, Davenport R (2003) Na^+ tolerance and Na^+ transport in higher plants. Ann Bot 91:503–527

Todorova D, Sergiev I, Alexieva V, Karanov E, Smith A, Hall M (2007) Polyamine content in Arabidopsis thaliana (L.) Heynh during recovery after low and high temperature treatments. Plant Growth Regul 51:185–191

Todorova D, Moskova I, Sergiev I, Alexieva V, Mapelli S (2008) Changes in endogenous polyamines and some stress markers content induced by drought, 4PU-30 and abscisic acid in wheat plants. In: Khan N, Singh S (eds) Abiotic stress and plant responses. I.K. International Publishing House, New Delhi/Bangalore/Mumbai, pp 205–215

Tonon G, Kevers C, Faivre-Rampant O, Graziani M, Gaspar T (2004) Effect of NaCl and mannitol iso-osmotic stresses on proline and free polyamine levels in embryogenic *Fraxinus angustifolia* callus. J Plant Physiol 161:701–708

Turkan I, Demiral T (2008) Salinity tolerance mechanisms of higher plants. In: Khan N, Singh S (eds) Abiotic stress and plant responses. I. K. International Publishing House, New Delhi/Bangalore/Mumbai, pp 105–123

Upreti KK, Murti GSR (2010) Response of grape rootstocks to salinity: changes in root growth, polyamines and abscisic acid. Biol Plant 54:730–734

Urano K, Yoshiba Y, Nanjo T, Igarashi Y, Seki M, Sekiguchi F, Yamaguchi-Shinozaki K, Shinozaki K (2003) Characterization of Arabidopsis genes involved in biosynthesis of polyamines in abiotic stress responses and developmental stages. Plant Cell Environ 26:1917–1926

Urano K, Yoshiba Y, Nanjo T, Ito Y, Seki M, Yamaguchi-Shinozaki K, Shinozaki K (2004) *Arabidopsis* stress-inducible gene for arginine decarboxylase *AtADC2* is required for accumulation of putrescine in salt tolerance. Biochem Biophys Res Comm 313:369–375

Verma S, Mishra SN (2005) Putrescine alleviation of growth in salt stressed *Brassica juncea* by inducing antioxidative defense system. J Plant Physiol 162:669–677

Waie B, Rajam MV (2003) Effect of increased polyamine biosynthesis on stress responses in transgenic tobacco by introduction of human S-adenosylmethionine gene. Plant Sci 164: 727–734

Wen X-P, Pang X-M, Matsuda N, Kita M, Inoue H, Hao Y-J, Honda C, Moriguchi T (2008) Overexpression of the apple spermidine synthase gene in pear confers multiple abiotic stress tolerance by altering polyamine titers. Transgenic Res 17:251–263

Wi SJ, Kim WT, Park KY (2006) Overexpression of carnation S-adenosylmethionine decarboxylase gene generates a broad-spectrum tolerance to abiotic stresses in transgenic tobacco plants. Plant Cell Rep 25:1111–1121

Willadino L, Camara T, Boget N, Claparols I, Santos M, Torné JM (1996) Polyamine and free amino acid variations in NaCl-treated embryogenic maize callus from sensitive and resistant cultivars. J Plant Physiol 149:179–185

Xing SG, Yu BJ, Zhang WH, Liu YL (2007) Higher accumulation of g-aminobutyric acid induced by salt stress through stimulating the activity of diamine oxidases in *Glycine max* (L.) Merr. roots. Plant Physiol Biochem 45:560–566

Yamaguchi K, Takahashi Y, Berberich T, Imai A, Takahashi T, Michael AJ, Kusano T (2007) A protective role for the polyamine spermine against drought stress in Arabidopsis. Biochem Biophys Res Commun 352:486–490

Yamamoto A, Sawada H, Shim IS, Usui K, Fujihara S (2011) Effect of salt stress on physiological response and leaf polyamine content in NERICA rice seedlings. Plant Soil Environ 57:571–576

Zacchini M, de Agazio M (2004) Spread of oxidative damage and antioxidative response through cell layers of tobacco callus after UV-C treatment. Plant Physiol Biochem 42:445–450

Zapata PJ, Serrano M, Pretel MT, Amoros A, Botella MA (2003) Changes in ethylene evolution and polyamine profiles of seedlings of nine cultivars *of Lactuca sativa* L. in response to salt stress during germination. Plant Sci 164:557–563

Zapata PJ, Serrano M, Pretel MT, Amoros A, Botella MA (2004) Polyamines and ethylene changes during germination of different plant species under salinity. Plant Sci 167:781–788

Zapata PJ, Serrano M, Pretel MT, Botella MA (2008) Changes in free polyamine concentration induced by salt stress in seedlings of different species. Plant Growth Regul 56:167–177

Zhao FG, Qin P (2004) Protective effect of exogenous polyamines on root tonoplast function against salt stress in barley seedlings. Plant Growth Regul 42:97–103

Zhao FG, Sun C, Liu YL, Zhang WH (2003) Relationship between polyamine metabolism in roots and salt tolerance of barley seedlings. Acta Bot Sin 45:295–300

Zhao J, Shi G, Qiuhong Y (2008) Polyamines content and physiological and biochemical responses to ladder concentration of nickel stress in *Hydrocharis dubia* (Bl.) Backer leaves. Biometals 21:665–674

Zhu J-K (2001) Plant salt tolerance. Trends Plant Sci 6:66–71

Zhu H, Ding GH, Fang K, Zhao FG, Qin P (2006) New perspective on the mechanism of alleviating salt stress by spermidine in barley seedlings. Plant Growth Regul 49:147–156

Chapter 14
Role of Jasmonates in Plant Adaptation to Stress

Losanka P. Popova

Abbreviations

ABA	Abscisic acid
GA	Gibberellic acid
JA	Jasmonic acid
JAZ proteins	Jasmonte-Zim proteins
MeJA	Methyl ester of jasmonic acid
JIPs	JA-induced proteins
PSII	Photosystem II
ROS	Reactive oxygen species
RuBPCase	Ribulose-1,5- bisphosphate carboxylase
RuBPOase	Ribulose-1,5- bisphosphate oxygenase

14.1 Introduction

During the past 20 years, a class of phytohormones derived from the metabolism of membrane fatty acids, collectively known as jasmonates, has attracted considerable attention. Jasmonates (jasmonic acid and related compounds) have been the only plant hormone class that is biosynthesized from fatty acid. The biological activity of MeJA, extracted from *Artemisia absinthium* L., was reported nearly 10 years later (Ueda and Kato 1980). Since then, jasmonates have been found in many species and are considered ubiquitous (Meyer et al. 1984; Hamberg and Gardner 1992).

L.P. Popova (✉)
Bulgarian Academy of Sciences, Institute of Plant Physiology and Genetics,
Acad. G. Bonchev str., bl. 21 1113, Sofia, Bulgaria
e-mail: lpopova@bio21.bas.bg

P. Ahmad et al. (eds.), *Ecophysiology and Responses of Plants under Salt Stress*,
DOI 10.1007/978-1-4614-4747-4_14,

The most abundant member of the group is jasmonic acid (JA). First identified and isolated from a culture of the fungus *Lasiodiplodia theobromae*, JA is the best-known and best-characterized member of the jasmonate family. Recent studies have shown that JA can be modified in various ways, such as conjugation to different amino acids such as leucine (JA-leucine) and isoleucine (JA-isoleucine), or hydroxylation, forming compounds of extreme importance for the activation of jasmonate-responsive defense genes (Creelman and Mullet 1995; Liechti and Farmer 2006; Thines et al. 2007).

Besides JA, another jasmonate of key importance is its methyl ester form, MeJA. MeJA was discovered in 1962 as a sweet-smelling compound in *Jasminium grandiflorum* L. flower extracts (Demole et al. 1962). After its discovery in jasmine flowers, JA was isolated from a pathogenic fungus, *Lasiodiplodia theobromae* (Aldridge et al. 1971).

First isolated from essential oil extracted from *Jasminum grandiflorum* petals, due to its volatile property, MeJA initially aroused considerable commercial interest from the perfume industry, stimulating studies focused on its structure and synthesis.

Other jasmonates with biological activity include tuberonic acid (from the leaves of potato *Solanum tuberosum* L., Yoshihara et al. 1989), dihydrojasmonic acid (from *Vicia faba* L., Miersch et al. 1989), and cucurbic acid (from seeds of *Cucurbita pepo* L., Fukui et al. 1977). Conjugation to isoleucine is necessary to elicit some jasmonate responses (Staswick and Tiryaki 2004), and many other conjugates exist (Gapper et al. 2002; reviewed by Hamberg and Gardner 1992).

14.2 Occurrence and Distribution of Jasmonates

Jasmonates seem to be ubiquitously distributed throughout the plant kingdom. They were detected in 206 plant species representing 150 family including ferns, mosses, and fungi (Meyer et al. 1984; Sembdner and Gross 1986). The level of JA in plant tissues varies as a function of tissue type, development, and external stimuli (for reviews, see Sembdner and Parthier 1993; Koda 1992; Vick and Zimmerman 1987; Hamberg and Gardner 1992). The highest levels of JA/JAMe are reported in flowers and reproductive tissues, whereas much lower levels are found in roots and mature leaves. In soybean seeds and seedlings, JA levels were highest in the youngest organs including the hypocotyl hook, plumule, and 12-h axis, compared to the zone of cell elongation and more mature regions of the stem, older leaves, and roots. JA levels are low in soybean seeds but increase to 2 mg g^{-1} fresh weight in developing axes within 12 h of imbibition. High levels of JA are also found in flowers and pericarp tissues of developing reproductive structures (Creelman and Mullet 1995, 1997; Lopez et al. 1987). The presence of JA and MeJA has been described in anthers and pollen of three species of *Camellia* (Yamane et al. 1982).

Levels of endogenous jasmonates increase upon wounding and are followed by activation of genes involved in plant defense responses such as those coding for proteinase inhibitors, enzymes of phytoalexin synthesis, vegetative storage proteins, thionins, and defensins (Creelman and Mullet 1997; Farmer et al. 1998; Ryan 2000). However, it is less well understood how the rise of jasmonates is regulated. The elevation of jasmonate levels is usually correlated with the activation of genes coding for JA biosynthetic enzymes (for review, see Wasternack and Hause 2002). Application of linolenic acid (LA) to plants resulted in accumulation of JA. Free LA levels doubled within 1 h after wounding while JA levels rose 10-fold. Therefore, the wound–induced increase in JA level could have resulted from release of LA from phospholipids, or the utilization of LA present before wounding for JA biosynthesis. When the time points of accumulation of different jasmonates were determined, JA levels were found to increase within 2–5 min of wounding. It is important that these changes occurred throughout the plant and were not restricted to wounded leaves. The speed of the stimulus leading to JA accumulation in leaves distal to a wound is at least 3 cm min^{-1}. The data give new insights into the spatial and temporal accumulation of jasmonates and have implications in the understanding of long-distance wound signaling in plants.

In soybean leaves that had been dehydrated to cause a 15% decrease in fresh weight, JA levels increased –5-fold within 2 h and declined to approximately control levels by 4 h (Creelman and Mullet 1995). Colonization of barley *(Hordeum vulgare* cv Salome) roots by an arbuscular mycorrhizal fungus, *Glomus intraradices* Schenck and Smith, leads to elevated levels of endogenous JA and its amino acid conjugate JA-isoleucine, whereas the level of the JA precursor, oxophytodienoic acid, remains constant. The rise in jasmonates is accompanied by the expression of genes coding for an enzyme of JA biosynthesis (allene oxide synthase) and of a jasmonate-induced protein (JIP23), (Hause et al. 2002).

14.3 Biosyntheses of Jasmonates

In 1984, Vick and Zimmermann provided one of the first insights into the biosynthesis of jasmonates.

Jasmonates are derived from the tri-unsaturated fatty acids – linolenic acid (18:3). The biosynthesis of JA starts with the insertion of oxygen at position 13 of α-linolenic acid catalyzed by 13-lipoxygenase. The resulting hydroperoxide is converted by allene oxide synthase into an unstable allene oxide that can rapidly be degraded in vitro by chemical hydrolysis. Under cellular conditions, the allene oxide is preferentially, if not exclusively, converted by allene oxide cyclase into (9S, 13S)-oxophytodienoic acid. Both the allene oxide synthase and the cyclase are located in the plastids, and they probably operate in concert or they are linked physically in some form of complex, although no direct evidence exists for this. The product is transferred to the peroxisomes, where a specific 12-oxo-phytodienoate reductase reduces the double bond in position 10, i.e. in the cyclopentenone ring, to 3-oxo-2-(pent-2′-enyl)-cyclopentane-1-octanoic acid. This is a key step in directing the metabolism towards jasmonic acid, as this compound only is able to undergo the three cycles of β-oxidation, catalysed by the multifunctional enzyme complex acyl-CoA oxidase, which are required to give the 12-carbon (−)-7-iso-jasmonic acid. This is the main isomer isolated from plant tissues and was long thought to be the active metabolite. However, it is now recognized that (+)-7-iso-jasmonic acid or cis-(epi)-jasmonic acid is in fact the active isomer. As both side chains are on the same side of the 3R,7S-cyclopentanone ring and the keto group at C-6 can tautomerize to an enol, the more stable (3R, 7R) isomer (−)- or trans-jasmonic acid is usually isolated as the main product (90% of the equilibrium mixture). According to the present knowledge, this enantiomer is the unique precursor for the naturally occurring (+)-7-iso-JA. The more stable (−)-JA is then formed by spontaneous isomerization. The metabolic pathway has been reviewed by Hause et al. (2002) and by Gfeller et al. (2010).

A number of amino acid conjugates of jasmonic acid have been found in plants, and those with leucine and isoleucine have especial importance. For example, jasmonoyl-isoleucine and hydroxylated and carboxylated analogues are formed in plants in response to wounding, and jasmonoyl-isoleucine specifically is now know to be a central element in hormone signaling by jasmonic acid. As with the free acid, the common isomer (−)-7-jasmonoyl-L-isoleucine is not the active isomer, but rather the much less abundant (+)-7-epimer. pH changes promote conversion of the (+)-7-epimer to the inactive (−)-7- form, suggesting that this may be a simple mechanism regulating the activity of the hormone through epimerization.

14.4 Physiological Functions of Jasmonates

14.4.1 Plant Growth and Development

During the 1980s the first reports of physiological effects attributed to JA and MeJA appeared, describing activities related to senescence in *Artemisia absinthium* (Ueda and Kato 1980), and growth inhibition in *Vicia faba* (Dathe et al. 1981). Since then, the number of metabolic and physiological activities attributed to jasmonates continues to increase. Jasmonates have been described as exerting a wide range of differing effects on virtually all plants, ranging from inhibition to promotion of plant processes. As described by Parthier (1991), the effect exhibited on the plant may even be concentration dependent, with some processes stimulated at lower concentrations but inhibited at higher concentrations. A few examples of activities influenced by jasmonates include inhibition of seed germination and seedling growth, stimulation of seed germination (at lower concentrations), promotion of seed dormancy breaking, and promotion of leaf senescence. The growth of vegetative tissues is also affected by treatment with jasmonates (Staswick et al. 1992; Huang et al. 2010). Plants treated with MeJA develop shorter petioles than do control plants (Cipollini 2005), and the short-petiole phenotype is observed in some genetically characterized mutants that accumulate higher than wild-type levels of JA in response to wounding (Bonaventure et al. 2007). The cultivation of barley

Photo 14.1 Effect of JA on the growth of barley seedling grown for 7 days on H_2O_2 (1), 2.5×10^{-7} JA (2), 2.5×10^{-6} JA (3), 2.5×10^{-5} JA (4), 2.5×10^{-4} JA (5)

plants in solution of JA exercised a considerable effect on the growth of the seedlings. Concentrations of the order 2.5×10^{-6}–2.5×10^{-4} M tend to inhibit growth. Lower concentrations had no effect on the growth of the seedlings. Similar in nature are also changes in the root system (Popova et al. 1988) (Photo 14.1).

In flowers, JA plays multiple roles that are related to general developmental processes (Maciejewska et al. 2004; Krajncic et al. 2006). Negative effects of jasmonates on flower opening and bud initiation have been reported for *Pharbitis nil* and *Nicotiana tabacum* (Barendse et al. 1985; Kesy et al. 2011). Moreover, a tissue-specific synthesis of JA in flowers has been described. On the other hand, JA appears to be necessary for pollen development and anther dehiscence in *Arabidopsis*. Much less is known on the role of JA for nectar secretion. The data presented by Radhika et al. (2010) showed that jasmonates have an influence on floral nectar secretion in *Brassica napus*. Pauwels et al. (2008) reported that high levels of MeJA repressed cell cycle progression by arresting cells in the G2 phase.

Elevated levels of JA has been shown to play a role in symbiotic processes such as arbuscular mycorrhiza and to control the induction of nitrogen-fixing root nodules. The involvement of jasmonates in tuberization in potato, yam and Jerusalem artichoke was inferred from their ability to induce tubers, and from changes in the levels of endogenous jasmonates during the growth of the plants, which can account for the initiation of tuberization. As to potato tuberization, JA and MeJA have strong tuber-inducing activity. tubers (Koda 1997).

JA can influence other aspects of plant growth and development, such as senescence and leaf abscission. At concentrations higher than 50 μM, jasmonates induced senescence in plant cell cultures and excised leaves. The senescence response includes a loss of chlorophyll, degradation of chloroplast proteins, and accumulation of new proteins. Jasmonates specifically inhibited translation of the large subunit of ribulose bisphosphate carboxylase (RuBPCase) by inducing cleavage of the rbcL transcript (Weidhase et al. 1987). While jasmonates can induce senescence when applied at high concentrations, it is not clear whether this regulator modulates senescence in vitro. Jasmonates can also induce tendril coiling (Weiler et al. 1993). Tendril coiling is induced by mechanical stimulation and roles for auxin and ethylene have been proposed. Although jasmonates can stimulate ethylene biosynthesis by inducing activity of the ethylene forming enzyme (Czapski and Saniewski 1992), studies on tendril coiling showed that jasmonates can mediate this response in the presence of an inhibitor of ethylene biosynthesis (Weiler et al. 1993). In early studies, JA, but not MeJA, was suggested to be the endogenous pollen germination regulator. More than 10 years later, two jasmonate compounds, N-[(−)-jasmonoyl]-(S)-isoleucine and N-[7-iso-cucurbinoyl]-(S)-isoleucine, were identified in pollen grains of *Pinus mugo* (Knöfel and Sembdner 1995). Jasmonates have been shown to stimulate fruit ripening, most likely through its action on ethylene biosynthesis (Czapski and Saniewski 1992). Kim et al. (2009) reported that transgenic overexpression of the *Arabidopsis* gene jasmonic acid carboxyl methyltransferase (*AtJMT*) in rice resulted in a large reduction in grain yield through increased MeJA and ABA levels in young panicles. Exposure of nontransgenic plants to drought conditions also increased MeJA and ABA levels in young panicles and significantly reduced grain yield. In both cases, the reduction in grain yield was due to lower numbers of spikelets and lower filling rates than were observed for nontransgenic controls. The ABA increase in *AtJMT* transgenic panicles grown in non-drought conditions suggests that MeJA, rather than drought stress, induces ABA biosynthesis under drought conditions. These results led the authors to postulate that plants produce MeJA during drought stress, which in turn stimulates the production of ABA, together leading to a loss of grain yield. Salt stress also increases the JA levels in roots of rice plants and in leaves of *Iris hexagona* (Moons et al. 1997; Wang et al. 2001). Application of 100 mmol MeJA to the intact bark of 30-year-old Norway spruce induced anatomical reactions related to defense. Within 30 days, a single MeJA treatment induced swelling of existing polyphenolic parenchyma cells and an increase in their phenolic contents and formation of additional parenchyma cells the cambial zone. Treatment enhanced resin flow and increased resistance to the blue-stain fungus, *Ceratocystis polonica* (Franceschi et al. 2002).

14.4.2 Jasmonates and Photosynthesis

The phytohormonal regulation of photosynthesis is an important aspect of the mechanism of photosynthetic regulation. A number of phytohormones (ABA, GA, kinetin, etc.) affect the rate of photosynthetic CO_2 fixation and photorespiration, and the activity of some enzymes of carbon and nitrogen metabolism (Popova et al. 1987, 1988). To a certain extent the action of JA is similar to the effect of ABA on a number of photosynthetic parameters. Two groups reported on the effect of jasmonates on photosynthesis- this of Weidhase in Germany and the group of Popova, in Bulgaria. We demonstrated that in barley (*Hordeum vulgare* L.) plants exogenous treatment with JA or MeJA revealed changes in a number of photosynthetic parameters, such as a decrease in the rate of photosynthetic CO_2 fixation and the activity of RuBPCase (Table 14.1).

A breakdown in the biosynthesis of Rubisco (Weidhase et al. 1987; Popova and Vaklinova 1988), an inhibition of the Hill reaction activity and some changes in the kinetic characteristics of the flash-induced O_2 evolution (Maslenkova et al. 1990, 1995) have been reported to occur as a result of JA and MeJA treatments. The observed inhibition in the Hill reaction activity and changes in the kinetic

Table 14.1 Effect of JA on the rate of photosynthetic CO_2 fixation, photorespiration, and the activity of RuBP carboxylase (RuBPCase) and RuBP oxygenase (RuBPOase). Data are of five experiments ± SE

JA (M)	CO_2 fixation $\mu M\ CO_2$ (mg chl h)$^{-1}$	Photorespiration $\mu M\ O_2$ (mg chl h)$^{-1}$	RuBPCase % of the control	RuBPOase % of the control
H_2O(control)	77.0 ± 7.6	3.34	100.0	100.0
2.5×10^{-7}	68.5 ± 8.5	5.79	74.6 ± 14.1	96.0 ± 3.4
2.5×10^{-6}	54.1 ± 8.5	5.32	72.3 ± 10.3	92.0 ± 4.8
2.5×10^{-5}	40.9 ± 5.7	4.24	65.7 ± 7.4	99.3 ± 3.7
2.5×10^{-4}	39.9 ± 3.0	3.95	7.4 ± 1.1	–

characteristics of the flash-induced O_2 evolution are hypothesized to be determined by alteration in chloroplast membrane during a prolonged treatment with JA. Treatment with JA resulted in the conversion of PSIIα to PSIIβ centers (Maslenkova et al. 1990) and the enhanced participation of a more resistant to stress cooperative mechanism for oxygen production (Zeinalov 1982).

In primary leaves of *Hordeum vulgare* L plants grown in the presence of JA alterations in the size of chloroplasts were induced that led to disorganization of chloroplast ultrastructure. The number of thylakoids per granum as well as the average length of granal and stromal thylakoids were lower in JA-treated plants (Popova and Uzunova 1996). MeJA affects plant transpiration (Lee et al. 1996; Wang 1999) by promoting stomatal closure (Raghavendra and Reddy 1987; Suhita et al. 2003). MeJA-induced stomatal closure is accompanied by an alkalization of the guard cell cytoplasm in *Paphiopedilum* spp. (Gehring et al. 1997).

Barley plant grown for 7 days in solutions of ABA and JA showed high increase in the activity of the cytoplasmic located carbonic anhydrase (Popova et al. 1991). The authors suggested that after exogenous treatment with ABA and JA (and probably during exposure of plants to stress) the internal concentration of CO_2 became limited for a normal activity of Rubisco and this led to appearance of additional mechanism for CO_2 concentration with the participation of carbonic anhydrase.

Metodiev et al. (1996) discussed the dual effect of JA on photosynthesis depending on the duration of treatment and JA concentration. They showed that long-term treatment (7 days) of barley plants led to a noticeable decrease in both the initial slope of the A/Ci curves and the maximum photosynthetic assimilation (A) at saturating C_i. The proportion of stomatal and nonstomatal factors in limitation of photosynthesis depended on the applied JA concentration. Short-term treatment (up to 2 h) with JA affected neither the stomatal conductivity for CO_2 nor the rate of photosynthetic CO_2 assimilation. The suggestion was that JA may affect photosynthesis indirectly, either as a stress-modulating substance, or through the alteration in gene expression.

MeJA vapors induced the biosynthesis of anthocyanin in light-grown soybean seedlings but inhibited anthocyanin accumulation in etiolated seedlings (Franceschi and Grimes 1991). Exogenously applied MeJA also induced anthocyanin accumulation in other plants (Tamari et al. 1995; Saniewski et al. 1998, 2003, 2006).

Table 14.2 Effect of JA on RubisCO protein level, the level of large and small subunits of Rubis CO and leaf soluble protein in barley leaves. Values are means ± SE for three experiments

JA (M)	Rubisco protein mg(mg sol pr)$^{-1}$	Large subunit mg(mg sol pr)$^{-1}$	Small subunit mg(mg sol pr)$^{-1}$	Soluble protein mg (g FW)$^{-1}$
H_2O (control)	538 ± 64.7	395.0 ± 17.4	162.23 ± 33.9	6.94 ± 2.04
2.5×10^{-7}	400 ± 28.9	265.9 ± 26.7	85.10 ± 14.6	6.00 ± 2.34
2.5×10^{-6}	266 ± 57.9	215.3 ± 33.0	69.33 ± 21.6	5.99 ± 2.21
2.5×10^{-5}	278 ± 105.9	211.8 ± 26.4	63.77 ± 18.7	5.40 ± 2.04
2.5×10^{-4}	123.4 ±61.6	172.5 ± 26.4	48.47 ± 16.0	3.82 ± 1.61

Work on the photosynthesis and transpiration effects of jasmonates application in conifers is limited, although many examples demonstrating the effect of exogenous MeJA application on other plants exist. Foliar applications of 1 mmol L^{-1} JA increased carbon export from the leaves to the stem and roots within hours of treatment in *Populus* spp. (Babst et al. 2005) and 100 mmol L^{-1} MeJA treatment reduced photosynthetic rate and growth in Scots pine *(Pinus sylvestris* L.), (Heijari et al. 2005).

14.4.3 Jasmonates and Protein Synthesis

Jasmonates induced accumulation of a number of proteins (JA-induced proteins, JIPs) in many plant species were first described in barley leaves (Weidhase et al. 1987), but since then have been reported for most plants. Maslenkova et al. (1992) induced experimental data showing an induction of JIPs mainly belonging to the thylakoid-bounded polypeptides. Most of the JIPs were identical to ABA- and NaCl-induced ones, leading to the assumption that exogenously applied jasmonates act as stress agents. Plant tissues treated with MeJA or exposed to stressors that induce *in planta* jasmonates accumulation synthesize high levels of JIPs (Parthier et al. 1992; Reinbothe et al. 1992; Mueller-Uri et al. 1988; Parthier 1989). Concomitantly, these tissues reduce or even shut-down the synthesis of most proteins that were present before MeJA or stress treatment (Mueller-Uri et al. 1988). Earlier experiments, obtained for detached leaf tissues of barley, have identified translation initiation as the site at which protein synthesis is downregulated by MeJA (Reinbothe et al. 1993). Reinbothe et al. (1994a) demonstrate that one of the cloned JIPs in barley, JIP60, is a ribosome-inactivating protein (RIP) that differs from previously characterized RIPs (18–23 kDa) in exhibiting a distinctive mode of action. Our data showed that treatment of barley plants with MeJA led to breakdown in the biosynthesis of Rubisco (Table 14.2 and Fig. 14.1)

It is important to note that jasmonates did not provoke the same or a similar set of effects in all plants. Some changes in gene expression such as the downregulation of proteins involved in the photosynthetic apparatus and the upregulation of some defence related proteins are commonly observed, but often the eventual outcome of the jasmonate-induced altered gene expression strongly depends on the species and

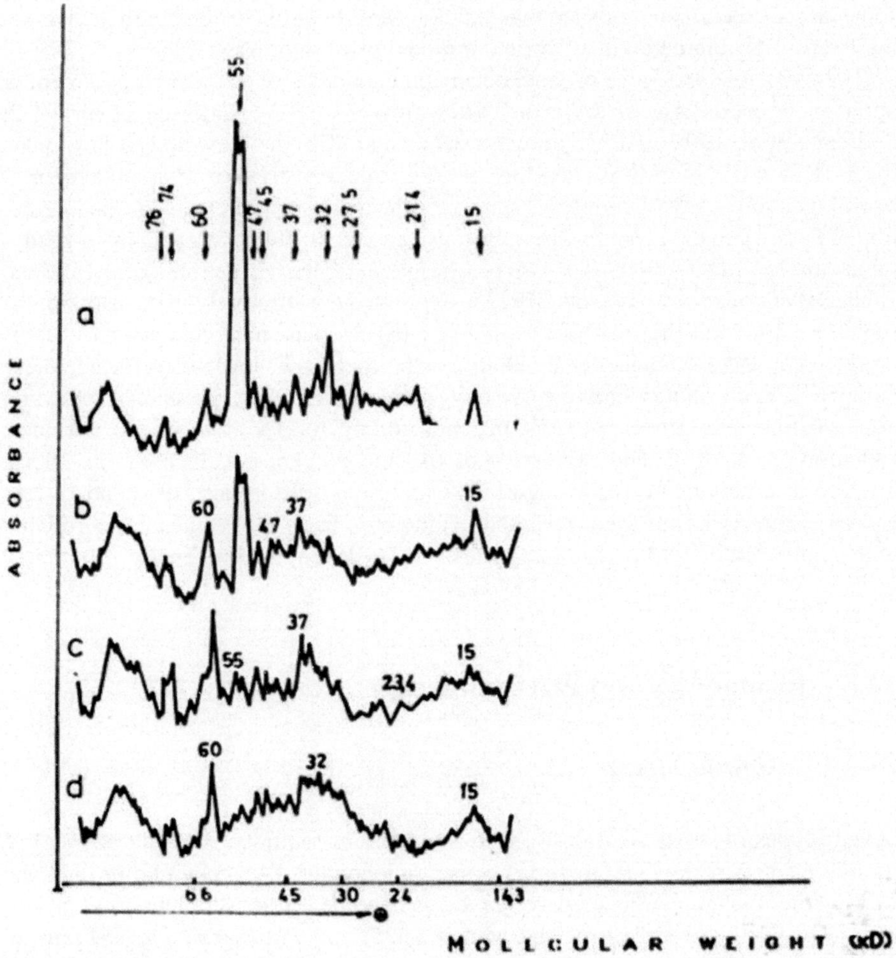

Fig. 14.1 Polypeptide profiles and gel scans of soluble proteins from barley leaves after 15% SDS-PAGE: *a* control (water), *b* 0. 25 μM JA, *c* 2.5 μM JA, *d* 25 μM JA, *e* 250 μM JA. The major protein differences among control and treated plants are indicated in *kDa* with *arrows*

tissue under investigation. Numerous JA-responsive genes have been identified that are clearly upregulated and lead to the synthesis of JIPs. Comparative analyses revealed that different plant species express different JIPs. Some types of JIPs have been identified in a single species. For example, some of the JIPs from barley seedlings have not been identified yet in any other plant. Many data indicate that plants developed, in addition to a set of common JA-responsive genes, a whole variety of specialized JA-responsive genes. At present, there is no simple genetic explanation for the origin of these specialized JA-responsive genes. The observations of Lannoo et al. (2006) provided the first firm evidence that even closely related species respond differently upon treatment with MeJA because of the presence/absence of some target gene(s), and indicate that extreme care should be taken before generalizing the results obtained with a particular model plant to other species.

The most abundant gene product occurring in barley leaves upon JA treatment or upon endogenous rise of JA is a 23-kD protein (JIP23, Andresen et al. 1992; Lehmann et al. 1995). JIP23 is always detectable after the elevation of jasmonate levels (Kramell et al. 2000). Therefore, the occurrence of JIP23 is a valuable reporter of endogenous rise of jasmonates as used for the analysis of the pathogenic interaction of barley leaves with powdery mildew (Hause et al. 1997). Also, in other tissues of the barley plant, there is a strict correlation of the expression of JIP23 and enhanced endogenous JA levels. JIP23 is expressed constitutively in the root tip, the scutellar node, and the leaf base, which are tissues that show enhanced JA levels (Hause et al. 1996; Maucher et al. 2000). Furthermore, the elevated JA level in these barley tissues correlates with reactive oxygen species (ROS) expression, suggesting a causal link between expression of genes coding for JA biosynthetic enzymes, elevation of JA levels, and expression of JA-induced genes (Maucher et al. 2000). Therefore, simultaneous recording of the expression of ROS and JIP23 and JA levels represents a tool for asking whether an increase in JA biosynthesis is correlated with expression of JA-biosynthetic genes and JA-dependent processes.

14.5 Jasmonates and Environmental Stresses

14.5.1 Abiotic Stress

Several types of stress conditions, such as wounding, salinity, drought, ozone, UV irradiation and pathogen infection, cause endogenous JA accumulation and the expression of jasmonate-responsive genes. Jasmonate levels were increased in soybean (*Glycine max,* Creelman and Mullet 1995) and *Pinus pinaster* (Pedranzani et al. 2007) upon plant exposure to drought and in tomato (*Solanum lycopersicum*, Pedranzani et al. 2003) and *Iris hexagona* (Wang et al. 2001) upon exposure to high salinity. In rice, both drought and high salinity increased jasmonate levels in the leaves and roots, resulting in the induction of stress-related proteins and JA biosynthetic genes (Moons et al. 1997; Kiribuchi et al. 2005; Tani et al. 2008).

14.5.1.1 Jasmonates and Salinity Stress

Among environmental stresses NaCl salinity is one of the main limitations to the growth and photosynthesis of nonhalophytic plants. The extent of this inhibition is correlated to the extent of NaCl salinity, the plant species, and the environment.

Barley is rated among the salt-tolerant crop species in the tribe *Triticeae* (Munns 2005). Response of barley to JA and osmotic/salinity stress treatment has been the focus of several studies (Maslenkova et al. 1992; Lehmann et al. 1995; Ortel et al. 1999).

Table 14.3 Effect of NaCl and JA on growth, photosynthetic CO_2 fixation and the activity of RuBPCase and PEPCase. Seedlings wet grown for 8 days. They were treated 4 days with 25 μM JA and then with 100 mM NaCl for the next 4 days. To obtain NaCl stepwise treated plants NaCl concentration was gradually increased by 25 mM every 2 days until 100 mM NaCl was reached

Treatment	Length of seedlings (cm)	CO_2 fixation (μmol CO_2) (mg chl h)$^{-1}$	RuBPCase (μmol CO_2) (mg min)$^{-1}$	PEPCase (μmol CO_2) (mg min)$^{-1}$	Chl. (a+b) (mg chl) (g FW)$^{-1}$
Control	9.94±0.65	64.9±6.7	0.38±0.09	0.064±0.09	2.30±0.07
100 mM NaCl	5.44±0.39	31.1±2.9	0.14±0.04	0.132±0.01	1.73±0.05
25–100 mM NaCl	8.76±0.81	48.8±6.3	0.33±0.10	0.119±0.03	1.78±0.05
25 μM JA	6.93±0.72	49.55±5.3	0.29±0.07	0.128±0.07	2.43±0.04
25 μM JA +100 mM NaCl	7.77±0.58	48.22±2.7	0.31±0.12	0.102±0.09	2.04±0.07

Table 14.4 Leaf gas exchange characteristics of barley plants after treatment with NaCl and JA. α, carboxylating efficiency; Γ, CO_2 compensation point; *A*, net CO_2 assimilation; *Ci*, intercellular CO_2 concentration; *Ls*, stomatal limitation of photosynthesis; r'_s, stomatal resistance to CO_2; *Tr*, transpiration rate. Values are means±SE for four experiments

Treatment	α (cm s^{-1})	Γ (μmol mol^{-1})	*A* (μmol CO_2 m^{-2} s^{-1})	*Ls* (%)	*Ci/Ca* (Ca=350)	r'_s (s cm^{-1})	*Tr* (mg H_2O m^{-2} s^{-1})
Control	0.068	92.2	8.75	27.5	0.72	7.8±0.5	29.5±2.0
100 mM NaCl	0.007	345.3	0.05	41.2a	0.95a	48.8±2.7	9.6±1.4
25–100 mM NaCl[b]	0.033	87.2	4.20	42.1	0.64	18.5±1.8	14.5±1.2
25 μM JA	0.037	135.6	3.79	47.0	0.77	15.8±0.2	19.8±0.8
25 μM JA+100 mM NaCl	0.054	77.7	4.71	39.7	0.69	17.5±1.3	16.9±0.9

[a]The data for L_s and C_i/C_a in the variant with 100 mM NaCl were calculated at C_a=400 ppm because of greater Γ

[b]Stepwise increased NaCl concentration from 25 to 10 mM within 8 days

Dombrowski (2003) reported a wounding–JA salinity interaction in tomato, where salt stress induced wound-related genes through the activation of the octadecanoid pathway. Most of these studies however, focus on the osmotic stress component of the salt stress usually involving the shock treatment with sudden increase in salt in the growth medium of whole plants or floating of barley leaf discs in salt solution. Popova and Maslenkova (1997) reported that pre-treatment of barley plants with JA for 4 days before salinization ameliorated the inhibitory effects of salinity on growth and photosynthesis when compared to a direct salt stress (Table 14.3).

Gas exchange measurements showed significant changes in the values of CO_2 compensation point, stomatal resistance and transpiration rate in plants treated with NaCl and JA. Pretreatment with JA or step-wise treatment with NaCl improved the values of these parameters and led to adaptation of barley seedlings to salinity stress (Tsonev et al. 1998, Table 14.4).

The data on structural and functional characteristics of chloroplasts in JA pretreated barley manifested that JA can improve photosynthetic performance of the NaCl stressed plants. Also, it was shown that JA pre-treatment prior to salinity stress resulted in lower Na^+ accumulation, therefore linking JA to Na^+ homeostasis. These results point to the suggestion of a possible role for JA in increasing the salt tolerance when it is applied to the plants for a short period before exposure to salinity stress, thus improving their non-vulnerability to unfavorable conditions, possibly involved in salinity tolerance mediated by JA (Popova and Maslenkova 1997).

Similar data have been reported by Walia et al. (2006). They observed that genes involved in JA biosynthesis showed increased abundance during salt stress. The authors hypothesized that JA plays a role in salt stress adaptation in barley. They tested this hypothesis by applying JA to barley plants and by observing its effect on salinity tolerance. In addition, they also examined the extent of interaction as indicated by gene expression overlap between JA and salinity stress response of barley using transcriptome profiling. Data confirmed that pre-treatment with JA partially alleviated photosynthetic inhibition caused by salinity stress. Expression profiling after a short-term exposure to salinity stress indicated a considerable overlap between genes regulated by salinity stress and JA application. Three JA-regulated genes, arginine decarboxylase, ribulose 1·5-bisphosphate carboxylase/oxygenase (Rubisco) activase and apoplastic invertase were identified. Rubisco activase transcript increased by seven-fold in JA-pre-treated salinized plants and 3.8-fold in JA treatment, but not differentially expressed at a significant level in salinity-stressed plants. The expression analysis results indicated an increase in the mRNA abundance of several photosynthesis-related genes in response to JA treatment (Walia et al. 2007).

Data by other authors indicated a decrease in levels of photosynthesis-related proteins such as Rubisco and light-harvesting cab-binding proteins upon treatment with jasmonates (Popova and Vaklinova 1988; Reinbothe et al. 1994). Although the abundance of Rubisco small subunit and Rubisco activase transcripts increased, the expression level of several other light-responsive genes including those encoding for cab-binding protein decreased in abundance upon JA treatment. There are reports indicating jasmonate regulation at post-transcriptional and translational levels for some genes (Reinbothe et al. 1993, 1994). Post-translational regulation is one possibility that may explain the differences between our expression results and previous reports on protein abundance for photosynthesis-related genes. Maslenkova et al. (1992) reported about effect of JA treatment and salinity stress on the polypeptide composition of soluble and membrane bounded polypeptide of barley seedlings. They showed that high JA concentration (250 μM) induced marked quantitative and qualitative changes concerning mainly the proteins with the approximately equal mobility as in NaCl-stressed plants. The most obvious increase in thylakoid polypeptide band intensity was at 55–56 kDa. The relative share of some polypeptides with apparent polypeptide masses above 66 kDa and polypeptide with lower molecular masses in the region 20.5–15 kDa was enhanced. At the same time one new band at 31 and 31.5 kDa was well expressed at 25 and 250 μM and became discernible in 100 μM NaCl-treated plants. The intensity of some polypeptides of soluble proteins of 60, 47, 37, 30, and 23.4 kDa increased with increasing JA concentration, whereas the intensities of other polypeptides bands (molecular masses 55, 21.4 and 15 kDa) decreased. The appearance of one new polypeptide, of 25.1 kDa, was observed only in NaCl-treated plants. The function(s) of jasmonate-induced polypeptides is not known. It was presumed that they participate in the modulation of aging reactions or of plant physiological responses to stress (Fig. 14.2).

Popova et al. (1991) observed a substantial increase of the cytoplasmic localized carbonic anhydrase in barley leaves after treatment with JA. The authors tempted to suggest that, under conditions of stress, there may an induction of CO_2 concentration mechanism and the observed increase in 30 kDa polypeptide corresponds to the same electrophoretic mobility and molecular mass of carbonic anhydrase. Various studies have also shown a relationship between salinity adaptation and an increase in CA activity (Latorella and Vadas 1973; Brown et al. 1987; Fisher et al. 1996). Therefore, the increased abundance of CA by JA pre-treatment 2 days before salinity stress can prime the subsequently stressed plants and make photosynthetic machinery more adaptive to salinity stress compared to directly salt-stressed plants. Less inhibited net photosynthetic rate in JA-pre-treated plants relative to salinity-stressed plants, coupled with increased transcript levels of Rubisco activase, and CA indicates a potential role of these genes in improved tolerance of barley as a consequence of JA pre-treatment.

Rice (*Oryza sativa* L.) is considered salt sensitive compared to other cereals. Data reported by Wilson (2007) showed that JA pretreatment in rice does not ameliorate salinity stress, possible due to increased accumulation of toxic ions and changes in ion homeostasis. Effects of 100 mM NaCl (osmotic stress) on shoot growth and biosynthesis of JA in two maize genotypes (*Zea mays* L.), salt-sensitive Across 8023 and resistant hybrid SR 03 were investigated (Shahzad et al. 2009). Exogenous treatment of JA, significantly reduced the maize seedling growth in terms of shoot extension, root length and root and shoot fresh weights. The differences in JA biosynthesis in genotypes of contrasting salt resistance indicated that jasmonic acid may have a role in plant adaptation to osmotic stress and increased JA in roots of the salt-sensitive genotype may be responsible for shoot growth inhibition. Similar data had been reported by Kang et al. (2005) for rice seedling differing in their sensitivity to salt stress and treated with JA. The obtained results indicated that application with exogenous JA can ameliorate salt-stressed rice seedlings, especially the salt-sensitive cultivar rather than the salt-tolerant cultivar. MeJA application greatly mitigated the adverse effects of NaCl on soybean growth and endogenous hormones. MeJA significantly increased ABA levels, while the endogenous amount of giberreline (GA_4) was reduced by the application of NaCl. The data revealed that MeJA counteracted the negative effects of NaCl stress on plant growth, chlorophyll content, leaf photosynthetic rate, leaf transpiration rate, and proline content (Yoon et al. 2009). Mansour et al. (2008) reported that treatment of broad been (*Vicia faba*) with JA led to reduction in vegetative growth, altered mineral composition of leaves and to partial alleviation of salinity stress. Wang (1999) showed that treatment of strawberry with MeJA alleviated the negative effect of water stress.

The influence of MeJA and NaCl on a *Catharanthus roseus* suspension cell culture with high production of ajmalicine was investigated. NaCl treatment increased

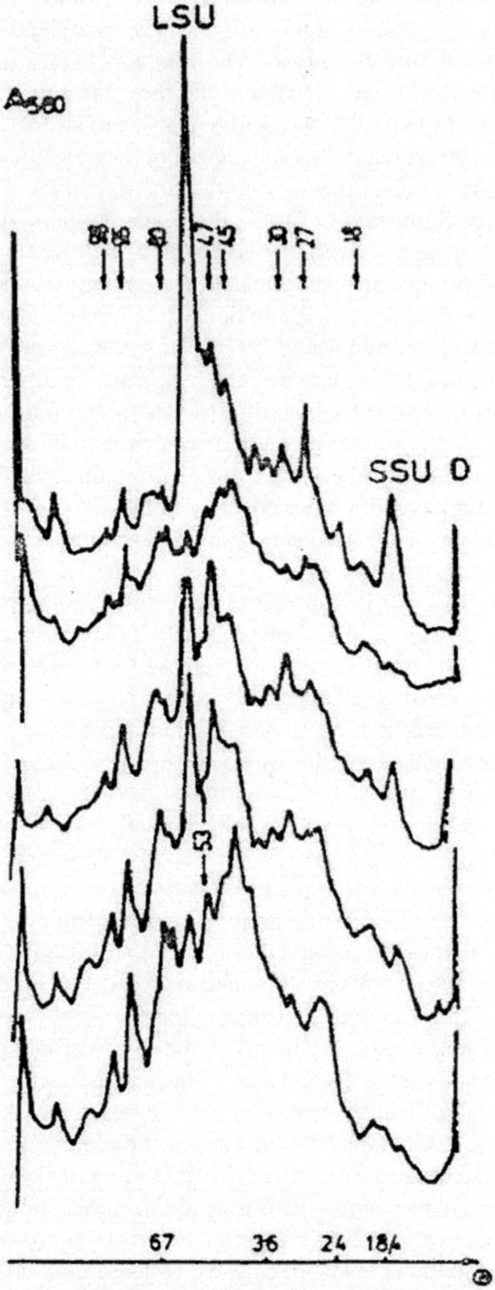

Fig. 14.2 Gel scans of soluble proteins from barley leaves after 15% SDS-PAGE. *a* control (water), *b* 100 mM NaCl, *c* 25–100 mM NaCl, *d* 25 μM JA, *e* 25 μM JA+100 mM NaCl. The major protein differences among control and treated plants are indicated in *kDa* with *arrows*

both the cell membrane permeability and ajmalicine accumulation. MeJA treatment promoted ajmalicine accumulation and maintained membrane permeability of cells under salt stress (Xiang et al. 2011).

14.5.1.2 Jasmonates and Heavy Metal Stress

A short time effects of 25 and 150 μM Cu^{2+} or 50 μM MeJA on growth of roots and leaves of *Phaseolus coccineus, Allium cepa* and *Zea mays* were investigated (Maksymek and Krupa 2007). The authors suggested that the leaf growth inhibition induced by excess Cu^{2+} was connected in *Z. mays* with jasmonate, and in *P. coccineus* with ethylene, NADPH oxidase and, to a minor degree, with jasmonate.

In *Arabidopsis thaliana* leaves, during the first hours of exposition to excess Cu or Cd, a considerable increase of hydrogen peroxide accumulation and superoxide radicals was observed. Increased superoxide radicals and hydrogen peroxide level was obtained after exogenous methyl jasmonate supply. Excess Cd, in contrast to Cu, increased in the first hour of SOD activity. The results indicated that the first effect of the investigated heavy metals stress was accumulation of superoxide radicals and hydrogen peroxide. That increase was mostly connected with the activity of NADPH oxidase, induction of jasmonate signaling pathways and partially, in the case of Cd, with superoxide dismutase activity (Maksymiek and Krupa 2002). Copper and cadmium in in vivo conditions induced accumulation of jasmonates in mature leaves of *Arabidopsis thaliana* and in young and oldest *Phaseolus coccineus* plants (Maksymiec et al. 2005, Maksymiek and Krupa 2006). The dynamics of jasmonate accumulation showed a biphasic character in both plants. In the first phase, after 7 h (*A. thaliana*) or 14 h (*P. coccineus*) of exposure to Cu or Cd, a rapid increase of JA level occurred, followed by a rapid decrease observed during seven successive hours. In the next phase, a repeated but slow increase of JA content occurred. The heavy metal stress induced in particular a more stable (3R,7R) form of jasmonates. The results indicated that JA is connected with the mechanism of toxic action of both heavy metals in plants, differentially reacting to exogenous JA and possessing variable dynamics depending on the plants studied as well as their growth stage.

The effect of MeJA at concentrations of 10 and 100 μM MeJA in alleviating Cd toxicity in and soybean plants were investigated. Pretreatment with lower MeJA concentration before Cd exposure significantly alleviated the Cd damage to shoot dry weight and total chlorophyll contents and also, reduced lipid peroxidation. The increase in catalase and polyphenol oxidase activities were also observed under cadmium stress. The activity of ascorbate peroxidase was enhanced by MeJA under no stress and cadmium stress. However, peroxidase activity increased under no stress conditions. The authors suggested that the application of MeJA caused significant alleviation of cadmium damages to soybean plants (Keramat et al. 2010).

Exposure of *Wolffia arrhiza* (*Lemnaceae*) to lead (Pb) resulted to decrease in fresh weight, chlorophyll *a*, carotenoid, monosaccharide and soluble protein content, and high accumulation of lipid peroxides. Exogenously applied 0.1 μM JA protected *W. arrhiza* fronds against Pb stress inhibiting heavy metal accumulation, restoring plant growth and primary metabolite level. It was also found that JA activated enzymatic (catalase, ascorbate peroxidase, NADH peroxidase) and non-enzymatic antioxidant (ascorbate, glutathione) system in *W. arrhiza*, and therefore, suppressed oxidative destruction of cellular components induced by heavy metal (Pijaotrowska et al. 2009).

14.5.1.3 Jasmonates and Extreme Temperature Stress

Jasmonate has been shown to increase the chilling tolerance of several plant species. Scientists found that in reducing chilling injury, jasmonates may play an integral role in the intracellular signal-transduction cascade that acts in inducible defense mechanisms that plants can develop against chilling stress.

Protection plant cells against environmental stress by jasmonates may also associate with involving special gene expression (Ding et al. 2001). Exposure of avocado (Gonzalez-Aguilar et al. 2003) and tomato (Ding et al. 2001) fruits to low concentration of jasmonates suppressed chilling injury and decay in low temperature storage fruits. Wang and Buta (1994) demonstrated that MeJA prevented the chilling injury of zucchini squash fruit (*Cucurbita pepo* L.) subjected to low temperature stress. In addition, jasmonates may prevent chilling injury by a mechanism that involves regulation of ABA and polyamine (PA) levels. ABA and PA have been found to stabilize cell membranes and maintain cellular structure. Therefore, regulation of ABA and PA levels by jasmonates can maintain cellular integrity and increase the tolerance of plant tissues to chilling injury. High temperature shocks has been useful in some peach (*Prunus persica*) cultivars to prevent chilling injury, although undesirable side effects have been determined. The young grape plants treated with 50 μmol L^{-1} JA solution and at 38 °C heat acclimation (HA) were studied. The result showed that SOD, POD, CAT, APX activity and soluble protein content in young grape plants become higher than under normal temperature. It was suggested that JA or HA enhances thermoprotection (Chen et al. 2006b)

The potential ability of stressed leaves to signal other leaves and induce the production of heat shock proteins (Hsps) suggests that a systemic nature of the heat-shock response might be adaptive for stress tolerance. It may prepare unstressed leaves for a stress that might occur in the near future. The systemic induction of Hsps may serve a protective function in unstressed leaves. For example, leaf temperatures can vary significantly within a single plant, throughout the day (Larcher 1995), which may stress only certain leaves at any given time.

14.5.1.4 Jasmonates and Oxidative Stress

There is no general consensus on the contribution of jasmonates to oxidative stress, besides inducing the production of ascorbic acid (AsA). AsA is the most abundant water-soluble antioxidant in plants. It provides the first line of defense against damaging reactive oxygen species (ROS), and helps protect plant cells from many factors that induce oxidative stress, including wounding, ozone, high salinity, and pathogen attack. Plant defenses against these stresses are also dependent upon jasmonates. MeJA also induces H_2O_2 production in plants (Orozco-Cárdenas and Ryan 1999; Orozco-Cárdenas et al. 2001; Wang and Wu 2005), although the increase in H_2O_2 may result in decreased AsA concentrations, which may in turn promote the imbalance between active oxygen species and antioxidants that leads to oxidative stress, (an increase in H_2O_2 is not always detrimental to plant function). MeJA promotes H_2O_2-induced stomatal closure (Suhita et al. 2004), thereby protecting plants from water stress. It has been shown in drought-resistant Mediterranean shrubs that an increase in H_2O_2 in the apoplast of mesophyll cells, xylem vessels and differentiating sclerenchyma cells might be more related to a structural role than to oxidative damage (Jubany-Marí et al. 2010). MeJA treatment led to enhanced production of H_2O_2 in *Arabidopsis* leaves and roots, indicating in vivo oxidative stress. With monobromobimane (mBBr) labeling to capture the oxidized sulfhydryl groups, 2D gel separation, and LC-MS/MS analysis Alvarez et al. (2009) isolated 35 protein spots that displayed significant redox and/or total protein expression changes. Thirty-three spots were identified in both control and MeJA-treated samples. Some of these proteins were redox responsive and others displayed abundance changes in response to MeJA. Many cysteine residues involved in the disulfide dynamics were mapped based on applied methods. Identification of redox proteins and their cysteine residues involved in the redox regulation is promising for a deeper understanding of the jasmonate signaling networks.

Pan et al. (1995) reported that growing peanut (*Arachis hypogaea* L.) seedlings in solution containing MeJA at 100, 200, and 400 mg L^{-1} resulted in increase the levels of protective enzymes SOD and CAT which were able to remove superoxide and H_2O_2 and leads to protection the integrity of plasma membrane from drought damage.

The application of MeJA (50 and 100 μM) in roots and shoots of maize seedlings subjected to paraquat decreased the lipid peroxidation with highest efficiency and meaningfully in roots and shoots as compared with controls (Norastehnia and Asghari 2006). Pretreatment of barley seedlings with 23 μM MeJA before exposure to paraquat fully blocked the inhibitory effect of Pq on photosynthesis and provided protection against subsequent Pq-induced oxidative damage (Hristova and Popova 2002).

Treatment of 40 day-old adventitious roots of *Panax ginseng* with 200 μM MeJA or SA caused an increase in the carbonyl and H_2O_2 contents. Total phenolic, flavonoid, ascorbic acid, non-protein thiol and cysteine contents and 1,1-diphenyl-2-picrylhydrazyl radical reducing activity were increased by MeJA and SA. The activities of glucose 6-phosphate dehydrogenase, phenylalanine ammonia lyase, substrate specific peroxidases (caffeic acid peroxidase, quercetin peroxidase and ferulic acid peroxidase) were higher in MeJA treated roots than the SA treated ones. Increased shikimate dehydrogenase, chlorogenic acid peroxidase and β-glucosidase activities and proline content were observed in SA treated roots than in MeJA ones. Cinnamyl alcohol dehydrogenase activity remained unaffected by both MeJA and SA.

These results strongly indicate that MeJA and SA induce the accumulation of phenolic compounds in ginseng root by altering the phenolic synthesis enzymes (Ali et al. 2007). Recent data showed that SA and MeJA influenced on the antioxidant systems in *Haematococcus pluviialis*. Both SA and MeJA at 500 μM concentration reduced the growth of alga with more pronounced effect of SA. Carotenoid and chlorophyll contents were decreased by SA and increased by MeJA (Vidhyavathi and Sarada 2011.) Para- Lobato et al. (2009) showed that treatment of sunflower (*Helianthus annuus* L.) seedlings with 50 μM MeJA led to a fast increase in ROS content, followed by a marked increase in the activity of H_2O_2-scavenging enzymes, guaiacol peroxidase, ascorbate peroxidase, and catalase. It was suggested that exogenous MeJA may be involved in the oxidative stress processes by regulating antioxidant enzyme activities.

JA and SA pretreatment enabled young grape plants to maintain higher content of soluble protein and activity of SOD, CAT, POD, APX and decrease electrolyte leakage to weaken oxidative stress under heat shock. It was suggested that JA or heat acclimation enhanced thermoprotection by similar mechanisms (Chen et al. 2006).

Suza et al. (2010) reported evidence showing that wounding and jasmonates influenced AsA accumulation in various plant species, and showed data from *Arabidopsis* and tomato testing the influence of jasmonates on AsA levels in wounded and unwounded plants. Their findings indicated that the effects of wounding and jasmonates on AsA accumulation differ between species, these factors both enhanced AsA accumulation in *Arabidopsis*, but depressed AsA levels in tomato. MeJA has been shown to cause H_2O_2 production in parsley suspension-cultured cells (Kauss et al. 1994) and to act as a signal molecule for the induction of defense genes in tomato plants (Orozco-Cárdenas et al. 2001).

Using molecular approaches Sasaki-Sekimoto et al. (2005) provided data that jasmonates activated the coordinated gene expression of factors involved in nine metabolic pathways belonging to two functionally related groups: (1) ascorbate and glutathione metabolic pathways, which are important in defence responses to oxidative stress, and (2) biosynthesis of indole glucosinolate, which is a defence compound occurring in the *Brassicaceae* family. They confirmed that jasmonates induced the accumulation of ascorbate, glutathione and cysteine and increased the activity of dehydroascorbate reductase, an enzyme in the ascorbate recycling pathway. They also showed that O_3 exposure caused the induction of several genes involved in antioxidant metabolism in the wild type *Arabidopsis*. The authors suggested that the coordinated activation of the metabolic pathways mediated by jasmonates provides resistance to environmental stresses.

Jubany-Mari et al. (2010) examined the contribution of MeJA to the acclimation of a Mediterranean shrub, *Cistus albidus* L., to water stress under natural climatic conditions. For this purpose, changes in MeJA, H_2O_2, AsA and the maximum efficiency of PSII photochemistry (F_v/F_m ratio) and lipid peroxidation were monitored in young leaves of two sets of plants: well-watered plants and plants exposed to water stress and then re-watered. MeJA accumulation in water-stressed plants showed a maximum increase after 11 weeks of water stress and a second increase after the re-watering, which was performed in week 17 of the experiment. Like MeJA, H_2O_2 variations showed a biphasic time course, reaching the first peak under mild water stress and the second peak during plant recovery. H_2O_2 accumulation was not associated with oxidative damage, and in addition to showing intact cell ultrastructure, water-stressed plants showed lower lipid peroxidation than well-watered plants. The F_v/F_m ratio remained above 0.75 throughout the experiment in both sets of plants. AsA concentrations began to increase at the beginning of water stress, before the increase in MeJA. AsA reached a steady maximum, which was maintained during water stress, and returned to initial values when plants were re-watered. On the basis of these results, the authors concluded that MeJA, H_2O_2 and AsA are involved in the mechanisms of plant resistance to water stress as follows: MeJA arrested the growth of young leaves and AsA prevented oxidative damage.

14.5.2 Biotic Stress

Plants respond to herbivores and pathogens attack with increased emission of volatile organic compounds. These molecules act as indirect defences when attracting natural enemies of herbivores and thus benefit the plant. It remains controversial whether undamaged plants capture chemicals released by damaged neighbouring plants and respond to them by increasing their defensive barriers against an imminent attack.

Within their environment, plants interact with a wide range of microorganisms, some of which are pathogenic and cause disease, and others that are beneficial and stimulate plant growth or activate natural defenses. Because of their agronomic importance, plant-microorganisms interrelations have been the major focus in many biology researches. To recognize and respond to this variety of pathogenic and beneficial microorganisms, plants have developed sophisticated strategies to perceive microorganisms and translate that perception into an appropriate adaptive response. This plant innate immune response is surprisingly complex and highly flexible in its capacity to recognize and respond to different invaders (Pozo et al. 2004). There are many data supporting that jasmonates are involved in plant defense responses to insect wounding, attack by various pathogens, and water deficit (Creelman and Mullet 1995; Wasternack and Parthier 1997; Seo et al. 2001; Rakwal et al. 2002). The levels of endogenous jasmonates were reported to be increased following pathogen exposure (Reymond and Farmer 1998; Thomma et al. 1998). Likewise, exogenous application of jasmonates to plants induced stress-related or pathogenesis-related (PR) genes (Moons et al. 1997; Mei et al. 2006). Thus, the role(s) of jasmonates in response to biotic stresses has been well documented.

In the absence of adaptive immunity displayed by animals, plants respond locally to biotic challenge via inducible basal defense networks activated through recognition and response to conserved pathogen-associated molecular patterns. In addition, immunity can be induced in tissues remote from infection sites by systemic acquired resistance (SAR), initiated after gene-for-gene recognition between plant resistance proteins and microbial effectors. The nature of the mobile signal and remotely activated networks responsible for establishing SAR remain unclear. Salicylic acid participates in the local and systemic response, but SAR does not require long-distance translocation of SA. Truman et al. (2007) showed that, despite the absence of pathogen-associated molecular pattern contact, systemically responding leaves rapidly activate a SAR transcriptional signature with strong similarity to local basal defense. They presented several lines of evidence that suggest jasmonates are central to systemic defense, possibly acting as the initiating signal for classic SAR. JA but not SA, rapidly accumulated in phloem exudates of leaves challenged with an avirulent strain of *Pseudomonas syringae*. In systemically responding leaves, transcripts associated with jasmonate biosynthesis were up-regulated within 4 h, and JA increased transiently. SAR can be mimicked by foliar JA application and is abrogated in mutants impaired in jasmonate synthesis or response. The authors concluded that jasmonate signaling appears to mediate long-distance information transmission.

While extensive research has examined plant defense responses to attack by herbivores and pathogens, plant responses to parasitism by other plants are not well characterized. The expression of induced plant defenses is mediated by complex signaling networks in which the plant hormones JA and SA play key roles. In general, JA-mediated signaling pathways are implicated in the regulation of antiherbivore defenses, while the SA pathway is associated with defense responses against pathogens. However, there are many exceptions to this basic framework, and recent work suggests that interactions between the JA and SA pathways may play important roles in fine-tuning defense responses. It is now clear that pathogen-induced modulation of signaling via other hormones contributes to virulence. A picture is emerging of complex crosstalk and induced hormonal changes that modulate disease and resistance, with outcomes dependent on pathogen lifestyles and the genetic constitution of the host. Robert-Seilaniantz et al. (2011) reported on recent advances, updating current knowledge on classical defense hormones SA, JA, and ET, and the roles of auxin, abscisic acid (ABA), cytokinins (CKs), and brassinosteroids in molding plant-pathogen interactions.

Treatment of cut freesia var. Cote d'Azur flowers with 0.1 μl MeJA L^{-1} vapour suppressed petal specking caused by *Botrytis cinerea* infection. MeJA did not exert direct antifungal activity in-vitro, suggesting that treatment in-vivo reduced *B. cinerea*-induced flower specking by induction of host defence responses. 0.1 μL MeJA L^{-1} significantly increased vase life of cut freesia flowers and delayed senescence judged by lower wilt scores and higher fresh weights as compared to untreated controls (Darras et al. 2005). Pretreatment of tomato by soaking the seeds in MeJA or treating them with gaseous MeJA efficiently reduced the development of disease caused by the necrotrophic fungus *Alternaria* (Kepczynska nd Krol 2011).

Mathew and Sankar (2012) studied the growth characteristics of *Ocimum basilicum* L., *Ocimum sanctum* L. and *Ocimum gratissimum* L. cell suspension cultures treated with elicitors, MeJA and chitosan individually and in combination. They found that when the elicitors were used individually they enhanced the accumulation of cell biomass, but when administered in combination there was no significant enhancement of cell biomass. The study demonstrated that both elicitors could effectively enhance cell biomass and can be used for effective induction of phytochemicals in *O. basilicum* L., *O. sanctum* L. and *O. gratissimum* L.

The effects of the application of the JA derivative n-propyl dihydrojasmonate (PDJ) on ET biosynthesis, volatile compounds, and endogenous JA and MeJA were examined in Japanese apricot (*Prunus mume* Sieb.) infected by a pathogen (*Colletotrichum gloeosporioides*). The inoculation with the pathogen-induced an increase in 1-aminocyclopropane-1-carboxylic acid (ACC), ET, JA, and MeJA. In contrast, PDJ application reduced the endogenous JA, MeJA, and ethylene production and expression of the ACC oxidase gene (PmACO1) caused by the pathogen infection. The authors suggested that PDJ application might influence ET production through PmACO1 and that aroma volatile emissions affected by pathogen infection can be correlated with the ET production, which is mediated by the levels of jasmonates (Nimitkeatkai et al. 2011)

14.6 Jasmonates and Secondary Metabolites

It is known that jasmonates have a central role in the regulation of the biosynthesis of several secondary metabolites, including terpenoids, alkaloids, phenylpropanoids, and antioxidants. The constitutive activation of the jasmonate-signaling pathway causes enhanced production of secondary metabolites in tomato plants. Exogenous application of MeJA promotes accumulation of caffeoylputrescine in tomato leaves, a metabolite derived from the phenylpropanoid and polyamine pathways; it has a putative function in the plant reproductive process (Chen et al. 2006). See et al. (2011) showed that the addition of different concentrations of MeJA showed different trends in pigment content and pigment production in *Melastoma melabathricum*. Lower concentrations of MeJA (2–5 mg L^{-1}) showed significantly higher pigment content and pigment production as compared to the higher concentrations of MeJA (12.5–50 mg L^{-1}). Since the addition of MeJA did not increase the cell biomass but pigment production, it was suggested that MeJA stimulated the synthesis of anthocyanin production via induction of particular enzymes that catalyzes the synthesis of anthocyanin. Synergistic and antagonistic effects on other hormones have been observed, too. The effects of two elicitors: JA and MeJA on cell growth as well as on rosmarinic acid accumulation in cell suspension cultures of *Mentha*×*piperita* were investigated. The highest rosmarinic acid accumulation 117.95 mg g^{-1} DW was measured 24 h after addition of 100 μM MeJA. There was no substantial influence of elicitors on rosmarinic acid secretion into the culture media. It was documented that suspension cultures of *M. piperita* treated with elicitors showed a decrease in biomass accumulation when compared to the control (Krzyzanowska et al. 2011).

Jasmonate derivatives induce the accumulation of so-called JIPs that were found in all plant species tested. Their accumulation can also be caused by desiccation or ABA effects. Jasmonate-induced-proteins are of varying molecular weights, and molecules of different size classes have immunologically been shown to be related. The major portion of these proteins is not glycosylated, has no proteolytic activity and is metabolically stable. Labelling with immunogold and electron microscopy showed that some of them are located within the nucleus, while others were detected in the vacuole. None have ever been found in mitochondria. JIPs are lacking in roots, in bleached leaves, and in leaves of chlorophyll-deficient *Hordeum vulgare* mutants. They exist in etiolated leaves, though. Jasmonates do not only regulate the transcription of these proteins, they do also influence the rate of translation of different groups of mRNA. They do, for example, decrease the production rate of several essential housekeeping proteins (Reinbothe et al. 1994).

14.7 Stress Signaling Role of Jasmonates

All terrestrial land plants are believed to be receptive to jasmonate. And while the effects of the jasmonate hormone are crucial to the ultimate survival of the organism, how the signaling pathway works has been a mystery.

Recently scientists at Washington State University discovered a key family of molecules, the Jasmonte-Zim (JAZ) proteins, which play a crucial role in the jasmonate pathway. "It is the last major hormone for which the central signaling components have not been described," explained Regents Professor and co-author of the study, John Browse. The study was published in the August 9th issue of Nature, the research paper was among the top ten most frequently downloaded from the Nature website that month (Chini et al. 2007).

JA is activated upon specific conjugation to the aminoacid L-isoleucine (Ile), which produces the highly bioactive hormonal signal (3R,7S)-jasmoyl-L-isoleucine (JA-Ile) that functionally and structurally mimicked by the *Pseudomonas syringae* phytotoxin coronatine (Staswick and Tiryaki 2004). JA responses typically depend on large-scale changes in gene expression. In unstressed cells containing low JA levels, transcription factors that promote expression of JA-responsive genes are repressed by members of the Jasmonate ZIM-domain (JAZ) protein family. Increased JA levels stimulate JAZ binding to coronatine insensitive1 (COI1), which is the F-box protein component of the E3 ubiquitin ligase SCFCOI1 (Thines et al. 2007).

Further research identified the JAZ proteins as repressor proteins, working to suppress the action of the jasmonate responsive genes. Jasmonate promotes the destruction of the JAZ repressor proteins, enabling the activation of genes involved in defense and development. Only when jasmonate is present the inhibition of these genes is released. Surprisingly, the genes encoding JAZ proteins are also activated by the jasmonate signal. In the study by Thines et al. (2007), eight members of the *JAZ* family were identified as rapid JA-responsive genes. Database searches identified five additional genes that encoded related proteins, bringing the total number of family members to 12.

The expression of the JAZ proteins in response to jasmonate is a safety feature for the plant. The presence of JAZ ensures that the jasmonate signaling will eventually be shut down again by initiating the formation of the proteins that it originally marked for destruction. This type of system is referred to as a negative feedback, a signal sets off a series of reactions that ultimately results in the cessation of the original signal.

In a review of the role of jasmonates in plants, Creelman and Mullet (1997) highlight the delicately balanced dual role of jasmonates in both plant defence and plant development. Resource allocation may be divided between growth and defence mechanisms, and models such as the "growth rate hypothesis" and the "growth-differentiation balance hypothesis" have been developed to describe the resource competition between growth and defence (Stamp 2003).

These discoveries will help the researchers to study and understand the mechanisms that plants use to defend themselves against insects and diseases and to address the fundamental question of how plants recognize and respond to tissue injury.

14.8 Jasmonates and Horticulture

The increasing demand for consumption of fresh horticultural crops, along with more restrictions on the use of synthetic chemicals to preserve produce quality, has encouraged scientific research to aim at developing new technologies based on natural products such as MeJA and its derivatives. MeJA treatment regulates diverse processes such as skin colour development (by promoting β-carotene synthesis and chlorophyll degradation), chilling injury and ion leakage. Treatment with MeJA reduces a number of stress-induced injuries that occur during the postharvest life of intact and fresh-cut fruits and vegetables. Studies show that fruits treated with MeJA maintain higher sugar, organic acid and vitamin C levels. MeJA has also been shown to stimulate ethylene production, by increasing 1-aminocyclopropane-1-carboxylatic acid (ACC) synthase and ACC oxidase activities, which in turn enhances fruit ripening. Furthermore, MeJA treatment inhibits the gray mould rot caused by *Botrytis cinerea* and reduces decay caused by the green mould *Penicillium digitatum.* Recently, it was reported that MeJA induced the expression of pathogenesis-related protein and alternative oxidase genes, increased the transcript accumulation of heat shock proteins, and enhanced antioxidant capacities and antioxidant enzyme activities. These findings help to explain the mode of action of MeJA in increasing chilling tolerance in fruits and vegetables (González-Aguilar et al. 2006).

The challenge for future research will be to develop treatments with optimal application and conditions for each crop; this will help to maintain quality and prevent deterioration of perishable crops for a period of time that is long enough to allow the commercialisation of produce in distant markets. In this regard, the use of MeJA in combination with packaging is a good alternative treatment, and has the potential to be used as a commercial treatment in the near future (González-Aguilar et al. 2006).

Scientists in Israel, Mexico, Spain and Taiwan were studying the practical use of MeJA in maintaining quality of fruits and vegetables (reviewed by Sevillano et al. 2010).

So far, MeJA appears effective in reducing chilling injury in avocados, cucumbers, grapefruit, mangoes, peppers, and zucchini squash, and in retarding physiological breakdown and decay in strawberries, pepper slices and celery sticks. Today, due to safety, application of these compounds for reducing environmental stress progressively increased and can be promising. However, the mechanism of MeJA treatment used to protect against decay and chilling injury is unclear. Inducing chilling tolerance make possible to storage fruits in low temperature for long time. Hadian and Zolfagharinasab (2007) reported that MeJA suppressed chilling injury and water loss and preserved external appearance in pomegranate fruits without abnormal effects on internal fruits quality.

Fan et al. (1998) reported that exogenous MeJA and JA promoted color change and improved the quality of 'Golden Delicious' and 'Fuji' apples (*Malus x domestica* Borkh.). González-Herranz et al. (2009) showed that the application of MeJA to grapes (*Vitis vinifera* L.) decreased fruit detachment force and promote the development of dry stem scars on the berries, both of which could improve the quality of machine-harvested raisin grapes. However, treatment with MeJA also promoted preharvest fruit drop, which is undesirable. They found that the optimization of the use of MeJA as a harvest aid for 'Thompson Seedless' requires application of between 2 and 10 mM MeJA followed by harvest within 3 days after treatment.

Tuber sprouting may be inhibited and/or melanization which occurs during processing may be reduced by exposure of the tubers to an effective amount of a jasmonate. With respect to the latter, treatment of tubers with jasmonate minimizes or controls the accumulation of reducing sugars, and hence reduces the development of undesirable dark pigments (melanization) which occurs during processing, such as during frying or cooking. By varying the amount of jasmonate applied to the tubers, either sprout inhibition or reduction of melanization (improvement of processed color quality) or both, may be preferentially selected. Exposure to low amounts of jasmonate provides greater sprout control but decreased improvement of processed color quality, while high amounts provide greater improvement in the processed color quality at the expense of poorer sprout inhibition. Exposure to intermediate amounts of jasmonate provides both tuber sprout control and processed color quality improvement (Patent 1995012311).

Transgenic fruits were used by Kausch et al. (2012) as a plant model to evaluate the effects of reduced endogenous MeJA on cellular metabolites in ripening tomato fruits. During on-shelf ripening, transgenic fruits were significantly reduced in the content of 19 out of 30 metabolites examined, including Ile, Val, Ala, Thr, Asn Tyr, Glu, Gln, His, Phe, Trp, GABA, citrate, succinate, myo-inositol, unidentified compound B, nucleic acid compound Nucl1, choline, and trigonelline as compared to the wild-type azygous counterparts. A significant increase in β-glucose levels in transgenic fruits was observed at the pink stage. The authors show that intracellular MeJA significantly regulates overall primary metabolism, especially aminome (amino acids and polyamines) of ripening fruits.

14.9 Conclusion and Future Perspectives

During the last few years the information on the nature of chemicals and the signaling pathways that they are involved against biotic and abiotic biotic stress factors has been generated. The mechanisms, by which the plants maintain their physiology and development under abiotic stress using different chemical signals at different times and in response to different stress conditions, and cross talk between each of them, still remained to be fully understood (Toteja and Sopory 2008).

Some research on horticultural uses of MeJA has focused on pre-harvest or post-harvest treatments to protect against microbial development on harvest tissues.

As biological control become more important and useful in horticultural crop production the used of jasmonates-induced defenses may provide valuable augmentation of integrated pest management strategies. Jasmonates have a crucial role in protecting ornamental and food crops from post-harvest disorders and diseases. Jasmonates may also enhance plant quality or to be used in propagation. Jasmonates can alter physiological processes in plants, to make plants more valuable for humans. Although only a few studies address these effects, the current ecological role of jasmonates remain very significant challenge. A major challenge for current biology is to integrate research approaches that address different levels of biological organization, from molecular to ecological systems.

Acknowledgements The author hereby thanks his colleagues Liliana Maslenkova, Tsonko Tsonev, Metodi Metodiev, and Zhivka Stoinova who have participated with their expert skills and helpful discussions through the numerous stages of the research over the years.

References

Aldridge DC, Galt S, Giles D, Turner WB (1971) Metabolites of Lasiodiplodia theobromae. J Chem Soc C Organ :1623–1627

Ali MB, Hah EJ, Paek YK (2007) Methyl jasmonate and salicylic acid induced oxidative stress and accumulation of phenolics in *Panax ginseng*. Bioreactor root suspension cultures. Mol 12(3):607–662

Alvarez S, Zhu M, Chen S (2009) Proteomics of *Arabidopsis* redox proteins in response to methyl jasmonates. J Proteomics 73(1):30–40

Andresen I, Becker W, Schlüter K, Burges J, Parthier B, Apel K (1992) The identification of leaf thionin as one of the main jasmonate-induced proteins of barley (*Hordeum vulgare*). Plant Mol Biol 19:193–204

Babst BA, Ferrieri RA, Gray DW, Lerdau M, Schlyer DJ, Schueller M, Thorpe MR, Orian CM (2005) Jasmonic acid induces rapid changes in carbon transport and partitioning in *Populus*. New Phytol 167(1):63–72

Barendse GWM, Croes AF, Vandenende G, Bosveld M, Creemers T (1985) Role of hormones on flower bud formation in thin-layer explants of Tobacco. Biol Plantar 27:408–412

Bonaventure G, Gfeller A, Proebsting WM, Hoerstensteiner S, Chételat A, Martinoia E, Farmer EE (2007) A gain of function allele of TPC1 activates oxylipin biogenesis after leaf wounding in *Arabidopsis*. Plant J 49:889–898

Brown AD, Goya IA, Larsen H, Lilley RMC (1987) A salt-sensitive mutant of *Dunaliella tertiolecta*. A role of carbonic anhydrase. Arch Microbiol 147:309–314

Chen H, Jones AD, Howe GA (2006a) Constitutive activation of the jasmonate signaling pathway enhances the production of secondary metabolites in tomato. FEBS Lett 580:2540–2546

Chen P, Yu S, Zhang Y (2006b) Effects of jasmonic acid and heat acclimation on thermotolerance and antioxidant enzymes of young grape plants. Life Sci Res 2006–03

Chini AS, Fonseca S, Fernández G, Adie B, Chico JM, Lorenzo O, García-Casado G, López-Vidriero I, Lozano FM, Ponce MR, Mico JL, Solano R (2007) The JAZ family of repressors is the missing link in jasmonate signalling. Nature 448:666–671

Cipollini D (2005) Interactive effects of lateral shading and jasmonic acid on morphology, physiology, seed production, and defense traits in *Arabidopsis thaliana*. Int J Plant Sci 166:955–959

Creelman RA, Mullet JE (1995) Jasmonic acid distribution and action in plants: regulation during development and response to biotic and abiotic stress. Proc Natl Acad Sci USA 92:4114–4119

Creelman RA, Mullet JE (1997) Biosynthesis and action of jasmonates in plants. Annu Rev Plant Physiol Plant Mol Biol 48:355–381

Czapski J, Saniewski M (1992) Stimulation of ethylene production and ethylene-forming enzyme activity in fruits of the non-ripening nor and rin tomato mutants by methyl jasmonate. J Plant Physiol 139:265–268

Darras AI, Terry LA, Joyce DC (2005) Methyl jasmonate vapour treatment suppresses specking caused by *Botrytis cinerea* on cut *Freesia hybrida* L. flowers. Postharvest Biol Technol 38(2):175–182

Dathe W, Rönsch H, Preiss A, Schade W (1981) Endogenous plant hormones of the broad bean, *Vicia faba* L. (−)-jasmonic acid, a plant growth inhibitor in pericarp. Planta 153:530–535

Demole E, Lederer E, Miercier DE (1962) Isolement et détermination de la structure du jasmonate de méthyle, constituant odorant charactéristique de l'essence de jasmin. Helv Chim Acta 45:675–685

Ding CK, Wang CY, Gross KC, Smith DL (2001) Reduction of chilling injury and transcript accumulation of heat shock proteins in tomato fruit by methyl jasmonate and methyl salicylate. Plant Sci 161:1153–1159

Dombrowski JE (2003) Salt stress activation of wound-induced gene expression in tomato plants. Plant Physiol 132(4):2098–2107

Fan XT, Mattheis JP, Fellman JK (1998) Responses of apples to postharvest jasmonate. J Am Soc Hortic Sci 123:421–425

Farmer EE, Weber H, Vollenweider S (1998) Fatty acid signaling in *Arabidopsis*. Planta 206:167–175

Fisher M, Gokhman I, Pick U, Zamir A (1996) A salt-resistant plasma membrane carbonic anhydrase is induced by salt in *Dunaliella salina*. J Biol Chem 271:17718–17723

Franceschi VR, Grimes HD (1991) Induction of soybean vegetative storage proteins and anthocyanins by low-level atmospheric methyl jasmonate. Proc Natl Acad Sci USA 88:6745–6749

Franceschi VR, Krekling T, Christiansen E (2002) Application of methyl jasmonate on *Picea abies (Pinaceae)* stems induces defense-related responses in phloem and xylem. Am J Bot 89:578–586

Fukui H, Koshimizu K, Usuda S, Yamazaki Y (1977) Isolation of plant growth regulators from seeds of *Cucurbita pepo* L. Agric Biol Chem 41:175–180

Gapper NE, Norris GE, Clarke SF, Lill RE, Jameson PE (2002) Novel jasmonate amino acid conjugates in *Asparagus officinalis* during harvest-induced and natural foliar senescence. Physiol Plant 114:116–124

Gehring CA, Irving HR, Mc Conchie R, Parish RW (1997) Jasmonates induce intracellular alkalization and closure of *Paphiopedilum* guard cell. Ann Bot 80:485–489

Gfeller A, Dubugnon L, Liechti R, Farmer EE (2010) Jasmonate biochemical pathway. Sci Signal 3:cm3

Gonzalez-Aguilar AB, Buta JG, Wang CY (2003) Methyl jasmonate and modified atmosphere packaging (MAP) reduce decay and maintain postharvest quality of papaya Sunrise. Postharvest Biol Technol 28:361–370

González-Aguilar GA, Tiznado-Hernández M, Wang CY (2006) Physiological and biochemical responses of horticultural products to methyl jasmonate. Stewart Postharvest Solut 2(1):1–9

González-Herranz R, Cathline KA, Fidelibus MW, Burns JK (2009) Potential of methyl jasmonate as a harvest aid for 'Thompson Seedless' grapes: concentration and time needed for consistent berry loosening. Hort Sci 44(5):1330–1333

Hadian J, Zolfagharinasab Z (2007) Influence of methyl jasmonate on inducing chilling tolerance in pomegranate fruits (*Malas Save*). Pak J Biol Sci 10:612–616

Hamberg M, Gardner HW (1992) Oxylipin pathway to jasmonates: biochemistry and biological significance. Biochim Biophys Acta 1165:1–18

Hause B, Demus U, Teichmann C, Parthier B, Wasternack C (1996) Developmental and tissue-specific expression of JIP-23, a jasmonate-inducible protein of barley. Plant Cell Physiol 37:641–649

Hause B, Kogel KH, Parthier B, Wasternack C (1997) In barley leaf cells, jasmonates do not act as a signal during compatible or incompatible interactions with the powdery mildew fungus (*Erisiphe graminis f.*, sp. Hordei). J Plant Physiol 150:127–132

Hause B, Maier W, Miersch O, Kramell R, Strack D (2002) Induction of jasmonate biosynthesis in *Arbuscular Mycorrhizal* barley roots. Plant Physiol 130:1213–1220

Heijari J, Nerg AM, Kainulainen P, Viiri H, Vuorinen M, Holopainen JK (2005) Application of methyl jasmonate reduces growth but increases chemical defence and resistance against *Hyloblus abietis* in Scots pine seedlings. In: Proceedings of 12th international symposium of insect-plant relationships 11, Berlin, 7–14 Aug 2004, p 283

Hristova VA, Popova LP (2002) Treatment with methyl jasmonate alleviates the effects of paraquat on photosynthesis in barley plants. Photosynthetica 40(4):567–574

Huang Y, Han C, Peng W, Peng Z, Xiong X, Zhu Q, Gao B, Xie D, Ren C (2010) Brassinosteroid negatively regulates jasmonate inhibition of root growth in *Arabidopsis*. Plant Signal Behav 5:140–142

Jubany-Marí T, Prinsen E, Munné-Bosch S, Alegre L (2010) The timing of methyl jasmonate, hydrogen peroxide and ascorbate accumulation during water deficit and subsequent recovery in the Mediterranean shrub *Cistus albidus* L. Environ Exp Bot 69:147–155

Kang DJ, Seo YJ, Lee JD, Ishii R, Kim KU, Shin DH, Park SK, Jang SW, Lee IJ (2005) Jasmonic acid differentially affects growth, ion uptake and abscisic acid concentration in salt-tolerant and salt-sensitive rice cultivars. J Agron Crop Sci 191(4):273–282

Kausch KD, Sobolev AP, Goyal RK, Fatima T, Beevi LR, Saftner RA, Handa AK, Matoo AK (2012) Methyl jasmonate deficiency alters cellular metabolome, including the aminome of tomato (*Solanum lycopersicum* L.) fruits. Amino Acids 42:843–885

Kauss H, Jublick W, Ziegler J, Krabler W (1994) Pretreatment of parsley (*Petroselinum crispum* L.) suspension cultures with methyl jasmonate enhances elicitation of activated oxygen species. Plant Physiol 105:89–94

Kępczyńska E, Paulina Król P (2011) The phytohormone methyl jasmonate as an activator of induced resistance against the necrotroph *Alternaria porri* f. sp. *solani* in tomato plants. J Plant Interac. doi:10.1080/17429145.2011.645169

Keramat B, Manouchehri KK, Arvin MJ (2010) Effect of methyl jasmonate treatment on alleviation of cadmium damages in soybean. J Plant Nutr 33(7):1016–1025

Kęsy J, Wilmowicz E, Maciejewska B, Frankowski K, Glazińska P, Kopcewicz J (2011) Independent effects of jasmonates and ethylene on inhibition of *Pharbitis nil* flowering. Acta Physiol Plant 33:1211–1216

Kim EH, Park SH, Kim JK (2009) Methyl jasmonate triggers loss of grain yield under drought stress. Plant Signal Behav 4(4):348–349

Kiribuchi K, Jikumaru Y, Kaku H, Minami E, Hasegawa M, Kodama O, Seto H, Okada K, Nojiri H, Yamane H (2005) Involvement of the basic helix-loop-helix transcription factor RERJ1 in wounding and drought stress responses in rice plants. Biosci Biotechnol Biochem 69:1042–1044

Knöfel HD, Sembdner G (1995) Jasmonates from pine pollen. Phytochemistry 38:569–571

Koda Y (1992) The role of jasmonic acid and related compounds in the regulation of plant development. Int Rev Cytol 135:155–199

Koda Y (1997) Possible involvement of jasmonates in various morphogenic events. Physiol Plant 100(3):639–646

Krajncic B, Kristl J, Janzekovic I (2006) Possible role of jasmonic acid in the regulation of floral induction, evocation and floral differentiation in *Lemna minor* L. Plant Physiol Biochem 44:752–758

Kramell R, Miersch O, Atzorn R, Parthier B, Wasternack C (2000) Octadecanoid-derived alteration of gene expression and the "oxylipin signature" in stressed barley leaves: implications for different signaling pathways. Plant Physiol 123:177–186

Krzyzanowska J, Czubacka A, Pecio L, Przybys M, Doroszewska T, Stochmal A, Oleszek W (2011) The effects of jasmonic acid and methyl jasmonate on rosmarinic acid production in *Mentha × piperita* cell suspension cultures. Plant Cell Tissue Org Cult 108:73–81

Lannoo N, Peumans WP, Van Damme EJM (2006) The presence of jasmonate-inducible lectin genes in some but not all *Nicotiana* species explains a marked intragenus difference in plant responses to hormone treatment. J Exp Bot 57(12):3145–3155

Larcher W (1995) Physiological plant ecology, 3rd edn. Springer, Berlin/Heidelberg/New York, pp 46–54

Latorella AH, Vadas RL (1973) Salinity adaptation by *Dunaliella tertiolecta*. I. Increases in carbonic anhydrase activity and evidence for a light-dependent Na^+/H^+ exchange. J Phycol 9:273–277

Lee TM, Lur HS, Lin YH, Chu C (1996) Physiological and biochemical changes related to methyl jasmonate-induced chilling tolerance of rice (*Oryza sativa* L.) seedlings. Plant Cell Environ 19:65–74

Lehmann J, Atzorn R, Brückner C, Reinbothe S, Leopold J, Wasternack C, Parthier B (1995) Accumulation of jasmonate, abscisic acid, specific transcripts and proteins in osmotically stressed barley leaf segments. Planta 197:156–162

Liechti R, Farmer EE (2006) Jasmonate biochemical pathway. Sci STKE 322:1–3

Lopez R, Dathe W, Bruckner C, Miersch O, Sembdner G (1987) Jasmonic acid in different parts of the developing soybean fruit. Biochem Physiol Pflanzen 182:195–201

Maciejewska BD, Kesy J, Zielinska M, Kopcewicz J (2004) Jasmonates inhibit flowering in short-day plant *Pharbitis nil*. Plant Growth Regul 43:1–8

Maksymiec W, Wianowska D, Dawidowicz AL, Radkiewicz S, Mardarowicz M, Krupa Z (2005) The level of jasmonic acid in *Arabidopsis thaliana* and *Phaseolus coccineus* plants under heavy metal stress. J Plant Physiol 162(12):1338–1346

Maksymiek W, Krupa Z (2002) Jasmonic acid and heavy metals in *Arabidopsis* plants: a similar physiological response to both stressors? J Plant Physiol 159(5):509–515

Maksymiek W, Krupa Z (2006) The effects of short-term exposition to Cd, excess Cu ions and jasmonate on oxidative stress appearing in *Arabidopsis thaliana*. Envirn Exp Bot 57(1–2):187–194

Maksymiek W, Krupa Z (2007) Effect of methyl jasmonate and excess cooper on leaf and root growth. Biol Plant 51(2):321–332

Mansour N, Mimi Z, Harb J (2008) Stress imposed on broad bean (*Vicia faba)* plants irrigated with reclaimed wastewater mixed with brackish water through exogenous application of jasmonic acid. In: Al Baz I et al (eds) Efficient management of wastewater. Springer, Berlin/Heidelberg, pp 91–102

Maslenkova LT, Zanev Y, Popova LP (1990) Oxygen-evolving activity of thylakoids from barley plants cultivated on different concentrations of jasmonic acid. Plant Physiol 93:1316–1321

Maslenkova LT, Miteva TS, Popova LP (1992) Changes in the polypeptide patterns of barley seedlings exposed to jasmonic acid and salinity. Plant Physiol 98:700–707

Maslenkova L, Toncheva S, Zeinalov Y (1995) Effect of abscisic acid and jasmonic acid (or MeJA) on photosynthetic electron transport and oxygen evolving reactions in pea plants. Bulg J Plant Physiol 21(4):48–55

Mathew R, Sankar PD (2012) Effect of methyl jasmonate and chitosan on growth characteristics of *Ocimum basilicum* L., *Ocimum sanctum* L. and *Ocimum gratissimum* L. cell suspension cultures. Afr J Biotechnol 11:4759–4766

Maucher H, Hause B, Ziegler J, Wasternack C (2000) Allene oxidase syntheses of barley-tissues specific regulation in seedling development. Plant J 21:199–213

Mei C, Qi M, Sheng G, Yang Y (2006) Inducible overexpression of a rice allene oxide synthase gene increases the endogenous jasmonic acid level, *PR* gene expression, and host resistance to fungal infection. Mol Plant Microbe Interact 19:1127–1137

Metodiev MV, Tsonev TD, Popova LP (1996) Effect of jasmonic acid on the stomatal and nonstomatal limitation of leaf photosynthesis in barley leaves. J Plant Growth Regul 15:75–80

Meyer M, Miersch O, Buttner C, Dathe W, Sembdner G (1984) Occurrence of the plant growth regulator jasmonic acid in plants. J Plant Growth Regul 3:1–8

Miersch O, Sembdner G, Schreiber K (1989) Occurrence of jasmonic acid analogues in *Vicia faba*. Phytochemistry 28:339–340

Moons A, Prinsen E, Bauw G, Montagu MV (1997) Antagonistic effects of abscisic acid and jasmonates on salt stress-inducible transcripts in rice roots. Plant Cell 9:2243–2259

Mueller-Uri F, Parthier B, Nover L (1988) Jasmonates induced alterations in gene expression in barley leaf segments analyzed *by* in vivo and in vitro protein synthesis. Planta 176:241–247

Munns R (2005) Genes and salt tolerance: bringing them together. New Phytol 3:645–663

Nimitkeatkai H, Shishido M, Okawa K, Ohara H, Ban Y, Kita M, Moriguchi T, Ikeura H, Hayata Y, Kondo S (2011) Effect of jasmonates on ethylene biosynthesis and aroma volatile emission in Japanese apricot infected by a pathogen (*Colletotrichum gloeosporioides*). J Agric Food Chem 59(12):6423–6429

Norastehnia A, Asghari MN (2006) Effects of methyl jasmonate on the enzymatic antioxidant defense system in maize seedlings subjected to paraquat. Asian J Plant Sci 5:17–23

Orozco-Cárdenas S, Ryan CA (1999) Hydrogen peroxide is generated systemically in plant leaves by wounding and systemin via the octadecanoic pathway. Proc Nat Acad Sci USA 96:6553–6557

Orozco-Cárdenas ML, Narváez-Vásquez J, Ryan CA (2001) Hydrogen peroxide acts as a second messenger for the induction of defense genes in tomato plants in response to wounding, systemin, and methyl jasmonate. Plant Cell 13:179–191

Ortel B, Atzorn R, Hause B, Feussner I, Miersh O, Wasternack C (1999) Jasmonate-induced gene expression in barley (*Hordeum vulgare*) leaves, the link between jasmonate and abscisic acid. Plant Growth Regul 29:113–122

Pan RC, Dou ZJ, Ye QS (1995) Effect of methyl jasmonate on SOD activity and membrane lipid peroxidation in peanut seedlings during water stress. Acta Phytophysiol Sinica 21(3):221–228

Parra-Lobato MC, Garcia NF, Olmos E, Alvarez-Tinaut AC, Jimenez MCG (2009) Methyl jasmonate-induced antioxidant defence in root apoplast from sunflower seedlings. Environ Exp Bot 66(1):9–17

Parthier B (1989) Hormone-induced alterations in gene expression. Physiol Pflanzen 185:289–314

Parthier B (1991) Jasmonates, new regulators of plant growth and development: many facts and few hypothesis of their action. Bot Acta 104:446–454

Parthier B, Bruckner C, Dathe W, Hause B, Herrmann HD, Knofel HM, Kramell J, Lehmann O, Miersch S, Reinbote G, Sembdner S, Vasternack U, Nieden Z (1992) Jasmonates: metabolism, biological activities and mode of action in senescence and stress responses. In: Regul Gr, Karssen CV, van Lon LC, Vreugdenhil D (eds) Progress in plant. Kluwer, Dordrecht, pp 276–285

Patent: 1995012311, Natural suppression of sprouting in stored potatoes using jasmonates

Pauwels L, Morreel K, Witte ED, Lammertyn F, Montagu MV, Boerjan W, Inzé D, Goossens A (2008) Mapping methyl jasmonate-mediated transcriptional reprogramming of metabolism and cell cycle progression in cultured *Arabidopsis* cells. Proc Natl Acad Sci USA 29:1380–1385

Pedranzani H, Racagni G, Alemano S, Miersch O, Ramírez I, Peña-Cortés H, Taleisnik E, Machado-Domenech E, Abdala G (2003) Salt tolerant tomato plants show increased levels of jasmonic acid. Plant Growth Regul 41:149–158

Pedranzani H, Sierra-de-Grado R, Vigliocco A, Miersch O, Abdala G (2007) Cold and water stresses produce changes in endogenous jasmonates in two populations of *Pinus pinaster* Ait. Plant Growth Regul 52:111–116

Pijaotrowska A, Bajgus A, Godlewska B, Caerpak Kaminska MR (2009) Jasmonic acid as modulator of lead toxicity in aquatic plant *Walfia arrhiza (Lemnaceae)*. Environ Exp Bot 63(3):507–513

Popova LP, Maslenkova LT (1997) Involvement of jasmonic acid in photosynthetic process in *Hordeum vulgare* L. during salinity stress. Recent Res Devel Plant Physiol 1:29–43

Popova LP, Uzunova AN (1996) Changes in the chloroplasts ultrastructure of barley leaves under treatment with jasmonic acid. Photosynthetica 32(4):635–639

Popova LP, Vaklinova SG (1988) Effect of jasmonic acid on the synthesis of ribulose-1,5-bisphosphate carboxylase-oxygenase in barley leaves. J Plant Physiol 133:210–215

Popova LP, Tsonev TD, Valklinova SG (1987) A possible role for abscisic acid in regulation of photosynthetic and photorespiratory carbon metabolism in barley leaves. Plant Physiol 83:824–828

Popova LP, Tsonev TD, Vaklinova SG (1988) Changes in some photosynthetic and photorespiratory properties in barley leaves after treatment with jasmonic acid. J Physiol Plant 69:161–166

Popova LP, Lazova GH, Miteva TS (1991) Carbonic anhydrase activity in barley leaves after treatment with abscisic acid and jasmonic acid. Com Rend ABS 44(5):51–54

Pozo MJ, van Loon LC, Pieterse MJ (2004) Jasmonates: signals in plant-microbe interactions. J Plant Growth Regul 23:211–222

Radhika V, Kost C, Bolland W, Heil M (2010) The role of jasmonates on floral nectar secretion. PLoS One 5:e9265

Raghavendra AS, Reddy KB (1987) Action of proline on stomata differs from that of abscisic acid, G-substances or methyl jasmonate. Plant Physiol 44:691–695

Rakwal R, Tamogami S, Agrawal GK, Iwahashi H (2002) Octadecanoid signaling component "burst" in rice (*Oryza sativa* L.) seedling leaves upon wounding by cut and treatment with fungal elicitor chitosan. Biochem Biophys Res Commun 295:1041–1045

Reinbothe S, Reinbothe C, Parthier B (1992) Differential accumulation of methyl jasmonate-induced mRNAs in response to abscisic acid and desiccation in barley (Hordeum vulgare). Physiol plant 86:49–56

Reinbothe S, Reinbothe C, Parthier B (1993) Methyl jasmonate represses translation initiation of a specific set of m RNAs in barley. Plant J 4:459–467

Reinbothe S, Reinbothe C, Lehmann J, Becker W, Apel K, Parthier B (1994a) JIP60, a methyl jasmonate-induced ribosome-inactivating protein involved in plant stress reactions. Proc Nat Acad Sci USA 91:7012–7016

Reinbothe S, Mollenhauer B, Reinbothe C (1994b) JIPs and RIPs: the regulation of plant gene expression by jasmonates in response to environmental cues and pathogens. Plant Cell 6:1197–1209

Reymond P, Farmer EE (1998) Jasmonate and salicylate as global signals for defense gene expression. Curr Opin Plant Biol 1:404–411

Robert-Seilaniantz A, Grant M, Jones JDG (2011) Hormone crosstalk in plant disease and defense: more than just jasmonate-salicylate antagonism. Annu Rev Phytopatol 49:317–343

Ryan CA (2000) The systemin signaling pathway: differential activation of plant defensive genes. Biochim Biophys Acta 1477:112–121

Saniewski M, Miszczak A, Kawa-Miszczak L, Wegrzynowicz-Lesiak E, Miyamoto K, Ueda J (1998) Effects of methyl jasmonate on anthocyanin accumulation, ethylene production, and CO_2 evolution in uncooled and cooled tulip bulbs. J Plant Growth Regul 17:33–37

Saniewski A, Horbowicz M, Puchalski J, Ueda J (2003) Methyl jasmonate stimulates the formation and the accumulation of anthocyanin in *Kalanchoe blossfeldiana*. Acta Physiol Plant 25:143–149

Saniewski A, Horbowicz M, Puchalski J (2006) Induction of anthocyanins accumulation by methyl jasmonate in shoots of *Crassula multicava* Lam. Acta Agrobot 59:43–50

Sasaki-Sekimoto Y, Taki N, Obayashi T, Aono M, Matsumoto F, Sakurai N, Suzuki H, Hirai MY, Noji M, Saito K, Masuda T, Takamiya K, Shibata D, Ohta H (2005) Coordinated activation of metabolic pathways for antioxidants and defence compounds by jasmonates and their roles in stress tolerance in *Arabidopsis*. Plant J 44:653–668

See KS, Bhatt A, Keng CL (2011) Effect of sucrose and methyl jasmonate on biomass and anthocyanin production in cell suspension culture of *Melastoma malabathricum* (*Melastomaceae*). Rev Biol Trop 59(2):597–606

Sembdner G, Gross D (1986) Plant growth substances of plant and microbial origin. In: Bopp M (ed) Plant growth substances 1985. Springer, Berlin, pp 139–147

Sembdner G, Parthier B (1993) The biochemistry and the physiological and molecular actions of jasmonates. Annu Rev Plant Physiol Plant Mol Biol 44:569–589

Seo HS, Song JT, Cheong JJ, Lee H, Lee YW, Hwang I, Lee JS, Choi YD (2001) Jasmonic acid carboxyl methyltransferase: a key enzyme for jasmonate-regulated plant response. Proc Natl Acad Sci USA 98:4788–4793

Sevillano L, Sanchez-Ballestra MT, Romojaro F, Flores FB (2010) Physiological, hormonal and molecular mechanisms regulating chilling injury in horticultural species. Postharvest technologies applied to reduce its impact. J Sci Food Agric 89:555–573

Shahzad AN, Pollman S, Schubert S (2009) Does jasmonic acid control the maize shoot growth during the first phase of salt stress? In: Proceedings of the international plant nutrition colloquium XVI, Department of Plant Sciences, UC Davis, pp 26–32

Stamp N (2003) Out of the quagmire of plant defense hypothesis. Q Rev Biol 78:23–55

Staswick PE, Tiryaki I (2004) The oxylipin signal jasmonic acid is activated by an enzyme that conjugates it to isoleucine in *Arabidopsis*. Plant Cell 16:2117–2127

Staswick PE, Su W, Howell SH (1992) Methyl jasmonate inhibition of root growth and induction of a leaf protein are decreased in a *Arabidopsis thaliana* mutants. Proc Natl Acad Sci USA 89:6837–6840

Suhita D, Kolla VA, Vavasseur A, Raghavendra AS (2003) Different signaling pathways involved during the suppression of stomatal opening by methyl jasmonate or abscisic acid. Plant Sci 164:481–488

Suhita D, Agepati S, Raghavendra J, Kwak JM, Vavasseur A (2004) Cytoplasmic alkalinization precedes reactive oxygen species production during methyl jasmonate- and abscisic acid-induced stomatal closure. Plant Physiol 134:1536–1545

Suza WP, Avila CA, Carruthers K, Kulkarni S, Goggin FL, Lorence A (2010) Exploring the impact of wounding and jasmonates on ascorbate metabolism. Plant Physiol Biochem 48(5):337–350

Tamari G, Borochov A, Atzorn R, Weiss D (1995) Methyl jasmonate induces pigmentation and flavonoid gene expression in petunia corollas: in possible role in wound response. Physiol Plant 94:45–50

Tani T, Sobajima H, Okada K, Chujo T, Arimura S, Tsutumi N, Nishimura M, Seto H, Nojiri H, Yamane H (2008) Identification of the OsOPR7 gene encoding 12-oxophytodienoate reductase involved in the biosynthesis of jasmonic acid in rice. Planta 227:517–526

Thines B, Katsir L, Melotto M, Niu Y (2007) JAZ repressor proteins are targets of the SCF (COI1) complex during jasmonate signalling. Nature 448:661–665

Thomma BPHJ, Eggermont K, Penninckk IAMA, Mauch-Mani B, Vogelsang R, Cammue BPA, Broekaert WF (1998) Separate jasmonate-dependent and salicylate-dependent defense-response pathways in *Arabidopsis* are essential for resistance to distinct microbial pathogens. Proc Natl Acad Sci USA 95:15107–15111

Toteja N, Sopory SK (2008) Chemical signaling under abiotic stress environment in plants. Plant Signal Behav 3(8):525–536

Truman W, Bennet MH, Kubigstelting I, Turnbull C, Grant M (2007) Arabidopsis systemic immunity uses conserved defense signaling pathways and is mediated by jasmonates. Proc Natl Acad Sci USA 104(3):1075–1080

Tsonev TD, Lazova GN, Stoinova ZhG, Popova LP (1998) A possible role for jasmonic in adaptation of barley seedlings to salinity stress. J Plant Growth Regul 17:153–159

Ueda J, Kato J (1980) Isolation and identification of a senescence-promoting substance from wormwood (*Artemisia absinthium* L.). Plant Physiol 66:246–249

Vick BA, Zimmerman DC (1984) Biosynthesis of jasmonic acid by several plant species. Plant Physiol 75:458–461

Vick BA, Zimmerman DC (1987) Oxidative systems for modifications of fatty acids: the lipoxygenase pathway. In: Stumpf PK, Con EE (eds) The biochemistry of plants: a comprehensive treatise, vol 9. Academic, Orlando, pp 53–90

Vidhyavathi R, Sarada R (2011) Effect of salicylic acid and methyl jasmonate on antioxidant systems of *Haematococcus pluvialis*. Acta Physiol Plant 33(3):1043–1049

Walia H, Wilson C, Wahid A, Condamine P, Cui X, Close TJ (2006) Expression analysis of barley (*Hordeum vulgare* L.) during salinity stress. Funct Integr Genomics 6:143–156

Walia H, Wilson C, Condamine P, Liu X, Ismail AM, Close TJ (2007) Large-scale expression profiling and physiological characterization of jasmonic acid-mediated adaptation of barley to salinity stress. Plant Cell Environ 30:410–421

Wang SY (1999) Methyl Jasmonates reduces water stress in strawberry. J Plant Growth Regul 18:127–134

Wang CY, Buta JG (1994) Methyl jasmonate reduces chilling injury in *Cucurbita pepo* through its regulation of abscisic acid and polyamine levels. Environ Exp Bot 34:427–432

Wang JW, Wu JY (2005) Nitric oxide is involved in methyl jasmonate-Induced defense responses and secondary metabolism activities of *Taxus* cells. Plant Cell Physiol 46:923–930

Wang Y, Mopper S, Hasenstein KH (2001) Effects of salinity on endogenous ABA, IAA, JA, and SA in *Iris hexagona*. J Chem Ecol 27:327–342

Wasternack C, Hause B (2002) Jasmonates and octadecanoids: signals in plant stress responses and development. Prog Nucleic Acids Res Mol Biol 72:165–221

Wasternack C, Parthier B (1997) Jasmonate-signalled plant gene expression. Trends Plant Sci 2:302–307

Weidhase RA, Lehmann J, Kramell HN, Sembdner G, Parthier B (1987) Degradation of ribulose-1, 5-bisphosphate carboxylase and chlorophyll in senescing barley leaf segments triggered by jasmonic acid and methyl ester and counteraction by cytokinin. Physiol Plant 69:161–166

Weiler EW, Albrecht T, Groth B, Xia ZQ, Luxem M, Li H, Andert L, Spengler P (1993) Evidence for the involvement of jasmonates and their octadecanoid precursors in the tendril coiling response of *Bryonia dioica*. Phytochemistry 32:591–600

Wilson C (2007) Effect of jasmonic acid on growth and ion relations of *Oryza sativa* L. grown under salinity stress. In: American Society of Agronomy Meetings, Paper number 3, pp 13–14

Xiang BB, Zhu YR, Wang WJ, Bai YL, Wang Y (2011) Influence of methyl jasmonate on cell membrane permeability and ajmalicine accumulation in salt-stressed *Catharanthus roseus* suspension cells. In: International conference on Bioinformatics and Biomedical Engineering, (iCBBE) 2011, Wuhan, 10–12 May 2011, pp 1–4

Yamane H, Abe H, Takahashi N (1982) Jasmonic acid and methyl jasmonate in pollens and anthers of three *Camellia* species. Plant Cell Physiol 23:1125–1127

Yoon JY, Hamayun M, Lee SKIJ (2009) Methyl Jasmonate alleviated salinity stress in soybean. J Crop Sci Biotechnol 12(2):63–68

Yoshihara T, Omer ESA, Koshino H, Sakamura S, Kikuta Y, Koda Y (1989) Structure of a tuber-inducing stimulus from potato leaves (*Solanum tuberosum* L.). Agric Biol Chem 53:2835–2837

Zeinalov Yu (1982) Existence of two different ways for oxygen evolution in photosynthesis and photosynthetic unit concept. Photosynthetica 16:27–35

Chapter 15
Role of Nitric Oxide in Improving Plant Resistance Against Salt Stress

Muhammad Farooq, Kadambot H.M. Siddique, and Sven Schubert

15.1 Introduction

Soil salinity is a serious threat to global crop production (Zhu 2001). More than 20% of cultivated land is affected by salinity worldwide (Zhu 2001; Rengasamy 2006) and owing to climate change, the area under salinity is expected to increase (Wassmann et al. 2009).

Salinity imposes both osmotic stress and ionic toxicity to plants disturbing the activities of cytosolic enzymes thereby causing nutritional disorders (Hernández et al. 1993, 1995; Serrano and Rodriguez 2002; Zhu 2003; Valderrama et al. 2006) and oxidative damage (Xiong and Zhu 2002; Xiong et al. 2002). Overproduction of reactive oxygen species (ROS) such as hydroxyl radical ($^{\bullet}OH$), single oxygen (O_2^{1}), superoxide (O_2^{-}), hydrogen peroxide (H_2O_2) and alkoxy radicals (RO) under salinity cause oxidative damage (Xiong and Zhu 2002; Xiong et al. 2002).

M. Farooq
Institute of Plant Nutrition, Justus-Liebig-University,
Heinrich-Buff-Ring 26-32, D-35392 Giessen, Germany

Department of Agronomy, University of Agriculture, Faisalabad 38040, Pakistan

The UWA Institute of Agriculture, The University of Western Australia,
Crawley, WA 6009, Australia

K.H.M. Siddique (✉)
The UWA Institute of Agriculture, The University of Western Australia,
Crawley, WA 6009, Australia
e-mail: kadambot.siddique@uwa.edu.au

S. Schubert
Institute of Plant Nutrition, Justus-Liebig-University,
Heinrich-Buff-Ring 26-32, D-35392 Giessen, Germany

P. Ahmad et al. (eds.), *Ecophysiology and Responses of Plants under Salt Stress*,
DOI 10.1007/978-1-4614-4747-4_15,

Nitric oxide (NO), an important bioactive molecule, is involved in the regulation of various physiological and biochemical processes in plants such as seed germination (Kopyra and Gwóźdź 2003; Arasimowicz and Floryszak-Wieczorek 2007), plant growth (Durner and Klessig 1999; Arasimowicz and Floryszak-Wieczorek 2007), stomatal oscillations (Neill et al. 2002; Bright et al. 2006), plant maturation and senescence (Leshem et al. 1998; Guo and Crawford 2005; Arasimowicz and Floryszak-Wieczorek 2007).

Nitric oxide is synthesized through both enzymatic and non-enzymatic reactions (Bethke et al. 2004), however its synthesis increases substantially under salinity. For instance, upon exposure to salt stress, endogenous level of NO was significantly increased in olive (*Olea europaea* L.; Valderrama et al. 2007), sunflower (*Helianthus annuus* L.; David et al. 2010) and Arabidopsis (*Arabidopsis thaliana* (L.) Heynh; Zhao et al. 2007a; Qiao et al. 2009). Through its involvement in the electron transport chain during respiration (Zottini et al. 2002), NO modulates the antioxidant defense system scavenging ROS under salinity in plants (Table 15.1; Kopyra and Gwóźdź 2003; Molassiotis et al. 2010). NO also helps to maintain a high K^+/Na^+ ratio in the cytosol (Table 15.1; Ruan et al. 2004; Siddiqui et al. 2011) through increased expression of plasma membrane and/or tonoplast H^+-ATPase and H^+-PPase under salinity (Zhao et al. 2004; Shi et al. 2007). Furthermore, NO triggers accumulation of organic osmolytes and compatible solutes, like proline, soluble sugars (Table 15.1; Guo et al. 2009), so as to maintain cell turgor and regulate water acquisition.

In this chapter, the role of NO in improving resistance against salinity in plants is being discussed through different methods.

15.2 Osmoregulation and Osmoprotection

Salinity causes osmotic stress which affects tissue water contents and water uptake in plants. Seed germination is the first step in the ontogeny of plants. Germination starts by water imbibition and ends with radical protrusion through resumption of metabolism (Bewleyl 1997). Owing to salinity-induced osmotic stress, seed germination significantly decreases under salinity (Misra and Dwivedi 2004) leading to poor crop stand establishment (Almansouri et al. 2001). However, NO application significantly increases germination and stand establishment in several crops under salinity (Table 15.2; Kopyra and Gwóźdź 2003; Li et al. 2005; Zheng et al. 2009).

In yellow lupin (*Lupinus luteus* L.), salinity significantly suppressed germination, nonetheless seed treatment with sodium nitroprusside (SNP), the NO donor, triggered germination under salinity (Kopyra and Gwóźdź 2003). Similarly, in the succulent shrub suaeda (*Suaeda salsa* (L.) Pall.), a halophyte, NO stimulated seed germination under salinity stress (Li et al. 2005). Pre-sowing seed treatment with 0.1 and 0.2 mM NO significantly improved the germination of rice (*Oryza sativa* L.) genotypes under salinity (Habib et al. 2010). Likewise in wheat (*Triticum aestivum* L.),

Table 15.1 Mechanisms of nitric-oxide-induced salinity resistance in different plant species

NO-mediated effect	Plant species	References
Increased expression of tonoplast H^+-ATPase and Na^+/H^+ antiporter gene	Maize	Zhang et al. (2006, 2007a)
Increased germination rate and root growth	Lupin	Kopyra and Gwóźdź (2003)
Survival of more green leaf tissue, and increased quantum yield for photosystem II	Rice	Uchida et al. (2002)
Expression of plasma membrane H^+-ATPase resulting in increased K^+/Na^+ ratio	Reed	Zhao et al. (2004)
Decreased Na^+ accumulation in cytosol	Maize	Zhang et al. (2004)
Increased K^+ content in cytosol	Wheat	Ruan et al. (2004)
Decreased Na^+ content with simultaneous increase in K^+ content	Seashore mallow	Guo et al. (2009)
Increased activity of catalase, superoxide dismutase, peroxidase and ascorbate peroxidase	Seashore mallow	Guo et al. (2009)
Increased activity of enzymatic (superoxide dismutase, guaiacol peroxidase, catalase, ascorbate peroxidase) and non-enzymatic (ascorbate, reduced glutathione) antioxidants	Tomato	Wu et al. (2011)
Increased proline accumulation	Seashore mallow	Guo et al. (2009)
	Wheat	Ruan et al. (2002)
	Tomato	Wu et al. (2011)
Increased accumulation of soluble sugars	Tomato	Wu et al. (2011)
Increased accumulation of putrescine, spermine and spermidine	Cucumber	Fan et al. (2012)
Increased activity of catalase, superoxide dismutase, dehydroascorbate reductase and ascorbate peroxidase, guaiacol peroxidase and glutathione reductase, and increased expression of plasma membrane and tonoplast H^+-ATPase and H^+-PPase	Cucumber	Shi et al. (2007)
Increased activity of catalase and superoxide dismutase, decrease in contents of malondialdehyde and hydrogen peroxide, and increased K^+/Na^+ ratio	Wheat	Zheng et al. (2009)
Increased activity of superoxide dismutase, peroxidase and ascorbate peroxidase and proline accumulation, and decrease in contents of malondialdehyde	Mustard	Zeng et al. (2011)

through improved water imbibition and metabolism, seed soaking in SNP solution increased the germination rate as well as coleoptile and radicle growth (Zheng et al. 2009). Apart from the accumulation of ions in vacuoles, plants synthesize low-molecular-mass organic compounds, the compatible solutes, which help to regulate turgor and water acquisition under osmotic stress (Yancey et al. 1982; Bohnert et al. 1995; Hasegawa et al. 2000). Several osmolytes including glycine-betaine, sugar alcohols, soluble sugars, proline, trehalose, polyols, etc., have been

Table 15.2 Influence of externally-applied nitric oxide on germination, and some morphological characteristics under salinity in different plant species

Plant species	Salinity level	NO treatment	Increase over salinity control	Reference
Yellow lupin	100 mM NaCl	100 μM SNP	88% increase in germination	Kopyra and Gwóźdź (2003)
Cucumber	100 mM NaCl	50 μM SNP	52% increase in shoot dry weight	Shi et al. (2007)
Cucumber	100 mM NaCl	50 μM SNP	48% increase in root dry weight	Shi et al. (2007)
Suaeda (brown seeds)	800 mM NaCl	50 μM SNP	62% increase in germination	Li et al. (2005)
Suaeda (brown seeds)	800 mM NaCl	200 μM SNP	27% increase in germination	Li et al. (2005)
Suaeda (black seeds)	400 mM NaCl	50 μM SNP	16% increase in germination	Li et al. (2005)
Suaeda (black seeds)	400 mM NaCl	200 μM SNP	82% increase in germination	Li et al. (2005)
Wheat	300 mM NaCl	100 μM SNP	20% increase in germination	Zheng et al. (2009)
Wheat	300 mM NaCl	100 μM SNP	18% increase in coleoptiles dry weight	Zheng et al. (2009)
Tomato (cv. Hufan148)	100 mM NaCl	100 μM SNP	11% increase in shoot dry weight	Wu et al. (2011)
Tomato (cv. Hufan2496)	100 mM NaCl	100 μM SNP	31% increase in shoot dry weight	Wu et al. (2011)
Tomato (cv. Hufan148)	100 mM NaCl	100 μM SNP	13% increase in root dry weight	Wu et al. (2011)
Tomato (cv. Hufan2496)	100 mM NaCl	100 μM SNP	40% increase in root dry weight	Wu et al. (2011)
Rice	100 mM NaCl	1 μM SNP	21% increase in surviving green parts	Uchida et al. (2002)
Rice	100 mM NaCl	10 μM SNP	17% increase in surviving green parts	Uchida et al. (2002)
Cucumber	50 mM NaCl	100 μM SNP	41% increase in plant height	Fan et al. (2012)
Cucumber	50 mM NaCl	100 μM SNP	67% increase in plant fresh weight	Fan et al. (2012)
Cucumber	50 mM NaCl	100 μM SNP	50% increase in plant dry weight	Fan et al. (2012)
Cucumber	100 mM NaCl	50 μM SNP	50% increase in shoot dry weight	Shi et al. (2007)
Cucumber	100 mM NaCl	50 μM SNP	33% increase in root dry weight	Shi et al. (2007)
Mustard	150 mM NaCl	100 μM SNP	51% increase in relative water contents	Zeng et al. (2011)

reported to accumulate in various plant species under salinity and drought (Yancey et al. 1982; Bohnert et al. 1995; Hasegawa et al. 2000; Farooq et al. 2009). In addition to their role in the maintenance of water balance in plant tissues, these osmolytes also act as osmoprotectants; for instance, proline scavenges free radicals (Chen and Murata 2000).

NO stimulates cytosolic synthesis of proline; for example, exogenous application of SNP significantly increased cytosolic proline accumulation in seashore mallow (*Kosteletzkya virginica* L.), conferring salinity resistance (Guo et al. 2009). Moreover NO application have found to increase proline accumulation in wheat, that scavenges ROS and stabilize the structure of the macromolecule (Ruan et al. 2002). Likewise in tomato (*Lycopersicom esculentum* Mill.), same treatment has shown to improve the accumulation of proline as well as soluble sugars under salt stress (Wu et al. 2011). Another study on tomato showed that exogenous NO significantly ameliorated the salinity induced decrease in photosynthetic pigments rate of photosynthesis (Wu et al. 2010).

15.3 Ion Homeostatis

In cytosol, a high K^+/Na^+ ratio is vital for normal cell functioning. However, significant enhancement of Na^+ uptake under salinity disrupts the physiological and biochemical activities in plant cells by disturbing K^+/Na^+ ratio that, eventually leads to death of cell, tissue or organ (Adams et al. 1992). Reduced Na^+ uptake, improved K^+ uptake and compartmentation of absorbed Na^+ are strategies for plants to cope with salinity-induced damages in this regard.

NO regulates these strategies responsible for salinity resistance in plants. Salt-inducible enzyme Na^+/H^+ antiporter is involved in removal of Na^+ from cytosol and/or its vacuolar compartmentation (Niu et al. 1993; Michelet and Boutry 1995; Apse et al. 1999; Chen et al. 2010). NO significantly increases the activity of vacuolar H^+-ATPase and H^+-Ppase – the driving forces for Na^+/H^+ exchange (Zhang et al. 2007a). In maize (*Zea mays* L.), for instance, exogenous application of NO substantially increased the activity of tonoplast H^+-ATPase and Na^+/H^+ antiport facilitating Na^+ compartmentation (Zhang et al. 2006, 2007a). In seashore mallow, however, NO application significantly decreased Na^+ contents and simultaneously increased K^+ contents in shoots under salt stress.

In the calluses of reed (*Phragmites communis* Trin.), NO application increased the expression of the plasma membrane H^+-ATPase, leading to a high K^+/Na^+ ratio in the cytosol (Zhao et al. 2004). Similarly in calluses from poplar (*Populus euphratica* L.), NO induced salt resistance by improving plasma membrane H^+-ATPase activity that increased the K^+/Na^+ ratio (Zhang et al. 2007b). In another study on maize, NO increased K^+ accumulation in roots, leaves and sheaths, and simultaneously decreased Na^+ accumulation contributing to salt resistance (Zhang et al. 2004). However in wheat, NO did not affect Na^+ content but K^+ content significantly increased thus increasing the K^+/Na^+ ratio in roots of wheat seedling under salinity (Ruan et al. 2004).

Exogenous application of SNP alleviated the salinity induced injury by maintaining a higher K^+/Na^+ ratio and an increased plasma membrane H^+-ATPase activity in callus of wild type *Arabidopsis* but not in callus of ethylene-insensitive mutant *inetr1-3* (Wang et al. 2009).

Several experiments using NO donors and inhibitors indicated plasma membrane H^+-ATPase and vacuolar H^+-ATPase and H^+-PPase dependent increase in K^+/Na^+ ratio in the cytosol. Therefore NO serves as a signal for inducing salt resistance in plants (Zhao et al. 2004; Zhang et al. 2006; Shi et al. 2007).

15.4 Antioxidative Defense System

Salinity-induced osmotic stress triggers the synthesis of reactive oxygen species (ROS) (Abogadallah 2010). Besides oxidative damage to lipids (Fridovich 1986; Wise and Naylor 1987), proteins and nucleic acids (Fridovich 1986; Imlay and Linn 1988), these cytotoxic ROS may also disrupt normal metabolism in plants (Farooq et al. 2009). However, NO substantially ameliorates oxidative stress damage by acting as an antioxidant, stimulating production of enzymatic and non-enzymatic antioxidants as well as osmoprotectants (Beligni et al. 2002; Guo et al. 2009; Tanou et al. 2009a; Molassiotis et al. 2010).

In mustard (*Brassica juncea* L.), NO application significantly increased the activity of superoxide dismutase (SOD), peroxidase (POD) and ascorbate peroxidase (APX), to prevent oxidative damage as indicated by the reduced value of malondialdehyde (MDA) (Zeng et al. 2011). Likewise in seashore mallow, exogenous application of NO increased the activity of catalase (CAT), SOD, POD and APX protecting plants from oxidative damage under salinity (Guo et al. 2009). Whereas in bitter orange (*Citrus aurantium* L.) trees, root pre-treatment with NO increased the activity of SOD, CAT, APX and glutathione reductase (GR), promoted maintenance of cellular redox homeostasis and mitigated oxidative damage under salt stress (Tanou et al. 2009a). This induction of the antioxidant system helps to resist salinity stress (Tanou et al. 2009a). Exogenous application of NO protected chickpea (*Cicer arietinum* L.) plants from salinity-induced oxidative damage through increased activity of antioxidant enzymes including SOD, CAT, APX, GR and dehydro-ascorbate reductase (DHAR) and higher ratios of glutathione/glutathione disulfide and ascorbate/dehydroascorbate (Sheokand et al. 2010). Similarly the exogenous application of NO significantly increased in the activity of antioxidant enzymes (superoxide dismutase, catalase, ascorbate peroxidase) in rice, thus increasing the resistance against salinity (Uchida et al. 2002).

NO can also act as an antioxidant to protect plants from oxidative damage (Beligni et al. 2002). In wheat, NO application substantially enhanced the activity of SOD and CAT (Ruan et al. 2002). Whereas in tomato, exogenous application of NO improved the antioxidant enzymes SOD, guaiacol peroxidase, CAT, APX, non-enzymatic antioxidant ascorbate and reduced glutathione under salinity thus helping to alleviate salt-induced oxidative damage (Wu et al. 2011).

15.5 Regulation of Signalling and Plant Hormones

Upon exposure to salinity stress, endogenous level of certain plant hormones such as abscisic acid (ABA) and cytokinins increase substantially (Thomas et al. 1992; Aldesuquy 1998; Vaidyanathan et al. 1999). Increased endogenous level of these hormones modulate salinity resistance in plants; for instance, ABA alters salt-stress-induced genes (de Bruxelles et al. 1996) regulating salinity resistance in plants (Gupta et al. 1998).

The protective role of NO during salinity-induced osmotic stress is dosage-dependent (Kopyra and Gwóźdź 2003). At low concentrations, the mechanism of NO in leaf water control is ABA-dependent but at high concentrations, NO can maintain leaf water by inducing stomatal closure independent of ABA accumulation (Xing et al. 2004).

NO mediates ABA signalling in guard cells to regulate stomatal oscillations; however with reduced endogenous NO, the degree of ABA-induced stomatal closure decreases (Neill et al. 2002). Exogenous NO application affirms its involvement in stomatal regulation in dose-dependent manner. For instance, externally-applied NO decreased the rate of transpiration through ABA-induced stomatal closure in several plant species including broad bean (*Vicia faba* L.), salpichroa (*Salpichroa organifolia* L.) and *Tradescantia* spp. (Garcia-Mata and Lamattina 2001). Nevertheless, application of NO inhibitors reverses NO-induced stomatal closure thus affirming its involvement in this process (Bright 2006).

15.6 Molecular Mechanism

Physiological and metabolic adaptations to salinity at the cellular level have led to the identification of several genes responsible for salt resistance in plants (Shinozaki et al. 1998). NO is also involved in the expression of certain genes under salinity (Chen et al. 2010). For example, expression of *NtGRAS1*, new member of the GRAS gene family was induced by NO application in tobacco (*Nicotiana tabacum* L.). *NtGRAS1* regulated the transcription of genes involved in salinity resistance in plants (Czikkel and Maxwell 2007).

Chen et al. (2010) reported that owing to NO-induced expression of the *HA1* and *SOS1*, salt secretion and net Na^+ efflux in salt glands was increased in mangrove plant (*Avicennia marina* (Forssk.) Vierh.) resulting in increased salinity resistance. Furthermore, increase in Na^+ sequestration into the vacuoles of the epidermis and hypodermal cells by NO-induced expression of *VHA-c1* and *NHX1* genes also confers salinity resistance in mangrove plant (Chen et al. 2010).

The protein kinases family is also involved in NO-mediated signaling cascades under salt stress. For example in tobacco cell suspensions exposed to salt stress, the osmotic stress-activated protein kinase (*NtOSAK*) activated by NO gets interacted with glyceraldehyde-3-phosphate dehydrogenase (GAPDH), the enzyme involved in glycolysis to provide energy and carbon molecules (Wawer et al. 2010).

In a recent proteomic study on citrus (*Citrus aurantium* L.), Tanou et al. (2009a) provided a global survey of the proteins controlled by salinity. Root pre-treatment with SNP prior to imposing salt stress reversed many of the NaCl-responsive proteins. These SNP/NaCl responsive proteins-mainly involved in photosynthesis (corresponding to enzymes of Calvin-Benson cycle), defense mechanism and energy/glycolysis are potentially important for salt resistance as these may exert a NO-mediated effect by acclimating citrus plants before the actual experience of salt stress. In another proteomic study on citrus, NO application modified accumulation levels of leaf S-nitrosylated proteins and decreased protein carbonylation (Tanou et al. 2009b). In maize, however NO application induced G-protein-associated protein accumulation (Bai et al. 2011).

Arabidopsis mutant *Atnoa1*, with an impaired *in vivo* NOS activity and low endogenous NO level, is more sensitive to salinity than its wild type (Zhao et al. 2007a). However, exogenous NO application to *Atnoa1* improved salinity resistance by mitigating salinity-induced oxidative damage (Zhao et al. 2007b).

15.7 Conclusion and Future Perspective

Nitric oxide, a small ubiquitous molecule, modulates an array of physiological and biochemical processes in plants. In plants, NO improves salinity resistance through production of osmolytes and specific proteins controlling water flux, ion homeostasis and antioxidant defense system scavenging free radicals. NO also regulates hormonal balance to improve salt resistance in plants.

Experimental evidence in supporting the above roles for NO has been obtained through the application of either gaseous NO or NO donors. However, different NO donors may release different molecular species of NO, which may have distinct effects on plants under salt stress; these differential effects should be monitored. Despite the exponential growth in NO research in plants under abiotic stresses, there is a need for much more – given the essential roles of NO in plant growth and development under salinity. Integration of NO functions with plant metabolism, growth and development, especially with plant hormones should be examined in more detail. A deeper understanding of the NO-induced transcription factors, regulating genes, the products of the major stress responsive genes and cross talk between different signaling components should also investigated.

References

Abogadallah GM (2010) Antioxidative defense under salt stress. Plant Signal Behav 5:369–374

Adams P, Thomas JC, Vernon DM, Bohnert HJ, Jensen RG (1992) Distinct cellular and organismic responses to salt stress. Plant Cell Physiol 33:1215–1223

Aldesuquy HS (1998) Effect of seawater salinity and gibberllic acid on abscisic acid, amino acids and water-use efficiency of wheat plants. Agrochimica 42:147–157

Almansouri M, Kinet JM, Lutts S (2001) Effect of salt and osmotic stresses on germination in durum wheat (*Triticum durum* Desf.). Plant Soil 231:243–254

Apse MP, Aharon GS, Snedden WA, Blumwald E (1999) Salt tolerance conferred by overexpression of a vacuolar Na^+/H^+ antiport in Arabidopsis. Science 285:1256–1258

Arasimowicz M, Floryszak-Wieczorek J (2007) Nitric oxide as a bioactive signalling molecule in plant stress responses. Plant Science 172:876–887

Bai X, Yang L, Yang Y, Ahmad P, Yang Y, Hu X (2011) Deciphering the protective role of nitric oxide against salt stress at the physiological and proteomic levels in maize. J Proteome Res 10:4349–4364

Beligni MV, Fath A, Bethke PC, Lamattina L, Jones RL (2002) Nitric oxide acts as an antioxidant and delays programmed cell death in barley aleurone layers. Plant Physiol 129:1642–1650

Bethke PC, Gubler F, Jacobsen JV, Jones RL (2004) Dormancy of Arabidopsis seeds and barley grains can be broken by nitric oxide. Planta 219:847–855

Bewleyl JD (1997) Seed germination and dormancy. Plant Cell 9:1055–1066

Bohnert HJ, Nelson DE, Jensen RG (1995) Adaptations to environmental stresses. Plant Cell 7:1099–1111

Bright J (2006) Nitric oxide signalling in *Arabidopsis thaliana* guard cells. PhD thesis, University of the West of England, Bristol

Bright J, Desikan R, Hancock JT, Weir IS, Neill SJ (2006) ABA-induced NO generation and stomatal closure in Arabidopsis are dependent on H_2O_2 synthesis. Plant J 45:113–122

Chen THH, Murata N (2000) Enhancement of tolerance of abiotic stress by metabolic engineering of betaines and other compatible solutes. Curr Opin Plant Biol 5:250–257

Chen J, Xiao Q, Wu F, Dong X, He J, Pei Z, Zheng H (2010) Nitric oxide enhances salt secretion and Na^+ sequestration in a mangrove plant, *Avicennia marina*, through increasing the expression of H^+-ATPase and Na^+/H^+ antiporter under high salinity. Tree Physiol 30:1570–1585

Czikkel BE, Maxwell DP (2007) *NtGRAS1*, a novel stress-induced member of the GRAS family in tobacco, localizes to the nucleus. J Plant Physiol 164:1220–1230

David A, Yadav S, Bhatla SC (2010) Sodium chloride stress induces nitric oxide accumulation in root tips and oil body surface accompanying slower oleosin degradation in sunflower seedlings. Physiol Plant 140:342–354

de Bruxelles GL, Peacock WJ, Dennies ES, Dolferus R (1996) Abscisic acid induces the alcohol dehydrogenase gene in Arabidopsis. Plant Physiol 111:381–391

Durner J, Klessig DF (1999) Nitric oxide as a signal in plants. Curr Opin Plant Biol 2:369–374

Fan H-F, Du C-X, Guo S-R (2012) Effect of nitric oxide on proline metabolism in cucumber seedlings under salinity stress. J Amer Soc Hort Sci 137:127–133

Farooq M, Wahid A, Kobayashi N, Fujita D, Basra SMA (2009) Plant drought stress: effects, mechanisms and management. Agron Sustain Dev 29:185–212

Fridovich I (1986) Biological effects of the superoxide radical. Arch Biochem Biophys 247:1–11

Garcia-Mata C, Lamattina L (2001) Nitric oxide induces stomatal closure and enhances the adaptive plant responses against drought stress. Plant Physiol 126:1196–1204

Guo FQ, Crawford NM (2005) Arabidopsis nitric oxide synthase1 is targeted to mitochondria and protects against oxidative damage and dark-induced senescence. Plant Cell 17:3436–3450

Guo Y, Tian Z, Yan D, Zhang J, Qin P (2009) Effects of nitric oxide on salt stress tolerance in *Kosteletzkya virginica*. Life Sci J 6:67–75

Gupta S, Chattopadhyay MK, Chatterjee P, Ghosh B, SenGupta DN (1998) Expression of abscisic acid-responsive element-binding protein in salt tolerant indica rice (*Oryza sativa* L. cv. Pokkali). Plant Mol Biol 137:629–637

Habib N, Ashraf M, Ahmad MSA (2010) Enhancement in seed germinability of rice (*Oryza sativa* L.) by pre-sowing seed treatment with nitric oxide (NO) under salt stress. Pak J Bot 42:4071–4078

Hasegawa PM, Bressan RA, Zhu JK, Bohnert HJ (2000) Plant cellular and molecular responses to high salinity. Annu Rev Plant Physiol Plant Mol Biol 51:463–499

Hernández JA, Corpas FJ, Gómez M, del Río LA, Sevilla F (1993) Salt-induced oxidative stress mediated by activated oxygen species in pea leaf mitochondria. Physiol Plant 89:103–110

Hernández JA, Olmos E, Corpas FJ, Sevilla F, del Río LA (1995) Salt-induced oxidative stress in chloroplasts of pea-plants. Plant Sci 105:151–167

Imlay JA, Linn S (1988) DNA damage and oxygen radical toxicity. Science 240:1302–1309

Kopyra M, Gwóźdź EA (2003) Nitric oxide stimulates seed germination and counteracts the inhibitory effect of heavy metals and salinity on root growth of *Lupinus luteus*. Plant Physiol Biochem 41:1011–1017

Leshem YY, Wills RBH, Ku VV (1998) Evidence for the function of the free radical gas-nitric oxide (NO) as an endogenous maturation and senescence regulating factor in higher plants. Plant Physiol Biochem 36:825–833

Li WQ, Liu XJ, Khan MA, Yamaguchi S (2005) The effect of plant growth regulators, nitric oxide, nitrate, nitrite and light on the germination of dimorphic seeds of *Suaeda salsa* under saline conditions. J Plant Res 118:207–214

Michelet B, Boutry M (1995) The plasma membrane H^+-ATPase: a highly regulated enzyme with multiple physiological functions. Plant Physiol 108:1–6

Misra N, Dwivedi UN (2004) Genotypic difference in salinity tolerance of green gram cultivars. Plant Sci 166:1135–1142

Molassiotis A, Tanou G, Diamantidis G (2010) NO says more than "yes" to salt tolerance – salt priming and systemic nitric oxide signalling in plants. Plant Signal Behav 5:209–212

Neill SJ, Desikan D, Clarke A, Hancock JT (2002) Nitric oxide is a novel component of abscisic acid signaling in stomatal guard cells. Plant Physiol 128:13–16

Niu X, Narasimhan ML, Salzman RA, Bressan RA, Hasegawa PM (1993) NaCl regulation of plasma membrane H^+-ATPase gene expression in a glycophyte and a halophyte. Plant Physiol 103:713–718

Qiao W, Xiao S, Yu L, Fan LM (2009) Expression of a rice gene OsNOA1 re-establishes nitric oxide synthesis and stress-related gene expression for salt tolerance in Arabidopsis nitric oxide-associated 1 mutant Atnoa1. Environ Exp Bot 65:90–98

Rengasamy P (2006) World salinization with emphasis on Australia. J Exp Bot 57:1017–1023

Ruan H-H, Shen W, Ye M, Xu L (2002) Protective effects of nitric oxide on salt-induced damages oxidative damages to wheat (*Triticum aestivum*) leaves. Chinese Sci Bull 47:677–681

Ruan H-H, Shen W, Xu L (2004) Nitric oxide modulates the activities of plasma membrane H^+-ATPase and PPase in wheat seedling roots and promotes the salt tolerance against salt stress. Acta Bot Sinica 46:415–422

Serrano R, Rodriguez PL (2002) Plants, genes and ions. EMBO Rep 3:116–119

Sheokand S, Bhankar V, Sawhney V (2010) Ameliorative effect of exogenous nitric oxide on oxidative metabolism in NaCl treated chickpea plants. Braz J Plant Physiol 22:81–90

Shi Q, Ding F, Wang X, Wei M (2007) Exogenous nitric oxide protect cucumber roots against oxidative stress induced by salt stress. Plant Physiol Biochem 45:542–550

Shinozaki K, Yamaguchi-Shinozaki K, Mizoguchi T, Uraro T, Katagiri T, Nakashima K, Abe H, Ichimura K, Liu QA, Nanjyo T, Uno Y, Luchi S, Srki M, Lto T, Hirayama T, Mikami K (1998) Molecular responses to water stress in *Arabidopsis thaliana*. J Plant Res 111:345–351

Siddiqui MH, Al-Whaibi MH, Basala MO (2011) Role of nitric oxide in tolerance of plants to abiotic stress. Protoplasma 248:447–455

Tanou G, Job C, Rajjou L, Arc E, Belghazi M, Diamantidis G, Molassiotis A, Job D (2009a) Proteomics reveals the overlapping roles of hydrogen peroxide and nitric oxide in the acclimation of citrus plants to salinity. Plant J 60:795–804

Tanou G, Molassiotis A, Diamantidis Gr (2009b) Hydrogen peroxide- and nitric oxide-induced systemic antioxidant prime-like activity under NaCl-stress and stress free conditions in citrus plants. J Plant Physiol 166:1904–1913

Thomas JC, McElwain EF, Bohnert HJ (1992) Convergent induction of osmotic stress responses. Plant Physiol 100:416–423

Uchida A, Jagendorf AT, Hibino T, Takabe T, Takabe T (2002) Effects of hydrogen peroxide and nitric oxide on both salt and heat stress tolerance in rice. Plant Sci 163:515–523

Vaidyanathan R, Kuruvilla S, Thomas G (1999) Characterization and expression pattern of an abscisic acid and osmotic stress responsive gene from rice. Plant Sci 140:21–30

Valderrama R, Corpas FJ, Carreras A, Gómez-Rodríguez MV, Chaki M, Pedrajas JR, Fernández-Ocaña A, del Río LA, Barroso JB (2006) The dehydrogenase mediated recycling of NADPH is a key antioxidant system against salt-induced oxidative stress in olive plants. Plant Cell Environ 29:1449–1459

Valderrama R, Corpas FJ, Carreras A, Fernandez-Ocana A, Chaki M, Luque F, Gómez-Rodríguez MV, Lmenero-Varea P, Del Río LA, Barroso JB (2007) Nitrosative stress in plants. FEBS Lett 581:453–461

Wang H, Liang X, Wan Q, Wang X, Bi Y (2009) Ethylene and nitric oxide are involved in maintaining ion homeostasis in Arabidopsis callus under salt stress. Planta 230:293–307

Wassmann R, Jagadish SVK, Heuer S, Ismail A, Redona E, Serraj R, Singh RK, Howell G, Pathak H, Sumfleth K (2009) Climate change affecting rice production: the physiological and agronomic basis for possible adaptation strategies. Adv Agron 101:59–122

Wawer I, Bucholc M, Astier J, Anielska-Mazur A, Dahan J, Kulik A, Wyslouch-Cieszynska A, Zareba-Koziol M, Krzywinska E, Dadlez M, Dobrowolska G, Wendehenne D (2010) Regulation of *Nicotiana tabacum* osmotic stress-activated protein kinase and its cellular partner GAPDH by nitric oxide in response to salinity. Biochem J 429:73–83

Wise RR, Naylor AW (1987) Chilling-enhanced photooxidation: evidence for the role of singlet oxygen and endogenous antioxidants. Plant Physiol 83:278–282

Wu X-X, Ding H-D, Chen J-L, Zhang H-J, Zhu W-M (2010) Attenuation of salt-induced changes in photosynthesis by exogenous nitric oxide in tomato (*Lycopersicon esculentum* Mill. L.) seedlings. Afr J Biotechnol 9:7837–7846

Wu X, Zhu W, Zhang H, Ding H, Zhang HJ (2011) Exogenous nitric oxide protects against salt-induced oxidative stress in the leaves from two genotypes of tomato (*Lycopersicom esculentum* Mill.). Acta Physiol Plant 33:1199–1209

Xing H, Tan LL, An LZ, Zhao ZG, Wang SM, Zhang CL (2004) Evidence for the involvement of nitric oxide and reactive oxygen species in osmotic stress tolerance of wheat seedlings: inverse correlation between leaf abscisic acid accumulation and leaf water loss. Plant Growth Regul 42:61–68

Xiong LM, Zhu JK (2002) Salt tolerance. In: Somerville CR, Meyerowitz EM (eds) The Arabidopsis book. American Society of Plant Biologists, Rockville

Xiong L, Schumaker KS, Zhu JK (2002) Cell signaling during cold, drought, and salt stress. Plant Cell 14:S165–S183

Yancey P, Clark ME, Had SC, Bowlus RD, Somero GN (1982) Living with water stress: evolution of osmolyte system. Science 217:1214–1222

Zeng C-L, Liu L, Wang B-R, Wu X-M, Zhou Y (2011) Physiological effects of exogenous nitric oxide on *Brassica juncea* seedlings under NaCl stress. Biologia Plant 55:345–348

Zhang YY, Liu J, Liu YL (2004) Nitric oxide alleviates the growth inhibition of maize seedlings under salt stress. J Plant Physiol Mol Biol 30:455–459

Zhang YY, Wang LL, Liu YL, Zhang Q, Wei QP, Zhang WH (2006) Nitric oxide enhances salt tolerance in maize seedlings through increasing activities of proton-pump and Na^+/H^+ antiport in the tonoplast. Planta 224:545–555

Zhang A, Jiang M, Zhang J, Ding H, Xu S, Hu X, Tan M (2007a) Nitric oxide induced by hydrogen peroxide mediates abscisic acid-induced activation of the mitogen activated protein kinase cascade involved in antioxidant defense in maize leaves. New Phytol 175:36–50

Zhang F, Wang YP, Yang YL, Wu H, Wang D, Liu JQ (2007b) Involvement of hydrogen peroxide and nitric oxide in salt resistance in the calluses from *Populus euphratica*. Plant Cell Environ 30:775–785

Zhao L, Zhang F, Guo J, Yang Y, Li B, Zhang L (2004) Nitric oxide functions as a signal in salt resistance in the calluses from two ecotypes of reed. Plant Physiol 134:849–857

Zhao MG, Tian QY, Zhang WH (2007a) Nitric oxide synthase-dependent nitric oxide production is associated with salt tolerance in Arabidopsis. Plant Physiol 144:206–217

Zhao MG, Zhao X, Wu YX, Zhang LX (2007b) Enhanced sensitivity to oxidative stress in Arabidopsis nitric oxide synthesis mutant. J Plant Physiol 164:737–745

Zheng C, Jiang D, Liu F, Dai T, Liu W, Jing Q, Cao W (2009) Exogenous nitric oxide improves seed germination in wheat against mitochondrial oxidative damage induced by high salinity. Environ Exp Bot 67:222–227

Zhu JK (2001) Plant salt tolerance. Trends Plant Sci 6:66–71

Zhu JK (2003) Regulation of ion homeostasis under salt stress. Curr Opin Plant Biol 6:441–445

Zottini M, Formentin E, Scattolin M, Carimi F, Schiavo FL, Terzi M (2002) Nitric oxide affects plant mitochondrial functionality *in vivo*. FEBS Lett 515:75–78

Chapter 16
Nitrogen and Phosphorus Nutrition Under Salinity Stress

Kenji Miura

16.1 Introduction

Environmental stresses, such as salinity, cold, and drought, influences crop productivity. High salinity affects approximately 30 % of the irrigated land and dry-land agriculture in the world (Pitman and Läuchli 2002). Salinity affects plant growth through the following factors; (1) water stress, which is caused by low osmotic potential of the solution, (2) imbalance of nutrition, and/or (3) specific ion effects. These factors are easily combined to influence plant growth (Marschner 1995; Shannon 1997; Munns and Tester 2008). Because plants obtains mineral nutrients including nitrogen (N) and phosphorus (P) from soil, almost all major living processes, such as growth, photosynthesis, and protein and lipid metabolism, are affected by accumulation of salts in soil (Evelin et al. 2009). Most crop plants have evolved to adapt low salinity condition, thus, these plants exhibit severe symptoms, including nutritional disorders, growth retardation, and senescence, under high salinity conditions. Nutrient availability is totally changed in high salt condition, because uptake of some minerals is competed with uptake of Na^+ or Cl^- (Bartels and Sunkar 2005; Munns and Tester 2008). For example, Na^+ induces K^+ deficiency and NO_3^- absorption is inhibited by excess Cl^- (Peuke and Jeschke 1999). Frequently, salinity causes a decrease in absorption of K^+, Ca^{2+}, NO_3^-, and phosphate (Pi) (Shokri and Maadi 2009). The ratios of Na^+/K^+, Na^+/Ca^{2+}, Ca^{2+}/Mg^{2+}, and Cl^-/NO_3^- are normally increased by salinity stresses, which provides high amount of Na^+ and Cl^-. Salinity stresses also damage macromolecules including chlorophyll, thus photosynthetic activity is reduced and senescence in leaves and growth retardation can be observed (Grattan and Grieve 1999; Feng et al. 2002; Wahid et al. 2004;

K. Miura (✉)
Faculty of Life and Environmental Sciences,
University of Tsukuba, Tsukuba, Japan
e-mail: kmiura@gene.tsukuba.ac.jp

P. Ahmad et al. (eds.), *Ecophysiology and Responses of Plants under Salt Stress*,
DOI 10.1007/978-1-4614-4747-4_16, © Springer Science+Business Media, LLC 2013

De Michele et al. 2009). To become salt tolerance in non-halophytic plants, these plants have developed the mechanism to exclude Na^+, leading to maintenance of a high K^+ /Na^+ ratio.

Several environmental factors are involved in nutrient availability and uptake by plants. In the soil solution, the activity of the nutrient ion is dependent on pH, concentration, and composition. For example, Pi can be easily insoluble in low pH and it is hard for plants to uptake Pi. The concentration and ratios of accompanying elements also affect the uptake and transport of the nutrient, such as K^+ and Na^+ as described above. Three possible types of salinity-fertility relationships have been proposed (Bernstein et al. 1974). (1) Salt tolerance could be increased by application of fertility when the soil fertility is low. (2) When the fertility level is the sub-optimal, salt tolerance is decreased even though the fertility is applied. (3) The impacts of salinity and fertility at the optimum and sub-optimum fertility levels are independent (no relationship), because the relative yield is the same for the two conditions. Generally, if the most limiting factor is relieved rather than the less limiting factors, plant growth can be improved, i.e. if the nutrient is severely limited, application of nutrition may increase plant growth even in salinity condition. Most soil in the world is suffered from nutrition deficiency. Thus, improving soil fertility to an adequate level would improve plant growth and crop yield in several fields, including salt stressed field. A clear understanding of how this interaction changes from low to high salinity is indispensable. In many cases, nutrient limitation is the limiting factor for plant growth more than salinity under low salinity stress condition, i.e. application of nutrition increases salt tolerance in such condition. Under moderate salinity condition, nutrient deficiency and salinity stress may be equally limiting to plant growth, and no clear interaction is observed. On the other hand, high salinity generally limits plant growth more than nutrient limitation (Grattan and Grieve 1992). In this review, we focus on the interaction between salinity and two nutrients, N and P, and how salinity affects homeostasis of N and P.

16.2 Salinity and N Uptake and Transport

Generally, N is the most limiting nutrient for plant growth in most soil, regardless of saline or non-saline. Consequently, the application of N fertility usually improves plant growth and crop yield. N is an important element to compose several cell components, such as amino acids and nucleic acids.

NO_3^- uptake is interfered with salinity, especially high Cl^-, in many plants, leading to decreasing NO_3^- contents (Khan and Srivastava 1998). High external Cl^- concentration causes salinity-induced deficiency of NO_3^-, resulting in reduction of yield and plant growth. Because of NO_3^-/Cl^- antagonism, fertilization of NO_3^- decreased concentration of Cl^- in leaves (Deane-Drummond 1986; Hu and Schmidhalter 1997). In another case, application of N enhanced the development of rice shoot biomass under salinity condition (Abdelgadir et al. 2005). NO_3^- concentration was reduced in tomato fruits when Cl^- concentration increased in the nutrient solution

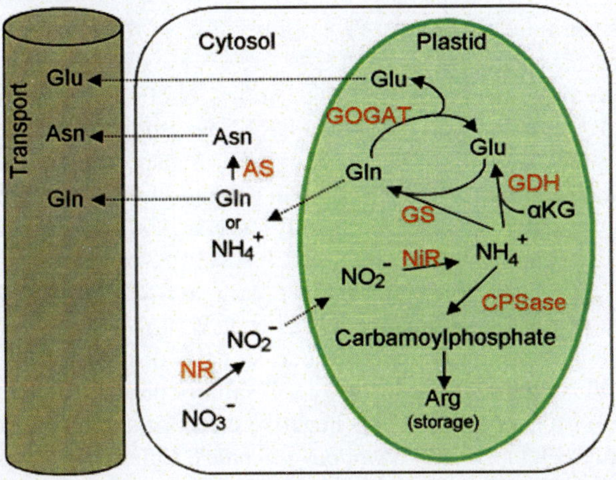

Fig. 16.1 Schematic pathway for nitrogen assimilation in plants. Nitrate reductase (NR) and asparagine synthetase (AS) are localized in cytosol, and nitrite reductase (NiR), glutamine synthase (GS), glutamate synthase (glutamine oxoglutarate aminotransferase; GOGAT), glutamate dehydrogenase (GDH), and carbamoylphosphate synthetase (CPSase) are localized within the plastids. The full names of each abbreviated amino acids are as follows: αKG, α-ketoglutarate; Arg, arginine; Asn, asparagine; Gln, glutamine; Glu, glutamate

(Chapagain et al. 2003). These results suggest a direct competitive effect between NO_3^- and Cl^-. Because many tree and vine plants are susceptible to Cl^- toxicity (Grattan and Grieve 1999), this type of competitive effect might be particularly important for these plants. In addition to direct competition of Cl^- with NO_3^-, the state of the membrane and/or membrane proteins also affects uptake of NO_3^- by changes in the plasmalemma integrity (Frechilla et al. 2001). The application of Ca^{2+} also increased NO_3^- uptake under salt stress (Dubey 1997), probably because Ca^{2+} is involved in maintaining membrane integrity and/or activity of NO_3^- transporters may be enhanced by Ca^{2+} under salinity conditions.

Inhibition of NH_4^+ uptake by salinity could be due to direct competition with Na^+ (NH_4^+/ Na^+ antagonism) and to the depolarizing effect of NaCl on the plasmalemma (Hawkins and Lewis 1993). These changes in uptake of N nutrient influence different steps of N transport and metabolism, such as uptake, reduction ($NO_3^- \rightarrow NO_2^- \rightarrow NH_4^+$), and protein synthesis (Fig. 16.1; Frechilla et al. 2001). Interestingly, the sensitivity of plant to salinity is changed by the N form (Speer and Kaiser 1994). Increase of NH_4^+/NO_3^- ratio makes wheat and maize more sensitive to salinity (Leidi et al. 1991; Botella et al. 1997) and cucumber accumulate more Na^+ and Cl^- and less Ca^{2+} and K^+ in leaves (Martinez and Cerda 1989). Application of NH_4^+ and 75 mM NaCl to salt-sensitive *Populus x canescens* significantly reduced N uptake, even though application of NO_3^-, instead of NH_4^+, did not show any significant change in the N uptake rate in salinity condition (Dluzniewska et al. 2006). On the other hand, the salinity-induced growth retardation of Madagascar periwinkle, *Catharanthus roseus* was largely improved by application of NH_4NO_3, compared to the application of only NO_3^- in (Zhonghua et al. 2011), possibly due to the competitive uptake of NH_4^+ and Na^+.

NO_3^- transporters are categorized into two families, NRT1 and NRT2. In *Arabidopsis thaliana,* 53 and 7 genes encode NRT1 and NRT2, respectively (Tsay et al. 2007). In addition, 5 genes are annotated to high-affinity NH_4^+ transports (AMT) in *Arabidopsis thaliana* (Loqué and Wirén 2004). The protein amount of the low-affinity transporter NRT1;1 was significantly decreased by salinity stress in *Arabidopsis,* whereas no alteration in the high-affinity transporters NRT2;1 AMT1;3, or AMT1;4 was observed (Monneuse et al. 2011). The transcript levels of tomato *NRT1.1*, *NRT1.2*, and *NRT2.1* are also down-regulated by salt treatment (Yao et al. 2008).

Negative effect of salt on N uptake is occasionally occurred. Reduction of total N uptake and soil N availability caused by soil salinity decreased in growth of grain legumes (van Hoorn et al. 2001). This negative effect of salt on N uptake was also observed in fenugreek (*Trigonella foenum-graecum*), a plant in the family Fabaceae (Evellin et al. 2012), lettuce, onion (Cantrell and Linderman 2001), cowpea (Silveira et al. 2001), and pea (Frechilla et al. 2001). Salt stress also inhibited nodule formation in legumes, leading to decreases in tissue N concentrations (Rao et al. 2002; Garg and Manchanda 2008; Garg and Manchanda 2009; Fahmi et al. 2011; Evellin et al. 2012). Comparing of leaf N content and growth in Mediterranean Fan Palm and Mexican Fan Palm under salt stress revealed that leaf nitrogen productivity (the rate of biomass gain per unit leaf N and time) was the major factor causing the differences in relative growth rate under salinity stress (Nieves et al. 2011)

Salinity often negatively affects plant growth and crop yield. But in some case, salt treatment improves quality of crops. NO_3^- has common harmful effects on human health (Anjana and Muhammad 2007), even though it is limiting factor for plant growth. Under saline stress, NO_3^- content was reduced in edible broccoli florets (López-Berenguer et al. 2009). Increased Cl^- concentration in the nutrient solution reduced NO_3^- content in tomato fruits, probably because of NO_3^-/Cl^- antagonism (Chapagain et al. 2003). The quality of fruits in terms of chemical constituents (mainly sugars and acids), pigments, and especially taste from saline-treated plants was improved in some case, compared to fruits from control plants. Decreased NO_3^-, as well as accumulation of sugars and several chemical constituents, may contribute to beneficial change in quality of fruits. Thus, saline treatment is already very popular for its nutritional and therapeutic properties.

16.3 Nitrogen Assimilation Affected by Salt Stress

In addition of N uptake, N metabolism, such as N assimilation, amino acid and protein biosynthesis is also affected by salt stress (Dluzniewska et al. 2006; Ehlting et al. 2007). Activities of NO_3^- assimilation enzymes, such as nitrate reductase (NR), nitrite reductase (NiR), glutamine synthetase (GS), and glutamate synthase (glutamine oxoglutarate aminotransferase; GOGAT) are altered under salt stress.

NR is the first enzyme in the NO_3^- assimilation pathway (Fig. 16.1) and is a limiting factor of plant growth and development (Skriver and Mundy 1990; Flores et al. 2002; Masclaux-Daubresse et al. 2010). NR activity is affected by several environmental stresses, including salt stresses (Solomonson and Barber 1990; Maighany and Ebrahimzadeh 2004; Tcherkez 2011). But the effect of salinity on NR in different plants have frequently been contradictory. NR activity increases with exogenous NO_3^- concentration in *Sprulina platensis* (Jha et al. 2007). In most cases, NR activity in leaves and roots is inhibited by salt stress, as reported for *Sorghum vulgare* (Rao and Gnaham 1990), bean, cotton (Gouia et al. 1994), maize (Abd-El Baki et al. 2000), and Algarrobo (*Prosopis alba*), a native legume from the semi-arid regions of north-western Argentina (Meloni et al. 2004). N concentration and NR and GS activities were higher in salt-tolerant genotypes of durum wheat, compared to salt-susceptible ones (Yousfi et al. 2012). In contrast, salinity stress stimulates NR activity in soybean roots (Bourgeais-Chaillou et al. 1992) and cucumber roots (Reda et al. 2011). In the case of a true mangrove *Bruguiera parviflora*, NR activity was increased with 100 mM NaCl treatment, but treatment with 200 and 400 mM NaCl decreased its activity (Parida and Das 2004). The modification of NR activity is very rapid. The NR protein is regulated by post-translational modifications, such as phosphorylation, which is catalyzed by specific protein kinase followed by association with the 14-3-3 protein (Campbell 1999; Kaiser et al. 2002; Shen and Huber 2006). This phosphorylation of NR leads to an inactive complex formation (Athwal and Huber 2002). Protein phosphatase type 1 and 2A dephosphorylates NR to reactivate (Kaiser and Huber 2001). Such post-translational modifications of NR may occur in response to salt stress because application of inhibitors for protein phosphatases prevents NR activity in cucumber roots (Reda et al. 2011). These data suggest that the effect of salinity on NR activity is dependent on plant species, tissues, and NaCl concentration.

The conversion of NO_2^- to NH_4^+ is catalyzed by NiR (Fig. 16.1). Generally, NH_4^+ is toxic to plants, because NH_4^+ triggers protein extrusion associated with NH_4^+ uptake, cytosolic pH disturbances, uncoupling of photophosphorylation, and so on (Kronzucker et al. 2001). Therefore, NH_4^+ is required to be assimilated into non-toxic organic nitrogen compounds, such as glutamine and glutamate. These amino acids are important precursors for almost all nitrogenous compounds. That is why the conversion of NH_4^+ to non-toxic organic nitrogen compounds is vital step for plants. NH_4^+ can be assimilated into nitrogenous organic compounds by the concerted action of two enzymes, GS-GOGAT cycle (Fig. 16.1). NH_4^+ may also be directly incorporated into glutamate by glutamate dehydrogenase (GDH) in the presence of the cofactor NAD(P)H (Fig. 16.1). These enzymes for N assimilation are also affected by salinity stress. Several key genes, such as *NR, NiR,* and *GS*, were significantly down-regulated by salinity in tomato (Ouyang et al. 2007). The enzyme activity of NR, NiR, and GS in tomato leaves was also reduced by salinity, whereas NR and GS activities were enhanced in salt-treated tomato roots (Debouba et al. 2006). Similarly, in potatoes, GS activity was down-regulated in leaves, but increases in roots after salinity stress (Teixeira and Fidalgo 2009). According to these results, it is suggested that NH_4^+ assimilation in roots is increased for providing a nitrogen source to shoots. In pearl millet (*Pennisetum typhoids* L.), the activities of NR, NiR, and GOGAT, but not GS, decreased after salt treatment (Albassam 2001). In *Brassica juncea* L., application of both N and sulfur was more effective in alleviating the adverse effect of NaCl on growth, and the activity of NR, NiR, GS, and GOGAT (Siddiqui et al. 2012). Because sulfur is also an important macronutrient and is required for constitution of coenzyme A, cysteine, enzymes, glutathione, and proteins (Marschner 2002), addition of both macronutrients may cumulatively affect reversal of growth and N-assimilation enzymes. Overexpression of a ferredoxin-NR gene has been shown to increase photorespiration and enhance salt and chilling stress tolerance in rice (Yan et al. 2006). Furthermore, high NO_3^- in the irrigation solution restores the activities of these N assimilation enzymes and ameliorates the retarded growth caused by salinity stress. The activity and transcript abundance of ferredoxin-dependent GOGAT is reduced in response to salinity in the leaves of the common iceplant (Popova et al. 2002). Interestingly, at lower salinity GS activity increased in leaves of wheat, but GS activity decreased at high salinity stress (Wang et al. 2007). Even in high salinity condition, GDH activity increased (Wang et al. 2007). At both lower and higher salinity conditions, glutamate was accumulated (Wang et al. 2007). It is suggested that under lower salinity GS-catalysis may be the main glutamate synthesis pathway and GDH-catalysis pathway may be activated at higher salinity conditions. In shoots of triticale seedlings, GS activity decreases under high salinity conditions, but no alteration of GDH activity has been observed (Kwinta and Cal 2005). Glutamate is a precursor of proline and proline is an important osmoprotectant as described below. Thus, plants need to keep a substantial supply of glutamate to maintain high-rate proline synthesis (Lutts et al. 1999; Hayashi et al. 2000). These findings indicate that salt stress has significant impacts on N reduction and assimilation in plants.

16.4 Photorespiration and Salt Stress

In C_3-plants, NH_4^+ is also produced through photorespiration as well as through primary assimilation from NR. And it is suggested that release of photorespiratory NH_4^+ is higher than the rate of production from NR (Wingler et al. 2000). Photorespiration lowers photosynthetic efficiency, because CO_2 and NH_4^+ should be re-assimilated with the concomitant consumption of both ATP and reducing power (Maurino and Peterhansel 2010). Water shortage and salinity lead to increases in photorespiration (Kriedemann and Downton 1981). Under stress conditions, high concentrations of polyamines, free amino acids, and ammonium ions are accumulated in plant tissues. In wheat, the flux of NH_4^+ from NR decreased and the contents of photorespiratory amino acids significantly increased under NaCl stress (Wang et al. 2007). Salinity stress increased glycine-NH_4^+ recycling from photorespiration, leading to the accumulation of glutamate, which is used for proline biosynthesis in cashew plants (Viégas and Silveira 1999).

16.5 Nitrogen Compounds and Salinity

Salinity stress stimulates protein degradation and inhibition of protein synthesis. And it also promotes accumulation of a number of metabolites, which are termed compatible solutes, because these solutes do not interfere with biochemical reactions. These metabolites include carbohydrates, such as mannitol, sucrose, and raffinose oligosaccharides, and N-containing compounds, such as amino acids and polyamines (Bohnert et al. 1995). Salt stress, like other abiotic stresses, can lead to oxidative stress through the increase in reactive oxygen species (ROS), such as superoxide (O_2^-), hydrogen peroxide (H_2O_2), and hydroxyl radicals (·OH). ROS may cause cellular damage through oxidation of lipids, proteins, and nucleic acids (Apel and Hirt 2004). Plants produce antioxidants such as glutathione and ascorbate, as well as activate ROS-scavenging enzymes, such as superoxide dismutase, catalase, ascorbate peroxidase, guaiacol peroxidase, and glutathione reductase to minimize the effects of oxidative stress caused by ROS (Alscher et al. 1997; Apel and Hirt 2004).

Amino acids are one form of nitrogen-containing compounds that have been shown to accumulate under salt stress (Mansour 2000). These amino acids include proline, arginine, alanine, glycine, serine, leucine, and valine, as well as the non-protein amino acids citrulline and ornithine (Rabe 1990). These amino acids could play a substantial role in mitigation of the effect of salt stress on K^+ homeostasis, leading to plant adaptation to salinity by decreasing the extent of the NaCl-induced K^+ efflux (Cuin and Shabala 2007). Among these amino acids, proline has a very important role and its content is remarkably increased in many plants (Ali et al. 1999). As described above, the main precursor of proline synthesis is glutamate. Glutamate is reduced to glutamate semialdehyde, which spontaneously cyclized to pyrroline 5-carboxylate. And pyrroline 5-carboxylate is converted to proline. The first step is catalyzed by P5CS (Δ^{-1}-pyrroline-5-carboxylate synthase), which is a key enzyme for proline synthesis, and expression of *P5CS* is induced under salt stress (Silva-Ortega et al. 2008). Proline acts as an osmoprotectant, can function as a protein stabilizer and a hydroxyl radical scavenger, stabilizes cell membranes by interacting with phospholipids, and serves as a source of carbon and nitrogen (Kavi Kishor et al. 2005). Overexpression of *P5CS* resulted in high accumulation of proline and increase in salt tolerance (Kishor et al. 1995; Sawahel and Hassan 2002). Exogenous application of proline also enhanced plant growth under salinity stress (Shaddad 1990). Because accumulation of proline is induced by several environmental stresses, including salt stress, it is suggested that proline accumulation is a universal stress response. Protein degradation occurs in response to environmental stresses. This degradation may provide amino acids, which functions in protection of ROS and is also required for synthesis of new proteins for survival (Mansour 2000).

In addition of proline, glycine betaine is also a major osmoprotectant, which is synthesized by many plants in response to stresses, such as salinity stress. Glycine betaine helps in maintaining the osmotic status of the cell to ameliorate the abiotic stress (Chinnusamy et al. 2005; Vinocur and Altman 2005). Glycine betaine is synthesized from choline catalyzed by choline monooxygenase and betaine aldehyde dehydrogenase. Overexpression of glycine betaine synthesizing genes resulted in the production of a high amount of glycine betaine, leading to tolerance to abiotic stresses, including salt stress (Rhodes and Hanson 1993; Vinocur and Altman 2005). Furthermore, application of glycine betaine by foliar spray improved plant growth under high salt conditions in rice and maize (Harinasut et al. 1996; Yang and Lu 2005). Proline and glycine betaine are N-containing compounds that are major osmolytes for the defensive machinery that copes with stress.

16.6 Interaction of Salinity and Phosphorus

P is an essential macronutrient for plants, because P is required for several key compounds, including the sugar-Pi intermediates of respiration and photosynthesis, the phospholipids of the plasma membrane, and nucleic acids (Taiz and Zeiger 2006). P can also be used for regulation of some enzymes (phosphorylation), transfer of energy, and the transport of carbohydrates. However, P is one of the most immobile nutrients in soils (Holford 1997). Thirty to forty percentage reduction in crop productivity in the world is estimated to be caused by P limitation (Vance et al. 2003).

Salinity stress decreases the uptake and concentration of P in plant tissues. Thus, plants exhibit reduced and stunted growth, dark green coloration of the leaves, production of slender stems, and death of older leaves (Taiz and Zeiger 2006). NaCl induces a high ionic strength in the soil, which reduces the activity of P. Uptake of P into plants under salt stress may be required for the maintenance of vacuolar membrane integrity, leading to facilitating the compartmentalization of Na^+ ions within vacuoles. This compartmentalization is important process to prevent the effect of Na^+ ions on metabolic pathways in the cytosol (Cantrell and Linderman 2001).

Unlike NO_3^-/Cl^- antagonism, uptake of $H_2PO_4^-$ and Cl- ions is not competitive. However, salinity-induced reductions in P concentrations in plant tissues have been found in several studies. Salinity stress reduces P availability, partly because the ionic strength caused by salinity reduces the activity of P. Salinity stress causes high concentration of $CaCl_2$ as well as NaCl in soil. Ca-P minerals are low solubility and hard for plants to absorb (Grattan and Grieve 1999). Because salinity decreases P availability to plants, the application of P to saline soils may improve plant growth in several cases. Even though both salinity and P deficiency reduce plant growth and crop yield, the more severe stress determines plant biomass. Because P availability for plants is limited in most land soil, application of P ameliorated growth of barley, corn, and pea (Shenker et al. 2003; Yousfi et al. 2007; Nenova 2008). However, it may be dependent on conditions. High external P increased Na uptake and reduced salt tolerance of soybean and barley (Phang et al. 2009; Zribi et al. 2011).

Salt stress results in not only Na^+ toxicity but also P-deficiency in plants. Interaction between the external P levels and the degree of salt tolerance has been investigated in various plants, but the conclusions are still contradictory. High external P could enhance growth of chickpea and the halophyte *Lavatera arborea* under salt stress (Okusanya and Fawole 1985; Saxena and Rewari 1991; Zaiter and Saade 1993). High external P also attenuated salt stress and increased the K^+/Na^+ ratio in wheat and barley (Al-Karaki 1997; Gibson 1988). P fertilization decreased the concentration of Na^+ in shoots, resulting in better survival, growth, and yield of rice (Qadar 1998) and wheat (Salim et al. 1999). Application of high P ameliorated damage of salinity stress in cucumber and pepper (Kaya et al. 2001). In *Anabaena doliolum,* chlorophyll and protein content decrease under salinity condition and the effects of NaCl were enhanced by P deficiency (Rai and Sharma 2006). Protective effects on the microculture of African violet were observed with P supplements even in high NaCl medium (Shibli et al. 2001). Contradictorily, in the case of corn and lupin, P supplements under salinity stress enhance salt injury (Nieman and Clark 1976; Treeby and van Steveninck 1988). Plant growth was decreased by P supplement in melon under salinity stress (Navarro et al. 2001). Application of P increased Na^+ uptake, reduced cell viability, and decreased the expression of the *SOS1* (putative plasma membrane-localized Na^+/H^+ antiporters; Shi et al. 2002) and *CNGC* (putative cyclic nucleotide-gated cation channel; Talke et al. 2003) homologs in soybean (Phang et al. 2009). Salt-treated barley exposed to P deficiency showed higher salt tolerance compared to plants grown with sufficient P supply (Zribi et al. 2011). In other cases, high P supplements exhibit no effect on the salt tolerance in lucerne and tepary bean (Rogers et al. 2003; Zaiter and Saade 1993). The interaction between salinity and P nutrition of plants depends to a large extent on the plant species, physiological developmental stage, environment and salt concentration, and P availability (Grattan and Grieve 1999). It was reported that if salinity stress is mainly related to chlorides, then crop yield could be improved through P fertilization (Kaya et al. 2001; Naheed et al. 2008).

16.7 Salinity and P Uptake and Transport

High concentration of salt inhibited Pi transport activity in barley roots and lettuce (Maas et al. 1979; Martinez et al. 1996) and induced P deficiency in a halophyte, a mangrove *Bruguiera parviflora* (Parida and Das 2004). P uptake was reduced by salinity stress under low P conditions, although no specific competition between Pi and Cl^- was observed in melon (Navarro et al. 2001). P is stored in vacuoles, but the mobility of P may be decreased by NaCl, resulting in inhibition of export from this storage compartment to other parts of plant. The Pi transporter PHT4;6 is localized to the Golgi apparatus and the *pht4;6* mutant exhibited salt sensitivity (Cubero et al. 2009). Phosphatase plays an important role in enhancement of P availability in soil. High salinity enhanced the enzyme activity in *Bruguiera parviflora* (Parida and Das 2004), the rhodophyte *Gracilaria tenuistipitata* (Lee et al. 1999), *Spinacia oleracea* leaves (Pan 1987), and wheat (Szabo-Nagy et al. 1992) to improve P uptake. The root tips of corn exposed to salt and high Pi have been shown to stimulate Pi uptake

(Roberts et al. 1984), but cotton subjected to high salinity stress resulted in reduced Pi uptake (Martinez and Lauchli 1994). On the other hand, P-overaccumulating mutants, *pho2* and *siz1,* which are impaired in a ubiquitin E2 (Aung et al. 2006) and a SUMO E3 ligase (Miura et al. 2005; Miura et al. 2010; Miura and Hasegawa 2010), respectively, have enhanced tolerance to salinity in *Arabidopsis* (Miura et al. 2011).

16.8 Na^+-Dependent Pi Transporters

In plants, H^+-coupled Pi transporters are dominant (Rausch and Bucher 2002), whereas Na^+-coupled Pi transporters (NaPi) are known to be active predominantly in mammals. Salinity induced changes in the H^+ electrochemical gradient across the plasma membrane and this change inhibits translocation of P from storage compartment to other parts. In mammals, Na^+-dependent Pi transport systems play a role in the Pi reabsorption process and the regulation of Pi homeostasis (Segawa et al. 2009). Three types of NaPi function in mammalian cells; NaPi-I represents a group of proteins for which the endogenous substrate, ionic coupling, and physiological function are still uncharacterized; NaPi-II plays a role in the intracellular Pi accumulation system; and NaPi-III has the characteristics of a housekeeping system. Recently, NaPi-type transporters have been reported in green algae and vascular plants (Mimura et al. 2002; Rubio et al. 2005; Pavón et al. 2008; Li et al. 2011). Biochemical and physiological studies have suggested an interaction between salt stress and phosphate metabolism in plants. In *Chara corallina*, external Na^+ is required for NaPi cotransport activity (Mimura et al. 2002). A NaPi at the plasma membrane was identified in the leaf and root cells of *Zostera marina* L. (Rubio et al. 2005). These transporters may contribute to Na and P homeostasis.

16.9 Conclusion and Perspective

Plants acquire mineral nutrients from the root in their native environment and several environmental factors affect uptake and availability of nutrients. Saline soils may be characterized by low activity of nutrient ions and by extreme ratios of Na^+/Ca^{2+}, Na^+/K^+, Ca^{2+}/Mg^{2+}, and Cl^-/NO_3^- in the soil solution. When plants are exposed to such saline soils, nutritional disorders may develop. But these disorders vary in their intensity and can differ among species as well as among cultivars within a species. Thus, several contradictory results have been shown in different species as described above. Salt tolerance also depends on the severity of salinity stress; thus, the addition of nutrients may increase, decrease, or have no effect on plant salt tolerance. But generally, relief of the more growth-limiting stress, such as salinity or nutrient deficiency, promotes growth more than the relief of a less limiting factor. Thus, soil conditions should be evaluated before applying nutrients. If nutrient deficiency is the severe growth-limiting factor, then growth may be improved by addition of proper nutrients. On the other hand, if salinity effects are more severe, then the application of nutrients would not improve plant biomass. In this case, decrease in salt concentration in soil should be required. Evaluation of intensity of salinity stress and nutrient deficiency is required for improvement of plant growth and crop productivity.

Acknowledgements This work was supported, in part, by Special Coordination Funds for Promoting Science and Technology, and by a Grant-in-Aid for Scientific Research on Innovative Areas on 'Environmental Sensing of Plants: Signal Perception, Processing Cellular Responses' from the Ministry of Education, Culture, Sports, Science and Technology.

References

Abd-El Baki GK, Siefritz F, Man HM, Welner H, Kaldenhoff R, Kaiser WM (2000) Nitrate reductase in *Zea mays* L. under salinity. Plant Cell Environ 23:515–521

Abdelgadir EM, Oka M, Fujiyama H (2005) Nitrogen nutrition of rice plants under salinity. Biol Plant 49:99–104

Albassam BA (2001) Effect of nitrate nutrition on growth and nitrogen assimilation of pearl millet exposed to sodium chloride stress. J Plant Nutr 24:1325–1335

Ali G, Srivastava PS, Iqbal M (1999) Proline accumulation protein pattern and photosynthesis in regenerants grown under NaCl stress. Biol Plant 42:89–95

Al-Karaki GN (1997) Barley response to salt stress at varied levels of phosphorus. J Plant Nutr 20:1635–1643

Alscher RG, Donahue JL, Cramer CL (1997) Reactive oxygen species and antioxidants: relationship in green cells. Physiol Plant 100:224–233

Anjana SU, Muhammad I (2007) Nitrate accumulation in plants, factors affecting the process, and human health implications. A Review Agron Sustain Dev 27:45–57

Apel K, Hirt H (2004) Reactive oxygen species: metabolism, oxidative stress and signal transduction. Annu Rev Plant Biol 55:1331–1341

Athwal GS, Huber SC (2002) Divalent cations and polyamines bind to loop8 of 14-3-3 proteins, modulating their interaction with phosphorylated nitrate reductase. Plant J 29:119–129

Aung K, Lin SI, Wu CC, Huang YT, Su CL, Chiou TJ (2006) *pho2*, a phosphate overaccumulator, is caused by nonsense mutation in a microRNA399 target gene. Plant Physiol 141:1000–1011

Bartels D, Sunkar R (2005) Drought and salt tolerance in plants. Crit Rev Plant Sci 24:23–58

Bernstein L, Francois LE, Clark RA (1974) Interactive effects of salinity and fertility on yields of grains and vegetables. Agron J 66:412–421

Bohnert H, Nelson D, Jensone R (1995) Adaptations to environmental stresses. Plant Cell 7: 1099–1111

Botella MA, Martinez V, Pardines J, Cerda A (1997) Salinity induced potassium deficiency in maize plants. J Plant Physiol 150:200–205

Bourgeais-Chaillou P, Perez-Alfocea F, Guerrier G (1992) Comparative effect on N-sources on growth and physiological responses of soybean exposed to NaCl-stress. J Exp Bot 254: 1225–1233

Campbell WH (1999) Nitrate reducatase structure, function and regulation: binding the gap between biochemistry and physiology. Annu Rev Plant Physiol Plant Mol Biol 50:277–303

Cantrell IC, Linderman RG (2001) Preinoculation of lettuce and onion with VA mycorrhizal fungi reduces deleterious effects of soil salinity. Plant Soil 233:269–281

Chapagain BP, Wiesman Z, Zaccai M, Imas P, Magen H (2003) Potassium chloride enhances fruit appearance and improves quality of fertigated greenhouse tomato as compared to potassium nitrate. J Plant Nutr 26:643–658

Chinnusamy V, Jagendorf A, Zhu JK (2005) Understanding and improving salt tolerance in plants. Crop Sci 45:437–448

Cubero B, Nakagawa Y, Jiang XY, Miura KJ, Li F, Raghothama KG, Bressan RA, Hasegawa PM, Pardo JM (2009) The phosphate transporter PHT4;6 is a determinant of salt tolerance that is localized to Golgi apparatus of *Arabidopsis*. Mol Plant 2:535–552

Cuin TA, Shabala S (2007) Amino acids regulate salinity-induced potassium efflux in barley root epidermis. Planta 225:753–761

De Michele R, Formentin E, Schiavo FL (2009) Legume leaf senescence. A transcriptional analysis. Plant Signal Behav 4:319–320

Deane-Drummond CE (1986) A comparison of regulatory effects of chloride on nitrate uptake, and of nitrate on chloride uptake into *Pisum sativum* seedlings. Physiol Plant 66:115–121

Debouba M, Gouia H, Suzuki A, Ghorbel MH (2006) NaCl stress effects on enzymes involved in nitrogen assimilation pathway in tomato *"Lycopersicon esculentum"* seedlings. J Plant Physiol 163:1247–1258

Dluzniewska P, Gessler A, Dietrich H, Schnitzler JP, Teuber M, Rennenberg H (2006) Nitrogen uptake and metabolism in *Populus* x *canescens* as affected by salinity. New Phytol 173:279–293

Dubey RS (1997) Nitrogen metabolism in plants under salt stress. In: Jaiwal PK, Singh RP, Gulati A (eds) Strategies for improving salt tolerance in higher plants. Oxford and IBH Publ. Co, New Delhi, pp 129–158

Ehlting B, Dluzniewska P, Dietrich H, Selle A, Teuber M, Haensch R, Nehls U, Polle A, Schnitzler JP, Rennenberg H, Gessler A (2007) Interaction of nitrogen nutrition and salinity in Grey poplar (*Populus tremula* x *alba*). Plant Cell Environ 30:796–811

Evelin H, Kapoor R, Giri B (2009) Arbuscular mycorrhizal fungi in alleviation of salt stress. Ann Bot 104:1263–1281

Evellin H, Giri B, Kapoor R (2012) Contribution of *Glomus intraradices* inoculation to nutrient acquisition and mitigation of ionic imbalance in NaCl-stressed *Trigonella foenum-graecum*. Mycorrhiza 22:203–217

Fahmi AI, Nagaty HH, Eissa RA, Hassan MM (2011) Effects of salt stress on some nitrogen fixation parameters in faba bean. Pak J Biol Sci 14:385–391

Feng G, Zhang FS, Li XL, Tian CY, Tang C, Rengel Z (2002) Improved tolerance of maize plants to slat stress by arbuscular mycorrhizal is related to higher accumulation of soluble sugars in roots. Mycorrhiza 12:185–190

Flores P, Botella MA, Martínez V, Cerdá A (2002) Response to salinity of tomato seedlings with a split-root system: nitrate uptake and reduction. J Plant Nutr 25:177–187

Frechilla S, Lasa B, Ibarretre L, Lamfus C, Aparico-Tejo P (2001) Pea responses to saline stress is affected by the source of nitrogen nutrition (ammonium or nitrate). Plant Growth Regul 35:171–179

Garg N, Manchanda G (2008) Effect of arbuscular mycorrhizal inoculation on salt-induced nodule senescence in *Cajanus cajan* (pigeonpea). J Plant Growth Regul 27:115–124

Garg N, Manchanda G (2009) Role of arbuscular mycorrhizae in the alleviation of ionic, osmotic and oxidative stresses induced by salinity in *Cajanus cajan* (L.) Millsp. (pigeonpea). J Agron Crop Sci 195:110–123

Gibson TS (1988) Carbohydrate metabolism and phosphorus/salinity interactions in wheat (*Triticum aestivum* L.). Plant Soil 111:25–35

Gouia H, Ghorbal MH, Touraine B (1994) Effects of NaCl on flows of N and mineral ions and NO_3^- reductase rate within whole plants of salt-sensitive bean and salt-tolerant cotton. Plant Physiol 105:1407–1418

Grattan SR, Grieve CM (1992) Mineral element acquisition and growth response of plants grown in saline environments. Agric Ecosys Environ 38:275–300

Grattan SR, Grieve CM (1999) Salinity, mineral nutrient relations in horticultural crops. Sci Hort 78:127–157

Harinasut P, Tsutsui K, Takabe T, Nomura M, Kishitani S (1996) Exogenous glycine betaine accumulation and increased salt tolerance in rice seedlings. Biosci Biotechnol Biochem 60:366–368

Hawkins HK, Lewis OAM (1993) Effect of NaCl salinity, nitrogen form, calcium and potassium concentration on nitrogen uptake and kinetics in *Triticum aestivum* L cv Gamtoos. New Phytol 124:171–177

Hayashi F, Ichino T, Osanai M, Wada K (2000) Oscillation and regulation of proline content by P5Cs and ProDH gene expressions in the light/dark cycles in *Arabidopsis thaliana* L. Plant Cell Physiol 41:1096–1101

Holford ICR (1997) Soil phosphorus: its measurement, and its uptake by plants. Aust J Soil Res 35:227–239

Hu Y, Schmidhalter U (1997) Interactive effects of salinity and macronutrient level on wheat. 2. Composition. J Plant Nutr 20:1169–1182

Jha P, Ali A, Raghuram N (2007) Nitrate-induction of nitrate reductase and its inhibition by nitrite and ammonium ions in *Spirulina platensis*. Physiol Mol Biol Plant 13:163–167

Kaiser WM, Huber SC (2001) Post-translational regulation of nitrate reductase: mechanism, physiological relevance and environmental triggers. J Exp Bot 52:1981–1989

Kaiser WM, Weiner H, Kandlbinder A, Tsai CB, Rockel P, Sonoda M, Planchet E (2002) Modulation of nitrate reductase: some new insights, an unusual case and a potentially important side reaction. J Exp Bot 53:875–882

Kavi Kishor PB, Sangam S, Amrutha RN, Sri Laxmi P, Naidu KR, Rao KRSS, Rao S, Reddy KJ, Theriappan P, Sreenivasulu N (2005) Regulation of proline biosynthesis, degradation, uptake and transport in higher plants: its implications in plant growth and abiotic stress tolerance. Curr Sci 88:424–438

Kaya C, Kirnak H, Higgs D (2001) Enhancement of growth and normal growth parameters by foliar application of potassium and phosphorus in tomato cultivars grown at high (NaCl) salinity. J Plant Nutr 24:357–367

Khan MG, Srivastava HS (1998) Changes in growth and nitrogen assimilation in maize plant induced by NaCl and growth regulators. Biol Plant 41:93–99

Kishor PBR, Hong Z, Miao GH, Hu CA, Verma DPB (1995) Overexpression of pyrroline-5-carboxylase synthetase increase proline production and confers osmotolerance in transgenic plants. Plant Physiol 108:1387–1394

Kriedemann PE, Downton WJS (1981) Photosynthesis. In: Paleg LG, Aspinall D (eds) The physiology and biochemistry of drought resistance in plants. Academic, London, pp 283–313

Kronzucker HJ, Britto DT, Davenport RJ, Tester M (2001) Ammonium toxicity and the real cost of transport. Trends Plant Sci 6:335–337

Kwinta J, Cal K (2005) Effects of salinity stress on the activity of glutamine synthetase and glutamate dehydrogenase in triticale seedlings. Polish J Environ Stud 14:125–130

Lee TM, Tsai CC, Shih MC (1999) Induction of phosphorus deficiency and phosphatase activity by salinity (NaCl) stress in *Gracilaria tenuistipitata* (Gigartinales, Rhodophyta). Phycologia 38:428–433

Leidi EO, Silberbush M, Lips SH (1991) Wheat growth as affected by nitrogen typ, pH and salinity. 1. Biomass production and mineral-composition. J Plant Nutr 14:235–246

Li SH, Xia BB, Zhang C, Cao J, Bai LH (2011) Cloning and characterization of a phosphate transporter gene in *Dunaliella salina*. J Basic Microbiol 51:1–8

López-Berenguer C, Martínez-Ballesta MC, Moreno DA, Carvajal M, García-Viguera C (2009) Growing hardier crops for better health: salinity tolerance and the nutritional value of broccoli. J Agric Food Chem 57:572–578

Loqué D, von Wirén N (2004) Regulatory levels for the transport of ammonium in plant roots. J Exp Bot 55:1293–1305

Lutts S, Majerus V, Kinet JM (1999) NaCl effects on proline metabolism in rice (*Oryza sativa* L.) seedlings. Physiol Plant 105:450–458

Maas EV, Ogata G, Finkel MH (1979) Salt-induced inhibition of phosphate transport and release of membrane protein from barley roots. Plant Physiol 64:139–143

Maighany F, Ebrahimzadeh H (2004) Intervarietal differences in nitrogen content and nitrate assimilation in wheat (*Triticum aestivum* L.) under salt stress. Pak J Bot 36:31–39

Mansour MMF (2000) Nitrogen containing compounds and adaptation of plants to salinity stress. Biol Plant 43:491–500

Marschner H (1995) Adaptation of plants to adverse chemical soil conditions. In: Marschner H (ed) Mineral nutrition of higher plants. Academic, London, pp 596–681

Marschner H (2002) Mineral nutrition of higher plants, 2nd edn. Academic, London

Martinez V, Cerda A (1989) Influence of N-source on rate of Cl, N, Na and K uptake by cucumber seedlings grown in saline condition. J Plant Nutr 12:971–983

Martinez B, Lauchli A (1994) Salt-induced inhibition of phosphate uptake in plants of cotton (*Gossypium hirsutum* L.). New Phytol 125:609–614

Martinez V, Bernstein N, Läuchli A (1996) Salt-induced inhibition of phosphorus transport in lettuce plants. Physiol Plant 97:118–122

Masclaux-Daubresse C, Daniel-Vedele F, Dechorgnat J, Chardon F, Gauflichon L, Suzuki A (2010) Nitrogen uptake, assimilation and remobilization in plants: challenges for sustainable and productive agriculture. Ann Bot 105:1141–1157

Maurino VG, Peterhansel C (2010) Photorespiration: current status and approaches for metabolic engineering. Curr Opin Plant Biol 13:249–256

Meloni DA, Gulotta MR, Martínez CA, Oliva MA (2004) The effects of salt stress on growth, nitrate reduction and proline and glycinebetaine accumulation in *Prosopis alba*. Braz J Plant Physiol 16:39–46

Mimura T, Reid RJ, Ohsumi Y, Smith FA (2002) Induction of the Na+/Pi cotransport system in the plasma membrane of Chara corallina requires external Na+ and low levels of Pi. Plant Cell Environ 25:1475–1481

Miura K, Hasegawa PM (2010) Sumoylation and other ubiquitin-like post-translational modifications in plants. Trends Cell Biol 20:223–232

Miura K, Rus A, Sharkhuu A, Yokoi S, Karthikeyan AS, Raghothama KG, Baek D, Koo YD, Jin JB, Bressan RA, Yun DJ, Hasegawa PM (2005) The *Arabidopsis* SUMO E3 ligase SIZ1 controls phosphate deficiency responses. Proc Natl Acad Sci USA 102:7760–7765

Miura K, Lee J, Miura T, Hasegawa PM (2010) SIZ1 controls cell growth and plant development in Arabidopsis through salicylic acid. Plant Cell Physiol 51:103–113

Miura K, Sato A, Ohta M, Furukawa J (2011) Increased tolerance to salt stress in the phosphate-accumulating *Arabidopsis* mutants *siz1* and *pho2*. Planta 234:1191–1199

Monneuse JM, Sugano M, Becue T, Santoni V, Hem S, Rossignol M (2011) Towards the profiling of the *Arabidopsis thaliana* plasma membrane transportome by targeted proteomics. Proteomics 11:1789–1797

Munns R, Tester M (2008) Mechanisms of salinity tolerance. Anuu Rev Plant Biol 59:651–681

Naheed G, Shahbaz M, Akram NA (2008) Interactive effects of rooting medium application of phosphorus and NaCl on plant biomass and mineral nutrients of rice (*Oryza sativa* L.). Pak J Bot 40:1601–1608

Navarro JM, Botella MA, Cerda A, Martinez V (2001) Phosphorus uptake and translocation in salt-stressed melon plants. J Plant Physiol 158:375–381

Nenova V (2008) Growth and mineral concentrations of pea plants under different salinity levels and iron supply. General Applied Plant Physiol 34:189–202

Nieman RH, Clark RA (1976) Interactive effects of salinity and phosphorus nutrition on the concentrations of phosphate and phosphate esters in mature photosynthesizing corn leaves. Plant Physiol 57:157–161

Nieves M, Nieves-Cordones M, Poorter H, Simón MD (2011) Leaf nitrogen productivity is the major factor behind the growth reduction in induced by long-term salt stress. Tree Physiol 31:92–101

Okusanya OT, Fawole T (1985) The possible role of phosphate in the salinity tolerance of *Lavatera arborea*. J Ecol 73:317–322

Ouyang B, Yang T, Li H, Zhang L, Zhang Y, Zhang J, Fei Z, Ye Z (2007) Identification of early salt stress response genes in tomato root by suppression subtractive hybridization and microarray analysis. J Exp Bot 58:507–520

Pan SM (1987) Characterization of multiple acid phosphatases in salt-stressed spinach leaves. Aust J Plant Physiol 14:117–124

Parida AK, Das AB (2004) Effects of NaCl stress on nitrogen and phosphorus metabolism in a true mangrove *Bruguiera parviflora* grown under hydroponic culture. J Plant Physiol 161: 921–928

Pavón LR, Lundh F, Lundin B, Mishra A, Persson BL, Spetea C (2008) Arabidopsis ANTR1 is a thylakoid Na+-dependent phosphate transporter: functional characterization in *Escherichia coli*. J Biol Chem 283:13520–13527

Peuke AD, Jeschke WD (1999) The characterization of inhibition of net nitrate uptake by salt in salt-tolerant barley (*Hordeum vulgare* L. cv. California Mariout). J Exp Bot 50: 1365–1372

Phang TH, Shao G, Liao H, Yan X, Lam HM (2009) High external phosphate (Pi) increases sodium ion uptake and reduces salt tolerance of 'Pi-tolerant' soybean. Physiol Plant 135:412–425

Pitman MG, Läuchli A (2002) Global impact of salinity and agricultural ecosystems. In: Läuchli A, Lüttge U (eds) Salinity: environment – plants – molecules. Kluwer, Dordrecht, pp 3–20

Popova OV, Ismailov SF, Popova TN, Dietz KJ, Golldack D (2002) Salt-induced expression of NADP-dependent isocitrate dehydrogenase and ferredoxin-dependent glutamate synthase in *Mesembryanthemum crystallinum*. Planta 215:906–913

Qadar A (1998) Alleviation of sodicity stressed rice genotypes by phosphorus fertilization. Plant Soil 203:269–277

Rabe B (1990) Stress physiology: the functional significance of the accumulation of nitrogen containing compounds. J Hort Sci 65:231–243

Rai AK, Sharma NK (2006) Phosphate metabolism in the cyanobacterium *Anabaena doliolum* under salt stress. Curr Microbiol 52:6–12

Rao KR, Gnaham A (1990) Inhibition of nitrate and nitrate reductase activity by salinity stress in *Sorghum vulgare*. Phytochemistry 29:1047–1049

Rao DLN, Giller KE, Yeo AR, Flowers TJ (2002) The effects of salinity and sodicity upon nodulation and nitrogen fixation in chickpea (*Cicer arietinum*). Ann Bot 89:563–570

Rausch C, Bucher M (2002) Molecular mechanisms of phosphate transport in plants. Planta 216:23–37

Reda M, Migocka M, Klobus G (2011) Effect of short-term salinity on the nitrate reductase activity in cucumber roots. Plant Sci 180:783–788

Rhodes D, Hanson AD (1993) Quaternary ammonium and tertiary sulfonium compounds in higher-plants. Annu Rev Plant Physiol Plant Mol Biol 44:357–384

Roberts JKM, Linker CS, Benoit AG, Jardetzky O, Nieman RH (1984) Salt stimulation of phosphate uptake in maize root tips studied by ^{31}P nuclear magnetic resonance. Plant Physiol 75: 947–950

Rogers ME, Grieve CM, Shannon MC (2003) Plant growth and ion relations in lucerne (*Medicago sativa* L.) in response to the combined effects of NaCl and P. Plant Soil 253:187–194

Rubio L, Linares-Rueda A, García-Sánchez MJ, Fernández JA (2005) Physiological evidence for a sodium-dependent high-affinity phosphate and nitrate transport at the plasma membrane of leaf and root cells of *Zostera marina* L. J Exp Bot 56:613–622

Sawahel WA, Hassan AH (2002) Generation of transgenic wheat plants producing high levels of the osmoprotectant proline. Biotech Lett 24:721–725

Saxena AK, Rewari RB (1991) Influence of phosphate and zinc on growth, nodulation and mineral composition of chickpea (*Cicer arietinum* L.) under salt stress. World J Microbiol Biotechnol 7:202–205

Segawa H, Onitsuka A, Kuwahata M, Hanabusa E, Furutani J, Kaneko I, Tomoe Y, Aranami F, Matsumoto N, Ito M, Matsumoto M, Li M, Amizuka N, Miyamoto K (2009) Type IIc sodium-dependent phosphate transporter regulates calcium metabolism. J Am Soc Nephrol 20:104–113

Shaddad MM (1990) The effect of proline application on the physiology of *Raphanus sativus* plants grown under stress. Biol Plant 32:104–112

Shannon MC (1997) Adaptation of plants to salinity. Adv Agron 60:75–119

Shen W, Huber SC (2006) Polycations globally enhance binding of 14-3-3ω to target proteins in spinach leaves. Plant Cell Physiol 47:764–771

Shenker M, Ben-Gal A, Shani U (2003) Corn growth and uptake under combined nitrogen and salinity environmental stresses. Plant Soil 256:139–147

Shi H, Quintero FJ, Pardo JM, Zhu JK (2002) The putative plasma membrane Na^+/H^+ antiporters SOS1 controls long-distance Na^+ transport in plants. Plant Cell 14:465–477

Shibli RA, Sawwan J, Swaidat I, Takat M (2001) Increased phosphorus mitigates the adverse effects of salinity in tissue culture. Common Soil Sci Plant Anal 32:429–440

Shokri S, Maadi B (2009) Effect of arbuscular mycorrhizal fungus on the mineral nutrition and yield of *Trifolium alexandrinum* plants under salinity stress. J Agron 8:79–83

Siddiqui MH, Mohammad F, Khan MMA, Al-Whaibi MH (2012) Cumulative effect of nitrogen and sulphur on *Brassica juncea* L. genotypes under NaCl stress. Protoplasma 249:139–153

Silva-Ortega CO, Ochoa-Alfaro AE, Reyes-Agüero JA, Aguado-Santacruz GA, Jiménez-Bremont JF (2008) Salt stress increases the expression of *p5cs* gene and induces proline accumulation in cactus pear. Plant Physiol Biochem 46:82–92

Silveira JAG, Melo ARB, Viegas RA, Oliveira JTA (2001) Salinity-induced effects on nitrogen assimilation related to growth in cowpea plants. Environ Exp Bot 46:171–179

Skriver K, Mundy J (1990) Gene expression in response to abscisic acid and osmotic stress. Plant Cell 2:503–512

Solomonson LP, Barber MJ (1990) Assimilatory nitrate reductase: functional properties and regulation. Annu Rev Plant Physiol Plant Mol Biol 41:225–253

Speer M, Kaiser WM (1994) Replacement of nitrate by ammonium as the nitrogen-source increases the salt sensitivity of pea plants. 2. Intercellular and intracellular solute compartmentation in leaflets. Plant Cell Environ 17:1223–1231

Szabo-Nagy A, Galiba G, Erdei L (1992) Induction of soluble phosphatases under ionic and non-ionic osmotic stresses in wheat. J Plant Physiol 140:629–633

Taiz L, Zeiger E (2006) Plant physiology, 4th edn. Sinauer Associates, Inc., Sunderland

Talke IN, Blaudez D, Maathuis FJM, Sanders D (2003) CNGCs: prime targets of plant cyclic nucleotide signaling? Trends Plant Sci 8:286–293

Tcherkez G (2011) Natural $^{15}N/^{14}N$ isotope composition in C_3 leaves: are enzymatic isotope effects informative for predicting the ^{15}N-abundance in key metabolites? Funct Plant Biol 38:1–12

Teixeira J, Fidalgo F (2009) Salt stress affects glutamine synthetase activity and mRNA accumulation on potato plants in an organ-dependent manner. Plant Physiol Biochem 47:807–813

Treeby MT, van Steveninck RFM (1988) Effects of salinity and phosphate on ion distribution in lupin leaflets. Physiol Plant 73:317–322

Tsay YF, Chiu CC, Tsai CB, Ho CH, Hsu PK (2007) Nitrate transporters and peptide transporters. FEBS Lett 581:2290–2300

van Hoorn JW, Katerji N, Hamdy A, Mastrorilli M (2001) Effect of salinity on yield and nitrogen uptake of four grain legumes and on biological nitrogen contribution from the soil. Agr Water Manag 51:87–98

Vance CP, Uhde-Stone C, Allan DL (2003) Phosphorus acquisition and use: critical adaptations by plants for securing a non-renewable resource. New Phytol 157:423–447

Viégas RA, Silveira JAG (1999) Ammonia assimilation and proline accumulation in young cashew plants during long term exposure to NaCl-salinity. Revista Brasileira de Fisiologia Vegetal 11:153–159

Vinocur B, Altman A (2005) Recent advances in engineering plant tolerance to abiotic stress: achievements and limitations. Curr Opin Biotech 16:123–132

Wahid A, Hameed M, Rasul E (2004) Salt-induced injury symptom, changes in nutrient and pigment composition and yield characteristics of mungbean. Int J Agric Biol 6:1143–1152

Wang ZQ, Yuan YZ, Ou JQ, Lin QH, Zhang CF (2007) Glutamine synthetase and glutamate dehydrogenase contribute differentially to proline accumulation in leaves of wheat (Triticum aestivum) seedlings exposed to different salinity. J Plant Physiol 164:695–701

Wingler A, Lea PJ, Quick WP, Leegood RC (2000) Photorespiration: metabolic pathways and their role in stress protection. Philos Trans R Soc Lond B 355:1517–1529

Yan SP, Zhang QY, Tang ZC, Su WA, Sun WN (2006) Comparative proteomic analysis provides new insights into chilling stress responses in rice. Mol Cell Proteomics 5:484–496

Yang X, Lu C (2005) Photosynthesis is improved by exogenous glycine-betaine in salt-stressed maize plants. Physiol Plant 124:343–352

Yao J, Shi WM, Xu WF (2008) Effects of salt stress on expression of nitrate transporter and assimilation-related genes in tomato roots. Russ J Plant Physiol 55:232–240

Yousfi S, M'sehli W, Mahmoudi H, Abdelly C, Gharsalli M (2007) Effect of salt on physiological responses of barley to iron deficiency. Plant Physiol Biochem 45:309–314

Yousfi S, Serret MD, Márquez AJ, Voltas J, Araus JL (2012) Combined use of $\delta^{13}C$, $\delta^{18}O$ and $\delta^{15}N$ tracks nitrogen metabolism and genotypic adaptation of durum wheat to salinity and water deficit. New Phytol 194:230–244

Zaiter HZ, Saade M (1993) Interactive effects of salinity and phosphorus-nutrition on tepary and common bean cultivars. Commun Soil Sci Plant Anal 24:109–123

Zhonghua T, Yanju L, Xiaorui G, Yuangang Z (2011) The combined effects of salinity and nitrogen forms on *Catharanthus roseus*: the role of internal ammonium and free amino acids during salt stress. J Plant Nutr Soil Sci 174:135–144

Zribi OT, Abdelly C, Debez A (2011) Interactive effects of salinity and phosphorus availability on growth, water relations, nutritional status and photosynthetic activity of barley (*Hordeum vulgare* L.). Plant Biol 13:872–880

Chapter 17
Potentiality of Sulphur-Containing Compounds in Salt Stress Tolerance

M. Iqbal R. Khan, M. Asgher, Noushina Iqbal, and Nafees A. Khan

17.1 Introduction

Among several abiotic stresses, salinity is one of the major abiotic stress that plants encounter. Soil salinity is a major abiotic stress affecting agricultural productivity (Ahmad 2010; Khan et al. 2012). It has been predicted that increased salinization of aerable land is expected to have devastating global effects resulting in 30% land loss within the next 25 years, and up to 50% by 2050 (Wang et al. 2003). It is estimated that abiotic stress such as, salinity, high temperature and drought leads to an average yield loss of >50% for most major crop plants (Boyer 1982; Bray et al. 2000). Soil fertility is adversely affected by salinity and has emerged as one of the most serious factors limiting plant growth and productivity including soil health (Khan et al. 2009b; Türkan and Demiral 2009). According to an estimate by Food and Agricultural Organization (2008), over 6% of the world's land is salt affected. In addition, out of 230 million hectares of irrigated land, 45 million hectares (~20%) are affected by salinity stress. Plants are affected in several ways by increasing salt concentrations. It causes osmotic stress, specific ion toxicity and nutrient deficiencies, thereby affecting a range of physiological processes involved in cell metabolism (Munns 2002; Ahmad and Umar 2011; Ahmad and Prasad 2011a, b; Ahmad et al. 2012; Iqbal et al. 2012; Katare et al. 2012). Salt stress influences all the major processes such as photosynthesis, protein synthesis, energy and lipid metabolism (Parida and Das 2005). According to Sreenivasulu (Sreenivasulu et al. 2007) the adverse affects of salt stress on plant growth are due to:

1. Reduction in the osmotic potential of the soil solution that reduces plant available water leading to water stress in plants

M.I.R. Khan (✉) • M. Asgher • N. Iqbal • N.A. Khan (✉)
Department of Botany, Aligarh Muslim University,
Aligarh 202002, India
e-mail: amu.iqbal@gmail.com; naf9@lycos.com

P. Ahmad et al. (eds.), *Ecophysiology and Responses of Plants under Salt Stress*,
DOI 10.1007/978-1-4614-4747-4_17,

2. Deteriotion of soil physical structure such that water permeability and soil aeration are diminished
3. Increase in the concentration of Na^+ ion causing severe ion toxicity, since Na^+ is not readily sequestered into vacuoles as in halophytes, and
4. Nutrient imbalances and deficiencies due to the interaction of salts with mineral nutrition.

These overall effects can ultimately lead to plant death.

Genotypic variation in plants may occur in response to salinity stress and production of oxidative stress (Khan and Panda 2008). The production of reactive oxygen species (ROS) is ubiquitous during metabolism and all plants can cope with them. However, considerable increase in ROS production can lead to a substantial cellular damage (Mittler 2002; Ahmad et al. 2010a, 2011; Ahmad and Umar 2011; Ahmad and Prasad 2011a, b). The question arises as to how a plant controls and speeds up its rate of ROS production and ROS scavenging when exposed to an abiotic stress particularly salinity.

Therefore, increasing the yield of crop plants in soils including salinized regions, is essential for feeding the world and strategies have to be evolved to improve salt stress tolerance (Blumwald et al. 2004). Any adaptation that regulates ROS generation in plants will provide efficient defense mechanism for tolerance against stress. The use of techniques and/or methods to alleviate adverse effects of salinity stress is expected to result in sustainable crop productivity. However, the management of salinity stress has been considered difficult because of its dynamic nature. Different strategies have been adopted to counteract the salinity effects (Ashraf 2009; Türkan and Demiral 2009), but the information on mineral nutrient status of plants and salinity tolerance is scarce (Choi et al. 2004; Al-Harbi et al. 2008; Khan et al. 2009b, 2010; Khorshidi et al. 2009).

Sulphur constitutes as one of the macronutrients necessary for the plant life cycle. Sulphur availability and sulphur content have a significant impact on plant productivity and utilization. Sulphur is an essential element for plant growth because it is present in major metabolic compounds such as amino acids (methionine and cysteine), antioxidant (glutathione), proteins, and sulpho-lipids (Abdallah et al. 2010) (Fig. 17.1). Sulphur is a ubiquitous and essential element for all living organisms. In plants besides these amino acids (i.e., cysteine and methionine), sulphur is also a component of iron-sulphur clusters, polysaccharides and lipids as well as in a broad variety of biomolecules such as vitamins (biotin and thiamine), cofactors (CoA and S-adenosyl- Methionine), peptides (glutathione and phytochelatins) and secondary products (allyl cysteine sulphoxides and glucosinolates) (Nocito et al. 2007).

Sulphur metabolism plays an important role in plant stress response (Rausch and Wachter 2005; North and Kopriva 2007). Since salinity is considered as one of the potential threats for agricultural productivity, the chapter focuses mainly to improve our understanding on the effects of salinity stress on plant physiology and metabolism and elucidate the potential of various sulphur-containing compounds in modulating salinity stress response.

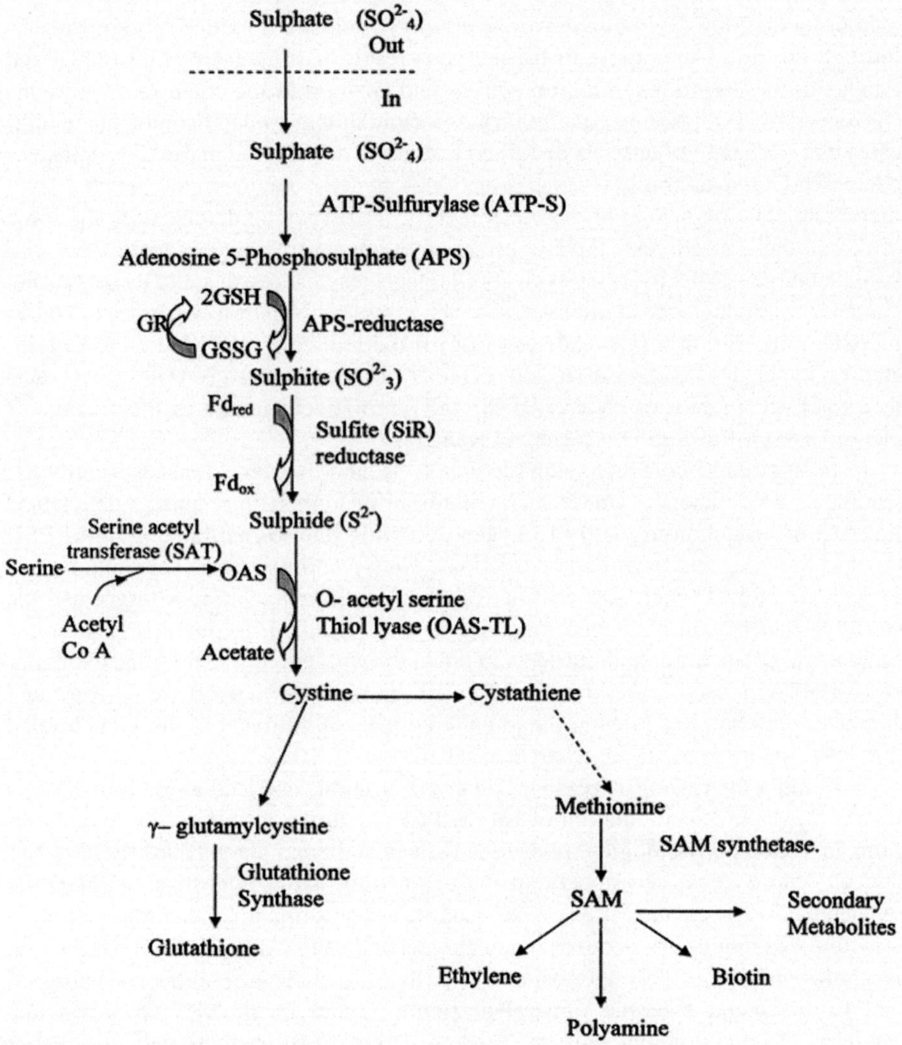

Fig. 17.1 Outline of sulphur assimilation and sulphur compounds derivation

17.2 Salt Stress: Consequences and Mechanism of Detoxification

Salt stress invariably affects photosynthesis and its related physiological variables are consistently affected by salinity stress in plants (Parida and Das 2005; Chaves et al. 2009; Nazar et al. 2011a; Ahmad et al. 2012; Shu et al. 2012). Photosynthesis is among the primary processes, together with cell growth, to be affected by salinity

(Munns et al. 2006). In view of available reports, salt stress reduces photosynthesis through limiting CO_2 supply to the leaf as a result of stomatal closure (Khan and Panda 2008), alterations in light reactions and changes in the contents of nutrients (Flexas et al. 2007; Khan et al. 2009b). A decline in the stomatal conductance indicates that exchange of gases is disturbed leading to a reduction in the CO_2 uptake or decreased CO_2 availability (Flexas et al. 2007), hence a decline in the net photosynthetic rate. The oxygen evolving machinery in PSII may be damaged by the ionic effects or these might have adverse effect on pigment protein complex of the thylakoid membrane (Allakhverdiev et al. 2000). Plants respond to salt stress by decreasing stomatal conductance to avoid excessive water loss. This in turn decreases the internal CO_2 concentration (Ci) and slows down the reduction of CO_2 by the Calvin-Benson cycle. This response leads to depletion of the oxidized $NADP^+$, which acts as a final acceptor of electrons in PSI, and alternatively increases the leakage of electrons to O_2 forming $O^-{}_2$ (Hsu and Kao 2003).

High salt uptake competes with the uptake of other nutrients ions, especially K^+, leading to K^+ deficiency. Under such conditions of high salinity and K^+ deficiency, a reduction in quantum yield of oxygen evolution due to malfunctioning of PSII occurs, showing photosynthetic reductions. Under salinity stress, accumulation of excess Na^+ and Cl^- decreases chemical activity causing cells to lose turgor and the stress caused by ion concentrations allows water gradient to decrease, making it more difficult for water and nutrients to move through the root membrane (Volkmar et al. 1998). Increasing salinity levels have been reported to decrease K^+, Ca^{2+} and N concentration in leaf resulting in impaired uptake of nutrients (Ashraf and Foolad 2007; Khorshidi et al. 2009; Khan et al. 2010).

In salinity, the rate of increase in salt concentrations can lead to ion toxicity primarily due to the accumulation of Na^+ and Cl^-, resulting loss in turgor and alterations in various physiological processes (Manchanda and Garg 2008). Excess Na^+ and Cl^- cause negative impacts on the acquisition and homeostasis of essential nutrients (Greenway and Munns 1980) leading to conformational changes in protein structure and membrane depolarization (Manchanda and Garg 2008). The reduction in photosynthesis has been associated with the disturbance in homeostasis of Na^+ and Cl^- ions and essential mineral nutrients (Gunes et al. 2007; Keutgen and Pawelzik 2009), stomatal closure (Steduto et al. 2000), reduction in leaf water potential (Silva et al. 2008), and the increased production of ROS in chloroplasts (Meneguzzo et al. 1999). It is also suggested that Cl^- ion accumulation adversely affects photosynthesis (Khayyat et al. 2009).

Furthermore, Na^+/Cl^- toxicity resulting from salt stress could disrupt the photosynthetic electron transport and provoke electron leakage to O_2 (Gossett et al. 1994; Borsani et al. 2001). This result in the decrease in Ci and slows down the reactions of Calvin cycle and induces photorespiration particularly in C3 plants, resulting in generation of more H_2O_2 in the peroxisome (Leegood et al. 1995). On other hand, the cell membrane-bound NADPH oxidase and the apoplastic diamine oxidase are supposed to be activated during salt stress and therefore, contribute to the generation of ROS (Mazel et al. 2004; Tsai et al. 2005). Salt stress increases the rates of respiration with the consequence of respiratory electron leakage to O_2 (Moser et al. 1991; Jeanjean et al. 1993).

Under saline environments, plant lipid metabolism is interrupted as a result of oxidative damage to membrane lipids by ROS and lipid peroxidation (Hernandez et al. 2002). NaCl in a dose-dependent manner increases the content of thiobarbituric acid reactive substances (TBARS) and H_2O_2 and lipid peroxidation resulting in oxidative stress in plant tissues (Jaleel et al. 2007).

17.2.1 Production of Reactive Oxygen Species (ROS) Under Salt Stress

Salt stress induces the production of ROS such as singlet oxygen ($^1O_2$), superoxide radical (O_2^-), hydrogen peroxide (H_2O_2), and hydroxyl radical (OH^-) (Ahmad et al. 2010a; Ahmad and Umar 2011). These ROS are necessary for inter- and intracellular signaling (Foyer and Noctor 1999; Sharma et al. 2012). However, at high concentrations they disrupt cell metabolism badly by oxidation of membrane lipids, proteins, and nucleic acids in plants (Noctor and Foyer 1998; Hernández et al. 2001).

Production of reactive oxygen species occurring in different cell organelles depends on, photosynthetic or non-photosynthetic tissues. In photosynthetic tissues, the chloroplast is the prime source of ROS having the capacity to produce high amounts of superoxide (O_2^-) and H_2O_2, especially during reduced rate of photosynthetic carbon fixation (Takahashi and Murata 2008). Whereas in non photosynthetic tissues, mitochondria are the major site for ROS production, but in a green cell of a plant, their contribution is considered small in comparison to chloroplasts (Navrot et al. 2007), although direct inference from non-photosynthetic organisms can underestimate their importance in ROS metabolism (Noctor et al. 2007).

Chloroplast is a major site of ROS production (Asada 2006) with H_2O_2 and O_2^- that arises as a consequence of the reduction of oxygen molecule (O_2) at photosystem I (PSI) (Asada 1999). $^1O_2$, a highly reactive excited state of O_2 (Halliwell and Gutteridge 1985) is also produced in PSII reaction centres that initiates fatty acid oxidation (Triantaphylides et al. 2008). Therefore, if one is to consider a role for ROS in retrograde signaling. H_2O_2, O_2^- and $^1O_2$ have been implicated in signaling (Apel and Hirt 2004; Mateo et al. 2004; Sattler et al. 2006) and consequently if produced in the chloroplast must be engaged in retrograde signaling from this organelle.

17.2.1.1 ROS Activation of Mitogen-Activated Protein Kinase (MAPK) Signaling Pathways Under Salt Stress

Mitogen-activated protein kinases (MAPKs) are a specific class of plant serine/threonine protein kinases that play a central role in the transduction of various extracellular and intracellular signals, including stress signals. These generally function as a cascade where MAPK is phosphorylated and activated by MAPK kinase (MAPKK), which itself is activated by MAPKK kinase (MAPKKK). All three of these kinases are interlinked together and are also called extracellular receptor kinases.

MAPK signaling modules are involved in eliciting responses to various stresses, H_2O_2 activates several MAPKs (Jonak et al. 2002). Sanan-Mishra et al. (2006) demonstrated the role of several plant MAP kinases in response to salinity. In *Arabidopsis*, H_2O_2 activates the MAPKs, MPK3, and MPK6 via MAPKKK ANP1, the over expression of ANP1 in transgenic plants resulted in increased tolerance to heat shock, freezing, and salt stress (Kovtun et al. 2000). A new MAPK gene (CbMAPK3) from *Chorispora bungeana* plant and its transcript level was up-regulated in response to salinity and cold stresses (Zhang et al. 2006a, b). It has been suggested that MAPK kinase signaling cascades activated by ROS helps in mediating salt stress tolerance in plants.

17.2.1.2 ROS as Signals Gene Expression

The microarray expression analysis has improved our knowledge regarding gene expression induced by ROS (Schena et al. 1995). Gene expression networks can be examined comprehensively by DNA microarrays during oxidative stress. Significant progress have been made in surveying gene expression in response to H_2O_2 in yeast (Gasch et al. 2000; Causton et al. 2001), animals (Finkel and Holbrook 2000), and higher plants (Desikan et al. 2000).

In plants, ROS-induced genes have been identified for receptor kinase (Desikan et al. 2000), and peroxisome biogenesis (Desikan et al. 2000). Approaches using cDNA profiling and DNA microarrays have analyzed large scale gene expression in response to ROS. Desikan et al. (2001) observed about 175 genes (i.e., 1–2% of the 11,000 genes on the microarray) among which 113 are induced and 62 are repressed. Of the total 113 induced genes, several encoded for proteins with antioxidant functions or were associated with defense responses or other stresses. Most of the genes out of 175 have no direct role in oxidative stress but may be linked indirectly, as a consequence of other biotic and abiotic stresses, explaining their sensitivity to H_2O_2. H_2O_2 induced genes encoding transcription factors, suggests that they may mediate downstream H_2O_2 responses. As in other organisms, expression of MAPKs induced by oxidative stress in *Arabidopsis,* in turn can mediate the induction of oxidative stress-responsive genes (Kovtun et al. 2000). Vranova et al. (2002) while analyzing the gene expression of tobacco plants treated with methyl viologen (MV) shows an altered expression of about 2% of the tobacco genes in acclimated leaves.

17.2.2 Detoxification of ROS Through SOS Pathway Under Salt Stress

Regulation of ion transport system is preliminary process to plant salt tolerance. Salt Overly Sensitive (SOS) pathway has a regulatory role under salt stress condition, as it regulates the ion homeostasis. In plants SOS signal-transduction pathway is extensively studied and is important for the maintenance of ion homeostasis and

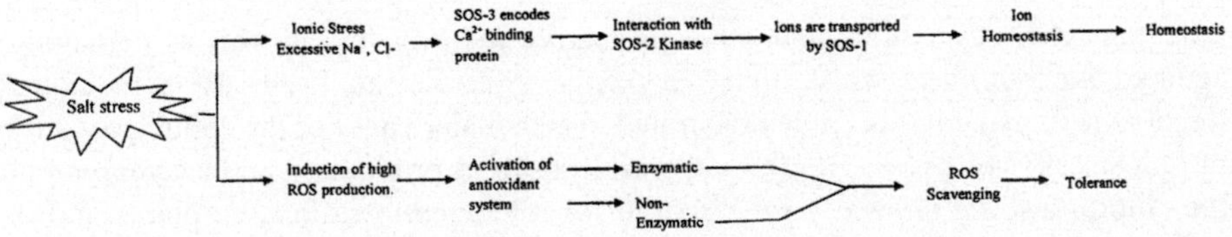

Fig. 17.2 Salt stress can be alleviated through these two pathways. Diagrammatic presentation of salt stress coping in plants through two separate systems

salt tolerance (Sanders 2000; Zhu 2003). In *Arabidopsis thaliana* SOS1 protein is the first putative plasma membrane Na^+/H^+ antiporter described in plants (Shi et al. 2000, 2002). The plasma membrane sited Na^+/H^+ antiporter SOS1 at the transcriptional and post-transcriptional level is controlled by the SOS pathway (Guo et al. 2001; Zhu 2001). It has been demonstrated that SOS mutants i.e. *sos1, sos2 and sos3*, could accumulate more proline under salt stress (Liu and Zhu 1997). Among the three SOS loci, SOS1 plays the greatest role in imparting salt-stress tolerance. *sos1* mutant plants are more sensitive to Na^+ stress as compared to *sos2* and *sos3* mutants (Zhu et al. 1998). SOS1 is located in plasma membrane responsible for extrusion of Na^+ out of the cell, while calcium activated SOS3–SOS2 protein complex is also involved in inhibiting HKT1, a low-affinity potassium transporter, that transports Na^+ ion under high-salt condition located in cytosol (Mahajan et al. 2008). Over expression of a plasma membrane Na^+/H^+ antiporter (SOS1) confers salt tolerance in *Arabidopsis*. It seems that extruding Na^+ ions from a higher plant is a feasible strategy to produce salt-tolerant crops (Shi et al. 2003). A loss of function mutation in the *Arabidopsis* SOS3 gene renders the mutant plants hypersensitive to NaCl. Interestingly, the salt-hypersensitive phenotype of *sos3* mutant plants can be partially rescued by increased concentrations of Ca^{2+} in growth media (Liu and Zhu 1997). SOS3 encodes an EF-hand Ca^{2+}-binding protein with a consensus sequence for N-myristoylation that functions as a calcium sensor for salt tolerance (Liu and Zhu 1998). SOS2 encodes a Ser/Thr protein kinase with an N-terminal kinase catalytic domain similar to SNF1/AMPK and a novel C-terminal regulatory domain (Liu et al. 2000). Salt stress elicits a transient increase of Ca^{2+} sensed by SOS3. Moreover SOS2 interacts as well a gets activated by SOS3 (Halfter et al. 2000). It has also been observed that SOS2/SOS3 kinase complex phosphorylates and activates SOS1 (Qiu et al. 2002) (Fig. 17.2).

Under salinity, Na^+ influx into root cells occurs via Na^+ permeable transporter (Amtmann et al. 1997; Tyerman et al. 1997), which in turn increases the sodium concentration and cause toxicity (Kingsbury and Epstein 1986).

17.2.3 Antioxidants in Salt Stress Detoxification

The toxic effect of salinity is through oxidative stress caused by enhanced production of ROS (Giraud et al. 2008; Ahmad et al. 2010a). These ROS may be signal inducing ROS scavengers and other protective mechanism, as well as damaging agents contributing to stress injury in plants (Prasad et al. 1994). ROS mediated membrane damage has been demonstrated to be a major cause of the cellular toxicity by salt stress (Mittova et al. 2004). Controlling ROS production and scavenging in the chloroplast are shown to be essential for tolerance to salinity in plants and in salinity-tolerant cultivars (Tseng et al. 2007). Singha and Choudhuri (1990) reported that H_2O_2 and O_2 are mainly responsible for NaCl-induced injury in *Vigna catjang* and *Oryza sativa* leaves. Antioxidant enzyme activities has been studied intensively and their significance in salt tolerance have also been proved, though these enzymes activities have shown to be associated with both salt tolerance as well as salt sensitivity due to differences in the ability to detoxify ROS (Munns and Tester 2008). Plants possess both enzymatic and non-enzymatic antioxidant defense systems to protect their cells against ROS (Ahmad et al. 2008, 2009, 2010a, 2011; Gill and Tuteja 2011). Gomathi and Rakkiyapan (2011) reported that the activity of antioxidant enzymes (ascorbate peroxidase, glutathione reductase, and superoxide dismutase) increased significantly under salt stress.

The scavenging of ROS by increased activation of antioxidant enzymes can improve salt tolerance (Alscher et al. 2002). A relationship between salt tolerance and increased activation of antioxidant enzymes has been demonstrated in several plants including mustard (Ahmad et al. 2010b, 2012; Syeed et al. 2011), mungbean (Nazar et al. 2011a), plantago (Sekmen et al. 2007), pea (Hernandez et al. 2000), *Arabidopsis*, rice (Dionisio-Sese and Tobita 1998), tomato, soyabean, maize (Azevedo-Neto et al. 2006) and *Vicia faba* (Azooz et al. 2011).

Salt stress is known to trigger oxidative stress in plant tissues through ROS production. To cope with oxidative stress-induced damage, ROS signals up regulation of antioxidants defense system. The increased expression of antioxidant system could be used as prospective selection criteria for breeding for salt tolerance in different crops (Ashraf 2009). Plants containing high levels of antioxidants can detoxify ROS thereby contributing to increased salt tolerance (Garratt et al. 2002). Considerable variations in the production of antioxidants, both enzymatic and non-enzymatic, in response to salt stress are evident at inter-specific or intra-specific level.

To control the level of ROS and to protect the cells under stress, plants possess the ability to detoxify ROS by producing different types of ROS scavenging compounds. The components of antioxidant defense system are enzymatic as well as non-enzymatic antioxidants (Fig. 17.2). Enzymatic antioxidants and non-enzymatic antioxidants in higher plants developed specific ROS-scavenging systems in different organelles to remove the ROS efficiently particularly under environmental stress such as salt stress. They coordinately work to provide plant cells with highly efficient machinery for detoxifying ROS.

Although manipulation of antioxidant genes seems to be a sound approach to counteract salt-induced oxidative stress. Attempts to improve oxidative stress tolerance particularly by manipulation of a single antioxidant gene has not always been successful, presumably because of the need for a balanced interaction of protective enzymes and other metabolites (Tseng et al. 2007). Thus, a balanced interaction of antioxidative enzymes as well as of other antioxidant metabolites may be vital to achieve a substantial improvement in plant stress tolerance (Ashraf 2009).

In plants over 150 genes encode for different ROS-detoxifying or ROS-producing enzymes forming well organized ROS gene web (Mittler et al. 2004). Several genes encoding for plant antioxidant enzymes have been cloned, characterized, and used in the construction of transgenic lines (Sarowar et al. 2005). Genetic analysis of the paraquet tolerant *Conyza bonariensis* indicated that all the three enzymes i.e. superoxide dismutase (SOD), ascorbate peroxidase (APX) and catalase (CAT) (of Halliwell-Asada pathway) co-segregate (Shaaltiel et al. 1988). Manipulation of genes that protect and maintain cellular functions or structure of cellular components has been targeted to produce transgenic plants enhancing stress tolerance. Molecular analysis of plants over expressing different enzymes of Halliwell-Asada cycle can provide us greater insights into the oxidative stress tolerance mechanism.

The chloroplast antioxidant system controls the cellular redox poise. Robert et al. (2009) demonstrated that accumulation of even apoplast superoxides is regulated by chloroplast redox signals. Redox imbalances are propagated over the chloroplast envelopes to the plasma membrane, where ROS-generation is activated. Vice versa, the chloroplast H_2O_2 detoxification system collapses if the cytosolic peroxidase capacity is limited (Pnueli et al. 2003; Mittler et al. 2006).

17.3 Regulation of Sulphate Uptake and Assimilation in Plants

17.3.1 Formation of Cysteine: The Key Compound

The source of sulphur in plants mainly the sulphate obtained from the soil in addition to sulphur oxide and hydrogen sulfide from the atmosphere are absorbed by the leaves through the stomata. Inorganic sulphur is taken up as sulphate from the soil through the root system and fixed into cysteine by the cysteine biosynthetic pathway (Saito 2000). The synthesis of cysteine represents the assimilation step of the reductive sulphate assimilation pathway. Serine acetyltransferase (SAT) and cysteine synthase (CSase) catalyze the sequential reactions leading to cysteine production which afterwards gets incorporated into proteins or antioxidant, or serves as sulphur donor of methionine or sulphur containing secondary products in plants.

Sulphate assimilation is highly regulated in a demand-driven manner (Lappartient and Touraine 1996; Leustek et al. 2000; Kopriva and Rennenberg 2004; Kopriva 2006). In plant sulphur assimilation, cysteine biosynthesis plays a central role in fixing inorganic sulphur from the environment into the metabolic precursor for cellular thiol-containing compounds. A key regulatory feature of this process is the physical association of the two enzymes involved in cysteine biosynthesis (serine acetyltransferase, SAT, and O-acetylserine sulfhydrylase, OASS) to form the cysteine synthase complex. Physiologically, this multienzyme complex acts as a molecular sensor in a regulatory circuit that coordinates sulphur assimilation and modulates cysteine production (Fig. 17.3).

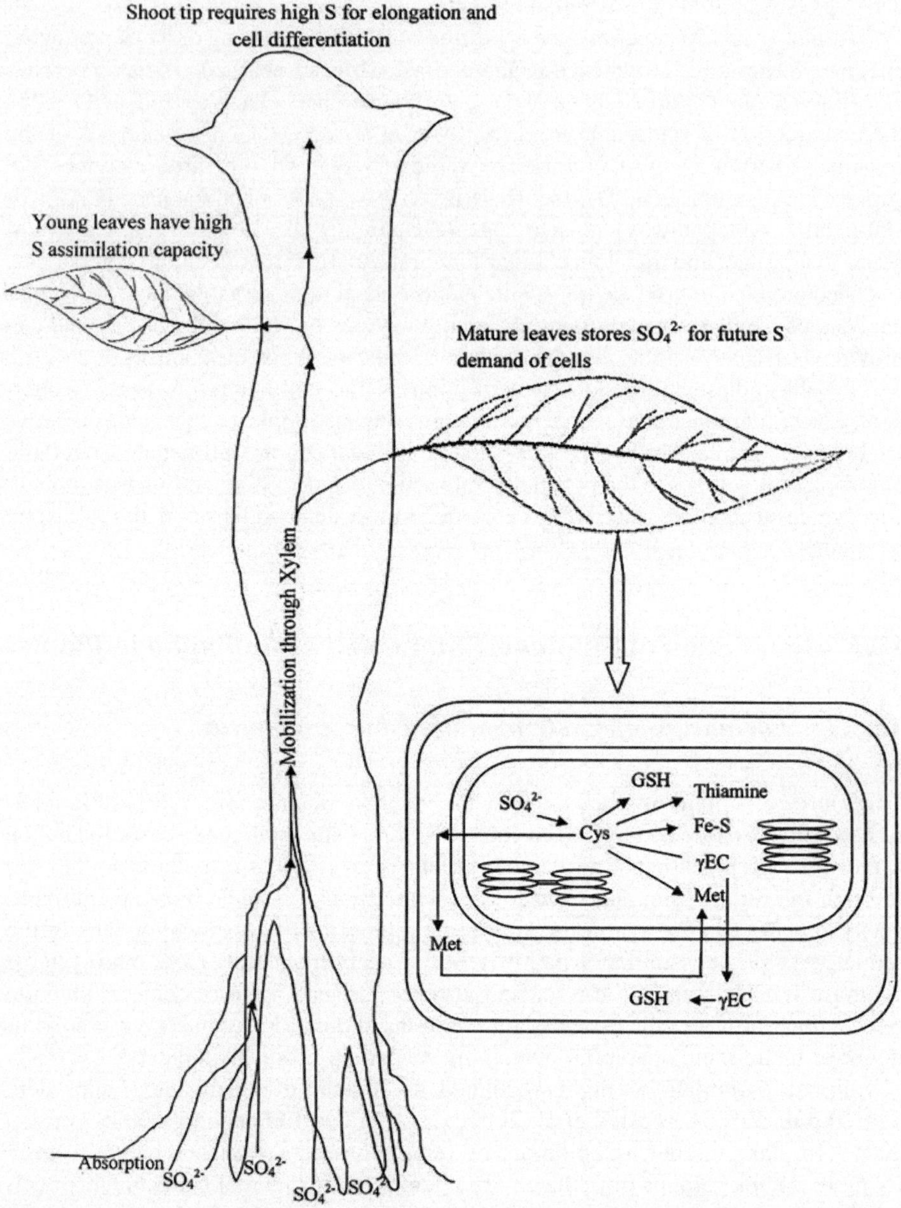

Fig. 17.3 Model of sulphate uptake and movement in plant and sulphate based on the function and tissue as well as developmental specific demand of sulphate. Absorption of sulphate completed by root of the plants, absorbed sulphate mobilized in plants through xylem and sink to leaves. Single cell illustrate the cysteine products and its presence in chloroplast. However methionine is synthesized in chloroplast but transported to cytosol of cell for next product of methionine

Metabolism of sulphate is initiated by its activation through the reaction of adenylation catalyzed by ATP sulphurylase (ATPS). ATPS is considered as the rate limiting enzyme of sulphur assimilation, as its over-expression enhances sulphur uptake by plants (Pilon-Smits et al. 1999). Activation of sulphate reduction is the dominant route for assimilation and is carried out in plastids (Leustek et al. 2000). Adenosine 5- phosphosulphate (APS) is reduced to sulfite by APS-reductase (APR). The key regulatory steps of sulphate assimilation are the transport of sulphate into the cells and the reduction of APS to sulfite by APR (Vauclare et al. 2002). Finally, sulfite is reduced to sulfide by sulfite reducatse (SiR). Sulfide is then transferred to activated serine by OASTL to form cystine.

An increased expression of sulphate transporters and enzymes of the assimilatory pathway are seen in response to stress (Schäfer et al. 1998; Heiss et al. 1999; Vanacker et al. 2000; Harada et al. 2002; Howarth et al. 2003a). Transgenic lines with increased fluxes to cysteine, and hence, protective and/or metal-chelating compounds such as glutathione and phytochelatins may contribute to an improved tolerance of plants to these stresses (Blaszczyk et al. 1999; Dominguez-Solis et al. 2001; Noji et al. 2001; Howarth et al. 2003b).

Cysteine synthesis in plants shows the final step of assimilatory sulphate reduction and almost an exclusive entry reaction of reduced sulphur into plant metabolism (Wirtz and Hell 2006). The importance of reduced sulphur is further illustrated by the multitude of functions that are directly or indirectly mediated by the major sulphur metabolites cysteine, methionine, and GSH. Further, increased rate of synthesis and accumulation of cysteine occur in order to form GSH during the abiotic defense response of plants to heavy metals or xenobiotics (Rüegsegger and Brunold 1992; Farago et al. 1994).

17.3.2 Compartmentation of Cysteine Metabolism

Intracellular and long-distance translocations are assumed to play an important role in plant sulphur metabolism. Transport of sulphate into chloroplasts is possibly mediated by a side-activity of the triose-phosphate translocator (Flugge Fischer et al. 1989). Cram (1990) has given little idea about the mechanism and specificity of tonoplast transport, although the vacuole presumably represents the main storage pool of sulphate.

It seems to be required in all cellular compartments that carry out protein biosynthesis, since serine acetyltransferase (SAT) and O-acetyl-serine (thiol) lyase (OAS-TL) have been found in cytosol, plastids, and mitochondria from various plants (Lunn et al. 1990; Rolland et al. 1992) possibly due to an inability of these organelles to transport cysteine across their membranes (Lunn et al.1990).

Studies of Heeg et al. (2008); Lopez-Martin et al. (2008) and Watanabe et al. (2008) have shown that mutants lacking cytosolic OAS TL A are fully viable, suggesting that plastid and possibly mitochondria are able to provide cysteine for cytosol requirements. Moreover, Heeg et al. (2008) reported the inability of mutant lacking plastidic OAS TL B, indicating that sulfide was able to exit the plastid to cytosol. Oastl AB double mutant, that lacks cysteine synthesis in the cytosol and plastids shows sulfide generated from sulphate in the plastid and diffuse through the cytosol in the mitochondria where it is integrated into cysteine.

17.3.3 Shoot to Root Signaling of Sulphur Under Salt Stress

Sulphate uptake is regulated by demand-driven control and this regulation is mediated by phloem-transported glutathione as a shoot-to-root signal. Exposure of the shoot to atmospheric H_2S enhanced the glutathione contents of leaves and roots and simultaneously, diminished sulphate uptake and xylem loading (Herschbach et al. 1995a, b; Rennenberg and Herschbach 1996; De Kok et al. 1997, 1998). It was shown that glutathione can signal the sulphur status of the shoot to the root (Rennenberg 1995; Rennenberg and Herschbach 1995). During S-deficiency, GSH content rapidly decreased in tobacco cell cultures but was rebuilt upon the resupply of sulphate and/or cysteine (Smith 1980). Sulphate transport consists of both constitutive and S nutrition- dependent regulated transport. A decreased intracellular content of sulphate, cysteine, and glutathione is concomitant with increasing transporter activity (Smith et al. 1997).

Salt stress induces the synthesis of cysteine and GSH and this may be a protective mechanism against salt-induced oxidative damage. Ruiz and Blumwald (2002) have studied the role of S assimilation and the biosynthesis of cysteine and GSH during the response to salinity stress of wild type and salt-tolerant transgenic *Brassica napus* L. (canola) plants over-expressing a vacuolar Na^+/H^+ antiporter. They have observed a threefold increase in cysteine and GSH contents in wild type plants exposed to salinity stress, but not in the transgenic plants. It has been shown that both APR activity and GSH level are increased in salt treated roots of plants and this occurs in a demand-driven manner where APR plays a key role in controlling the pathway (Lappartient and Touraine 1996; Kocsy et al. 2004; Kopriva 2006). This shows that S-assimilation is up-regulated in salt-stressed plants and this up-regulation is a demand driven.

Plants under salt stress increase their demand for GSH synthesis and this demand is transported downstream to the root to enhance S uptake and assimilation essential for GSH synthesis. Nazar et al. (2011b) reported GSH synthesis occurs in a demand-driven manner under salinity stress to scavenge ROS. Glutathione synthesized under salt stress is correlated with tolerance to salinity (Noctor et al. 1998; Kocsy et al. 2004; Ruiz and Blumwald 2002) and the synthesis of GSH depends upon S supply and S assimilation. Cysteine biosynthesis and hence GSH biosynthesis is linked to the pathway of reductive sulphate assimilation, a pathway regulated not only in response to the availability of sulphur, but also to environmental conditions that influence plant growth and development (Rausch and Wachter 2005).

Regulatory mechanism for sulphate uptake and assimilation results in direct sensing of the plant nutritional status rather than of the composition of the external solution supplied to the roots (Lappartient and Touraine 1996; Lappartient et al. 1999). This control necessary involves an inter-organ signaling mechanism in which a terminal product of the assimilatory pathway may act as long distance repressor signal. Experimental evidences suggest GSH as a phloem-translocated signal responsible for this regulation (Herschbach and Rennenberg 1991; Lappartient and Touraine 1996; Lappartient et al. 1999) generating a typical demand-driven coordinate transcriptional regulation of genes involved in sulphate assimilation and GSH biosynthesis. Sulphate withdraws from the growing medium decreases the levels of sulphate, cysteine, and GSH in plant tissues leading to the induction of sulphate transporter systems and key enzymes along the assimilatory pathway (Lappartient and Touraine 1996; Lappartient et al. 1999). Higher expression of ATPS activity has been shown necessary for the maintenance of optimal GSH levels required for the proper functioning of ascorbate-glutathione (AsA-GSH) cycle in mustard (Khan et al. 2009a). ATPS, the first enzyme in sulphur assimilation, is sensitive to glutathione concentrations and its activity may be controlled by the enzyme's ability to sense available thiol stores, not redox state (Yi et al. 2010).

17.4 Involvement of Sulphur Containing Compounds in Salt Tolerance

Sulphur does not play any specific role in biological systems directly; in many cases the presence of sulphur atoms in biomolecules is responsible for their catalytic or electrochemical properties and thus for their involvement in specific biochemical mechanisms. For example, the extreme nucleophilicity of the sulfhydryl group of cystine residues confers to many thiols -the capacity to react with a broad spectrum of agents, ranging from free radicals, active oxygen species, and cytotoxic electrophilic organic xenobiotics, to stress (Rabenstein 1989; Leustek et al. 2000), and accounting for the unique roles that amino acid plays in biological systems. Two cystine residues interact with each other in an oxidation reaction to form a disulfide bound which participates in maintaining protein structure and, sometimes, in the regulation of protein activity (Åslund and Beckwith 1999). The interconversion of two thiols in disulfide can also be involved in redox cycles, in maintaining the cell redox status and in the physiological responses to the oxidative stress. Such a simple cycle represents the chemical base making glutathione, a powerful cell redox buffer. Finally, GSH is the direct precursor of phytochelatins (PCs), a class of enzymatically synthesized cysteine rich peptides where the sulfhydryl groups of cysteine residues are involved in buffering potentially toxic elements (Zenk 1996; Na and Salt 2011).

17.4.1 Cysteine

The biosynthesis of cysteine constitutes the final step of the sulphur-reduction pathway in plants (Leustek and Saito 1999). Consequently, the formation of cysteine is the crucial step for assimilation of reduced sulphur into organic compounds (Noctor et al. 1998). The integration of reduced sulphur into cysteine is catalyzed by O-acetylserine (thiol) lyase (OASTL) and requires two substrates: free sulfide, provided by the sulphate-reduction pathway, and O acetylserine, an energetically-activated derivative of serine that is metabolically unique to cysteine synthesis (Wirtz et al. 2004). Studies have shown that exposure to salt-stress conditions induced higher rates of cysteine synthesis as a result of an increased expression of the cytosolic form of OASTL (Barroso et al. 1999; Fediuc et al. 2005; Romero et al. 2001) also it could be that this enzyme is related with salt tolerance (Fediuc et al. 2005; Romero et al. 2001).

During cysteine biosynthesis, an acetyl-group from acetyl-CoA is transferred to L-serine, creating the highly reactive compound O-acetyl-L-serine catalyzed by SAT. The subsequent synthesis of cysteine in plants is accomplished by the sulfihydrilation of O acetyl-L-serine in the presence of sulfide mediated by OAS-TL. Studies have shown that SAT activity is the limiting step in the regulation of cysteine synthesis (Blaszczyk et al. 1999; Hofgen et al. 2001).

Ruiz and Blumwald (2002) showed that the increase in NaCl in the growth medium led to an increase in SAT and OAS-TL activities both in wild-type and transgenic canola plants. Increase in SAT and OAS-TL (10% and 20%, respectively) was found in the transgenic plants growing in the presence of 150 mM NaCl, whereas wild type plants displayed a marked increase in SAT and OAS-TL activities (236% and 181%, respectively). In the same experiment, the salt-stress-induced differences in ATPS, SAT and OAS-TL activities between wild-type and transgenic plants correlated with the leaf cysteine concentrations. No differences in cysteine content was seen in the transgenic plants, whereas 150% increase in cysteine content was observed in wild-type plants.

17.4.2 Glutathione

Glutathione (GSH) – a tripeptide (cysteine, glutamic acid and glycine) acts as a storage and transport form of reduced S. It controls S assimilation and plays important roles including cell differentiation, cell death, control of redox status, protection against biotic and abiotic stresses, protein folding, precursor of phytochelatins and detoxification of xenobiotics (Foyer et al. 2001; Mullineaux and Rausch 2005). Furthermore, GSH acts as an antioxidant, quenching the ROS generated in response to various stresses. Cysteine is the precursor of GSH, a low molecular weight, water soluble non-protein thiol compound that functions in protection of plants against varied environmental stresses (De Kok et al. 2005).

Studies support the notion that cysteine availability in plants also plays an important role in determining cellular GSH concentrations through the kinetic restriction of the reaction catalyzed by γ-ECS (Noctor et al. 1997). Environmental conditions that stimulate synthesis of GSH will therefore place an increased demand upon S-assimilation into cysteine. Cysteine, the initial product of sulphate assimilation in plants, and its immediate metabolite GSH are both cellular constituents that originate from the three most important pathways of plants primary metabolism, i.e. photosynthesis, nitrogen assimilation, and sulphate assimilation (Leustek et al. 2000; Brunold et al. 2003; Kopriva and Rennenberg 2004). During salt stress, there is a high demand for cysteine, which is in agreement with the role of sulphuric compounds as antioxidants (Barroso et al. 1999).

Interconversion of two thiols disulfide involved in redox cycle is required for maintaining the cell redox status and the physiological responses to the oxidative stress. Disulfide/thiol exchange reactions involving the GSH pool and the production of H_2O_2 are believed to be crucial for regulation of stress adaptation at molecular level (Foyer et al. 1997). The oxidative stress induces a remarkable increase in the transcripts of two ascorbate peroxidase genes. These enzymes are cytoplasmic and are involved in regulating the redox level of the GSH pool. SODs, ascorbate peroxidase, dehydroascorbate reductase (DHAR) and GSH are associated with defence mechanism against stress. It is important to note that sulphur deprivation leads to oxidative stress in plants (Pilon-Smits and Pilon 2006). Such a simple cycle represents the chemical base making GSH a powerful cell redox buffer. The sulfhydryl group in the GSH structure arises from its capability to readily react with a wide range of electrophilic compounds to form covalently bound glutathione S-conjugates (Leustek et al. 2000) catalyzed by glutathione S-transferases (GSTs). GST has been shown to be essential in the detoxification processes of toxins, xenobiotics and catabolic products (Marrs et al. 1995; Marrs 1996; Leustek et al. 2000). Finally, GSH is the direct precursor of phytochelatins (PCs), a class of enzymatically synthesized cystine rich peptides in which the sulfhydryl groups of cystine residues are involved in buffering potentially toxic elements (Zenk 1996).

To maintain the protection of the cell against ROS and other free radicals, oxidized glutathione (GSSG) is permanently reduced by glutathione reductase to GSH by using NADPH ($GSSG + NADPH + H^+ \Rightarrow 2\,GSH + NADP^+$) (Wonisch and Schaur 2001). Similarly, Zagorchev et al. (2012) shows that total glutathione (GSH + GSSG) concentrations and redox state were associated with growth and development in control cultures and in moderately salt-stressed cultures and were affected by severe salt stress and also shows that the comparatively high contribution of ECySS/2 Cys to Ethiol–disulphide in cultures exposed to severe salt stress suggests that Cys and CySS may be important intracellular redox regulators with a potential role in stress signaling to maintain ROS scavenging and oxidative stress signaling.

In non-stress situations glutathione occurs to 90% in its reduced form. Deficiency of GSH within plant cells during oxidative stress situations demonstrates the need of the plant for cellular protection against ROS or other radicals whereas high levels of GSSG indicate oxidative stress within the plant. However, GSH levels within cells are not just influenced by oxidation and reduction processes, but are even more the product of equilibrium between synthesis, degradation, use, and short- and long-distance transport of glutathione and its precursors (Foyer et al. 2001). Especially, the availability of GSH precursors in plastids and the cytosol are interfering with GSH synthesis. Cysteine for example is supposed to be the rate-limiting factor during glutathione synthesis in non-stressed plants (Kopriva and Rennenberg 2004; Kopriva and Koprivova 2005) leading to a decreased ability of the plant to fight oxidative stress.

17.4.3 Thioredoxin Systems

Thioredoxins are small 12–13-kDa proteins with a redox active disulfide bridge and are widely distributed in all types of organisms (Schürmann and Jacquot 2000). Thioredoxins are heat-stable, ubiquitous proteins with a conserved pair of vicinal cysteines (–Trp-Cys-Gly-Pro-Cys- Lys-) (Holmgren 1989). Plants contain more types of thioredoxin systems than any other organism. The cytosolic system in plants is composed of type h thioredoxins reduced by NADPH in a reaction catalyzed by NADPH thioredoxin reductase (NTR) 1 (Schürmann and Jacquot 2000). In chloroplasts, the thioredoxin system includes type f, m, and x thioredoxins (Collin et al. 2003), which are reduced by ferredoxin in a reaction mediated by ferredoxin thioredoxin reductase (Schürmann and Jacquot 2000). This system is involved in light/dark regulation of chloroplast enzymes (Schürmann and Jacquot 2000; Ruelland and Miginiac-Maslow 1999; Buchanan 1991; Levings and Siedow 1995) and in oxidative stress responses (Motohashi et al. 2001; Balmer et al. 2003). The plant mitochondrial system is formed by a different type of thioredoxin (type o), which is reduced by an NADPH-dependent thioredoxin reductase similar to the cytosolic enzyme (Laloi et al. 2001). Thioredoxin is also believed to be involved in defense against oxidative stress through its ability to reduce hydrogen peroxide by acting as a hydrogen donor for a yeast peroxidase (Chae et al. 1994).

In addition, a thioredoxin/peroxiredoxin (Trx/Prx) system comprising of NADPH-dependent thioredoxin reductase (NTR) and glutaredoxin has been described in plant mitochondria similar to that in chloroplasts (Rouhier et al. 2005; Finkemeier et al. 2005). This system is involved in redox homeostasis and may also act as an antioxidant by eliminating hydroperoxides, including H_2O_2 (Laloi et al. 2001; Barranco-Medina et al. 2007). Recently, the presence of a new PsTrxo1 in pea leaves localized in mitochondria and the nucleus was reported. Moreover, alternative oxidase (AOX) has been shown to be regulated by the mitochondrial PsTrxo1 (Marti et al. 2009). The possible role of Trx and Prx in response of plants to abiotic stress including salinity and their involvement in plant tolerance have not been studied in detail (Barranco-Medina et al. 2007; Pulido et al. 2009; Tripathi et al. 2009), although a role in redox sensing and signal transduction have been proposed by Rouhier and Jaquot (2005).

Marti et al. (2011) examined the physiological function of PsTrxo1, transcript and protein levels of the PsTrxo1/PsPrxII F system in pea plants under short- (5 days) and long-term (14 days) salt stress conditions. It was found that short term (i.e., 5 days) provoked no change in the Trx activity, whereas a significant increase in Trx activity (20% increases as compared to the control plants) was observed in mitochondria isolated from plants grown in the presence of 150 mM NaCl for 14 days. Recently, Zhang et al. (2012) showed that OsTRXh1 regulates the redox state of the apoplast and influences plant development and salt stress responses.

17.4.4 Methionine and Its Derivatives

Methionine plays a multiple levels role in cellular metabolism: as a protein constituent, in the initiation of mRNA translation, and as a regulatory molecule in the form of S-adenosylmethionine (SAM). Methionine synthesis, accumulation, and consumption are under stringent regulatory control (Matthews 1999; Hesse and Höfgen 2003). The importance of methionine and/or its derivatives for promoting germination and seedling growth has been demonstrated in *Arabidopsis* (Gallardo et al. 2002). Overexpression of a gene involved in methionine biosynthesis have shown to increase salt tolerance suggests the methionine pathway is sensitive to salt stress and that methionine supplementation could improve salt tolerance (Glaser et al. 1993).

A relationship between the levels of SAM synthase (an enzyme in SAM synthesis from methionine) content was also observed in tomato plants, in which the level of SAM-S increased significantly under salt stress (Sanchez-Aguayo et al. 2004). Ogawa and Mitsuya (2012) demonstrated that the utilization of S-methylmethionine is important for the salinity tolerance of *Arabidopsis* plants at the germination and early growth stages.

Ethylene and polyamine phytohormones share same precursor SAM (Wang et al. 2002) and mediate salt tolerance (Cao et al. 2007; Liu et al. 2006). Ethylene receptor is the first component of ethylene signaling that regulates plant growth, development and stress responses. In *Arabidopsis*, ETR1 and ERS1 belong to the group I subfamily, while ETR2, EIN4, and ERS2 are subfamily II receptors. The expression of ETR1 gene seems to be down regulated by salt and osmotic stress at both transcription and protein levels in *Arabidopsis* (Zhao and Schaller 2004). The absence of spermine (polyamine) causes an imbalance in Ca^{2+} homeostasis in the mutant plant that determines the role of spermine in high salt stress responses (Yamaguchi et al. 2006).

17.4.5 Vitamins

Thiamine (Vitamin B1), in the form of thiamine diphosphate (TPP), acts as a cofactor for several enzymes in key cellular metabolic pathways such as glycolysis, the pentose phosphate pathway and the tricarboxylic acid cycle (TCA), in addition to amino acid and non mevalonate isoprenoid biosynthesis. It has been shown to involve in tolerance to DNA damage and as an activator of disease resistance in plants (Ahn et al. 2005, 2007).

Another thiamine precursor i.e., thiamine diphosphate (TDP) serves as a coenzyme in a number of the major metabolic pathways, including acetyl-CoA synthesis, the TCA cycle, anaerobic ethanolic fermentation, the oxidative pentose phosphate pathway, the Calvin cycle, the branched–chain amino acid pathway, and plant pigment biosynthesis (Friedrich 1987). Bacteria, yeasts, and higher plants can synthesize TDP from simple universal precursors (Begley et al. 1999; Nosaka 2006; Roje 2007). El-Shintinawy and El-Shourbagy (2001) observed that the addition of 2 μM thiamine in 100 mM NaCl concentration alleviates the reduction of growth in plants. However, the alleviation was greater in roots than shoots. In the same report, the alleviation of salt stress with thiamine was described to be related with the induction of certain low molecular mass proteins and increased contents of cystine and methionine.

Biotin (Vit. H) acts as a coenzyme, covalently bound to a lysine residue of a group of enzymes that catalyze carboxylation, decarboxylation or transcarboxylation reactions (Moss and Lane 1971). An essential role for biotin is its attachment to the active site of carboxylases such as acetyl CoA carboxylase (ACCase) and methycrotonyl CoA carboxylase (MCCase), that mediate carboxylation reactions in crucial metabolic processes such as the synthesis and catabolism of amino acids, fatty acids and isoprenoids (Nikolau et al. 2003; Alban et al. 2000).

In addition to its catalytic function as an enzyme bound prosthetic group, biotin may have a role in regulating gene expression (Che et al. 2003). Li et al. (2012) reported that biotin deficiency results in light-dependent spontaneous cell death and modulates defense gene expression. Hamdia (2000) observed that biotin or pyridoxine ameliorates salinity stress in lupine plants and on growth and metabolites of zea mays plant. However, more research is needed to evaluate the role of biotin in salt stress tolerance.

17.4.6 Coenzyme A

Coenzyme A is a cofactor for a multitude of enzymatic reactions, including the oxidation of fatty acids, carbohydrates, and amino acids, as well as many synthetic reactions (Begley et al. 2001). Rubio et al. (2006) identified and characterized *Arabidopsis* T-DNA knockout mutants of two CoA biosynthetic genes, HAL3A and HAL3B. The HAL3A gene encodes a 4′-phosphopantothenoyl-cysteine decarboxylase that generates 4′-phosphopantetheine. HAL3B, the gene product of which is 86% identical to that of HAL3A, present in the *Arabidopsis* genome. HAL3A appears to have a predominant role over HAL3B according to their respective mRNA expression levels. Furthermore over expression of HAL3A shows improved plant tolerance to salt stress (Espinosa-Ruiz et al. 1999; Yonamine et al. 2004).

Rubio et al. (2008) characterized a viable *Arabidopsis* T-DNA mutant affected in the penultimate step of the CoA biosynthesis pathway catalyzed by the enzyme phosphopantetheine adenylyltransferase (PPAT). They used ppat-1 knockdown mutant and found reduction in PPAT transcript levels (approximately 90%) impairing plant growth and seed production. The sum of CoA and acetyl-CoA levels was severely reduced (60–80%) in ppat-1 seedlings compared to wild type, and catabolism of storage lipids was delayed during seedling establishment. Conversely, PPAT overexpressing lines showed, on average, approximately 1.6-fold higher levels of CoA+acetyl-CoA levels, as well as enhanced vegetative and reproductive growth and salt/osmotic stress resistance. However, dry seeds of over expressing lines contained 35–50% more fatty acids than wild type, implying an essential role of CoA biosynthesis in storage oil accumulation. Biochemical analysis of the recombinant PPAT enzyme revealed an inhibitory effect of CoA on PPAT activity. These results suggest that the reaction catalyzed by PPAT is a regulatory step in the CoA biosynthetic pathway that plays a key role for plant growth, stress resistance, and seed lipid storage.

17.5 Conclusions and Future Perspective

Salinity is one of the dangerous threats to agriculture reducing soil fertility and crop productivity. Using sulphur or sulphur-containing compounds in agriculture can be a good tool in alleviation of salinity stress. In plants sulphur assimilates play dual role; they provide structural components of essential cellular molecules besides acting as signaling molecules for environmental communication. Sulphur-containing compounds enhance tolerance by modulating physiological processes and by up regulating specific gene for stress tolerance proteins. It is important to carry out detailed studies on sulphur assimilation pathway, its control through internal and external factors and to gain more physiologically relevant insight into plant responses to their environment.

Currently vitamins gained attention for research but still study on vitamins especially biotin is still in its beginning and it needs much attention for future research as biotin is an essential cofactor in many physiological processes and its deficiency can lead to plants death whereas its presence can alleviate salt stress. Regulation and expression of the majority of sulfate transporters are controlled by the sulfur nutritional status of the plant. It is important to carry out detailed studies on thiols, their subcellular compartmentation and the regulation of differential enzyme activities in different plant tissues. Future strategies are focused to modulate steps of S assimilation pathways leading to the production of thiols and their products in plants through manipulating enzymes in S-assimilation pathway under salinity stress.

Acknowledgement Authors gratefully acknowledge financial assistance received by the University Grants Commission, New Delhi.

References

Abdallah M, Dubousset L, Meuriot F, Etienne P, Avice JC, Ourry A (2010) Effect of mineral sulphur availability on nitrogen and sulphur uptake and remobilization during the vegetative growth of *Brassica napus* L. J Exp Bot 61:2335–2346

Ahmad P (2010) Growth and antioxidant responses in mustard (*Brassica juncea* L.) plants subjected to combined effect of gibberellic acid and salinity. Arch Agro Soil Sci 56(5):575–588

Ahmad P, Prasad MNV (2011a) Environmental adaptations and stress tolerance in plants in the era of climate change. Springer Science+Business Media, New York

Ahmad P, Prasad MNV (2011b) Abiotic stress responses in Plants: metabolism, productivity and sustainability. Springer Science+Business Media, New York

Ahmad P, Umar S (2011) Oxidative stress: role of antioxidants in plants. Studium Press, New Delhi

Ahmad P, Sarwat M, Sharma S (2008) Reactive oxygen species, antioxidants and signaling in plants. J Plant Biol 51(3):167–173

Ahmad P, Jeleel CA, Azooz MM, Nabi G (2009) Generation of ROS and non-enzymatic antioxidants during abiotic stress in plants. Bot Res Intern 2(1):11–20

Ahmad P, Jaleel CA, Salem MA, Nabi G, Sharma S (2010a) Roles of Enzymatic and non-enzymatic antioxidants in plants during abiotic stress. Crit Rev Biotechnol 30(3):161–175

Ahmad P, Jaleel CA, Sharma S (2010b) Antioxidative defence system, lipid peroxidation, proline metabolizing enzymes and Biochemical activity in two genotypes of *Morus alba* L. subjected to NaCl stress. Russ J Plant Physiol 57(4):509–517

Ahmad P, Nabi G, Ashraf M (2011) Cadmium-induced oxidative damage in mustard [*Brassica juncea* (L.) Czern. & Coss.] plants can be alleviated by salicylic acid. South Afr J Bot 77:36–44

Ahmad P, Hakeem KR, Kumar A, Ashraf M, Akram NA (2012) Salt-induced changes in photosynthetic activity and oxidative defense system of three cultivars of mustard (*Brassica juncea* L.). Afr J Biotechnol 11(11):2694–2703

Ahn IP, Kim S, Lee YH (2005) Vitamin B1 functions as an activator of plant disease resistance. Plant Physiol 138:1505–1515

Ahn IP, Kim S, Lee YH, Suh SC (2007) Vitamin B1-induced priming is dependent on hydrogen peroxide and the NPR1 gene in Arabidopsis. Plant Physiol 143:838–848

Alban C, Job D, Douce R (2000) Biotin metabolism in plants. Annu Rev Plant Physiol Plant Mol Biol 51:17–47

Al-Harbi AR, Wahb-Allah MA, Abu-Muriefah SS (2008) Salinity and nitrogen level affects germination, emergence, and seedling growth of tomato. Int J Veg Sci 14:380–392

Allakhverdiev SI, Sakamoto A, Nishiyama Y, Inaba M, Murata N (2000) Ionic and osmotic effects of NaCl-induced inactivation of photosystems I and II in Synechococcus sp. Plant Physiol 123:1047–1056

Alscher RG, Erurk N, Heath LS (2002) Role of superoxide dismutases (SODs) in controlling oxidative stress in plants. J Exp Bot 53:1331–1341

Amtmann A, Laurie S, Leigh R, Sanders D (1997) Multiple inward channels provide flexibility in Na^+/K^+ discrimination at the plasma membrane of barley suspension culture cells. J Exp Bot 48:481–497

Apel K, Hirt H (2004) Reactive oxygen species: metabolism, oxidative stress, and signal transduction. Annu Rev Plant Biol 55:373–399

Asada K (1999) The water-water cycle in chloroplasts: Scavenging of active oxygen and dissipation of excess photons. Ann Rev Plant Physiol Plant Mol Biol 50:601–639

Asada K (2006) Production and scavenging of reactive oxygen species in chloroplasts and their functions. Plant Physiol 141:391–396

Ashraf M (2009) Biotechnological approach of improving plant salt tolerance using antioxidants as markers. Biotechnol Adv 27:84–93

Ashraf M, Foolad MR (2007) Roles of glycine betaine and proline in improving plant abiotic stress tolerance. Environ Exp Bot 59:206–216

Åslund F, Beckwith J (1999) Bridge over troubled waters: sensing stress by disulfide bond formation. Cell 96:751–753

Azevedo-Neto AD, Prisco JT, Enéas-Filho J, Braga-de-Abreu CE, Gomes-Filho E (2006) Effect of salt stress on antioxidative enzymes and lipid peroxidation in leaves and roots of salt-tolerant and salt-sensitive maize genotypes. J Environ Exp Bot 56:87–94

Azooz MM, Youssef AM, Ahmad P (2011) Evaluation of salicylic acid (SA) application on growth, osmotic solutes and antioxidant enzyme activities on broad bean seedlings grown under diluted seawater. Inter J Plant Physiol Biochem 3:253–264

Balmer Y, Koller A, del Val G, Manieri W, Schürmann P, Buchanan BB (2003) Proteomics gives insight into the regulatory function of chloroplast thioredoxins. PNAS USA 100:370–375

Barranco-Medina S, Krell T, Finkemeier I, Sevilla F, Lazaro JJ, Dietz KJ (2007) Biochemical and molecular characterization of the mitochondrial peroxiredoxin PsPrxII F from *Pisum sativum*. Plant Physiol Biochem 45:729–739

Barroso C, Romero LC, Cejudo FJ, Vega JM, Gotor C (1999) Salt-specific regulation of the cytosolic O-acetylserine(thiol)lyase gene from Arabidopsis thaliana is dependent on abscisic acid. Plant Mol Biol 40:729–736

Begley TP, Downs DM, Ealick SE (1999) Thiamine biosynthesis in prokaryotes. Arch Microbiol 171:292–300

Begley TP, Kinsland C, Strauss E (2001) The biosynthesis of coenzyme A in bacteria. Vitam Horm 61:157–171

Blaszczyk A, Brodzik R, Sirko A (1999) Increased resistance to oxidative stress in transgenic tobacco plants over expressing bacterial serine acetyl transferase. Plant J 20:237–243

Blumwald E, Grover A, Good AG (2004) Breeding for abiotic stress resistance: challenges and opportunities. New directions for a diverse planet. In: Proceeding of the 4th international crop science congress, Brisbane, 26 Sep–1 Oct 2004

Borsani O, Valpuesta V, Botella MA (2001) Evidence for a role of salicylic acid in the oxidative damage generated by NACL and osmotic stress in Arabidopsis seedling. J Plant Physiol 126:1024–1030

Boyer JS (1982) Plant productivity and environment. Science 218:443–448

Bray EA, Bailey-Serres J, Weretilnyk E (2000) Responses to abiotic stresses. In: Gruissem W, Buchannan B, Jones R (eds) Biochemistry and molecular biology of plants. ASPP, Rockville, pp 1158–1249

Brunold C, Von Ballmoos P, Hesse H, Fell D, Kopriva S (2003) Interactions between sulphur, nitrogen and carbon metabolisms. In: Davidian JC, Grill D, De Kok LJ, Stulen I, Hawkesford MJ, Schnug E, Rennenberg H (eds) Sulphur transport and assimilation in plants: regulation, interaction, and signaling. Backhuys, Leiden, pp 45–56

Buchanan BB (1991) Regulation of CO_2 assimilation in oxygenic photosynthesis: the ferredoxin/thioredoxin system. Perspective on its discovery, present status, and future development. Arch Biochem Biophy 288:1–9

Cao WH, Liu J, He XJ, Mu RL, Zhou HL, Chen SY, Zhang JS (2007) Modulation of ethylene responses affects plant salt-stress responses. Plant Physiol 143:707–719

Causton HC, Ren B, Koh SS, Harbison CT, Kanin E, Jennings EG, Lee TI, True HI, Lander ES, Young RA (2001) Remodeling of yeast genome expression in response to environmental changes. Mol Biol Cell 12:323–337

Chae HZ, Chung SJ, Rhee SG (1994) Thioredoxin dependent peroxide reductase from yeast. J Biol Chem 269:27670–27678

Chaves MM, Flexas J, Pinheiro C (2009) Photosynthesis under drought and salt stress: regulation mechanisms from whole plant to cell. Ann Bot 103:551–560-580

Che P, Weaver LM, Wurtele ES, Nikolau BJ (2003) The role of biotin in regulating 3-methylcrotonyl-coenzyme A carboxylase expression in Arabidopsis. Plant Physiol 131:1479–1486

Choi WY, Lee KS, Ko JC, Park HK, Kim SS, Kim BK, Kim CK (2004) Nitrogen fertilizer management for improving rice quality under different salinity conditions in tidal reclaimed area. Korean J Crop Sci 49:194–198

Collin V, Issakidis-Bourguet E, Marchand C, Hirasawa M, Lancelin JM, Knaff DB, Miginiac-Maslow M (2003) The Arabidopsis plastidial thioredoxins: new functions and new insights into specificity. J Biol Chem 278:23747–23752

Cram WJ (1990) Uptake and transport of sulphate. In: Rennenberg H, Brunold C, De Kok LJ, Stulen I (eds) Sulphur nutrient and sulphur assimilation in higher plants. SPB Academic Publishing, The Hague, pp 3–11

De Kok LJ, Stuiver CEE, Rubinigg M, Westerman S, Grill D (1997) Impact of atmospheric sulphur deposition on sulphur metabolism in plants: H_2S as sulphur source for sulphur deprived *Brassica oleracea* L. Bot Acta 110:411–419

De Kok LJ, Stuiver CEE, Stulen I (1998) The impact of elevated levels of atmospheric H_2S on plants. In: De Kok LJ, Stulen I (eds) Responses of plant metabolism to air pollution and global change. Backhuys, Leiden, pp 51–63

De Kok LJ, Castro A, Durenkamp M, Kralewska A, Posthumus FS, Elisabeth Stuiver E, Yang L, Stulen I (2005) Pathways of plant sulphur uptake and metabolism – an overview. Landbaufor Volken 283:5–13

Desikan R, Neill SJ, Hancock JT (2000) Hydrogen peroxide-induced gene expression in *Arabidopsis thaliana*. Free Rad Biol Med 28:773–778

Desikan R, Mackerness S, Hancock JT, Neill SJ (2001) Regulation of the Arabidopsis transcriptome by oxidative stress. Plant Physiol 127:159–172

Dionisio-Sese ML, Tobita S (1998) Antioxidant responses of rice seedlings to salinity stress. Plant Sci 135:1–9

Dominguez-Solis JR, Gutierrez-Alcala G, Romero LC, Gotor C (2001) The cytosolic O-acetylserine (thiol) lyase gene is regulated by heavy metals and can function in cadmium tolerance. J Biol Chem 276:9297–9302

El-Shintinawy F, El-Shourbagy MN (2001) Alleviation of changes in protein metabolism in NaCl-stressed wheat seed by thiamine. Biol Plant 44:541–545

Espinosa-Ruiz A, Belles JM, Serrano R, Culianez-Macia FA (1999) Arabidopsis thaliana AtHAL3: a flavoprotein related to salt and osmotic tolerance and plant growth. Plant J 20:529–539

Farago S, Brunold C, Kreuz K (1994) Herbicide safeners and glutathione metabolism. Physiol Plant 91:537–542

Fediuc E, Lips SH, Erdei L (2005) O-acetylserine (thiol) lyase activity in Phragmites and Typha plants under cadmium and NaCl stress conditions and the involvement of ABA in the stress response. J Plant Physiol 162:865–872

Finkel T, Holbrook NJ (2000) Oxidants, oxidative stress and the biology of aging. Nature 408:239–247

Finkemeier I, Goodman M, Lamkemeyer P, Kandlbinder A, Sweetlove LJ, Dietz KJ (2005) The mitochondrial type II peroxiredoxin F is essential for redox homeostasis and root growth of *Arabidopsis thaliana* under stress. J Biol Chem 280:12168–12180

Flexas J, Diaz-Espejo A, Galmes J, Kaldenhoff R, Medrano H, Ribas-Carbo M (2007) Rapid variations of mesophyll conductance in response to changes in CO_2 concentration around leaves. Plant Cell Environ 30:1284–1298

Flugge Fischer K, Gross A, Sebald W, Lottspeich F, Eckerskorn C (1989) The triose phosphate phosphoglycerate-phosphate translocator from spinach chloroplast: nucleotide sequence of a full-length cDNA clone and import of the in vitro synthesized precursor protein into chloroplasts. Embo J 8:39–46

Food and Agricultural Organization, FAO (2008) Land and plant nutrition management service. http://www.fao.org/ag/agl/agll/spush/. Accessed 25 Apr 2012

Foyer CH, Noctor G (1999) Leaves in the dark see the light. Science 284:599–601

Foyer CH, Lopez-Delgado H, Dat JF, Scott IM (1997) Hydrogen peroxide and glutathione-associated mechanisms of acclamatory stress tolerance and signalling. Physiol Plant 100:241–254

Foyer CH, Theodoulou FL, Delrot S (2001) The functions of inter- and intracellular glutathione transport systems in plants. Trends Plant Sci 6:486–492

Friedrich W (1987) Thiamin (Vitamin B1, aneurin). In: Handbuchder vitamine. Urban and Schwartzenberg, Munich, pp 240–258

Gallardo K, Job C, Groot SPC, Puype M, Demol H, Vandekerckhove J, Job D (2002) Importance of methionine biosynthesis for Arabidopsis seed germination and seedling growth. Physiol Plant 116:238–247

Garratt LC, Janagoudar BS, Lowe KC, Anthony P, Power JB, Davey MR (2002) Salinity tolerance and antioxidant status in cotton cultures. Free Rad Biol Med 33:502–511

Gasch AP, Spellman PT, Kao CM, Carmel-Harel O, Eisen MB, Storz G, Botstein D, Brown PO (2000) Genomic expression programs in the response of yeast cells to environmental changes. Mol Biol Cell 11:4241–4257

Gill SS, Tuteja N (2011) Cadmium stress tolerance in crop plants probing the role of sulphur. Plant Signal Behvi 6:215–222

Giraud E, Ho LH, Clifton R (2008) The absence of alternative oxidase 1a in Arabidopsis results in acute sensitivity to combined light and drought stress. Plant Physiol 147:595–610

Glaser HV, Thomas D, Gaxiola H (1993) Salt tolerance and methionine biosynthesis in Saccharomyces cerevisiae involves a putative phosphatase gene. EMBO J 12:3105–3110

Gomathi R, Rakkiyapan P (2011) Comparative lipid peroxidation, leaf membrane thermostability, and antioxidant system in four sugarcane genotypes differing in salt tolerance. Int J Plant Physiol Biochem 3:67–74

Gossett DR, Millhollon EP, Lucas MC (1994) Antioxidant response to NaCl stress in salt-tolerant and salt-sensitive cultivars of cotton. Crop Sci 34:706–714

Greenway H, Munns R (1980) Mechanism of salt tolerance in non halophytes. Annu Rev Plant Physiol 31:149–190

Gunes A, Inal A, Alpaslan M, Eraslan F, Bagci EG, Cicek N (2007) Salicylic acid induced changes on some physiological parameters symptomatic for oxidative stress and mineral nutrition in maize (Zea mays L.) grown under salinity. J Plant Physiol 164:728–736

Guo Y, Halfter U, Ishitani M, Zhu JK (2001) Molecular characterization of functional domains in the protein kinase SOS2 that is required for plant salt tolerance. Plant Cell 13:1383–1400

Halfter U, Ishitani M, Zhu JK (2000) The Arabidopsis SOS2 protein kinase physically interacts with and is activated by the calcium-binding protein SOS3. Proc Natl Acad Sci USA 97:3735–3740

Halliwell B, Gutteridge JMC (1985) Free radicals in biology and medicine, 3rd edn. Oxford University Press, Oxford

Hamdia MA (2000) Physiological studies on the influence of biotin or pyridoxine in amelioration the effect of salinity in lupine plants. Bull Fac Sci Assiut Univ 29:283–291

Harada K, Imamura F, Hiraishi T (2002) Experimental study on the effect in reducing Tsunami by the coastal permeable structures. In: Final proceedings of the international offshore polar engineering conference, San Francisco, pp 652–658

Heeg C, Kruse C, Jost R, Gutensohn M, Ruppert T, Wirtz M, Hell R (2008) Analysis of the Arabidopsis O-acetylserine(thiol)lyase gene family demonstrates compartment-specific differences in the regulation of cysteine synthesis. Plant Cell 20:168–185

Heiss S, Schäfer HJ, Haag-Kerwer A, Rausch T (1999) Cloning sulphur assimilation genes of *Brassica juncea* L.: cadmium differentially affects the expression of a putative low-affinity sulphate transporter and isoforms of ATP sulphurylase and APS reductase. Plant Mol Biol 39:847–857

Hernandez JA, Jimenez A, Mullineaux P, Sevilla F (2000) Tolerance of pea (Pisum sativum L.) to long-term salt stress is associated with induction of antioxidant defenses. Plant Cell Environ 23:853–862

Hernández JA, Ferrer MA, Jiménez A, Ros-Barceló A, Sevilla F (2001) Antioxidant systems and $O_2^{\bullet-}$/H_2O_2 production in the apoplast of *Pisum sativum* L. leaves: its relation with NaCl induced necrotic lesions in minor veins. Plant Physiol 127:817–831

Hernandez JMJ, Lucas MM, de Felipe MR (2002) Antioxidant defense and damage in senescing lupin nodules. Plant Physiol Biochem 40:645–657

Herschbach C, Rennenberg H (1991) Influence of glutathione (GSH) on sulphate influx, xylem loading and exudation in excised tobacco roots. J Exp Bot 42:1021–1029

Herschbach C, De Kok LJ, Rennenberg H (1995a) Net uptake of sulphate and its transport to the shoot in spinach plants fumigated with H_2S or SO_2: does atmospheric sulphur affect the 'inter-organ' regulation of sulphur nutrition. Botanica Acta 108:41–46

Herschbach C, De Kok LJ, Rennenberg H (1995b) Net uptake of sulphate and its transport to the shoot in tobacco plants fumigated with H_2S or SO_2. Plant Soil 175:75–84

Hesse H, Höfgen R (2003) Molecular aspects of methionine biosynthesis in Arabidopsis and potato. Trends Plant Sci 8:259–262

Hofgen R, Kreft O, Willmitzer L, Hesse H (2001) Manipulation of thiol contents in plants. Amino Acids 20:291–299

Holmgren A (1989) Thioredoxin and glutaredoxin systems. J Biol Chem 264:13963–13966

Howarth JR, Domínguez-Solís JR, Gutiérrez-Alcalá G, Wray JL, Gotor C, Romero LC (2003a) The serine acetyltransferase gene family in Arabidopsis thaliana and the regulation of its expression by cadmium. Plant Mol Biol 51:589–598

Howarth JR, Fourcroy P, Davidian JC, Smith FW, Hawkesford MJ (2003b) Cloning of two contrasting high-affinity sulphate transporters from tomato induced by low sulphate and infection by the vascular pathogen Verticillium dahliae. Planta 218:58–64

Hsu SY, Kao CH (2003) The protective effect of free radical scavengers and metal chelators on polyethylene glycol-treated leaves. Biol Plant 46:617–619

Iqbal N, Masood A, Khan NA (2012) Phytohormones in salinity tolerance: ethylene and gibberellins cross talk. In: Khan NA, Nazar R, Iqbal N, Anjum NA (eds) Phytohormones and abiotic stress tolerance in plants. Springer, Berlin, pp 77–98

Jaleel CA, Manivannan P, Lakshmanan GMA, Sridharan R, Panneerselvam R (2007) NaCl as a physiological modulator of proline metabolism and antioxidant potential in *Phyllanthus amarus*. C R Biologies 330:806–813

Jeanjean R, Matthijs HCP, Onana B, Havaux M, Joset F (1993) Exposure of the cyanobacterium Synechocystis PCC 6803 to salt stress induces concerted changes in respiration and photosynthesis. Plant Cell Physiol 34:1073–1079

Jonak C, Ökresz L, Bögre L, Hirt H (2002) Complexity, cross talk and integration of plant MAP kinase signalling. Curr Opin Plant Biol 5:415–424

Katare DP, Nabi G, Azooz MM, Aeri V, Ahmad P (2012) Biochemical modifications and enhancement of psoralen content in salt-stressed seedlings of *Psoralea corylifolian* Linn. J Funct Environ Bot 2:65–74

Keutgen AJ, Pawelzik E (2009) Impact of NaCl stress on plant growth and mineral nutrient assimilation in two cultivars of strawberry. Environ Exp Bot 65:170–176

Khan MH, Panda SK (2008) Alterations in root lipid peroxidation and antioxidative responses in two rice cultivars under NaCl-salinity stress. Acta Physiol Plant 30:81–89

Khan NA, Anjum NA, Nazar R, Iqbal N (2009a) Increased activity of ATP-sulphurylase, contents of cysteine and glutathione reduce high cadmium-induced oxidative stress in high photosynthetic potential mustard (*Brassica juncea* L.) cultivar. Russ J Plant Physiol 56:670–677

Khan NA, Nazar R, Anjum NA (2009b) Growth, photosynthesis and antioxidant metabolism in mustard (*Brassica juncea* L.) cultivars differing in ATP-sulphurylase activity under salinity stress. Sci Hort 122:455–460

Khan NA, Syeed S, Masood A, Nazar R, Iqbal N (2010) Application of salicylic acid increases contents of nutrients and antioxidative metabolism in mungbean and alleviates adverse effects of salinity stress. Int J Plant Biol 1:e1

Khan MIR, Syeed S, Nazar R, Anjum NA (2012) An insight into the role of salicylic acid and jasmonic acid in salt stress tolerance. In: Khan NA, Nazar R, Iqbal N, Anjum NA (eds) Phytohormones and abiotic stress tolerance in plants. Springer, Berlin, pp 277–300

Khayyat M, Rajaee S, Sajjadinia A, Eshghi S, Tafazoli E (2009) Calcium effects on changes in chlorophyll contents, dry weight and micronutrients of strawberry (*Fragaria ananassa* Duch.) plants under salt stress conditions. Fruits 64:1–10

Khorshidi MB, Yarnia M, Hassanpanah D (2009) Salinity effect on nutrients accumulation in alfalfa shoots in hydroponic condition. J Food Agric Environ 7:787–790

Kingsbury RW, Epstein E (1986) Salt sensitivity in wheat. Plant Physiol 80:651–654

Kocsy G, Szalai G, Galiba G (2004) Effect of osmotic stress on glutathione and hydroxyl methyl glutathione accumulation in wheat. J Plant Physiol 161:785–794

Kopriva S (2006) Regulation of sulphate assimilation in Arabidopsis and beyond. Ann Bot 97:479–495

Kopriva S, Koprivova A (2005) Sulphate assimilation and glutathione synthesis in C4 plants. Photosynth Res 86:363–373

Kopriva S, Rennenberg H (2004) Control of sulphate assimilation and glutathione synthesis: interaction with N and C metabolism. J Exp Bot 55:1831–1842

Kovtun Y, Chiu WL, Tena G, Sheen J (2000) Functional analysis of oxidative stress-activated mitogen-activated protein kinase cascade in plants. Proc Natl Acad Sci USA 97:2940–2945

Laloi C, Rayapuram N, Chartier Y, Grienenberger JM, Bonnard G, Meyer Y (2001) Identification and characterization of a mitochondrial thioredoxin system in plants. Proc Nat Acad Sci USA 98:14144–14149

Lappartient AG, Touraine B (1996) Demand-driven control of root ATP sulphurylase activity and SO_4^{2-} uptake in intact canola. Plant Physiol 111:147–157

Lappartient AG, Vidmar JJ, Leustek T, Glass ADM, Touraine B (1999) Inter-organ signaling in plants: regulation of ATP sulphurylase and sulphate transporter genes expression in roots mediated by phloem-translocated compound. Plant J 18:89–95

Leegood RC, Lea PJ, Adcock MD, Hausler RE (1995) The regulation and control of photorespiration. J Exp Bot 46:1397–1414

Leustek T, Saito K (1999) Sulfate transport and assimilation in plants. Plant Physiol 120:637–643

Leustek T, Martin MN, Bick JA, Davies JP (2000) Pathways and regulation of sulphur metabolism revealed through molecular and genetic studies. Annu Rev Plant Physiol Plant Mol Biol 51:141–165

Levings CS, Siedow JN (1995) Regulation by redox poise in chloroplasts. Science 268:695–696

Li J, Gunter B, Helenius E, Kariola T, Tapio Palva E (2012) Biotin deficiency causes spontaneous cell death and activation of defense signaling. Plant J 70: 315–326

Liu J, Zhu JK (1997) An Arabidopsis mutant that requires increased calcium for potassium nutrition and salt tolerance. Proc Natl Acad Sci USA 94:14960–14964

Liu J, Zhu, JK (1998) A calcium sensor homolog required for plant salt tolerance. Science 280:1943–1945

Liu J, Ishitani M, Halfter U, Kim CS, Zhu JK (2000) The *Arabidopsis thaliana* SOS2 gene encodes a protein kinase that is required for salt tolerance. Proc Nat Acad Sci USA 97:3730–3734

Liu GF, Cheng GL, Jiang J, Bai S, Yu Y, Cai ZJ, Dong JX, Li SJ (2006) The transformation of betA gene into the pollen plantlets of *Populus simonii* and *P. nigra*. J Plant Physiol Mol Biol 32:163–168

Lopez-Martin MC, Becana M, Romero LC, Gotor C (2008) Knocking out cytosolic cysteine synthesis compromises the antioxidant capacity of the cytosol to discrete concentrations of hydrogen peroxide in Arabidopsis. Plant Physiol 147:562–572

Lunn JE, Droux M, Martin J, Douce R (1990) Localization of ATP sulphurylase and Oacetylserine(thio1)lyase in spinach leaves. Plant Physiol 94:1345–1352

Mahajan S, Pandey GK, Tuteja N (2008) Calcium- and salt-stress signaling in plants: shedding light on SOS pathway. Arch Biochem Biophys 471:146–158

Manchanda G, Garg N (2008) Salinity and its effect on the functional biology of legumes. Acta Physiol Plant 30:595–618

Marrs K (1996) The functions and regulation of glutathione S-transferases in plants. Annu Rev Plant Physiol Plant Mol Biol 47:127–158

Marrs KA, Alfenito MR, Lloyd AM, Walbot V (1995) A glutathione transferase involved in vacuolar transfer encoded by the maize gene Bronze-2. Nature 375:397–400

Marti MC, Olmos E, Calvete JJ, Diaz I, Barranco-Medina S, Whelan J, Lazaro JJ, Sevilla F, Jimenez A (2009) Mitochondrial and nuclear localization of a novel pea thioredoxin: identification of its mitochondrial target proteins. Plant Physiol 150:646–657

Marti MC, Florez-Sarasa I, Camejo D, Ribas-Carbo M, Lázaro JJ, Francisca Sevilla F, Jiménez A (2011) Response of mitochondrial thioredoxin PsTrxo1, antioxidant enzymes, and respiration to salinity in pea (*Pisum sativum* L.) leaves. J Exp Bot 10:1093

Mateo A, Mühlenbock P, Rustérucci C, Chang CCC, Miszalski Z, Karpinska B, Parker JE, Mullineaux PM, Karpinski S (2004) Lesion simulating disease 1 is required for acclimation to conditions that promote excess excitation energy. Plant Physiol 136:2818–2830

Matthews BF (1999) Lysine, threonine and methionine biosynthesis. In: Singh BK (ed) Plant amino acids: biochemistry and biotechnology. Marcel Dekker, New York, pp 205–225

Mazel A, Leshem Y, Tiwari BS, Levine A (2004) Induction of salt and osmotic stress tolerance by overexpression of an intracellular vesicle trafficking protein AtRab7 (AtRabG3e). Plant Physiol 134:118–128

Meneguzzo S, Navarri-Izzo F, Izzo R (1999) Antioxidative responses of shoots and roots of wheat to increasing NaCl concentrations. J Plant Physiol 155:274–280

Mittler R (2002) Oxidative stress, antioxidants and stress tolerance. Trends Plant Sci 7:405–410

Mittler R, Vanderauwera S, Gollery M, Van Breusegem F (2004) Reactive oxygen gene network of plants. Trends Plant Sci 9:490–498

Mittler R, Kim Y, Song LH, Coutu J, Coutu A, Ciftci-Yilmaz S, Lee H, Stevenson B, Zhu JK (2006) Gain- and loss-of function mutations in Zat10 enhance the tolerance of plants to abiotic stress. FEBS Lett 580:6537–6542

Mittova V, Guy M, Tal M, Volokita M (2004) Salinity up-regulates the antioxidative system in root mitochondria and peroxisomes of the wild salt-tolerant tomato species *Lycopersicon pennellii*. J Exp Bot 55:1105–1113

Moser D, Nicholls P, Wastyn M, Peschek GA (1991) Acidic cytochrome c6 of unicellular cyanobacteria is an indispensable and kinetically competent electron donor to cytochrome oxidase in plasma and thylakoid membranes. Biochem Int 24:757–768

Moss J, Lane MD (1971) The biotin-dependent enzymes. Adv Enzymol 35:321–442

Motohashi K, Kondoh A, Stumpp MT, Hisabori T (2001) Comprehensive survey of proteins targeted by chloroplast thioredoxin. Pro Nati Acad Sci USA 98:11224–11229

Mullineaux PM, Rausch T (2005) Glutathione, photosynthesis and the redox regulation of stress responsive gene expression. Photosynth research 47:459–474

Munns R (2002) Comparative physiology of salt and water stress. Plant Cell Environ 25:239–250

Munns R, Tester M (2008) Mechanisms of salinity tolerance. Ann Rev Plant Biol 59:651–681

Munns R, James RA, Lauchli A (2006) Approaches to increasing the salt tolerance of wheat and other cereals. Exp J Bot 57:1025–1043

Na GN, Salt DE (2011) The role of sulfur assimilation and sulfur-containing compounds in trace element homeostasis in plants. Environ Exp Bot 72:18–25

Navrot N, Rouhier N, Gelhaye E, Jacquot JP (2007) Reactive oxygen species generation and antioxidant systems in plant mitochondria. Physiol Plant 129:185–195

Nazar R, Iqbal N, Masood A, Syeed S, Khan NA (2011a) Understanding the significance of sulfur in improving salinity tolerance in plants. Environ Exp Bot 70:80–87

Nazar R, Iqbal N, Syeed S, Khan NA (2011b) Salicylic acid alleviates decreases in photosynthesis under salt stress by enhancing nitrogen and sulphur assimilation and antioxidant metabolism differentially in two mungbean cultivars. J Plant Physiol 168:807–815

Nikolau BJ, Ohlrogge JB, Wurtele ES (2003) Plant biotin-containing carboxylases. Arch Biochem Biophys 414:211–222

Nocito FF, Lancilli C, Giacomini B, Sacchi GA (2007) Sulphur metabolism and cadmium stress in higher plants. Plant Stress 1:142–156

Noctor G, Foyer CH (1998) Ascorbate and glutathione: keeping active oxygen under control. Ann Rev Plant Physiol Plant Mol Biol 49:249–279

Noctor G, Arisi ACM, Jouanin L, Valadier MH, Roux Y, Foyer CH (1997) Light-dependent modulation of foliar glutathione synthesis and associated amino acid metabolism in poplar over expressing gamma-glutamyl cysteine synthetase. Planta 202:357–369

Noctor G, Arisi ACM, Jouanin L, Kunert KJ, Rennenberg H, Foyer CH (1998) Glutathione: biosynthesis, metabolism and relationship to stress tolerance explored in transformed plants. J Exp Bot 49:623–647

Noctor G, De Paepe R, Foyer CH (2007) Mitochondrial redox biology and homeostasis in plants. Trends Plant Sci 12:125–134

Noji M, Saito M, Nakamura M, Aono M, Saji H, Saito K (2001) Cysteine synthase over expression in tobacco confers tolerance to sulphur-containing environment pollutants. Plant Physiol 126:973–980

North KA, Kopriva S (2007) Sulphur in resistance to environmental stresses. In: Hawkesford MJ, De Kok LJ (eds) Sulphur in plants – an ecological perspective. Springer, New York, pp 143–168

Nosaka K (2006) Recent progress in understanding thiamin biosynthesis and its genetic regulation in Saccharomyces cerevisiae. Appl Microbiol Biotech 72:30–40

Ogawa S, Mitsuya S (2012) *S*-methylmethionine is involved in the salinity tolerance of *Arabidopsis thaliana* plants at germination and early growth stages. Physiol Plant 144:13–19

Parida AK, Das AB (2005) Salt tolerance and salinity effects on plants: a review. Eco toxico Environ Saf 60:324–349

Pilon-Smits EAH, Pilon M (2006) Sulfur metabolism in plastids. In: Wise RR, Hoober JK (eds) Advances in photosynthesis and respiration – the structure and function of plastids. Kluwer, Dordrecht, pp 387–402

Pilon-Smits EAH, Hwang S, Lytle CM, Zhu Y, Tai JC, Bravo RC, Chen Y, Leustek T, Terry N (1999) Overexpression of ATP sulphurylase in Indian mustard leads to increased selenate uptake, reduction and tolerance. Plant Physiol 119:123–132

Pnueli L, Liang H, Rozenberg M, Mittler R (2003) Growth suppression, altered stomatal responses, and augmented induction of heat shock proteins in cytosolic ascorbate peroxidase (Apx1)-deficient Arabidopsis plants. Plant J 34:187–203

Prasad TK, Anderson MD, Martin BA, Steward CR (1994) Evidence for chilling-induced oxidative stress in maize seedlings and a regulatory role for hydrogen peroxide. Plant Cell 6:65–74

Pulido P, Cazalis R, Cejudo FJ (2009) An antioxidant redox system in the nucleus of wheat seed cells suffering oxidative stress. The Plant J 57:132–145

Qiu QS, Guo Y, Dietrich MA, Schumaker KS, Zhu JK (2002) Regulation of SOS1, a plasma membrane Na+/H+exchanger in *Arabidopsis thaliana*, by SOS2 and SOS3. Proc Nat Acad Sci USA 99:8436–8441

Rabenstein DL (1989) Metal complexes of glutathione and their biological significance. In: Dolphin D, Poulson R, Avramovic O (eds) Glutathione: chemical, biochemical and medical aspects. Wiley, New York, pp 147–186

Rausch T, Wachter A (2005) Sulphur metabolism: a versatile platform for launching defence operations. Trends Plant Sci 10:503–509

Rennenberg H (1995) Processes involved in glutathione metabolism. In: Wallsgrove RM (ed) Amino acids and their derivatives in higher plants: biosynthesis and metabolism. Cambridge University Press, Cambridge, pp 155–171

Rennenberg H, Herschbach C (1995) Sulphur nutrition of trees: a comparison of spruce (*Picea abies* L.) and beech (*Fagus sylvatica* L.). Z Pflanzenernähr Bodenk 158:513–517

Rennenberg H, Herschbach C (1996) Responses of plants to atmospheric sulphur. In: Iqbal JM, Yunus M (eds) Plant responses to air pollution. Wiley, New York, pp 285–293

Robert EMR, Koedam N, Beeckman H, Schmitz N (2009) A safe hydraulic architecture as wood anatomical explanation for the difference in distribution of the mangroves *Avicennia* and *Rhizophora*. Functional Ecol 23:649–657

Roje S (2007) Vitamin B biosynthesis in plants. Phytochemistry 68:1904–1921

Rolland N, Droux M, Douce R (1992) Sub cellular distribution of O-acetylserine (thiol) lyase in cauliflower (*Brassica olerucea* L.) inflorescence. Plant Physiol 98:927–935

Romero LC, Domínguez-Solís JR, Gutiérrez-Alcalá G, Gotor C (2001) Salt regulation of O-acetylserine(thiol)lyase in Arabidopsis thaliana and increased tolerance in yeast. Plant Physiol Biochem 39:643–647

Rouhier N,Villarejo A, Srivastava M, Gelhaye E, Keech O,Droux M, Finkemeier I, Samuelsson G, Dietz KJ, Jaquot JP (2005). Identification of plant glutaredoxin targets. Antiox Redox Signal 7: 919–929

Rouhier N, Jaquot JP (2005) The plant multigenic family of thiol peroxidases. Free Rad Biol Med 38:1413–1421

Rubio S, Larson TR, Gonzalez-Guzman M, Alejandro S, Graham IA, Serrano R, Rodriguez PL (2006) An Arabidopsis mutant impaired in coenzyme A biosynthesis is sugar dependent for seedling establishment. Plant Physiol 140:830–843

Rubio S, Whitehead L, Larson TR, Graham LA, Rodriguez PL (2008) The Coenzyme A biosynthetic enzyme phosphopantetheine adenylyltransferase plays a crucial role in plant growth, salt/osmotic stress resistance, and seed lipid storage. Plant Physiol 148:546–556

Rüegsegger A, Brunold C (1992) Effect of cadmium on c- glutamylcysteine synthesis in maize seedlings. Plant Physiol 99:428–433

Ruelland E, Miginiac-Maslow M (1999) Regulation of chloroplast enzyme activities by thioredoxins: activation or relief from inhibition. Trends Plant Sci 4:136–141

Ruiz JM, Blumwald E (2002) Salinity-induced glutathione synthesis in *Brassica napus*. Planta 214:965–969

Saito K (2000) Regulation of sulphate transport and synthesis of sulphur-containing amino acids. Curr Opin Plant Biol 3:188–195

Sanan-Mishra N, Tuteja R, Tuteja N (2006) Signaling through MAP kinase networks in plants. Arch Biochem Biophys 452:55–68

Sanchez-Aguayo I, Rodriguez-Galan JM, Garcia R, Torreblanca J, Pardo JM (2004) Salt stress enhances xylem development and expression of S-adenosyl-L-methionine synthase in lignifying tissues of tomato plants. Planta 220:278–285

Sanders D (2000) Plant biology: the salty tale of Arabidopsis. Curr Biol 10:486–488

Sarowar S, Kim EN, Kim YJ, Ok SH, Kim KD, Hwang BK, Shin JS (2005) Over expression of a pepper ascorbate peroxidase-like1 gene in tobacco plants enhances tolerance to oxidative stress and pathogens. Plant Sci 169:55–63

Sattler SE, Mène-Saffrané L, Farmer EE, Krischke M, Mueller MJ, DellaPenna D (2006) Nonenzymatic lipid peroxidation reprograms gene expression and activates defense markers in Arabidopsis tocopherol deficient mutants. Plant Cell 18:3706–3720

Schäfer HJ, Haag-Kerwer A, Rausch T (1998) cDNA cloning and expression analysis of genes encoding GSH synthesis in roots of the heavy metal accumulator *Brassica juncea* L: evidence for Cd induction of a putative mitochondrial g-glutamylcysteine synthetase isoform. Plant Mol Biol 37:87–97

Schena M, Shalon D, Davis R, Brown P (1995) Quantitative monitoring of gene expression patterns with complementary DNA microarray. Science 270:467–470

Schürmann P, Jacquot JP (2000) Plant thioredoxin systems revisited. Ann Revi Plant Physiol Plant Mol Biol 51:371–400

Sekmen AH, Türkan I, Takio S (2007) Differential responses of antioxidative enzymes and lipid peroxidation to salt stress in salt-tolerant Plantago maritime and salt-sensitive *Plantago media*. Physiol Plant 131:399–411

Shaaltiel Y, Chua NH, Gepstein S, Gressel J (1988) Dominant pleiotropy controls enzymes cosegregating with paraquet resistance in Conyza bonariensis. Theor Appl Genet 75:850–856

Sharma P, Jha AB, Dubey RS, Pessarakli M (2012) Reactive oxygen species, oxidative damage, and antioxidative defense mechanism in plants under stressful conditions. J Bot. doi:10.1155/2012/217037

Shi H, Ishitani M, Wu SJ, Kim CS, Zhu JK (2000) The *Arabidopsis thaliana* salt tolerance gene SOS1 encodes a putative Na1/H1 antiporter. Proc Natl Acad Sci USA 97:6896–6901

Shi H, Xiong L, Stevenson B, Lu T, Zhu JK (2002) The Arabidopsis salt overly sensitive 4 mutants uncover a critical role for vitamin B6 in plant salt tolerance. Plant Cell 14:575–588

Shi H, Kim YS, Guo Y, Stevenson B, Zhu JK (2003) The Arabidopsis SOS5 locus encodes a putative cell surface adhesion protein and is required for normal cell expansion. Plant Cell 15:19–32

Shu S, Guo SR, Sun J, Yuan LY (2012) Effects of salt stress on the structure and function of the photosynthetic apparatus in *Cucumis sativus* and its protection by exogenous putrescine. Physiol Plant. doi:10.1111/j.1399-3054.2012.01623

Silva C, Martínez V, Carvajal M (2008) Osmotic versus toxic effects of NaCl on pepper plants. Biol Plant 52:72–79

Singha S, Choudhuri MA (1990) Effect of salinity (NaCl) stress on H_2O_2 metabolism in Vigna and Oryza seedlings. Biochem Physiol Pflanz 186:69–74

Smith IK (1980) Regulation of sulphate assimilation in tobacco cells. Effect of nitrogen and sulphur nutrition on sulphate permease and oacetylserine sulphydrylase. Plant Physiol 66:877–883

Smith FW, Hawkesford MJ, Ealing PM, Clarkson DT, van den Berg PJ, Belcher AR, Warrilow AGS (1997) Regulation of expression of a cDNA from barley roots encoding a high affinity sulphate transporter. Plant J 12:875–884

Sreenivasulu N, Sopory SK, Kavi Kishor PB (2007) Deciphering the regulatory mechanisms of abiotic stress tolerance in plants by genomic approaches. Gene 388:1–13

Steduto P, Albrizio R, Giorio P, Sorrentino G (2000) Gas exchange response and stomatal and non-stomatal limitations to carbon assimilation of sunflower under salinity. Environ Exp Bot 44:243–255

Syeed S, Anjum NA, Nazar R, Iqbal N, Masood A, Khan NA (2011) Salicylic acid-mediated changes in photosynthesis, nutrients content and antioxidant metabolism in two mustard (*Brassica juncea* L.) cultivars differing in salt tolerance. Acta Physiol Plant 33:877–886

Takahashi S, Murata N (2008) How do environmental stresses accelerate photoinhibition? Trends in Plant Sci 13:178–182

Triantaphylides C, Hoeberichts FA, Krischke M, Ksas B, Gresser G, Havaux M, Van Breusegem F, Mueller MJ (2008) Singlet oxygen is the major reactive oxygen species involved in photooxidative damage to plants. Plant Physiol 148:960–968

Tripathi BN, Bhatt I, Dietz KJ (2009) Peroxiredoxins: a less studied component of hydrogen peroxide detoxification in photosynthetic organisms. Protoplasma 235:3–15

Tsai YC, Hong CY, Liu LF, Kao CH (2005) Expression of ascorbate peroxidase and glutathione reductase in roots of rice seedlings in response to NaCl and H_2O_2. J Plant Physiol 162:291–299

Tseng MJ, Liu CW, Yiu JC (2007) Enhanced tolerance to sulphur dioxide and salt stress of transgenic Chinese cabbage plants expressing both superoxide dismutase and catalase in chloroplasts. Plant Physiol Biochem 45:822–833

Türkan I, Demiral T (2009) Recent developments in understanding salinity tolerance. Environ Exp Bot 1:2–9, Special issue

Tyerman SD, Skerrett M, Garrill A, Findlay GP, Leigh RA (1997) Pathways for the permeation of Na^+ and Cl^- into protoplasts derived from the cortex of wheat roots. J Exp Bot 48:459–480

Vanacker H, Carver TLW, Foyer CH (2000) Early H_2O_2 accumulation in mesophyll cells leads to induction of glutathione during the hypersensitive response in the barley-powdery mildew interaction. Plant Physiol 123:1289–1300

Vauclare P, Kopriva S, Fell D, Suter M, Sticher L, von Ballmoos O, Krähenbuhl U, Op den Camp R, Brunold C (2002) Flux control of sulphate assimilation in *Arabidopsis thaliana*: adenosine 5 phosphosulphate reductase is more susceptible to negative control by thiols than ATP sulphurylase. Plant J 31:729–740

Volkmar KM, Hu Y, Steppuhn H (1998) Physiological responses of plants to salinity: a review. Can J Plant Sci 78:19–27

Vranova E, Inze D, Van Brensegem F (2002) Signal transduction during oxidative stress. J Exp Bot 53:1227–1236

Wang ZY, Nakano T, Gendron J, He J, Chen M, Vafeados D, Yang Y, Fujioka S, Yoshida S, Asami T (2002) Nuclear-localized BZR1 mediates brassinosteroid-induced growth and feedback suppression of brassinosteroid biosynthesis. Dev Cell 2:505–513

Wang W, Vinocur B, Altman A (2003) Plant responses to drought, salinity and extreme temperatures: toward genetic engineering for stress tolerance. Planta 218:1–14

Watanabe M, Kusano M, Oikawa A, Fukushima A, Noji M, Saito K (2008) Physiological roles of the b-substituted alanine synthase gene family in Arabidopsis. Plant Physiol 146:310–320

Wirtz M, Hell R (2006) Functional analysis of the cysteine synthase protein complex from plants: structural, biochemical and regulatory properties. J Plant Physiol 163:273–286

Wirtz M, Droux M, Hell R (2004) O-Acetylserine(thiol)lyase: an enigmatic enzyme of plant cysteine biosynthesis revisited in *Arabidopsis thaliana*. J Exp Bot 55:1785–1798

Wonisch W, Schaur RJ (2001) Chemistry of glutathione. In: Grill D, Tausz M, De Kok LJ (eds) Significance of glutathione to plant adaptation to the environment. Kluwer, Dordrecht/ Boston /London, pp 13–26

Yamaguchi K, Takahashi Y, Berberich T, Imai A, Miyazaki A, Takahashi T (2006) The polyamine spermine protects against high salt stress in *Arabidopsis thaliana*. FEBS Lett 580:6783–6788

Yi H, Galant A, Ravilious GE, Preuss ML, Jez JM (2010) Sensing sulfur conditions: simple to complex biochemical regulatory mechanisms in plant thiol metabolism. Mol Plant 3:269–279

Yonamine I, Yoshida K, Kido K, Nakagawa A, Nakayama H, Shinmyo A (2004) Overexpression of NtHAL3 genes confers increased levels of proline biosynthesis and the enhancement of salt tolerance in cultured tobacco cells. J Exp Bot 55:387–395

Zagorchev L, Seal CE, Kranner I, Odjakova M (2012) Redox state of low-molecular-weight thiols and disulphides during somatic embryogenesis of salt-treated suspension cultures of *Dactylis glomerata* L. Free Radi Res 46:656–664

Zenk MH (1996) Heavy metal detoxification in higher plants: a review Gene 179:21–30

Zhang A, Jiang M, Zhang J, Tan M, Hu X (2006a) Mitogen-activated protein kinase is involved in abscisic acid-induced antioxidant defense and acts downstream of reactive oxygen species production in leaves of maize plants. Plant Physiol 141:475–487

Zhang T, Liu Y, Xue L, Xu S, Chen T, Yang T, Zhang L, An L (2006b) Molecular cloning and characterization of a novel MAP kinase gene in Chorispora bungeana. Plant Physiol Biochem 44:78–84

Zhang CJ, Zhao BC, Ge Wei-Na, Zhang YF, Song Y, Sun DY, Guo Yi (2012) An apoplastic H-type thioredoxin is involved in the stress response through regulation of the apoplastic reactive oxygen species in rice. Plant Physiol 157:1884–1899

Zhao XC, Schaller GE (2004) Effect of salt and osmotic stress upon expression of the ethylene receptor ETR1 in *Arabidopsis thaliana*. FEBS Lett 562:189–192

Zhu JK (2001) Plant salt tolerance. Trends Plant Sci 6:66–71

Zhu JK (2003) Regulation of ion homeostasis under salt stress. Curr Opin Plant Biol 6:441–445

Zhu JK, Liu J, Xiong L (1998) Genetic analysis of salt tolerance in *Arabidopsis thaliana*: evidence of a critical role for potassium nutrition. Plant Cell 10:1181–1192

Chapter 18
Soil and Water Management for Sustained Agriculture in Alluvial Plains and Flood Plains Exposed to Salinity: A Case of Neretva River Valley

Monika Zovko, Davor Romić, Marija Romić, and Gabrijel Ondrašek

18.1 Introduction

Flood plains and alluvial plains contain deposits formed by periodical deposition of suspended sediments from river water during flood events. Regardless of the river size (i.e., large – such as the Nile, the Mississippi, Ebro river; or smaller streams), they all provide parent material for creating deep and fertile soils suitable for agricultural production. On the other hand, majority of these areas are globally the most densely populated (Stuyfzand 1993). Intensively altered by land reclamation, flow regulation, channelisation – drainage and irrigation networks and over-pumping of freshwater, plains are not balanced systems. They are rather dynamic endpoint sites of fluvio-ecological processes, socio-economic activities and wildlife population dynamics (Leuven and Poudevigne 2002).

All these alterations made this river landscape prone to sea water intrusion and developing of salt-affected soils. Seawater intrusion takes place either at typical coastal aquifers due to over-pumping of freshwater (Kallioras et al. 2006) or within delta regions due to drainage of the deltaic areas and reduction in the river flow, due to capillary rise of saline water in the soil profile or due to irrigation with saline water (Romić et al. 2008).

The problem of soil salinisation is present on all continents, in more than 100 countries, and in Europe it is most pronounced in the Mediterranean basin region (Acosta et al. 2011a). Salinisation in the Mediterranean area is related largely to climatic factors and irrigation development (Walter et al. 2001). The salt-affected soils cannot support vegetation and consequences include salt-induced water deficit, ion toxicity, nutrient imbalance, yield volatility and yield reduction. This leads to

M. Zovko (✉) • D. Romić • M. Romić • G. Ondrašek
Faculty of Agriculture, University of Zagreb, Zagreb, Croatia
e-mail: mzovko@agr.hr

P. Ahmad et al. (eds.), *Ecophysiology and Responses of Plants under Salt Stress*, DOI 10.1007/978-1-4614-4747-4_18, © Springer Science+Business Media, LLC 2013 some very important indirect effects of salinisation: a change in the biodiversity of wetlands (Lyons et al. 2007), a change in the agricultural production structure and transformation of socio-economic conditions (Romić et al. 2010a).

The Neretva river valley in the Mediterranean part of Croatia faces similar situation. Within this region there are around 6,000 ha of intensively used agricultural land made up by reclamation of flood plain and drained with open channels. Agricultural production is becoming more endangered because of periodical or temporal soil and water salinisation (Ondrašek et al. 2011). Salinisation in the area naturally occurs by sea water intrusion through river mouth and by costal aquifer through underground. In addition, the changes in hydrological conditions affected by numerous water engineering schemes and facilities within the Neretva basin contribute to intensified sea-water intrusion, causing severe groundwater salinisation. Whereas agricultural production in this area is unfeasible without irrigation, farmers are forced to use saline water. The extraction of water for irrigation can by itself enhance seawater intrusion, thus increasing the salinity of sources. This problem is both temporarily and spatially varied, being most pronounced in the summer time when the sea water intrusion to the land culminates. Summer irrigation of crops may cause variable salinisation of the root zone that will dynamically evolve throughout the growing season (Maggio et al. 2011). In spite of the fact that seasonal salinisation can be partially controlled by fulfilling appropriate leaching requirements (Rhoades et al. 1992), at advanced salinisation, permanent modifications of the soil physical-chemical properties will occur (De Pascale et al. 2003; Romić et al. 2008).

Meeting the challenge of managing salt-affected areas requires efficient and accurate quantifying, inventorying and mapping of soil salinity hazard (Pisinaras et al. 2010). An analysis of spatial variability explains the characteristics and direction of the salinisation process, identifies the main factors of salinity and spatial trends, and it can also indicate areas at high risk which require specific management in order to mitigate adverse effects of increased salt concentrations. It is therefore particularly important to monitor and determine salinisation of soils in order to timely propose management measures which will slow down this process and/or reduce its adverse effects. The knowledge gathered on the physiological and biochemical basis, is necessary for achieving maximum crop yield under saline conditions which is the principal objective of all agriculturists (Koyro et al. 2012).

Based on all mentioned, the research program was set out to develop and validate appropriate knowledge and technologies to sustain and improve the production capacity of the available soil and water resources of the Neretva river valley. The hypothesis underlying this research is that the environmental risk of soil salinisation in coastal river valleys is determined by the spatial characteristics of the area that can favour saltwater intrusions, by the physical and chemical properties of the soils, and by the type of agricultural land use. To achieve this goal, research in soil and water management is carried out under two main research program components, namely: (i) land resources surveys and inventory and (ii) soil and water management in saline conditions. In this chapter, the ones concerning soil salinity survey, water quality monitoring, field and greenhouse experiments have been considered.

18.2 Regional Conditions of the Study Area

18.2.1 The Neretva River Basin

The research was carried out in floodplain of the Neretva River Valley in the Mediterranean part of Croatia (43°00□N, 17°30□E) covering 5,216 ha of agricultural land (Fig. 18.1). The Neretva River sources in eastern Bosnia and Herzegovina, at elevation of 1,200 m above sea level. It flows into the Adriatic Sea close to Ploče (Croatia) about 220 km downstream. The river is subject to tidal fluctuations in the estuarine part of the basin, which extends from the river mouth to the city of Gabela, about 25 km inland (Ljubenkov 2006). The Neretva River flow can be divided into two parts. In the upper part the river flows swiftly through the mountainous landscape, while in the last 30 km of its course, in Croatia, the river spreads into an alluvial delta covering approximately 12,000 ha, before emptying into the Adriatic Sea.

The Neretva's lower course and delta were shaped by high waters that periodically washed down from the mountains, bringing dissolved organic substrate, whose sedimentation created fertile soils. The delta area in Croatia has been reduced by extensive land reclamation projects, and the river now flows in only 3 out of 12 former distributaries channels. The marshes, lagoons and lakes that used to characterize this fluvial plain have all disappeared, and only fragments of the old Mediterranean wetlands have survived. These parts are today protected as ornithological and ichthyological reserves and a natural landscape.

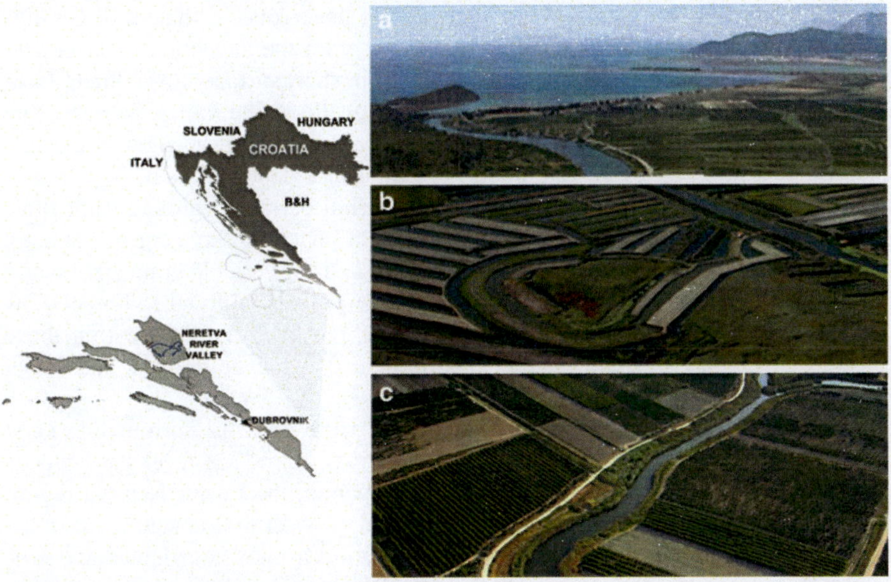

Fig. 18.1 Geographic location of the case study area, and panoramic views of the (**a**) Neretva river estuary; (**b**) polder type land; and (**c**) land on fluvial terraces along watercourses

Climatically, the Neretva basin marks a transition from continental into Mediterranean climate. The Lower Neretva region is semi-arid with hot, dry summers and wet winters. Most rainfall occurs from October to April, with an annual average (1980–2000) of 1,230 mm. The long-term mean annual air temperature is 15.7°C, with the highest (25.2°C) in July. Annual Penman-Montheith reference evapotranspiration amounts to 1,196 mm and the greatest values (191 mm) also occur in July. During the growing season (April–September) there is an apparent water deficit of up to 503 mm; this amount can only meet the needs of evapotranspiration. The greatest demand for water is in July when the difference between reference evapotranspiration and effective rainfall is at its highest and reaches up to nearly 160 mm.

18.2.2 Geology and Hydrogeology

The lower Neretva River Valley was formed on two very different groups of deposits: (1) Mesozoic and Paleogene carbonate rocks with some Paleogene flysch; and (2) Quaternary sediment and poorly lithified deposits. Carbonate rocks make up the basement of the valley, its flanks and smaller isolated hills within the valley. These rocks are intensely fractured and deeply karstified. Numerous faults, fissures, joints and sinkholes make the carbonate rocks very permeable to water and allow accumulation and circulation of groundwater. The permeability of these rocks, however, can be highly variable, depending on the level of fracturing and karstification. One of the main characteristics of these rocks is that all direct precipitation immediately percolates underground, with the exception of large and long-lasting precipitation events that can result in short surface flows. Such flows commonly also turn into sinking streams or can flood the valley surface. The flysch deposits, present along the northwestern and southeastern edges of the valley, are of limited thickness and lateral extent. These deposits are mainly impermeable, but are cut by faults and serve as a partial barrier to groundwater flow. Surficial Quaternary sediments consist mainly of peat and clay (organic marshy deposits), which are underlain by clayey sands, sands of variable grain size, sandy clay, gravely sands, sandy gravel, and Holocene gravels (alluvial sediments), as well as Pleistocene conglomerates. There are three main aquifers within these sediments: (1) in Holocene sands; (2) in middle Pleistocene gravels, and (3) in lower Pleistocene conglomerates and gravels.

Above sea level, fresh water with free-water table from the surrounding karst massif directly feeds the valley, while at elevations below sea level the water is saline and enters the valley under pressure. In winter, the Lower Neretva region must be protected from flood waters and the level of water in drainage canals must be regulated, while during the dry, vegetation period it must be irrigated and protected from sea-water intrusions. The Neretva River discharge varies considerably with minimal amounts occurring in summer and autumn and maximum ones in winter and spring (Romić and Vranjes 2010). Owing to many underground karst streams in the basin, there are many springs carrying large amounts of water, especially in winter. The groundwater coming from this surrounding karst area supplies numerous streams, lakes and cavities. Such a great variability of karst underground geology and morphology favours water flow and mixing of sea water with fresh water. Sea water intrudes into the surface water flows, particularly into the Neretva River. Salinisation of smaller streams is caused by mixing with waters from brackish karst springs on the limestone edges of the valley (Bonacci and Roje-Bonacci 1997), and from the aquifer with highly saline groundwater in the deeper layers.

18.2.3 Characteristics of Salt-Affected Soils and Agricultural Land Uses

The soils in the Neretva delta were formed by the deposition of sediment washed out from the hilly karst region into the basin. The eroded material was transported by stream and flood waters and deposited on limestone bedrock over geological time. The alluvial plain with diverse deposits was formed by a gradual sea level rise and by flooding of the river delta. Terrain elevations were gradually raised along the stream banks, from which flood waters were bringing mud and sand in the past. Fresh mud and sand deposition was reduced by a hydrological regime regulation and by the consequent cessation in frequent valley flooding.

Soils within the Neretva river valley belonging mostly to the division of hydromorphic soils are characterized by the excessive water presence caused primarily by the shallow groundwater. Soil types within the soil class differ regarding the depth of the glayic horizon in the soil profile, soil texture, organic matter content and salinity level. All soils with the salt content >0.13% (w/w) (or $EC_e > 2$ dS m^{-1}) were considered as salinized. Based on these criteria, and using WRB classification (FAO 1998), Fluvisols Eutric Calcaric, partly hydroameliorated by channels, consisting of floodplain loam, dominate on fluvial terraces. Dominating soils in the low-land area are Molic Fluvisols Calcaric, hydroameliorated by channels, differing greatly regarding the depth of glayisation and salinity level, consisting of floodplain silt, and Molic Glaysols, hydroameliorated, partly salinized with the shallow glayic horizon, and mostly calcaric, consisting of floodplain silt-clay. Soils in the part of the study area are classified as Histic Glaysols being developed from maritime sediments. The region is protected of flooding by the pumping system. These soils are still subject to dynamic transformations due to variations in water levels and regimes, the relief position, and the proximity to the water courses and the sea.

The first form of land reclamation in the Lower Neretva region was a traditional method of manual excavation and heaping the sediments up perpendicularly to the river flow at higher elevations along watercourses. Sediment was dug up manually and piled to about 0.8 m high to form small terraces and avoid flooding during the high tide. With mostly hydro-ameliorated soils from the alluvium and without excess water stagnating in the rhizosphere, these terraces are used traditionally to grow vegetables and fruits.

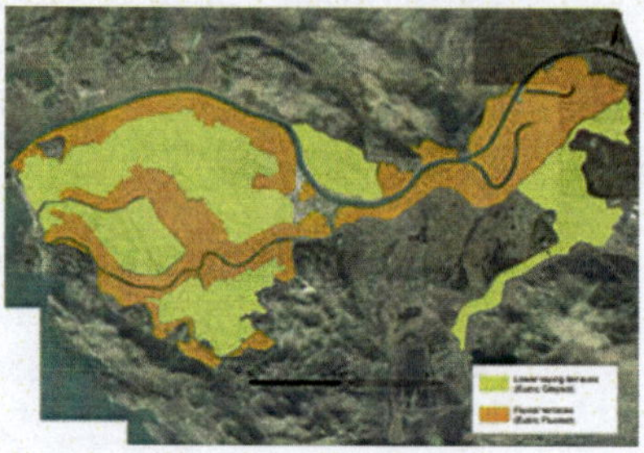

Fig. 18.2 Land zones within the study area divided according to dominant soil type and land use (Adopted from Romić et al. 2012)

The first available historical records about organized land reclamation of the Neretva delta wetland date back to the end of the seventeenth century. In the mid-twentieth century, modern land-reclamation projects had been carried out in the region with the expert and technical help of FAO (UN). Polder-type parcels were formed in the areas of low-elevation terrain where land surface was still above the low-tide level, and the water was pumped continuously over a dike. Agricultural production on these parcels was managed by the state agricultural company and had been mainly used for citrus growth and intensive vegetables production. After the company was closed, most of the parcels were used by private enterprises, but the property ownership problem has not been solved so far.

Based on all mentioned, it is possible to delineate two different land zones according to the different land reclamation and hydroamelioration practices as well as the past and present land use. In this way, the lower-laying terraces comprising mostly polder-type parcels, were separated from the land on fluvial terraces traditionally used by small farmers (Fig. 18.2).

18.3 Soil Salinity in the Neretva River Valley: Spatial and Temporal Variations

18.3.1 Spatial Variations of Soil Salinity

An analysis of spatial variability explains the characteristics and direction of the salinisation process, identifies the main factors of salinity and spatial trends, and it can also indicate areas at high risk which require specific management in order to mitigate adverse effects of increased salt concentrations. Apart from identifying salt-affected soils and monitoring soil salinity, it is important to assess the trend in salinity development and map the salinity hazard that might threaten a given area (Shrestha and Farshad 2009).

The identification, mapping, and further monitoring of the process of salinisation of agricultural soils represent the basic, but also the most costly part of the saline soil management. This is especially pronounced in the Mediterranean region because

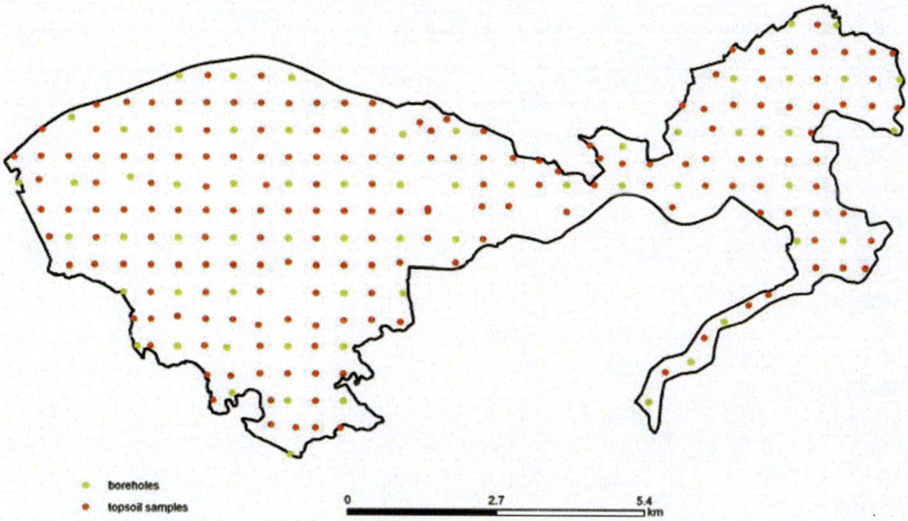

Fig. 18.3 The locations of the soil sampling site in the Neretva river valley. Red circles are the locations of the topsoil (0–25 cm) samples and yellow circles are locations of the borehole (0–100 cm) samples.

of the variability of soils, climate, small size of farm holdings and soil management practices (Zalidis et al. 2002). In the Neretva River valley the size of the farm parcels varies from 0.01 to 7 ha (Romić et al. 2010b). In order to ensure uniform coverage of the Neretva river valley area, a grid sampling scheme from the collection of agricultural topsoil samples (0–25 cm) and from boreholes was selected (Fig. 18.3). The topsoil samples were taken using regular rectangular grid with a 500-m distance between the points, and from boreholes with a 1,000-m distance between the points. Total 183 grid points were obtained for surface soil sampling and 63 grid points for borehole soil sampling from four depths (0–25, 25–50, 50–75 and 75–100 cm). The observation sites were spatially referenced using GPS and data were stored in different GIS layers. The field investigations were launched in August 2010 when the accumulation of salts in the surface soil horizon is the greatest.

The soil samples were air-dried and passed through a 2-mm sieve. For the soil salinity assessment, soil saturation extract (EC_e) was analyzed (Rhoades et al. 1989). Sodium and potassium concentrations were determined by atomic emission spectrometry (AAS PerkinElmer 3110). Chloride and sulfate concentration were determined by spectrophotometry using a segmented flow method with a Skalar San+Analyzer (HRN/EN ISO 10304-1/1998); Quality control procedure consisted of reagent blanks, duplicate samples and several referenced soil and sediment samples of a similar matrix from the inter-laboratory calibration program (Houba et al. 1996).

According to the standard criteria soil is considered saline if the value of electrical conductivity of the saturated aqueous extract is >4 dS m^{-1} (SSSA 1997). Although this is only 10% of the seawater salinity, from the agronomic point of view this level

Table 18.1 Statistical parameters for soil saturation extract from the four soil layers in the study area of the Neretva river valley

	EC_e (dS m^{-1})	Ca^{2+} (mg l^{-1})	Mg^{2+}	HCO_3^-	Na^+	SO_4^{2-}	Cl^-
0–25 cm							
Median	1.0	208	31	232	13	242	45
Min	0.3	48	1.9	98	3.2	4.0	10
Max	10.6	1,337	445	604	2,040	2,658	6,044
25–50 cm							
Median	1.1	176	29	183	33	222	74
Min	0.2	35	3.9	79	3.1	3.2	7.3
Max	12.0	988	533	744	3,136	3,123	7,534
50–75 cm							
Median	2.0	285	37	153	76	437	134
Min	0.19	35	3.9	67	2.9	4.3	5.9
Max	18	1,828	1,264	689	7,870	4,734	16,320
75–100 cm							
Median	2.4	468	53	159	113	1,261	159
Min	0.2	32	9.7	61	4.4	3.1	5.1
Max	60.3	1,315	2,217	1,220	11,200	5,218	19,900

is sufficiently high so that only cultures relatively salt-resistant can tolerate it. Medium salt-sensitive cultures tolerate concentrations of soil solution, which correspond to the value of EC_e 2 dS m^{-1}. Agricultural production in the Lower Neretva region is based on the cultivation of vegetable crops and citrus fruit, generally sensitive to salt-affected soils. Furthermore, farmers still haven't started with introduction of more salinity tolerant cultivars of vegetables, nor with the selection of more resistant rootstock for citrus fruit (Romić et al. 2007b). A tolerable degree of soil solution salinity also depends on the salt distribution in the soil profile, soil water content and overall agricultural practice including irrigation. This is why the general boundary level of EC between saline and non-saline soils has to be adjusted in terms of specific agri-environmental conditions of the particular area (Eynard et al. 2005). Based on knowledge gathered in the study area, the value of electrical conductivity of the soil saturation extract >2 dS m^{-1} was taken as a boundary level between saline and non-saline soil.

Compared with proposed boundary level, 24% of surface (0–25 cm) and subsurface (25–50 cm) samples is considered saline. In deeper layers this percent absolutely increase, from 46% of saline samples at the depth of 50–75 cm to 59% at the depth of 75–100 cm. The lowest range of EC is observed in the surface layer ranging from 0.3 to 10.6 dS m^{-1}, and a median value of 1 dS m^{-1} (Table 18.1). Values of EC_e > 1 dS m^{-1} are typical for soils of semiarid and arid areas. Climatic conditions, typical for the Mediterranean area, cause that the annual evaporation exceeds precipitation. Ions released into the soil solution, whether by weathering of minerals under the influence of salinized ground water or by irrigation with salinized water, accumulate in the secondary minerals formed by reducing moisture content of these soils. Na, K, Ca and Mg contribute the most to the salinity of the soil. Ionic species relationship in non-saline soil solution of moderate climate is Ca>Mg>K>Na (Bohn et al. 1985). In salt-affected and alkaline soil solution the concentration and the relationship of ionic species changes significantly. The predominant salt type is important because of the effect of individual ions on the soil properties and the possibility that they may be toxic to plants (Pisinaras et al. 2010). The principal cations in soil saturation extracts are Ca^{2+}, Mg^{2+} and Na^{+} while SO_4^{2-} and Cl^{-} are the dominant anions in the study area (Table 18.1). A noticeable increase of individual ions in the subsurface layer in relation to the surface soil layer is mostly the result of the transport of salt from the surface soil layer. The concentration range of dominant anions is very wide, from the lowest 5.1 mg l^{-1} Cl^{-} to the maximum 19,900 mg l^{-1} Cl. The highest concentrations of sodium, sulphate, and chloride indicate the accumulation of water-soluble salts as the result of direct impact of saline groundwater, i.e. sea water intrusion. Due to the changed natural regime of surface water and groundwater and their interaction, there is an increased inflow of saline water from the Neretva and deep quaternary layers into the network of drainage canals. That means that saline water intensively rises toward the surface, while at the same time the drainage system interrupts the inflow of such water to the arable surface soil layer. Still, certain amount of salt does reach the soil within a 1-m depth by capillary rise and other intermolecular forces in clayey materials.

In order to quantify soil salinity hazard in the Neretva river valley spatial distribution maps for EC_e of soil saturate extracts were produced using distance weight (IDW) method of interpolation. An interpolated map for each sampled depth was generated in the ArcGIS extension of the Geostatistical Analyst (Fig. 18.4).

The spatial intensity of salinity evidently changes moving away from the Neretva River mouth. Interpolated maps of EC_e show that high salinity hazard spreads within about 2 km inland from the sea (Fig. 18.4). The salinity increased with the depth, as well. The processes of primary salinisation of soil from groundwater become more expressed in the deeper layers. In the soil layer at a depth of 50–75 cm salts come from saline groundwater, and this is particularly marked in the area of the Lower Neretva closest to the sea. The highest value of electrical conductivity of 60 dS m^{-1} was measured in the soil solution at a depth of 75–100 cm. It indicates that main soil salinity pattern in the Neretva river valley is induced.

It is also evident that corridor of high soil salinity hazard exists in sub-area Luke along the narrow part of the valley. To fully understand the differences between subareas and the spatial trends in salinity, topographic and lithologic conditions, land use and quality of irrigation water have to be considered (Walter et al. 2001; Romić et al. 2012). Previous intensive citrus production in Luke area (gross around 300 ha) has been completely abounded 15 years ago after the state agricultural company was closed. Thus, in consequence, there has been failure to maintain expensive hydro-ameliorative facilities which have deteriorated as a result. Maintaining the appropriate regime of flow in drainage canals is directly related with groundwater flow. Relatively low elevations (partly even below the sea level) and relatively high groundwater levels require the application of constant pumping in order to achieve agricultural production, as well as to protect agricultural areas from salinisation.

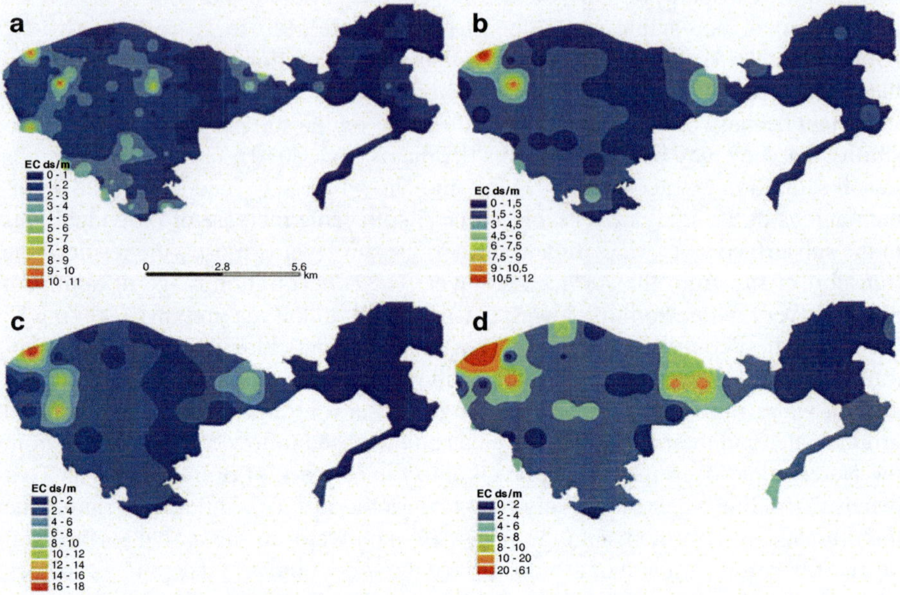

Fig. 18.4 Interpolated maps of EC_e in soil of the study area at (**a**) 0–25 cm depth (**b**) 25–50 cm depth (**c**) 50–75 cm depth (**d**) 75–100 cm depth (Adopted from Romić et al. 2010b)

All these changes that have taken place in recent decades evidently increase the salinity risk of the area. In an artificially created environment like polder-system land within the Neretva River valley, every change alters highly sensitive equilibrium which can have the long term consequences on further developing of salt-affected soils. The situation is different with plots formed at higher elevations along watercourses through a traditional method of manual excavation and heaping up of sediments, which was traditionally used only by agricultural producers. These are mostly hydro-ameliorated soils from the alluvium; they have a more favourable water-air regime, without excess water stagnating in the rhizosphere. The increased total salt content in the surface soil is, however, also related to the application of saline water for irrigation. The salinity of irrigation water significantly depends on the source of water intake, and that is the main reason for an uneven distribution of soil salinity in the study area.

18.3.2 Seasonal Variations of Soil Salinity

The processes of salinisation spatially highly variable are also subject to seasonal changes. It is well known that there is close relationship between degree of salts accumulation in the root zone and the groundwater table depth/regional hydrology as well as the topography that can significantly affect water movement and storage.

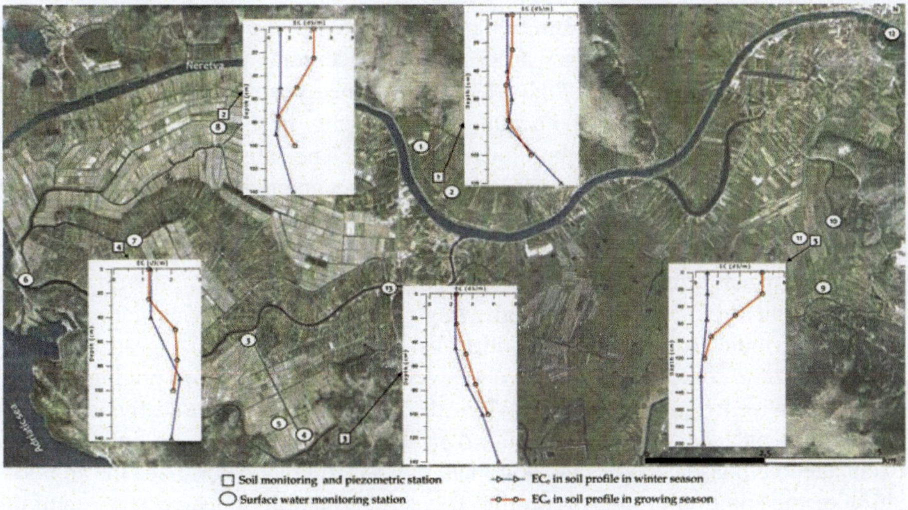

Fig. 18.5 Soil, surface and ground-water monitoring station and seasonal changes in soil salinity along profile in the Neretva river valley

Furthermore, hydraulic properties of the soils can be the key factor for soil salinity fluctuation (Akramkhanov et al. 2011). The irrigation practice and rainfall between irrigation events are also important factors which affect the distribution of salt within the soil profile (Ayars 2003). Knowing where the highest fluctuation in salinity will take place allows the identification of vulnerable areas and there site specific management (Acosta et al. 2011b).

Five soil monitoring stations were established in 2009 (Fig. 18.5) within the National project of monitoring soil and water salinisation in the Neretva river valley. Soils monitoring stations are distributed at different agricultural land and soil type within the vicinity of ground water monitoring wells. In the first year of the 5-year monitoring cycle soil pits were opened and detailed physical and chemical analyses of samples from each horizon were performed. Two sampling campaigns per year are carried out: the first one in winter season (February) and the second one in dry growing season (August). The soil profiles were sampled in 0.25 m increments to a depth of 1 m.

Comparison between winter and dry growing season showed that EC_e levels were constant throughout the profiles in monitoring stations 1, 3 and 4. However, significantly higher EC_e levels were determined in the subsurface horizons then at the surface in both sampling period at the stations 1, 3 and 4. Short term salinity seasonal fluctuation is detected at the soil monitoring stations 2 and 5. In the winter season EC_e was less than 2 dS m^{-1} throughout the profiles, while in the dry growing season the EC_e rose up to 4.4 dS m^{-1} at station 2 and up to 6 dS m^{-1} at station 5 in the top 25 cm.

Seasonal and vertical soil salinity pattern observed at the monitoring stations 1, 3 and 4 indicates that the primary process of salts accumulations in the root zone is through the capillary rise with high water tables. The dynamics of the groundwater level, has a direct effect on the accumulation of salts in the rhizosphere. This is particularly visible at the soil monitoring station 1. The salinity profile within this monitoring station indicates accumulative effect accentuated by clay texture (up to 50% of clay, data not shown) at the subsurface horizons (below 60 cm). The higher percentage of sand texture in surface horizons inhibits further capillary rise, and consequently salt accumulation in the soil surface.

The extent of EC_e increase in dry growing season at the monitoring stations 2 and 5 can be attributed to irrigation with degraded quality water. This suggests that salts in the root-zone could be accumulated to levels deleterious for crops predominantly cultivated with irrigation practice. When the water which was used for irrigation comes into balance with the soil, the distribution of ions between solution and exchangeable phase becomes the dominant mechanism that determines the composition of the solution in the soil profile. It is evident that in the short term, rainfall events during the winter time in the Neretva valley could be large enough to provide salt leaching from surface soils horizons. Studies conducted in Mediterranean conditions (Romić and Romić 1997) found that natural leaching from the soil can remove significant amounts of salt which entered the soil through irrigation with salinized water. Namely, in the Mediterranean climate the majority of precipitation falls in autumn-winter period. In this very period are the salts leached which accumulated during the irrigation season. Romić and Romić (1997) also found that in natural conditions the amounts of salt leached from the soil depend on the type of soil. The research has shown that a larger quantity of salt was leached from rendzina than from vertisol. However Adam et al. (2012) showed that desalination of vertic clay soils induced by repetitive ponding and flushing cycles can remove 35.7% of the initial salt content in the topsoil layer.

18.4 Water Quality in the Neretva River Valley

The decline in availability of freshwater for irrigation due to its allocation to other sectors (urban and industry), especially in arid and semi-arid regions such as the Mediterranean basin, has resulted in intensive use of waters of poor quality. Furthermore, inappropriate management of a coastal aquifer, highly sensitive to disturbance, may lead to its destruction as a source of fresh water much earlier than other aquifers that are not connected to the sea. This irrational and systematic groundwater abstraction for irrigation purposes of the area has resulted in the aggressive intrusion of seawater wedge, particularly during the last decade (Kallioras et al. 2006). Besides, large hydrotechnical interventions such as construction of dams, hydroelectric power plants and other hydrotechnical structures can also change the water regime within a catchment area, and consequentially lower the quality of water for different purposes.

An important impact of agriculture on the quality of soil and water resources in the Mediterranean is the increasing conductivity and associated salinisation of soils and the intrusion of seawater into the groundwater aquifers near the coast (Zalidis et al. 2002). The profitable cultivation of vegetables and fruit in the Lower Neretva region is practically unfeasible without irrigation. The water requirement of the selected cultivated crops in the Lower Neretva region was calculated in the Preliminary Design "Irrigation in the Lower Neretva Region" (Romić et al. 2007a). According to these calculations, irrigation requirements for the Lower Neretva region range from 50 to 60 mm in the production of cruciferous vegetables to as much as 561 mm in the growing of apples with an inter-row cover cropping. However, not even today is irrigation used to the extent required by climatic and hydro-pedological conditions of the region. Although the whole of the Neretva valley abounds in water, the provision of quality water for irrigation, especially in summer, could represent a serious problem. Water quality for irrigation in the region varies significantly depending on the source of water (Romić et al. 2008). In order to use irrigation water without adverse effects to agricultural crops, as well as to soil, its chemical composition has to be known.

Monitoring of the water quality in addition to the adequate management may certainly prevent the damages on the agricultural land and crops. A system of monitoring developed up to this point and measurements conducted, both in surface waters and groundwater in the alluvium provide a good starting point for further research (Romić et al. 2010a). The surface water monitoring program was established in 2009 as a part of National project of monitoring soil and water salinisation in the Neretva valley. A total of 13 surface water monitoring stations and 5 groundwater monitoring stations were established covering the whole Neretva river valley (Fig. 18.5). Four double piezometers for monitoring the state in shallow and deep aquifer serve for the measurement of water level and quality. Standard protocols for sampling, sample stabilization and analysis were adopted for all water quality variables, pH, electrical conductivity (EC dS m^{-1}) and ionic composition (APHA 1995). Water samples for lab analysis have been taken in 13 locations from open waterways and from 5 shallow piezometers once per month.

Measurements obtained by the water quality monitoring showed that together with electrical conductivity (Fig. 18.6), the concentrations of chloride and sulphate, bicarbonate as well as of calcium and sodium in surface and groundwater samples (Fig. 18.7) are variable. This variability occurs not only with regard to their spatial distribution but also over time. It is observed that the salt front is pushed towards the sea in the winter period (January), while the summer period brings significant salinisation (August). The salt wedge goes furthest upriver in mid-summer, while it was pushed out of the riverbed in the end.

In general at all surface water monitoring stations water can be classified as slightly to moderately saline according to FAO guidelines for irrigation water quality (Ayers and Westcot 1994). The highest level of EC in surface water was detected in August in the pumping station (SW 9) and the drainage channel (SW 2). Drainage network plays an important role in protection from floods and salt. During the rainy season, its role is primarily to collect significant amounts of rainfall during the

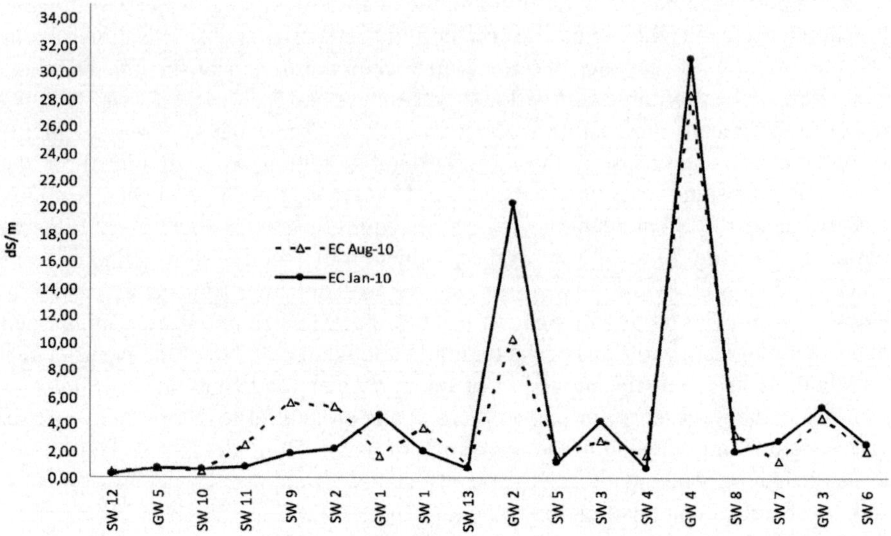

Fig. 18.6 Electrical conductivity (dS m^{-1}) in water samples collected in January and August 2010 at each surface (SW) and groundwater (GW) monitoring stations. All the data are reported with regard the distance from the Neretva river mouth (For geographical references see Fig. 18.5)

heavy downpours, and in the dry season to drain all brackish water which would salinize upper 2–3 m necessary for agricultural activity. In order to provide protection against floods within ameliorated areas of the Lower Neretva region, a total of eight pumping stations were constructed. The effect of maintaining the appropriate regime of flow in drainage canals by water repumping is evident during irrigation period.

A pick of EC (~32 dS m^{-1}) is observed in groundwater (GW 4) in January as well as in August. This groundwater monitoring station is located close to the river mouth. Numerical modelling has shown significantly different characteristics of shallow and deep aquifer in the lower Neretva River (Gotovac 2005). The deep aquifer has proven communication with the sea, so the salinity does not change significantly in both regimes, the dry and the rainy. Since there is a constant hydraulic gradient in the vertical direction, the model shows that 10 million cubic meters a year of highly salinized brackish water flows into the shallow aquifer through confining clay layer (Gotovac 2005).

The water quality also shows variability depending on their hydrological integration (Schmalz et al. 2009). In general far from the river mouth surface and groundwater quality are in most parameters much more alike than at the close to the river mouth (Fig. 18.7). Close to the river mouth groundwater is more mineralized than surface water (Fig. 18.6). The exception is in April 2010 when markedly higher concentrations of Cl, and SO_4 were in the surface water close to the river mouth and in the June far from the river mouth. The observed higher concentrations can also point on the effect of collecting brackish water by drainage channel (Romić et al. 2009).

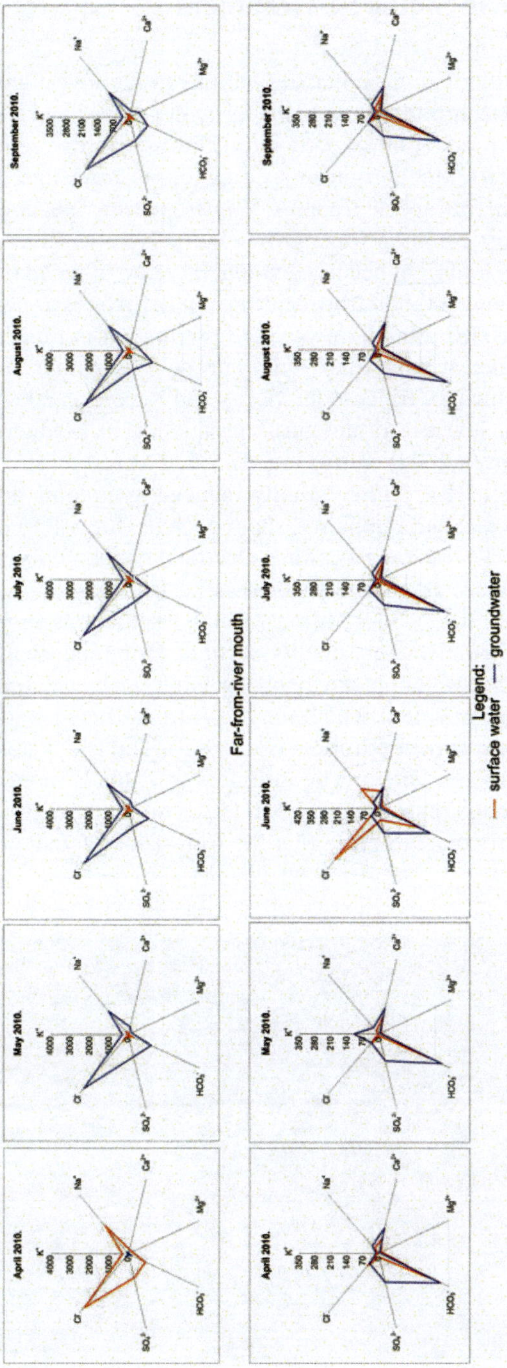

Fig. 18.7 Amoeba diagrams showing comparison of the chemical parameters during growing season in surface water from lateral drainage channel and groundwater close-to-the-river mouth (SW 4 and GW 3) and far-from-the-river mouth (SW 10 and GW 5) in the Neretva river valley (For geographical references see Fig. 18.5)

18.5 Agriculture in Saline Environment

Due to specific agricultural/environmental qualities in the investigated area of the Neretva valley, the agricultural production is fairly developed and it can be defined as fruit-and-vegetable production. Woody (fruit) crops are represented with crops ranging from exclusively Mediterranean species (citrus fruit, figs, olives, etc.) to traditionally continental fruit species (apples, plumes, peaches, pears, etc.). However, citrus fruits are the most common fruits grown in the Neretva River valley. This is especially important to highlight because evident increase of the mandarin oranges. Nowadays there are about 400,000 fruit-bearing mandarin trees as against 16 535 in 1967 (Romić et al. 2007b). Mandarin porange production is limited to relatively small belt of the Mediterranean climate with not too cold winter but adequate for flowering and development of fruit colour (Levy and Syversten 2004). On the other hand, soil salinity can severely limits citrus production in Mediterranean coastal areas (Levy and Syvertsen 2004). Citrus are classified as salt sensitive species with soil salinity threshold of 1.4 dS/m. Salinity-induced symptoms are non-specific chlorosis, smaller leaf size, and impaired shoot growth (Fig. 18.8) as well as fruit yields decrease up to 13% for each 1 dS m^{-1} electrical conductivity increase (Maas 1993). The yield decrease is probably attributable to osmotic stress rather than toxicity (Pedrero et al. 2012). In addition to high oscillations in yields, another problem is achieving the satisfactory quality of fruits. Difference in the water potential, is the driving force of water uptake by plants. High salt concentration in the root zone, give a high water potential which may lead to restricted water uptake. The most prominent problem is fruit puffiness (Fig. 18.8). The less water available to a plant, the problem is more marked. The puffiness of fruits is more marked in the years when high air temperatures and high relative humidity occur in the colour-changing phase, in particular if before that the fruit had suffered a drought stress.

Fig. 18.8 Chlorosis symptoms in mandarins leaves induced by salinity stress, and fruit puffiness

The juicy fruit analysis has shown that puffy fruits in statistical terms are significantly less juicy, have less soluble dry matter and a lower juice pH (Romić et al. 2007b). All the problems related with the soil water deficiency are more pronounced if a mandarin tree had been grafted to the hydrophilic stock of *Poncirus trifoliata* most frequently used by the farmers in the Neretva valley due to its resistance to low temperatures. Since these rootstocks are also very susceptible to salinity, it increases the hazard of salinity damage if leaching is reduced with reduced irrigation (Levy and Syvertsen 2004). It is important to note that mandarin trees recover very slowly from the water stress. The selection of an optimal system and method of irrigation is undoubtedly one of the most efficient agricultural engineering measures which would reduce an adverse effect of unequally distributed rainfall on the yields and quality of mandarin fruits (Romić et al. 2007b).

The vegetable production pattern is fairly developed, and due to specific – particularly geographic and climatic – conditions, the selection of vegetable crops and the duration of their vegetation, two harvests per year are possible in the field. Unlike the fruit production, the vegetable production is subject to relatively rapid changes and to the intensification of production, which also has to be taken into account when planning the future model of the development of agricultural production and assessing the application of agricultural engineering measures. In line with the projection of future yields in the study "Irrigation in the Lower Neretva Region" (Romić et al. 2007a), which is based on the reduction of yields under conditions without irrigation, with the introduction of irrigation yield will increase the most for tomato, paprika, eggplant, water melon, musk melon, leek and Swiss chard, ranging from as much as 1.5 to 2.2 times.

The experiment was set up with the aim to determine the influence of increasing salt concentrations in irrigation water using different irrigation systems (sprinkler and drip) on: crop growth, changes in soil solution composition, changes in leaf tissue mineral content, and changes in crop yield. Saline irrigation reduced the crop growth and thus had detrimental effects on yield (Fig. 18.9). Application of saline water rapidly changed the ion composition of soil solution. Significant changes in Ca, Na and Cl concentrations in soil solution were determined parallel to increasing the irrigation water salinity. The experiment has shown as well that salinity affects nutrient uptake and accumulation and also nutrient partitioning within the plant. In the case of drip irrigation, increasing water salinity reduced potassium uptake by plants, but enhanced sodium and chloride uptake. As the ion concentration in the cell wall rapidly escalate, and the cell rapidly dehydrate (Munns 2002). Crops with perturbed nutrients relations are more susceptible to invasion of different pathogenic (micro)organisms and/or physiological dysfunctions; their edible parts have markedly less economic and nutritional value due to reduced fruit size and shelf life, non-uniform fruit shape, decreased vitamin content (Ondrašek 2008).

Despite the salt-induced yield reductions, the implication for applying water of poor quality using drip irrigation is clears (Ondrašek et al. 2006). This is especially valid in the regions with high annual precipitations, occurring mostly during the winter. Besides the usual agricultural management, the possible harmful effects of salt accumulation in soils may thus be prevented also by natural leaching. Promising

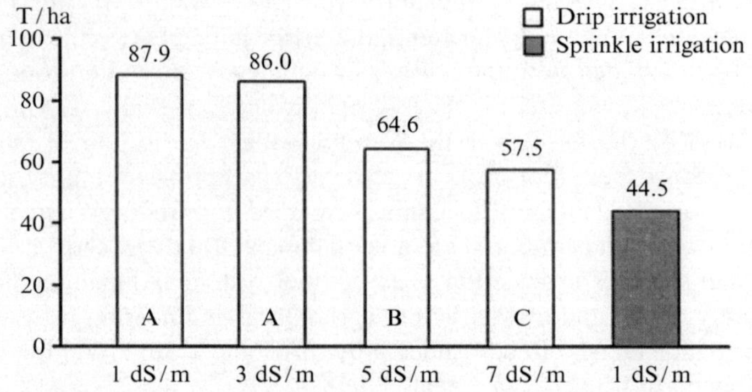

Fig. 18.9 Marketable yield decline of watermelon affected by water salinity and irrigation method (Adopted form Romić et al. 2008)

option of leaching management has to be linked to specific crops mechanisms operating in saline soils as leaching the soil can restore the water relation of the plant but not affect salt levels in leaves (Jacobsen et al. 2012).

The problem of soil salinisation is very complex and the assessment of salinity risks need to evaluate not only water stress to plants and potential toxic impacts of salt but also the potential increase in the bioavailability of metals (Du Laing et al. 2009; Romić et al. 2011b) as well as increased organic matter decomposition (Weston et al. 2011). Estuarine sediments are frequently rich in contaminants transported from land. Also well-known is the ability of sediments to faithfully record environmental impact, including the heavy metal contamination, on fluvial systems over time. For the Neretva estuary, the main source of contaminants is industries in the upper part of the catchments area. The lower part is an intensively agricultural region, producing generally vegetables and fruits, and salinity poses an increasing threat in the Neretva estuary, since the major route of human exposure to toxic elements is via consumption of vegetables grown on contaminated soil. Matrix of saline (sodic) soil is naturally poor and/or depleted in metal-binding interfaces (Ondrašek et al. 2010) Soil salinity, especially increased concentration of dissolved Cl-ligands, significantly influences solubility of some trace elements like cadmium (Ondrašek et al. 2009). The fate of metals and potential toxic risk depend primarily on the sorption-desorption balance and how this balance changes in regard to changes in environmental conditions (Cerquiera et al. 2011). If desorption processes are slow, and thus pollution permanent, it can jeopardize plant production. On the other hand, fast desorption processes trigger potentially toxic metals and thus may affect the quality of groundwater (Romić 2009).

Two experiments were set up in a polyethylene greenhouse located at the experimental station on the University of Zagreb Faculty of Agriculture (Croatia) in 2004 and 2006. The first one was aimed to test the hypothesis that organic matter decreases the bioavailable Cd^{2+} pool and therefore restricts its phytoextraction. The effect of four salinity levels (0, 20, 40 and 60 mM NaCl) and three Cd levels (0.3, 5.5 and

10.4 mg kg^{-1}) in peat soil on mineral accumulation and distribution as well as vegetative growth and fruit yield parameters of muskmelon (Cucumis melo L.) were assessed. The second experiment was done on radish (*Raphanus sativus* L. var. sativus, cv. Tarzan). Results of the greenhouse experiment with muskmelon grown on Cd-enriched peat provide evidence that cadmium transfer from saline and contaminated organic soil to edible fruity tissue is low and not mediated by NaCl salinity. These results suggest that (i) muskmelon appears to have low capacity to take up and transport Cd from roots to fruits, and/or (ii) to translocate Cd via phloem from developed leaves to fruits, suggesting that phloem mobility of Cd in muskmelon is relatively poor and unaffected by NaCl salinity (Ondrašek et al. 2009). When radish plants were tested, it was shown that cadmium uptake and leaf deposition was markedly enhanced by NaCl, whereas Cd translocation and hypocotyl accumulation was up to sixfold lower and not influenced by NaCl salinity. Despite low concentration of trace elements in topsoil, it was still necessary to monitor Cd and Cu concentration in floodplain soils in the Neretva river valley since most of them might be phytoaccumulated especially in excessive concentration of dissolved Cl^- ligands (Romić et al. 2012).

18.6 Conclusion and Future Perspective

In this chapter an overview of recent and current research program being carried out within the Neretva River valley has been presented. The result considering land resources surveys showed that the degree of salinisation is strongly related to characteristics of the soils and their location in the catchments area. Salinity of soils within the root zone can be highly variable and in order to estimate the risk of salinisation it is important to examine soil salinity as a function of depth.

Analyses of the data obtained by the water quality monitoring showed that salt concentrations in surface waters of the Neretva River valley changed substantially during the year as a result of the hydrological regime, demonstrating a spatial as well as a temporal variability of the water electrical conductivity and sodium and chlorine concentrations.

Agricultural production in salt-affected Mediterranean environments is unfeasible without irrigation. The overall sustainability of agricultural production is dependent not only of good quality water, but also on adequately maintaining reclaimed areas, site-specific management of soil and the overall practices in agriculture.

The salinisation problem may enhance trace metals mobilization, its bioavailability and phytoaccumulation. Hence, more holistic approach has to be applied in mitigation adverse effects of salinity on plant growth in saline environments.

Since salinity–induced land degradation is related to complex environmental dynamics, the challenge is to carry out an integrated and multidisciplinary study aimed to identify the main impact of agriculture on soil and water quality degradation and their reflections on the socio-economical conditions. A case of Neretva river valley is good example how an appropriate management for saline soils and crop production in saline environments should bring together different studies and disciplines. Scientific partners with different sectorial expertise should provide guide lines for the land management based on the knowledge of the different environmental component in the involved territories and of the degradation processes affecting them.

By detecting and modelling the main soil degradation processes it will be possible to propose possible alternative scenarios of agricultural management in order to reach a trade of between environmental protection income in terms of costs and benefits of a sustainable agricultural management.

References

Acosta JA, Faz A, Jansen B, Kalbitz K, Martinez-Martinez S (2011a) Assessment of salinity status in intensively cultivated soils under semiarid climate, Murcia, SE Spain. J Arid Environs 75:1056–1066

Acosta JA, Jansen B, Kalbitz K, Faz A, Martinez-Martinez S (2011b) Salinity increases mobility of heavy metals in soils. Chemosphere 85:1318–1324

Adam I, Michot D, Guero Y, Soubega B, Moussa I, Dutin G, Walter C (2012) Detecting soil salinity changes in irrigated Vertisols by electrical resistivity prospection during a desalinisation experiment. Agr Water Manage 109:1–10

Akramkhanov A, Martius C, Park SJ, Hendrickx JMH (2011) Environmental factors of spatial distribution of soil salinity on flat irrigated terrain. Geoderma 163:55–62

APHA (1995) Standard methods for the examination of water and wastewater. 19th ed. APHA, AWWA and WEF, Washington, DC

Ayars EJ (2003) Filed crop production in areas with saline soils and shallow saline groundwater in the San Joaquin Valley of California. In: Goyal SS, Sharma SK, Rains DW (eds) Crop production in saline environments: global and integrative perspectives. The Haworth Press, Binghamton/New York, pp 353–386

Ayers RR, Westcot DW (1994) Water quality for agriculture. FAO, Rome

Bohn HL, McNeal BL, O'Connor GA (1985) Soil chemistry, 2nd edn. Wiley, New York

Bonacci O, Roje-Bonacci T (1997) Sea water intrusion in coastal karst springs: example of the Blaz Spring (Croatia). Hydro Sci J 42(1):89–100

Cerquiera B, Covelo FE, Andrade L, Vega FA (2011) The influence of soil properties on the individual and competitive sorption and desorption of Cu and Cd. Geoderma 162:20–26

De Pascale S, Ruggiero C, Barbieri G, Maggio A (2003) Physiological response of pepper (*Capsicum annuum* L.) to salinity and drought. J Am Soc Hort Sci 128:48–54

Du Laing G, Rinklebe J, Vandecasteele B, Meers E, Tack FMG (2009) Trace metal behaviour in estuarine and riverine floodplain soils and sediments: a review. Sci Total Environ 407:3972–3985

Eynard A, Lal R, Wiebe K (2005) Crop response in salt-affected soils. J Sustain Agr 27(1):5–50

FAO/ISRIC/ISSS (1998) World reference base for soil resources. World Soil Resources Report, 84, FAO, Rome

Gotovac H (2005) Variable density flow and transmits in aquifers. MS thesis, University of Split

Houba VJG, Uittenbogaard J, Pellen P (1996) Wageningen evaluating programmes for analytical laboratories (WEPAL), organization and purpose. Comm Soil Sci Plant 27(3–4):421–431

Jacobsen SE, Jensen CR, Liu F (2012) Improving crop production in the arid Mediterranean climate. Field Crops Res 128:34–47

Kallioras A, Pliakas F, Diamantis I (2006) Conceptual model of a coastal aquifer system in northern Greece and assessment of saline vulnerability due to seawater intrusion conditions. Environ Geol 51:349–361

Koyro HW, Ahmad P, Geissler N (2012) Abiotic stress responses in plants: an overview. In: Ahmad P, Prasad MNV (eds) Environmental adaptations and stress tolerance of plants in the era of climate change. Springer, New York

Leuven RSEW, Poudevigne I (2002) Riverine landscape dynamics and ecological risk assessment. Freshwater Biol 47:845–865

Levy Y, Syvertsen JP (2004) Irrigation water quality and salinity effects in citrus trees. Hortic Rev 30:37–82

Ljubenkov I (2006) Numerical model of stratified flow in river mouth. MS thesis, University of Split

Lyons MN, Halse SA, Gibson N, Cale DJ, Lane JAK, Walker CD, Mickle DA, Froend RH (2007) Monitoring wetlands in a salinizing landscape: case studies from the Wheatbelt region of Western Australia. Hydrobiologia 591:147–164

Maas EV (1993) Salinity and citriculture. Tree Physiol 12:195–216

Maggio A, De Pascale S, Fagnano M, Barbieri G (2011) Saline agriculture in Mediterranean environments. Ital J Agron 6(7):36–43

Munns R (2002) Comparative physiology of salt and water stress. Plant Cell Environ 25:239–250

Ondrašek G (2008) Salt stress and rhizosphere contamination by cadmium: phytoaccumulation of nutrients and toxic metal. Dissertation, University of Zagreb

Ondrašek G, Romić D, Romić M, Duralija B, Mustać I (2006) Strawberry growth and fruit yield in a saline environment. Agric Conspec Sci 71(4):155–158

Ondrašek G, Romić D, Rengel Z, Romić M, Zovko M (2009) Cadmium accumulation by muskmelon under salt stress in contaminated organic soil. Sci Total Environ 407(7):2175–2182

Ondrašek G, Rengel Z, Romić D, Savić R (2010) Environmental salinisation processes in agroecosystem of Neretva River estuary. Novenytermeles 59:223–226

Ondrašek G, Rengel Z, Veres S (2011) Soil salinisation and salt stress in crop production. In: Shanker A (ed) Abiotic stress in plants – mechanisms and adaptations. InTech INtech, Rijeka, pp 171–190

Pedrero F, Allende A, Maria IG, Alacron JJ (2012) Soil chemical properties, leaf mineral status and crop production in a lemon tree orchard irrigated with two types of wastewater. Agr Water Manage 109:54–60

Pisinaras V, Tsihrintzis VA, Petalas C, Ouzounis K (2010) Soil salinisation in the agricultural lands of Rhodope District, northeastern Greece. Environ Monit Assess 166:79–94

Rhoades JD, Manteghi NA, Shouse PJ, Alves WJ (1989) Estimating soil salinity from saturated soil-paste electrical conductivity. Soil Sci Sot Am J 53:428–433

Rhoades JD, Kandiah A, Mashali AM (1992) The use of saline waters for crop production. FAO Irrigation and drainage paper 48, Rome

Romić M (2009) Gospodarenje tlom i vodom za održivu poljoprivredu (soil and water management for sustained agriculture). University of Zagreb, Faculty of Agriculture

Romić D, Romić M (1997) Ratio of salt content added by irrigation with saline water and percolated from the root zone. In: International conference on water management, salinity and pollution control towards sustainable irrigation in the Mediterranean region, Vol. IV: Wastewater re-use in irrigated agriculture, Technomack, Bari

Romić D, Vranjes M (2010) Monitoring of soil and water salinisation in Neretva River delta. Report for 2009, University of Zagreb & University of Split

Romić D, Romić M, Dolanjski D, Ondrašek G, Salopek Z, Husnjak S, Borošić J (2007a) Irrigation in the Lower Neretva region. Report, University of Zagreb

Romić D, Romić M, Ondrašek G, Zovko M, Jemrić T, Dobričević N, Voća S (2007b) Mineral composition and sensory evaluation of Satsuma fruits (*Citrus unshiu* cv. Chahara) in the Neretva River Valley. Report, University of Zagreb

Romić D, Ondrašek G, Romić M, Borošić J, Vranjes M, Petosic D (2008) Salinity and irrigation method affect crop yield and soil quality in watermelon (*Citrullus lanatus* L.) growing. Irrig Drain 57(4):463–469

Romić D, Zovko M, Romić M, Ondrašek G (2009) Quality aspects of the surface water used for irrigation in the Neretva delta (Croatia). J Water Land Develop 2:3–13

Romić D, Romić M, Zovko M, Juračak J, Ondrašek G, Bakić H (2010a) Harmonization of agricultural intensification with long-term ecological integrity in Croatia. In: Wiggering H, Knierim A, Ende HP, Pintar M (eds) Innovations in European rural landscapes. Springer-Verlag Berlin Heidelberg, pp 97–114

Romić D, Romić M, Husnjak S, Ondrašek G, Zovko M, Bakić H, Filipović V, Stričević I (2010b) Level of salinity and sustainable use of soil in the lower Neretva region, GEF Neretva and Trebišnjica Management. Report, University of Zagreb

Romić D, Romić M, Zovko M, Bakic H, Ondrašek G (2012) Trace metals in the coastal soils developed from estuarine floodplain sediments in the Croatian Mediterranean region. Environ Geochem Health. doi:10.1007/s10653-012-9449-z

Schmalz B, Springer P, Fohrer NP (2009) Variability of water quality in a riparian wetland with interacting shallow groundwater and surface water. J of Plant Nutr Soil Sci 172:757–768

Shrestha DP, Farshad A (2009) Mapping salinity hazard: an integration application of remote sensing and modelling based techniques. In: Zinck AJ, Metternich G (eds) Remote sensing of soil salinisation: impact on land management. CRC Press, Boca Raton

Soil Science Society of America (SSSA) (1997) Glossary of soil science terms. SSSA, Madison, p 134

Stuyfzand PJ (1993) Hydrochemistry and hydrology of the costal dune area of the Western Netherlands. Dissertation, de Vrije Universiteit van Amsterdam

Walter C, McBratney AB, Douaoui A, Minasny B (2001) Spatial prediction of topsoil salinity in the Chelif Valley, Algeria, using local ordinary kriging with local variograms versus whole-area variogram. Aust J Soil Res 39:259–272

Weston NB, Vile MA, Neubauer SC, Velinsk DJ (2011) Accelerated microbial organic matter mineralization following salt-water intrusion into tidal freshwater marsh soils. Biogeochemistry 102:135–151

Zalidis G, Stamatiadis S, Takavakoglou V, Eskridge K, Misopolinos N (2002) Impacts of agricultural practices on soil and water quality in the Mediterranean region and proposed assessment methodology. Agr Ecosyst Environ 88:137–146

Brief Biodata of Editor and Co-editor

Dr. Parvaiz Ahmad (Editor)

Dr. Parvaiz is Assistant professor in Botany at A.S. College, Srinagar, Jammu and Kashmir, India. He has completed his post-graduation in Botany in 2000 from Jamia Hamdard New Delhi India. After his Ph.D. from Indian Institute of Technology (IIT) Delhi, India in 2007 he joined International Centre for Genetic Engineering and Biotechnology, New Delhi, India. His main research area is stress physiology and molecular biology. He has published more than 30 research papers in peer reviewed journals and 16 book chapters. He has also edited three volume two with Springer NY USA and one with Studium Press Pvt. India Ltd., New Delhi, India. He have received Junior Research Fellowship and Senior Research Fellowship by CSIR, New Delhi, India, during his Ph.D. Dr. Parvaiz has been awarded Young Scientist Award under Fast Track scheme in 2007, by Department of Science and Technology, Govt. of India, New Delhi, India. Dr. parvaiz is actively engaged in studying the molecular and physio-biochemical responses of different agricultural and horticultural plants under environmental stress.

Prof. Mohamed M. Azooz (Co-editor)

Prof. Dr. M.M. Azooz is a Professor of Plant Physiology in department of Botany, Faculty of Science, South Valley University, Qena, Egypt. He received the B.Sc. in 1984 and M. Sc. in 1990 from Assiut University, Egypt; and Ph.D. in 1997 from South Valley University, Qena, Egypt in collaboration with Tübingen University, Germany. He has published more than 70 research articles in peer reviewed journals and five book chapters in international books. He has supervised many M.Sc. students and Ph.D. scholars till date and many are in progress. He is a member of editorial board and reviewer of many international journals. Prof. Azooz has been

P. Ahmad et al. (eds.), *Ecophysiology and Responses of Plants under Salt Stress*,
DOI 10.1007/978-1-4614-4747-4, © Springer Science+Business Media, LLC 2013

included in Marquis Who's in the World (Biography, 28th Edition, 2011), and in the "Top 100 Scientists" list, International Biographical Centre, Cambridge-UK, 2011. Major focus of his research has been on the physiological, biochemical and molecular responses of plants against various biotic and abiotic stresses and the mechanisms of tolerance.

Prof. M.N.V. Prasad (Co-editor)

Prof. M.N.V. Prasad is a Professor in Department of Plant Sciences at the University of Hyderabad, India. He received the B.Sc. (1973) and M.Sc. (1975) degrees from Andhra University, India, and the Ph.D. degree (1979) in botany from the University of Lucknow, India. Prasad had published 138 articles in peer reviewed journals and 84 book chapters and conference proceedings in the broad area of environmental botany and heavy metal stress in plants. He is the author, co-author, editor, or co-editor for 12 books (John Wiley and Sons Inc. New York; CRC Press, Boca Raton; Springer-Verlag Heidelberg; Narosa Publishing House, New Delhi; Kluwer Academic Publishers, Dordrecht; Marcel Dekker, New York, Fizmatlit Publishers. Moscow, Ministry of Environment and Forests, Govt. of India). He is the recipient of Pitamber Pant National Environment Fellowship of 2007 awarded by the Ministry of Environment and Forests, Government of India. He has served as visiting professor/guest researcher in many universities in different parts of the world viz., Australia, Belgium, Canada, Finland, Kazakhstan, Poland, Portugal and Sweden.

Index

P. Ahmad et al. (eds.), *Ecophysiology and Responses of Plants under Salt Stress*,
DOI 10.1007/978-1-4614-4747-4, © Springer Science+Business Media, LLC 2013

P

T